The Speciation and Biogeography
of Birds

The Speciation and Biogeography of Birds

Ian Newton

Monks Wood Research Station
Cambridgeshire, UK

Illustrations by Keith Brockie

ACADEMIC PRESS

An imprint of Elsevier Science

Amsterdam • Boston • London • New York • Oxford • Paris
San Diego • San Francisco • Singapore • Sydney • Tokyo

Copyright 2003, Elsevier Science Ltd.

Academic Press
An Imprint of Elsevier Science
84 Theobald's Road, London WC1X 8RR, UK
http://www.academicpress.com

Academic Press
An Imprint of Elsevier Science
525 B Street, Suite 1900, San Diego, California 92101-4495, USA
http://www.academicpress.com

ISBN 0-12-517375-X

Library of Congress Catalog Number: 2002115609

A catalogue record for this book is available from the British Library

Typeset by J&L Composition, Filey, North Yorkshire
Printed and bound in Italy

03 04 05 06 07 PT 9 8 7 6 5 4 3 2 1

Contents

Preface

This book is concerned mainly with the process of species formation in birds, and with the factors that affect their geographical ranges. The current distributions of species depend not just on prevailing conditions, but also on past conditions which have enabled species to reach the areas they have or eliminated them from areas formerly occupied. Important processes involved include: on the longest time scales, the movements of the continents and the formation and loss of volcanic islands; on shorter time scales, the changes in glacial and other climatic conditions; and on yet shorter time scales, the effects of human activities on habitats and biodiversity.

Any attempt to understand these processes places us at the interface between several traditional scientific disciplines: palaeontology, ecology, genetics and evolutionary biology, in addition to the 'earth sciences' of geology, oceanography and climatology. These various sciences form the basis of biogeography, which itself underlies many aspects of evolutionary biology. Some of the founders of modern evolutionary theory, notably Charles Darwin and Alfred Russel Wallace, were biogeographers, although they may not have thought of themselves as such.

Research in all the relevant sciences has made enormous advances in recent years, supporting some earlier ideas but changing others, as well as providing fresh insights. These advances have shed new light on speciation and extinction events, and on the shifts in geographical ranges that have shaped the composition of current avifaunas. Particularly influential has been the development of molecular biology: the study of the structure of DNA, the chemical responsible for genetic inheritance. Such study has enabled us to trace the history of species in ways not previously imaginable.

In this book, I have attempted to integrate new information from these various fields of science in order to gain more understanding of speciation in birds, and of the histories and distributions of populations. The book is intended primarily for advanced students of population and evolutionary biology, but jargon has been cut to a minimum in the hope that the text will be enjoyed by a wider readership, including interested bird-watchers. A glossary is provided, and no background in mathematics is necessary.

After an introductory chapter, giving some necessary background information, the book is arranged in five main sections. These deal respectively with: (1) species formation and diversity in birds; (2) major distribution patterns on continents, islands and seas; (3) effects of past glacial and other climatic changes in shaping current distribution and diversity patterns; (4) contemporary factors limiting geographical ranges; and (5) the influence of dispersal and migration on bird distribution patterns. The book ends with a concluding chapter summarising the main findings. With this treatment, some topics inevitably arise in more than one chapter, but I have tried to keep repetition to a minimum, and to cross-refer wherever possible. Because many species are mentioned from different parts of the globe, scientific names are given with common names throughout.

Much of this book was written while I was the grateful recipient of a grant from the Leverhulme Trust. Many of the maps and diagrams in the book have been redrawn from previously published sources, as acknowledged in the captions. For permission to use this material, I thank Academic Press, the American Ornithologists' Union, Birdlife Finland, Birdlife South Africa, Blackwell Science Limited, the British Ecological Society, the British Ornithologists' Union, the Ecological Society of America, *Evolution* Editorial Office, the National Academy of Sciences, *Nature*, the Royal Society, the Smithsonian Institution, Oxford University Press, Springer-Verlag and University of Chicago Press, as well as those of the original authors that I managed to track down.

I am grateful to Mrs Doreen Wade for typing the several drafts of each chapter and for helping to put the book together; also to Linda Birch of the Alexander Library in Oxford for helping me find some relevant literature, and to Angela Turner for meticulous copy-editing. Throughout, I have benefited from discussion with many friends and colleagues too numerous to mention individually. Among them, Anthony Cheke, Mike Brooke, Ian Owens, David Parkin and Pertti Saurola kindly allowed me to read their manuscripts before publication. For commenting on particular chapters, I thank Mark Avery (14), Bill Bourne (8 and parts of 9), Mike Brooke (8), Anthony Cheke (6 & 7), Fred Cooke (part of 2), Jeremy Greenwood (2 & 3), Shelley Hinsley (14), Peter Jones (11), Mick Marquiss (2 & 3) and Iain Taylor (part of 5). Finally, for commenting helpfully on the whole book in draft I am grateful to David Jenkins and my wife, Halina. Any errors that remain are entirely my own.

Ian Newton

Osprey *Pandion haliaetus*, one of the most widely distributed of bird species.

Chapter 1
Introduction

This book is about the formation and diversity of bird species, their geographical distributions and their migration patterns. It is an attempt at a fresh synthesis which draws from recent developments in the biological sciences, as well as in the earth sciences of geology and climatology. For in order to understand the diversity and distribution patterns of birds (or any other organisms), we have to take account both of past events in the earth's history that enabled particular species to evolve, persist and reach the areas they have, and of present conditions that limit their current ranges. The three main questions addressed in this book are therefore (1) how have new bird species arisen, (2) how did they come to be distributed in the way they are, and (3) how are current patterns of distribution and diversity maintained.

Scientific interest in such matters dates back to the nineteenth century, and some of the ideas developed then, notably by Charles Darwin and Alfred Russel Wallace, are just as valid today. Nonetheless, great strides in our understanding of bird biogeography have been made in recent years, with new developments in the study of bird evolution arising from the growth of genetic (DNA) technology; in palaeo-ecology arising from research in geology, climatology and plant and animal fossils; and in knowledge of current bird distribution and migration patterns, arising from bird mapping and ringing schemes. These developments enable us to reappraise earlier ideas, modify and extend some of them, and discard others.

Birds offer particular advantages for biogeographical research. Their evolutionary relationships, one species to another, are relatively well understood, benefiting from years of work on museum skins, from observational studies in the field, and, most recently, from analyses of their DNA. In consequence, the whole classification of birds has been reappraised from DNA studies, at least at the levels of orders and families (Sibley & Ahlquist 1990), and for many groups also at the levels of species and subspecies (Chapter 2). Throughout this book, I have mostly adopted the DNA-based classification and listing of birds laid out by Sibley & Monroe (1990), but have sometimes deviated from it, especially where more recent reappraisal has suggested changes.

Secondly, compared with most other animals, the geographical patterns of morphological variation found within bird species have been extremely well studied. Every regional handbook on birds can now give details, not only of the size and colour differences between similar species, but also of the size and colour variation found across the geographical range of a single species, and from one subspecies to another. Such details can often also be given on song and other characteristics. Hence, for birds more than for other animals, the geographically definable population has become the customary taxonomic unit of study.

Thirdly, because of the popularity of birds, and the comparative ease with which they can be found and identified, their distributions over much of the world have been well mapped. This process was started by explorer-naturalists and collectors of museum skins, and for some tropical regions their findings still provide the main source of information. Over much of the world, however, distributional records are increasingly obtained by bird-watchers. In fact, the spatial study of bird distributions has now entered a new era, with the development of grid-based 'Atlas' surveys, involving the collective efforts of hundreds of bird-watchers. In the last 30 years, such detailed species-by-species range maps have become available for Europe, North America, Australia and Southern Africa, as well as for several smaller regions (Sharrock 1976, Blakers *et al.* 1984, P. Lack 1986, Root 1988a, Gibbons *et al.* 1993, Price *et al.* 1995, Hagemeijer & Blair 1997, Harrison *et al.* 1997).

Compared with most other organisms, birds are highly mobile, and some are able to fly long distances over seas, deserts or other inhospitable terrain. Not surprisingly, therefore, birds are some of the most widely distributed of organisms. Unlike many other kinds of animals, they are well represented, not only on continents and nearby islands, but also on remote oceanic islands, on some of which the original colonists have undergone major radiations to produce a diversity of different species. Much of the theory of island biogeography, discussed in Chapter 6, is based on studies of birds (MacArthur & Wilson 1967, Diamond & May 1976, Lack 1976).

Another consequence of the high mobility of birds is the development of seasonal migration, through which the same populations can occupy different parts of the world at different times of year. Migration between fixed breeding and non-breeding areas is not by tradition regarded as an aspect of biogeography, yet it occurs in many groups of animals, from insects to fish and mammals. It reaches its greatest development in birds, however, causing large-scale seasonal changes in bird distributions over the earth's surface. It occurs in some degree in most bird species that live in seasonal environments, from the arctic tundras to the tropical savannahs and grasslands. Only in the relatively stable conditions of tropical rain forest do the majority of bird species remain resident year-round, but even these forest areas receive a seasonal influx of wintering migrants from higher latitudes. Overall, more than 1,000 million birds are estimated to fly each year on return migrations between breeding and non-breeding areas, which for many species lie on different continents. Hence, in attempting to understand the factors that affect bird distributions, we cannot ignore the remarkable and spectacular phenomenon of migration. It adds an extra dimension to the distributional ecology of birds.

Research on bird movements has always been popular with ornithologists. It received a major boost about 100 years ago with the start of scientific bird ringing, and a further boost in the 1950s with the development of more efficient trapping methods, including mist nets and cannon nets. Since then more than 200 million birds have been individually ringed worldwide, giving hundreds of thousands of recoveries that reveal the movement patterns of different populations. Ring recoveries are biased to some extent towards areas with high human population and literacy. Nonetheless, together with observations (made directly or with radar), these ring recoveries have revealed a network of bird migration routes that encompass all habitable parts of the globe (for European birds, see Zink 1973–85, Zink & Bairlein 1995, Wernham *et al.* 2002). In recent years, it has become possible to attach radio-transmitters to individual migrants and track from satellites their day-to-day movements, thus gaining further information on routes and stop-over sites.

With their great mobility, birds might be considered less suitable than other, less dispersive organisms for addressing some types of biogeographical questions. Using their powers of flight, birds could in theory reach almost anywhere on earth, a facility that could greatly reduce their value in studies of the history and evolution of regional faunas. In fact, not all birds are strong fliers, and many are reluctant to cross water or other hostile terrain (Chapter 16). Moreover, the majority of species show great site fidelity: individuals of some species stay in the same localities throughout their lives, and many long-distance migrants return year after year to the same breeding and wintering areas. So birds are less free in their wanderings than their powers of flight might suggest, and they also often have difficulty in establishing themselves in new areas (Chapter 16). In consequence, birds still show sufficient regional structure in their distribution patterns to reveal much about their geographical histories and evolution.

ORDERS, FAMILIES AND SPECIES

On the Sibley–Monroe classification, the world's living birds are grouped into 23 orders, comprising 146 families and more than 9,700 species. Eleven out of the 23 extant orders of birds are cosmopolitan, being represented on all the main land-masses, except Antarctica (Chapter 5, **Table 5.1**). They include the game birds (Galliformes), waterfowl (Anseriformes), rollers and others (Coraciiformes), cuckoos (Cuculiformes), parrots (Psittaciformes), swifts (Apodiformes), owls (Strigiformes), pigeons (Columbiformes), cranes and others (Gruiformes), storks and others (Ciconiiformes) and songbirds (Passeriformes). Three other orders are represented on four continents, three on three continents, two on two continents, and four on only one continent (counting Europe and Asia as a single continent). Many of these orders are also represented on oceanic islands. In general, therefore, the majority of bird orders are widely represented around the world.

At the level of the family, cosmopolitanism is less common (**Table 5.1**). The only landbird families that are well represented on all continents are mainly aquatic types, namely herons (Ardeidae), ibises (Threskiornithidae), pelicans (Pelecanidae), rails (Rallidae), ducks and geese (Anatidae), plus the owls (Tytonidae and Strigidae) and raptors (Accipitridae and Falconidae). These nine types, which comprise 16% of all bird families, all have ancient lineages, good dispersive powers or both. About 18 other families are represented on all continents but on some by only one or two species. Many other families are restricted to either the Old World or the New World, or to a particular continent or even a small part of it. The family Opisthocomidae contains only a single species, the Hoatzin *Opisthocomus hoazin*, a nearly flightless leaf-eating bird that occurs only in a small region of northern South America. Other bird families are confined to certain islands or island groups, such as the Rhynochetidae (represented by the Kagu *Rhynochetos jubatus*) in New Caledonia, the Todidae (todies) in the West Indies, and the Apterygidae (kiwis), Acanthisittidae (wrens) and Callaeatidae (wattlebirds) in New Zealand.

At the level of the species, cosmopolitanism is relatively even rarer and most bird species are restricted to a single continent or island group (**Table 5.12**). Of the nearly 8,000 landbird species that live on continents, only 0.04% breed on all five continents, 0.06% on four continents, 0.3% on three continents, 6.0% on two continents and about 93.6% on only one continent. Many of these species are also found on oceanic islands, but in addition more than 1,600 landbird species are found only on such islands (Chapter 6).

For obvious reasons, as one works through the taxonomic hierarchy, from orders through families to species, distributions become increasingly restricted; but considering the potential dispersive powers of birds, the level of range restriction shown by most species is especially striking (for a different rendering of the same data by biogeographical region, see **Figure 5.5**).

Like most other kinds of plant or animal, therefore, each species of bird lives only in certain parts of the world, its geographical range. We take this for granted, but it is not always obvious why particular species are found in one part of the world and not in another, why some are found over wide areas and others over small areas, why on continents and not on islands, or why on some islands and not on others. Because of their propensity to perform long-distance migrations, such questions seem more apposite for birds than for most other organisms. Yet whether resident or migrant, bird species vary enormously in the geographical extents of their distributions.

On a world scale, the six most widely distributed landbird species, which breed on every continent except Antarctica, include the Barn Owl *Tyto alba*, Osprey *Pandion haliaetus*, Peregrine Falcon *Falco peregrinus*, Great Egret *Ardea alba*, Cattle Egret *Bubulcus ibis* and Glossy Ibis *Plegadis falcinellus*. Five other species breed on four continents, namely the Fulvous Whistling Duck *Dendrocygna bicolor*, Common Moorhen *Gallinula chloropus*, Kentish Plover *Charadrius alexandrinus*,

Black-crowned Night Heron *Nycticorax nycticorax* and Striated Heron *Butorides striatus*. Some other species, such as the House Sparrow *Passer domesticus*, occur naturally on a single land-mass but have been widely introduced elsewhere, while certain shorebirds nest across the arctic, but when not breeding extend to coast-lines over much of the world (again excluding Antarctica).

At the other extreme, some bird species are found year-round only on a single small oceanic island and may occupy no more than a few square kilometres. Probably the smallest range of any living bird is held by the Laysan Teal *Anas laysanensis*, now restricted to the 3.6 km^2 Pacific island of Laysan, and centred on a single saline pool. The maximum population that Laysan Island could support at the best of times is probably around 500 individuals, but numbers have often been much lower. Bone remains indicate that the species once occurred on other Hawaiian Islands too (Cooper *et al.* 1996). At least eight endemic species once lived on the 13 km^2 Lord Howe Island situated about 550 km east of Australia and some still survive there. In addition, a flightless wren *Xenicus lyalli*, now extinct, once occurred only on the 2.6 km^2 of Stevens Island, situated between the North and South Islands of New Zealand. Of course, species reduced by human action to a handful of pairs can occupy even smaller areas, but this is usually a temporary step en route to extinction or recovery.

Interest in bird distributions has increased greatly in recent years, stemming partly from conservation concerns. The total population size of any species depends on the area over which it occurs, the amount of suitable habitat within that area, and its average density within that habitat. Species that occur over small areas may be at greater risk than those that occur over wide areas, so range size is one measure of vulnerability that is useful in setting conservation priorities. Knowledge of bird distributions has also been used in ranking areas according to their conservation value, areas of high priority being those that fall within the current ranges of the largest number of species (or species of interest), as exemplified in the concept of 'endemic bird areas' (Bibby *et al.* 1992, Stattersfield *et al.* 1998). The term endemic means occurring nowhere else. Birds can be endemic to geographical regions at different taxonomic levels (orders, families, genera and species) and on different spatial scales (continents, regions, islands). They can be endemic to a location either because they originated there and never dispersed elsewhere (like many island species) or because they now survive in only a small part of their former range, not necessarily the part where they evolved. These two categories are sometimes separated as 'neo-endemics' and 'palaeo-endemics', respectively.

THE HISTORICAL CONTEXT

As indicated at the outset, the study of geographical ranges may take a static view, recording the facts of distribution as they appear now and relating them to prevailing conditions. Or it may take a historical view, seeking to derive the present situation from knowledge of past events in an ever-changing world. Whereas the static approach is largely descriptive and factual, the historical approach, concerned with processes, is analytical but often speculative. An understanding of bird distributions can come only from a blending of the two approaches. At best, the study of present conditions may reveal what limits the current distributions of birds to particular regions, while the study of past events can reveal what has led those species to occur where they do. The inclusion of the historical context makes parts of biogeographical theory untestable in the strict sense, but if we ignore past events, we can gain no appreciation of the factors that have led to present distribution patterns, and hence are likely to promote future changes.

Past influences on bird distributions include the movements of land-masses (on time scales of millions of years), glacial and other climatic vicissitudes (on scales of tens of thousands of years), or shorter-term climatic and habitat changes (on scales of hundreds or tens of years). Over and above these natural events, the most important influence of recent centuries is human impact. Many species of birds now occur over much smaller areas than in the recent past, or are extinct altogether as a result of habitat destruction or other human action, while other species occur over larger areas, as a result of habitat provision or introductions. Through both natural and human-induced causes, then, bird ranges are not static, but are continually shifting in position, expanding or contracting, on time scales that vary from many millions of years to individual years. Inevitably, the range of any species must begin small, wherever the species evolves, and then over time it may expand and contract in size, finally shrinking as the species declines to extinction, only to be replaced by later-evolving forms.

Movements of land-masses

On the longest geological time scale (**Table 1.1**), the most important biogeographical event affecting the evolution and distributions of organisms has been the slow alteration of the geography of the earth through plate tectonics (formerly called continental drift). Throughout the earth's history, land-masses have moved over the earth's surface, continually joining and splitting, and changing the numbers and configurations of continents and oceans. The evidence for such movements comes partly from the shapes of continents (reflecting how they could once have fitted together), but also from the identical rock-types on opposite sides of oceans (reflecting the positions of splits), and several other lines of geological evidence, including discovery of the mechanism responsible (for summary see Brown & Lomolino 1998).

Understanding the modern distributions of birds or other organisms depends on some knowledge of the past movements of continents. The splitting of the single ancient land-mass, Pangea, into two smaller ones, Gondwanaland in the southern hemisphere and Laurasia in the northern, occurred soon after the earliest birds appeared, as judged by the dating of the fossilised *Archaeopteryx* at 150 million years ago to the mid Jurassic Period. As fossils of this species were found in Germany, we know for sure that the earliest birds were in Laurasia, and it seems likely that they were also in Gondwanaland.

Soon after its formation, Laurasia itself began to fragment, and by the mid Cretaceous Period, about 100 million years ago, it was subdivided by shallow seas into three main land-masses, which subsequently formed western North America, eastern North America and most of Eurasia. These land-masses combined and split in various ways to give the two main northern land-masses of today, with Greenland off eastern North America. Land connections existed at times between

Table 1.1   Geological time scale, from the Jurassic Period when birds first evolved to the present. From Lamb & Sington 1998.

	Time before present (millions of years)
CENOZOIC ERA	
Quaternary Period	**2–present**
Holocene	0.01–present
Pleistocene	2–0.01
Tertiary Period	**65–2**
Pliocene	5–2
Miocene	24–5
Oligocene	37–24
Eocene	58–37
Palaeocene	65–58
MESOZOIC ERA	
Cretaceous Period	**144–65**
Jurassic Period	**213–144**

The geological time scale is hierarchical, with each division among eons, eras, periods and epochs marking transitions among geological strata and particularly in their embedded fossil assemblages. The Phanerozoic Eon covers the time of life on earth from the start of the Cambrian Period about 590 million years ago. Traditionally, the Cenozoic Era was divided into the Tertiary Period which lasted 63 million years and the Quaternary Period which lasted only two million years. A newer scheme divides the Cenozoic into only two periods of more equal length, the Palaeogene (65–24 million years before present) and the Neogene (24 to present), the division between them being set at the start of the Miocene.

eastern North America and Europe and between western North America and Asia, providing corridors for the movements of plants and animals. North America and Eurasia share many species of plants and animals, and are often coupled as a single biogeographical region, the Holarctic.

Roughly coincident with the break-up of Laurasia, Gondwanaland also began to fragment. It gave rise to South America, Africa, Arabia, Madagascar, India, Australia–New Guinea, New Zealand and New Caledonia, most of which drifted northwards, while the remaining fragment of Antarctica, originally much warmer than today, moved south to take up its position over the pole (**Figure 1.1**). India and other smaller fragments moved north to form some southern parts of Eurasia, in the process pushing up the Alps and Himalayas. The timing of these major events, according to recent estimates, is given in **Table 1.2**. Current evidence suggests that the land that gave rise to Africa was first to break away from Gondwana, about 100 million years ago, followed by India–Madagascar (100–80 million years ago), New Zealand and New Caledonia (80 million years ago) and then Australia–New Guinea (64–45 million years ago) and finally South America at its southern tip (35 million years ago), leaving Antarctica to drift southwards.

We can envisage the drifting continents like giant rafts, carrying their cargoes of plants and animals which, after separation, were free to evolve independently of those on other land-masses. In a sense then, each land area provided a separate experiment in evolution. Subsequent connections between continents enabled the mixing of these mainly independent biotas, notably between South and North America about 3.5 million years ago, between Africa and southern Eurasia about

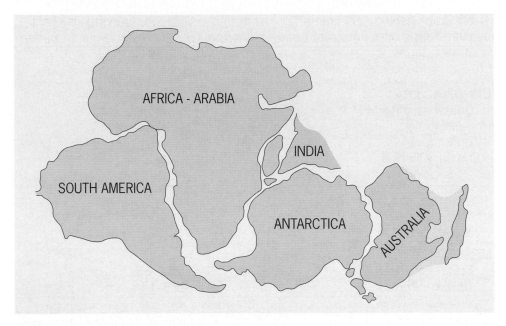

Figure 1.1 The break-up of Gondwanaland.

12 million years ago, and between Eurasia and North America on several occasions, most recently only 10,000 years ago. It remains uncertain whether Australia, on reaching its present position, ever had any continuous land connection with southeast Asia, but the many intervening islands must always have provided stepping stones for dispersing organisms. As biotas intermingled, species that exploited similar habitats and foods are likely to have competed, leading to further differentiation and to many extinctions, as confirmed in the fossil record for the North–South American connection, though more for mammals than for birds (Simpson 1965; Chapter 5).

As the continents split, drifted and regrouped, seas expanded and contracted, mountain ranges rose and were later eroded away, and islands appeared and disappeared. New mountains, oceans and land areas would have altered atmospheric and oceanic circulations, continually changing regional climates. New coastlines and shallow seas greatly increased the habitat for coastal birds, and separated it into different regions isolated by sea or land, thus allowing for independent regional evolution of coastal species. Knowledge of such past physical changes to the earth's surface has proved especially helpful in explaining disjunct distributions, in which organisms from the same families now occur in two or more widely separated parts of the world (Chapter 5). The current distributions of many plants and animals provide clear evidence of their past connections: for example, many species on the southern continents are much more closely related to one another than to any forms on the northern continents.

Table 1.2 Timing of major geological events that affected the flora and fauna of different regions. Gondwanaland refers to the great southern continent from which some major modern land-masses were derived, and which at the time of its fragmentation was mainly covered by humid forest. One remaining fragment, Antarctica, is now devoid of landbirds. Mainly from Norton & Sclater 1979, Smith *et al.* 1981, Audley-Charles 1983 and Besse & Courtillot 1988, in addition to various other sources.

Africa	Severance from South America, and hence from Gondwanaland, in the mid Cretaceous about 100–80 million years ago. By the end of the Cretaceous, the two continents were separated by over 800 km of sea, but now-submerged volcanic islands occurred between them. Joining to Eurasia at the western end of the Mediterranean in the Miocene, 9–8 million years ago, and to Arabia–Turkey 20 million years ago. Severance from western Eurasia at the western end (Gibraltar) about five million years ago and at Arabia–Turkey in the mid Miocene, 16 million years ago, recreating the Mediterranean Sea which had largely evaporated in the interim. Separation from the Indomalayan Region by development of the Red Sea and extension of deserts in the Middle East in the late Pliocene, three million years ago.
Australia–New Guinea	Severance from Gondwanaland in the early Eocene, about 55 million years ago. Present position by the Pliocene, adjacent to the Asian plate, five million years ago, producing new collision and volcanic islands between the two.
New Zealand and New Caledonia	Severance from Gondwanaland in the mid Cretaceous, 85–82 million years ago, perhaps with subsequent land connections with New Caledonia and Australia until 75 million years ago.
South America	Final severance from Gondwanaland, possibly as late as 35 million years ago. Severance from Africa in the mid Cretaceous 100–80 million years ago. Joining to North America, in the Pliocene about 3.5 million years ago, after separation throughout the Tertiary.
India	Severance from Gondwanaland, in the mid Cretaceous, more than 100 million years ago, but some estimates give 140 million years ago; retained a connection with Madagascar until about 85–80 million years ago, when it broke away and drifted north, leaving behind fragments as the Seychelles Islands, 65–60 million years ago. A late Cretaceous land-bridge has been postulated from India to Madagascar to Antarctica via the Kerguelen plateau (see Cracraft 2001). Joining to Asia, in the early Eocene, 55–40 million years ago (forming the Himalayan mountains), but some estimates put this event in the late Cretaceous, 70 million years ago.
Eurasia	Severance from North America at the Atlantic (Greenland) connection in the Eocene, 45 million years ago at high latitudes, the split having occurred earlier at lower latitudes. Joined to eastern Asia at the Pacific (Bering) connection in the Eocene, but as exposed land only during dry periods, the last of which was 13,000–10,000 years ago.
North America	Severance from Eurasia at the Atlantic (Greenland) connection, in the Eocene, 45 million years ago.

	Severance from Eurasia at the Pacific (Bering) connection 13,000–10,000 years ago. Joining to South America at the Panamanian connection in the late Miocene, six million years ago.
Madagascar	Severance from Africa in the mid Jurassic, 160–155 million years ago, or even later, and attained its present position relative to Africa by 120 million years ago; split from India about 85–80 million years ago when India broke away and drifted north, leaving behind the Seychelles Islands 65–60 million years ago.

About 180 million years ago all land-masses were joined as a single giant continent called Pangea, which broke up into a northern supercontinent, Laurasia, and a southern one, Gondwanaland. These in turn became progressively subdivided by the formation of new oceans over the last 125 million years, and at the same time migrated to different parts of the globe.

The importance of Gondwanaland

The gradual fragmentation of the single southern land-mass (Gondwanaland) into separate continents was one of the most significant events in avian biogeography. At the end of the Cretaceous Period (65 million years ago) or later, the southern continents were still more or less interconnected, but by then some modern bird orders had probably evolved (Chapter 2; Cracraft 1973, 2001, Cooper & Penny 1997, Fedducia 1999). Such birds could therefore have reached their present distributions on different continents by direct descent from Gondwanaland. The fragmentation of this single land-mass, followed by drift of the fragments to their present positions as continents, accounts for many otherwise puzzling distribution patterns, for example the concentration of several distinct types of birds (including the large flightless ratites, Struthioniformes) on all three southern continents, as well as on Madagascar and New Zealand (**Figure 1.2**; Cracraft 1973)[1]. An alternative view is that different species of ratites arose independently by convergence, each evolving flightlessness separately from winged ancestors on different land-masses. This issue is still not totally resolved, but relationships within the Struthioniformes (based on anatomy and DNA analyses) point to these birds having a single common ancestor, with their present distributions resulting mainly from the break-up of a former single land area, though not necessarily before the development of flightlessness (Haddrath & Baker 2001; for dissenting views see Härlid *et al.* 1998, Feduccia 1999; for further discussion see Chapter 2).

Early Gondwanan forms that are still represented among living birds are likely to have included the Passeriformes (songbirds), Gruiformes (cranes and rails), Galliformes–Anseriformes (game birds–waterfowl), Columbiformes (pigeons),

---

[1]*Among flightless ratites, the rheas (Rheidae) and a fossil family (Opisthodactylidae) are associated with southern South America, the ostriches (Struthionidae) are found in Africa but in the late Tertiary were also widely distributed across Eurasia, the recently extinct elephant-birds (Aepyornithidae) were confined to Madagascar, the cassowaries (Casuariiadae) are found in Australia, New Guinea and some nearby islands, and the kiwis (Apterygidae) and the recently extinct moas (Dinornithidae) are found only on New Zealand, and a fossil ostrich was also found in India (Olson 1985b). Marsupial mammals also became isolated in South America and Australia with the break-up of Gondwanaland. During the early Tertiary they radiated on these two continents, while placental mammals were replacing them on other land-masses.*

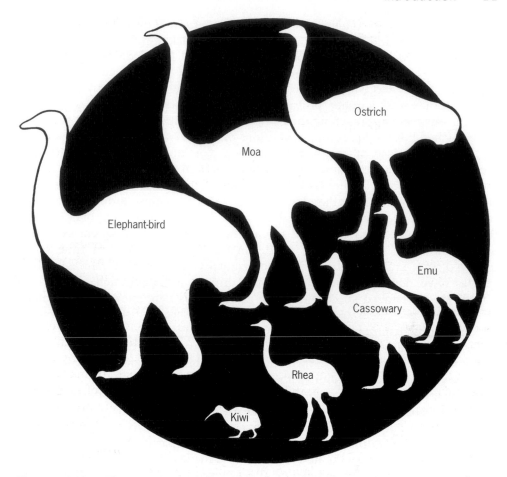

Figure 1.2 The various modern species of flightless ratites, found on land areas derived from the ancient supercontinent of Gondwanaland.

Apodiformes–Caprimulgiformes–Strigiformes (swifts–nightjars–owls), Psittaciformes (parrots), Sphenisciformes (penguins), as well as the Struthioniformes–Tinamiformes (ratites–tinamous). The evidence on their ages is still inconclusive (Chapter 4), but these groups may have been present before the final break-up of Gondwanaland, which began in the late Cretaceous and ended in the Eocene (Cracraft 2001)[2]. The majority of these groups have since spread widely around the world, but the penguins and others are still almost restricted to land areas derived from Gondwana. Such areas now hold what seem to be some of the most primitive of birds. Examples include seriemas (Cariamidae), screamers

---

[2]*Within these orders, Cracraft (1973) listed 28 bird families as 'groups for which reasonable arguments can be advanced for southern hemisphere dispersal'. These include the Apterygidae, Dinornithidae, Casuariidae, Columbidae, Psittacidae, Cuculidae, Aegothelidae, Podargidae, Megapodiidae, Furnariidae, Rhinocryptidae, Menuridae, Atrichornithidae, Tyrannidae, Acanthisittidae and Phylotomidae, among others. More recent findings have not altered this view, even though some of these groups also spread widely within the northern continents.*

(Anhimidae), finfoots (Heliornithidae) and sunbittern (Eurypygidae) of South America; the ground-rollers (Brachypteraciidae), cuckoo-roller (Leptosomidae) and mesites (Mesitornithidae) of Madagascar; the Freckled Duck *Stictonetta naevosa* and Magpie Goose *Anseranas semipalmata* of Australia, and the wrens (Acanthisittidae) and wattlebirds (Callaeatidae) of New Zealand.

Some other bird families probably arose much later, when the continental land-masses were approaching their present configuration. They could therefore have reached their full present distributions on several continents only by dispersal from their place of origin. Hence, any major land-mass is likely to have received its current complement of bird families from two sources, one by inheritance from its parent land-mass (autochthonous families) and another by successive colonisations from other land-masses (allochthonous families), some of which would have involved overwater flights. Both avifaunal components could then have changed and expanded to varying degrees by subsequent evolution and radiation on the land-mass itself. On the 35–80 million year time scales involved, it is unlikely that any of the original Gondwanan or Laurasian bird species have remained unchanged to the present, although such ancient and apparently unchanged species (dubbed 'living fossils') are known among many other organisms, celebrated examples being the coelacanth fish from the Indian Ocean and the tuatara lizard from New Zealand.

Colonists on any land-mass are likely to have arrived at different times, and from one or more source areas, their success in establishment depending partly on how they fared in competition with species already there. If the invaders were successful, they could have eliminated and replaced a large proportion of the original inhabitants. Hence, on any land area, the proportion of the avifauna derived from autochthonous families is likely to depend partly on the distance of that land-mass from other source areas (isolation) and partly on the differential in competitiveness between the new invaders and the existing inhabitants.

Distinguishing autochthonous from allochthonous families on any land-mass is seldom straightforward. Families now endemic to a land-mass did not necessarily evolve there, for they may represent relicts from a once much wider distribution. Thus, many families that are now restricted to Australia or South America are represented by fossils in the northern continents (examples given later; Chapter 5). Likewise, the flightless Kagu *Rhynochetos jubatus* of New Caledonia has as its closest relative the volant Sunbittern *Eurypyga helias* of tropical South America. Both these species are isolated relicts of a once widespread group, fossils of which have been found in Eocene deposits of Wyoming and Germany (Houde *et al.* 1997, Miyaki *et al.* 1998, Cracraft 2001). Conversely, families that were originally endemic to a region may have since spread so widely that it is now impossible to discern their region of origin. The fossil record, which carries the history of past distribution patterns, has proved useful in confirming the former presence of certain families on continents from which they are now absent. Moreover, fresh finds of fossils are constantly being made, continually altering the picture of former bird distribution patterns.

The present distributions of bird families between continents may thus bear only limited relationship to their earlier distributions. Such range changes could have followed partly from the climate and vegetation changes that occurred as continents drifted over the earth's surface, and partly perhaps from the outcome of competition from new invaders. Any proposed explanation of the current and past distribution patterns of birds and other organisms is likely to be valid only if it is consistent with the known geological and climatic history of the earth, and with the fossil record, insofar as it is known.

Climatic changes

The evidence for past climate changes comes mainly from cores taken from sea and lake beds and from ice sheets, which can be analysed for carbon and oxygen isotopes, for plant and animal remains (especially pollen) and other physical and biological clues. The annual layers in ice cores reflect past precipitation levels, and the carbon dioxide levels in air bubbles trapped within the ice reflect the composition of past atmospheres. On shorter and more recent time scales, coral and tree growth patterns provide indices of past growing conditions. All these indicators can be dated and checked for consistency, one with another, and from different parts of the world.

At the start of the Tertiary Period, some 65 million years ago, the global climate was much warmer than now, and subtropical conditions prevailed as far as 50°N. Since then, there has been a net cooling and drying of climates. The temperature difference between the poles and equator has gradually steepened, in turn increasing sea currents and winds, and causing greater seasonality of climate and greater latitudinal zonation in vegetation. The antarctic ice cap began to grow from the end of the Eocene (37 million years ago) onwards, starting on the highest mountains and gradually spreading over the whole continent. It eventually locked up so much ice that it caused a general reduction in global sea-levels. The arctic ice cap (lying over sea water) was apparently formed much later, beginning only about 2.4 million years ago (Shackleton *et al.* 1984). Until then, the Arctic Ocean was so warm that it supported little or no sea ice even in winter, and a boreal forest of conifers grew in northernmost Greenland. We must assume that tundra, if it existed at all as a distinct northern habitat, was restricted to high-latitude areas too dry or high for trees.

Superimposed on the overall downward temperature trend were frequent temperature oscillations which gradually increased in amplitude, culminating in the glacial cycles of the Quaternary Period, the last 2.4 million years (de Menocal & Bloemendal 1995). Since then glaciers in both hemispheres have fluctuated in size fairly regularly, giving a succession of glacial and interglacial periods at high latitudes. Glacial periodicity shifted about 900,000 years ago from cycles of about 41,000 years to cycles of about 100,000 years. These glaciations, which are thought to have numbered more than 20, varied in severity, but had profound effects on the distributions of plants and animals, including birds, as is evident from fossil and other evidence (Chapter 9).

Changes in climate during the Quaternary led to continual changes in sea-level, which in turn altered the sizes and configurations of land areas. Rises tended to reduce and fragment land areas, leaving the higher ground as islands, whereas falls tended to enlarge and join together land areas, but also exposed many new islands, previously submerged. These processes in turn had large effects on the dispersal, distributions and evolutionary histories of plant and animal populations (Chapter 9).

Some 26,000–14,000 years ago, the most recent glaciation covered high-latitude regions in both hemispheres, cooled and dried climates worldwide, and lowered sea-levels by at least 120 m. At this time, about one-third of land in the northern hemisphere was covered with ice, compared with about one-tenth today, and because of lowered sea-levels, many areas that are now islands were joined to nearby mainland (such as Britain to Europe, Ceylon to India, or the Falklands to South America). These areas became cut off from the mainland around 10,000 years ago when ice melt caused a general rise in sea-level. In addition, during the last glaciation, the Bering Straits were ice-free and dry, so that northeastern Siberia was connected to northwestern North America, enabling boreal and tundra plants and animals to disperse between these two major land-masses. The fact that this huge high-latitude area escaped the ice increased its importance as a refuge and as a dispersal route for plants and animals.

The end of the last glaciation around 10,000 years ago marked the start of the current interglacial period, the Holocene. Added together, the various cold periods characterised about 90% of the last 2.4 million years, warm interglacial conditions, like those prevailing now, being unusual in this longer term setting. The various measures of past climates mentioned above have given surprising indications of the speed of climate change, showing that transitions between fundamentally different climates have sometimes occurred within periods as short as a few decades.

Human impacts

It is impossible to assess the recent distributional history of plants and animals without taking into account the effects of human activities, which over much of the world have become the dominant influence on where species live, and on whether they live or die. Around two million years ago, from among several hominid lineages in Africa, *Homo erectus* emerged and soon spread to other regions, including the Middle East and Asia where fossils as old as 1.7 million years have been found. A second wave of expansion of *H. erectus* from Africa occurred in the mid Pleistocene, by which time cranial capacity had increased. On the basis of both fossil and DNA evidence, modern-type humans (with high rounded skull, prominent chin and small brow ridges) first appeared in Africa about 130 thousand years ago, and are thought to have spread out as a third wave around 100 thousand years ago, interbreeding with the original *H. erectus* (Templeton 2002). Since then, the resulting *H. sapiens* has gradually colonised almost every habitable land area on earth, invariably with adverse effects on other species. The known period since *H. erectus* first appeared in the fossil record is about as long it takes two populations of the same songbird species to diverge into two different species (Chapter 3), and modern-type humans have existed for less than one-tenth of this period. Hence, most of the birds that these three waves of early people encountered as they spread out of Africa were the same as those we see today. Modern people spread first through the Middle East across southern Asia, and then much later from Eurasia to Australasia some 60,000–40,000 years ago, and from the Middle East into Europe about 41,000 years ago. They reached North America from Asia only 14,000–12,000 years ago, spreading gradually from the northwest southwards, so that South America was the last continent to experience human impact (for summaries, see Avise 2000, Hewitt 2000, Templeton 2002).

Our species is thought to have taken most of its evolutionary history to reach a total population level of ten million individuals, only ten thousand years ago. With the advent of agriculture, this number grew to about 100 million about two thousand years ago, then to 2.5 billion by 1950. Within less than the space of a single human lifetime it more than doubled to nearly 6.0 billion in 1995. If recent rates of population growth continue, the world human population is projected to reach 12 billion before 2100. The actual outcome will have enormous implications for the fate of life on earth, including human.

Perhaps the first major effect of humanity on the natural world, from early in human history, came from the extensive use of fire which, in warm regions, extended and created grassland at the expense of forest and scrub. This in turn led to soil erosion and desertification in many regions, but especially in parts of Africa, the Middle East and Australia. The second suspected major impact, as human numbers and hunting skills grew, was the extinction of dozens of species of large mammals and birds, which disappeared from region after region as modern-type humans spread (Chapter 9). It culminated, in the last few thousand years, in the extinction of many bird species following the discovery and colonisation by people of increasingly remote oceanic islands (Chapter 7). The third major impact was widescale deforestation by cutting and burning, which has occurred at an increasing pace over the past few thousand years and especially in the last few hundred years, in line with the destruction or modification of many other natural vegetation types, mainly in the interests of a progressively developing agriculture.

Over this whole period of human history, three activities have been influential in reducing biodiversity: (1) overhunting, which has affected increasing proportions of species, as people have spread and technology improved; (2) habitat destruction, degradation and fragmentation, leading to piecemeal loss of populations; and (3) translocations of species, enabling those released in new areas to eliminate local species through competition, predation or infection with alien parasites and pathogens. In addition, pollution now has increasing impacts on the natural world, especially greenhouse gas emissions which threaten climate changes as great as in any glacial–interglacial cycle, with a human-induced increase in mean global temperatures of 0.6°C in the twentieth century alone (Chapter 9). In fact, all these factors are now acting at unprecedented speed to reduce the numbers and distributions of wild species, as human numbers, resource demands and technological skills continue to grow.

During our own lifetimes many of us have witnessed great changes in the abundance and distributions of various bird species. They have given us some idea of the rate at which individual species have retracted, or have colonised islands or spread over continents, as well as the rate at which genetically controlled changes in bird morphology, physiology and migratory behaviour have occurred. On this short time scale, the role of humanity, it seems, has come to supersede all natural processes in influencing the rest of terrestrial plant and animal life. Nowadays, almost every place on earth, from the polar ice caps to the equator, and from the upper atmosphere to the deep oceans, has been altered by human action, mainly in the past few centuries. We cannot hope to understand the current distributions of birds or any other organisms without taking these facts into account.

THE VALUE OF FOSSILS

We can gain some idea of the evolutionary history of birds, and of their past distribution patterns, from the study of fossils. Given enough dated fossil remains from all the families of birds from all the main land-masses, we could work out the evolutionary and distributional histories of each group of birds, and map them on the changing surface of the earth. If we knew the precise birthplace of each taxon, we could then tell whether certain geographical regions have been more important than others as 'cradles of evolution', and then go on to examine the underlying factors that promote evolutionary innovation. We could see how new groups increase in numbers of species, radiate to exploit different opportunities, spread out to occupy new terrain, take over from older groups, and finally shrink to extinction. Sadly, the fossil record for birds is far too incomplete and patchy to attempt anything close to this, but it has still given some astonishing insights, relevant to the current distributional scene.

Because the weak pneumatised bones of birds do not fossilise easily, the fossil record may be less extensive for birds than for other vertebrates, and it seems safe to conclude that only a minute proportion of the bird species that have ever lived has been found as fossils. Of course, the larger species with bigger bones fossilise more readily than small species with finer bones, and species that die in wetlands fossilise most readily because of the ease with which their carcasses are buried and protected from decomposers and scavengers. However, animals that die in water can be transported long distances, and in this way different species that never co-existed in life can end up in the same deposits. More recent subfossil remains are also found in caves, in sand-dunes, swamps, bogs and localised tar pits and at archaeological sites where birds were eaten, and their bones rejected by people.

The fossil record is geographically and temporally very uneven, and some identifications are far from certain. At best, a fossil can give us a name with varying levels of precision from order to species, a date and a location. Fossils are usually dated from the strata in which they lie, making sure that no disturbance events might have caused their displacement from one stratum to another. With more recent material, for which more precise dates are often required, the radiocarbon technique has proved especially helpful. This method is based on measuring the ratio of a radioactive isotope of carbon (called C^{14}) to normal carbon (C^{12}), and is most useful for specimens less than 40,000 years old[3].

---

[3]*Throughout recent time, C^{14} has formed in the atmosphere at about the same rate as it has decayed, so that its concentration in air and water has remained roughly constant through time. While an organism is alive, the ratio of C^{14} to C^{12} within it stays the same because carbon is constantly exchanged between the organism and the atmosphere, but when an organism dies, no new carbon is taken in and the C^{14} gradually decays. Hence, provided that the sample has not been contaminated with more recent organic matter, the ratio of C^{14} to C^{12} within the organism provides an estimate of time since death. After 5,700 years, approximately half the C^{14} originally present in the organism has disappeared, and after 38,000 years only about 1% remains. So, over this time span, the technique provides a potentially sound basis for dating specimens, but older ones become progressively more problematic.*

Given the biases resulting from differential preservability, burial and subsequent alteration, together with potential errors in the dating, identification and interpretation of fossils, caution is needed in drawing conclusions. Fossil species are inevitably distinguished primarily on skeletal features. They are in no way equivalent to living bird species, which are distinguished primarily on external features, such as plumage colours. Many closely related living species would be inseparable on skeletal features, especially from the fragmentary material usually available as fossils. The implication is that different specimens of a single type of fossil might in reality represent more than one species. Nonetheless, the past 30 years have seen great advances in the study of fossil birds, in part because of an increase in the number of qualified researchers, the rapid rate of discovery of new specimens, the development of new preparation techniques, and the considerable enlargement of museum collections of comparative skeletal material from modern birds (Olson & Kurochin 1988).

In our present context, fossil finds are important in three main respects. Firstly, they indicate that some extant bird orders and families have existed at least since the early Tertiary Period, at a time when some of today's land-masses were joined together, or in very different relative positions from those they occupy now. This fact has been used to explain some otherwise puzzling disjunct distributions, such as that of the Struthioniformes mentioned earlier. Secondly, it indicates that some bird families, now considered endemic to particular regions, were formerly much more widely distributed. For example, the Accipitrid vultures are now restricted to the Old World and the Cathartid vultures to the New World. The two types are often cited as an example of convergent evolution, of different types of birds evolving independently to fill the same niche in different places. Yet fossil remains reveal that representatives of both families were once found together in both the Old and New Worlds. Similarly, some families now restricted to tropical low latitudes were once also found at higher latitudes (Chapter 5). Whereas the presence of correctly identified fossils can confirm the former presence of a taxon in a particular region, the absence of fossils cannot be taken to imply absence of that taxon. It may be that the crucial bones have still to be unearthed. Moreover, the fact that many bird species are migratory (and perhaps always were) means that the distributional records do not necessarily imply breeding.

Thirdly, recent subfossil finds on various islands around the world have shown that many bird species disappeared soon after human colonisation, and hence presumably under the influence of human impact (Chapter 7). They imply that only a thousand years ago, up to several thousand more bird species occurred on earth than are present now. These various finds have made it necessary to revise many long-established ideas in biogeography, and on the palaeo-histories of birds, casting new light on their present distributions. Some biologists doubt the reliability of some of the fossil identifications. Some specimens, however, are so well preserved or distinctive that identification (at least to family) is certain; others based on damaged or sparse material (sometimes single bones) are clearly much less certain.

Despite the limitations, a great deal of information about the history of life has been obtained from fossils. Only fossils document with certainty the kinds of organisms that occurred at particular times and places in the past. Fossils provide minimum estimates of the ages of different taxa and reveal some of the enormous diversity of prehistoric life, some of the history of lineages, and their expansions and contractions in range. In addition, the fossil record provides the only long-term measures of the rate of evolutionary change that are based on direct evidence. Indirect measures, based on molecular (DNA) divergence among living species (Chapter 2), necessarily ignore untold numbers of extinct species scattered through the evolutionary tree, and their unknown relationships to those alive today. When a series of correctly dated fossils is available for a particular group, it is sometimes possible to see how the lineage evolved, and hence to distinguish ancient from derived features. This in turn can help in distinguishing the most primitive among the surviving species.

Although geological processes preserve the record of life, they do not do so indefinitely, for they also destroy that record. Volcanoes can melt sedimentary rock and obliterate the record within it. Ocean floors, formed over eons, eventually get sucked into the earth's interior and melted for recycling. The longer a fossil-bearing stratum is exposed to such risks, the more likely is its disappearance. In consequence, the most recent past is much better represented among the fossil record than is the more distant past (Raup 1976).

Recent fossil finds from China and Mongolia have thrown new light on the origin of birds. They have helped to confirm that birds arose directly from bipedal theropod dinosaurs (rather than from thecodont reptiles as previously thought). The theropods shared many derived features with birds, including many aspects of skeletal anatomy, hollow bones, bird-like reproductive features and even feathers, which evolved in these dinosaurs before one lineage gave rise to birds (Prum 2002). Today, we regard feathers as the most important defining avian feature, distinguishing birds from all other animals. Yet ancient theropod fossils clearly show the remains of feathers, varying in different types from simple filaments to complex structures, with rachis, vane and barbs indistinguishable from those of modern birds (Norell *et al.* 2002, Prum 2002). For many millions of years, feathered theropods and feathered flying birds lived concurrently on earth. Even egg-brooding, another distinguishing feature of birds, may have evolved first in theropod dinosaurs. One striking fossil shows an *Oviraptor* lying on top of a terrestrial clutch of eggs in typical brooding posture, with forelimbs held apart from the body, as if covering eggs with its wing feathers. In effect, then, it seems that birds are simply another lineage of dinosaurs, the only one that has survived to the present. From this impressive beginning, birds evolved to become the most species-rich vertebrates on earth, apart from fish. Living types represent the tip of an evolutionary iceberg that may have run to hundreds of thousands of species. Birds have been flying across the skies for well over 100 million years, while humans have been walking the ground below for only around two million, and modern *Homo sapiens* for less than 150,000 years.

SUMMARY

The geographical range of any bird species results from an interaction between the bird itself and its environment. Because species evolve and environments change, ranges are dynamic, in process of continual change. Ranges depend on past events which influenced where each species evolved and reached, and on present conditions which limit current distributions. Unlike many less mobile animals, birds occur commonly on oceanic islands, as well as on continents, and perform migrations on a vast scale, which result in massive seasonal changes in their distributions over the earth's surface.

On the longest time scales, bird distributions have been influenced by plate tectonics, which have altered the positions and configurations of the major land-masses. On somewhat shorter time scales, they have been influenced by long-term climatic and vegetational changes, as during past glaciations, and on more recent time scales, by short-term climatic fluctuations, and especially by human impact.

When correctly identified and dated, fossils can give information on the minimal ages and past distributions of certain types of large birds, and can also be used to document the evolution and extinction of particular lineages. They have revealed that many families of birds were much more widely distributed in the past than they are now, and that many species have disappeared from oceanic islands following human colonisation. They have also shown that birds are likely to have evolved directly from theropod dinosaurs, some of which had many typical avian features, including feathers.

Part One
Evolution and Diversity of Birds

Ostrich *Struthio camelus*, a member of the flightless ratites.

Chapter 2
Bird species and their relationships

Consideration of the origin of bird species may seem far removed from the theme of a book concerned primarily with bird distributions, but it is relevant for two reasons. Firstly, species formation in birds, as we understand it, is very much a geographically based process which mainly occurs through the divergence of geographically isolated populations from a common ancestor; and secondly, the place where a species originates can affect its ultimate distribution. The three chapters that comprise this section are therefore concerned with the delimitation and classification of bird species, with the process of species formation, and with the various factors that influence the numbers and diversity of species present at one time. As this subject area in itself embraces a huge field of research, I can do no more than outline the link between bird distribution patterns on the one hand, and species formation and diversity on the other, in order to provide a basis for some later chapters.

Most of our understanding of the speciation process in birds was worked out in the first half of the twentieth century, mainly by inference from current morphological and distribution patterns, so it has now become a familiar part of evolutionary biology (Mayr 1942, 1963). Nonetheless, great progress on the delimitation of species, and on estimating their relationships and dates of origin, has been made in recent decades from study of the structure of DNA — the molecule that encodes and transmits genetic inheritance. This fast-developing field of molecular genetics is providing further insight into the evolutionary history of birds and other organisms, extending and modifying the findings from traditional approaches.

Everyone has some idea of what is meant by a species, and in all parts of the world people have given different species characteristic names. Each bird species normally looks and sounds somewhat different from all other species, mates only with others of its kind and not with other species, and breeds 'true', producing offspring that look like their parents. In addition, each species is found in only particular parts of the world and is associated with particular habitats and feeding behaviour. To put it in biological terms, each species forms an independent and closed genetic system, reproductively isolated from other species, occupying a distinct geographical range and a specific ecological niche. It also occupies a discrete temporal frame in geological time, running from its formation to its extinction or evolutionary change to another species. Thus defined, species can be considered as the most significant units of biodiversity.

Most problems in the recognition and delimitation of species arise because evolution is a continuing process, so species exist in all stages of formation, vary in their level of distinctiveness, and show exceptions to most of the general features mentioned above. In particular, it is often hard to decide whether two geographically separated but related populations should be considered as two subspecies of the same species, two semispecies or as two allospecies of a superspecies. Subspecies are simply distinctive geographical subdivisions of a species, often called races; semispecies are populations that are almost distinct enough to be regarded as 'good' species, but not quite; and superspecies are arrays of geographically replacing forms (allospecies) clearly derived from the same parental form but too different from one another to be regarded as races of the same species (Chapter 3). It is thus the intermediate stages in speciation that reveal most about the process, but that cause most difficulty in classification. Moreover, some species may look alike but differ in non-visible features that prevent them from interbreeding. We can recognise such species as separate only by more detailed study, taking account of characters other than overall appearance. The important point in a distributional context is that different populations of many closely related kinds of birds, whether classed as subspecies or as species, occupy distinct geographical areas.

To signify the relationships between different species, it is not enough merely to delimit and name them. We also need to arrange them in some system of classification which groups similar types together and keeps more distant ones apart, in a way that reflects their evolutionary history. Thus, similar species are arranged into genera, similar genera into families, similar families into tribes, similar tribes into orders, and so on, to produce a nested hierarchical system that reflects the genealogical relationships among the groups. Ideally, the different taxonomic ranks, from species to order, should reflect the increasing time, youngest to oldest, since their formation (or geological ages). The classification in use at any one time reflects our current knowledge, and modifications are continually being made in light of fresh information. Taxonomy is the practice of classification and naming, and phylogeny (or systematics) is the study of evolutionary relationships.

Together these two sciences create some measure of order in our understanding of an otherwise bewildering variety of life.

DELIMITATION OF SPECIES

Assessing the degree of genetic relatedness among species is not easy. For many years, birds were designated as distinct species solely on the basis of their morphology — on size, anatomical structure or plumage pattern — the very features that could be examined by eye most readily in museum specimens. The conventional definition of a species was thus the 'morphological species concept', which recognised that each species usually looks different from other species, including its closest relatives. All individuals of at least one sex or age group of a species should be distinguishable on one or more characters from the same sex and age group of all other species. Based on the judgement of the classifier, taxa were then arranged, depending on how much they differed, into the hierarchical system mentioned above, from subspecies and species, through genus and family to order, which was supposed to reflect their evolutionary relationships and histories, and hence the branching (cladistic) pattern of their phylogeny.

Convergence

In constructing any phylogeny, one of the main problems with using morphological criteria alone is convergence, when species evolve to resemble one another because they have the same way of life, rather than because they are closely related. This has been a major cause of uncertainty and misclassification in the past. Examples of convergence are common among birds, and include various nectar-feeding species which in different parts of the world represent four different families whose members look alike, namely the hummingbirds (Trochilidae) in the New World, the sunbirds (Nectariniidae) in the Old World, the honeyeaters (Meliphagidae) in Australasia and the honeycreepers (Fringillidae) in the Hawaiian Islands (**Figure 2.1**). Despite their different ancestries, all these species look similar to one another, because they exploit the same type of food in the same way. Other examples of convergence include New World vultures (Cathartinae) and Old World vultures (Accipitrinae), northern hemisphere auks (Alcinae) and southern hemisphere penguins (Spheniscidae), New World warblers (Parulini) and Old World warblers (Sylviidae), and New World flycatchers (Tyrannidae) and Old World flycatchers (Muscicapidae). It is as though evolution has offered similar specific 'niches' (or job vacancies) in different parts of the world, and whatever appropriate stock was available locally evolved to fill them.

Some remarkable examples of convergence exist among particular pairs of species that, although different in ancestry, look and behave alike. For example, the Little Auk *Alle alle* of high northern latitudes and the Magellanic Diving Petrel *Pelecanoides magellani* of high southern latitudes belong to different orders of birds, but they occupy equivalent niches in opposite hemispheres (Warham 1990). They have almost identical proportions, plumages, flight and feeding behaviour; both have throat pouches for storing food and both lay single white eggs in burrows or rock piles. Both have rapid whirring flight on short stubby wings, which is

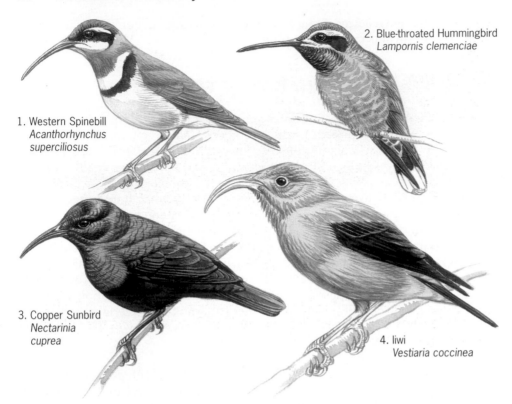

Figure 2.1   Four species of nectar-feeding birds representing four different families. In geographically separate regions convergent evolution has produced, from different ancestral stock, birds of similar morphology (with similar structure of bill and tongue) adapted for feeding on nectar from long tubular flowers. 1. Honeyeater, Meliphagidae; 2. Hummingbird, Trochilidae; 3. Sunbird, Nectariniidae; 4. Honeycreeper, Fringillidae.

unusual for a petrel, but ideal for underwater pursuit of prey **(Figure 2.2)**. Other similarities extend to many features of the skeleton, such as rather flattened wing bones, free finger joints allowing rotation and extension for underwater propulsion, rearwardly placed feet, the same number and arrangement of vertebrae, as well as similar ratios of weight to wing area. In both species, the sterna and relatively long ribs with the elongated rib-cage are seen as adaptations for diving and underwater movement. Both species also become flightless during moult, as they shed all their flight and tail feathers simultaneously. Were it not for their tubular nostrils, which mark them as petrels (Procellariiformes), the Magellanic and other Diving Petrels would probably have been classed along with the auks (Charadriiformes). If one includes extinct birds along with living ones, many more examples of convergence are apparent, as different types of birds have evolved the same designs to fill the same niches at different periods in geological time **(Box 2.1)**.

Figure 2.2   Convergent evolution as illustrated by the Magellanic Diving Petrel *Pelecanoides magellani* (left) of the Southern Ocean and the Little Auk *Alle alle* (right) of the North Atlantic.

Box 2.1   Examples of convergent evolution, from comparisons of extinct birds with living ones.

If we include extinct birds, other impressive examples of convergence come to light. For example, among raptorial birds, the ancient teratorns derived from pelecaniform birds existed alongside the three other lineages of raptors that persist today, namely the Falconidae, Accipitridae and Cathartinae, the latter derived from storks. Similarly, in the late Oligocene and early Miocene, some giant penguin-like birds occupied the northern Pacific at the same time as true giant penguins occupied the southern seas. These northern forms belonged to an extinct family, the Plotopteridae, placed within the Order Pelecaniformes. They were thus unrelated to penguins, but similar by convergence. They were some of the largest flightless aquatic birds yet discovered but, like the auks and penguins, they used their wings as underwater paddles. There were several species with body lengths up to 2 m (Olson & Hasegawa 1979). Another example of convergence involving extinct aquatic birds includes the Cretaceous bird *Hesperornis*, which had an almost identical body shape to the modern loons *Gavia*, adapted for underwater travel.

Convergence can also occur at the level of entire communities of many species, resulting in geographically separated assemblages consisting of similar-looking species, with similar ecology. Many Australian birds, for example, have their equivalents in Eurasia and North America, so that all three continents have look-alike species occupying equivalent niches, but with totally different ancestries. Again, the same characteristics have evolved in different species independently, presumably as a result of selection from using the same types of habitat and food-sources in different places. Such precise morphological and behavioural resemblance between unrelated species implies that certain niches are discrete and predictable, and can be exploited most efficiently by birds with specific characteristics. In fact, however, close species-to-species convergence is unusual among birds as a whole, and the majority of species have no close equivalents elsewhere. This in turn implies that most types of resource are subdivided differently on different continents, a consequence of different continental avifaunas having evolved largely independently of one another (Chapter 5). Almost always, some potential avian niches seem to be unfilled on one continent or another, as exemplified by the lack of nectar-feeding birds in Europe and the lack of woodpeckers in Australia.

The basis of decision making

A second problem with traditional taxonomy is the subjective nature of decisions on similarities and differences, leading to the old adage that a species is a species when a competent taxonomist says so. The lack of an objective system of decision making meant that even supposedly competent taxonomists sometimes could not agree on the delimitation of species. It also meant that, because the birds of different regions were often described and classified by different taxonomists, inconsistencies arose in the way that birds were treated from one part of the world to another. In addition, traditional taxonomy used only some of the more obvious criteria of classification, and ignored others.

Later, with the development of field ornithology, other features of behaviour and voice began to be used in the separation of closely related bird species. This not only clarified some previously puzzling relationships, but also led some apparently single species to be split into two or more separate ones. Thus, the Willow Warbler *Phylloscopus trochilus* and Eurasian Chiffchaff *P. collybita* were originally classed together as one species because in the museum they look the same, but field-recognition of their strikingly different songs immediately marked them as separate species which did not interbreed. In more recent years, greater study (including the use of songs and calls) in other *Phylloscopus* warblers has increased the numbers of Eurasian species recognised by more than 25% over those delimited by morphology alone (Price 1996, Irwin *et al*. 2001a). These warblers provide examples of 'cryptic' species: taxa that are barely distinguishable on appearance, but differ in behaviour and other ways, and do not interbreed. Similarly, the Brazilian Pygmy Owl *Glaucidium minutissimum* was on morphological grounds considered as a single species with numerous subspecies, but study of vocal characters revealed that four subspecies would be more appropriately ranked as separate species, each occupying a different region (Howell & Robbins 1995). The use of tapes and sonograms of bird voices has been a helpful development because vocalisations are clearly used by the birds themselves in mate recognition. Typically, individuals respond to recorded calls of their own species, but ignore those of other species. This is true even though some aspects of bird vocalisations, at least in some species, are culturally influenced, copied from other individuals, like human dialects (Thorpe 1961, Wright & Wilkinson 2001).

However many characters of this type are used to delimit species, decision making is still partly subjective, which could lead to incorrect groupings of unrelated birds. Gradually, therefore, taxonomists turned to alternative, more quantitative measures of affinity: based on numerical taxonomy (proportion of characters shared), or on biochemical gene products, such as egg-white and tissue proteins, polymorphic enzyme systems (allozymes) and so on, aimed to reduce the subjectivity in decisions. More recently, however, the development of molecular biology has provided other ways to examine the relationships between taxa, by measuring the degree of difference in DNA structure between them. Such methods enable quantification of the actual genetic differences between populations, and hence are assumed to be immune to the effects of convergence. If this assumption is valid, DNA methods have the potential to provide more objective measures than any previous method of the relationships between populations. In the last 20 years, the findings from DNA studies have led to modification of earlier classifications, and provided further clarification in the relationships between some 'difficult' species. One such difficult species on New Guinea was MacGregor's Bird-of-Paradise *Macgregoria pulchra* which, on analysis of its DNA, seemed not to be a bird-of-paradise but a honeyeater, and yet another example of morphological convergence (Cracraft & Feinstein 2000)[1].

Phylogenetic research is now at a transitional stage: with some systematists working primarily with morphological and other obvious features, others with biochemical 'molecular' methods and yet others with both. In cases of dispute, genetic methods seem likely to prevail in the long run, if only because they give more objective and measurable differences between species, and are assumed to circumvent the problem of convergence. The introduction of molecular methods has thus revitalised taxonomic research, although such methods are not without their own problems. Also, they have not altogether removed subjectivity from decision making, because it is still a matter of judgement on how different two populations have to be before they are considered as species, rather than as sub-species. Particularly in recently evolved sister species, genetic and phenotypic (or morphological) differences do not correlate well, as species can differ much more in one respect than in the other (see later). Also, while convergence seems less likely to occur in the DNA than in body form, DNA has been insufficiently studied in this respect. So despite the flood of new information, we still have no generally acceptable set of criteria to define a species (Chapter 4). Nor do we have a consistent methodology for decision making. This is not as bad as it sounds, however, because most species are usually so distinct from one another in several respects that no one argues over the label.

---

[1]*The alternative explanation was divergence in the DNA of* Macgregoria *away from bird-of-paradise towards honeyeater, at least in the small piece that was analysed. The larger the segment used, the less likely this becomes.*

PHYLOGENETIC CLASSIFICATIONS

Phylogenies are based on the principle of common ancestry and, because evolution proceeds mainly by a branching process, as one species splits into two or more, and these species in turn split (see later), the supposed evolutionary history of a group of organisms is often depicted as a tree. The apparent primitive ancestral form can be placed at the base, forming the trunk of the tree, while the main groups derived from it are represented by the main branches, and so on through to the twigs that represent the individual species. To pursue the analogy further, the height of the phylogenetic tree is related to time, but the tree has a ragged top, the varying lengths of the branches reflecting evolution that has occurred at different rates. Each branch node records a divergence event when one species splits into two. The horizontal spread of the branches represents the diversity of the group concerned, and the topmost twigs are the individual species living today. The branches below the living twigs at the top are dead; they represent the ancestors that, except for fossil remains, are unknown to us. The majority of the dead branches are broken off below the top; they are the extinct lineages.

The aim of any phylogeny is to reconstruct the branching pattern of the tree and, if possible, to date each branching event. If we could do these things for all organisms, we would have reconstructed the phylogeny of life on earth. If we could do it for birds alone, we would have reconstructed the relatively small cluster of branches and twigs that represents the phylogeny of the 10,000 or so living bird species. The resulting 'cladogram' would depict the sequence of past speciation events through which the common ancestor of all birds gave rise to all its descendent species alive today.

Any proposed phylogeny is only as valid as the information and assumptions on which it is based. New information is continually being added, but there is usually scope for debate, particularly over alternative structures within the overall hierarchical system, the arrangement of groups within groups. The classification adopted at any one time dictates the order in which different kinds of birds are listed in books, or arranged in museum drawers. Ideally again, species should be listed in such a way that closely related forms are put close together, and unrelated ones far apart. As knowledge and fashion have changed over the years, different textbooks and checklists have arranged the same bird species in markedly different sequences. One gets used to one system and then it is changed. However, it will always be difficult to arrange what is essentially a multi-dimensional tree-canopy into a one-dimensional list without discarding a lot of information, and it is clearly important to take account of new information as it emerges.

The use of DNA in the construction of phylogenies

Phylogenetic relationships among birds are increasingly studied through analyses of DNA which, as stated above, can provide measures of the 'genetic distances' between species. The DNA molecule consists of a double helix, composed of two long and complementary chains. Each chain is made up of four types of small nucleotide bases whose sequence within the chain carries all the genetic information transmitted from parents to offspring. For analysis, the DNA is usually extracted from blood, but can be obtained from most other tissues too. One of the earliest techniques applied is that of DNA–DNA hybridisation. This method takes advantage of the double-strand structure of the DNA molecule. Chemical bonds normally hold these strands together, but high temperature causes the bonds to break, the strands to separate, and the DNA to 'melt'. After separation, two strands from the same genome will subsequently re-bind together perfectly on cooling, because all their component base pairs are complementary, but strands from different genomes will bind less well because not all their base pairs will fit together. It is possible to adjust the conditions in a test tube so that single strands are produced by dissociation of the DNA from two separate individuals. Relaxing the conditions allows the strands to re-associate and form a hybrid molecule. If the individuals are taxonomically divergent (say from different families), the strands will be sufficiently different for alignment to be poor and the melting point to be lower than for species more closely related (say from the same family). Thus, by analysing the melting points of hybrid DNA from a wide diversity of species, it is possible to estimate their similarity levels based upon the structure and composition of their DNA. In essence, DNA–DNA hybridisation assesses the average percentage difference in base-pair complementarity between pairs of taxa, and thus gives a measure of the overall genetic distance between them. As it turns out, a melting point reduced by 1°C is equivalent to a 1% difference in the DNA of the two species compared.

The assumptions are that, the more alike genetically two species are, the more closely related they are, and hence, the more recently in geological time they shared a common ancestor. If one accepts these assumptions, the big advantage of such a procedure is that the classification resulting from it is totally objective: there is no subjective guess as to the relative importance of different characters, and classification can be based entirely on a single measurable parameter reflecting genetic relationship. Moreover, this measurable parameter is based on the entire genome.

The method of DNA–DNA hybridisation is most useful in examining the relationships between higher taxa such as orders and families. The first attempt to apply it on a large scale to any group of organisms was by Sibley & Ahlquist (1990), who produced the first coarse-grained DNA-based classification of modern birds into orders, families and tribes, with dendrograms showing relationships between them. These data were in turn used to produce a revised ordering of the world's birds based upon the gross similarity of their DNA (Sibley & Monroe 1990). The method proved much less useful for lower taxa, such as genera and species. This was because the DNA of closely allied species is so similar that any differences that emerge from hybridisation of their DNA approach the level of experimental error. In particular, the standard errors on the DNA–DNA hybridisation distances tend to be large relative to the actual distance measurements. For the lower taxonomic levels, therefore, methods of greater resolving power were needed.

Initial research with DNA thus provided a new and independent hypothesis about bird relationships. To a large degree it confirmed earlier phylogenies based on mainly morphological criteria. More than 75% of the Sibley–Ahlquist data concur with phylogenies derived by other methods. This in itself provides strong reassurance that the DNA–DNA hybridisation method defines natural groups. However, findings from DNA–DNA hybridisation also resulted in some substantial discrepancies with earlier phylogenies. The most surprising, and still controversial, finding was the large number of apparently disparate families that, on DNA–DNA hybridisation data, seemed to fall within the single order Ciconiiformes, including the storks, shorebirds, raptors, grebes, shags, herons, penguins and many others. Clearly, further investigation of these and other suspect findings was required before they could be settled one way or another.

The technique proved important in identifying the closest relatives of previously puzzling groups, such as the New Zealand Wrens (Acanthisittidae) and the Hawaiian Honeycreepers (Drepanidini). It also helped to reveal similarities due to convergence rather than to common ancestry. For example, Australian songbirds were once classified with their Eurasian equivalents which they resembled but, because their DNA revealed that they have a quite different ancestry from any Eurasian species, they are now placed in separate families. They have evidently gone through an evolutionary radiation in Australia analogous to that of the marsupial mammals there (Chapter 5).

This early DNA research gave several other pointers, including the close relationship of starlings to mockingbirds, of New World vultures to storks, and of the African Shoebill *Balaeniceps rex* to pelicans. It indicated that grebes and loons, formerly placed close together, are in fact phylogenetically far apart, and that the traditional order Pelecaniformes consisted of four distinct groups of birds. The majority of these relationships had been suggested before, but the evidence for them had been considered insufficient to change a pre-existing classification. Moreover, most of the more substantial changes to earlier classifications suggested by the DNA–DNA hybridisation work have been supported by subsequent studies by different researchers using different molecular methods (see later). The main underlying assumption in all these methods, mentioned earlier, is that large-scale convergence does not occur in the DNA of unrelated species.

Molecular clocks

If measurements of the genetic distance between species could be converted to a time scale, the technique of DNA–DNA hybridisation could be used to construct phylogenetic trees; that is, to separate taxa with a common origin, and to determine the branching pattern of their divergences in relative time. To convert relative time to absolute time, it is necessary to calibrate the DNA–DNA hybridisation measurements against fossil evidence of divergence dates, or against dated geological events (such as the fragmentation of land areas), which can be assumed to have split ancestral taxa into two or more living taxa. The validity of such dating procedures, which in theory could give the ages of different taxa, depends upon (1) the ability of the technique to measure with reasonable accuracy the 'genetic distances' between taxa, and (2) an assumption of the same average rate of DNA evolution in all bird lineages (meaning that the average rate of nucleotide substitution, measured across the whole genome over time, is the same in all lineages). If this were so, the degree of genetic difference between two existing species would reflect the time since they diverged from a common ancestor.

The assumption of an equal rate of DNA change in all lineages is clearly important, and a potential source of error, accepting at the very least that rates are likely to vary between different types of birds, according to their generation times (early-maturing short-lived species get in more generations per unit time than do late-maturing long-lived ones and hence may have more opportunities for mutational changes in DNA to accumulate; for evidence, see Wu & Li 1985, Martin & Palumbi 1993, Nunn & Stanley 1998). Not all parts of the DNA molecule seem to be of equal evolutionary significance. About 90% of the molecule in higher organisms is thought to be neutral (or non-coding), in that changes have no effect on fitness or phenotype in the conditions prevailing, so are neither selected for nor against[2]. The assumption is that nucleotide substitutions (mutations) that are neutral occur through chance at a more or less standard average rate per generation (**Box 2.2**). On this basis, any differences in nucleotide sequences between related species mainly reflects the time since those species diverged from a common ancestor. In contrast, those small parts of the molecule that control phenotypic characters are not neutral, but respond strongly to natural and sexual selection (Chapter 3). It is these parts that cause species to diverge in morphology and other features, and are thus important in speciation. In higher organisms, if the neutral part does indeed comprise around 90% of the whole genome, even in the absence of knowledge on which part this is, it should have most influence on any time estimates that are made.

Box 2.2   Molecular evolutionary clocks.

The idea that molecular structure might change at consistent average rates through evolutionary time first arose in the 1960s, as a result of studies of the sequence of amino acids in proteins. When the sequences were compared in equivalent proteins from different species, it was noted that the numbers of amino acid differences along the protein chain were correlated with the length of time since the lineages had diverged from one another, as judged from fossil evidence. The greater the number of differences, the older were the lineages. It seemed that proteins evolved at constant average rates, and might therefore be used as 'molecular clocks' to date the divergence events in phylogenetic trees. The idea was not that the individual mutations occurred at regular intervals, only that the mean rates, averaged over long time periods, were reasonably constant from one lineage to another. Also, the average mutation rates were found to vary from one protein to another, by up to several hundred-fold.

Because proteins are produced by specific genes, the implications were that equivalent genes from different related species might also evolve at constant

---

[2]*The idea of neutrality in part of the DNA is based largely on assumption rather than on critical evidence. The evidence known to me provides consistency with the idea rather than testing of it (which would anyway be difficult). It echoes an earlier view about the small morphological differences between closely related species. As they were studied by museum taxonomists, they were thought to result from chance and have no function. Only later did field studies reveal their importance in behaviour and ecology, and hence their responsiveness to natural selection. As the late Professor A. J. Cain remarked: 'An animal is not in the best position to demonstrate the function of its characteristics when lying on its back in a museum drawer.' How much more true this may be when the creature is represented only by molecules in a test tube.*

average rates, but again that these rates might differ from one gene to another. These new ideas seemed to cut across the theory of natural selection on which beneficial mutations were favoured and deleterious ones removed.

This led Kimura (1983) to propose the 'neutral theory of molecular evolution', which asserts that 'the great majority of evolutionary changes, at the molecular level, as revealed by the comparative study of protein and DNA sequences, are caused not by Darwinian selection, but by random drift of selectively neutral mutants'. The theory does not deny the role of natural selection in determining the course of evolution, but it assumes that only a minute fraction of DNA mutational changes are adaptive in nature, and hence favoured by natural selection. Others are deleterious and so are selected against. However, the great majority of mutations may be phenotypically silent nucleotide substitutions that, in the environmental conditions prevailing, exert no significant influence on survival or reproduction. Such neutral mutations could penetrate the population, or be lost from it, by random drift; but they may become selected for or against, if change in the environment so dictated.

One line of evidence in favour of the neutral theory is the well-established fact that less than 10% of the DNA in the genome of living organisms codes for a detectable protein product. The other 90% or more may store variability that Sibley & Ahlquist (1990) called 'money in the bank', for use in future if environmental conditions change. In other words, neutral mutations have no immediate phenotypic effect, and remain silent until activated and selected by environmental change. In this way, natural selection has produced a mechanism that promotes the survival of lineages that have the ability to incorporate high levels of genetic variability, only part of which is obviously functional at one time. On this basis, the neutral mutation system, otherwise hard to explain, is like any other adaptive product of natural selection. The neutral theory also explained a previously puzzling phenomenon found in many different organisms, namely the high incidence of molecular polymorphisms at a large number of genetic loci that were accompanied by no obvious phenotypic effects and no obvious correlation with environmental conditions.

The rate of accession of neutral mutations determines the rate of molecular evolution. It is dependent on generation time (or age of first breeding), or more likely on the number of DNA replications (cell cycles) per unit time in the germ line. If the rate of genetic evolution, whether in the whole genome or in particular genes, is to be converted to an absolute time scale, it must be calibrated against the known dates of geological events which would have split populations, or against the divergence of dated fossils (see text). Some of the first attempts to estimate by molecular methods the timing of branching events in phylogenetic trees used the whole nuclear genome to provide a measure of the genetic distances between species, as in the Sibley–Ahlquist phylogeny of birds. Later attempts used the nucleotide sequences of particular genes or other parts of the genome.

Any molecular clock must be calibrated on independent evidence, based on dated fossils or geological events. Sibley & Ahlquist used dated geological events, mainly the break-up of Gondwanaland (Chapter 1), to calibrate the clock, measuring by DNA–DNA hybridisation the degree of genetic difference between lineages assumed to have split at that time.[3] On this basis, the estimated divergence dates of the different bird orders, based on DNA differences, was largely consistent with the fossil evidence, such as it is, with most modern orders estimated to have arisen in the Cretaceous Period. In other words, the DNA method gave dates that pre-dated or coincided with the earliest relevant fossils but did not post-date them (but see Chapter 4). At present, it is hard to predict how future work will modify the concept of a molecular clock, or alter the estimates of divergence dates already made. Clearly, however, any method that allows us to transform units of genetic distance into units of time could be of great significance in reconstructing the history and relationships of birds and other organisms.

Nucleotide sequence analysis

Soon after publication of the Sibley–Ahlquist avian phylogeny, a more sophisticated technology came on stream. Instead of analysing the structure of 'total' DNA by the hybridisation technique, it became possible to identify the nucleotide bases that comprise a strand of DNA itself. It was then feasible to select an equivalent fragment of DNA from a wide range of species, and to determine the sequence of nucleotide bases within the fragment from each species. A comparison of the proportion of bases that differs between species gives an alternative measure of genetic divergence. The longer the fragments analysed, and the greater the number of nucleotide bases, the more reliable the result. Furthermore, careful comparison of the nucleotide sequence in different species allows reconstruction of the order in which the substitutions occurred, and hence the evolutionary history of the group. In some regions of the DNA, this rate of substitution again seems to have been fairly constant through time (see later), so that differences between species could be translated directly into divergence dates. The more distantly related two taxa are, the greater the difference they are likely to show in any particular stretch of their DNA. Finally and perhaps most usefully, the method can be applied not only to higher taxa, such as orders and families, but also to lower taxa, such as species and subspecies. Once an equivalent stretch of DNA has been sequenced for several different taxa, the application of various mathematical algorithms reveals the most likely (parsimonious) phylogenetic relationship between those taxa, from which an evolutionary tree can be drawn. Such trees usually

---

[3]*They used, in particular, the separation of Africa from South America 80 million years ago against the genetic distance between the Ostrich* Struthio camelus *and Rhea* Rhea americana, *both non-flying ratite birds, but which may not in fact be valid sister taxa (Cooper et al. 1992). But on this basis, the overall mean rate of DNA divergence between these two species was estimated at 0.22% per million years. This compared with a similar estimate based on the split between the Old World cuckoos (Centropodidae and Cuculidae) and the New World cuckoos (Crotophagidae, Neomorphidae and Opisthocomidae) of 0.22% per million years (Sibley & Ahlquist 1990), and with an estimate of 0.37% (range 0.34–0.40%) for galliform birds that was calibrated against six dated fossils (Helm-Bychowski & Wilson 1986).*

accord well with conclusions drawn from morphological analyses, but where they do not, they indicate the need for further investigation.

Although the Sibley–Ahlquist phylogeny, when first published, met with substantial criticism, subsequent studies using DNA sequencing have largely confirmed its more controversial findings (Sheldon & Bledsoe 1993, Avise 1994)[4]. Important support came from work by Bleiweiss *et al.* (1994a, b, 1995), who studied selected parts of the sequence using analytical techniques that overcome the problems of assuming a uniform rate of molecular evolution. Their data, for both controversial and non-controversial parts of the classification, produce the same relationships as those inferred by Sibley & Ahlquist. No other previous bird classification has been exposed to such critical examination, but no classification, however derived, can be 'proved correct' in a phylogenetic sense.

As the years pass, we will have more and more comparisons of relationships found by DNA–DNA hybridisation and those produced by direct sequence analysis. Both procedures have strengths and weaknesses, but in combination they offer big improvements in our understanding of the phylogeny of birds and other organisms. The technique of DNA–DNA hybridisation remains useful in the resolution of the older branches because it averages the whole genome. Sequencing methods have proved the best available for measuring genetic differences between species and subspecies. Their reliability depends partly on the length of the DNA fragment analysed, so they are likely to become even more influential when analyses of long nucleotide sequences become routine. This will give greater confidence in the results, and enable older as well as younger branches to be resolved.

The fact that different parts of the DNA molecule have different degrees of importance in evolutionary change may explain the poor correlation (mentioned earlier) between the extents of genetic difference between taxa and the visible (morphological) differences between them. A relatively small genetic change might be associated with a small or a large change in phenotype, as for example when a small DNA change may cause a major change in the size or colour of a species.

---

[4]*They have, for example, supported the views that (1) grebes and loons are genetically very distant from one another, and that the loons cluster with the penguins and petrels (Hedges & Sibley 1994); (2) that the gamebirds (Galliformes) and waterfowl (Anseriformes) are each other's closest living relatives (Mindell et al. 1997), diverging from one another an estimated 90 mya (van Tuinen & Hedges 2001); (3) that the New World quails (Odontophoridae) are a sister clade to the pheasants (Phasianidae), and not part of the same clade (Kornegay et al. 1993); (4) that the New World barbets are closer to toucans than to Old World barbets (Lanyon & Hall 1994); (5) that New World vultures are more closely related to storks than to Old World vultures (Avise et al., 1994); (6) that the different pelecaniform birds are not closely related to one another, as previously thought, but that they consist of three or four distinct groups, with the pelicans closest to the Shoebill* Balaeniceps rex *(Hedges & Sibley 1994, Siegel-Causey 1997a); (7) that the finfoots (Heliornithidae) are monophyletic, with limpkins closely related; (8) that the American Wrentit* Chamaea fasciata *and the Old World* Sylvia *warblers are closely related (Shirihai et al. 2001); and (9) that the endemic passerine groups of Australasia are the results of adaptive radiation within that area, not the products of a series of invasions from Asia (Baverstock et al. 1991, Christidis 1991, Christidis & Schodde 1991). Purported relationships within the Australasian passerines were changed to some extent in light of further work; and some of the relationships proposed by Sibley & Alquist (1990) within the Gruiformes have not withstood the test of sequence analyses (Houde et al. 1997).*

Evidently, among closely related species, morphological characters are not good indices of genome evolution, even though they may be crucial in speciation.

Nucleotide sequence analysis in studies of species and subspecies

The mitochondrial genome has been much used for taxonomic studies at the lower levels of species and subspecies. It consists of about 37 functionally distinct genes arranged in a circle of single-copy DNA composed of about 16,000–17,000 base pairs in birds. The entire nucleotide sequence of mitochondrial (mt) DNA is known for several bird species, and the sequences of particular mitochondrial genes are known for many more. Because mitochondria occur only in the cytoplasm of egg cells, and not in sperm, they are inherited by male and female offspring only from their mothers, and only the female offspring can pass on mt DNA to future generations. The haploid mitochondrial genome is thus passed down the generations for the most part clonally, without the complications of recombination that result from the fusing at fertilisation of nuclear genetic material from male and female. In addition, parts of the mitochondrial genome show high intraspecific polymorphism and are subject to high mutation rates. Mitochondrial DNA therefore yields high resolution of the differences between closely related taxa, likely to have diverged within the last few million years. Its analysis has also revealed so-called cryptic phylogroups — geographical populations that are recognisable as distinct on biochemical but not on morphological evidence (for kiwis *Apteryx*, see A. J. Baker *et al.* 1995).

Usually fewer than 1,000 bases are sequenced. This may seem to be a large number, but it is only about 6% of the total mitochondrial genome, and equivalent to a negligible fraction of the nuclear genome, which consists of an estimated 2,000 million bases in birds. Moreover, different individuals may be carrying different versions (alleles) of the gene under study. For DNA analysis to be meaningful, enough individuals of each species must be examined to be sure whether the range of variation found between populations is significantly greater than that within them.

Studies of mt DNA have helped in the separation of some closely related species pairs. One example concerns the eastern and western populations of Bonelli's Warbler *Phylloscopus bonelli*, which occur in separate regions of the western Palaearctic, and have long been treated as two subspecies. Helbig *et al.* (1995) showed that 8.6% of the bases in a stretch of the mt DNA cytochrome *b* gene differed between the two populations. This was as great as the difference between Bonelli's Warbler *Phylloscopus bonelli* and Wood Warbler *Phylloscopus sibilatrix* which have long been classed as different species, and greater than the differences shown by any other subspecies yet analysed. Variation within the three taxa was substantially less. In conjunction with consistent morphological differences, and their distinctive songs which were known to function in mate choice, the authors concluded that Western and Eastern Bonelli's Warblers would be more appropriately classed as separate species, *P. bonelli* and *P. orientalis*.

By the same procedure, together with song and behavioural differences, the original Eurasian Chiffchaff *Phylloscopus collybita* was split into three separate allospecies, namely *P. brehmii* of Iberia and *P. canariensis* of the Canary Islands, as well as *P. collybita* which occupies most of the rest of western Europe (Helbig *et al.*

1996). This latter species overlaps with a fourth, previously recognised species, the Mountain Chiffchaff *P. sindianus* in Asia. Other examples of species recommended for splitting into two as a result partly of DNA analyses include Meller's Duck *Anas melleri* of Madagascar, formerly classed as a subspecies of the Mallard *A. platyrhynchos* (Young & Rhymer 1998), the Atlas Flycatcher *Ficedula speculigera* of North Africa, formerly classed as a subspecies of the Pied Flycatcher *F. hypoleuca* (Saetre *et al.* 2001), and Barlow's Lark *Certhilauda barlowi* of southwest Africa, formerly classed as a subspecies of the Karoo Lark *C. albescens* (Ryan *et al.* 1998). These are just a few of many possible examples, and further subdivisions of existing taxa are likely in future as molecular and other new approaches are brought to bear.

The underlying problem is again the poor correlation between physical appearance and evolutionary relationship. As in other organisms, one can find in birds examples of species that are closely related yet look very different, and conversely of others that appear almost identical, yet behave as separate species. However, it is important that molecular data are not used on their own to support a change in taxonomic status, but only along with morphological, ecological and other data. This is because genetic divergence between taxa mainly reflects the time since they became separated, and the taxa need not necessarily have diverged morphologically, behaviourally or ecologically in that time (for examples see later). It is these latter changes that lead to speciation, and they result mainly from divergence in selection pressures rather than from neutral mutations (Chapter 3).

Related species, classified on morphological grounds, usually show 6–10% sequence divergence in the mt DNA cytochrome *b* gene, while subspecies of the same species typically show 0.2–2.5% divergence. There is, however, considerable overlap in the range of mitochondrial DNA differences found among recognised (closely related) species, and the range of differences found among subspecies. In several genera, species pairs with mt DNA differences of less than 2% have been documented, as in *Gyps* vultures (Seibold & Helbig 1995), *Anser* geese (Shields & Wilson 1987a), *Anas* ducks (Kessler & Avise 1984) and *Rallus* rails (Avise & Zink 1988). In each of these pairs, taxa are morphologically distinct, breed in separate regions, and would probably be ranked as species on any criterion. In fact, bird taxa diagnosed as species can differ by as little as 0.1% in mt DNA cytochrome *b*, and subspecies by as much as 6.4%, and this span of overlap is likely to increase as more birds are examined (Sangster 2000). Such overlap should not surprise us because, as mentioned above, DNA distances measure accumulated mutations (reflecting time since split) and not taxonomic divergence. Once again, we see a poor relationship between genetic and phenotypic characters. There can be no fixed degree of genetic divergence that defines a speciation event, especially as the genes being sequenced are unlikely to be the ones that control the reproductive characteristics of species. The main advantage of measuring mt DNA differences is that they infuse taxonomic decisions with a more explicit genealogical perspective.

More molecular clock calibrations

As mentioned earlier, any molecular clock must be calibrated either with respect to dated geological events, such as the splitting of land-masses or the emergence of volcanic islands, or with respect to dated fossils. As an example of the use of geological evidence, the volcanic islands of the Hawaiian chain are arranged sequentially by age, with the oldest surviving islands in the northwest (e.g. Kauai at 5.1 million years) and the youngest in the southeast (e.g. Hawaii at 0.43 million years). On the assumptions that the islands had been correctly aged by geologists (using the potassium–argon (K–Ar) method[5]) and that each island was colonised soon after its formation by birds from the nearest (next youngest) island, the maximum possible ages of endemic bird taxa could be calculated (**Figure 2.3**). On this basis, the mean rate of sequence divergence in the mt DNA of various types of Hawaiian honeycreepers (Drepanidini) was estimated at 2.0% of the total bases per million years for restriction fragments and at 1.6% per million years for the cytochrome *b* gene (Tarr & Fleischer 1993, Fleischer *et al*. 1998).

As an example of the use of fossil evidence, Shields & Wilson (1987a) examined the mt DNA (restriction fragments) from five species of North American geese.

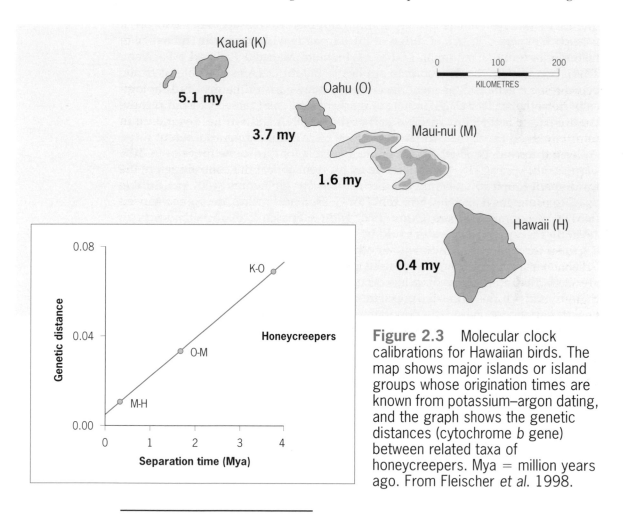

Figure 2.3 Molecular clock calibrations for Hawaiian birds. The map shows major islands or island groups whose origination times are known from potassium–argon dating, and the graph shows the genetic distances (cytochrome *b* gene) between related taxa of honeycreepers. Mya = million years ago. From Fleischer *et al*. 1998.

[5]*The potassium–argon method is based on the knowledge that radioactive potassium ($^{40}K^{19}$) decays to stable calcium ($^{40}Ca^{20}$) and the inert gas argon ($^{40}Ar^{18}$); the half-life of potassium^{40} is 1.31 billion years. Measurement of the decay of rubidium 87 (^{87}Rb) to strontium 86 (^{86}Sr) is used mainly for dating rocks older than 100 million years.*

These birds are well represented in the fossil record which suggests that the *Anser* and *Branta* geese diverged from a common ancestor 4–5 million years ago, fossils before this date being inseparable. The average extent of nucleotide sequence divergence between various *Anser* and *Branta* species was calculated as 9%. On this basis, the mean rate of sequence divergence in goose mt DNA was estimated at roughly 2% per million years (range 1.8–2.3%). Estimates for other types of birds are of similar order (**Table 2.1**), so a mean rate of 2% mt DNA sequence divergence per million years (i.e. 1% per lineage) is often taken as a working figure to apply to other species pairs in which no independent calibration has been made (e.g. Bermingham *et al*. 1992, Zink 1996a, Klicka & Zink 1997, Avise *et al*.1998).

With so few calibrations yet available for birds, the 2% estimate may prove not to be generally applicable. In particular, birds with longer generation times would be expected to show slower rates of nucleotide change, assuming a constant mean rate of mutational change per generation. In other words, the molecular clock would be expected to tick at different rates in birds with markedly different life histories, as found in mammals (**Table 2.1**, footnote; Martin & Palumbi 1993, Avise 2000). In addition, the 2% estimate applies to the mt DNA as a whole or to the cytochrome *b* gene and, as mentioned above, some parts of the mt DNA, as well as parts of the nuclear DNA, mutate at markedly different rates. For example, the control region is the most variable part of the mt DNA and evolves several times more rapidly than the rest: different parts of the control region at different rates. At least three independent estimates are available for birds: one for Snow Geese *Chen caerulescens* at 21% per million years for domain I of the control region, 9% for domain II and 15% overall; another for *Cepphus* guillemots at 2% per million years for domains II and III, and a third for Geospizine Finches (domains unspecified) of 5% per million years (Quinn 1992, Kidd & Friesen 1998a, Freeland & Boag 1999). In theory, mutation rates could be calculated for any part of the genome of a species through comparison with another part for which the rate was known.

Whatever parts of the genome are used to calculate divergence dates, assumptions are made that no interbreeding has occurred between the two populations since their divergence, that no severe population bottlenecks have occurred to reduce the genetic variance of one or both populations, and that both have accumulated mutational differences in a clock-like manner. This is a lot to ask, and deviations from these assumptions would be expected to influence the estimated divergence dates.

Studies on extinct birds

Modern genetic techniques can be used to extract and examine the DNA of individuals that have been dead for many years, and even of species that have been extinct for hundreds or thousands of years, if the material is well preserved. In this way, analysis of ancient DNA extracted and amplified[6] from the bones of

---

[6]*In this context, amplify means making more copies of the DNA to give a larger amount to work with. In vitro amplification is done using the polymerase chain reaction which, since its discovery, has formed the basis of nearly all ancient-DNA research because it permits a designated sequence to be amplified from a small number of damaged templates amidst a background of non-specific DNA. Because ancient DNA is invariably damaged, studies have generally been restricted to sequences of less than 500 base pairs. Preserved bone has often proved to be a better source of DNA than surrounding tissue.*

Table 2.1 Calibrations of molecular clocks, as calculated for the mt DNA of birds. The figures show the estimated rate of nucleotide sequence divergence per million years, from calibration of genetic differences between species against independently dated events.

Divergence between	DNA material used	Mean rate of sequence divergence per million years	Calibration source	Source
Anser and Branta geese	mt DNA (restriction fragments)	2%	Age of fossils	Shields & Wilson 1987a
Hawaiian honeycreepers	mt DNA (cytochrome b gene)	1.6%	Age of volcanic islands	Fleischer et al. 1998
	mt DNA (restriction fragments)	2.0%	Age of volcanic islands	Tarr & Fleischer 1993
Phoebastria and Diomedia albatrosses	mt DNA (cytochrome b)	1.58%	Age of fossils	Nunn et al. 1996
Thallassarche and Phoebastria albatrosses	mt DNA (cytochrome b)	2.86%	Age of fossils	Nunn et al. 1996
Balearica and Grus cranes	mt DNA (cytochrome b)	0.7%	Age of fossils	Krajewski & King 1996

Because so few independent calibrations of molecular clocks are available for birds, it is helpful to look at the figures found for other vertebrates. Among mammals, estimates clearly vary with body size and associated generation times. For mt DNA, divergence rates of 3.8–11.3% per million years have been estimated for small rodents, of around 2.1% for larger species (primates, bears, wolves and horses), and of 1.0–0.4% for whales (Avise 2000). Compared to these rates, the bird rates seem low, especially the 2% figure taken for passerines. Estimated rates are much lower for cold-blooded amphibia and reptiles (Martin & Palumbi 1993, Avise 2000). These animals have markedly slower metabolic rates than birds and mammals, and some larger reptiles also have longer generation times. These various data have given rise to the views that nucleotide substitutions are negatively correlated with body size and generation time and positively correlated with metabolic rate, all of which are themselves intercorrelated. They emphasise the potential for error in rate estimates that are based on very few reliable calibrations from fossil or geological evidence. Occasionally, pronounced rate differences have been suggested for closely related lineages.

extinct birds has revealed their relationships to living species. For example, DNA from various extinct moas, which were found on New Zealand, was compared with DNA from various living ratites found in other parts of the world (Cooper *et al.* 1992, Haddrath & Baker 2001). It was found that the kiwis are more closely allied to the modern ratites (emus, ostriches and cassowaries), and that the moas represented a distinct earlier lineage, basal to other ratites. On the basis of a molecular clock, the moa lineage emerged roughly at the time when New Zealand became separated from Gondwanaland, around 80 million years ago, but the kiwi lineage may have evolved more recently, possibly from a flying form; an alternative view that both moas and kiwis diverged from a common ancestor on New Zealand gained no support from the DNA. Possibly other ratites once occurred on New Zealand, but became extinct, leaving only the moas and kiwis, and now only the kiwis.

Similarly, DNA from the bones of extinct moa-nalos (*Thambetochen, Ptaiochen, Chelychelynechen*) revealed that these large flightless waterfowl, which disappeared from the Hawaiian Islands after human settlement, were more closely related to dabbling ducks than to geese. They represent an ancient lineage that colonised the Hawaiian Islands and became flightless long before the emergence of the youngest island, Hawaii, about 0.43 million years ago (Sorenson *et al.* 1999). In the absence of mammals, they evolved to become important terrestrial herbivores, feeding on leafy vegetation in a wide range of habitats.

Subfossil material has provided further calibration of the rates of molecular evolution. It is useful only for parts of the DNA which evolve extremely rapidly, within the time frame of thousands of years. The method has been applied to well preserved bones of Adélie Penguins *Pygoscelis adeliae* found beneath Antarctic colonies, and dating back more than 7,000 years (Lambert *et al.* 2002). Comparison of DNA from these ancient bones with that from the blood of living Adélie Penguins indicated that part of the mt DNA (hypervariable control region I) evolved at a rate equivalent to 96% change per million years. This figure from a single species is about 2–7 times greater than the phylogenetic estimates obtained from equivalent parts of the DNA from different species (as above); at the time of writing, it stands as an outlier, but it is based on the changes over a relatively very short time period.

The fact that DNA can be extracted from the tissues of extinct animals means that we must now distinguish between biological extinction, the death of the species as we know it, and molecular extinction, the point at which the genetic code (or blueprint) for that species is also lost from preserved and subfossil material. Experience with the moas and others shows that parts of the genetic code can last unchanged for at least a few hundred years beyond death, and in organisms frozen in ice this period could extend to thousands of years (more than 60 thousand years in mammals, Barnes *et al.* 2002). While the genetic code persists, some geneticists believe that one day advances in technology will enable us to use this blueprint to resurrect the whole animal again, thus rescuing long extinct species from archaeological and palaeontological remains residing in museums. If this seems far-fetched, perhaps it is, but remember that most of what has been achieved with DNA since 1970 was previously unimaginable.

Review of molecular methods

Molecular systematics initially used relatively simple assessments of genetic similarity, such as electrophoresis which detects differences in the products of specific genes, such as allozymes[7], or DNA–DNA hybridisation which provides a single quantitative measure of the genetic distance between two organisms. Nowadays, with further advances, most studies are based on comparisons of nucleotide sequences. Since different genes are now known to evolve (mutate) at different rates, one can select a gene likely to give the desired degree of resolution: a fast-evolving gene for working out relationships among recently derived, closely related species, and a slower-evolving gene for resolving the more basal branches of more distantly related lineages. The nuclear DNA fibrinogen intron 4 gene is good for dating separations that occurred up to 40 million years ago; the mt DNA cytochrome *b* gene is useful for separations that occurred within the past 4–5 million years; and the even faster-evolving mt DNA control region sequences or nuclear microsatellite[8] sequences are useful for dating separations that occurred in the past 10,000 years, since the end of the last glaciation. It would be unwise to use fast-evolving genes to date more ancient splits, because on long time scales more than one mutation might have occurred at the same nucleotide site, thus underestimating the number of changes that have occurred since divergence, and the resulting time span involved. Not only is the technology for sequencing genes improving all the time, so are the computer algorithms and statistical methods for constructing likely cladograms (phylogenetic trees) from molecular data.

In recent years, the impact of biochemical approaches on systematics and biogeography has been profound, partly because such characters are thought to evolve in an essentially time-dependent manner, as a result of stochastic processes. They circumvent the problems associated with morphological characters, such as variable rates of evolution and convergence, all of which confound phylogenetic inference. In biochemical measures, the expected magnitude of change in sister lineages should be the same, depending entirely on the time since independent evolution began and on rate of change in the characters involved. Sister taxa should have more character states in common than should non-sister taxa.

---

[7]*The genetic structure of bird populations can be examined through studies of the DNA itself or of the specific chemicals (allozymes) that DNA produces and that exist in more than one form, depending on the structure of the DNA itself.*

[8]*Microsatellites are short regions of nuclear DNA in which a motif usually of only 2–5 bases is repeated, head to tail, a number of times. These regions often gain or lose repeat motifs, a process that causes extensive length polymorphism. The advantages of microsatellite polymorphism over other genetic markers are that (1) the levels of polymorphism are high, and for the most part selectively neutral, and (2) they are abundant in the genomes of all higher organisms. They can be used to distinguish individuals in the same population, to determine gender, parentage and relatedness, the genetic structure of populations, and for comparisons between species. Moreover, they are locus specific and inherited in Mendelian fashion. They show some of the fastest mutation rates recorded. Minisatellites differ from microsatellites in being larger with more tandem repeats.*

All methods have limitations, however, and a molecular-based cladogram should be considered as no more than a hypothesis about relationships, heavily dependent on the assumptions, data and analytical methods used to construct it. Most cladistic methods assume a branching system of diversification, and make no allowance for possible reticulate evolution (hybridisation). They usually also have no way of including fossil material in the construction (but see above). Hence, not all speciation events, and none of the extinction events, are likely to be represented in a cladogram, and their omission could considerably distort its structure.

Integration of traditional and molecular methods in addressing biogeographical questions

Sometimes the evidence from new morphological studies, from fossil finds and from DNA analyses come together, in a consistent manner, to elucidate earlier ideas on the origins and biogeographical histories of particular groups (Härlid & Arnason 1999). Take the Passeriformes, for example, which, with about 5,700 species, form 58% of all living bird species. They are divided into two main groups: the largely South American suboscines and the nearly worldwide oscines. Owing to the primitive structure of both the syringeal muscles and the middle ear, the New Zealand Wrens (Acanthisittidae) were definable on morphological grounds as neither oscines nor suboscines (Feduccia & Olson 1982). They were thus recognised as primitive within the order Passeriformes, pre-dating the evolution of the oscines and suboscines. The same conclusion was reached on the basis of DNA–DNA hybridisation (Sibley & Ahlquist 1990).

Because the New Zealand wrens are endemic to New Zealand, it was proposed that Passeriformes originated in the Australasian (or Gondwanan) region (Feduccia & Olson 1982, Sibley & Ahlquist 1990). As it happens, the oldest known putative passerine fossils are also Australian, from the early Eocene (Tertiary) Period, 54 million years ago (Boles 1995), around the time that Australia began to separate from Gondwanaland. In the northern hemisphere, the earliest passerine fossils date from the Oligocene Period (28 million years ago), although small non-passerine fossils were abundant well before then. During the mid Tertiary Period, the northern radiation of passerines appears to have been extremely rapid, and passerine fossils in sediments from this time exceed the total number of all other fossils (Ballmann 1969). A passerine origin in the Cretaceous Period would support the ideas of a southern origin of Passeriformes, an isolation of suboscine passerines in South America and a mid Tertiary radiation in the northern hemisphere. As yet, however, no passerine fossils have been found from the Cretaceous.

More recent analyses of nuclear DNA largely supported the idea (Barker *et al.* 2002, Ericson *et al.* 2002). They gave findings consistent with the views that: (1) the acanthasittids were isolated when New Zealand separated from Gondwanaland about 82–85 million years ago; (2) suboscines were in turn derived from an ancestral lineage that inhabited at least the western part of Gondwanaland (that gave rise to South America); and (3) the ancestors of the oscines (songbirds) were subsequently isolated by the separation of Australia from Antarctica (the parvorder Corvidi of Australia–New Guinea being basal to all other extant oscine passerines). The later movement of oscine passerines from Australia into the northern hemisphere thus reflects the northward spread of these former Gondwanan lineages (Ericson *et al.* 2002).

Another example of the integration of molecular with traditional methods to resolve phylogenetic and biogeographical problems involves the ratite birds (Struthioniformes). These birds were formerly considered as having evolved earlier than most other living birds, because of their primitive (palaeognath) palate structure, which distinguished them from most other modern birds, with a more recent (neognath) palate structure. However, all modern birds go through a palaeognath stage during their embryonic development, and on the basis of their mitochondrial DNA, the Struthioniformes are now considered to have evolved from neognath birds, and to have secondarily acquired their palate structure, probably by neoteny (the retention of juvenile characters into adult life) (Härlid & Arnason 1999). Analyses of nuclear and mt DNA from living ratites, as well as from the bones of extinct moas, have indicated that all these birds are monophyletic, derived from a single common ancestor, yet are genetically very distinct, which suggests that they separated from one another long ago (Sibley & Ahlquist 1990, Cooper *et al.* 1992, Haddrath & Baker 2001). The Struthioniformes were estimated to have diverged from the Galliformes (game birds) 92 million years ago, the moas from other ratites 79 million years ago (roughly coinciding with the separation of New Zealand from Gondwanaland), the emus and cassowaries from other ratites 64–54 million years ago (before the separation of Australia), and the rheas from the emus 45 million years ago (following the severance of South America). Most of these dates are consistent with the separation of the various southern land areas from Gondwanaland, thus supporting the land fragmentation (vicariance) model of ratite origins. However, the Ostrich–Rhea split at 65 million years ago cannot be reconciled with the separation of Africa from South America about 100–80 million years ago and the kiwi–Australian ratite split at 62 million years ago cannot be reconciled with the separation of New Zealand about 80 million years ago (Haddrath & Baker 2001). The ancestors of these species may therefore have dispersed from elsewhere to their present ranges, either as flying or flightless forms, and this stage of their histories remains unresolved. The possibility that the ancestors of the Ostrich may have travelled overland, across the northern hemisphere to reach Africa, is consistent with the presence of an ostrich-like ratite in middle-Eocene Europe (Houde 1987).

As another example, take the parrots, which reach their greatest diversity on the three southern continents derived from Gondwanaland. Analyses of DNA (by hybridisation and sequencing methods) were consistent with the view that these birds also evolved on Gondwanaland, and were carried on the southern continents as they drifted apart, thereafter evolving independently on each continent (Sibley & Ahlquist 1990, Miyaki *et al.* 1998). In other words, as with the ratites, the genetic evidence supports the land fragmentation (vicariance) hypothesis of present distributions, rather than dispersal. On an independently derived molecular clock, Australian and Neotropical parrot lineages were estimated to have diverged around the end of the Cretaceous period, coinciding with the break-up of Gondwanaland (Miyaki *et al.* 1998). Molecular phylogenies have thus provided an additional source of information on current biogeographical patterns and the evolutionary processes that have moulded them (Avise 1994). They support the view that Gondwanaland, and its fragmentation, played a crucial role in the early evolution and radiation of modern birds (Cracraft 2001).

GEOGRAPHICAL VARIATION IN SPECIES

To understand species formation in birds, it will help first to mention the geographical variation that occurs within species. Those species that occur over large geographical areas often show gradual (clinal) variation across the range, supposedly developed in response to gradients in climatic or other conditions (**Table 2.2**). Such clines within species may affect morphological characters, such as body size (larger in colder climates, known as Bergman's rule), coloration (darker in warmer and wetter climates, known as Gloger's rule), the proportion of different colour morphs (greater proportion of pale morphs in colder climates) or wing shape (longer and more pointed at higher latitudes). They may also affect internal features, such as relative heart size (larger at high latitudes), life-history features, such as migration (more prevalent at high latitudes) and clutch size (larger at higher latitudes), aspects of physiology, such as cold tolerance (greater at higher latitudes), and many other features (Huxley 1942). Moreover, different species occurring across the same region often show parallel variation in the same feature, presumably because they are all exposed to similar conditions, thus providing further circumstantial evidence for the adaptive nature of such variation.

While most known character gradients in birds are latitudinal, this is not true of all, and in some species different character gradients may run in different directions. In some North American bird species, body size declines from north to south across the range, while plumage colour weakens from east to west (for shrikes *Lanius*, see Miller 1931; for Fox Sparrow *Passerella iliaca*, see Swarth 1920). In some Australian species, body sizes decrease from south to north (high to low latitudes, as in the northern hemisphere), whereas plumages often become paler from the humid edge to the arid interior (Ford 1974b). Some birds also show altitudinal clines, notably in body size (for wrens, see Chapman & Griscom 1924), while in some desert species plumage colours change across the range to match the substrate.

Clines are presumed to result from local adaptation which is impeded to some extent by gene flow from neighbouring populations so as to give a gradual and continuous trend in certain characters across the range. Only where a part of the population is isolated from the rest, as on an island, so that gene flow is effectively absent, can that subpopulation develop a truly distinctive character (forming a well-marked subspecies). Sometimes on continental areas, however, superimposed on the normal gradual variation are 'steps' of abrupt change in various characters. Such steps have sometimes been put down to an unusually steep change in selection pressures over a short distance, but they also occur in fairly uniform terrain, and more likely mark the place where two subspecies met after differentiating in isolation, giving a narrow zone of intergradation (examples in North America include the various races of Fox Sparrow *Passerella iliaca*, Swarth 1920; and in Europe the different races of Long-tailed Tit *Aegithalos caudatus*, see Chapter 10). Typically, such subspecies are less distinct than those on islands because they intergrade with one another.

Table 2.2   Taxonomic expression of various types of morphological variation in bird species.

Polymorphic	Differentiated into two or more distinct types, such as light and dark colour phases, which are independent of sex differences, but interbreed freely in the same area. Examples: light and dark morphs of Grey Goshawk *Accipiter novaehollandiae* in Australia, Gyr Falcon *Falco rusticolus* and various jaegers *Stercorarius* across the arctic; bridled and unbridled Common Murre *Uria aalge* in the North Atlantic.
Polytypic	Differentiated into subspecies, each in a different region. The subspecies may fall into clines.
Clines	Continuous and gradual variation across the range which may or may not be differentiated into subspecies.
Monotypic	Not differentiated into subspecies or into any other array of well-marked morphological forms.

The importance of clines in our present context is that they reveal the adaptations of bird populations to local conditions. By emphasising gradational changes, and the orderly interconnections of populations, they help to provide a fuller picture of species diversity, and the origin of local adaptations that, given sufficient isolation, can lead to speciation. The subspecies category, although based primarily on morphological features, is useful because it enables us to make geographical subdivisions within species, while at the same time retaining a meaningful species concept. A subspecies is thus defined as a local form that will interbreed with other forms of its own species wherever they come into contact. Subspecies tell us much about the biology and biogeographic history of the species concerned, and about how selection pressures change across continents and between continents and islands. Subspecies are maintained by spatial segregation, each occupying a different region, but blending with others in neighbouring regions.

Clines, subspecies and dispersal

How can we explain the fact that some bird species show greater geographical variation in morphology than others that occur over much the same area? In some small bird species, more than 15 different races have been described, each from a different part of the continental range, whereas some larger birds show little or no geographical variation in size or colour across a continent or even across the whole northern hemisphere (though they may vary in other non-visible characters). Clearly, past or present population fragmentation might have some influence, for anything that reduces gene flow might allow more marked adaptation to local conditions. However, whether distribution is fragmented or continuous, another factor of importance is individual dispersal on which the spread of genes depends. In the majority of birds pair formation occurs on the breeding areas, and the crucial movements include shifts between natal site and subsequent breeding site, and between the breeding sites of different years (Chapter 17).

In some passerine species, individuals disperse over relatively short distances (often less than 10 km), which must limit the spatial extent of gene flow. This in turn could allow local adaptations to develop, and hence local populations to diverge to some extent, giving clines in behaviour, physiology and morphology across the range. Such species typically show marked geographical variation, sometimes with many subspecies (**Table 2.3**). In contrast, some larger birds, such as many herons, can move long distances between hatch sites and breeding sites (sometimes more than 500 km). Such birds show little or no geographical variation in morphology across the range, presumably reflecting a more uniform genetic structure. If genes move long distances with every generation, then the differentiation of local populations is likely to be poor to non-existent (Endler 1977). In fact, the degree of clinal or subspecific differentiation among birds is clearly linked with their dispersal behaviour. Moreover, among related species, migrants typically show less geographical variation in size and colour than non-migrants (**Table 2.3**). These are generalisations, however, and many species provide exceptions (for further discussion, see Chapter 17).

In conclusion, the main lessons drawn from studies of geographical variation in birds include the importance of natural selection in causing local adaptation and the role of population fragmentation in promoting differentiation. Out of earlier research on geographical variation (mainly in birds) arose the concepts of allopatric speciation and the 'biological species', discussed in Chapters 3 and 4.

Geographical variation in DNA structure

The study of geographical variation thus provides a glimpse of evolution at its most basic spatial level. In recent years, the emphasis has shifted from morphology to molecular features in a search for spatial patterns of genetic variation that can be interpreted in evolutionary terms. It has thrown new light on the phylogenetic and biogeographical history of birds, both within and between species. The whole area of research has been termed phylogeography, defined as the study of the geographical distributions of genealogical lineages (Avise 1989, 2000). Both nuclear and mt DNA have been used for this purpose, but mt DNA has proved the most useful, owing to its maternal inheritance and relative immunity to the effects of recombination and hybridisation. It can thus provide an unadulterated

Table 2.3   Geographical variation in different types of Palaearctic birds, as indicated by the proportions of monotypic species and by the numbers of subspecies per polytypic species. From Rensch 1933.

	Per cent monotypic of total number of species	Mean number of subspecies per polytypic species
Large birds[1]	54.5	1.6
Small birds, migratory[2]	39.9	3.2
Small birds, non-migratory[3]	29.6	7.2

[1] Five families — Ardeidae (herons), Ciconiidae (storks), Threskiornithidae (ibises), Otidae (bustards) and Gruidae (cranes), with 44 species.
[2] Nine families or subfamilies — including Laniidae (shrikes), Silviidae (warblers), Turdidae (thrushes), Hirundinidae (swallows), Muscicapidae (flycatchers) and Motacillinae (wagtails), with 288 species.
[3] Six families or sub-families — Corvidae (crows), Certhiidae (treecreepers), Sittidae (nuthatches), Paridae (tits), Troglodytinae (wrens) and Picidae (woodpeckers), with 115 species.

history of the female line. These features also give mt DNA an effective population size only one-fourth that of nuclear DNA, and thus make it more useful than nuclear DNA for revealing past population bottlenecks and founder events, both of which give rise to current populations derived from small numbers of individuals (and hence of limited genetic variance). From studies of spatial patterns of mt DNA variation, one can thus evaluate in the history of species the relative roles of dispersal (gene flow), bottlenecks and past barriers to movements.

In general, the degree of genetic differentiation between two populations of a species increases with the distance that separates them (Peterson 1992, Nesje *et al.* 2000). This is expected because, the further apart two populations lie, the less the interchange of individuals (and resulting gene flow) between them. Little genetic variation in a species across a wide area is taken to imply a past bottleneck or recent expansion from a small area, while lack of geographical structure is taken to indicate high gene flow across the range or again recent expansion from a small area. Most studies have been based on the mt DNA cytochrome *b* gene which is thought to diverge in separated bird populations at a rate of about 2% per million years (see earlier). It is thus hard to detect from this gene very recent separations, such as those occurring within the last 10,000 years.

Some species show no spatial structure in either morphological or genetic features over wide areas. This is true of the Northern Pintail *Anas acuta*, for example, in which individuals show substantial variation in mt DNA but no obvious differentiation between well-separated North American breeding areas (Cronin *et al.* 1996). The implication is that the Pintail has maintained a large population size throughout its history, and a high rate of gene flow between widely separated parts of its range (panmixia). This fits what is known of Pintail behaviour, with birds often nesting at long distances from their natal and former breeding sites, and settling each year wherever local habitat is favourable (Chapter 17). Similar high variation in mt DNA, with lack of geographical structure, has been found in the Short-tailed Shearwater *Puffinus tenuirostris*; however, because this species shows high site fidelity, with the majority of young returning to their natal colonies to breed, lack of geographical structure was attributed to recent rapid range expansion from a smaller area (Austin *et al.* 1994).

In some other species, mt DNA has been found to vary between populations in different parts of the breeding range, as reflected in the relative frequencies of different haplotypes. Thus in the Fox Sparrow *Passerella iliaca*, four types of mt DNA haplotypes have been found, each corresponding to a different morphological subspecies, relatively uniform within itself (*iliaca*, *unalaschensis*, *schistacea* and *megarhyncha*) (Zink 1994). In other species divergence in mt DNA was found, with no equivalent divergence in morphology, as exemplified by eastern and western European populations of the Great Reed Warbler *Acrocephalus arundinaceus* (Bensch & Hasselquist 1999). The interpretation was that eastern and western populations were separated at some time in their distant past, which enabled divergence in their mt DNA, but this was not accompanied by any striking morphological differentiation. The two populations have since come together in morphologically seamless contact. Other examples of such 'cryptic phylogroups' are given in Chapter 10 (and for an Australian example, see Miura & Edwards 2001).

Yet other species have shown little evidence for mt DNA differentiation across the range, despite marked subspeciation. In the Song Sparrow *Melospiza melodia*, which breeds over much of North America, variation in size and colour is extreme, ranging from dark birds weighing 46 g that breed in the Aleutian Islands to 20 g pale birds found at the Salton Sea in southern California. The most recent taxonomic review (American Ornithologists' Union 1957) partitions this variation into 34 subspecies, making the Song Sparrow one of the most polytypic bird species in North America. Interestingly, though, subspecies of Song Sparrows could not be identified from mt DNA (restriction fragment) analyses. The suggested explanation was that, after the post-glacial dispersal of Song Sparrows across the continent, the subsequent evolution of size and plumage coloration, which was under the control of part of the nuclear DNA, proceeded faster than the evolution of mt DNA (Zink & Dittman 1993a).

Other bird species studied across wide regions of North America also show little or no evidence of mt DNA differentiation, including the Northern Flicker *Colaptes auratus* which has 'red-shafted' and 'yellow-shafted' colour types (Moore *et al.* 1991), the Red-winged Blackbird *Agelaius phoeniceus* which has 23 described subspecies (Ball *et al.* 1988), the Downy Woodpecker *Picoides pubescens* which has seven subspecies (Ball & Avise 1992) and the Chipping Sparrow *Spizella passerina* which has three subspecies (Zink & Dittman 1993b). Again, the implication is that change in morphology has evolved more rapidly than change in mt DNA. In such species, genetic structuring may be detected in other parts of the genome that evolve more rapidly than those parts of the mt DNA so far examined.

Another striking disparity between morphological and mt DNA differentiation is provided by the redpolls, among which two largely sympatric species are usually recognised: the dark-rumped *Carduelis flammea* with four well-marked subspecies (*C. f. flammea, cabaret, islandica* and *rostrata*) and the white-rumped *C. hornemanni* with two subspecies (*C. h. hornemanni* and *exilipes*). In these birds, the divergence in mt DNA (restriction fragments) between populations is small and unrelated to their morphology. Seutin *et al.* (1995) estimated the genetic divergence between the two 'species' to be at most 0.25%, very much smaller than the usual divergence between other bird species. This could be due to either very recent morphological divergence in the redpolls or to extensive hybridisation following the meeting of previously isolated populations. Hybridisation between dark-rumped and white-rumped birds has been well documented in many areas (Williamson 1975). In contrast, the west European Lesser Redpoll *C. (f.) cabaret* has recently spread north into the breeding range of *C. f. flammea*, and in the one area where both were studied no hybridisation was found, suggesting that both might warrant specific status (Knox *et al.* 2001). The two forms also differ slightly in voice and behaviour.

The relatively small number of species studied so far have thus given four main types of pattern, each with its own interpretation: (1) no geographical pattern in either mt DNA or in morphology — attributed to recent expansion from a limited area too small to permit geographical structuring or to panmixia across a wide area; (2) geographical structuring in mt DNA matching the geographical variation in morphology, as reflected in clines and subspecies — attributed to long residence in the same range or in a number of core areas within it; (3) geographical structuring apparent in mt DNA but no geographical variation in morphology — attributed to past population fragmentation but no morphological differentiation, followed by subsequent expansion and joining; and (4) no detected geographical structuring in mt DNA, but obvious clinal or subspecific variation in morphology — attributed to recent expansion from a single limited area, followed by rapid morphological adaptation to local conditions (which may subsequently be affected by hybridisation). The interpretations offered of these various patterns are, of course, little more than plausible guesses, and we normally have no way of testing them. They also depend on the validity of the assumptions mentioned earlier. Clearly, however, the combination of new molecular and traditional morphological studies could provide insights into the biogeographical history of populations that are not discernible from morphology alone.

Of particular value are studies that examine different species occupying the same broad range to find whether they show parallel patterns of genetic structure. This might be expected if they had shared a long history together, and had been exposed to the same environmental influences. Zink (1996b) reviewed studies of mt DNA variation in five co-distributed species of widespread North American birds (**Figure 2.4**). Three species, the Song Sparrow *Melospiza melodia*, Chipping Sparrow *Spizella passerina* and Red-winged Blackbird *Agelaius phoeniceus*, showed no geographically structured mt DNA patterns, with little variation in haplotypes. This was consistent with recent population expansion from a single limited area, and thus with the relatively recent establishment of their current distributions. The other two species, the Canada Goose *Branta canadensis* and Fox Sparrow *Passerella iliaca*, each showed marked but different spatial patterns in mt DNA variation. So these two species have evidently had different biogeographical histories from the other three, both involving past isolating events that produced genetic divisions in their populations. These and other studies help to confirm additional indications that any regional avifauna is now composed of a mixture of species with different biogeographical histories, but including some species that may have shared the same histories. It emphasises the past mobility of species, and the continual changes that are likely to have occurred in the species composition of particular regions (Chapters 5, 9–11).

Inference on past bottlenecks

Genetic theory predicts that populations that suffer substantial reduction in size will thereby lose some genetic variance. On this basis, current low levels of genetic variability have often been used to infer past bottlenecks (although alternative explanations are sometimes possible).[9] A clear-cut example was provided for the Greater Prairie Chicken *Tympanuchus cupido* by comparing pre-bottleneck and post-bottleneck populations represented in museum collections. In the State of Illinois the species numbered millions of individuals in the 1860s but, through human activity, it had been reduced to fewer than 50 individuals in 1993 (though

---

[9]*One measure of genetic diversity is the number of alleles (variants of an equivalent gene) in the population. Another measure is the amount of heterozygosity (where an individual has different alleles at the corresponding loci of the two parental chromosomes). Heterozygosity is influenced by the number of different alleles in a population and their relative frequencies; high levels are found in populations with many different alleles, none of which are common. This measure is somewhat less sensitive to sample size than is the number of alleles alone.*

Figure 2.4   Approximate breeding distributions and diagrammatic haplotype phylogenies for five species of North American birds. Not all haplotypes are shown. The two Canada Goose *Branta canadensis* lineages correspond to small-bodied (I) and large-bodied (II) geese, respectively. A haplotype divergence of 2% is thought to correspond to a distributional split about one million years ago. From Zink 1996b.

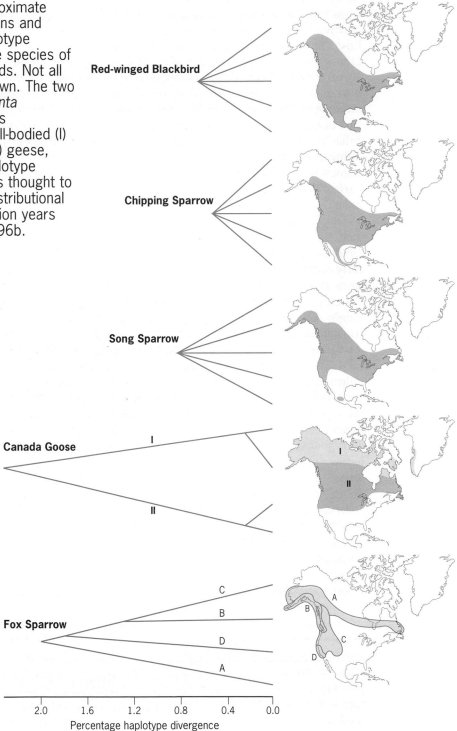

many more birds remained elsewhere). Analysis of nuclear DNA microsatellites from museum specimens from central Illinois revealed the loss of specific alleles (known to have been present earlier this century) following population collapse. Lost alleles included common ones present in all other populations sampled and others unique to the Illinois population (Bouzat *et al*. 1998).

As another example, the numbers of Mauritius Kestrels *Falco punctatus* are thought to have recovered from a single wild pair in 1974 to more than 200 pairs today. An attempt was made to assess the loss of genetic variation resulting from this bottleneck by comparing 12 microsatellite DNA loci in museum specimens collected long before the population crash with those in birds taken from the living population. Ancestral variation was high and more or less as expected in a continental kestrel population occupying a similar sized area, but current variation was much lower: across all loci allelic diversity had fallen by 55% and heterozygosity by 57% (Groombridge *et al*. 2000).

The role of mt DNA in pointing to more ancient population bottlenecks can be illustrated by findings from several New Zealand groups (Cooper & Millener 1993, Cooper & Cooper 1995). Five extinct moa genera, which were morphologically quite diverse, possess a surprisingly limited amount of genetic diversity. The same is true of the three living kiwi species and of the Acanthisittid wrens, one of which is extinct. Evidently some event reduced the diversity of each group to a single mitochondrial lineage, from which there was later radiation. This would conceal any previous diversity, and the three groups would appear to have radiated simultaneously. Reduced tectonic activity in the Palaeocene reduced New Zealand to a broad flat lowland. During the large sea-level changes of the late Oligocene (29–23 million years ago), New Zealand was inundated for six million years, reducing the land area to a string of islands, totalling less than one-fifth of the current land area. This fits the reduction in population size, and hence in genetic diversity, estimated to have occurred then.

In many bird species, reduction in population size most often results from habitat shrinkage and fragmentation, but the colonisation of islands and other areas by a small number of individuals is also likely to give populations with low genetic variance. In the absence of further input to the gene pool, genetic variance is likely to remain low for many generations even though numbers may rise greatly in that time, because genetic variance can be increased only by mutation which is slow. Further examples of species that have apparently passed through demographic bottlenecks include the Red Knot *Calidris canutus* and Ruddy Turnstone *Arenaria interpres* discussed in Chapter 10, and a colony of the Fairy Prion *Pachyptila turtur* discussed in Chapter 17.

*   *   *   *

In conclusion, phylogeographic analyses involving DNA have added greatly to our understanding of the spatial structure of bird populations. This structure is often concordant with differentiation in other characters or with past or present geographical isolation. However, many species that show no geographical structure in parts of their mt DNA still show marked subspecific differentiation, attributed to recent evolution following range expansion from a smaller area. Splits in populations have occurred at various times in the past, giving deep or shallow divergence in mt DNA, followed by subsequent re-contact. Not all past population splits have resulted in morphological divergence, and in several species, past splits (like past bottlenecks) have been identified only from DNA divergence (Chapter 10). So not only has the study of DNA increased the information available on the history of particular bird species, it has enabled approximate dates to be put on significant events in this history, and hence has helped to decipher the factors responsible for current population genetic structure. This aspect is considered further in Chapters 10–11, in relation to effects of glacial and other long-term climatic change on bird populations.

SUMMARY

Species are often defined as interbreeding natural populations that are more or less reproductively isolated from other such populations. Each species differs from other species in geographical range, habitat or diet, or in more than one of these respects. Initially, most species were identified and classified solely on morphological and anatomical features, but now other criteria are also used, including DNA structure.

In recent years changes have been made to the taxonomic assignments of many types of bird largely as a result of analyses of DNA, and, on the basis of molecular clocks, estimates have been made of the ages of various bird taxa, from subspecies to orders. No one could doubt the value of DNA in phylogenetic research, but findings are beset by three crucial, but inadequately tested, assumptions: (1) that convergent evolution does not apply to DNA (or occurs on only a negligible scale); (2) that a large part of the DNA is non-functional (neutral) and does not change through selection; (3) that at least the neutral part of the DNA changes (through mutation) at a constant average rate over time, which is consistent for equivalent parts of the DNA molecule in closely related species. Only further research will reveal the validity or otherwise of these assumptions and of the time estimates that depend on them.

Some species show clinal variation across their geographical range in morphological, physiological or other characters, in adaptation to gradients in local conditions. Such polytypic species can often be subdivided into distinct subspecies, each occupying a different part of the range. Subspecies tend to be most distinct on islands or other areas that limit genetic interchange with other populations of the same species. Studies of geographical patterns in mt DNA, along with similar patterns in morphology, have revealed much about the history of particular species, of the occurrence of genetic bottlenecks in their history, and of the dates at which different subspecies began to diverge.

Blue Goose-Snow Goose *Chen caerulescens* hybridising pair.

Chapter 3
Species formation

For a population to become distinct, as a new species, it must be more or less reproductively isolated from other populations, either by geography or by other means, such as mating preference. Only then can it pursue its own independent evolutionary track, free from substantial genetic input from other populations. New bird species are thought to arise in two main ways, namely by the transformation of one species into another over time (anagenesis) or by the splitting of one species into two or more (cladogenesis). Both processes involve gradual change; but the first process gives a single product of evolutionary change, perhaps through a succession of different species over time, while the second gives an increase in the number of species existing at the same time. A third possible speciation process, involving the hybridisation of two existing species to produce a new species, is probably rare or non-existent in birds (although frequent in some other kinds of organism such as higher plants).

ANAGENESIS

To detect anagenesis, we must usually depend on a good fossil record, sufficient to reveal how a given stock has gradually changed through time, giving a succession of forms (chronospecies) that merit different specific titles. Although in various animals some trends have emerged from the fossil record, we can never be sure that they were unilinear, and that the chronospecies form an unbroken series, rather than products of successive branching events that were followed by rapid extinction of all but one lineage. In general, the fossil record of birds is too poor to reveal such trends, but one can imagine some of the more distinct species on oceanic islands (say) passing through intermediate stages in their evolution that, if still extant, would be regarded as specifically distinct from both the parental and the present-day forms. It is perhaps in this type of confined situation that anagenesis is most likely to occur.

CLADOGENESIS AND ALLOPATRIC SPECIATION

In contrast to anagenesis, the operation of species splitting (cladogenesis) is much easier to detect and very apparent in birds, research on which has played a key role in working out the theory of geographical (allopatric) speciation (beginning with the observations of Darwin (1859) and Wallace (1876), and extended in the twentieth century by the work of Stresemann (1919) and Rensch (1933), followed by Mayr (1942, 1963) and Lack (1947)). As explained in the previous chapter, many bird species show marked geographical variants (often given formal recognition as subspecies), the boundaries between which may be gradual, but often coincide with physical barriers, such as sea channels or mountain ranges which the birds concerned seldom cross. Many species that breed in Britain differ slightly in size and colour from their continental equivalents, and species on archipelagos often differ between islands, either slightly so are classed as subspecies, or more so and sufficiently to be classed as species. Such barriers to dispersal split species into separate breeding units which, being reproductively isolated, can then evolve independently, gradually diverging from one another. This type of speciation can thus be viewed as a process of lineage splitting, leading to differentiation and the formation of two or more non-interbreeding populations (or species) from an original one.

The importance of geographical isolation in the speciation of birds is particularly evident in 'allospecies', defined as closely related species, all clearly and recently derived from the same ancestral form, but each occupying a different area from all others. Many species on archipelagos, or in extensive mainland habitats, occur as allospecies, having arisen by geographical isolation from the same formerly widespread parental species. Any group of allospecies is sometimes referred to collectively as a 'superspecies' to designate its connections and common origin. An example of such a superspecies group is provided by three African buzzards, with Archer's Buzzard *Buteo archeri* occupying the highlands of northern Somalia, the Augur Buzzard *B. augur* extending from Ethiopia to Namibia and Angola and the Jackal Buzzard *B. rufofuscus* in southern Africa.

Populations living in different areas inevitably diverge from one another for several reasons. Firstly, selection for some characteristics might have incidental effects on others, altering the genetic structure of each population imperceptibly from generation to generation, by a process called random genetic drift (random because of its non-directional nature and drift because of its slowness). Secondly, mutations arise independently in each population, but do not spread to the others. Many such mutations may be deleterious and so be selected against, but some may be neutral, while others may improve the survival of individuals that carry them, and so be favoured by selection. Because mutation is also random, different variants may arise in each population, further contributing to their divergence. These processes give accident, as well as adaptation, a place in evolution.

The third process leading to the differentiation of allopatric populations is natural selection, which occurs when some individuals in a population survive or reproduce better than others because they possess traits that enable them to perform better in that particular environment. If those traits have a heritable basis, the genes governing them will be passed on to the next generation, and differential gene transmission leads to further evolutionary change. This process of adaptation by natural selection gradually alters the characteristics of each population through time, and because environmental conditions inevitably differ from region to region, populations in different regions gradually diverge. It is in this way that environmental (ecological) factors have their greatest influence on speciation.

Fourthly, sexual selection could also promote the differentiation of allopatric populations. This could occur if the separated populations diverged in their mate choice or other social behaviour in a way that affected morphology or other mate recognition characters (West-Eberhard 1983). Provided that the females in different areas came to differ in their mate choice, the argument runs, morphological divergence would follow, especially in the males. By this process, divergence in plumage or other features could occur extremely rapidly, especially in polygynous species where a small percentage of favoured males fertilise all the females. Evidence for the power of sexual selection in species multiplication is indirect. It stems from correlations between the numbers of species in families on the one hand and the prevailing mating system (Mitra *et al.* 1996) or degree of sexual dimorphism on the other (Barraclough *et al.* 1995, Møller & Cuervo 1998, Owens *et al.* 1999). The two latter characteristics are assumed to reflect the strength of female choice of mate in birds, and hence the strength of sexual selection (Owens & Hartley 1998). Such selection is held to account for the wide variety of bright colour patterns found among birds, which are often shown to advantage in ritualised display movements. In many species, the females are duller than the males, because of their greater role in incubation and brooding; the resulting greater vulnerability of the females to predation selects for more cryptic plumage colours (Martin & Badyaev 1996).

Any factor or combination of factors that promotes divergence of isolated populations, whether from random genetic processes and mutation or from selection through ecological or social processes, could lead to speciation. But whereas the action of natural and sexual selection has been confirmed repeatedly from studies of various wild bird populations (see later), the evidence for random genetic processes is much less firmly based. Such processes are theoretically possible, and have apparently occurred in laboratory populations of some organisms; but whether they frequently affect phenotypic features in wild bird populations remains an open question. So-called founder effects, discussed below, are in the same category.

Vicariance and dispersal

Population fragmentation, enabling speciation, can occur in two main ways. The first process (vicariance) happens when geological or climatic events impose large discontinuities in a previously continuous distribution, creating separate subpopulations no longer in contact. The rates at which the different subpopulations then diverge, given little or no genetic interchange between them, is likely to depend partly on local selection pressures, in turn governed partly by the extent to which environmental conditions differ between them. On much the grandest and longest time scale, the tectonic divisions resulting in continental drift led long ago to large-scale fragmentation of populations, promoting the emergence of distinctive continental faunas (Chapter 5). On a smaller and shorter time scale within a continent, physical changes, such as the raising of mountain ranges, have also acted to split populations. Perhaps the most frequent factor, however, which operates worldwide on still shorter time scales, results from the marked climatic changes that have repeatedly led to the fragmentation of major vegetation belts. Because most bird taxa are tied to particular vegetation types, their populations have been continually fragmented, allowing differentiation and speciation to occur. Climatic changes also led to changes in sea-levels, with rises dissecting land areas and severing islands from mainlands. All such fragmentation processes are grouped under the one category of vicariance, but the climatic-induced ones are reversible on time scales of tens of thousands of years, as climate swings from warm to cold or wet to dry. Many examples of the latter type of speciation are given in Chapters 10 and 11.

The second fragmentation process that can lead to speciation results from the dispersal of individuals across a pre-existing barrier (such as a sea, a wide river or a mountain range) to found a new population, sufficiently isolated from the original. This type of process is held to account for the divergence of the many bird species that have colonised remote oceanic islands (examples in Chapter 6). A new small population might diverge rapidly from the parental one for two reasons (Mayr 1954). Firstly, the parental population with a large geographical range constitutes a kind of compromise between diverse selection pressures operating in different parts of the range, but a small isolate is suddenly freed from such conflicting pressures and can adapt more precisely to its local environment. Secondly, the small number of colonising individuals usually involved might by chance have a restricted and non-random selection of genetic material from the parental population. This in turn could result in rapid divergence of the new population by genetic drift, involving the chance fixation of some new mutation or gene combination which could set the population on a different evolutionary trajectory. This model of speciation, termed peripatric or founder speciation, was developed from the observation that peripheral (often insular) populations of a polytypic species tend to be the most distinct. The theory is hard to test because it is merely a variant of allopatric speciation in which the founders are genetically limited or atypical compared to the population from which they are drawn. It will remain conjectural until the genetic changes involved in speciation have been measured. Nonetheless, some island populations have reduced genetic variance, relative to mainland forms, as shown for example in Loggerhead Shrikes *Lanius ludovicianus* on San Clemente Island, or in various vireos on Caribbean islands compared with their counterparts on the North American mainland (Mundy *et al.* 1997, Zwartjes 1999).

The rate at which any population can adapt to changed conditions presumably depends partly on the rate at which novel beneficial mutations arise, and partly on the rate at which these mutations can be fixed in the population. This latter rate is likely to be more rapid in smaller than in larger populations because of the smaller number of individuals that must acquire the beneficial mutation and the reduced chance of outbreeding. A similar effect might occur when a population is reduced in numbers by some environmental catastrophe. Such a crash may result in a 'genetic bottleneck' that again causes random changes in the genetic structure of the population so that its subsequent evolution follows a different course from before. This is partly because future adaptation depends on the genetic variation residing within the population. Any loss of variation reduces the number of ways that genetically-restricted populations might respond to selective pressures; in extreme situations, they might be unable to respond at all, thus increasing their risk of extinction.

It is not always easy to tell whether two sister species, derived directly from the same parental form, arose as a result of vicariance or dispersal. The vicariance model is supported if the separating barrier has been intermittent through time or if it is younger than the estimated age of the parent population. In contrast, the dispersal model is supported if the sister species have differentiated across a permanent barrier that is judged to be older than the species pair. Through analyses of DNA, it has become possible not only to estimate the ages of different sister taxa, but also to assess the relative frequencies of vicariance and dispersal in species formation in particular continental regions. The assumption is that vicariant processes were likely to affect a wide range of species simultaneously, so all the new taxa resulting from a particular isolating event will be of similar age. In contrast, dispersal over major barriers is likely to be a rare and idiosyncratic event, affecting species individually, and at different times, so that any new taxa resulting from major dispersal events are likely to be of widely different ages. On this basis, Zink *et al.* (2000) examined the mt DNA sequence data for six lineages of birds that occur mostly in the aridlands of western North America, namely towhees (genus *Pipilo*), gnatcatchers (*Polioptila*), quails (*Callipepla*), warblers (*Vermivora*) and two groups of thrashers (*Toxostoma*). From estimates of species' ages, using a standard molecular clock, it was concluded that up to 75% of these species could have been formed by vicariant processes, and that the remaining species probably resulted from occasional dispersal events. The effects of history seemed still apparent in the mt DNA of those species, even though they were dated at more than one million years old. In contrast, molecular data for 38 pairs of vertebrates from the West Indies islands proved mostly consistent with the dispersalist explanation of origin, as would be expected on islands (Hedges *et al.* 1992).

Multiple invasions

Multiple invasions present a special case of speciation following dispersal, and involve successive immigrations of the same form to the same area at different times. This seems to be a frequent event on some islands. If the stock resulting from one wave of immigration has diverged sufficiently by the time the next wave of the same species arrives, the two stocks remain distinct, behaving as separate species. Apparent examples include the Blue Chaffinch *Fringilla teydea* (earliest colonist) and Common Chaffinch *F. coelebs* (latest colonist) on the Canary Islands, and the Takahe *Porphyrio mantelli* (earliest colonist) and Pukeko *P. porphyrio* on New Zealand. If, on the other hand, differentiation has been slight, the new colonists will blend with the old, and the new immigration will pass unnoticed. This partly explains why islands far from a source area more often hold endemic species than do closer islands, subject to more frequent immigration (Chapter 6). Remote islands might also offer more unusual conditions, promoting greater divergence through natural selection.

Examples exist of quite well-differentiated forms breaking down under recent immigration. The darkly coloured New Zealand Variable Oystercatcher *Haematopus unicolor* most likely arose from a double invasion of the Pied Oystercatcher *H. longirostris* from Australia (Baker 1991). Descendants of the first invasion became melanic in New Zealand, and these birds now hybridise extensively with pied birds from the second invasion. In this example, the time between successive invasions was apparently too short to allow the evolution of reproductive isolation. The same is apparently true of the Black and Pied Stilts (*Himantopus novaezelandiae* and *H. leucocephalus*) in New Zealand, the former of which is now practically unrecognisable as a result of hybridisation. It shows again that marked phenotypic divergence does not inevitably lead to speciation.

Examples also exist of two races of a species reaching an island at more or less the same time, and forming a hybrid population. For example, on Dampier Island such a population of the Dusky Megapode *Megapodius freycinet* was formed by the interbreeding of two subspecies, one from the mainland of New Guinea and the other from the Bismarck Archipelago (Mayr 1963). If further immigration ceased, and the population in time developed phenotypic uniformity, this may be another way in which distinct island forms could evolve, whether classed as new subspecies or as new species. Examples of distinct forms produced by hybridisation are given later.

Ring species

Another isolation process can affect some species that have wide geographical ranges. Such species may form a single interbreeding population unbroken by topographic barriers, but sheer distance prevents the mingling of individuals from the remoter parts of the range. Clinal divergence may then occur in relation to environmental features, such as climate. Populations at the two ends may be as different from one another as separate species and may indeed behave as separate species if they meet, even though they are connected by a graded series of geographical forms. This is evident in the phenomenon of ring species, in which a series of races form what amounts to a geographical loop. Neighbouring forms on the loop interbreed but the forms at either end are reproductively isolated from one another where they overlap in the same region. Thus the Greenish Warbler *Phylloscopus trochiloides* is connected around the Tibetan Massif by a series of slightly different races to the Two-barred Greenish Warbler *P. plumbeitarsus*

(Figure 3.1). Where *P. trochiloides* and *P. plumbeitarsus* meet, they overlap in similar habitat but do not hybridise. They do not react to one another's songs and their mt DNA also indicates long separation (Irwin *et al.* 2001a). Similarly, the Herring Gull *Larus argentatus* in the North Atlantic is connected westward by a series of slightly different forms that encircle the northern hemisphere. The far end form is the Lesser Black-backed Gull *Larus fuscus*, which does not interbreed with the Herring Gull where the two co-exist around the eastern North Atlantic[1]. Other examples from Eurasia are provided by the shrikes *Lanius schach/tephronotus*, and the warblers *Acrocephalus arundinaceus/stentoreus*. They all clearly illustrate the speciation process because intermediate stages are still represented within the ring, although this does not exclude the possibility that the different forms were periodically isolated from one another, and rejoined at various stages in their history. They also illustrate the relativity of the terms species and subspecies, in that the two ends of a chain of subspecies may be much more different from one another than might other species living side by side.

THE RE-MEETING OF PREVIOUSLY ISOLATED POPULATIONS

When two populations have diverged morphologically, but remain geographically separate, it is not usually possible to decide with certainty whether they are subspecies or 'good' species, which would not interbreed if they met. Any decision on this point is to some extent subjective, but can be made according to considered guidelines (Mayr 1969b). However, when two previously separated populations expand and meet, a more definite conclusion may be reached, depending on what happens. Either (1) the two populations might fuse into a single hybrid swarm, effectively behaving as a single species (e.g. Variable Oystercatcher *Haematopus unicolor* and Pied Oystercatcher *H. longirostris* in New Zealand, Baker 1991), or (2) they may hybridise along a limited zone of contact, in which case the process of speciation has advanced further (e.g. 'Red-shafted' and 'Yellow-shafted' Flickers *Colaptes auratus* in North America, Moore & Koenig 1986), or (3) they may persist as distinct breeding populations, either (a) remaining in separate areas (parapatry, e.g. Melodious Warbler *Hippolais polyglotta* and Icterine Warbler *H. icterina* in Europe), (b) overlapping to a small extent (partial sympatry, e.g. Common Nightingale *Luscinia megarhynchos* and Thrush Nightingale *Luscinia luscinia* in Europe), or (c) spreading through one another's ranges to live together as separate species (almost complete sympatry, e.g. Garden Warbler *Sylvia borin* and Blackcap *S. atricapilla* in western Eurasia). All taxa that remain distinct despite breeding in close proximity to one another are usually labelled as different species. They would include all those in category 3 above, and some or all of those in category 2 (depending mainly on prevailing taxonomic fashion). The important point is that, in allopatry, lack of gene flow between taxa may be due solely to their geographical isolation. However, in sympatry and

---

[1]*This oft-quoted example needs further assessment, as it is more complicated than originally envisaged, and data on gene flow between the different taxa are largely lacking.*

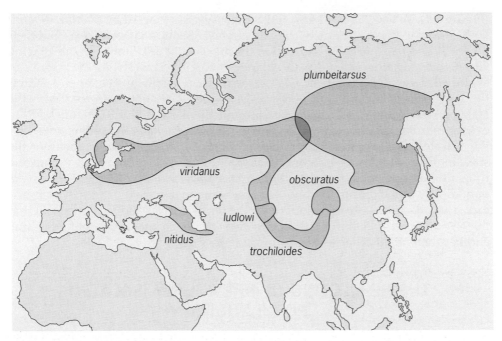

Figure 3.1 Distribution of races of the Greenish Warbler *Phylloscopus trochiloides*, a ring species. Racial variation is clinal through the intergradation of the subspecies *viridanus* (eastern Europe to western Asia and Afghanistan), *ludlowi* (southeast Afghanistan to Kumaon), *trochiloides* (central and eastern Himalayas), *obscuratus* (central China) and *plumbeitarsus* (eastern Russia); the gap in the distribution ring is attributed to deforestation. All other subspecies interbreed with adjacent subspecies except for *viridanus* and *plumbeitarsus* which do not intergrade where their distributions overlap, and warrant specific status, as the Greenish Warbler *P. trochiloides* and the Two-barred Greenish Warbler *P. plumbeitarsus*. The closely allied Green Warbler *P. t. nitidus* is fully allopatric in a separate forest area. From Irwin *et al.* 2001b.

parapatry, gene flow can no longer be prevented by geographical isolation, but depends instead on the behaviour and other features of the species themselves.

Range overlap between two expanding sister taxa is most likely to occur if they have diverged enough to have somewhat different ecological requirements, preferring different habitats or food sources. Only then would competition between them be reduced sufficiently for them to live together in the same area. As they continue to co-exist and compete, further ecological and reproductive isolation may develop between them. Often two species that utilise a range of habitats or foods where they occur alone subdivide them where they overlap, each using different parts of the total habitat or food resource (Lack 1971). Hence, while geographical isolation is important as the mechanism that generates diversity, subsequent ecological and reproductive isolation are the mechanisms that maintain it. Sympatry arises from range changes that occur after speciation has been completed or almost completed in allopatry.

Typically, then, closely related sympatric bird species differ in several kinds of characters that reflect their differing ecologies and mate selection. In Europe, some of the most similar-looking sympatric species include the Crested Lark *Galerida cristata* and Thekla Lark *G. theklae*, the Marsh Tit *Parus palustris* and Willow Tit *P. montanus*, the Eurasian Treecreeper *Certhia familiaris* and Short-toed Treecreeper *C. brachydactyla*. In North America, some of the most similar include the Grey Flycatcher *Empidonax wrightii* and Dusky Flycatcher *E. oberholseri*, the Summer Tanager *Piranga rubra* and Hepatic Tanager *P. flava*. All these species pairs are unusual, and the majority of closely related species that live together in the same area differ much more markedly in appearance and habits.

Hybridisation[2]

Hybrid zones between sister taxa are of great interest to the evolutionist but rather awkward for the taxonomist, because it is largely an arbitrary decision whether to call the hybridising forms subspecies (because they interbreed) or species (because hybridisation is limited). The ornithological literature contains many examples of birds classed as separate species and then being reclassed as sub-species after a hybrid zone was discovered. It also contains examples of birds classed as subspecies of the same species being proposed as separate species, after they were found breeding in the same area without hybridising (e.g. Lesser Redpoll *Carduelis cabaret* and Common (Mealy) Redpoll *C. flammea* in northern Europe (Knox *et al.* 2001); Pale-bellied Brant *Branta hrota* and Black Brant *B. nigricans* in northern Canada (Sangster *et al.* 1999)). There is always the risk that such deci-sions are premature, and that the same taxa may be found interbreeding freely in a later year or a different area. In addition, inconsistencies abound, with some quite distinct forms separated by a hybrid zone being considered as subspecies (such as Carrion Crow *Corvus c. corone* and Hooded Crow *C. c. cornix*) and others showing equally distinct features being classed as species (such as Red-headed Bunting *Emberiza bruniceps* and Black-headed Bunting *E. melanocephala*).

In many well-studied animal hybrid zones, hybrids have reduced fitness (sur-vival or breeding success), and the stability of the zone persists because this reduced fitness is balanced by continual movement of pure-bred individuals into the zone from either side (Barton & Hewitt 1989). Each zone is thus maintained as a dynamic equilibrium (balance) between gene flow via immigration and natural selection against the hybrids (for examples of stable hybrid zones, see Chapter 10). Alternatively, transient hybridisation, which results from range expansion and recent contact between reproductively compatible taxa, can cause the fusion of two previously distinct forms or the gradual swamping of one form by another. It can thus be an agent of extinction. For example, the Blue-winged Warbler *Vermivora pinus* and Golden-winged Warbler *V. chrysoptera* in North America were

---

[2]*Hybridisation is defined here as interbreeding between individuals of distinct taxa, and a hybrid zone is defined as an area where two related taxa overlap and where pure phenotypes of both parent species can be found alongside obvious hybrids and backcrosses. Introgression is gene flow between populations that hybridise, acknowledged when hybrids backcross to one or both parental forms. A hybrid swarm is a population in which introgression has occurred to varying degrees.*

once allopatric, but have come together through mutual range expansion. About 50 years after contact in any given locality, the Golden-winged and hybrid phenotypes disappear, leaving only the Blue-winged (Gill 1980, 1997). Similarly, interbreeding with the Pied-billed Grebe *Podilymbus podiceps* may have contributed to the extinction of the Atitlán Grebe *P. gigas* in Guatemala (Cade & Temple 1995), and currently the Madagascar Grebe *Tachybaptus pelzelnii* and the Alaotra Grebe *T. rufolavatus* are both thought to be disappearing from Madagascar through hybridisation with the Little Grebe *T. ruficollis* (Collar & Andrew 1988).

Other examples of extinction by hybridisation concern species that have reached new areas through human action (Rhymer & Simberloff 1996). Thus, the Seychelles Turtle Dove *Streptopelia picturata rostrata* is disappearing as a distinct subspecies through hybridisation with the introduced nominate race from Madagascar (Long 1981); the Grey (Pacific Black) Duck *Anas superciliosa* is declining in New Zealand through hybridisation with the introduced Mallard *A. platyrhynchos* (Haddon 1984); and the White-headed Duck *Oxyura leucocephala* in Spain is threatened by hybridisation with the introduced Ruddy Duck *O. jamaicensis* (Hughes 1993; see Newton 1998a for other examples). In some such cases, resulting from recent contact, hybridisation can be expected to become rarer if hybrids have lower fitness than parental types. This is because parental individuals that do not hybridise leave most descendants, promoting the evolution of reproductive isolation. In other cases, especially if non-hybridising genotypes are initially rare, the two forms may merge before reproductive isolation can evolve.

Among wild birds, every gradation in the hybridisation process can be found, from (1) complete free interbreeding of parental forms, with no obvious fitness penalties; (2) occasional interbreeding between parental forms, but with fitness penalties; (3) regular interbreeding along a hybrid zone only, with no fitness penalties; (4) regular interbreeding along a hybrid zone, but with fitness penalties; (5) interbreeding along a moving hybrid zone, resulting in the gradual replacement of one phenotype by another; and (6) no known hybridisation, and hence no means of assessing the fitness consequences (**Table 3.1**; Grant & Grant 1992). Hybrid zones may be narrow or broad, stable over many decades or changing unidirectionally, while mating of birds within the zone may be random or assortative with respect to phenotype. Hybrids may be at a disadvantage with respect to parent types, partly or totally sterile, or they may breed as well as or better than parent forms. Reduced fertility and viability of hybrids are often expressed in the heterogamic sex (females in birds, males in mammals) (Haldane 1922). This so–called 'Haldane's rule' was shown to hold in hybridising flycatchers (*Ficedula albicollis* and *F. hypoleuca*), in which hybrid females were infertile while hybrid males were not (Gelter *et al.* 1992). Other sex differences in fertility are evident in hybrids between the Indigo Bunting *Passerina cyanea* and Lazuli Bunting *P. amoena* (Baker & Boylan 1999) and between the Carrion Crow *Corvus c. corone* and Hooded Crow *C. c. cornix* (Saino & Villa 1992). But among hybrids between the Eastern Meadowlark *Sturnella magna* and Western Meadowlark *S.neglecta* both sexes were sterile (Lanyon 1979). Given this spectrum of natural variation, the difficulties of attempting to delimit species on reproductive–genetic criteria are self-evident. Because hybridisation usually involves close taxonomic relatives, however, it tells us something about genetic relatedness and about how species evolve.

Table 3.1 Differences in the types and consequences of hybridisation in birds.

Complete free interbreeding of parental forms with no obvious fitness penalties.

Examples: Snow Goose-Blue Goose *Chen caerulescens* (Cooke *et al.* 1995)[1].
             Mallard *Anas platyrhynchos* and New Zealand Grey Duck
             A. superciliosa (Haddon 1984).

Occasional interbreeding between parental forms, but with fitness penalties.

Examples: Western Meadowlark *Sturnella neglecta* and Eastern Meadowlark
             S. magna (Lanyon 1979).
             European Pied Flycatcher *Ficedula hypoleuca* and Collared Flycatcher
             F. albicollis (Gelter *et al.* 1992, Saetre *et al.* 1999)[1].
             Various taxa of Galapagos finches *Geospiza* (Grant & Grant 1996).

Regular interbreeding along a hybrid zone, with no obvious fitness penalties.

Examples: Red-shafted Flicker *Colaptes auratus cafer* and Yellow-shafted Flicker
             C. a. auratus (Moore & Koenig 1986).
             Western Gull *Larus occidentalis* and Glaucous-winged Gull
             L. glaucescens (Good *et al.* 2000)[2].

Regular interbreeding along a hybrid zone, but with fitness penalties expressed as reduced hybrid viability, fertility or competitiveness.

Examples: Red-breasted Sapsucker *Sphyrapicus ruber* and Red-naped Sapsucker
             S. nuchalis (Johnson & Johnson 1985)[1].
             Carrion Crow *Corvus c. corone* and Hooded Crow *C. c. cornix* (Saino &
             Villa 1992, Rolando 1993)[1].

Interbreeding resulting in the gradual replacement of one phenotype by another, as manifest in a moving hybrid zone.

Examples: Blue-winged Warbler *Vermivora pinus* replacing Golden-winged Warbler
             V. chrysoptera (Gill 1980)[1].
             Indigo Bunting *Passerina cyanea* replacing Lazuli Bunting *P. amoena*
             (Emlen *et al.* 1975, Baker & Boylan 1999)[1].
             Baltimore (Northern) Oriole *Icterus galbula* replacing Bullock's Oriole
             I. g. bullockii (Rhymer & Simberloff 1996)[1].

[1] Assortative mating observed.
[2] In this study, hybrids performed better than pure-bred birds in the overlap zone, a situation called 'bounded hybrid superiority'.

A particularly well-studied example of hybridisation following the recent meeting of two formerly separated populations concerns the arctic-nesting Snow Goose *Chen caerulescens* in North America (Cooke *et al.* 1995). This species exists in two phases, white from the western part of the range and blue from the eastern part. The two forms have recently met; they interbreed freely and are in the process of merging. Historical evidence implies that very little intermingling occurred prior to 1920, when the two forms were allopatric in both breeding and wintering areas (where pair formation occurs). A major change occurred subsequently, probably triggered by human impact, which increased and expanded potential food-supplies on the wintering areas in the Gulf States. This enabled the white and blue phases, previously isolated, to come together and form mixed pairs, the males then accompanying the females to their particular breeding areas.

Because the blue form is genetically dominant and the white recessive, there is now a gradually changing ratio of white to blue birds across both the wintering and the breeding ranges. In view of these events, current taxonomists class the two morphs and their intermediates as the same species, but earlier taxonomists (before 1930), operating before hybridisation began, classed the two forms as separate species, as they looked strikingly different and were allopatric. In this example, visible differences were again misleading and a change in taxonomic status reflected a natural event resulting from range expansion. Although the two colour phases are inter-fertile, they show some degree of assortative mating (white preferring white and blue preferring blue), and this, together with known rates of dispersal (gene flow) and the dominance of the blue form, can explain the present ratios of white to blue birds found in different colonies across the breeding range.

Approximately 9% of bird species have been recorded hybridising with at least one other species, some frequently and others rarely. However, most species in the 9% have provided only one or a few records, resulting from chance observations, so the true percentage of species that occasionally hybridise could well be much higher (Grant & Grant 1992). Hybridisation seems much more frequent in some types of birds, such as waterfowl, than in others, but this could be due to greater difficulty in recognising hybrids in some groups. Most known hybridisation occurs between species in the same genus. This immediately cuts down its incidence because many species do not occur in the same region as a congener, while other species have no congeners, being the sole members of their respective genera. The ability to produce fertile young in a hybrid cross is partly a consequence of the chromosomal number and composition of the two parents. Closely related species may have quite different chromosomal characteristics; but equally two more distantly related species whose morphology and behaviour are very different may be able to produce fertile young because events such as chromosome and gene rearrangements happen not to have occurred. Among North American warblers, many cases of intergeneric hybrids are known. These and other examples point to relatively widespread genomic compatibility and potential for hybridisation among birds that look strikingly different (Grant & Grant 1992). The fact that hybrids remain rare may lie largely in the powers of avian social recognition, which ensures that individuals normally pair only with others of their own species (see above), but such acuity in recognition may in turn have evolved partly through the selective disadvantage of hybridisation.

Individual birds may mate with a member of a different species either (1) because of their inability to discriminate between two potential types of mate, (2) because they show the wrong preference (perhaps through mis-imprinting), (3) because of the lack of a conspecific mate, (4) because of some other constraint in mate choice, or (5) because of indiscriminating extra-pair copulation forced by the male of one species on the female of another. As shown by experiment, the young of some bird species become imprinted on their parents, and select birds with parental features as mates. If, through cross-fostering experiments, they are raised by another closely related species, mixed pairs often result (for flycatchers see Löhrl 1955, for gulls see Harris 1970). Other species are genetically programmed to respond to the colour patterns and displays of their own species, and do so even when raised by another species (for finches see Immelmann 1965). In many bird species, whatever the mechanism of mate choice, females are more discriminating than males (for ducks see Schutz 1965, for gulls see Harris 1970), and many known examples of hybridisation can be attributed to lack of choice caused by shortage of conspecifics (Randler 2000).

Evolution of reproductive isolation

As new bird species evolve from isolated populations of the same parental form, they might diverge sufficiently so as to become reproductively incompatible, perhaps unrecognisable to each other as potential mates. They would then not interbreed if they later met and the entire speciation process would have been completed in allopatry. Alternatively, they may still be reproductively compatible, and start to hybridise when they meet. Hybridisation might then result in the complete merging of gene pools or the swamping of one gene pool by another; but if the hybrids are inferior in some way, more specific isolating mechanisms might evolve, gradually lessening the amount of interbreeding. The speciation process would then have been started in allopatry and completed in sympatry. Hence, where we find regular hybridisation in natural populations, despite hybrid inferiority, it may be a temporary phenomenon, as reproductive isolation develops (for example by mate preference), although this may be hard to prove without prolonged study. The chance of anti-hybridisation mechanisms developing fortuitously while the two forms are living apart is presumably small compared to the rapid selection for reproductive isolation that can occur once they come into contact. Some species pairs that come together without hybridising may, of course, have had contact in the past, with isolating mechanisms evolved then persisting to the present. In addition, isolating mechanisms involved through contact with one species might well function against another at a later date. The achievement of full reproductive isolation in sympatry marks a point of no return in the differentiation of the lineages involved, in that merging is not thereafter possible.

Some isolating mechanisms may act before fertilisation (pre-zygotic) and others after fertilisation (post-zygotic), so as to prevent the production of hybrid offspring. With pre-zygotic isolation, successful fertilisation does not happen because mating does not occur, or because egg and sperm are incompatible. With post-zygotic isolation, successful fertilisation occurs, but the resulting hybrids are unviable, infertile or unfit in other ways. In these circumstances, individuals that hybridise are disadvantaged, because they leave few or no offspring, so would be expected eventually to disappear from the population. Pre-zygotic mechanisms include all those morphological and behavioural differences between species that affect individual mate choice and prevent errors in mate recognition (such as colour patterns and display). This mode of speciation may be achieved by female choice selecting for male traits that facilitate species recognition, and prevent confusion with related species. Because of reduced hybrid fitness, females benefit from choosing a mate that is conspicuously of their own species. This could lead to gradual divergence in the male plumage patterns of different species, which in turn reduces the frequency of hybridisation, promoting reproductive isolation. Such reinforcement has been shown to occur in *Ficedula* flycatchers, for example, in areas where two species occur together (Saetre *et al.* 1997).

Closely related bird species usually differ in song and plumage, and these traits appear often to have evolved by sexual selection (Andersson 1994, Petrie 1999).

However, in many birds, especially passerines, songs and other recognition features are partly learnt and not entirely under genetic control (Harris 1970, Fleischer & Rothstein 1988, Catchpole & Slater 1995). This implies that pre-mating reproductive isolation in emerging species could happen quickly, without much genetic change. Both natural and sexual selection could generate morphological, behavioural and vocal differences between populations, after which sexual imprinting could lead to these traits being used in mate recognition. Each species may, of course, respond to more than one feature, the combination of features giving more precise mate discrimination than one alone.

Sexual selection sometimes leads to the development of extreme features in the males of some species, such as the train of the peacock (Indian Peafowl) *Pavo cristatus*. In such polygynous species, successful males can fertilise many females, which intensifies the selection for whatever characters form the basis of female choice. In peacocks, a relationship between mating success and train length has been shown by observation (Petrie *et al.* 1991). In other bird species the experimental modification of pronounced secondary sexual characters has confirmed that such characters influence female choice, and that males with the most extreme characters are preferred (Andersson 1994).

Hybrids might be inferior on genetic grounds if they produce fewer offspring than the parent forms, or on ecological grounds if they are less successful in competition with the parent forms for food and other resources (Lack 1947, Grant & Grant 1996). In the latter case, sympatry might encourage further divergence in morphology through competition. To judge from studies on the Galapagos Finches, some degree of hybridisation can occur for a long time (possibly more than a million years) after species first come together in secondary contact (Grant & Grant 1996). All six *Geospiza* finches have been found to hybridise with at least one other congeneric species, producing fertile offspring. Evidently, in these birds speciation was not completed in allopatry. The hybrids were not noticeably less fertile or viable than pure-bred offspring, but they suffered greater mortality in competition for food, being less efficient than either parent form. It was thus this ecological differential that was likely to provide the selection pressure to finish the speciation process, by leading to the gradual elimination of hybridising individuals from the sympatric populations of both species. From comparisons among these and other species, it is clear that the degree of morphological differentiation between populations does not necessarily reflect their sexual incompatibility. The Eurasian Chiffchaff *Phylloscopus collybita* and Willow Warbler *P. trochilus* look alike but do not appear to hybridise, while the North American Blue-winged Warbler *Vermivora pinus* and Golden-winged Warbler *V. chrysoptera* look quite different, but interbreed freely. Evidently, reproductive isolation does not necessarily evolve concomitantly with morphological divergence; and, in the absence of selection favouring such isolation, reproductive compatibility between two congeneric taxa can occur as a retained trait.

Creative role of hybridisation

Hybridisation can lead to the extinction of species, or at least of their phenotypes, as described above, but it also produces novel combinations of genes, as well as new alleles. It could therefore play a creative role in speciation, either directly by producing a new compound form which stabilises to a new species, or indirectly by increasing genetic variation in each of the interbreeding species and thereby allowing further evolutionary change in one or both. The first process is well known in plants (in which it usually involves polyploidy: a doubling of the chromosome number), but is at most rare and poorly documented in birds (see below). In the second process, hybridisation might produce new species, not by the passive fusing of genes, but by the action of natural selection on the expanded gene pool, which could facilitate further evolutionary change, to culminate in a new species. In this way, hybridisation could contribute to the speciation process without being solely responsible for it, by enhancing genetic variation and thereby relaxing the genetic constraints on particular directions of evolutionary change (Grant & Grant 1996). This could be a more significant process in bird evolution than recognised so far, given the many opportunities that populations have to fragment and diverge and then to come together again and interbreed (see Chapter 10). Such reticulate patterns in evolution, involving the splitting and rejoining of populations, are probably commonplace on time scales of thousands of years. Genetic variation could be added by hybridisation much more rapidly than by mutation, increasing the raw material on which selection can act.

Appreciation of this possibility has already helped to clarify some otherwise puzzling relationships. For example, the close similarity in mt DNA between the Pomarine Skua *Stercorarius pomarinus* and the Great Skua *Catharacta skua* is most plausibly explained on the assumption of past transfer of mt DNA by hybridisation from *S. pomarinus* to *C. skua*, with the latter retaining its own nuclear genome (Andersson 1999). The two species are also similar in their ectoparasites. Although usually placed in separate genera, they might now be better classed together even though this enlarged genus would contain species from two lineages. One species may have obtained genetic material from another while retaining its own independent lineage. This is quite different from saying that the two species have fused to form a new species. It is more akin to a language acquiring some words from another language than to two languages being mixed together to form a completely new one.

The close similarity of the mt DNA between the White-crowned Sparrow *Zonotrichia leucophrys* and Golden-crowned Sparrow *Z. atricapilla* in North America has also been attributed to past hybridisation (Weckstein *et al.* 2001). How might this situation arise, where a species could have nuclear DNA from one lineage and mt DNA from another? Because the mt DNA is maternally inherited, the first-generation hybrid would retain the mt DNA of its mother, but acquire a nuclear genome half from the mother and half from the father. Provided that the hybrid female was fertile, she could breed with a male of her father's species. Repeated backcrossing by the female line to males of the original father species would result in the gradual conversion of the nuclear DNA to the original father species, while the mt DNA of the original mother species would be retained. This is one mechanism through which mt DNA could be permanently transferred from one species to another, a process that could be helped by small population size, by selection favouring individuals of mixed lineage or by females imprinting on the song of their fathers (Weckstein *et al.* 2001).

Supposed examples of hybridisation leading to the development of phenotypically stable populations have been occasionally recorded in wild birds, following the cessation of gene flow from the parental populations. The so-called 'Italian Sparrow' *Passer hispaniolensis italiae* is purported to have originated as a hybrid population between the House Sparrow *P. domesticus* and the Spanish Sparrow *P. hispaniolensis*, and the same supposedly holds for the sparrow populations on some Mediterranean islands and at some Saharan oases (Meise 1936). Similarly in western North America, the Junco race *J. hyemalis cismonatus* apparently originated as a hybrid population between *J. h. hyemalis* and *J. h. oreganus* (Miller 1941). In these examples, the resulting populations have achieved phenotypic uniformity (more or less), but to my knowledge none has been subjected to modern DNA analysis, or been shown to be reproductively isolated from parental forms. Hence, while hybridogenic speciation may occur in birds, it has not been convincingly demonstrated.

Recent studies of hybridisation have done more than point to a creative role in evolution, for they have also highlighted the need for caution in attempts to use genetic data to reconstruct phylogenies and calculate divergence dates (Avise 1989, Degnan 1993, Weckstein *et al.* 2001). As a result of introgressive hybridisation, sympatric species may be genetically more similar than allopatric species. In fact, the Galapagos finches *Geospiza fuliginosa* and *G. fortis* were more similar to each other in allozymes at two localities on Santiago and Santa Cruz Islands where they occurred together, than either was to a conspecific elsewhere (Yang & Patton 1981). Hence, in phylogenetic studies of these species, the branching points may be incorrectly determined; but how general a problem this is among birds remains to be seen. As a general point, we can expect that the more extensive has been the reproductive interaction between two lineages, the more difficult it becomes to decipher their separate histories from analyses of DNA.

SYMPATRIC SPECIATION

Apart from hybridisation, all the means of speciation discussed above involve geographical isolation, at least in the early stages. At various times, preferential mating (sexual selection) has been suggested as a means of sympatric speciation: that is, species formation without the need for geographical isolation. If individuals tended to mate selectively, perhaps choosing mates of the same type as themselves, this process could split a population of individuals living in the same area into subsidiary populations which, though mingling with one another spatially, would function as separate breeding populations and so be able to evolve separately. They could only persist, however, if they exploited different resources or different parts of a habitat; otherwise one would be likely to outcompete and replace the other. Although assortative mating has been found to occur in birds (as between light and dark forms of the Parasitic Skua *Stercorarius parasiticus* (O'Donald 1959) and between light and dark forms of the Snow Goose *Chen caerulescens* (Cooke *et al.* 1995)), it never seems complete or continuous enough to split a species into two or more types. In fact, no evidence has yet emerged from birds that such sexual selection could be important in causing sympatric speciation, nor that the different types differ in ecology sufficiently to co-exist in the long term.

Further evidence pointing to the lack (or extreme rarity) of sympatric speciation in birds comes from studies of mt DNA. If sympatric speciation were frequent, one would expect to find sympatric species showing all degrees of mt DNA divergence from nil upwards. In fact, none of the closely related non-interbreeding sympatric species yet analysed (and known to me) shows less than about 3% divergence. This implies that they came together only after a long period in allopatry. Allopatric populations, in contrast, whether species or subspecies, can show much lower or much higher degrees of mt DNA divergence.

Some sympatric species do indeed show little or no consistent differentiation in particular segments of their DNA, but this cannot be ascribed unequivocally to recent sympatric speciation, because they are also known or strongly suspected to interbreed (making them inseparable on DNA). This is the case in various Galapagos finches, redpolls and crossbills (Seutin *et al.* 1995, Freeland & Boag 1999, Piertney *et al.* 2001). The three taxa of crossbills found in Britain (Common Crossbill *Loxia curvirostra*, Scottish Crossbill *L. scotica* and Parrot Crossbill *L. pytyopsittacus*) show no consistent differences in either mt DNA or nuclear microsatellite sequences, and substantial overlap in the main morphological character (bill depth) that supposedly separates them (Piertney *et al.* 2001). It is therefore uncertain how or when they originated, how often they hybridise, and whether they will remain distinct.

Nonetheless, sympatric speciation remains theoretically possible in birds, the main requirements being (1) ecological variation within a population, (2) assortative mating linked with this variation, and (3) strong selection against the intermediates. Thus if a population exploited two distinct niches, and if mate preference was linked with niche preference, this could eventually lead to the splitting of one population into two. One might imagine such a process acting in crossbills (*Loxia*), for example, which feed on the seeds from conifer cones. Individuals vary in bill size; small-billed individuals feed most efficiently on conifer species that have small, soft cones, while large-billed individuals feed mainly from conifers that have large hard cones (Newton 1972, Benkman 1993). As the birds form pairs while in feeding flocks, small-billed birds could tend to pair together, as could large-billed birds, thus producing two distinct types from a single variable population. Intermediates are likely to feed less efficiently on both small and large cones, so be selected against. This in turn could enhance selection for assortative mating, in which like paired with like. Crossbills do indeed occur as different ecotypes, which differ in body size, call notes, bill size and food preferences; they are usually found in different parts of the range, but occasionally intermix in the same areas (Benkman 1987, Groth 1988, Marquiss & Rae 2002). However, no one knows for sure whether the different forms evolved in this way rather than allopatrically. Among birds, assortative mating might more readily occur if the different 'ecotypes' also look or sound different (as is true of crossbills). The important point is that, for sympatric speciation to occur, gene flow need not cease completely, provided that selection overrides its effects through the continual elimination of intermediates.

If sympatric speciation occurs at all in birds, it is likely to affect relatively few species, because few show discontinuity in niche preference linked with mating behaviour. In most bird species, feeding and mating are independent activities. Some of the most species-rich families of birds are also some of the most sexually dichromatic (assumed to reflect strong sexual selection, Owens *et al.* 1999), but this does not rule out allopatric speciation or a two-step process starting in allopatry and finishing in sympatry (which is probably the norm). In various types of animals, sympatric speciation involving intense sexual selection has been conjectured to be a key factor in explosive radiations in which many species are produced in a short time from a single ancestral species. The cichlid fishes of Lake Victoria have often been cited as an example, but the evidence is unconvincing (e.g. Fryer 2001).

DESIGNATION OF SPECIES

It will be evident by now that the term species can embrace populations at very different stages of their evolution. It helps to designate these different stages in nomenclature, with terms such as:

(1) semispecies: between subspecies and species, and perhaps including all populations that form clear and stable hybrid zones;
(2) allospecies: populations derived from the same common ancestor whose ranges differ and do not touch, but where the different allospecies may be so similar to one another as to leave doubt whether they would remain distinct if they came into contact;
(3) paraspecies: populations derived from a common ancestor whose ranges touch, but which exclude each other geographically or altitudinally; they are similar ecologically and may interbreed occasionally, but do not merge or spread through one another's ranges;
(4) superspecies: an over-arching term for all the semispecies, allospecies and paraspecies recently derived from the same parental form, and occupying different areas from one another;
(5) sympatric species: taxa that overlap in at least parts of their ranges without interbreeding, gene flow being prevented by intrinsic isolating mechanisms.

Because evolution is an ongoing process, there are of course no sharp boundaries between these categories, and, in the case of allospecies, the splitting of the ancestral stock may have occurred in stages, so that the products are of different ages, showing different degrees of divergence.

The term species group is often used to designate sympatric or partly sympatric species that are closely related. Because they can persist together in the same area, they represent the next step in evolution from allospecies, and can no longer be regarded as members of superspecies. It is the successional process from allopatry, through parapatry (abutting ranges) to sympatry that leads to increased species richness in particular areas with fully compatible species living together. Allopatric species pairs are sometimes called 'sister species', while sympatric relatives which are alike in appearance and habits are sometimes called 'sibling species'.

ADAPTIVE RADIATION

When a species colonises different areas, with each population isolated from the others, each may adapt to somewhat different local habitats and food-types. If these separate populations later meet and do not interbreed, competition between them may encourage further specialisation, eventually producing an array of species each adapted to different niches. This cycle of segregation and differentiation through local adaptation, followed by re-contact and further differentiation, could in theory be repeated again and again, each time increasing species numbers and diversity. The original colonists could thus undergo a process of 'adaptive radiation', defined as the diversification of a single lineage into a range of different species in the same general region, each with distinct morphology and ecology. The essential driving forces of this process are in turn: geographical isolation, local adaptation, occasional dispersal and competition, leading to further divergence (**Table 3.2**). Some of the most impressive and rapid examples of adaptive radiation are found on the most isolated of island groups, which receive immigrants from mainland areas only extremely rarely.

Table 3.2   Steps envisaged in the adaptive radiation of island birds.

Step 1: an island in an archipelago is colonised from a continent by a species that establishes a local population;

Step 2: the species spreads to a second island and establishes a new population, in a process that could be repeated several times on different islands, giving two or more allopatric populations that differentiate from one another partly in response to local conditions;

Step 3: contact through dispersal of members of two populations that, if differentiated enough, remain as separate species, perhaps diverging further in sympatry;

Step 4: repetition of the process of dispersal and differentiation, eventually to give a range of different species exploiting different niches.

The Hawaiian Honeycreepers (Drepanidini) are endemic to the Hawaiian Islands. Some species have the short heavy bills of seed-eaters, others have the short thin bills of insectivores, and yet others have the long curved bills of nectar-feeders, all present in a range of different sizes in adaptation to different foods and feeding niches (**Figure 3.2**). It is envisaged that, long ago, a finch-like bird colonised the Hawaiian chain and, with few other birds present, was able to take advantage through its descendants of various ecological opportunities. As new islands were colonised, new species were formed that could then reinvade the original island. Since each new species had to adapt to a slightly different niche in order to co-exist with an older species, an array of varying forms developed that eventually covered almost the entire spectrum of passerine bird adaptations. Unfortunately, we cannot know the full extent of this radiation, because many species disappeared soon after the islands were colonised by Polynesian people more than a thousand years ago, and yet others are known to have become extinct following the arrival of European people more than 200 years ago (Chapter 7). Thirty-three species are known from historical collections and more than 17 others from subfossil remains, making more than 50 in all (Fleischer & McIntosh 2001). The living forms show much subspecific variation between islands (Pratt *et al.* 1987). The honeycreepers are not alone in showing adaptive radiation on the Hawaiian Islands, however, for the same process has occurred in other birds (Chapter 7), as well as in insects, snails and other animals and plants, in each case from one or a few original colonist species.

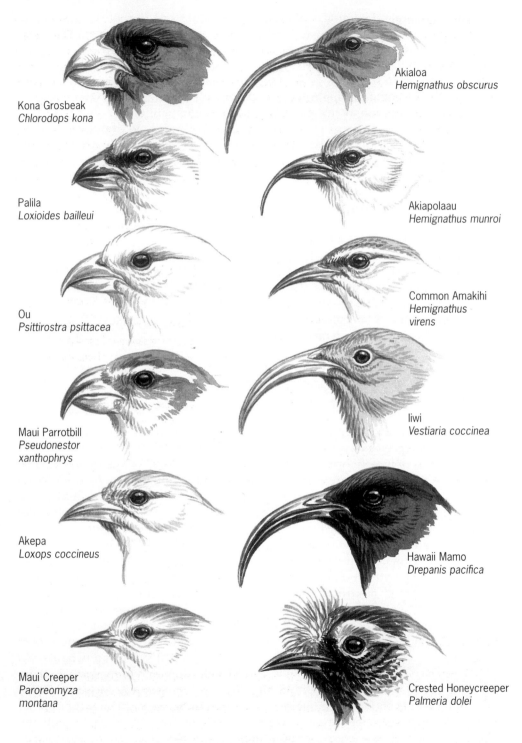

Kona Grosbeak
Chlorodops kona

Akialoa
Hemignathus obscurus

Palila
Loxioides bailleui

Akiapolaau
Hemignathus munroi

Ou
Psittirostra psittacea

Common Amakihi
Hemignathus virens

Maui Parrotbill
Pseudonestor xanthophrys

Iiwi
Vestiaria coccinea

Akepa
Loxops coccineus

Hawaii Mamo
Drepanis pacifica

Maui Creeper
Paroreomyza montana

Crested Honeycreeper
Palmeria dolei

Figure 3.2   Adaptive radiation as exemplified by some of the honeycreepers of the Hawaiian Islands. For further details see text.

Using starch-gel electrophoresis of proteins, Johnson *et al.* (1989) examined the relationships among various species of honeycreepers (**Figure 3.3**). They concluded that (1) these birds are indeed derived from a single ancestor (i.e. are monophyletic); (2) they are more similar genetically to emberizine than to cardueline finches; (3) based on a molecular clock for proteins, the ancestral species colonised the islands 7–8 million years ago (this date agrees with the emergence of Nihoa, now largely submerged, but antedates the appearance of Kauai 5 million years ago, the oldest of the present high-elevation islands); (4) the oldest and most divergent lineage of living drepanidines includes the creepers *Oreomystis* and *Paroreomyza*; (5) the youngest lineages are represented by the nectar-feeders *Himatione* and *Vestiaria*, the thick-billed finches *Loxioides* and *Telespiza* and a

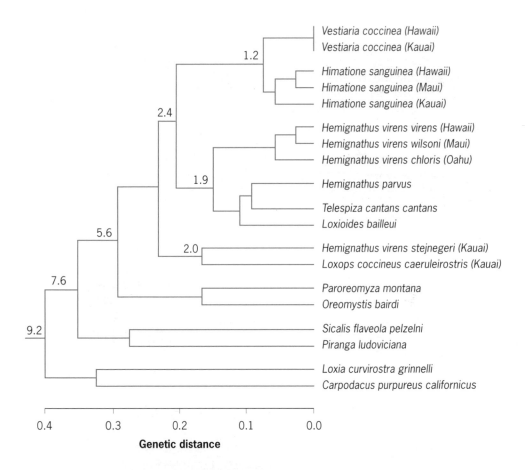

Figure 3.3   Phylogeny of the Hawaiian honeycreepers (Drepanidini) based on analyses of 36 protein-encoding loci. The figures at the branching points are approximate times in millions of years for divergence of the taxa. The most recently evolved species diverged in the last 1–2 million years and subspecies within the last 0.5 million years approximately. The four species at the bottom are not drepanidines, but different kinds of emberizine and cardueline finches. Redrawn from Johnson *et al.* 1989.

diverse array of other forms (*Loxops* and *Hemignathus*). Further analyses, involving mt DNA, confirmed the monophyly of the honeycreepers, but suggested a closer relationship to the carduelines than the emberizines, and that divergence from a North American ancestor occurred about 6.4 million years ago (again predating the formation of Kauai) (Fleischer & McIntosh 2001). Once again, despite some inconsistencies requiring further study, biochemical methods provided insight into the evolutionary history of an interesting group of birds.

Another well-studied example of adaptive radiation is provided by the Galapagos finches, famous because (along with the giant tortoises of the same islands) they started Darwin on his study of evolution, and later led Lack (1944, 1947) to develop his ideas on competition and ecological segregation in birds. Unlike the Hawaiian honeyeaters, none of the Galapagos finches is known to have been eliminated by human action. About 14 species are recognised, again with a range of different sized and shaped bills, adapted to different foods, some species having distinct subspecies on different islands (**Figure 3.4**). These birds include the ground finches, *Geospiza*, some of which feed mainly on seeds and others on cactus plants, and the tree finches, placed in three genera *Camarhynchus*, *Platyspiza* and *Cactospiza*, which include one vegetarian and five mainly insectivorous species. One of the latter, *Camarhynchus pallidus*, clambers on trees like a woodpecker and excavates insects from bark with a stick. A third type, *Certhidea*, looks like a warbler, has a long slender beak and eats small soft insects. The fourth type is *Pinaroloxias inornata*, the Cocos Finch, found on Cocos Island 900 km away, which also looks like a warbler. From a single ancestral colonist, these birds have thus evolved on the islands into seed-eaters, nectar-eaters, fruit-eaters, cactus-feeders, wood-borers and insect-eaters, some foraging on the ground, and others in trees (Lack 1947). Some even squeeze the blood from the growing feathers of incubating seabirds, break seabird eggs to eat the contents or pick ticks off reptiles. Up to ten different species now co-exist on particular islands.

Another striking feature is that these finches are very variable within species. The outlying islands lack certain species and their place is taken by variants of others, with corresponding modification of the beak. This process of 'character displacement', which apparently occurs through competition with related species, has resulted in species being more different where they co-exist than where they do not. In the finches, character displacement is most apparent in the bill, which enables species to specialise on different foods. It is striking that so many distinct types appear on islands separated by distances mostly less than 50 km. Geographical separation is clearly crucial, for while the central islands, lying close together, have no endemic subspecies, the outlying islands have mostly distinct forms.

Studies of nuclear and mt DNA agree that the Galapagos finches had a single common ancestral form, and are now most appropriately divided into three groups, comprising ground finches, tree finches and the Warbler Finch *Certhidea olivacea*, with the latter as the basal taxon (Freeland & Boag 1999, Petren *et al.* 1999). The Warbler Finch complex (eight island subspecies) is paraphyletic, because it embraces two divergent genetic lineages. Hybridisation or incomplete speciation are two likely explanations for the sharing of haplotypes among taxa. It is estimated that the Warbler Finch and tree finch lineages split about 750,000 years ago (Freeland & Boag 1999), an estimate not too dissimilar from that of 570,000 years

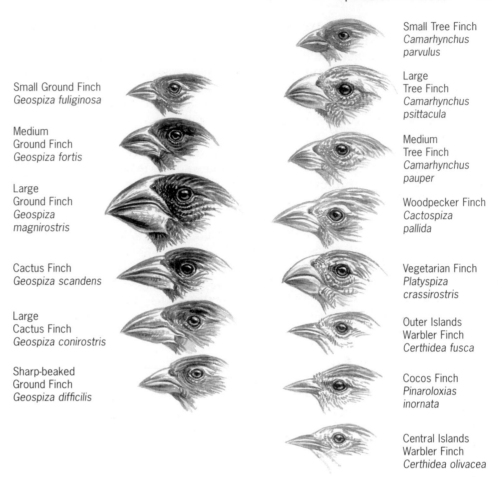

Small Tree Finch
*Camarhynchus
parvulus*

Large
Tree Finch
*Camarhynchus
psittacula*

Medium
Tree Finch
*Camarhynchus
pauper*

Woodpecker Finch
*Cactospiza
pallida*

Vegetarian Finch
*Platyspiza
crassirostris*

Outer Islands
Warbler Finch
Certhidea fusca

Cocos Finch
*Pinaroloxias
inornata*

Central Islands
Warbler Finch
Certhidea olivacea

Small Ground Finch
Geospiza fuliginosa

Medium
Ground Finch
Geospiza fortis

Large
Ground Finch
*Geospiza
magnirostris*

Cactus Finch
Geospiza scandens

Large
Cactus Finch
Geospiza conirostris

Sharp-beaked
Ground Finch
Geospiza difficilis

Figure 3.4   Adaptive radiation in beak structure as exemplified by some of the finches of the Galapagos Islands. For further details see text.

since the divergence of *Certhidea* from other geospizines based on allozyme data (Yang & Patton 1981). In addition, genetic divergence among the various Galapagos Island populations of the Warbler Finch pre-dates the radiation of all other Darwin's finches, and implies that the original colonist species may have been warbler-like in appearance. The Cocos Island Finch evolved after the Galapagos Finch radiation was under way, supporting the hypothesis that this distant island was colonised from Galapagos (Petren *et al.* 1999).

   The genetic data suggest that the radiation of these birds took around 2.8 million years (Grant & Grant 1996, correcting Yang & Patton 1981), and that the diversity of bill types and feeding behaviour was achieved with only minor structural genetic change. On this time schedule, the radiation would have started when there were fewer islands than now, which could explain the slow initial diversification, compared to the more recent diversification of the group (Yang & Patton 1981). Rather than radiating slowly within a pre-existing group of islands, the birds radiated progressively as more islands became available. In other words, speciation may for a time have been limited by opportunities (Grant & Grant 1996). The whole radiation is much younger and more recent than that of the honeycreepers on Hawaii, and the species themselves are less distinctive looking. Their nearest living relatives are the South American grassquits *Tiaris*, from which they were presumably derived (Sato *et al*. 2001).

While the most striking examples of adaptive radiation are from islands, the process also occurs on continents but, presumably because of the greater range of established species there, it tends to occur in a more restricted manner. Finches radiate to produce a greater variety of finches, each adapted to different finch niches, thrushes to fill different thrush niches and warblers to fill different warbler niches. It is rare for one group to move into the ecological space of several others, as has happened repeatedly on islands in the absence of potential competitors. However, without DNA or other special studies, cases of marked divergence may go unsuspected on continents, an example being *Pseudopodoces* of the Tibetan plateau, formerly regarded as a type of crow, but now recognised as a highly divergent tit (James *et al*. 2003). Other well-known examples of regional adaptive radiation among birds include the honeyeaters (Meliphagidae) of Australia–New Guinea and the vanga shrikes (Vanginae) of Madagascar.

Examples of natural selection in action

Studies of the Medium Ground Finch *Geospiza fortis* on the Galapagos Island of Daphne have provided some remarkable observations of natural and sexual selection in action (Grant 1986). During a drought on Daphne Island, large individuals with large beaks survived best because they were able to deal with the large and hard seeds that remained after most of the smaller and softer seeds had been eaten. The mean bill size of the surviving population was thus altered within a period of a few months and, as bill size was inherited, this represented an evolutionary change caused by natural selection. Another result of the drought was that males survived better than females, giving a sex ratio of about 5:1. At the next breeding season, females selected as mates the largest (= largest billed) from the available males, so that the same traits were further accentuated by sexual selection. In other years, advantages in small size were shown, and evolutionary changes away from the average tended to be short term.

These findings showed that evolutionary changes within a population could occur extremely rapidly, within a few months, given sufficient initial variation and strong enough selection pressure. They also implied that competition within and between species was the main determinant of the ecological niche and distribution of each species, and that small inherited variations in traits such as beak size could lead the individuals in a population to differ in their ecology, and the selective forces they experience.

Another example of rapid micro-evolutionary change in a wild population is provided by the House Sparrows *Passer domesticus* from Europe that were released in New York City in 1852. In the next 100 years these birds spread over the whole continent, and local populations evolved distinctive traits. Birds from the northeastern States became large and dark, whereas those from the southwestern deserts became small and pale (Johnston & Klitz 1977). Because such trends parallel those in other birds, they are assumed to represent adaptations to local conditions. House Sparrows changed recognisably in morphological features within 20 years of colonising new areas. Yet other examples of rapid morphological change involve House Sparrows and Common Mynas *Acridotheres tristis* introduced to New Zealand (Baker & Moeed 1979, Baker 1980), Laysan Finches *Telespiza cantans* introduced to other islands (Freed *et al.* 1987), House Finches *Carpodacus mexicanus* in newly colonised parts of North America (Badyaev *et al.* 2000), and Song Sparrows *Melospiza melodia* from year to year in British Columbia (Schluter & Smith 1986). All provide examples of change in the genetic structure of a population through the differential survival and reproduction of its members. In some of these studies, natural and sexual selection were actually witnessed, not merely inferred.

The capacity of animal and plant populations to respond rapidly to selective pressure is forcefully illustrated, as Darwin realised, in the origin of domestic strains; and more recently laboratory experiments on various short-lived organisms (notably fruit-flies) have achieved genetic changes within a few generations under strong artificial selection. Yet the rapid rates of evolution occasionally seen over short time scales in the wild and in experiments in laboratories seem at odds with the slow rates generally recorded on geological time scales. To some extent this apparent paradox was resolved by analysing evolutionary rates measured over different time scales (Gingerich 1993). Fast rates were maintained for only short periods, and the longer the observed time span, the lower the measured mean rate. Long evolutionary periods may have included short episodes with rates of change as fast as the modern examples, but over time these short episodes were diluted by long periods with little or no change. Moreover, over the existence of most species, the intensity and direction of selection are likely to change repeatedly, so the species simply wobbles around its phenotypic mean, with no consistent long-term trend. The relatively rare events involving the origin of a major new taxon require much greater than usual consistency in directional selection.

Temporal variation in rates of evolutionary change has given rise to the 'gradualist' and 'punctuated equilibrium' models of evolutionary change. The rapid changes that occur during speciation, as a result of selection to promote local adaptation, reproductive isolation and ecological divergence from related forms, are near the fast end of the range. At the other extreme, some organisms have remained unchanged, at least in fossil features, for millions of years. At the level of the lineage, some have shown gradual and continuous change over millions of years (the gradualist model), while others have shown short periods of rapid change associated with early radiation, interspersed with long periods of stability (the punctuated equilibrium model). In addition, within lineages, different characteristics evolve at different rates (a pattern called mosaic evolution), so that every organism is a patchwork of features, some of which have changed substantially in the recent past and others little for many millions of years. This is true of both DNA sequences and phenotypic features.

TAXON CYCLES

Each lineage, during its evolution, may pass through distinct and recognisable stages, involving expansion, differentiation and then contraction, the so-called 'taxon cycle' (Wilson 1961). The evidence for this view is based mainly on the observation that, at any one time, species with different levels of taxonomic distinctiveness, and hence of different ages, have different distributions. On islands, in particular, species evolve through a series of stages from newly arrived colonists, indistinguishable from their mainland relatives, to highly differentiated single-island endemics that ultimately become extinct. Imagine that a species arrives on an island (the 'birth' stage). It then spreads to become widespread, occurring on many islands in the same region, but is still taxonomically undifferentiated (Stage 1). Over time, the populations on different islands diverge, perhaps forming different subspecies on different islands (Stage 2). Eventually, some of the island populations die out, perhaps through competition with later colonists, while remaining populations continue to diverge and become increasingly specialised, forming distinct species (Stage 3). With further passage of time, other populations die out, leaving one remaining population, as an endemic species on a single island (Stage 4). Eventually, that too dies out, bringing the entire lineage to an end.

The bird faunas of some island groups can be interpreted as representing different stages in this temporal cycle (**Table 3.3, Figure 3.5**, Ricklefs & Cox 1972, 1978, Ricklefs & Bermingham 1999). On the Lesser Antilles in the Caribbean, species purportedly representing Stage 1 in their evolution from South American ancestors occur on most islands; their populations show little or no differentiation between islands or between islands and mainland. Species representing Stage 2 often have irregular distributions, occurring on some islands but not on others, and are distinguished to the level of endemic subspecies. Species representing Stage 3 consist of scattered endemic subspecies and species that still show obvious affinities to related populations on other islands and on the mainland. Species representing Stage 4, the last stage of the cycle, form highly differentiated endemic species or genera persisting as relicts on single islands. Similar patterns are apparent in the avifaunas of other island groups, and also in other types of animals (see Wilson 1961 for ants).

Species at different stages of the cycle also tend to occupy different habitats. Newly arrived colonists typically occur in coastal or disturbed habitats (as do introduced species). As they differentiate in isolation, they also expand into

Table 3.3   Criteria for assigning species to stages of the taxon cycle, according to Ricklefs & Cox 1972.

Stage	Taxonomic differentiation	Geographical distribution
1	None	Widespread
2	Differentiated	Widespread
3	Differentiated	Fragmented
4	Endemic	Single Island

The logic of the temporal sequence is as follows. Because no undifferentiated taxa have gaps in their distributions (being present on all islands in an archipelago), Stage 3 species can only have been derived from Stage 1 by differentiation (Stage 2), followed by local extinctions, and not by haphazard long-distance colonisation. Because parallel evolution of island populations is unlikely, Stage 1 species can only have originated by rapid colonisation (relative to differentiation) from a single source. Taxa could remain at Stage 1 indefinitely through continued immigration from the mainland, and migration between islands.

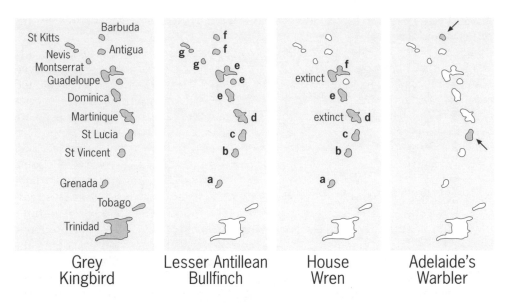

Figure 3.5   Representative distributions of passerine landbirds across the major islands of the Lesser Antilles, on Tobago and on the continental island of Trinidad. Letters next to island populations of the Lesser Antillean Bullfinch *Loxigilla noctis* and the House Wren *Troglodytes aedon* indicate named subspecies. The populations of Adelaide's Warbler *Dendroica adelaidae* on St Lucia and Barbuda are also strongly differentiated. The Grey Kingbird *Tyrannus dominicensis* is distributed throughout the West Indies; Adelaide's Warbler also occurs on Puerto Rico. From Ricklefs & Bermingham 1999.

other habitats, such as native forest. Under the continuing influence of interspecific competition, they then become increasingly specialised and restricted in distribution. Highly differentiated endemic forms are typically confined to a narrow range of habitats, usually rain forest or montane forest in the interior of islands. As the cycle proceeds, the older taxa show increasingly restricted habitat distribution, reduced population size and hence increased probability of extinction (**Figure 3.6**). Meanwhile, earlier colonists have been replaced by a wave of new colonists occupying the coastal and other disturbed habitats. Because immigration rates, local extinction rates and habitat preferences vary from stage to stage in this cycle, new and old colonists are likely to be concentrated in different habitats and geographical areas, with the older colonists in the older, more stable habitats.

The taxon cycle, as described above, was inferred from current distribution patterns, but studies of mt DNA of landbirds in the Lesser Antilles have now provided relative ages for taxa based on genetic differentiation among island populations (**Figure 3.7**). Stage 1 and Stage 2 species revealed similar genetic distances (sequence divergences), implying that morphological (taxonomic) differentiation appeared relatively quickly on these islands. The average genetic distance

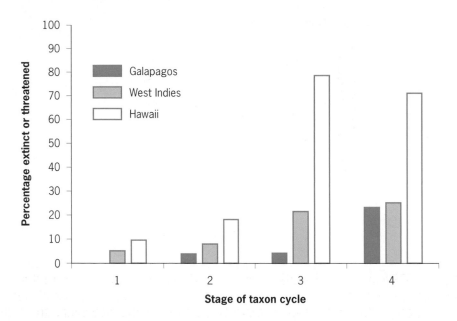

Figure 3.6 Proportions of island populations that are either extinct or threatened by extinction among taxa at each stage of the taxon cycle in the West Indies, Galapagos and Hawaiian Islands. The status of many of the Hawaiian species is uncertain because of human-caused extinctions prior to scientific collecting in the islands. However, the pattern with respect to stage in taxon cycle is robust to different interpretations of taxonomy and distribution. From Ricklefs & Bermingham 1999.

(in mt DNA) for Stage 2 species was about 1.1%, suggesting an average age of about half a million years (to reach subspecies stage). The average genetic distance for Stage 3 species, for which distribution gaps indicated at least one extinction event, was about 4.5%, indicating an average age of over two million years. Few Stage 4 species were analysed, but the oldest appear to have colonised the Lesser Antilles as long as five million years ago. These age estimates are consistent with the idea that the older species are those that tend to have restricted geographical and ecological distributions (Ricklefs & Bermingham 1999). Because ecology and distribution are strongly correlated with age across taxa, the rate of evolutionary change through the taxon cycle seemed relatively consistent among independently evolving populations. Moreover, because young taxa have continuous distributions within the West Indies, gaps in the ranges of older taxa indicate extinctions of island populations. Starting from this premise, Ricklefs & Bermingham (1999) estimated exponential extinction rates of about 50% per million years for island populations in the Lesser Antilles. This indicates an average population life span of about two million years. This was also the mean period required for speciation.

There is an apparent paradox in this proposed taxon cycle in that, as island populations adapt to the local environment and differentiate in isolation, they are still replaced by colonising species (that have never before experienced the local

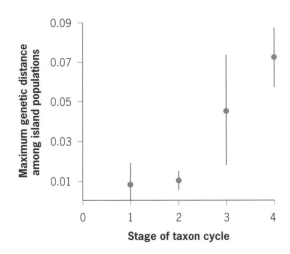

Figure 3.7   Genetic distances (means and standard deviations), based on mt DNA (ATPase 6 and 8 genes), among birds grouped according to the stage of the taxon cycle. In Stage 4 species, the distance is to the closest sister taxon. Data are presented as means and standard deviations to indicate the range of variation. Means differ significantly between stages: $F_{3,21} = 8.35$, $P=0.0008$, $R^2 =0.54$. From Ricklefs & Bermingham 1999.

conditions), and eventually decline to extinction. On the hypothesis proposed, the cycle is driven by fresh colonising species, which themselves undergo evolutionary differentiation and habitat shifts, and thereby also affect the evolution and ecology of longer-established species. The most remote islands receive fewest colonists, which prolongs the cycle and thereby permits the evolution of more highly differentiated endemics, adapted to the peculiar conditions found on remote islands. At the other extreme, islands close to mainlands have such high rates of immigration and associated gene flow that differentiation of populations is limited or prevented altogether. Taxon cycles should therefore be regarded as neither inevitable nor as routine evolutionary mechanisms; they are merely descriptions of some associated evolutionary and ecological processes.

On a larger scale, the relationship between global geographical range size and species age was examined for six monophyletic groups of extant birds by Webb & Gaston (2000). On the basis of molecular-derived estimates of age, broad relationships emerged between the ages of individual species and their global geographical range sizes. There was some suggestion of an increase in range sizes immediately post-speciation, followed by a subsequent decline towards extinction. These relationships among several groups of dissimilar birds provided further indication that bird range sizes may change in a consistent manner through evolutionary time. If generally true, the expectation of small range size at both ends of a species' lifetime may help to explain the fact that most extant bird species have small ranges (Chapter 14).

The concept of a taxon cycle thus provides a useful paradigm for understanding variation among species in geographical range, ecological distribution and vulnerability to extinction. However, the concept has not escaped criticism, mainly on the grounds that the patterns could be explained in other ways (e.g. Pielou 1979, Pregill & Olson 1981). However, the ageing of taxa from DNA analyses, and the relationship between taxon age and distribution, has largely squashed this criticism. There has also been resistance to the idea that distributions might have an evolutionary component. Nevertheless, taxon cycles may well be widespread, though most apparent on island archipelagos (Ricklefs & Cox 1972).

Consequences of specialisation

One major lesson from the taxon cycle is consistent with the fossil evidence from a wide range of organisms: although a particular lineage may have flourished for a period of time and radiated to produce a diverse range of species, this success is usually ephemeral on a geological time scale. Eventually, the rate of extinction exceeds the rate of speciation, and the group diminishes and finally becomes extinct. As revealed by the fossil record, episodes of speciation, radiation and extinction have occurred many times in the history of birds, and very many more times in the history of life, to produce a huge variety of extinct and living forms (Sepkoski 1989, Jablonski 1995, Rosenzweig 1995, Feduccia 1999). However, total life span may vary greatly from lineage to lineage.

Specialisation in any form is often regarded as something advanced and efficient, in contrast to more mundane characters which are often labelled as 'primitive'. Yet a high degree of specialisation, while it confers short-term advantages, can bring long-term vulnerability. It is useless being the only bird in the world that can open a certain nut if conditions change and the nut trees perish (Hall 1974). Often, therefore, specialisation might become a prelude to extinction, the terminal stage in a taxon cycle, however long that stage may last.

Occasionally, however, an evolutionary change that at first appearance represents an extreme specialisation enables an organism to exploit new opportunities and radiate to fill a new adaptive zone. The first bird was a very specialised reptile, but its modified scales led to the feathers that provided endothermy and flight, and thereby opened up a whole new way of life, so that a new group of vertebrates could radiate and colonise the world (Feduccia 1999). The most striking features of such large-scale evolution, of other animals as well as of birds, are the extremely rapid divergence of lineages near the time of their origin, followed by long periods in which basic body plans and ways of life are retained. The extreme speed of evolutionary changes that follow the evolution of each new type of animal is consistent with the difficulty of finding intermediate forms in the fossil record.

So-called key innovations, which lead to further adaptive radiation, can of course occur at any stage in the evolution of a group. Take, for example, the habit of seed storage by titmice (Sheldon & Gill 1996). The genus *Parus* (including *Poecile)* is divided into two lineages: one consists of seven species none of which is known to cache seeds, and the other consists of 23 species all of which are seed cachers. The benefits of seed stores for winter survival could have helped to promote the greater diversification of the seed-caching lineage. Other examples of key innovations associated with greater diversification of taxa are given by Heard & Hauser (1995).

SPECIATION TIMES

Estimates from land-bridge islands

The question of how long subspecies and species take to evolve was first addressed mainly for birds found on islands that were once connected to continents (Mayr 1963). Many such islands have subspecies and species that were clearly derived from mainland forms. On the assumption that such subspecies and species developed in the period since their separation, their maximum ages can be calculated, provided that the date of severance of the island from the mainland is known. Thus the British Isles began to split from mainland Europe about 7,500 years ago. Out of about 155 regular breeding species, they now have 33 kinds which have been subspecifically differentiated from their continental counterparts, differing in size and colour, and two other taxa, the Scottish Crossbill *Loxia scotica* and the Red Grouse *Lagopus l. scoticus*, which are at an intermediate stage between subspecies and full species (semispecies). On this basis, one might conclude that subspeciation takes less than 7,500 years, and full speciation much longer than this.

More rapid rates can be calculated in this way for Tasmania, which now lies 252 km from mainland Australia, from which it was severed by rising sea-levels about 12,000 years ago, having been joined for 29,000 years previously. It now has no fewer than 14 endemic species and 27 endemic subspecies out of a total of 104 breeding native landbirds (**Table 3.4**). The endemic species include two monotypic genera now confined to the island, ten members of superspecies that have sister species on the mainland, and two members of double invasions, the most recent of which are undifferentiated or subspecifically differentiated from the mainland forms. While the more distinctive endemics in monotypic genera may once have occurred more widely, but have since become extinct on the mainland, almost certainly the remaining ones evolved their current level of differentiation on Tasmania (Ridpath & Moreau 1966, Keast 1981a). Additional endemic species or races occur on most of the other islands off Australia separated at the same time as Tasmania (Keast 1981a).

The problem with this type of evidence is that, unless the species was absent from an island altogether during the glaciation (and period of low sea-level), one can never be sure that its population was wholly undifferentiated then. Tasmania was never completely glaciated and, lying further south than mainland Australia, it is likely always to have had some distinct habitats, as it does today. In other words, at least some of the endemics might have been confined to Tasmania for much longer than the last 12,000 years, including times when the island was merely a southern peninsula of Australia. The same argument could apply to endemic species and subspecies on many other land-bridge islands (including Britain) and, in other species, rising sea-levels could have merely broken a previously clinal gradient. Moreover, while some species have changed morphologically in this time, many others have not.

The fact that some continental islands may have held differentiated taxa before they were last separated from continents limits their value in assessing rates of evolution. The safest figures derive from islands that were largely or wholly glaciated and received most or all of their avifauna in the last 12,000 years or less.

Table 3.4 Endemic bird species on Tasmania and their nearest Australian equivalents. Modified from Ridpath & Moreau 1966.

	Member of	Australian counterparts
Tasmanian Native Hen *Gallinula mortierii*	Superspecies	Black-tailed Native Hen *Gallinula ventralis* (allospecies)
Green Rosella *Platycercus caledonicus*	Superspecies	Crimson Rosella *Platycercus elegans*, Adelaide Rosella *P.e. adelaididae* and Yellow Rosella *Platycercus flaveolus* (allospecies)
Orange-bellied Parrot *Neophema chrysogaster*	Superspecies	Blue-winged Parrot *Neophema chrysostoma* (allospecies)
Swift Parrot *Lathamus discolor*	Monotypic genus	—
Tasmanian Scrubwren *Sericornis humilis*	Superspecies	White-browed Scrubwren *Sericornis frontalis* (allospecies)
Scrubtit *Acanthornis magnus*	Monotypic genus	—
Forty-spotted Pardalote *Pardalotus quadragintus*	Double invasion	Spotted Pardalote *Pardalotus punctatus* (Second invader)
Tasmanian Thornbill *Acanthiza ewingii*	Double invasion	Brown Thornbill *Acanthiza pusilla* (Second invader)
Yellow Wattlebird *Anthochaera paradoxa*	Superspecies	Red Wattlebird *Anthochaera carunculata* (allospecies)
Yellow-throated Honeyeater *Lichenostomus flavicollis*	Superspecies	White-eared Honeyeater *Lichenostomus leucotis* (allospecies)
Strong-billed Honeyeater *Melithreptus validirostris*	Superspecies	Black-chinned Honeyeater *Melithreptus gularis* (allospecies)
Black-headed Honeyeater *Melithreptus affinis*	Superspecies	White-naped Honeyeater *Melithreptus lunatus* (allospecies)
Dusky Robin *Melanodryas vittata*	Superspecies	Hooded Robin *Melanodryas cucullata* (allospecies)
Black Currawong *Strepera fuliginosa*	Superspecies	Pied Currawong *Strepera graculina* (allospecies)

In addition to the endemic species listed above, Ridpath & Moreau (1966) listed 27 distinct subspecies but some are doubtfully differentiated: Wedge-tailed Eagle *Aquila auda*, Australian Owlet-Nightjar *Aegotheles cristatus*, Emu *Dromaius novaehollandiae* (now extinct), Brown Quail *Coturnix ypsilophora*, Tawny Frogmouth *Podargus strigoides*, Ground Parrot *Pezoporus wallicus*, Eastern Rosella *Platycercus eximius*, Lewin's Rail *Lewinia pectoralis*, Morepork *Ninox novaeseelandiae*, Australian Masked Owl *Tyto novaehollandiae*, Black-faced Cuckoo-shrike *Coracina novaehollandiae*, Australian Raven *Corvus coronoides*, Grey Butcherbird *Cracticus torquatus*, White-backed Magpie *Gymnorhina hypoleuca*, Eastern Spinebill *Acanthorhynchus tenuirostris*, Tawny-crowned Honeyeater *Phylidonyris melanops*, Noisy Miner *Manorina melanocephala*, Grey Shrike-thrush *Colluricincla harmonica*, Golden Whistler *Pachycephala pectoralis*, Grey Fantail *Rhipidura fuliginosa*, Brown Thornbill *Acanthiza pusilla*, Superb Fairywren *Malurus cyaneus*, Little Grassbird *Megalurus gramineus*, Southern Emuwren *Stipiturus malachurus*, Spotted Quail-thrush *Cinclosoma punctatum*, White-fronted Chat *Epthianura albifrons*, White-eye *Zosterops lateralis*.

One such island is probably Iceland which has ten endemic subspecies, among about 50 breeding landbirds, most of which could not have been present in glacial times. This again implies that at least several thousand years are needed for subspeciation, assuming that these species colonised soon after conditions became suitable.

Estimates from oceanic islands

The majority of volcanic islands that arose from the sea bed are less than five million years old, and the ages of many have been estimated using potassium–argon or other chemical methods (Chapter 2). On this basis, all the endemic bird species must have formed well within this period, but without more precise knowledge of when their ancestors arrived, the range of possible ages is wide. Moreover, most volcanic islands once had others nearby that have since become eroded and submerged. Hence, birds that colonised the older islands may have moved to the younger ones as they became available, so that the lineages may be older than the islands they now occupy. This is not true of the 35 km² Norfolk Island, however, which lies nearly 1,400 km east of Australia, and has been dated at three million years old (Schodde *et al.* 1983). At the time of its discovery by European sailors, it held five endemic species, nine endemic subspecies and ten other native birds of land and freshwater habitats. Its bird taxa therefore showed varying degrees of differentiation from Australian relatives, and presumably evolved from ancestral colonists within this period, giving the oldest endemics a maximum age of up to three million years.

Flightless birds on islands are particularly revealing, because they must have lost their powers of flight only after having got there. The extinct ibises from Hawaii are known only from islands less than 1.8 million years old (Olson & James 1982) and in this time their flightlessness had become evident in their skeletal remains (Olson & James 1982). Aldabra Island in the Indian Ocean was submerged several times in the Pleistocene, and most recently re-emerged about 80,000 years ago. All the birds must have colonised after this date, including the Aldabra White-throated Rail *Dryolimnas cuvieri aldabranus* whose skeleton has also been considerably modified. Its living ancestor on Madagascar can still fly. The fact that many mainland species have flightless subspecies on islands provides further indication that flightlessness can evolve rapidly (Chapter 6).

An even more remarkable situation exists on two islands (Long and Tolokiwa) in northern Melanesia (Mayr & Diamond 2001). A volcanic eruption on Long in the seventeenth century wiped out the entire fauna of these islands, but they now hold 4–5 recognisably distinct bird populations (classed as subspecies), which must therefore have developed from colonists within about 300 years. Their formation may, however, have involved founder events and the stabilisation of hybrid populations, rather than selection alone. Nonetheless, they serve to illustrate how rapidly distinct forms can arise in the wild given appropriate conditions. More estimates of the time needed for speciation are likely to become available as more volcanic islands are precisely aged, especially those that appeared above the waves in the last few million years.

Estimates from continents

The rise in sea-level that led to the severance of islands from continents was caused partly by the melting of glaciers at the end of the last ice age. Before that time, forests in both Eurasia and North America were much more restricted than today, being confined to a small number of refuges, each well separated from the others (Chapter 9). The same bird species in each of these refuges could therefore have evolved independently, giving rise to distinct subspecies and species. Many pairs of closely related species in North America are thought to have arisen in distinct east and west forest refuges, as are similar pairs of European birds (Chapter 10). The assumption was formerly made that these various species pairs arose during the last glaciation, which peaked around 26–14,000 years ago. They must therefore have formed within a period of at least 12,000 years. This is a minimum estimate, because the forest probably fragmented long before the glaciation reached its peak, and remained so for long afterwards.

These estimates suffer from the same problem as those for island birds, namely that the taxa concerned may have differentiated to some extent well before the last glaciation. There was not just one glacial cycle, but many, running one after another (Chapter 9). The birds of land-bridge islands, and of fragmented mainland habitats, would therefore have been exposed to repeated periods of separation, when divergence could have occurred, broken by periods of fresh contact, when diverged populations could have expanded and interbred. This raises the possibility that speciation could have been an extended process in which progressive divergence during periods of separation received periodic set-backs during the shorter periods of re-contact (Chapter 10). On such a system, speciation of particular birds may have begun in any one of many glaciations, and without knowing which, taxa cannot be aged precisely in this way.

Estimates from DNA analyses

Analyses of DNA have provided an alternative means of estimating the ages of species and subspecies, that is by dating the population separations that started the speciation process. The majority have been based on nucleotide sequencing of the mt DNA (often the cytochrome *b* gene). This procedure gives the date of divergence of pairs of related taxa from the number of independent mutations that have arisen since their separations (Chapter 2). If several subspecies and sister species thought to have been split by the same geological or climatic event can be compared, a range of values can be obtained for different lineages, which puts some limits on the duration of the speciation process. However, estimates are again made on the inadequately tested assumption of a consistent rate of mutation in equivalent pieces of DNA from closely related taxa.

The ageing of species by molecular methods has often given results that differed greatly from those obtained by more traditional methods, and that vary greatly between species pairs supposedly affected by the same event. For example, divergence dates for North American subspecies and sister species appear to range from less than 200,000 to more than five million years, with typical values for sister species of around 2.5 million years (Klicka & Zink 1999, correcting Avise *et al.* 1998). About 80% of all phylogeographic separations so far studied in North

American sister species are estimated from DNA to have started more than one million years ago (Klicka & Zink 1999). Similar estimates for mammals, made in the same way, gave a median figure of 2.2 million years, close to that for birds (Avise 2000). If these estimates are correct, many of the population fragmentations that led to full speciation in birds and mammals must have begun earlier than the late Pleistocene, perhaps in the preceding period (the Pliocene), though speciation might have been an extended process as described above, starting in the Pliocene and completing in the late Pleistocene. In contrast, fragmentations beginning in the late Pleistocene may by now have differentiated only to the level of sub-species. Other estimates from mt DNA suggest an average divergence time for non-sister species in the same genus of about 3.9 million years (Johns & Avise 1998). Clearly, the more divergent the taxa, the greater the time since their separation.

Other estimates based on DNA analyses of the time taken for species formation have been made for island birds, clearly derived from a particular mainland ancestor. For example, the change from the Common Chaffinch *Fringilla coelebs* to Blue Chaffinch *F. teydea* on the Canary Islands seems to have occurred within the last million years, judging from the degree of divergence in mt DNA and in protein-encoding genes (Baker *et al.* 1990). These two chaffinches look different and occur together without interbreeding. Among different races on various Atlantic Islands, the ancestors of the Azores subspecies *F.c. moreletti* were estimated to have colonised these islands (presumably from Iberia) about 600,000 years ago, this being followed by rapid colonisation of Madeira and (for the second time) the Canary Islands (Marshall & Baker 1999). These various chaffinches are regarded as subspecies, but are very distinctive looking. Their estimated ages lie within the estimated geological ages of the volcanic Canary Islands on which they now occur, which range between one and 16 million years (Juan *et al.* 2000).

Other estimates from islands include the Mauritius Parakeet *Psittacula echo* which is estimated to have split from Asian *Psittacula* stock 1.5 million years ago, and the very distinctive Mauritius Kestrel *Falco punctatus* which is estimated to have diverged directly from Common Kestrel *F. tinnunculus* stock about 1.7 million years ago, rather than via Madagascar which was colonised somewhat earlier (Groombridge *et al.* 2000). These estimates are well within the estimated age of Mauritius at less than 20 million years. Another island example is the Hawaiian Goose *Branta sandvicensis* which, on the basis of mt DNA (cytochrome b gene), is estimated to have split from the North American Canada Goose *Branta canadensis* following a colonisation event less than three million years ago — most probably 0.9 million years ago (Quinn *et al.* 1991). This event preceded the divergence of the large- and small-bodied forms of Canada Geese on the North American continent, also dated within the last three million years, but most probably around 0.6 million years ago. Those other estimates for island birds, mentioned earlier in this chapter, mostly fall in the range 0.5–5.0 million years for species and up to one million years for subspecies. Moreover, many distinctive subspecies that show no differentiation in their mt DNA and occupy recently deglaciated areas can be assumed to have formed in as little as 10,000 years (Chapters 2 and 10). As stressed above, errors in all these estimates could arise from faulty dating of the event used to calibrate the 'molecular clock', or from non-uniform rates of molecular evolution between related species. Nevertheless, we must accept that, while rates of divergence may have varied between species, many of the distinctive species that we know today have evolved in the last few million years, and some well within the last million years. As speciation times, they are maximum estimates because, while mt DNA may have continued to mutate to the present day, the general appearance and behaviour of the species may have stabilised long ago.

CONCLUDING REMARKS

As stressed at the outset, an understanding of how species are formed is central to understanding their distributions. Moreover, the way in which species are defined and delineated has a major influence on how we assess the locations and sizes of their geographical ranges. In particular, geographical forms classed as subspecies would each occupy particular areas which, added together, would give the range of the species as a whole. Yet if these same geographical forms were classed instead as allospecies, then we would have several smaller ranges in place of one large one. The way in which species are defined also has a major influence on the numbers of species that are recognised, both overall and in particular regions. This in turn influences all regional comparisons that depend on assessments of species numbers. Evolutionary reality and taxonomic practice thus underpin most of the issues discussed in this book and in fact most of biogeography, an aspect discussed further in Chapter 4.

Taxonomic reality involves recognition of the continuum of variation found in nature. This continuum holds both through time as lineages evolve, diversify and die out, and through space as populations adapt to local conditions, fragment and diverge. The continuity of the evolutionary process means that decisions on the limits of species, and hence on their distributions, are inevitably to some extent arbitrary. The important point is that species should be defined and delimited in a consistent manner from family to family and from region to region, incorporating changes in understanding as they occur.

Taxonomy and distribution are linked in yet another way. Taxonomic categories are hierarchical, so that an order contains within it sets of families, which in turn contain sets of genera and then species that supposedly represent the evolutionary lineage. In the same way, the geographical range of an order contains within its boundaries the ranges of all its families, which in turn contains the ranges of all its genera and then all its species. In this sense, too, genealogy and distribution are not separate issues: they are simply different sides of the same biological coin.

SUMMARY

New species can be formed by the gradual transformation of one species into another through time, but most evidence for birds points to allopatric speciation — to the fragmentation of a single parental species to give different geographically isolated subpopulations which, lacking genetic input from other subpopulations, can then diverge from one another, eventually forming distinct species. Divergence could occur through genetic drift and mutation, as well as through natural and sexual selection. The process can be seen most clearly in the occurrence of allospecies — different species each occupying a different area but clearly derived from the same parental form. Thus speciation in birds is essentially a geographically based process, and the place of origin of a species can do much to influence its eventual distribution.

Further differentiation involving the development of reproductive and ecological isolation, enables taxa developed in separate areas to expand their ranges and co-exist as sympatric species. Reproductive isolation between two taxa develops if: (1) interbreeding is prevented by mate recognition or incompatibility of egg and sperm, or (2) interbreeding occurs, but hybrids are less viable or fertile than parental forms or are unable to attract mates. Ecological segregation develops if the taxa diverge sufficiently in habitat and food needs to enable them to co-exist without competing to the level that one eliminates the other. Overall, the most significant forces in the speciation process in birds are selection, non-random mating and gene-flow (or lack of it).

Sympatric speciation, involving the simultaneous formation of two or more species within the same area from the same parental species, is probably rare in birds, if it occurs at all.

Hybridisation between different taxa is potentially important in birds, mainly in merging previously distinct taxa, but also perhaps in triggering the formation of new taxa distinct from either parental form (although its frequency cannot yet be reliably assessed). Apparently stable hybrid zones between allopatric species and subspecies are fairly common in birds.

Among related bird species, the degrees of phenotypic differentiation, reproductive isolation and genetic differentiation are poorly correlated with one another.

From analysis of mt DNA, species formation in birds is estimated to take around 2.5 million years but varying between 0.2 and 5.5 million years, while sub-speciation is estimated to take less, sometimes less than 10,000 years. These estimates are based on the assumption that the molecular clock ticks at a consistent rate between related lineages (often taken as around 2% per million years for the mt DNA cytochrome b gene). In general, they are not inconsistent with estimates from other evidence.

Striking examples of adaptive radiation in birds, and of taxon cycles, are found on islands. Some adaptive radiations, involving the diversification of a single lineage into several widely different species, are estimated to have taken less than eight million years in the Hawaiian honeycreepers and less than three million years in the Galapagos finches. The central idea of the taxon cycle is that each lineage, during its evolution, passes through recognisable stages, involving expansion, differentiation, contraction, and then extinction. Taxa at different stages of this cycle show characteristic differences in distribution patterns, habitat preferences and taxonomic distinctiveness. In some birds of the Lesser Antilles, speciation is estimated to have taken about two million years, and the full taxon cycle from colonisation to single island endemic more than five million years.

The continuity of the evolutionary process, through space and time, means that decisions on the limits of species and their distributional boundaries are to some extent arbitrary, but the way in which species are defined and delineated can influence our perceptions of distribution patterns and diversity.

Willow Warbler *Phylloscopus trochilus*, a member of a large genus of similar-looking species.

Chapter 4
Species numbers

Any consideration of the delimitation and classification of bird species raises the question of how many there are. In fact, the number of bird species in the world is known with much greater accuracy than are the numbers of species of any other types of organism, except perhaps mammals. The most recent worldwide listing of birds, by Sibley & Monroe (1990), follows the DNA-based classification for higher taxonomic categories (tribes, families and orders), but mainly other information for species (and to a lesser degree genera), which were delimited primarily on the basis of the most recent studies available at the time, taking account of vocalisations and other potential isolating mechanisms. This classification recognises 23 orders of living birds, divided into 146 families and 9,736 species (based on the updated 1996 CD version), applying the biological species concept (see later).

The species number is larger than in the preceding list, and is still under constant revision in light of new findings. Each year, a handful of previously undescribed bird species is discovered (mean 5.5 per year during 1938–90, Vuilleumier *et al.*

1992), mainly through further exploration of little-known areas. These newly found species usually occur in very restricted areas, and many are also small and inconspicuous, so it is not surprising that they were previously missed (Blackburn & Gaston 1995). In addition, each year, some other 'already known' species are each split into two or more separate species, either through reassessment of existing museum skins or from use of new criteria, such as vocalisations or DNA analyses. To set against these additions, each year one or more known species may become extinct (usually through human action), and in light of new information or reassessment, other birds previously considered as different species may be combined as single species.

Clearly, those changes resulting from further research could continue almost indefinitely, as could those resulting from swings in taxonomic fashion. Because the arbitrary dividing line between subspecies and species has fluctuated over the years, in response to shifts in prevailing opinion, the number of recognised species has changed accordingly. For example, the decision by Sibley & Monroe (1990) to promote many geographically very distinct 'subspecies' to the status of allospecies immediately added about 1,000 'new' species to previous lists. Ever since organisms were first scientifically classified in the nineteenth century, taxonomists have been designated as 'lumpers or splitters', according to their preference to combine or subdivide closely related taxa. About a century ago, the splitters prevailed in ornithology, followed by the lumpers, but now the splitters are again gaining influence, largely under the rubric of the phylogenetic species concept (see later). Hence, bird species lists fluctuated by two-fold during the twentieth century according to whether species were defined broadly or narrowly, and are likely to lengthen in the coming years, again through greater subdivision of already known taxa.

The number of recognised species has, of course, increased relatively more in some families than in others. The owls are an extreme example: Peters (1940) listed 141 known species in the families Tytonidae and Strigidae; less than 60 years later, del Hoyo et al. (1999) listed 195 species in these families and König et al. (1999) recognised 'at least 212 species'. This represents a 38% increase over the period concerned. The genus Otus experienced the greatest growth, from 36 to 62 species, partly through the discovery of 12 previously unknown species and partly through the elevation of 14 previously designated subspecies to species.

Extinctions are a different matter. Those resulting from the destruction of tropical forests are likely to involve many undescribed species of plants and animals. They will almost certainly include some birds, so that known avian extinction rates (on average less than one bird species per year, Newton 1998a) are probably underestimates of the real values. Over the past 400 years, at least 127 named bird species have become extinct. About 116 of these were island endemics (Newton 1998a). To these recent extinctions of named species we could add up to a few thousand other bird species known or surmised to have disappeared from islands in the past 1,500 years, owing to human action, and now represented only as subfossil bone remains (Olson & James 1991, Steadman 1995; Chapter 7). The implication is that, in the absence of people, the number of bird species on earth would be at least 20% greater than it is now. In the last few hundred years, human-induced extinction rates have become very much greater than the average background rate calculated from the fossil record (Lawton & May 1995). In addition to birds, recent extinctions include at least 58 species of mammals, 100 reptiles and 64 amphibia.

In terms of the loss they represent to biodiversity, not all species are equal: one without close relatives represents a more unique evolutionary heritage than a species that forms part of a large family. For example, the loss of unique species such as the Kagu *Rhynochetos jubatus* or the Hoatzin *Opisthocomus hoazin*, both of which are the sole remaining representatives of their families, would constitute a more significant loss of genetic material than would the extinction of any two species from a large and relatively uniform group, such as the *Phylloscopus* warblers. Known extinctions of birds over the past thousand years or so have included a much greater proportion of unusual species from distinct lineages than expected by chance (Chapter 7). They therefore represent a greater loss of evolutionary history than expected from the number of species involved. In terms of conservation, we may increasingly in future have to assign quantitative value to the taxonomic distinctiveness of different species, for use in setting priorities and in the selection and management of protected areas (May 1990).

Species richness of families

With around 9,736 extant bird species distributed among 146 taxonomic families, each family contains an average of 67 species. In practice, however, over half the species are contained in just 12 large families, each of which holds more than 250 species. The largest family on the Sibley–Monroe classification is the Fringillidae, with at least 996 species. At the other end of the spectrum, almost half of bird families contain fewer than ten species each, and collectively they hold fewer than 280 species (**Figure 4.1**). This pattern of species distribution in families differs significantly from the pattern expected by chance, mainly because the numbers of

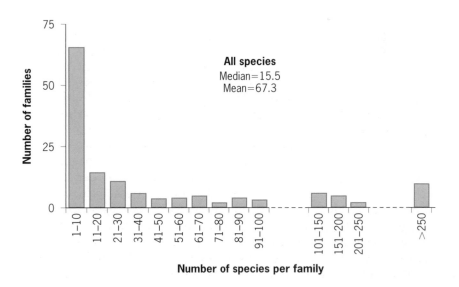

Figure 4.1   Species numbers in the 146 families of living birds. Based on information in Sibley & Monroe 1990.

species-rich and of species-poor families greatly exceed random expectation (Nee *et al.* 1992, Owens *et al.* 1999). It is not simply a matter of family age, for no correlation is apparent between the supposed age of families and the numbers of species they contain (Owens *et al.* 1999). Obviously other factors are at work.

The first point to make is that such patterns of phylogenetic inequality are not unusual. In most kinds of plants and animals, whatever the taxonomic level considered, from phyla through orders and families to genera, a small proportion of taxa contain far more species than the average (Marzluff & Dial 1991). Secondly, whatever the type of animal, high species richness is usually associated with relatively small body size and short generation time (Maurer *et al.* 1992, Brown 1995). These relationships hold for birds, with the small-bodied, short-generation Passeriformes containing 57% of all bird species (Owens *et al.* 1999). This may be because smaller birds (or other animals) can subdivide a habitat into more niches than can large ones: small passerines can split the feeding sites and food-resources of a forest more finely than can larger pigeons or raptors, say. Moreover, small birds generally have shorter generation times than larger ones (as reflected in age at first breeding, annual reproductive rate and life span). Short generation time may itself help to increase diversity because it could promote rapid speciation (through increasing the maximum rate of mutation and divergence) and reduce extinction risk because it enables rapid recovery from vulnerable low numbers (Marzluff & Dial 1991). Small birds also tend to disperse over shorter distances than large ones, which in turn facilitates the isolation of populations that promotes speciation (Chapter 17). Passerines themselves seem to have more neural capacity, and a greater ability to learn and experiment than many other birds, which may give them a lead in exploiting new environments. These and other relationships suggest that interaction between properties of both the animals themselves and of their environment influence species richness, but the matter is far from resolved (Barraclough *et al.* 1995, 1998, Rosenzweig 1995).

COMMENTS ON SPECIES CONCEPTS

As already indicated, the number of species that are recognised depends not only on the processes that lead to their formation and extinction, but also on the way we choose to subdivide them. Species could be delimited in any one of several apparently reasonable ways, according to different 'species concepts'. Some kinds of birds are so distinctive that they would be classed as species on any of the accepted concepts, but others could be classed as species on some concepts but as subspecies or lower on others. Hence, whichever concept is in general use affects the numbers of species that are recognised. We must of course accept that the birds themselves may be able to detect differences between taxa that are not obvious to us, and that we can detect differences (such as protein or DNA structure) that are not necessarily obvious to the birds.

For many years, birds throughout the world have been subdivided according to the so-called 'biological species concept' (BSC, also called 'isolation species concept'), which was derived from study of the geographical variation and distribution patterns of birds (Mayr 1942). On this concept, species are viewed as 'groups of actually or potentially interbreeding natural populations which are reproductively isolated from other such groups' (Mayr 1963). Species are thus regarded as more or less closed genetic systems, such that, if hybridisation occurs, it is limited in some way, and does not result in the fusion of two separate gene pools. Lack of effective interbreeding is the key element in this concept, so that populations normally maintain their distinctiveness through space and time. This gives species a special status: they are distinct entities because they do not interbreed and merge. It also gives the BSC a solid biological foundation, based on current distribution and reproductive behaviour.

Hybridisation presents problems in species delimitation, because of the wide variation in its frequency and in its consequences. For bird species that occur together in the same area when they are breeding, we can be certain from observation that they do not normally hybridise, and have retained their integrity through time. However, for the vast majority of species which do not occur sympatrically when breeding, we have to make a judgement on whether they would be likely to retain their distinctiveness if they met. In this respect, the BSC is prospective, in that only future events could show whether many currently recognised species that now have different ranges would remain distinct if they came together.

Problems with the BSC have arisen from too strict an application of the criterion of reproductive isolation. If individuals of different species mated and produced either no progeny or infertile progeny, we would be fully justified in calling them separate species, but if they produced some fertile progeny, say 10%, could we still do so? In fact, as explained in Chapter 3, many birds classed as separate species interbreed occasionally, and in some cases the hybrids have some degree of fertility. Moreover, hybridisation occurs not only between closely related sister species, but occasionally also between more distantly related ones, some of which show distinct hybrid zones (Moore *et al.* 1991, Zink 1994, Freeman & Zink 1995). The problems are even greater in other organisms, such as flowering plants in which hybridisation often produces new species.

The main criticisms of the BSC are therefore that: (1) it is process-based, derived from an inferred speciation process rather than from a rigorous analysis of taxonomic characters (even though the two would usually give the same result); (2) the criterion of effective reproductive isolation is often untestable, leading to informed guesses on the status of some closely related allopatric populations; and (3) it overplays the significance of hybridisation between different taxa, and in particular, through uniting paraphyletic taxa because they interbreed along a contact zone, it might misrepresent their separate evolutionary histories (Cracraft 1983, Zink & McKitrick 1995). The BSC is also restricted to sexually reproducing organisms, and is inappropriate for asexually reproducing ones, which form a large proportion of all organisms. Other objections seem less crucial because they result largely from misapplication of the BSC through insufficient knowledge or inadequate analyses. Nonetheless, despite the problems associated with the term 'reproductive isolation', the biological species concept has been widely accepted by ornithologists, partly because in general it reflects the distributional and phylogenetic realities well. It has also stood the test of time, has given a period of relative stability to nomenclature, and seems to present fewer problems than other species concepts, discussed below. It has been much easier to criticise the BSC than to improve on it.

One variant of the BSC is the 'recognition species concept' (RSC), which defines species as 'groups of individuals that share a common fertilisation system or specific mate recognition system' (Paterson 1985). This concept thus emphasises the factors that ensure interbreeding within a population of individuals, rather than the factors that prevent or reduce interbreeding between members of different populations, as does the BSC. It also assumes that speciation results from changes in fertilisation or mate recognition systems, and thus depends on detail that may not be available. Another variant of the BSC is the 'cohesion species concept' (CSC), which emphasises the genetic and ecological cohesion that is a fundamental characteristic of species, rather than the causes of that cohesion, as does the BSC and RSC (Templeton 1989). The CSC is also appropriate for both sexual and asexual organisms.

The main current alternative to the BSC and related concepts is the 'phylogenetic species concept' (PSC), in which species are defined as recognisable lineages: a species is 'the smallest diagnosable cluster of individual organisms within which there is a parental pattern of ancestry and descent' (Cracraft 1983). Any criterion — whether morphological, behavioural or genetic — can be used to identify individuals as belonging to the same 'diagnosably different' lineage provided that the criteria are genetically controlled, and result from evolution[1]. Whereas biological species are defined primarily by lack of interbreeding, for phylogenetic species this is merely one of several possible criteria, allowing for the possibility that interbreeding may be a retained or transient trait. On this basis, the Red-shafted Flicker *Colaptes auratus cafer* and Yellow-shafted Flicker *C. a. auratus* in North America would be considered as separate phylogenetic species (rather than geographical races of a single species), as would the Carrion Crow *Corvus c. corone* and Hooded Crow *C. c. cornix* in Eurasia. Although the members of each pair interbreed in a zone of overlap (and for this reason are usually classed together as a single species on the BSC), the parental forms are nevertheless separate lineages that maintain their genetic and morphological integrity. In a nutshell then, whereas the BSC is a 'horizontal' concept which is based on current reproductive status, the PSC is a 'vertical' concept which is based on phylogenetic history. It thus defines species in a manner that is consistent with the theory and practice of phylogenetics. It does not matter whether two species are prevented from merging back into each other by biological factors (such as reproductive isolation) or by geographical factors (giving spatial isolation). Several versions of the PSC have been put forward by researchers working on different kinds of organisms, but all emphasise phylogenetic relationships (descent) and not current reproductive relationships.

Full-scale application of the PSC, if it ever became generally accepted, would greatly increase the numbers of geographically distinct bird populations that were classed as species rather than as subspecies. One such estimate put the numbers of living bird species based on the PSC at 20,000, almost double the number accepted on the BSC (Snow 1997). This is not very different from the 18,939 species listed by Sharpe (1899–1909) a century ago, who also classed many currently

---

[1] *Defining species solely on their DNA is hardly a practical proposition, because the method could not be used in the field, but would depend on a retrospective judgement based on a biochemical test of a tissue sample.*

recognised subspecies as species. The numbers themselves are unimportant, but they highlight one potential problem with the PSC as applied to birds, namely that it could become a runaway process, leading to repeated subdivision of existing species, especially if the subdivision were based not just on morphological but also on physiological and other non-visible criteria. We would then lose in nomenclature the reflection of degree of difference that the terms species, allospecies and subspecies denote, for the term species would be used to embrace all three categories of distinctiveness. Furthermore, subspecies that intergrade with other subspecies or for any other reason are not safely diagnosable would be abolished in nomenclature, and large amounts of genetic diversity would receive no taxonomic recognition. The category of subspecies is useful because it indicates nearest relatives that differ less from one another than from the various species in that genus, and also that these nearest relatives are allopatric. Both the BSC and the PSC attempt to distinguish the terminal twigs in the evolutionary tree, but whereas the BSC distinguishes in its nomenclature the different types of twigs, the PSC as currently defined, does not. There are, in addition, more practical problems in applying the PSC than the BSC, particularly in little known avifaunas (Snow 1997)[2].

A precursor to the PSC is the 'evolutionary species concept' (ESC), which defines species as populations of distinct lineage with 'their own evolutionary tendencies and historical fates' (Simpson 1951, Wiley 1981). It emphasises the idea of a species as representing a distinct lineage that maintains its integrity (with respect to other lineages) through both time and space, not just one of these aspects, as do other concepts. On the ESC, every population is considered a species if it is sufficiently isolated by either genetic or geographical barriers from other populations that it can diverge. This concept admits specific rank to taxa that hybridise strongly, so long as the hybridisation occurs within well-defined contact zones and does not break down the genetic integrity of the taxa involved. The ESC takes account of the fact that populations can diverge in isolation and then come together again (reticulate evolution). Such populations would be recognised as independent units while they were isolated and as one species when they re-joined, as in the Snow–Blue Goose *Chen caerulescens* example mentioned in Chapter 3. The ESC can accommodate all kinds of species, including those that reproduce asexually; it also acknowledges subspecies. With these characteristics, the ESC is particularly useful in biogeographical research. Its application would increase the number of taxa classed as species, but not to the same extent as the PSC. Its main drawback is that it gives no simple criteria on which species can be defined. It is in fact a theoretical or 'primary' concept that encompasses the other concepts (Mayden & Wood 1995, Mayden 1997). They can all be viewed as operational or 'secondary' concepts, which provide working criteria for the determination of specific status under particular circumstances. Without such criteria, we could not separate species one from another. These various secondary concepts can thus be regarded as operational arms of the ESC (Parkin 2002). In effect, the

---

[2]*Applying these concepts to* Homo sapiens, *some of the different coloured races of people could be classed as different species on the PSC (because they are diagnosably distinct on colour and appearance), rather than as essentially allopatric subspecies of the same species, as on the BSC.*

ESC tells us what a species is, and the other concepts how to recognise it, based on heritable characteristics. In their application, all concepts have problems with hybridisation, all face arbitrary decisions over borderline cases, and all carry assumptions about the past and expectations about the future.

Yet other species concepts have been proposed from time to time, but they do not reflect the distributional and phylogenetic realities of birds well and are more appropriate to other organisms. Also, it is hard to apply any existing species concept satisfactorily to extinct forms. Fossil birds are distinguished on skeletal features and are in no way equivalent to living birds, which are separated largely on plumage or behaviour. Many extant closely related species would be indistinguishable from one another on skeletal features. Palaeontologists also have the problem of dividing continuous slowly changing evolutionary lineages into discrete taxonomic entities equivalent to species.

The fact that different species concepts give different numbers of species from the same range of natural diversity emphasises the difficulty of attempting to force the continuum of natural variation into a human-designed simplified system of categorisation. Many of the entities that are currently called species are not such discrete natural units as any of the species concepts imply. Changing from one species concept to another does not remove problems; it merely moves the boundaries between species, and alters the taxonomic decisions that emerge as problematic. Different concepts may indeed be best for different purposes; all have areas of agreement, notably in requiring meaningful reproductive isolation, and none is likely to satisfy everyone. Moreover, no labelling system can reflect adequately what is going on all the time in the natural world. The words of Darwin (1859) ring as true today as when they were written: 'No one definition has yet satisfied all naturalists; yet every naturalist knows vaguely what he means when he speaks of a species.'

Whatever one thinks of these different concepts, for the most part they divide up the natural diversity of birds in essentially the same way. Another type of question is whether people from totally different cultures perceive the natural diversity of birds in broadly the same way as we do, and subdivide it into the same named units. In general, this seems to be so, at least among those types of organisms that are important to them. Thus pre-literate peoples of New Guinea had vernacular names for 136 of the 137 native birds recognised as separate species by trained zoologists (Mayr 1963). This held even when the different species were extremely similar to one another, and separated by the most minor of morphological features.

SPECIES THROUGH TIME

Because the evolution of any group of organisms is assumed to begin with one or a few species, and lead to the many alive today, species numbers must increase through time, at least during the early development of a group. The increase is likely to be rapid initially, as the new group expands to take advantage of opportunities, and then to slow down, as openings become filled. So what determines how many species could co-exist at one time? Addressing this question involves consideration of the ultimate factors that limit the capacity of the earth to support species, and the proximate factors that govern the formation of new species and the extinction of existing ones.

Two of the most important ultimate limiting factors for landbirds are the numbers and sizes of existing land areas, which together influence the opportunities for allopatry. Because species evolve and persist in land areas that give some security from other species that might interbreed with or outcompete them, the number of bird species at one time must depend partly on the degree of fragmentation of the earth's land areas, and of the various habitats that those areas support. The more numerous the separated areas, the more species they could collectively hold. Thus, if all the islands in the world were joined together, yet retained the species they have now, this would give a species to land area ratio much greater than that in any equivalent continental land area (Chapter 6). This is testimony to the effect of land fragmentation in enhancing species numbers. Other important factors on particular land areas include the variety of habitats present (in turn dependent largely on spatial variation in topography and climate), and the structural diversity of those habitats, which affects the numbers of niches they offer for birds, and hence the opportunities for sympatry (Chapter 11).

At the proximate level, the number of species alive at one time, either worldwide or in particular areas, depends on the balance between speciation and extinction rates. This is analogous to the way that the number of individuals in a population depends on the balance between birth and death rates. Because most speciation is a branching (multiplicative) process, the number of species could in theory increase exponentially, as could the number of individuals in a population. However, constraints to this process come from the limited opportunities for speciation (analogous to opportunities for reproduction) and from ecological limits to the number of species that can co-exist (analogous to population carrying capacity). One can imagine that environmental factors might limit growth in species numbers, in the same way that they limit the growth of populations, but the factors involved are not necessarily the same in both cases.

While the fragmentation of land areas and their habitats can promote the formation of new species, the same process carried to extreme can promote their extinction, as habitat areas disappear or become too small to support their populations long term. External abiotic influences, such as climate change or volcanic eruptions, can destroy species in this way. In addition, biotic changes, including the immigration of superior competitors, predators or virulent parasites, can cause species extinctions, as documented many times on oceanic islands (Chapter 7). However, it is not the rates of speciation and extinction that determine equilibrial species numbers, but the balance between them. Lineages could combine high extinction with high speciation, and still maintain high (or low) numbers of species at any one time.

Extinction is assumed to be the ultimate fate of every species. At each stage in the history of life, the earth was populated by different kinds of plants and animals, most of which disappeared, to be replaced by others in an evolutionary succession. The numbers of fossil types known from different periods must be adjusted to allow for inequalities in the stratigraphic record: that is, for the relative amounts of different-aged sedimentary rock available at the surface for exploration. Allowing for these inequalities, extinctions have apparently occurred continuously, but with occasional brief episodes of catastrophic mass extinction, when large proportions of existing plants and animals disappeared, apparently through rapid and drastic environmental change. Only the last of these events, at the end of the Cretaceous Period (65 million years ago), occurred within the evolutionary history of birds. It removed 75–80% of the fossilised organisms known to have occurred then, including the large dinosaurs.

In addition to this major event, five smaller extinction episodes are known to have occurred subsequently, at fairly regular intervals of around 26 million years (Sepkoski 1989). Although they broke the usual regime of slow background extinction, their effects did not last long, as the diversity of many groups recovered within a few million years, as new species evolved to take up the available openings (Rosenzweig 1995). For much of geological time, therefore, diversity may have been unaffected by periodic mass extinctions. To judge from fossils of various non-avian taxa, species numbers fluctuated within relatively narrow limits, perhaps two-fold over periods of 5–10 million years, with no clear upward or downward trend. This is consistent with the idea of an equilibrium in species numbers, broken by periodic catastrophes (**Figure 4.2**). However, it is unlikely that species richness has ever exactly attained equilibrial levels, firstly because the environment itself is always changing (through alterations in land area and configuration and in climate), and secondly because there has probably been a gradual improvement in the ability of organisms collectively to use the resources of the earth. This is suggested by the increase in the types of organisms present over geological time, and the occasional expansion to previously unused parts of the environment, from sea to fresh water to land to air. To judge from the fossil record, the diversity of birds has increased progressively through most of their evolutionary history, decreasing in the last 2–3 million years or so, and especially since human influence became felt (Chapters 7 and 9). It would, of course, be interesting to know how many bird species had lived at different times in the past, and how this number had changed through the ages. Some biologists have made such estimates, but we really have no useful bases for them, except perhaps for the last few thousand years (Chapter 7).

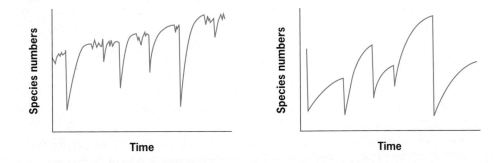

Figure 4.2   Models representing changes in species numbers through time. (Right) Catastrophic environmental events occur at long and irregular intervals, each time causing marked reductions in species numbers from which gradual recovery occurs, reaching a fairly stable equilibrium level. (Left) Severe environmental events occur so frequently that before numbers have recovered to an equilibrium level, they are again reduced.

Early birds

The earliest known bird fossil, *Archaeopteryx*, dates from the upper Jurassic, 150 million years ago, a time when the dinosaurs still prevailed, but when mammals and flowering plants were already in existence. So far, seven specimens of *Archaeopteryx* have been found, plus a single feather, all from the same period, and all from central Europe. This primitive creature was about the size of a crow, with teeth and clawed fingers and the long bony tail of a reptile, but with the wishbone and feathered wings of a bird, and the wing feathers divided into primaries and secondaries. The tail supported a row of feathers down each side. The feathers themselves resembled those of modern birds, not just in number and arrangement, but also in their asymmetrical shape, with narrow outer vanes.

Until only a few years ago, the fossil record for the next 50 million years of bird evolution was depressingly meagre. From the late Jurassic *Archaeopteryx* to the late Cretaceous loon-like *Hesperornis* and tern-like *Ichthyornis*, only skimpy avian remains were known. This situation changed dramatically with the discovery of bird remains in early Cretaceous lake-bed deposits in Spain and particularly in China, where thousands of fossils have been discovered in recent years, including several different kinds of birds. These new fossils have helped to fill in the picture of early bird evolution from dinosaurs, with several intermediate types (Ackerman 1998). One called *Protarchaeopteryx* was a turkey-sized creature with strong legs, but more primitive symmetrical feathers; it was actually less advanced than *Archaeopteryx*, but more recent, dated at 120 million years old. Another of similar age, called *Ambiortus*, was recovered from early Cretaceous deposits in Mongolia, and is clearly a modern-type bird, as shown by the shape of the wishbone, the fused carpometacarpus, keeled sternum and other features (Kurochkin 1985). Evidently birds representing different stages of evolution persisted on earth at the same time, as they do today. Another fossil of an ancient bird possessing a long bony tail like *Archaeopteryx* was recently found in Madagascar, confirming the presence of such creatures in Gondwanaland.

When did the different types of birds that we know today first appear on earth? At a gross scale, particular types of birds can be aged from the fossil record but, because of the sketchiness of this record, the estimated ages can be no more than minimal values. It is always possible that older fossils exist but have not been found. Another source of information comes from studies of living birds, in which the degree of difference in DNA between any two species is taken to reflect the time since their divergence (Chapter 2). On this basis, and using the DNA–DNA hybridisation technique, Sibley & Ahlquist (1990) assigned ages to the various orders and families of birds that, in general, were not inconsistent with dated fossil remains: that is, where the dates did not match, the DNA dates generally pre-dated the fossil dates rather than vice versa.

The fossil evidence, such as it is, shows that at least four living orders of birds had relatives living as long ago as the Cretaceous Period (namely Charadriiformes, Gaviiformes, Anseriformes and Procellariiformes), but suggests that most modern orders appeared in the early Tertiary (Padian & Chiappe 1998, Feduccia 1999). They arose as an 'explosive radiation' within a time frame of 5–10 million years, paralleling a similar radiation in mammals, both following the mass extinction event at the end of the Cretaceous. These conclusions are drawn from limited evidence, however, because relatively few avian fossils of any kind are available from the Cretaceous, particularly from the southern continents that formed part of Gondwanaland. The molecular evidence, in contrast, which is based on both nuclear and mitochondrial DNA, suggests that most modern bird orders arose well before the Tertiary, and that at least 22 lineages in 15 orders crossed the Cretaceous–Tertiary (K–T) boundary, 65 million years ago (Cooper & Penny 1997, van Tuinen & Hedges 2001). These 15 orders form 65% of the total orders represented by living birds. They included a wide range of types, from passerines through parrots to ratites (namely the Struthioniformes, Tinamiformes, Galliformes, Anseriformes, Psittaciformes, Pelecaniformes, Charadriiformes, Passeriformes, Strigiformes, Falconiformes, Threskiornithiformes, Gruiformes, Gaviiformes, Podicipediformes and Procellariiformes). The molecular evidence further implies that different types were added gradually during the Cretaceous Period rather than rapidly by an explosive radiation in the early Tertiary, as the fossils suggest (Cooper & Penny 1997, Paton *et al*. 2002). However, like all DNA evidence, these conclusions are based on inadequately tested assumptions about molecular clocks, and insufficient independent calibrations of rates of molecular evolution.

The third line of evidence comes from the current southern hemisphere distribution patterns of many bird families, each of which is found only on two or more land areas derived from Gondwanaland. Their distributions are therefore consistent with the view that these families arose on Gondwanaland, and reached their present distributions on the drifting continents (Cracraft 2001). For this to have happened, they must have arisen in the Cretaceous Period, and the break-up of the southern continents would have been a major force in bird evolution and diversification. However, an alternative possibility is that current southern hemisphere families reached their present distributions via the northern continents, where they have since died out. With the current scarcity of appropriately dated fossils, the idea of a Gondwanan origin for many bird families may be undermined by future fossil discoveries. Further information is required to resolve this central issue in bird evolution. The fossil evidence also suggests a mass extinction of birds at the end of the Cretaceous, along with the dinosaurs. At least 20 types of birds, classed as distinct orders, did not survive into the Tertiary Period, including the Enantiornithes, Hesperornithiformes and Ichthyornithiformes (Feduccia 1999). They lived alongside some surviving orders of birds, but may have pre-dated others. Virtually all modern families of birds were present by the end of the Pliocene, and are represented in the fossil record of that time.

More detail on fossil history of birds is provided in other other texts, such as Olson (1985b) and Feduccia (1999). My aim here is merely to make the points that, in the 150 million years of avian evolution, birds in general have become progressively more distinct from their reptilian ancestors, that some totally distinctive orders of birds present at the end of the Cretaceous Period have since died out completely, while others from that time have survived to the present day. Yet other orders may have evolved since then. The present avifauna therefore consists of a mixture of orders and families of greatly varying ages, as well as of greatly varying species richness. Knowledge of fossil specimens from these major taxa provides no way in which we can reliably estimate the numbers of species present at any time in the distant past.

SUMMARY

The latest worldwide estimate lists 9,736 current bird species, but this figure is under continual modification as new species are discovered and as known species go extinct, and also as new research or taxonomic fashion leads to the re-evaluation of existing known species and subspecies.

The number of bird species recognised at one time depends partly on the way in which species are defined and delimited. Species definitions can be divided into two groups. One cuts the evolutionary tree across the horizontal, with species being regarded as the different groups of extant organisms at the tips of the branches: that is, individuals from the same species look alike, share a common gene pool and recognise each other as mates, but are different from individuals of other species. Variants of this group include the Biological (Isolation), Cohesion and Recognition Species Concepts. These definitions more or less ignore the history of the different lineages, and delimit species by their present-day reproductive and ecological features.

The other group of species definitions looks at the species in the vertical (time) axis, as organisms that are evolving independently as branching lines in the evolutionary tree. Variants of this group include the Phylogenetic and Evolutionary Species Concepts. They place little weight on behavioural and ecological relationships as they are now, but concentrate instead on historical development with respect to related forms. The version of the PSC applied in ornithology also upgrades many subspecies to the level of species.

Because modern genetic techniques yield information on the history of lineages that was previously unattainable, genealogy has come to play an increasing role in decisions on species delimitation and classification. This is as it should be, for it makes use of the new information and helps to ensure that history is represented at all levels of classification.

Currently, there is more instability than usual in bird classification and nomenclature. This has followed partly from the introduction of molecular techniques, still in active development, partly to a swing in taxonomic fashion from lumping towards splitting, and also from attempts to harmonise variations in common name usage in different parts of the world.

Fossils have given minimal estimates of the ages, and dates of origin, of the different orders and families of birds. In recent years, other estimates have been obtained from analyses of DNA structure, for orders, families or species, assuming that the degree of difference between any two species is proportional to the time since their divergence. For most extant orders of birds, estimates from DNA pre-date those obtained from fossils, and suggest that many of them arose in the Cretaceous period, more than 65 million years ago. Other bird orders present at that time have long since disappeared. The modern avifauna consists of orders, families and species of greatly varying ages.

Part Two

Major Distribution Patterns

Raggiana Bird-of-Paradise *Paradisaea raggiana*, a member of a group centred on New Guinea.

Chapter 5
Continental birds: biogeographical regions

Many types of birds and other organisms are found only on a single continental land-mass, while others are found on more than one. The level of distinctiveness in the plants and animals of any land-mass reflects not only the position of that land-mass with respect to others, but also its geological history: its movements over the globe, the duration and degree of its isolation, and the opportunities for separate evolution that it has provided. We can therefore use the present distribution patterns of birds or other organisms, along with palaeo-geological evidence, to make inferences about how such patterns came about, and the time scales involved, as well as about the general dispersive abilities of birds.

In the nineteenth century, naturalists became aware of large-scale patterns in the distributions of plants and animals, and categorised parts of the globe according to the distinctiveness of their flora and fauna. The global system developed for passerine birds by Philip Lutley Sclater (1858) was later modified by Alfred Russel Wallace (1876) to apply to animals in general, and is still in use today. It forms not only one of the most fundamental descriptions of animal distribution patterns, but also one of the main empirical foundations of biogeography.

In this scheme, the land-masses of the earth are divided into six main biogeographical realms, namely the Palaearctic Region (most of the Eurasian landmass), the Indomalayan (or Oriental) Region (southeast Asia and nearby islands), the Afrotropical (or Ethiopian) Region (Africa south of the Sahara but including southern Arabia), the Australasian Region (mainly New Guinea, Australia and New Zealand), the Nearctic Region (North America–Greenland) and the Neotropical Region (Central and South America). The crucial barriers between these regions, past or present, are large stretches of sea or desert, long high mountain ranges, or big climatic–vegetation differences, the relative importance of each type of barrier varying from one region to another (Darlington 1957). There are of course transitional areas, such as Arabia, southern China, Mexico and the islands between Asia and Australia, where faunas intermingle and where the boundaries between realms can be defined only in an arbitrary way.

Because of the earth's tectonic history, not all the boundaries between these various realms coincide exactly with the continents of conventional geography (**Figure 5.1**). For biogeographical purposes, the land-mass of Europe and northern Asia is best treated as a unit, northern and southern Asia are best treated as separate units divided by the Himalayas, while the Sahara desert rather than the Mediterranean Sea forms the chief latitudinal barrier to dispersal between Europe and Africa. In the New World, the northern limit of the tropics is a more important faunal boundary than any provided by the configuration of the continents.

The six originally proposed regions have been variously combined and rearranged or subdivided by subsequent authors. For example, the Malagasy Region (Madagascar and associated islands) is sometimes separated from the Afrotropical as a separate region or subregion, as is New Zealand from the rest of Australasia, while the Palaearctic and Nearctic Regions are sometimes combined as a single Holarctic Region. Moreover, each region can be divided even more finely to reflect the different climatic and habitat zones that it contains. The islands of the central Pacific are usually incorporated into a seventh region, Oceania. And the Antarctic, which holds only seasonally present marine birds (and not landbirds), is usually excluded from the scheme together.

The proportions of bird families, genera and species that different regions share depend largely on the effectiveness of the barriers between them to bird dispersal, not least the distances between the regions, both now and in the geological past. Through continued tectonic movements, some land-masses that are now far apart were close together or connected in the past (Chapter 1). Such regions are likely now to share a greater proportion of families, evolved before the break, than of genera and species evolved after the break. Conversely, regions that have become close or connected only in the geologically recent past are likely to share many genera and species as well as families. Nonetheless, some types of birds, notably waterbirds, show remarkable dispersive powers, and have been known to cross oceans in historical times, the establishment of the Old World Cattle Egret *Bubulcus ibis* in the New World during the late nineteenth century providing a striking example (Chapter 16). In general, however, the longer an area has been isolated,

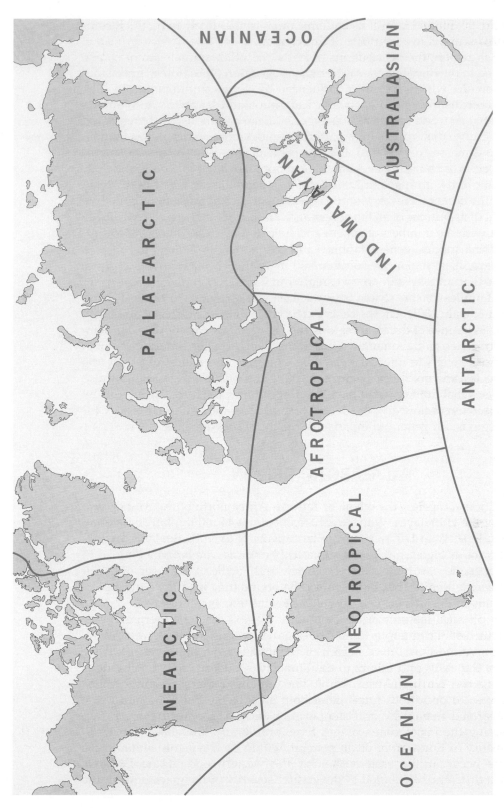

Figure 5.1 The main biogeographical realms of the world.

the higher the taxonomic rank of its endemic organisms is likely to be. Such comparisons are discussed in a later part of this chapter.

Within each region, the distributions of birds and other animals are of course closely related to the distributions of the major vegetation types (forest, grassland, desert, and so on), which are in turn influenced by climate and topography. The most important distinctions are between waterbirds and landbirds and, among the latter, between forest and non-forest species. Seabirds are usually considered separately, because their distribution patterns depend more on the oceans than on the major land-masses (Chapter 8). For our present purposes, then, exclusion of about 320 species of seabirds leaves a total of about 9,416 landbird species (including freshwater birds) arranged in 2,002 genera, 140 families and 23 orders, based on the classification of Sibley & Monroe (1990). As a basis for discussion, the biogeographical distributions of all the orders and families of birds are given in **Table 5.1**, together with the numbers of genera and species they each contain. The numbers of landbird species, genera, families and orders that are known to breed in each biogeographical region are summarised in **Table 5.2**, the numbers of these various taxa that are endemic to each region in **Table 5.3**, and the numbers of species and families that are shared between each pair of regions in **Tables 5.4** and **5.5** (Newton & Dale 2001). On the basis of the latter figures, indices of similarity between the avifaunas of different regions can be calculated, as shown later. Subsequent discoveries and taxonomic revision are likely to alter the figures somewhat, but not enough to affect the main conclusions. The tables give only the breeding species, and not the passage migrants or non-breeding visitors; they also exclude species that are known to have been introduced to particular regions by people. The sections below give the details of land form and avifaunal history for particular regions; for general comparisons, see the Discussion later.

PALAEARCTIC REGION

This vast region comprises the whole of Europe, Africa north of the Sahara, and Asia north of the Himalayas. With a total area of about 46 million km², extending from longitude 10°W to 170°W and from latitude 25°N to 80°N, the Palaearctic is more than twice as big as any other region. Its boundaries are broadly defined in the west, north and east by the Atlantic, Arctic and Pacific Oceans, separating it from the Nearctic Region, but in the south the boundaries with the Afrotropical and Indomalayan Regions are less clearly defined as faunas intermingle.

The main physical features of the Palaearctic Region are a northern arctic or tundra zone, a belt of coniferous forest (the taiga) extending from Norway to eastern Siberia, an almost continuous chain of mountains from the Pyrenees, Alps, Carpathians, Caucasus and Elburz to the Himalayas, and an almost continuous belt of desert across northern Africa, Arabia, and most of southern Asia (**Figure 5.2**). Extensive areas of deciduous forest lie at both ends of the Eurasian land-mass, and are separated mainly by arid steppe grasslands. These features and their changes during the Pleistocene ice ages have profoundly influenced the distributional history of Eurasian birds. In general, within each type of habitat, more bird species occur at the eastern than at the western end of the Eurasian land-mass, a difference attributed to the greater severity of the glaciations in the

Table 5.1 The order and families of landbirds, numbers of genera and species and biogeographical distribution patterns. based on Sibley & Monroe 1990, updated on CD-rom by Sibley 1994. Note that in other recent lists, the numbers of genera and species may differ from this list, as may the classification of families.

Orders and families	Numbers of:		Breeding in different zoogeographical regions							Total number of regions
	Genera	Species	Palaearctic	Indomalayan	Afrotropical	Australasian	Nearctic	Neotropical	Oceania	
Order STRUTHIONIFORMES	5	10			*	*		*		3
Family Struthionidae	1	1			*					1
Family Rheidae	1	2						*		1
Family Casuariidae	2	4				*				1
Family Apterygidae	1	3				*				1
Order TINAMIFORMES	9	47						*		1
Family Tinamidae	9	47						*		1
Order CRACIFORMES	17	69		*		*	*	*	*	5
Family Cracidae	11	50					*	*		2
Family Megapodiidae	6	19		*		*			*	3
Order GALLIFORMES	62	215	*	*	*	*	*	*		6
Family Phasianidae	49	178	*	*	*	*	*	*		6
Family Numididae	4	6			*					1
Family Odontophoridae	9	31					*	*		2
Order ANSERIFORMES	50	163	*	*	*	*	*	*	*	7
Family Anhimidae	2	3						*		1
Family Anseranatidae	1	1				*				1
Family Dendrocygnidae	2	9		*	*	*	*	*		5
Family Anatidae	45	150	*	*	*	*	*	*	*	7
Order TURNICIFORMES	2	17	*	*	*	*				4
Family Turnicidae	2	17	*	*	*	*				4
Order PICIFORMES	51	354	*	*	*		*	*		5
Family Indicatoridae	4	17		*	*					2
Family Picidae	28	214	*	*	*	*	*	*		6
Family Megalaimidae	3	27	*	*						2
Family Lybiidae	7	41			*					1
Family Ramphastidae	9	55						*		1
Order GALBULIFORMES	15	51						*		1
Family Galbulidae	5	18						*		1

Table 5.1   continued

Orders and families	Numbers of:		Breeding in different zoogeographical regions							Total number of regions
	Genera	Species	Palaearctic	Indomalayan	Afrotropical	Australasian	Nearctic	Neotropical	Oceania	
Family Bucconidae	10	33						*		1
Order BUCEROTIFORMES	9	54	*	*	*	*				4
Family Bucerotidae	8	52	*	*	*	*				4
Family Bucorvidae	1	2			*					1
Order UPUPIFORMES	3	12	*	*	*					3
Family Upupidae	1	4	*	*	*					3
Family Phoeniculidae	1	5			*					1
Family Rhinopomastidae	1	3			*					1
Order TROGONIFORMES	6	39		*	*		*	*		4
Family Trogonidae	6	39		*	*		*	*		4
Order CORACIIFORMES	34	153	*	*	*	*	*	*	*	7
Family Coraciidae	2	12	*	*	*	*				4
Family Brachypteraciidae	3	5			*					1
Family Leptosomidae	1	1			*					1
Family Momotidae	6	9						*		1
Family Todidae	1	5						*		1
Family Alcedinidae	3	25	*	*	*	*				4
Family Halcyonidae	12	61	*	*	*	*			*	5
Family Cerylidae	3	9	*	*	*		*	*		5
Family Nyctyornithidae	1	2		*						1
Family Meropidae	2	24	*	*	*	*				4
Order COLIIFORMES	2	6			*					1
Family Coliidae	2	6			*					1
Order CUCULIFORMES	29	145	*	*	*	*	*	*		6
Family Cuculidae	16	81	*	*	*	*				4
Family Centropodidae	1	30		*	*	*				3
Family Coccyzidae	4	18					*	*		2
Family Crotophagidae	2	4					*	*		2
Family Neomorphidae	5	11					*	*		2
Family Opisthocomidae	1	1						*		1
Order PSITTACIFORMES	81	360		*	*	*	*	*	*	6

Taxon	Genera	Species								Regions
Family Psittacidae	81	360	*	*	*	*		*	*	6
Order APODIFORMES	19	105	*	*	*	*	*	*	*	7
Family Apodidae	18	101	*	*	*	*	*	*	*	7
Family Hemiprocnidae	1	4		*		*				2
Order TROCHILIFORMES	108	326					*	*		2
Family Trochilidae	108	326					*	*		2
Order MUSOPHAGIFORMES	5	23			*					1
Family Musophagidae	5	23			*					1
Order STRIGIFORMES	45	300	*	*	*	*	*	*	*	7
Family Tytonidae	2	17	*	*	*	*	*	*	*	7
Family Strigidae	23	168	*	*	*	*	*	*	*	7
Family Aegothelidae	1	8				*				1
Family Podargidae	1	3				*				1
Family Batrachostomidae	1	11		*						1
Family Steatornithidae	1	1						*		1
Family Nyctibiidae	1	7						*		1
Family Eurostopodidae	1	7		*		*				2
Family Caprimulgidae	14	78	*	*	*	*	*	*		6
Order COLUMBIFORMES	44	315	*	*	*	*	*	*	*	7
Family Raphidae	2	3			*					1
Family Columbidae	42	312	*	*	*	*	*	*	*	7
Order GRUIFORMES	53	197	*	*	*	*	*	*	*	7
Family Eurypygidae	1	1						*		1
Family Otididae	6	25	*	*	*	*				4
Family Gruidae	2	15	*	*	*	*	*			5
Family Aramidae	1	1					*	*		2
Family Heliornithidae	3	3		*	*			*		3
Family Psophiidae	1	3						*		1
Family Cariamidae	2	2						*		1
Family Rhynochetidae	1	1				*				1
Family Rallidae	34	143	*	*	*	*	*	*	*	7
Family Mestornithidae	2	3			*					1
Order CICONIIFORMES	251	1,024	*	*	*	*	*	*	*	7
Family Pteroclidae	2	16	*	*	*					3
Family Thinocoridae	2	4						*		1
Family Pedionomidae	1	1				*				1

Table 5.1 continued

Orders and families	Numbers of:		Breeding in different zoogeographical regions							Total number of regions
	Genera	Species	Palaearctic	Indomalayan	Afrotropical	Australasian	Nearctic	Neotropical	Oceania	
Family Scolopacidae	20	88	*	*	*	*	*	*	*	7
Family Rostratulidae	1	2	*	*	*	*		*		5
Family Jacanidae	6	8		*	*	*	*	*		5
Family Chionidae	1	2			*			*		2
Family Pluvianellidae	1	1						*		1
Family Burhinidae	1	9	*	*	*	*		*		5
Family Charadriidae	16	88	*	*	*	*	*	*	*	7
Family Glareolidae	6	19	*	*	*	*				4
Family Laridae	27	129	*	*	*	*	*	*	*	7
Family Accipitridae	65	237	*	*	*	*	*	*	*	7
Family Sagittariidae	1	1			*					1
Family Falconidae	10	63	*	*	*	*	*	*	*	7
Family Podicipedidae	6	22	*	*	*	*	*	*		6
Family Phaethontidae	1	3	*	*	*	*	*	*	*	7
Family Sulidae	3	9	*	*	*	*	*	*	*	7
Family Anhingidae	1	4	*	*	*	*	*	*		6
Family Phalacrocoracidae	1	38	*	*	*	*	*	*	*	7
Family Ardeidae	19	65	*	*	*	*	*	*	*	7
Family Scopidae	1	1			*					1
Family Phoenicopteridae	1	5	*	*	*			*		4
Family Threskiornithidae	14	33	*	*	*	*	*	*		6
Family Pelecanidae	2	9	*	*	*	*	*	*		6
Family Ciconiidae	11	26	*	*	*	*	*	*		6
Family Fregatidae	1	5		*	*	*		*	*	5
Family Spheniscidae	6	17			*	*		*		3
Family Gaviidae	1	5	*				*			2
Family Procellariidae	23	114	*	*	*	*	*	*	*	7
Order PASSERIFORMES	1,165	5,752	*	*	*	*	*	*	*	7
Family Acanthisittidae	2	4				*				1
Family Pittidae	1	31	*	*	*	*		*		4
Family Eurylaimidae	8	14		*	*			*		2

Family			1	2	3	4	5	6	7	
Family Philepittidae	2	4			*					1
Family Incertae sedis	1	1						*		1
Family Tyrannidae	146	544					*	*		2
Family Thamnophilidae	44	191					*	*		2
Family Furnariidae	66	280					*	*		2
Family Formicariidae	7	60						*		1
Family Conopophagidae	1	8						*		1
Family Rhinocryptidae	12	33						*		1
Family Climacteridae	2	7				*				1
Family Menuridae	2	4				*				1
Family Ptilonorhynchidae	8	20				*				1
Family Maluridae	5	26				*				1
Family Meliphagidae	41	181		*		*			*	3
Family Pardalotidae	16	68		*		*				2
Family Petroicidae	13	44				*				1
Family Irenidae	2	10		*						1
Family Orthonychidae	1	2				*				1
Family Pomatostomidae	1	5				*				1
Family Laniidae	3	30	*	*	*	*	*			5
Family Vireonidae	4	54					*	*		2
Family Corvidae	128	648	*	*	*	*	*	*	*	7
Family Callaeatidae	3	3				*				1
Family Incertae sedis	2	4				*				1
Family Bombycillidae	5	8	*				*	*		3
Family Cinclidae	1	5	*	*			*	*		4
Family Muscicapidae	70	450	*	*	*	*	*	*	*	7
Family Sturnidae	37	147	*	*	*	*	*	*	*	7
Family Sittidae	2	25	*	*			*			3
Family Certhiidae	23	98	*	*	*		*	*		5
Family Paridae	7	63	*	*	*		*	*		5
Family Aegithalidae	3	8	*	*			*	*		4
Family Hirundinidae	14	89	*	*	*	*	*	*	*	7
Family Regulidae	1	6	*	*			*			3
Family Pycnonotidae	22	137	*	*	*				*	4
Family Hypocoliidae	1	1	*							1
Family Cisticolidae	20	122	*	*	*	*				4

Table 5.1 continued

Orders and families	Numbers of: Genera	Species	Breeding in different zoogeographical regions Palaearctic	Indomalayan	Afrotropical	Australasian	Nearctic	Neotropical	Oceania	Total number of regions
Family Zosteropidae	14	96	*	*	*	*			*	5
Family Sylviidae	95	560	*	*	*	*	*		*	6
Family Alaudidae	19	92	*	*	*	*	*	*		6
Family Nectariniidae	8	172	*	*	*	*				4
Family Melanocharitidae	3	10				*				1
Family Paramythiidae	2	2				*				1
Family Passeridae	57	389	*	*	*	*	*	*	*	7
Family Fringillidae	240	996	*	*	*	*	*	*	*	7

Table 5.2 Numbers of landbird taxa that breed in different biogeographical regions, with the numbers listed separately for (a) each region as a whole, (b) the main continental part of each region together with its land-bridge islands, and (c) the oceanic islands associated with each region. Species that occur on both continent and oceanic islands of the same region are listed under continental (b). Figures in brackets are percentages of the total known numbers of extant landbird taxa in the world (taken as 9,416 for species, 2,002 for genera, 140 for families and 23 for orders, see text). + = <1%. Based on information in Sibley & Monroe 1990, updated on CD-rom by Sibley 1996.

	Palaearctic	Indomalayan	Afrotropical	Australasian	Nearctic	Neotropical	Oceania	Overall
(a) All birds								
Species	937(10)	1,697(18)	1,950(21)	1,592(17)	732(8)	3,370(36)	187(2)	9,416
Genera	288(14)	431(22)	473(24)	457(23)	302(15)	893(45)	82(4)	2,002
Families	58(41)	73(52)	75(54)	73(52)	52(37)	71(51)	23(16)	140
Orders	14(61)	17(74)	19(83)	16(70)	15(65)	18(78)	10(43)	23
(b) Continental birds								
Species	903(10)	1,259(13)	1,714(18)	926(10)	723(8)	3,170(34)	0(0)	7,789
Genera	286(14)	398(20)	412(21)	340(17)	301(15)	847(42)	0(0)	1,784
Families	58(41)	70(50)	70(51)	65(46)	52(37)	70(50)	0(0)	130
Orders	14(61)	16(70)	19(83)	15(65)	15(65)	18(78)	0(0)	23
(c) Island birds								
Species	34(+)	438(5)	236(3)	666(7)	9(+)	200(2)	187(2)	1,627
Genera	29(1)	168(8)	121(6)	224(11)	9(+)	104(5)	82(4)	532
Families	17(12)	47(34)	42(30)	49(35)	8(6)	29(21)	23(16)	81
Orders	8(35)	16(70)	13(57)	16(70)	6(26)	14(61)	10(43)	19

Areas of different continental regions are taken as ($\times 1,000,000$ km^2): Palaearctic 46, Indomalayan 9.6, Afrotropical 21, Australasian 8.9, Nearctic 21 and Neotropical 18.2.

Table 5.3 Numbers of landbird taxa that are endemic as breeders in different biogeographical regions, with the numbers listed separately for (a) each region as a whole, (b) the main continental part of each region together with its land-bridge islands, and (c) the oceanic islands associated with each region. Species that occur on both continent and oceanic islands are listed under continental (b). Figures in brackets are percentages of the total known taxa in that region or subregion (from **Table 5.1**). Based on information in Sibley & Monroe 1990, updated on CD-rom by Sibley 1996.

	Palaearctic	Indomalayan	Afrotropical	Australasian	Nearctic	Neotropical	Oceania
(a) All birds							
Species	442(47)	1,184(70)	1,807(93)	1,415(89)	395(54)	3,121(93)	163(87)
Genera	26(9)	126(29)	293(62)	280(61)	58(19)	686(77)	31(38)
Families	0(0)	3(4)	16(21)	18(25)	0(0)	20(28)	0(0)
Orders	0(0)	0(0)	2(11)	0(0)	0(0)	2(11)	0(0)
(b) Continental birds							
Species	411(46)	801(64)	1,572(92)	848(92)	387(54)	2,925(92)	0(0)
Genera	25(9)	108(27)	240(58)	221(65)	58(19)	642(76)	0(0)
Families	0(0)	3(4)	11(15)	14(22)	0(0)	19(27)	0(0)
Orders	0(0)	2(11)	0(0)	0(0)	2(11)	0(0)	0(0)
(c) Island birds							
Species	31(91)	383(88)	235(100)	567(85)	8(89)	196(98)	163(87)
Genera	1(3)	18(11)	53(43)	59(26)	0(0)	44(42)	31(38)
Families	0(0)	0(0)	5(12)	4(8)	0(0)	1(3)	0(0)
Orders	0(0)	0(0)	0(0)	0(0)	0(0)	0(0)	0(0)

Table 5.4 Numbers of species shared between different pairs of regions, and indices of similarity* in avifaunal composition (see text).

Number of species shared

	Palaearctic	Indomalayan	Afrotropical	Australasian	Nearctic	Neotropical
Palaearctic	*	361	116	33	107	15
Indomalayan		*	83	167	29	16
Afrotropical			*	24	15	17
Australasian				*	8	9
Nearctic					*	239
Neotropical						*

Simpson index = 100. N_c/N_1

	Palaearctic	Indomalayan	Afrotropical	Australasian	Nearctic	Neotropical
Palaearctic	*	3.85	12.27	3.52	14.62	1.60
Indomalayan		*	4.89	10.42	3.96	0.94
Afrotropical			*	1.44	2.19	0.87
Australasian				*	1.09	0.56
Nearctic					*	33.33
Neotropical						*

Jaccard index = 100. $N_c/(N_1+N_2-N_c)$

	Palaearctic	Indomalayan	Afrotropical	Australasian	Nearctic	Neotropical
Palaearctic	*	15.9	4.2	1.3	6.9	0.3
Indomalayan		*	2.3	5.3	1.2	0.3
Afrotropical			*	0.7	0.6	0.3
Australasian				*	0.3	0.2
Nearctic					*	5.8
Neotropical						*

N_c = number of taxa shared by areas
N_1 = number of taxa in first area (with the smallest number)
N_2 = number of taxa in second area (with the largest number)

*Indices of similarity are calculated using the Simpson and Jaccard indices which take account of the total number of taxa in each region and the proportion of the total that are shared. To give a worked example on the Simpson index, 107 of 732 species found in the Nearctic Region are also found in the Palaearctic; the coefficient for this comparison is therefore $100 \times 107/732 = 14.6$. On the Jaccard index, the coefficient of similarity is $100 \times 107/(732 + 937 - 107) = 6.9$. On both indices, a value of 0 would indicate no overlap and 100 would indicate complete overlap.

Table 5.5 Numbers of families shared between different pairs of regions, and indices of similarity in avifaunal composition. For definitions of Simpson and Jaccard indices, see Table 5.4, footnote.

Number of families shared

	Palaearctic	Indomalayan	Afrotropical	Australasian	Nearctic	Neotropical
Palaearctic	*	56	51	46	37	34
Indomalayan		*	59	55	40	39
Afrotropical			*	50	36	36
Australasian				*	33	31
Nearctic					*	46
Neotropical						*

Simpson index $= 100.\ N_c/N_1$

	Palaearctic	Indomalayan	Afrotropical	Australasian	Nearctic	Neotropical
Palaearctic	*	96.55	87.93	79.31	71.15	58.62
Indomalayan		*	80.82	75.34	76.92	53.42
Afrotropical			*	68.49	69.23	48.00
Australasian				*	63.46	42.47
Nearctic					*	88.46
Neotropical						*

Jaccard index $= 100.\ N_c/(N_1 + N_2 - N_c)$

	Palaearctic	Indomalayan	Afrotropical	Australasian	Nearctic	Neotropical
Palaearctic	*	0.75	0.62	0.54	0.51	0.36
Indomalayan		*	0.66	0.60	0.47	0.37
Afrotropical			*	0.51	0.40	0.33
Australasian				*	0.36	0.27
Nearctic					*	0.60
Neotropical						*

N_c = number of taxa shared by areas
N_1 = number of taxa in first area (with the smallest number)
N_2 = number of taxa in second area (with the largest number)

west (Chapter 9). As expected from its geographical position, the Palaearctic climate is markedly seasonal, the difference between summer and winter temperatures increasing with latitude. Northern parts of the continent lie under ice and snow for more than half the year. Not surprisingly, therefore, the number of breeding bird species per unit area tends to decrease northwards and the proportion of those breeding species that leave for the winter increases (Chapter 18; Newton & Dale 1996b).

The Atlantic has proved an effective barrier to range extension in landbirds. No purely Palaearctic passerine species has established itself in eastern North America in the last 200 years, except perhaps the Northern Wheatear *Oenanthe oenanthe* in northeastern Canada, whilst only a single non-passerine — the Cattle Egret *Bubulcus ibis* — has recently established itself in the New World by crossing the Atlantic at wider tropical latitudes. In addition, the Little Gull *Larus minutus* has gained a foot-hold in at least two parts of eastern Canada and the Fieldfare *Turdus pilaris* has established itself in Greenland. These birds form less than 1% of all Palaearctic bird species. Likewise, despite the many North American birds blown across the Atlantic or assisted by ships every year, not one species has yet established itself in Europe, although the Spotted Sandpiper *Actitis macularia* and Wilson's Phalarope *Phalaropus tricolor* have bred on one or more occasions.

In contrast, the narrow Bering Strait, which separates eastern Asia from western North America, was dry land during each of the glacial periods (most recently 10,000 years ago), and today holds a series of islands separated by short sea crossings. Several Palaearctic species have spread eastwards into Alaska (e.g. Yellow Wagtail *Motacilla flava*, Northern Wheatear *Oenanthe oenanthe* and Arctic Warbler *Phylloscopus borealis*), and conversely, several Nearctic species have spread westwards into eastern Siberia (e.g. Snow Goose *Chen caerulescens*, Grey-cheeked Thrush *Catharus minimus*, Yellow-rumped Warbler *Dendroica coronata*). All these species inhabit tundra or boreal forest, which form the predominant vegetation types of Alaska and northeast Siberia at the closest points between the continents (for other examples, see Chapter 9).

In southern Europe, the Mediterranean Sea dates back about five million years to the early Pliocene, when the Straits of Gibraltar opened and Atlantic waters poured in. With its numerous islands, this sea seems to form a much less effective barrier to bird dispersal than the Sahara Desert which extends from one side of Africa to the other, and is as large as the United States or Australia. The birds of North Africa (north of the Sahara) are overwhelmingly Palaearctic in affinity, with 165 such species breeding regularly but not extending into sub-Saharan Africa even as migrants. They include endemics such as Dupont's Lark *Chersophilus duponti*, Moussier's Redstart *Phoenicurus moussieri*, Tristram's Warbler *Sylvia deserticola* and Algerian Nuthatch *Sitta ledanti*. However, several birds of sub-Saharan provenance also reside there, such as White-rumped Swift *Apus caffer* and Black-crowned Tchagra *Tchagra senegala* (Moreau 1966). The birds of the northern and central Sahara are still predominantly Palaearctic, occurring mainly at oases; and at both ends Palaearctic species extend southwards (e.g. Cirl Bunting *Emberiza cirlus* in the west and Eurasian Spoonbill *Platalea leucorodia* and Little Owl *Athene noctua* in both west and east) and Afrotropical species extend northwards (e.g. Small Buttonquail *Turnix sylvatica* and Red-knobbed Coot *Fulica cristata* in the west and Senegal Thick-knee *Burhinus senegalensis*, Little Green Bee-eater *Merops*

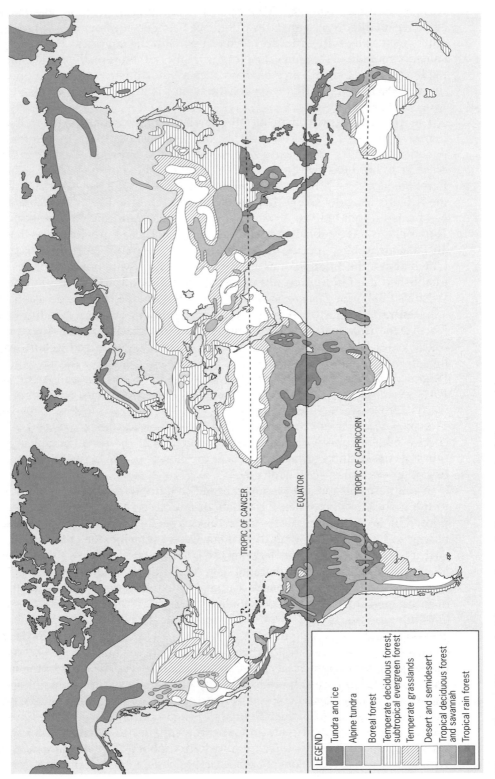

Figure 5.2 The main biomes (vegetation zones) of the world.

LEGEND

Tundra and ice

Alpine tundra

Boreal forest

Temperate deciduous forest, subtropical evergreen forest

Temperate grasslands

Desert and semidesert

Tropical deciduous forest and savannah

Tropical rain forest

TROPIC OF CANCER

EQUATOR

TROPIC OF CAPRICORN

orientalis and Nile Valley Sunbird *Anthreptes metallicus* in the east). Even in the driest of times, the Nile valley would always have provided a link between the Palaearctic and Afrotropical faunas. The landbirds of the Azores, Madeira and the Canary Islands are Palaearctic in origin, as are most species on the Cape Verde Islands much further south.

Many Palaearctic species reach southern Arabia, as do a few Indomalayan ones from further east, but mountainous southwest Arabia is usually included in the Afrotropical Region. The birds of the Persian Gulf and Iran are predominantly Palaearctic, with some Afrotropical and Indomalayan species; but further east a more exact boundary separates Palaearctic and Indomalayan species in the mountains of the Afghan–Pakistan border, and then east along the higher Himalayas into China about 32°N, where the River Yangtze is usually considered a faunal boundary, but where in fact lies another transition zone.

At least 937 landbird species breed regularly in the Palaearctic Region as defined above (or 903 excluding island forms, **Table 5.2**), of which more than half are passerines. The total number is low, considering the great size of this region, which also has a low level of endemism, with no families, and only 9% of genera and 47% of species being restricted to the region (**Table 5.3**). Overall, the Palaearctic Region shares 39% of its species and 97% of its families with the adjoining Indomalayan Region, 12% of its species and 88% of its families with the Afrotropical Region, and 11% of its species and 64% of its families with the Nearctic Region, plus smaller percentages with the other regions (**Tables 5.4** and **5.5**). Localised endemics occur on Madeira and the Canary Islands (nine species), Cyprus (two species), the Azores (two species), Corsica (one species) and the Caucasus mountains (three species), and other mountains and islands of eastern Asia. The volcanic island of Iceland in the north Atlantic was largely ice-covered until about 9,000 years ago; it now holds about 50 breeding landbird species, only ten of which have endemic subspecies (Lack 1969).

For much of their history, the northern land-masses of Eurasia and North America were connected, either across the North Atlantic via Greenland (until the Eocene), or across the North Pacific, via the Bering land-bridge (last broken 10,000 years ago) (**Table 1.2**). Shared landbird species represent 11% and 15%, respectively, of the total breeding landbird species in these regions (**Table 5.4**). However, within the two regions the proportions of shared species increase northwards, and many of the species of boreal forest, tundra and arctic ocean are circumpolar, making them Holarctic in distribution (**Table 5.6**). Such species include some waterfowl (Anatidae), grouse (Tetraoninae), raptors (Accipitridae and Falconidae), owls (Tytonidae and Strigidae) and finches (Carduelinae) among landbirds, plus divers (Gaviidae), auks and gulls (Laridae) among seabirds. Specific examples include Red Crossbill *Loxia curvirostra*, Northern Goshawk *Accipiter gentilis* and Common Goldeneye *Bucephala clangula* of boreal forest; Rock Ptarmigan *Lagopus mutus*, Gyr Falcon *Falco rusticolus* and Lapland Longspur *Calcarius lapponicus* of tundra; and King Eider *Somateria spectabilis*, Sabine's Gull *Xema sabini* and Glaucous Gull *Larus hyperboreus* of the arctic ocean. Moreover, some other northern species, although different between the two land-masses, are very closely related, presumably having diverged in recent geological times, though which continent was the ancestral home may have differed between pairs (**Table 5.7**).

Table 5.6 Numbers of bird species (including marine birds) that breed naturally in both Eurasia and North America. From Udvardy 1958.

Habitat	Species found in both Eurasia and North America	% of Eurasian birds	% of North American birds
Landbirds			
Tundra	36	67	75
Taiga	28	53	44
Forest	7	7[1]	3
Open grassland	2	8	6
Mediterranean scrub	0	0	0
Desert scrub	0	0	0
Montane	1	14	17
Various	10	62	55
Freshwater birds			
Open water and marsh	23	29	26
Seabirds			
Arctic	32	100	87
Temperate	18	47	62
Totals	157	15	26

[1] Although more typical of temperate forests, all seven also occur in Taiga.

Studies of mt DNA (cytochrome *b*) have revealed different degrees of genetic differentiation between counterpart taxa, and hence point to markedly different divergence dates. Among various diurnal raptors and owls, divergence dates were estimated at about 0.5 million years ago for the Northern Harriers *Circus c. cyaneus* and *C. c. hudsoni*, at 1.5 million years for the Merlins *Falco c. columbarius* and *F.c. aesalon*, 4.3 million years for the large owls *Bubo bubo* and *Bubo virginianus*, 6.3 million years for the Boreal and Saw-whet Owls *Aegolius funereus* and *A. acadicus*, 7.7 million years for the scops owls *Otus scops* and *O. hoyi*, and 7.9 million years for the Pygmy Owls *Glaucidium passerinum* and *G. gnoma* (Heidrich & Wink 1998, Wink *et al.* 1998). The dates indicate that species have moved from one continent to another, and then evolved independently, at a range of different dates in the past eight million years.

On the other hand, each land-mass now has certain families of birds not represented on the other, and certain groups now found only on one are represented by fossils on the other (see later). Furthermore, both the Palaearctic and Nearctic are poorly represented in species from some otherwise widespread families. For example, of 86 species of kingfishers (Alcedinidae and Halcyonide) worldwide, only seven breed in the Palaearctic and two in the Nearctic; and of 360 species of parrots (Psittacidae), only one breeds naturally in the Palaearctic and two (including one recently extinct) in the Nearctic.

Old World families barely represented in North America include the Old World warblers (Sylviidae, two of 560 species), nuthatches (Sittidae, four of 25 species), bush tits (Aegithalidae, one of eight species), dippers (Cinclidae, one of five species), shrikes (Laniidae, two of 30 species), treecreepers (Certhiidae, one of 98 species) and larks (Alaudidae, one of 92 species). The North American repre-

Table 5.7 Examples of ecologically equivalent and closely related species that occur in Eurasia and North America, respectively. Most (but not all) the pairs listed are 'sister species' apparently derived from the same parental form.

Eurasia	North America
Grey Heron *Ardea cinerea*	Great Blue Heron *Ardea herodias*
Whooper Swan *Cygnus cygnus*	Trumpeter Swan *Cygnus buccinator*
Eurasian Wigeon *Anas penelope*	American Wigeon *Anas americana*
Eurasian Teal *Anas crecca*	Green-winged Teal *Anas carolinensis*
White-headed Duck *Oxyura leucocephala*	Ruddy Duck *Oxyura jamaicensis*
Common Pochard *Aythya ferina*	Canvasback *Aythya valisineria*
Tufted Duck *Aythya fuligula*	Ring-necked Duck *Aythya collaris*
Common Coot *Fulica atra*	American Coot *Fulica americana*
Eurasian Oystercatcher *Haematopus ostralegus*	American Oystercatcher *Haematopus palliatus*
Pied Avocet *Recurvirostra avosetta*	American Avocet *Recurvirostra americana*
Black-winged Stilt *Himantopus himantopus*	Black-necked Stilt *Himantopus mexicanus*
Little Ringed Plover *Charadrius dubius*	Semi-palmated Plover *Charadrius semipalmatus*
European Golden Plover *Pluvialis apricaria*	American Golden Plover *Pluvialis dominica*
Eurasian Curlew *Numenius arquata*	Long-billed Curlew *Numenius americanus*
Hazel Grouse *Bonasa bonasia*	Ruffed Grouse *Bonasa umbellus*
Eurasian Woodcock *Scolopax rusticola*	American Woodcock *Scolopax minor*
Common Buzzard *Buteo buteo*	Red-tailed Hawk *Buteo jamaicensis*
Common Kestrel *Falco tinnunculus*	American Kestrel *Falco sparverius*
Eurasian Eagle Owl *Bubo bubo*	Great Horned Owl *Bubo virginianus*
Eurasian Pygmy Owl *Glaucidium passerinum*	Mountain Pygmy Owl *Glaucidium gnoma*
Grey-headed Chickadee *Poecile cincta*	Boreal Chickadee *Poecile hudsonicus*
Willow Tit *Parus montanus*	Black-capped Chickadee *Parus atricapillus*
Eurasian Treecreeper *Certhia familiaris*	Brown Creeper *Certhia americana*
Goldcrest *Regulus regulus*	Ruby-crowned Kinglet *Regulus calendula*
Eurasian Siskin *Carduelis spinus*	Pine Siskin *Carduelis pinus*
Siberian Jay *Perisoreus infaustus*	Grey Jay *Perisoreus canadensis*

sentatives of these families are often more closely related to species in eastern Asia than in Europe, again suggesting a Beringian route of entry. Similarly, several large New World groups, such as the tyrant flycatchers (Tyrannidae), vireos (Vireonidae), wood warblers (Parulinae), and emberizine finches (Emberizinae), are absent or have many fewer species in Eurasia. Given that different passerine groups prevail in North America and Eurasia, only limited exchange of passerines has occurred, and mainly through Beringia.

In both the Palaearctic and Nearctic regions, some bird species leave in autumn to winter in the tropics. On the whole, this sector of the passerine avifauna is phylogenetically more different between the two land-masses than is the remaining passerine sector, consisting of residents and short-distance migrants (Böhning-Gaese *et al.* 1998). This fact has led to the view that many long-distance passerine migrants on both land-masses derive from essentially tropical families that have spread north to breed at higher latitudes. They include mainly insectivorous species, including the Old World and New World warblers and flycatchers. The same is not true, however, for non-passerines, including many shorebirds and waterfowl, among which the same species of long-distance migrants occur on both land-masses.

Compared with some other regions, the avifauna of the Palaearctic is relatively well known. However, parts of China, especially some of the remaining mountain forests, may still hold undescribed species. Large parts of the region, notably western Europe, India and China, have long had high-density human populations, so that little remains of the natural vegetation. In these parts, bird distributions are likely to have greatly changed from their natural state, and some species may have disappeared in earlier centuries before ornithologists could record them.

INDOMALAYAN REGION

This region comprises a large part of southeast Asia and adjacent islands, so lies mainly within the tropics. Its northern boundary, dividing it from the Palaearctic Region, is a climatic-vegetation one, much of which corresponds with the Himalayan mountain chain. South of the Asian mainland, the Indomalayan Region includes the continental islands of Taiwan and Hainan, as well as the greater part of the Indonesian Archipelago, the Philippines, Borneo, Sulawesi, Sumatra, Java and Bali, south to the Chagos and Cocos Islands. In total, the Indomalayan Region covers about 9.6 million km^2. Between the Indomalayan and Australian Regions is the transition zone (called 'Wallacea' after Alfred Russel Wallace). Many of the islands in this zone reveal clear-cut divisions, as well as close intermingling, between the two faunas. In the nineteenth century, Wallace (1876) proposed a dividing line between the two regions, running northeast between Bali and Lombok, and then between Borneo and Celebes (Sulawesi) (**Figure 5.3**). This made sense because, when the sea level was 120 m lower than today, as during glaciations, all islands from Bali and Borneo westwards through Java and Sumatra lay on the Sunda shelf and would have been joined to continental Asia, whereas those from Lombok and Celebes eastwards would not. This partly explains the division of so many plants and animals at these points, even though Bali and Lombok are separated by only 40 km. To the east of the Sunda shelf lies a deep water channel containing other (oceanic) islands, including the Moluccas and Lesser Sundas, and then the Sahul shelf which supports New Guinea and Australia. However, later researchers drew lines between the Indomalayan and Australasian regions at somewhat different positions, partly because the point of the 50:50 balance varies between different types of plants and animals, as expected from their different rates of cross-water spread. While Wallace's Line follows the edge of the Sunda Shelf, Lyddeker's Line follows the edge of the Sahul Shelf, the islands between the two lines forming the transition zone of Wallacea.

Uncertainty surrounds the date that the Philippine Islands were last attached to southeast Asia, for they are separated by water much deeper than 120 m. Their birds are generally more distinctive than are those of other islands that were joined to the mainland as recently as 10,000 years ago. Sulawesi is also separated by deep water and may never have been connected to mainland Asia. Although it has land mammals, they form a restricted selection whose ancestors could have

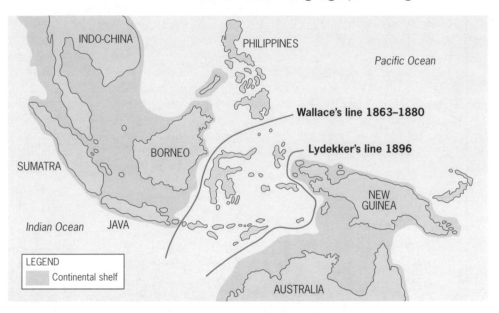

Figure 5.3   The area in southeast Asia (Wallacea) that marks the boundary between the Indomalayan and Australasian faunas, showing the dividing lines suggested by Wallace (1876) and Lydekker (1896).

arrived by swimming (Musser 1986). Its birds are more similar to those of the Philippines and Lesser Sunda Islands than to those of Sumatra and Borneo.

Much of the Indomalayan Region, before human intervention, was covered with rain forest, but extensive areas of savannah, grassland and desert also occurred, especially in India (**Figure 5.2**). Sizeable stretches of all these habitats still remain, while various montane habitats, with rhododendron, bamboo, coniferous forest and high tundra occur in the Himalayas.

In numbers of landbird species, the Indomalayan Region is the third richest, with about 1,697 species in total, or 1,259 excluding island forms (**Table 5.2**). Three (4%) of the bird families are peculiar to the Indomalayan Region, namely the leaf-birds (Irenidae, with 14 species), Asian frogmouths (Batrachostomidae, 11 species) and bearded bee-eaters (Nyctyornithidae, two species), but 126 (29%) genera and 1,184 (70%) species are also endemic **(Tables 5.3 and 5.8)**. Within the region, major areas of species endemism occur in some mountain areas and on the Philippines and other islands. The region shares 21% of its species and 97% of its families with the adjoining Palaearctic region, 10% of its species and 75% of its families with the Australasian Region, 5% of its species and 81% of its families with the Afrotropical Region, and smaller percentages with other regions **(Tables 5.4 and 5.5)**. It is also a major wintering area for east Palaearctic breeding birds.

The Indomalayan Region shares more bird families (but not more species) with Africa than with any other region probably because, as recently as the Miocene, these two regions were connected by a broad stretch of tropical habitats through the Middle East. Only later did the African and Indian floras and faunas become separated by deserts and seas. Moreover, while zoogeographers separate the Indomalayan and Afrotropical Regions, plant geographers often combine them as

Table 5.8 Families of landbirds endemic to each zoogeographical region.

	Scientific name	Family English name	Number of species
Afrotropical Region			
Mainland Africa	Struthionidae	Ostrich	1
	Scopidae	Hamerkop	1
	Sagittariidae	Secretary-bird	1
	Musophagidae	Turacos	23
	Coliidae	Mousebirds	6
	Phoeniculidae	Woodhoopoes	5
	Bucorvidae	Ground–hornbills	2
	Lybiidae	African Barbets, Tinkerbirds	41
	Numididae	Guineafowl	6
	Picathartidae	Rockfowl, Rockjumpers	4
	Rhinopomastidae	Scimitar-bills	3
Madagascar or other islands	Aepyornithidae[1]	Elephant-birds	
	Mesitornithidae	Mesites	3
	Leptosomidae	Cuckoo-rollers	1
	Brachypteraciidae	Ground-rollers	5
	Philepittidae	Asities	4
	Raphidae[1]	Dodo, Solitaire	3
Indomalayan Region			
	Batrachostomidae	Asian frogmouths	11
	Irenidae	Fairy-bluebirds, Leafbirds	14
	Nyctyornithidae	Bearded Bee-eaters	2
Neotropical Region[2]			
	Tinamidae	Tinamous	47
	Rheidae	Rheas	2
	Anhimidae	Screamers	3
	Opisthocomidae	Hoatzin	1
	Psophiidae	Trumpeters	3
	Cariamidae	Seriemas	2

Family	Common name	No.
Eurypygidae	Sunbittern	1
Thinocoridae	Seedsnipes	4
Steatornithidae	Oilbird	1
Nyctibiidae	Potoos	7
Galbulidae	Jacamars	18
Bucconidae	Puffbirds	33
Ramphastidae	New World Barbets, Toucans	55
Todidae	Todies	5
Momotidae	Motmots	9
Formicariidae	Ground antbirds	60
Conopophagidae	Gnateaters	8
Rhinocryptidae	Tapaculos	30

Australasian Region

Australia–New Guinea–Tasmania

Family	Common name	No.
Casuariidae	Emus, Cassowaries	4
Pedionomidae	Plains-wanderer	1
Aegothelidae	Owlet-nightjars	7
Menuridae	Lyrebirds, Scrub-birds	2
Climacteridae	Treecreepers	7
Ptilonorhynchidae	Bowerbirds	20
Petroicidae	Australo-New Guinean robins, Scrub-Robins, Jacky-winter, etc.	44
Anseranatidae	Magpie Goose	1
Maluridae	Fairywrens	26
Melanochartitidae	Berrypickers, Longbills	10
Orthonychidae	Logrunner, Chowchilla	2
Paramythiidae	Painted Berrypickers	2
Podargidae	Australian Frogmouths	3
Pomatostomidae	Australo-New Guinean babblers	5

New Zealand

Family	Common name	No.
Apterygidae	Kiwis	3
Acanthisittidae	New Zealand Wrens	4
Callaeatidae	New Zealand Wattlebirds	3

New Caledonia

Family	Common name	No.
Rhynochetidae	Kagu	1

The Nearctic and Palaearctic Regions have no endemic families.

[1] Recently extinct.

[2] The list excludes the Furnariidae, which is effectively endemic in the Neotropical Region, because only one of its 280 species extends north into northern Mexico which is here classed as in the Nearctic Region.

a single Palaeotropical Region. Among birds, the Indomalayan Region shares with the Afrotropical Region, either exclusively or overwhelmingly, the passerine families Eurylaimidae (broadbills), Pycnonotidae (bulbuls), Nectariniidae (sunbirds) and the subfamily Ploceinae (weaver birds). The typical Afrotropical family of the honeyguides (Indicatoridae) is represented in the Indomalayan (but in no other) Region by the genus *Indicator*.

The family Phasianidae is particularly well represented in the Indomalayan Region, and the spectacular genera *Pavo*, *Gallus*, *Lophura*, *Pucrasia*, *Catreus*, *Argusianus*, *Polyplectron* and *Rollulus* are centred here. The Indomalayan Region is the centre of radiation for many other distinctive birds, including pittas, laughing thrushes, drongos, parrotbills and flower-peckers.

The richness of the Indomalayan fauna may be partly because it results from the fusion of three separate faunas, each originating on a separate land-mass, with infiltrations from at least two others. Parts of southeast Asia may have been the first to unite with the rest of Asia, in the late Jurassic or early Cretaceous about 100 million years ago, while India collided with southern Asia in the early Eocene about 55 million years ago, after a long period of isolation as an island continent (Chapter 1). Finally, Africa became connected to southern Asia about 20 million years ago in the early Miocene and, as mentioned above, its fauna achieved a direct route into India, via a broad belt of tropical vegetation through the Middle East. In addition, further birds could have moved in directly from Eurasia to the north, and, with greater difficulty, from Australasia to the southeast. Three of these faunas were ultimately of Gondwanan origin, but would have undergone long independent evolution on separate land-masses, while the northern Eurasian fauna would have had an even more independent history. A further factor contributing to the richness of the Indomalayan avifauna is the inclusion of many islands, isolated enough to encourage speciation. The islands of Indonesia are extraordinarily rich in bird species, of which an exceptionally large proportion are endemic, either to the region as a whole or to particular islands (Andrew 1992, Stattersfield *et al.* 1998). The Philippines alone comprise some 7,100 islands of greatly varying sizes, scattered over 3.2 million km² of the tropical Pacific and China seas. They hold a great wealth of species, with one of the highest degrees of endemism in the world. Of 403 species known to breed on the Philippines, 172 (43%) are found nowhere else (Kennedy *et al.* 2000). These figures compare with 88 (36%) out of 247 breeding species on Sulawesi and 39 (9%) out of 434 breeding species on Borneo.

It may seem strange that the southeastern part of Eurasia should hold a bird fauna so different from that on the rest of the land-mass. This can be attributed partly to the effectiveness of the Himalayan chain and Tibetan plateau as a combined barrier to north–south range extension, partly to the abundance of islands, and partly to the fact that the Indomalayan Region (with the boundary as drawn in **Figure 5.1**) holds the only tropical forest in Eurasia.

Knowledge of the avifaunas of the Indomalayan Region is patchy, especially in parts of Indonesia and the Philippines. Some islands have been visited by ornithologists on only a few short occasions, so some endemic species probably remain undiscovered. Parts of the region contain high-density and rapidly growing human populations, and deforestation rates are among the highest in the world. If recent trends continue, many restricted-range species are likely to disappear in the coming decades, including some as yet undescribed.

AFROTROPICAL (OR ETHIOPIAN) REGION

This mainly tropical region comprises Africa south of the Sahara, together with Madagascar, the Comoro, Seychelles and Mascarene Islands, but as mentioned above, the islands are sometimes separated as a distinct 'Malagasy' Region. Also as explained above, the northern border with the Palaearctic Region is poorly defined, especially across Arabia. Extending between latitudes 20°N and 35°S, the total area of the Afrotropical Region is about 21 million km² (against 30 million km² for the whole of Africa), of which only about 1.25 million km² lie south of the tropics. Africa has had connections with other land-masses to the north for the past 20 million years. The high areas of the east were uplifted only about 12 million years ago, and the Rift valley, with its large lakes, developed only within the last million years.

In Africa, the main vegetation types run chiefly in latitudinal belts which become progressively more arid and open from the equator both northwards and southwards (**Figure 5.2**). Rain forest occurs mainly within 10° on either side of the equator in west Africa. To the north and south, forest gives way to progressively more open habitats, from woodland through savannah and grassland to desert. This basic pattern is complicated by topography: most of the southern and eastern parts of the region lie more than 1,000 m above sea-level and are too cool and dry for the forest found at lower altitudes. These raised plateaux are therefore covered with open woodland, scrub, savannah or grassland. The majority of the high mountains occur in the east of the region. Over most of Africa, the climate is markedly seasonal, with distinct wet and dry seasons; and with some regional exceptions, rainfall decreases with increasing distance north and south from the equator.

Although the Afrotropical Region occupies similar latitudes to the Neotropical Region, it contains less than half as many landbird species, with about 1,950 regular breeders (or 1,714 excluding island forms) (**Table 5.2**). This may be partly because tropical and montane forests cover much smaller areas in Africa than in South America, while high arid plateaux are more extensive. Well over half of all Afrotropical species are passerines, among which weaver finches (Ploceinae), waxbills (Estrildinae), starlings (Sturnidae), larks (Alaudidae), shrikes (Laniidae) and sunbirds (Nectariniidae) are especially well represented. The region is also rich in francolins, bustards, barbets, honeyguides and grass-warblers *Cisticola*, but is relatively poor in parrots and woodpeckers. The two species of oxpecker (Buphaginae), which are found only in Africa, have evolved in association with large herbivorous mammals.

Some 21% of all Afrotropical bird families, 62% of genera and 93% of species are restricted to the region (**Table 5.3**). The 11 endemic families of Africa and six others (including two recently extinct) of the Malagasy subregion mostly contain small numbers of species (**Table 5.8**). Some of these endemic families are represented among fossils from Europe or North Africa (**Table 5.9**, later), indicating that their present distributions are relictual, in that they have survived in the Afrotropics but died out elsewhere.

The main areas of local endemicity within Africa include the Ethiopian massif, the Cameroon mountains, the East African mountains, the Somali and the Namibian arid zones, and the Gulf of Guinea Islands, particularly Principé and

São Tomé (Moreau 1966). The Ethiopian Highlands form the largest mountain massif in Africa, totalling some 70,000 km² above 1,500 m, bisected by the Rift Valley. Of 25 endemic species, 21 are montane, and most of these occur commonly over the upland massif. The highlands of Eastern Africa are also extensive but more divided by lowland barriers. The highest peaks in Africa, Mount Kilimanjaro (5,894 m), Mount Kenya (3,199 m) and the Ruwenzori Range (5,119 m), all support alpine vegetation and glaciers.

The Afrotropical Region shares 6% of its species and 68% of its families with the Palaearctic to the north, and 4% of its species and 79% of its families with the Indomalayan Region (**Tables 5.4** and **5.5**). For many of these shared species, it is hard to decide in which direction colonisation took place, though some seem fairly certain: (1) from Eurasia to Africa — three wagtails and the pipit *Anthus similis*; (2) from southeast Asia to Africa — the pittas (Pittidae, two species in Africa), broadbills (Eurylaimidae, four species in Africa), drongos (tribe Dicrurini, eight species in Africa), the *Terpsiphone* flycatchers (tribe Monarchini, six species in Africa) and the African Tailorbird *Orthotomus metopias*, all of which belong to families or genera better represented in Asia than in Africa; (3) from Africa to southeast Asia — *Cisticola juncidis*, the *Ploceus* weavers and probably the Estrildinae (Hall 1972). Without DNA analyses, however, for some of these groups it is hard to tell whether the African or Asian representatives are the most primitive.

One of the most dramatic features of the Afrotropical avifauna is the enormous influx of migrants from the Palaearctic Region in the northern winter (Moreau 1972). About a third of all Palaearctic species winter wholly or mainly in the region, chiefly in savannah and scrub habitats and some reach the southern tip of the continent (Chapter 18). Several species occasionally breed in their African 'winter' quarters, and the close affinity of many species of the southern grasslands with Palaearctic species to the north suggests that other African taxa may have originated in this way (Chapter 15; Snow 1978b).

Malagasy subregion. This subregion comprises Madagascar and its outlying islands, east to the Mascarenes and north to the Seychelles. At 587,000 km², Madagascar is the fourth largest island on earth, and contains both forest and open land. It is only 400 km from Africa, from which its distinctive avifauna is mainly derived. Madagascar is thought to have broken away from Gondwanaland as a unit with India more than 100 million years ago, and then separated from India 80–85 million years ago, as the latter drifted northwards as a separate land-mass. Some of Madagascar's endemic bird families may date back to this time, notably the suboscine asities (Philepittidae), which have primitive passerine features (Cracraft 1981), the roatelos (Mesitornithidae) (Houde *et al.* 1997), the now extinct elephant-birds (Aepyornithidae), and perhaps also the ancestors of some of the extant parrots, cuckoos and pigeons. Most of the rest of the avifauna could have arrived later by cross-water flights. The various levels of endemism, from families to subspecies, attest to the continuation of the colonising process over a long period of time. Fossils of birds similar to Madagascan ground-rollers (Brachypteraciidae) are known from North America (Eocene), while the Madagascan vanga shrikes (Vanginae) are similar in DNA to the African helmet shrikes. This part of the Madagascan avifauna is relictual, consisting of groups that have died out elsewhere. Another part includes Asian elements, including three genera unknown in Africa, namely *Ninox*, *Hypsipetes* and *Copsychus*.

The avifauna of Madagascar appears somewhat poor and unbalanced, compared with that of Africa. Despite its great size and tropical position, it has only about 198 breeding landbird species, but about one-fourth of the genera and more than half of species are endemic (Langrand 1990). The majority of genera contain only one to several species (average 1.4, compared to 4.1 in the Afrotropics as a whole) and the largest genus (*Coua*) has only ten species. The small average number of species per genus implies a low rate of recent speciation (or immigration) or a high rate of recent extinction, giving a low speciation:extinction ratio. One adaptive radiation has given rise to a diverse range of songbird species, resembling bulbuls, babblers and warblers (Cibois *et al.* 2001); and another has given rise to 14 species of vanga shrikes (Vanginae), whose differences are so great that they are classified in 12 different genera. Compared with Africa, fruit-eaters, seed-eaters and raptors seem poorly represented.

The associated islands of the Mascarenes, Seychelles and Comoros seem to have derived their birds mainly from Madagascar (and hence Africa), with which they share about 22 landbird species, plus some additional species from Asia, and perhaps elsewhere. Many now extinct flightless species (such as the dodo and solitaires) once lived on the Mascarenes, and many of the remaining landbird species are endemic (Chapter 7).

In general, the birds of the Afrotropical region are relatively well known, but further exploration is especially needed of some regions, notably Angola, Mozambique, Congo, Ethiopia and Somalia and elsewhere of some of the mountain forests and islands. Previously unknown species continue to be described, but at a slower rate than for tropical parts of South America and southeast Asia. In many parts of Africa, human populations are growing rapidly, and deforestation and desertification are occurring at an unprecedented rate. Some small forest areas could soon disappear altogether, causing the loss of some endemic species, possibly including some undescribed ones.

AUSTRALASIAN REGION

This region comprises some eastern Indonesian Islands, New Guinea, Australia and New Zealand, and neighbouring Pacific islands out to the Solomons, New Hebrides, New Caledonia and Chathams. The heart of the region is the island continent of Australia, with an area of about 7.7 million km^2 (including Tasmania). New Guinea covers an area of 809,000 km^2, and New Zealand and its sub-antarctic islands about 271,000 km^2. The total land area of the region, including the smaller island groups, is about 8.9 million km^2. Habitats cover almost the full global range, from alpine grassland to rain forest, but a large proportion of the total area (in Australia) is semi-arid or arid.

Australia extends between 11° and 43° S, nearly 40% lying within the tropics, the rest largely in the subtropics, with a temperate southern fringe. It is a land of low relief, with large expanses of plain which impart uniformity in landscape and vegetation. It is the driest of all the continents, two-thirds having an annual rainfall of less than 50 cm, and one–third less than 25 cm. It has a hot, arid centre with concentric zones of more favoured country, culminating in restricted humid forest areas in interrupted coastal strips, particularly in the east and southwest (**Figure**

5.2). The northwest is mostly hot and dry. Down the east side runs a range of low forested mountains, the 'Great Dividing Range'. These mountains are too low to form a significant barrier to bird movements, which are instead constrained mostly by expanses of desert. Both New Guinea and New Zealand, with their high mountain cordilleras, high rainfall and lush vegetation — the one tropical, the other temperate — present marked physiographic and vegetational contrasts.

Australasia is so distinctive and biogeographically illuminating that it warrants much more text space than any other region. For more than 55 million years Australia itself has been physically isolated from the rest of the world's landmasses, although for a time ancient islands (now joined to southeast Asia) occurred in the sea between Asia and Australia. This long period of isolation gave rise to a unique flora and fauna largely derived from Gondwanan heritage. A long period of geological quiet, coupled with lack of glaciations, means that the majority of the soils are old and poor, having long been leached of most of their nutrients, a process hastened by frequent fire. Few nutrients flow down the rivers, so that Australia round parts of its coastline has some of the least productive seas on earth.

Several other features set Australia apart from other continents. The annual variations in climate, notably rainfall, are far greater than the within-year seasonal variations. The climate is governed mainly by the El Niño–Southern Oscillation, so that over much of the continent, droughts occur for up to several years at a time, followed by heavy rain and devastating floods which create large temporary inland lakes: hence, plant productivity and bird populations fluctuate greatly, and many bird species of the interior breed well only at intervals of several years. Fire has long influenced the vegetation of much of Australia, and increased in frequency following human colonisation, some 40–60,000 years ago. In consequence, many of the most widespread vegetation types are fire-resistant or fire-dependent. Associated with variable rainfall and fire regimes, about a third of all bird species have been described as nomadic, in that they are not necessarily found in the same areas in successive years. Some of the species that breed in the centre during wet periods move to the edges during droughts, and may not breed for several years at a time. Another curious feature of the Australian bird and mammal fauna is the relatively large proportion of species that nest or roost in tree hollows. About 11% of all Australian birds are obligate cavity nesters, about twice as many as on any other continent (Newton 1998a).

Australia, New Guinea and Tasmania have repeatedly been connected to one another at times of low sea-level, most recently in the last glaciation, until about 10,000 years ago. They broke from Gondwanaland as a single unit (Meganesia) more than 55 million years ago, so that some of the bird lineages are old (as judged by anatomical and DNA features). The isolation of the Australasian avifauna was probably helped by the fact that relatively few species migrate there from Asia (and those that do are mostly arctic-nesting shorebirds, Chapter 18), and likewise very few Australasian landbirds migrate northwards to winter in Asia. For much of its history, Australia was wetter than today, and much more forested, but as it drifted northwards extensive dry areas appeared from the Miocene, and especially in the Pleistocene. The region now holds about 1,592 breeding landbird species, or 926 on New Guinea–Australia–Tasmania together (**Table 5.2**). Although Australia (with about 538 species) and New Guinea (with about 578 species) each has many landbird species that the other lacks, this can be mainly attributed to the difference in habitat, rather than the distance between them.

In terms of landbird species per unit area, Australasia is the second richest of all regions, mainly because of the fragmented nature of forests in Australia, and the many islands in the region, both of which have favoured speciation. In addition, some 18 (25% of the total) families are endemic to the region, as are 280 (61%) of genera and 1,415 (89%) of species (**Tables 5.3** and **5.8**). This is a greater degree of endemism than in any other region, except the Neotropical. Australasia shares 10% of its species and 75% of its families with the Indomalayan Region to the north, and smaller percentages with other regions (**Tables 5.4** and **5.5**).

Within Australasia, species of parrots (Psittacidae), pigeons and doves (Columbidae), honeyeaters (Meliphagidae) and kingfishers (Alcedinidae, Halcyonidae) are especially well represented. In Australia and New Guinea, the honeyeaters are the largest and arguably the most diverse group of birds, occupying every available habitat from montane forest and tropical rain forest to the arid interior of Australia. The region also has more than its share of curiosities, including the spectacular birds-of-paradise on New Guinea, the bowerbirds of Australia which construct bowers of sticks decorated with colourful ornaments, the megapodes which lay their eggs to incubate in sand, decaying vegetation or volcanically heated soil, and the nocturnal kiwis of New Zealand which have nostrils at the end of the beak.

The discovery of fossil bird feathers from Lower Cretaceous deposits in Victoria confirmed that some birds existed in the land-mass that became Australia early in avian evolution. The earliest known passerine in the world came from early Eocene deposits in Australia, dated about 55 million years old, around the time of the separation of Australia from Gondwanaland (Boles 1995). Thus one major component of the Australasian avifauna could have been inherited, *in situ*, directly from Gondwanaland (Serventy 1972). In addition to passerines (Atrichornithidae, Grallinidae, Meliphagidae, Ptilonorhynchidae and Corvidae), this autochthonous component included ratites and penguins, and probably also parrots, halcyonine kingfishers, bronze-cuckoos (*Chrysococcyx*), lyrebirds, pigeons, galliform birds, grebes and waterfowl (Cracraft 1972, Marchant 1972, Serventy 1972, Schodde 1991, Härlid *et al.* 1998). After the separation of Australia, some of these groups clearly underwent additional evolution and radiation, as is evident from their present distribution patterns.

Some of the groups just mentioned have relatives (in the same families) in Africa and South America, perhaps reflecting common Gondwanan origins. Some families, such as ratites, parrots and pigeons, are shared between all three southern continents, while others, such as trogons, are shared only between South America and Africa. However, some families currently restricted to the Australasian Region, or more widely within the southern hemisphere, are represented among Eocene or Oligocene fossils from Eurasia, including ratites, megapodes, frogmouths and owlet-nightjars (**Table 5.9**). Their modern distributions may therefore be relictual, like the African families mentioned earlier (Olson 1988). Other distinctive Australian birds probably disappeared long ago in competition with newer colonists from the north.

To judge from DNA–DNA hybridisation studies, many Australo-Papuan passerines result from an endemic radiation within Australia rather than from

Table 5.9 Families that are now considered endemic to one of the southern continents that are apparently represented in the fossil record of the northern continents. From Brodkorb 1971, Olson 1982, 1988, Bochenski 1985 and Mourer-Chauviré 1982, 1985, 1988.

Afrotropical families formerly represented in Europe	Epoch
Parrots (Psittacidae)	Miocene
Trogons (Trogonidae)	Oligocene–Miocene
Barbets (Capitonidae)	Miocene
Mousebirds (Coliidae)	Eocene–Miocene
Hornbills (Bucerotidae)	Eocene–Miocene
Turacos (Musophagidae)	Eocene–Miocene
Woodhoopoes (Phoeniculidae)	Eocene–Miocene
Broadbills (Eurylaimidae)	Eocene–Miocene
Secretary-birds (Sagittariidae)	Miocene
Guineafowl (Numididae)	Pleistocene
Neotropical families formerly represented in Europe	
Seriemas (Cariamidae)	Eocene–Oligocene
Parrots (Psittacidae)	Eocene–Oligocene
Cracids (Cracidae)	Oligocene
Capitonids (Capitonidae)	Miocene
Potoos (Nyctibiidae)	Eocene–Oligocene
Motmots (Momotidae)	Oligocene
New World vultures (Cathartidae)	Eocene
Trogons (Trogonidae)	Oligocene–Miocene
Todies (Todidae)	Eocene–Oligocene
Neotropical families formerly represented in North America	
Cracids (Cracidae)	Eocene–Miocene
Seriemas (Cariamidae)	Eocene–Oligocene
Oilbirds (Steatornithidae)	Eocene
Motmots (Momotidae)	Miocene
Todies (Todidae)	Oligocene
Jacamars (Galbulidae)	Eocene
Puffbirds (Bucconidae)	Eocene
Old World families formerly represented in North America	
Pratincoles (Glareolidae)	Miocene
Button-quails (Turnicidae)	Miocene
Australasian families formerly represented in Europe	
Megapodes (Megapodiidae)	Eocene–Oligocene
Frogmouths (Podargidae)	Eocene–Oligocene
Owlet-nightjars (Aegothelidae)	Eocene–Oligocene

In addition, flighted and flightless palaeognathous birds (ratites and tinamous) are represented in the fossil record of both Europe and North America in the Palaeocene–Eocene.

successive colonisations from Eurasia, as previously supposed (Sibley & Ahlquist 1985). Such birds comprise the modern Parvorder Corvida which long ago (30–40 million years ago on DNA evidence) separated from the Parvorder Passerida. The majority of these birds now fall into three distinct assemblages (**Figure 5.4**). They include some remarkable cases of convergence in form and habit towards unrelated birds in the northern hemisphere. The *Gerygone* warblers look and behave like sylviid and parulid warblers of Eurasia and North America, the sittellas (*Daphoenositta*) climb trees and have dagger-shaped beaks like nuthatches (Sittidae), and the small *Myzomela* honeyeaters resemble sunbirds. The evidence suggests an adaptive radiation of a few early forms, which parallels that of the Australian marsupials. All these features combine to give Australasia one of the most interesting and distinctive avifaunas in the world.

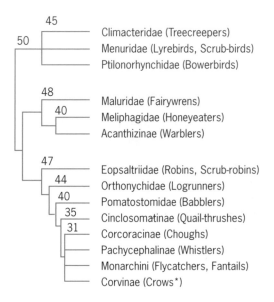

Figure 5.4   Relationships and approximate dates of divergence (millions of years ago) of Australian passerines. Modified from Sibley & Ahlquist (1985).

The second major component of the Australasian avifauna is assumed to derive from the Palaearctic–Indomalayan Regions to the north, where close relatives are now found. Although 12 families fall into this category, the majority are represented in Australia by only one or two species. These birds probably arrived over a long period by cross-water colonisation, followed in some cases by secondary radiation within the region (Mayr 1944). They include members of several near-cosmopolitan waterbird and raptor genera, a few strigids, swifts and coraciids, and among passerines, pittas (*Pitta*), swallows (*Hirundo*), thrushes (*Zoothera*), silver-eyes (*Zosterops*) and others. Together, these various species form almost one-fifth of Australia's current land and freshwater bird species. They show varying degrees of differentiation, some having evolved into distinct genera (e.g. some estrildine finches and sylviid warblers), while others are conspecific with their

Eurasian ancestors (e.g. Black Kite *Milvus migrans*, Richard's (Australasian) Pipit *Anthus novaeseelandiae*). Evidently, they have arrived successively over a long period (Mayr 1944).

The progressive attenuation of some dominant African bird groups, through Asia, towards Australia gives further indication that most avian relationships between Africa and Australia do not extend back to Gondwanaland, but are instead the result of more recent colonisation (Keast 1981a). The majority of groups that are represented in the Afrotropical, Indomalyan and Australasian regions have fewest genera and species in Australasia (**Table 5.10**). Some, such as bulbuls, hornbills and shrikes, have reached the Australasian Region, but not Australia itself. A few groups limited to the Afrotropical–Indomalayan–Australasian Regions have undergone independent radiations in Africa and Australia, including the turnicid quails, emerald cuckoos and estrildine finches.

Table 5.10   Decline in species numbers and diversity between the Afrotropical–Indomalayan Regions and Australasia among some bird families represented in all three regions. G – genera, S – species. Modified from Keast 1981a.

	Africa		Asia		Australia	
	G	S	G	S	G	S
Sunbirds (Nectariniidae)	2	75	5	24	1	2
Bee-eaters (Meropidae)	2	15	2	5	1	2
Hornbills (Bucerotidae)	4	20	8	13	1	1
Shrikes (Laniidae)	9	55	1	2	1	1
Larks (Alaudidae)	10	49	3	3	1	1
Rollers (Coraciidae)	2	6	2	2	1	1
Wagtails (Motacillinae)	3	26	1	4	1	1

In contrast to this apparent colonisation of Australia from Indomalaya and beyond, very few characteristic Australian genera have breached Wallace's line to reach the Indomalayan Region, and none has reached Africa (Keast 1981a). Pachycephaline flycatchers, *Monarcha* flycatchers and *Gerygone* warblers occur on the Malay Peninsula, while *Artamus* woodswallows have also reached India and China. It seems that the Indomalayan forms again have a better colonisation record than Australasian ones. Similarly, Palaearctic species show a marked attenuation in numbers of genera and species through Indonesia to Australia. This situation provides an example of the unequal exchange of genera and species between continents, usually attributed to the competitive superiority of the members of one avifauna over the members of another (see later). Nevertheless, several groups of Australian birds, including the Pachycephalinae (whistlers), Meliphagidae (honeyeaters) and Artamini (woodswallows) have colonised Pacific Islands.

Several Eurasian bird families have no representatives in Australia, further emphasising the distinctiveness of the avifauna. Notable absentees include Old World vultures (Accipitrinae), pheasants (*Phasianus*), skimmers (Rynchopini), sandgrouse (Pteroclidae), trogons (Trogonidae), rollers (Coraciidae), woodpeckers (Picidae), broadbills (Eurylaimidae), finches and buntings (Fringillidae). The flamingos were formerly placed in this category too, until the finding of fossil remains proved that several species were present around Australia's inland lakes to the late Pleistocene, disappearing in subsequent dry periods (Serventy 1972).

New Guinea is tropical, mountainous (highest peak >5,000 m and glaciated) and wet, mostly covered in rain forest and close to the Indomalayan Region, so not surprisingly it has a rich avifauna. It has more bird species than Australia, but in only one-tenth of the area (Beehler *et al.* 1986). It is the centre of distribution of birds-of-paradise, bowerbirds, cassowaries, megapodes and crowned pigeons, and is also rich in other pigeons, parrots, cuckoos, kingfishers, honeyeaters and Australian warblers and flycatchers, with numerous endemic genera. It shares many other taxa with Australia but, as expected, it has many more rain forest specialists and a generally higher proportion of birds of Asian origin. Many of the forest species now found only on New Guinea may well have occurred also in Australia in the Pliocene when forest was more extensive (Pratt 1991). New Guinea has also acted as a source of re-invasion for Australia, notably of honeyeaters (*Meliphaga*) and fruit-doves (*Ptilinopus*). It is not only rich in birds, but in other animals and plants too, making it one of the most species-rich islands on earth.

New Zealand now has three components to its avifauna, an ancient group, together with more recent colonists, some from Australia and others from the Holarctic (Baker 1991). It evidently had birds when it broke away from Gondwanaland, about 82 million years ago and preceding the separation of Australia, notably the New Zealand wrens (Acanthisittidae) which on DNA evidence pre-date this separation, and the extinct moas (Dinornithidae) and possibly the kiwis (Apterygidae, but see Chapter 2). The wattlebirds (Callaeatidae), comprising the Saddleback *Philesturnus carunculatus*, the Kokako *Callaeas cinerea* and the extinct Huia *Heteralocha acutirostris*, may be equally ancient, as may be the extinct New Zealand Thrushes (Turnagridae). All these taxa are endemic to New Zealand, and have no close relatives elsewhere. They may be some of the most primitive of all surviving landbirds. Endemic genera usually included in this ancient component are the parrots *Nestor* and *Strigops*, the snipe *Coenocorypha*, the plovers *Thinornis* and *Anarhynchus*, the Blue Duck *Hymenolaimus*, the giant extinct rail-like birds *Dicaphorapteryx* and *Aptornis* and the rails *Gallirallus*. Some of these genera may yet be found through DNA analyses to be related to taxa in other parts of the world, but at the moment they seem to stand alone. Insular influences on the evolution of New Zealand birds are seen in the flightlessness of several species (**Table 6.5**), in the lack of sexual dimorphism and in the occurrence of melanistic forms, a common cause of which is assumed to be lack of selection pressure in an environment free of mammalian predators (Chapter 6).

Although New Zealand has been separated from Australia for more than 80 million years, many other of its landbirds are more obviously derived from Australian families, including the honeyeaters (*Meliphaga*), warblers (*Gerygone*), robins (*Petroica*) and fantails (*Rhipidura*). This colonization is of long standing, as revealed by the varying degrees of differentiation of the New Zealand from the ancestral Australian forms. At least three species are unrecorded in deposits older than 1,000 years old, but had colonised by the time of European settlement, namely Southern Shoveler *Anas rhynchotis*, Pukeko *Porphyrio p. melanotus* and Australian (Swamp) Harrier *Circus approximans*. Many others arrived after European colonisation, including eight Australian species in the last 150 years

(Chapter 16). In total, at least 33 species of landbirds are in this category. Recent human modification of the landscape in this time may have increased the numbers of vagrant species that could establish themselves, compared to earlier times when the range of habitats was more restricted. The third component of the New Zealand avifauna, derived from Holarctic species which apparently did not colonise via Australia, include the South Island (Pied) Oystercatcher *Haematopus finschi*, closely related to the Palaearctic *H. ostralegus*, the New Zealand Scaup *Aythya novaeseelandiae*, closely related to the Holarctic *A. fuligula*, *A. marila* and *A. affinis*, the extinct Auckland Islands Merganser *Mergus australis* related to the Scaly-sided Merganser *M. squamatus*, and the extinct Chatham Island Eagle *Haliaeetus australis*, related to sea-eagles of the northern hemisphere (Baker 1991).

Compared with other areas of similar size, New Zealand is now generally poor in breeding birds, with only 65 surviving species of native land and freshwater birds, plus 38 introduced species (Baker 1991, Bell 1991). In contrast, it is exceptionally rich in seabirds, with no fewer than 84 species breeding on the main islands and their outliers, out to Chatham and Macquarie (Baker 1991). This gives 187 current breeding species in total (Bell 1991). Several factors have contributed to the poverty of New Zealand's landbirds. Like Madagascar, it has an unusually low species per genus ratio (1.4, compared to 3.5 for the Australasian Region as a whole), again suggesting recent extinctions and little recent speciation. Second, in the Oligocene period, about 28 million years ago, New Zealand suffered a huge reduction and fragmentation of land area, to about 20% of the current level, distributed as five main islands and other smaller ones, all of mostly low relief. Coinciding with a marked temperature drop, this event is likely to have extinguished many species extant at the time, and reduced the genetic diversity of those remaining. Studies of mt DNA in museum specimens of three ancient lineages (kiwi, moa and acanthisittid wrens) gave results consistent with a major genetic bottleneck (= greatly reduced population sizes) at that time, and with subsequent radiation from one of a few mitochondrial lineages within each group later on (Cooper & Cooper 1995).

Thirdly, New Zealand has experienced a very different history from the rest of Australasia. Lying much further south, and 1,600 km from Australia, it has always been colder and subject to frequent glaciation which, together with volcanic activity, has given it more fertile soils. It is among the most isolated of the world's land-masses, and its biota have pursued different evolutionary paths. The lack of native land mammals (apart from bats) is surprising because monotreme mammals were almost certainly present when New Zealand split from Gondwanaland. It has few reptiles. To set against this, until colonised by humans about 1,000 years ago, New Zealand had an extraordinary assemblage of birds, some of which had radiated to occupy niches filled elsewhere by herbivorous mammals (Chapter 7). At least 44 species have become extinct in this period, including about 11 species of moas, which ranged from the towering 3-m high 270-kg giant *Dinornis maximus* to smaller, 1-m high, 20-kg forms (Chapter 7).

The island of New Caledonia separated from Gondwana about the same time as New Zealand, about 80 million years ago. At only 16,000 km^2, one-fifth the size of New Zealand, it has a mountainous spine and poor soils, with rain forest, dry forest and maquis heath as the main natural vegetation types. Not surprisingly, it has a poor avifauna, with about 71 species. The most notable include the strange flightless Kagu *Rhynochetos jubatus*, in a family to itself, as well as four other endemic genera and 14 other endemic species. As on New Zealand, many interesting species were lost following human colonisation (Chapter 7).

In general, the Australasian avifauna is well known, but parts of New Guinea are in particular need of further exploration. Many species of this island seem to have curiously disjunct distributions (Chapter 12), but in some species this may be partly a consequence of insufficient exploration. The steep terrain, difficult forest habitats and hostile natives have made parts of this island almost inaccessible to ornithologists. In Australasia as a whole, many bird species have naturally small geographical ranges, and others are threatened by introduced predators or by continuing forest and scrub destruction. Considering the immense biogeographical importance of the region, it would be a pity to lose yet more species to human impact.

NEARCTIC REGION

This region embraces North America north of the tropics and the large island of Greenland. It is bordered by the Arctic Ocean in the north, the Bering Strait and the Pacific Ocean in the west, and the Atlantic Ocean in the east. The southern border is usually placed through Mexico, along the northern edge of the tropical rain forest and across a major lowland gap in the north–south mountain chain; it is therefore a climatic-vegetation boundary, which does not correspond with the ancient sea water barrier near Panama that until 3.5 million years ago separated North from South America. Thus defined, the Nearctic Region extends from about 20°N to about 85°N, and from about 168°W to 15°W, covering around 21 million km^2, plus another two million km^2 (mostly ice-covered) in Greenland.

In this large region, a wide range of climates is encountered, but the winters become progressively colder with increasing latitude and altitude. The major topographical features of North America, in contrast to Europe, extend longitudinally. The west consists largely of long lines of mountains, the Rocky Mountains and related chains, extending from Alaska southwards into Central America and continuing into the Andes of South America. The interior of North America is occupied by plains extending from the Gulf of Mexico northwards into Canada. In the east, a minor chain of mountains, the Appalachians, extends from Georgia and Alabama into Pennsylvania and continues in several more or less isolated mountain ranges to the Maritime Provinces of Canada. This essentially longitudinal arrangement of the mountains is to a large extent responsible for some otherwise puzzling features of North American bird life. The area east of the Rockies has about 90 breeding landbird species not shared with the west, while the smaller area west of the Rockies has more than 175 species not shared with the east, despite the similarity in habitats between west and east. Many other species have distinct eastern and western forms, divided either by the mountains or by the central grasslands (Chapter 10). Smaller areas of endemicity occur on the mountains themselves, or in lowland areas between the different western mountain ranges or between the Pacific coastline and the mountains.

Apart from the effects of the longitudinal mountain ranges, landscape and bird life change latitudinally, with the climatic and vegetational belts becoming increasingly better defined northwards (**Figure 5.2**). Circumpolar tundra and coniferous forest belts are well defined, with both extending southwards along the mountains. Further south lie less well-defined belts of deciduous forest, limited to areas of higher rainfall. Between the Mississippi Valley and the Rocky Mountains lie extensive grasslands ('prairies'); and in northern Mexico and the southwestern states are extensive arid areas, some of them true deserts. The bird life of each of these vegetational areas differs more or less markedly from that of other areas.

In all, around 732 bird species breed regularly in the Nearctic Region (or 723 excluding island forms), less than in any other major biogeographical region (**Table 5.2**). The region has no endemic bird families, but about 19% of the genera and 54% of the species are found nowhere else (**Table 5.3**). Moreover, some sub-families, including the Emberizinae, are represented by more genera and species in the Nearctic than in any other region. The Nearctic Region shares 33% of its breeding species and 88% of its families with the Neotropical Region immediately to the south, 15% of its species and 71% of its families with the Palaearctic Region, and smaller percentages with other regions **(Tables 5.4** and **5.5)**.

The composition of the North American bird fauna is again best understood in the light of its history. The North American continent was connected with Europe (via Greenland) in the early Tertiary when the Atlantic Ocean was much narrower than today, and has also had intermittent connections with Asia across the Bering Strait bridge (the last only 10,000 years ago). Not surprisingly, therefore, it has strong faunal affinities with Eurasia (see above). The connection between North and South America occurred about 3.5 million years ago in the Pliocene, although a number of 'stepping stone' islands would have facilitated some bird interchange long before that. In general, because Central America is well vegetated, and links North and South America, there is much less separation of the Nearctic and Neotropical avifaunas than of the Palaearctic and Afrotropical ones which are separated by desert.

During the first half of the Tertiary, the southern half of North America was humid and tropical as far north as latitudes 38°–40° (roughly the level of San Francisco to Washington, DC), as shown by the palaeo-botanical record. This seems to have permitted the evolution of a tropical North American fauna rather distinct from that of South America (Mayr 1985). The two tropical faunas later intermingled, mainly after the Panamanian land-bridge was established in the Pliocene. In the meantime, there had been a steady process of cooling and reduction of rainfall in North America, in part caused by the rising of the mountain ranges in the west. This resulted in the development of deserts, populated by colonists from the adjacent, more humid habitats.

In addition to seabirds and worldwide taxa, the North American avifauna consists essentially of four elements, according to Mayr (1946, 1985): (1) an old indigenous one that is presumed to have developed during the Tertiary isolation; (2) a younger Holarctic element dating back to the Eocene trans-Atlantic connection with Europe; (3) a more recent Holarctic element resulting from recent immigration from Asia across the Beringian connection; and (4) an immigrant element from South America. As in any faunal analysis, there is some ambiguity, but it is easy to guess to which of the four stated components most families and genera of North American birds might belong.

Some of the groups that Mayr (1946, 1985) classed as indigenous to North America include the wrens (Troglodytinae), dippers (Cinclidae), gnatcatchers

(Polioptilinae), silky flycatchers (*Ptilogonys*), American sparrows (Emberizinae), the mostly tropical motmots (Momotidae) and the Palmchat *Dulus dominicus* (Dulini). Several families of non-passerine birds, such as the New World quails (Odontophoridae), may also have evolved in North America.

Because North America and Europe were a single continent up to the Eocene, it is impossible to determine the exact area of origin of the older Holarctic element, such as cranes (Gruidae), grouse (Tetraoninae) and certain thrushes (Turdidae). The present distributions of these species give little clue as to whether they originated in the Old World or the New. The recent (and hence less differentiated) Palaearctic element evidently came across the Bering Straits; and many such species are still confined to the northwest of North America. These recent immigrants include the Horned Lark *Eremophila alpestris*, the Brown Creeper *Certhia americana*, pipits (*Anthus*), nuthatches (*Sitta*), Wrentit (*Chamaea fasciata*), various corvids (*Perisoreus, Pica, Corvus, Nucifraga*), various tits (*Parus (Poecile), Auriparus, Psaltriparus*), kinglets (*Regulus*), Barn Owl *Tyto alba* and some hirundines (*Riparia, Hirundo*), with other, even more recent, colonists listed in Chapter 9. Some likely colonists of the Nearctic Region from the Neotropical Region are mentioned in the next section. The most notable are the tyrannid flycatchers (Tyrannidae), which comprise the only family of suboscine birds to have undergone a major radiation in North America.

Because of its proximity to North America, Greenland is usually classed as part of the Nearctic Region. Of 59 bird species that breed there regularly, 38 are circumpolar in distribution, eight clearly derive from North America (being absent from Eurasia), while 13 derive from Eurasia (being absent from North America) (Salomonsen 1967). The majority of North American species are found only on the west side, and most European species on the east side, being separated by the ice cap. As almost the whole of Greenland was ice-covered until 6,000 years ago, much of its avifauna results from recent immigration. Despite this, its redpolls are classed as subspecifically endemic.

The Nearctic avifauna is the best known of all, mainly because all parts of the region have remained open to ornithological exploration for more than 200 years. In this time, however, extensive habitat destruction has occurred, as forests have been felled, grasslands cultivated and wetlands drained. Over this period more species are known to have become extinct in North America than in any other continental land-mass, namely the Great Auk *Pinguinus impennis* (last seen 1846), Labrador Duck *Camptorhynchus labradorius* (1875), Carolina Parakeet *Conuropsis carolinensis* (1914), Passenger Pigeon *Ectopistes migratorius* (1914) and Imperial Woodpecker *Campephilus imperialis* (1950s), plus the Ivory-billed Woodpecker *Campephilus principalis* which may still survive on Cuba. Over the past 20 years, taxonomic splitting into separate species of taxa previously regarded as subspecies has probably occurred to a greater extent in North America than elsewhere, thus increasing species numbers relative to those of other regions.

NEOTROPICAL REGION

This Region embraces Central and South America, together with the West Indies and other islands near South America, out to the Galapagos in the Pacific and the

Falklands in the South Atlantic. The Region extends from the northern edge of the tropical rain forest in Mexico, at about 20°N, southwards to Cape Horn, at about 57°S, and from about 112°W to 35°W, covering around 18.2 million km². It is essentially a tropical area, mainly forest, but also includes desert, grassland ('pampas'), and other temperate and high montane habitats. Through the transition zone in Central America, there is a gradual northwards replacement of Neotropical by Nearctic species. A few tropical birds, such as a trogon and a chachalaca, extend north into the United States, while some Nearctic ones, such as turkeys, extend into South America. In consequence, Central America contains a strongly mixed fauna — an old indigenous element and numerous more recent invaders from South America, including tinamous, jacamars, puffbirds, toucans, woodcreepers, ovenbirds, antbirds, manakins and cotingas, all of which now occur in Central America in diminishing numbers from south to north. The southern continent continues to contribute to the Central American avifauna, for with the destruction of forest barriers chiefly in recent decades, some open-country birds have also spread north.

South America is a continent of extremes. The Andes lie near the western edge throughout its length (**Figure 5.2**). They form the longest mountain range in the world, extending through nearly 50° of latitude and covering 1.8 million km² of land, with a great diversity of habitats, both forested and open. The central part supports a high desert, the Altiplano, which holds the largest expanse of salt on earth, with large saline lakes supporting flamingos. Two areas of mountains lie to the east of the Andes — the isolated Guiana–Venezuela highlands (with Mount Roraima and Mount Duida), and the eastern Brazilian highlands. The Amazon, at more than 5,000 km long, drains 40% of the continent, and is the largest river in the world, supporting more than 3,000 species of fish, but acting as a significant barrier to the movements of birds. The Amazon Basin receives more than 3 m of rain per year which provides 57,000 km² of seasonally flooded forest. There are some extensive savannahs ('llanos') north of the Amazon, particularly in the upper Orinoco basin of Venezuela and northeastern Colombia, and more extensive ones from the Matto Grosso south into Patagonia. The latter include the greatest area of seasonal swamp on earth, the 60,000 km² of flood, drought and fire that forms the Pantanal. Very arid country, some of it true desert (the Atacama), extends along the Pacific coastal plain from Ecuador south to Chile near Valparaiso. Some parts can go for years without rain. As Wallace (1876) said, the Neotropical Region 'is distinguished from all the other great zoological divisions of the globe by the small proportion of its surface occupied by deserts, by the large proportion of its lowlands, and by the altogether unequalled extent and luxuriance of its tropical forests'. For sheer variety of landscape and vegetation, the region is without equal.

The Neotropical Region is also the richest of all regions in the number and diversity of its breeding bird species. It holds a total of at least 3,370 breeding landbird species, or 3,170 species excluding island forms, which comprise more than one-third of all land and freshwater bird species on earth (**Table 5.2**). These species fall into 71 families and 893 genera. Two (11%) orders are endemic to the region, together with 20 (28%) families, 686 (77%) genera and 3,121 (93%) species (**Table 5.3**). This represents the greatest level of endemism shown by any region, reflecting the ancient and distinctive evolutionary history of the South American land-mass. The Neotropical Region shares only 7% of its species but 65% of its families with the Nearctic Region immediately to the north, and smaller percentages with other regions (**Tables 5.4** and **5.5**). In addition to its year-round residents, the Neotropical Region receives many migrant species from North America during the northern winter. However, in contrast to the situation in Africa, where most Palaearctic migrants live in woodland and savannah, the majority of the Nearctic migrants to Central and South America live in forest.

Many of the most characteristic families have undergone extensive recent evolutionary radiation to produce a large number of species. They include the guans (Cracidae, 50 species), ovenbirds and woodcreepers (Furnariidae, 280 species), ground antbirds (Formicariidae, 60 species), tyrant flycatchers, cotingas and manakins (Tyrannidae, 544 species), puffbirds (Bucconidae, 33 species), gnateaters (Conopophagidae, eight species), jacamars (Galbulidae, 18 species), motmots (Momotidae, nine species), New World barbets and toucans (Ramphastidae, 55 species), tapaculos (Rhinocryptidae, 33 species) and tinamous (Tinamidae, 47 species). These families together comprise 1,137 species, more than 10% of the world's extant birds. Much of this diversity has been attributed to events in the Pliocene and Pleistocene, which supposedly promoted much allopatric speciation (Chapter 11). Other endemic families, however, contain only one or a few species, as shown in **Table 5.8**. Again, not all the bird families now endemic to South America necessarily evolved there, but instead could be relicts from former widespread distributions (see later).

The South American continent also has a rich fauna of freshwater birds, consisting essentially of cosmopolitan families but with many endemic genera and species. The more characteristic of these include the torrent ducks *Merganetta*, the Coscoroba Swan *Coscoroba coscoroba*, the South American shield-geese *Chloephaga* and the steamer ducks *Tachyeres* (mostly salt water). The latter are reminiscent of the Musk Duck *Biziura lobata* of Australia.

The absence from the Neotropics of otherwise widespread groups also emphasises the distinctiveness of the region's bird life. There are no button quails (Turnicidae), cranes (Gruidae), bustards (Otididae), hornbills (Bucerotidae), broadbills (Eurylaimidae), corvids of the genus *Corvus*, titmice (Paridae), nuthatches (Sittidae), treecreepers (Certhiidae) or shrikes (Laniidae). Compared with other southern continents, South America lacks six families that are found in Australia, four found in Africa, and eight others that occur in both Africa and Australia.

How could South America, isolated as a giant island continent for more than 30 million years, achieve such a great diversity of life? Keast (1972) attributed the richness of the bird life to: (1) the great extent of rain forests, which cover about 30% of the continent; (2) the fact that the contemporary fauna is essentially a double one, the result of the joining of North and South America at the end of the Pliocene; (3) a high degree of regional endemism, and allopatric duplication; (4) presumed fewer Pleistocene extinctions than in the northern continents where glacial effects were more drastic; and (5) marked speciation within South America, which has a high degree of familial endemism (28% of families), attributed to its long isolation as a separate land-mass. Some of these features apply equally or more so to other regions but no other region shows the combination of all features. Furthermore, the Andes separate east and west, with many species found one side or the other, especially in the forest areas of the north. These high mountains add to the diversity of habitats by providing altitudinal zones of vegetation, each of which has a distinct avifauna, and by providing separate high-altitude islands of vegetation, also with partly distinct avifaunas. In fact, the Andes mountain chain holds most of the main centres of species endemism on the continent (Chapter 11). An additional factor contributing to the current overall richness of the avifauna might have been the lateness at which the continent was colonised by humans (about 12,000 years ago), and the reduced human impact compared with most other continents.

About 90% of all suboscine birds occur in the neotropics. In contrast, the largest group of living birds, the songbirds or oscines (over 4,000 species worldwide), are proportionately less abundant in the neotropics than in other regions. Suboscines were probably isolated in South America during their major radiation, and many genera have not penetrated Central America and the West Indies (except for Trinidad, once connected to the mainland), possibly because of competition with the resident oscines. Several other families in mainland tropical America and the West Indies have pan-tropical distributions, including the parrots (Psittacidae) and trogons (Trogonidae). The nearly cosmopolitan families of pigeons (Columbidae) and woodpeckers (Picidae) are also abundant in South America.

Most of South America lies within the tropics. The south temperate avifauna occupies only a small area (because of the narrowness of the southern part of the continent), and is characterised more by its paucity than by its distinctiveness. There are, however, a number of endemic genera, some of them, such as the seed-snipe (Thinocoridae), forming endemic families. In addition, the tapaculos (Rhinocryptidae) are of largely temperate zone distribution.

When North and South America joined, substantial faunal exchange occurred between the two continents (the so-called 'Great American Interchange'). On the basis of present distribution patterns, various authors have speculated on which bird families moved from south to north and which from north to south, but in general more types of birds seem to have colonised South America from North America than vice versa (Mayr 1985). The main likely immigrants from North America include the tanagers (Thraupinae), cardinals (Cardinalinae), and thrushes (Turdidae), among others. The Andean chain, extending through Central America in almost continuous mountainous connection with the North American Rocky Mountains, has permitted the immigration of some typical Holarctic species into South America, such as pipits *Anthus*, the Horned Lark *Eremophila alpestris* and the Short-eared Owl *Asio flammeus*.

The flightless ratites in Africa, South America and Australia are probably remnants of the former Gondwanan connection, and the same may be true for other families shared between South America and Africa, such as parrots, trogons, snake-birds, sungrebes, jacanas, painted snipes and waterfowl, some of which are pan-tropical in distribution. However, the two continents share only 17 species, representing only 1% of the landbird species found in the Afrotropics and 0.5% of those found in the Neotropics. The assumptions are that some of these shared species made the ocean crossing directly, including the Cattle Egret *Bubulcus ibis* in the nineteenth century, and the Striated Heron *Butorides striatus*, two whistling duck *Dendrocygna* species and a pochard *Netta erythrophthalma* somewhat earlier, and that other shared species may have reached one or other region via the

Holarctic. Long after the separation of Africa from South America, the two continents were close enough to permit the easy exchange of some bird species, and until the Tertiary more volcanic islands than remain today may have existed in the south Atlantic, providing 'stepping stones'. Apart from ratites and a few others, the contemporary avifaunas of South America and Australia have even less in common, despite their shared Gondwanan origin, but some older plant and insect groups are shared.

The West Indies have an impoverished avifauna, evidently received over a long period (several geological epochs) by over-water dispersal. About 280 landbird species now breed in the West Indies and adjacent islands, including 31 endemic genera and 150 endemic species. This is a surprising number of endemics considering the closeness of these islands to both North and South America. The islands were formed in the Cretaceous, more than 110 million years ago, and presumably supported many kinds of birds then. However, the ancestors of most of the current avian inhabitants came from nearby continental areas. The Greater Antilles have avifaunal similarities with North and Central America, and the Lesser Antilles with South America, but speciation has resulted in an endemic avifauna that is distinct from that of mainland America (Lack 1976). The most distinctive species include the Palmchat *Dulus dominicus*, which is endemic to Hispaniola and the only member of its family, and several todies (Todidae), which form a family endemic to various Caribbean islands. The todies are coraciiform birds, related to the motmots, and represented in North America and Europe by fossils (**Table 5.9**). Other islands, such as the Galapagos, the Falklands and the Juan Fernandez group, received most of their fauna from the adjacent parts of South America. The fauna of the Galapagos Islands is distinctive, supporting an adaptive radiation of finch-like birds, and many other distinctive animals (Chapters 3 and 6).

Knowledge of the South American avifauna is relatively good, but more field surveys are needed in some regions. New species continue to be discovered, and more than 25 were described in the 1990s alone. Habitat destruction is less advanced than in other continents, but is now proceeding rapidly, so that many localised endemics may disappear in the coming decades. The eastern coastal forests of Brazil are now almost gone, and many plants and animals unknown to science are likely already to have become extinct. With such high levels of allopatry, the total species numbers calculated for the Neotropics are more responsive to trends in taxonomic fashion (splitting or lumping) than are those for other regions, but even on the strictest of lumping regimes, the Neotropical Region would still emerge as exceptional in the richness of its avifauna.

OCEANIA

The Pacific Ocean occupies over one-third of the earth's surface, at more than 166 million km^2, an area greater than all land areas combined. Within this vast expanse of water are scattered more than 23,000 relatively small islands. They are concentrated in the central and southwestern parts of the ocean, leaving a broad band in the northern and eastern sectors entirely devoid of land. Thus the majority of the islands are tropical or subtropical. They are divided into three regions: Micronesia and Melanesia in the west and Polynesia in the central Pacific. Despite the vastness of the region, the total land area of all the islands is only 46,632 km^2, over half of which is contributed by the three largest. Hawaii and Viti Levu each cover more than 10,000 km^2 and Vanua Levi covers more than 5,000 km^2; six islands lie in the range 1,000–2,000 km^2 and all the rest extend to less than 1,000 km^2. Most of the islands are volcanic and some are mountainous. Seventeen have peaks extending to over 1,000 m, and two of these extend to over 4,000 m. They have some of the most dramatic rainfall gradients in the world and a wide range of natural habitats, from forested to open, but with wetlands relatively small and scarce. However, the majority of islands are small and low-lying, the coral atolls of the Tuomota archipelago, for example, being mostly below 7 m altitude.

All islands have received their birds entirely by cross-water colonisation. They hold only a relatively small number of taxa with affinities on the nearest continents. The largest families are the Pachycephalidae (whistlers, 40 species), Columbidae (doves, 34 species), Muscicapidae (flycatchers, 28 species), Rallidae (rails, 21 species), Psittacidae (parrots, 19 species) and Fringillidae (honeycreepers, 19 species) (Cornutt & Pimm 2001). All these families have lost many other species since their islands were first colonised by people (Chapter 7), but endemic species are still found in nearly all island groups.

The islands fall into five biogeographical regions, but those of the central Pacific are usually treated separately from those of the major continental regions, as a distinct region, Oceania. With the boundaries as shown in **Figure 5.1**, Oceania now holds about 187 naturally occurring landbird species, which represent 23 families (**Table 5.2**). At 2% of the world's bird species, this is a trivial number compared to the main continental regions. However, considering the total land area involved, it represents a species density more than 20 times greater than on the richest of continents. Moreover, as expected of oceanic islands, the levels of endemism are high, with 38% of genera and 87% of species being found nowhere else (**Table 5.3**). The birds of Oceania provide striking examples of recent species formation and of faunal attenuation — of successive families, genera and species dropping out with increasing distance from the continental source area (see **Figure 6.5**; Williamson 1981, Keast 1991). As in the case of the other regions, these figures exclude introduced species, which now form a substantial part of the total on many Pacific Islands (Chapter 7).

Lying more than 4,000 km from North America and even further from Asia, the Hawaiian islands form the world's most isolated archipelago, which still contains the world's most diverse island fauna, even though a large proportion of its species have now gone (Chapter 7). The Hawaiian islands are unusual among Pacific Islands in having derived about half of their avian (but not the other) colonisers from North America (to which they are closest), rather than from the Indomalayan–Australasian Regions. Of approximately 92 land and freshwater species that currently breed on the Hawaiian Islands, 33 are endemic, six are non-endemic natural colonists and 53 are introduced. Non-endemic natural colonists include some cosmopolitan species, such as the Black-crowned Night Heron *Nycticorax nycticorax*, Common Moorhen *Gallinula chloropus* and Short-eared Owl *Asio flammeus*. As these species show little or no morphological divergence from their continental counterparts, they have presumably arrived sufficiently recently or frequently to prevent any obvious differentiation developing, a view supported by DNA analyses (Fleischer & McIntosh 2001). The owl may have established itself only after rodents were introduced by the original Polynesian colonists some 1,700 years ago. Including recently extinct forms, native bird species on the Hawaiian Islands could have evolved from no more than 21 different colonising species, including representatives of six passerine and seven non-passerine families (James 1991, Fleischer & McIntosh 2001).

Although the islands of Oceania have mostly been well explored, it is hard to form a good impression of their natural avifaunas because so many species became extinct following human colonisation (Chapter 7). Moreover, species continue to disappear under the influence of continuing deforestation and introduced predators. Because of the small land areas involved, and the vulnerability of their birds to human activities, the avifauna of Oceania has been impacted more severely than that of any other biogeographical region, and this shows every sign of continuing. The few endemic species confined to low-lying islands may be further affected by rise in sea-level, associated with global warming.

ANTARCTICA

This region, excluded from the usual biogeographical division of the world, is usually defined as the continent of Antarctica, together with all the islands, sea ice and ocean northwards to the Antarctic Convergence, the circumpolar boundary where north-flowing antarctic surface water sinks beneath warmer, south-flowing sub-antarctic water. The mean position of this convergence is usually constant to within 100 km, roughly following the 50°S parallel in the Atlantic and Indian Oceans, but lying between 55°S and 62°S in the Pacific Ocean.

The antarctic continent is the remaining fragment of the ancient land-mass of Gondwanaland, which was once covered mainly by forest, but after the other continents broke away, it drifted to its present position over the south pole. It is colder than the arctic because of its greater average elevation, with mountains extending to 5,000 m elevation. Most of its 14.3 million km² are now covered with thick glacial ice, with only about 8,000 km² of rock and soil exposed in summer, mainly on the Antarctic Peninsula and around the edge of the continent. Much of this exposed land is either too inhospitable or too far from the open sea for birds, which mainly use rock ledges and rock debris, shingle beaches and fixed sea ice for breeding. At the end of summer, the sea freezes from the shoreline outwards, advancing at up to 3 km per day, so that, by the end of winter, it stretches in places up to several hundred kilometres from land, effectively doubling the size of the continent. In spring the ice begins to melt from the outer edge inwards, so non-flying penguins gain easy access to onshore nesting areas for only a few months each year.

No landbirds live on the main antarctic land-mass, and seasonally present seabirds are dominated by representatives of the two most marine types, namely penguins and petrels. However, landbirds occur on sub-antarctic islands, including four duck *Anas* species (two at South Georgia, one each at Kerguelen and Macquarie), a pipit *Anthus antarcticus* (South Georgia only), Common Starling *Sturnus vulgaris* (self-introduced from New Zealand), Lesser Redpoll *Carduelis flammea cabaret* (also self-introduced from New Zealand) and an introduced rail *Gallirallus australis* at Macquarie Island (where another rail *Hypotaemidia macquariensis* and a parakeet *Cyanorhamphus novaezelandiae erythrotis* became extinct by 1894 and 1911, respectively). Farther north on the temperate sub-antarctic islands, a number of other birds occur, principally shearwaters *Puffinus*, gadfly petrels *Pterodroma* and additional (mostly endemic) landbirds, particularly on the Tristan da Cunha group and the Chatham Islands.

DISCUSSION

Species numbers and diversity

As will be clear by now, biogeographical regions differ greatly in the numbers and densities of breeding bird taxa they support, whether species, genera or families (**Table 5.2** and **5.11**). Such differences have been attributed to: (1) the different past opportunities for avifaunal build-up, by a combination of inheritance (autochthony), colonisation and *in situ* speciation (relative to extinction); and (2) differences in the past and present capacities of the regions to support large numbers of different taxa (Keast 1972). The current carrying capacity of the regions for landbirds depends largely on topographic and climatic variation, which influences the vegetation, and hence the variety and fragmentation of the habitats available.

Table 5.11   Number of species and other taxa per 1,000,000 km^2 that breed in the continental part of different biogeographical regions. Based on data in **Table 5.2**.

	Palaearctic	Indomalayan	Afrotropical	Australasian	Nearctic	Neotropical
Species	20.4	176.8	92.9	178.9	34.9	185.2
Genera	6.3	44.9	22.5	51.3	14.4	49.1
Families	1.3	7.6	3.6	8.2	2.5	3.9
Orders	0.3	1.8	0.9	1.8	0.7	1.0

Of the three southern regions, the Neotropical is by far the richest, with about 3,370 species (or 185 per million km^2), while the Afrotropical has 1,950 species (93 per million km^2) and the Australasian has 1,592 (179 per million km^2). Of the three northern realms, the mainly tropical Indomalayan Region is by far the richest, with 1,697 species (or 177 per million km^2), while the Palaearctic (937 species or 20 per million km^2) and Nearctic (732 species or 35 per million km^2) realms are relatively poor, a fact attributed largely to latitude (climate) and to recent glaciations which eliminated many species, as their habitats were reduced and displaced (Chapter 9). The major continental regions thus show 4.6-fold variation in the numbers of recognised landbird species they hold, or 9.1-fold variation in the numbers of species per unit area. The Region of Oceania, comprising a large number of Pacific Islands, now holds only 187 landbird species, some 2% of the world's total, placed in 4% of genera and 16% of families; but in the recent past, before human colonisation, it held many more (Chapter 7). Future discovery of currently unknown species, together with further research on known taxa, is likely to alter these figures. Because the most species-rich regions are the least known, however, future work is likely to increase, rather than lessen, the differences in species numbers between regions. Moreover, the above figures refer to breeding birds alone, and exclude those present only as non-breeding migrants. Because many more landbird migrants move from the northern to the southern continents for the non-breeding season than vice versa (Chapter 18), inclusion of non-breeding visitors in the totals would further increase the stated differences in species richness between the northern and southern realms. The totals also exclude 320 species classed as seabirds, which have different distribution patterns (Chapter 8).

At higher taxonomic levels, the differences between regions are less marked, with 1.4-fold variation in numbers of both families and orders between the main continental regions (**Table 5.12**). Also, at these higher taxonomic levels, the ranking of regions is slightly different, with the Afrotropical Region emerging as the richest, as it holds breeding species from 54% of all bird families and from 83% of all bird orders. Again Oceania falls well below the others, with only 16% of families and 43% of orders represented.

Table 5.12   Number of landbird taxa that breed in different numbers of zoogeographical regions (total 7, including Oceania). Numbers listed separately for (a) all birds, (b) continental birds, and (c) oceanic island birds. Species that breed on both continents and oceanic islands are listed under continental (b). + = <0.05%. Figures in brackets are percentages of total known numbers of landbird taxa in the World, or on continents or islands, as appropriate (from **Table 5.1**).

(a) All birds

	1	2	3	4	5	6	7
Species	8,527(91)	784(8)	76(+)	19(+)	1(+)	5(+)	4(+)
Genera	1,500(75)	304(15)	86(4)	49(2)	28(1)	20(1)	15(+)
Families	57(41)	19(14)	9(6)	16(11)	11(8)	11(8)	17(12)
Orders	4(17)	1(4)	2(9)	3(13)	1(4)	4(17)	8(35)

(b) Continental birds

	1	2	3	4	5	6	7
Species	6,944(89)	740(10)	76(1)	19(+)	1(+)	5(+)	4(+)
Genera	1,294(73)	292(16)	86(5)	49(3)	28(2)	20(1)	15(1)
Families	47(36)	19(15)	9(7)	16(12)	11(9)	11(9)	17(13)
Orders	4(17)	1(4)	2(9)	3(13)	1(4)	4(17)	8(35)

(c) Island birds

	1	2	3	4	5	6	7
Species	1,583(97)	44(3)	0(0)	0(0)	0(0)	0(0)	0(0)
Genera	206 (39)	12(2)	0(0)	0(0)	0(0)	0(0)	0(0)
Families	10(12)	0(0)	0(0)	0(0)	0(0)	0(0)	0(0)
Orders	0(0)	0(0)	0(0)	0(0)	0(0)	0(0)	0(0)

Perhaps surprisingly, the numbers of different bird taxa do not increase with increasing size of biogeographical region (**Appendix 5.1**). Instead, the numbers decrease slightly (but not significantly), a relationship that is greatly influenced by the small size of the Australasian and Indomalayan Regions (which are rich in species) and the very large size of the Palaearctic Region (which is poor in species). This negative relationship contrasts with the situation in mammals, in which the numbers of genera and families correlate positively with continental areas (Flessa 1975). Evidently, the greater dispersal powers of birds, together with the differences between continents in their capacities to produce and maintain bird species, has overridden any effects of area *per se*. By implication, while the numbers and distributions of habitable land areas are a major factor influencing the diversity of birds, the actual sizes of land areas (at least at the level of the continents) are much less important. In particular, the four regions that have tropical forests have far more bird species overall than the two that do not. On smaller land areas with equivalent habitats, however, as on many island archipelagos, species–area relationships are well established in birds (Chapter 6).

Number of species per genus and per family

Regions that hold the largest numbers of landbird species also hold the largest numbers of genera and families (**Appendix 5.1**). They also hold noticeably more species per family than other regions, the Neotropical Region being extreme, with an average of 47.5 species per family, and two families (Tyrannidae and Furnariidae) containing 544 and 280 species respectively (**Figure 5.5**). Moreover, the richest regions also hold the greatest proportions of endemic species and families. They have evidently offered greater speciation opportunities (relative to extinctions) than other regions, and because diversity extends beyond species to families, they have provided conditions suitable for diverse avifaunas for much of their history.

Avifaunal distinctiveness

The bird faunas of the different realms also vary in degree of endemism, and in the extent to which the endemism extends from species through genera to families. Each realm has its own dominant groups, containing large numbers of species, which commonly show great morphological and ecological diversity. Take, for example, the great variety of parulid warblers in North America, of tyrannid flycatchers in South America, of silviid warblers in Eurasia, of weaver finches in Africa or of honeyeaters in Australia. Within each region, endemic species are most numerous, as expected, on mountains and islands which present the most isolated habitat areas.

The converse of endemism is reflected in the proportions of taxa that each region shares with other regions. Overall, about 91% of all the earth's landbird species are found in only one biogeographical region, 8% are found in two regions, 0.8% in three regions, and 0.3% in more than three (**Table 5.12**)[1]. In other words, the majority of species are extremely restricted in biogeographical terms. Without such restriction, it would not have been possible for Sclater (1858) and Wallace (1876) to distinguish the regions in the first place, but it re-emphasises that, despite the great mobility of birds, intercontinental colonisation is relatively uncommon.

Moving through the taxonomic hierarchy, from species through genera to families and orders, distributions change progressively from restricted to widespread

---

[1]*If the Palaearctic and Indomalayan Regions were combined, to give the whole Eurasian land-mass, the equivalent figures become 93% in one region, 6% in two regions and the remaining 1% in three or more regions.*

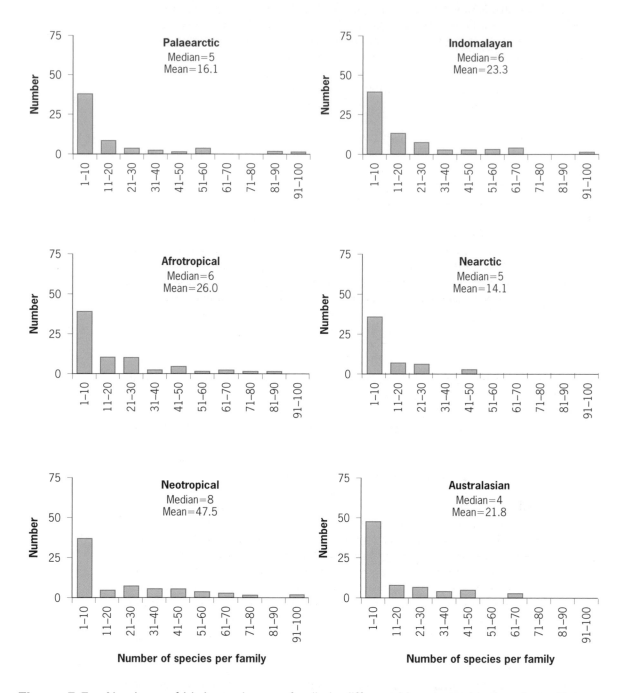

Figure 5.5   Numbers of bird species per family in different biogeographical regions. Note the low mean numbers of species per family in the Palaearctic and Nearctic compared with other regions, and the exceptional richness of the Neotropical Region. From Newton & Dale 2001.

(**Table 5.12, Figure 5.6**). Only four (17%) out of the 23 landbird orders are represented in a single biogeographical region, another four (17%) are represented in six of the seven regions, and eight (35%) are represented in all seven regions. This difference between taxonomic levels is a common finding in other organisms, and can be attributed to the greater age of the higher taxa, compared to species, and the greater opportunities through time for wider dispersal to have occurred. When the majority of the bird orders and many of the families evolved, the continental land-masses were in very different relative positions from those of today. This is held to account for the presence of some families on all three southern continents, despite their current wide separation (Cracraft 1973; Chapters 1 and 2).

Not surprisingly, each region shares the greatest number of landbird species with the closest other region, and the fewest species with the most remote region (**Tables 5.4** and **5.5**). For example, the Palaearctic Region shares 361 species with the adjoining Indomalayan Region and only 15 species with the remote Neotropical Region, while the Nearctic Region shares 239 species with the Neotropical Region and only nine with the remote Australasian Region.

Although much greater proportions of families than of species extend to more than one region, the distributions of families follow the same general trends as shown by species (**Table 5.5**). Thus the Palaearctic Region still shares most landbird families (56) with the adjoining Indomalayan Region and the fewest (34) with the remote Neotropical Region, while the Nearctic Region shares 46 families with the Neotropical Region and 31 with the remote Australasian Region. The correlation coefficient (r) between the degree of sharing at species and family levels in the six main regions is 0.631 ($P = 0.0117$). However, whereas at the species level, the Afrotropical and Indomalayan Regions each share most species with the Palaearctic, at the family level they share more families with one another than either does with the Palaearctic (**Table 5.5**). This can be attributed to the fact that the families were mostly in existence at the time when the Afrotropical and Indomalayan Regions were in unbroken contact across the Middle East, whereas many of the species evolved after the separation of these regions.

On the basis of such comparisons, the most distinctive bird fauna, as measured by the proportion of taxa that are endemic, is the Neotropical one, followed by the Afrotropical and Australasian ones, and then the Indomalayan, Nearctic and Palaearctic ones. In general, the level of distinctiveness in the bird life is matched by similar distinctiveness in other fauna and flora (Wallace 1876), and again reflects the geological and climatic history of the different regions and their isolation from other land areas. Only where the different realms are close together or adjoining do their faunas conspicuously intergrade. This re-emphasises the effectiveness for most birds of oceans and deserts as barriers to dispersal, and the difficulties for most species of penetrating far into another avifauna.

Recent range extensions

Scoring species merely by presence or absence in particular biogeographical regions gives a generous assessment of their distributions because the majority occur in only small parts of only one region (Chapter 13). The exaggeration is especially great for those species that occur primarily in one or two continents,

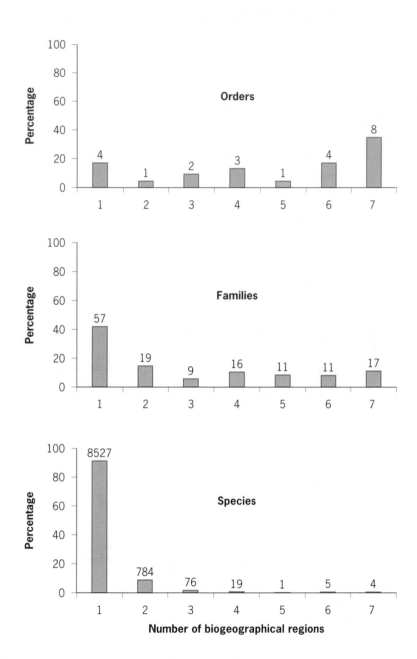

Figure 5.6 Percentage of different landbird orders (total 23), families (total 140) and species (total 9,416) that breed in 1–7 different biogeographical regions (including Oceania). Moving from orders through families to species, the different taxa show increasingly restricted distributions. Numbers of orders, families and species given above columns, based on information in Sibley (1996).

and occupy only a small adjacent part of another. Examples of the latter include the several species that extend from Eurasia across the Bering Strait into Alaska and the others that extend in the opposite direction from North America into northeast Siberia. In fact, every continent has some species that occupy a small area of an adjacent continent to which they are presumed to have recently spread. If such apparently recent range extensions were excluded, species distributions would appear much more biogeographically restricted than indicated in **Table 5.2**, and even more species than the stated 91% would be allocated to a single region.

The same holds for higher taxa. Several families are confined to a single region, apart from one or a few species that extend to another region. For example, the 280 species of Furnariidae are all confined to the Neotropics as here defined, apart from two that extend north into the southern Nearctic Region; and similarly the 68 species of Pardalotidae are confined to Australasia, apart from one that extends into the Indomalayan Region. Other examples occur among families centred on other regions, which again tends to give a generous impression of familial distributions, and a minimal measure of familial endemism.

The meeting of faunas

After the joining of North and South America, birds could move more freely north and south than before. This situation contrasts with that in the Old World where massive east–west barriers of mountain, sea and desert have long isolated temperate avifaunas from tropical ones. This may explain why continental interchange has been more marked within the New World than within the Old World (Snow 1978a, Mönkkönen *et al.* 1992). Many North American birds, including some of the three largest groups of insectivores breeding in North America (Tyrannidae, Vireonidae and Parulini), are thought to be of tropical origin. This is in contrast to the situation found in Europe, where the equivalent insectivores (Sylviidae, Muscicapidae, Turdidae and Oriolini) are probably of Palaearctic origin, even though they all have representatives south of the Sahara. This contrast between the New and Old World situations again attests to the effectiveness of barriers separating the Palaearctic and Afrotropical faunas, and particularly in reducing the northward spread of Afrotropical species.

Whenever land areas have come together in the past, many more species seem to have moved one way than the other. Much presumably depends on the relative number of species in each land area, and on their relative competitiveness. The most competitive species in one avifauna are likely to eliminate their closest counterparts in the other if they come together, thus influencing the extent to which biotas remain geographically distinct, intermingle or replace one another. Such asymmetry in faunal interchange was well recognised by Darwin (1859) and Wallace (1876). It is apparent in the disproportionate colonisation of South America by North American mammals (Simpson 1965), and by the almost one-way interchange of birds from Asia to Australia (for further discussion, see Chapter 9).

Despite the possibilities of interchange, most regions have several families, numerous genera and hundreds of species that have not penetrated the adjoining land-mass. The most tempting explanation, virtually impossible to prove, is that the distinctiveness of biotas is maintained mainly by their resistance to invasion, notably by competition. This could explain the virtual restriction of the Old World warblers and flycatchers to the Old World, and the virtual restriction of their New World equivalents to the New World. Other groups have less obvious counterparts, and, if competition is involved, it must include many species (Chapter 14). Perhaps in some way species in entire biotas, while co-existing with each other, collectively provide a strong barrier to invaders from outside. If the species already present in an area are exploiting to the full the ecological opportunities offered by the available habitats, it may be difficult for new species to establish themselves, however often they reach the area, especially if these species are adapted to different floristic and faunal backgrounds. The introduced species that have become established in all main continental regions would seem to argue against this view, but in fact most such species are restricted to man-made habitats, and have seldom penetrated the remaining natural habitats which continue to hold mainly native birds. Nonetheless, it is still hard to dismiss an alternative explanation of only limited intermingling, namely that unique adaptations to features of their own regions, rather than direct competition, prevents the spread of more species across regional boundaries. Factors constraining the spread of particular species are discussed further in Chapter 14.

While some southern hemisphere bird taxa, such as parrots, passerines and pigeons, might be traced back to common origins in Gondwanaland, the role of Greater India in the biogeographical history of birds remains a mystery. This continental land-mass was once much bigger than is peninsular India today, and was tropical in climate throughout the late Cretaceous and early Tertiary. The absence of obvious relictual species in India is especially puzzling. Could it be that the Indian biota derived from Gondwanaland were almost entirely eliminated by interactions with a more advanced biota from the large Eurasian land-mass when the two areas joined? Unfortunately, the fossil record for the Indian subcontinent is still poorly known, giving insufficient evidence to check this possibility. It is a region crying out for more palaeontological studies.

Implications from fossils

The present distribution patterns of birds, on which the biogeographical division of the world is based, may bear only limited likeness to previous patterns (Olson 1988). The fossil record of birds is now sufficient to indicate how the distributions of certain bird families have changed with time. However, this record is much stronger for the northern than for the southern continents, and on all continents there are large regional gaps in the record (as for India noted above).

Perhaps the most striking finding to emerge is that many major taxa that are now endemic to the southern hemisphere are actually relics of once more widespread taxa that also occurred in the northern hemisphere during the Tertiary (**Table 5.9**). Hence, the majority of the faunal realms have been defined, not only by their autochthonous faunas, but also by the persistence there of taxa that have died out elsewhere. For example, in the late Oligocene and Miocene Periods, 20–40 million years ago, when climates were warmer than today, many families now restricted to the tropics were widespread in what is now the temperate zone. In western Europe, the avifauna included representatives of the parrots (Psittacidae), trogons (Trogonidae), barbets (Lybiidae), mousebirds (Coliidae), hornbills (Bucerotidae), touracos (Musophagidae), woodhoopoes (Phoeniculidae), broadbills (Eurylaimidae) and others. Some were present in Europe until the end of the Pliocene, about two million years ago (Brodkorb 1971, Bochenski 1985, Olson 1988). In addition, several families now restricted to the New World were also represented in Europe in Eocene–Oligocene times, including the cracids (Cracidae), capitonids (Capitonidae), potoos (Nyctibiidae), motmots (Momotidae) and New World vultures (Cathartidae). Similarly in the New World, several families that are now restricted to the neotropics once occurred more widely (Brodkorb 1971, Olson 1988). The cracids (Cracidae) and motmots (Momotidae) occurred throughout the Tertiary in North America, while the seriemas (Cariamidae) were abundant in the early Tertiary and the oilbirds (Steatornithidae) are known from Eocene deposits in Wyoming. Another example is provided by the current restriction of palaeognathous birds (Struthioniformes and Tinamiformes) to the southern continents, because such birds are known as fossils from the Palaeocene and Eocene of North America and Europe (and the Ostrich *Struthio camelus* occurred in southern Asia into historical times, Olson 1985b).

However, certain other bird groups that are now widespread still remain unrecorded in North American or Eurasian fossil avifaunas dated earlier than the early Miocene, about 24 million years ago. These groups include the pigeons (Columbiformes), parrots (Psittaciformes), passerines (Passeriformes), grebes (Podicipediformes) and waterfowl (Anseriformes), all of which are likely to have evolved in the southern continents at an earlier period, and spread into northern regions later. Each of these groups still has its greatest distribution, and the largest number of distinctive genera, in the southern hemisphere and is likely to be of Gondwanan origin (Olson 1988).

From their present distributions, some bird families were thought to have evolved in one region and later spread to another, but the fossil record has sometimes turned this around. Australia appears to have been completely isolated from Asia until it approximated its present position in the Miocene. If the earliest interchanges between Australia and Asia date from that time, then the megapodes (Megapodiidae), present in Eurasia in the Eocene and Oligocene, did not reach Australia until then. Hence, rather than being the place of origin of the megapodes, Australia was secondarily colonised by them. What all this implies, then, is that we have little knowledge of what the truly autochthonous avifaunas of either Australia or other continents might have been like. All these statements, however, depend on the crucial fossils having been correctly identified, at least to the level of family. Clearly, mistakes in identification are possible (for discussion of caprimulgids, see G. Mayr 1999), but collectively the fossils involve so many specimens, and include such distinctive birds, that we must at the least accept that: (1) bird distributions have changed greatly since the start of the Tertiary (the past 65 million years); and that (2) many families that are now considered characteristic of particular regions once occurred in other parts of the world. Changing climates, or the movements of the land-masses themselves through different climatic zones, could account for many of the changes.

CONCLUDING REMARKS

The same biogeographical patterns as described above for birds are also evident in other organisms, both animal and plant. This in itself provides a strong pointer that these patterns are meaningful, especially as some otherwise puzzling features can be explained in terms of past tectonic events. All the regions now offer a diversity of environments. For example, all have mountains more than 3,000 m in elevation, all except the Nearctic and Palaearctic contain lowland tropical rain forest, and all have other forest, grassland and desert habitats. If the distributions of birds were limited solely by ecological factors, regardless of geography, we would expect a different pattern of provincialism. The same taxa would be found around the world in areas of similar climate and habitat, rather than distinct taxa on each continent, which have radiated to occupy the diversity of habitats available there. This again reflects the importance of past events in earth history in shaping present species diversity and distribution patterns.

Although generalised and simplified, current geographical patterns in bird faunas provide a useful foundation for avian biogeography, but they are essentially descriptions of current patterns for which the explanations are no more than inferences. The boundaries between regions were described in the nineteenth century, mainly on subjective or arbitrary criteria, and more thorough statistically based analyses may change them somewhat, as has happened several times with the Indomalayan–Australasian boundary. Moreover, these broad patterns depend on prevailing knowledge of taxonomic relationships, and, as this improves, the supposed boundaries between regions and relationships between taxa may be further modified. Nonetheless, any such changes are unlikely to alter the main patterns that have emerged, or the main conclusions that have been drawn.

SUMMARY

The different biogeographical realms vary greatly in the numbers and types of bird taxa they hold. By far the richest region ornithologically is the Neotropical, which holds 36% of all known landbird species. This is followed by the Afrotropical Region (21% of species), the Indomalayan Region (18% of species) and the Australasian Region (17% of species), and then by the Palaearctic Region (10% of species) and the Nearctic Region (8% of species). In general, regions with the largest numbers of species also have the greatest proportions of endemic species, genera and families, as well as the largest numbers of species per family.

Each region shares the greatest number of landbird species with the closest other region, and the fewest species with the most remote region. No relationship is apparent between the size of each biogeographical region and the numbers of species, genera and families it holds; rather those regions with tropical forest have many more bird taxa overall than those without.

Species numbers in each realm are influenced by the range of habitats present (tropical forest being especially rich), and by the geological and climatic history of that realm, with its opportunities for adaptive radiation and immigration on the one hand and for extinctions on the other. Some types of birds were present on the great southern continent of Gondwanaland before its break-up, probably accounting for the presence of the same families on two or three of the southern continents, which are now widely separated.

The avifaunas of each realm are presumed to consist of a mixture of (1) autochthons, whose ancestors were present on that land-mass since early in the evolution of birds, and (2) allochthons, whose ancestors colonised from other land areas at varying dates in the past. Some families now endemic to particular realms were once more widely distributed, as shown by fossils, so they represent distributional relics which have died out elsewhere. In the absence of a better fossil record, it is impossible to say with any confidence where in the world most bird orders evolved, but some (notably the Passeriformes, Columbiformes and Psittaciformes) almost certainly arose in the southern hemisphere, before the fragmentation of Gondwanaland.

Despite some distant common ancestry, to a large extent the current bird faunas of each region have evolved independently of one another, and radiated to occupy the various habitats and niches available. Most major groups of birds have their ecological counterparts on other continents, often from widely different families.

Appendix 5.1   Results of regression analyses (r values) of number of taxa in relation to (a) area, (b) number of genera, (c) number of families, (d) number of species per genus, and (e) number of species per family in different zoogeographical regions (including Oceania). Figures show Pearson's coefficients. Asterisks indicate the individual significance levels for each pairwise test. *P<0.05, **P<0.01, ***P<0.001. For a total of 14 independent tests (as in species comparisons), and an overall experimentwise significance level of 5%, P values for individual comparisons should be 0.0036 or less. For a total of 11 independent tests (as in genera comparisons) the equivalent figure should be 0.0046 or less, and for 8 independent tests (as in families comparisons) the equivalent figure should be 0.0064 (Sokal & Rohlf 1981:241–242). Correlations that are significant on this basis are shown in bold.

	Area (km^2)[1]	Number of genera	Number of families	Number of species per genus	Number of species per family
	(a)	(b)	(c)	(d)	(e)
Overall number of species	−0.355	**0.981*****	0.650	0.645	**0.989*****
Continental species	−0.162	**0.979*****	0.635	0.698	**0.995*****
Island species	–	**0.981*****	0.892**	0.933**	0.947**
Overall number of genera	−0.346	–	0.520	0.487	**0.993*****
Continental genera	−0.234	–	0.553	0.538	**0.983*****
Island genera[2]	–	–	**−0.953*****	0.946**	0.943**
Overall number of families	−0.574	–	–	0.926**	0.532
Continental families	−0.472	–	–	0.761	0.552
Island families[3]	–	–	–	0.871*	0.826*

[1] Analysis excludes Oceania.
[2] Genera containing island species.
[3] Families containing island species.

Brown Kiwi *Apteryx australis*, a flightless endemic of New Zealand.

Chapter 6
Island birds: general features

On the most recent tally, some 1,627 extant landbird species occur only on islands, about 17% of all non-marine bird species. Individual islands usually hold fewer landbird species than equivalent areas on continents, but different islands often hold some different species from one another. So adding together all island bird species, and dividing by all land occupied by islands, gives an overall species density nearly four times greater than the average for continents. Moreover, the differential would be increased further by the inclusion of three other categories of birds, namely: (1) island endemics obliterated in recent centuries by human action, which would at least double the number (see later); (2) island species that also occur on continents; and (3) seabirds that nest on islands but mostly forage at sea (about 320 species worldwide). It is at once apparent, therefore, that, allowing for their small total area, islands overall hold a disproportionately high percentage of all bird species. The percentage is probably far higher than in most other kinds of organisms because of the superior dispersal powers of birds, which have enabled them to reach remote islands denied to many other organisms.

It is the tropical Pacific Islands of Oceania that collectively hold the greatest density of bird species on earth. The many small islands of this region have a total surface area of only 46,632 km² (Pratt *et al.* 1987), yet they now support a total of 187 endemic landbird species (Chapter 5). This gives a density of 40 breeding species per 10,000 km², which is 53 times greater than the average species density for continents, and 21 times greater than the average density for South America, which is the richest of all continents (Chapter 5). Moreover, the island figure should be more than doubled to take account of the various other groups mentioned above. In whatever way the figures are expressed, they provide ample testimony to the importance of islands in promoting species formation and diversity. This is due primarily to the geographical (and hence genetic) isolation that they provide.

Studies on the fauna and flora of islands have had a major influence on evolutionary and biogeographical theory from the time of Darwin (1859). This is largely because islands represent replicated natural systems, rather like experiments, in which some factors are relatively constant while others vary from island to island. One can therefore investigate, in a way not possible on continents, the factors that might influence the presence or absence of particular species or the richness and composition of communities. Most studies of island biogeography, however, were made before the scale of human-induced extinctions was known. They were therefore based on bird communities depleted to varying degrees by human action, and it is not certain to what extent the same patterns would hold if the missing species could have been included in the analyses. I shall return to this point later.

DIFFERENT TYPES OF ISLANDS

In biogeography, it is useful to distinguish two main types of islands (**Table 6.1**). Oceanic islands originate mostly from volcanos or coral. They have never been connected to a continent, so must have received their entire flora and fauna by cross-water colonisation. Many are separated from continents by immense stretches of ocean that present formidable barriers to the movements of most species. As a result, oceanic islands have received only a subset of continental species, those best able to get there, albeit on rare occasions. They tend to lack non-flying mammals, amphibia and freshwater fishes, unless put there by people. Rarity of immigration has in turn allowed much independent evolution, when the colonists, being isolated from mainland ancestors, adapt to local conditions. Hence, oceanic islands typically hold large proportions of endemic species, well differentiated from their closest continental relatives. Remote archipelagos, notably the Galapagos and Hawaiian Islands, provide some of the best known examples of adaptive radiation in birds and other organisms (Chapter 3). These various features make oceanic islands unique in their ecology, and not simply scaled-down versions of continental ecosystems. Their disposition around the world is shown in **Figure 6.1**.

An important feature of volcanic or coralline islands is that they are often relatively short lived in geological time. Volcanic islands arise from the sea bed: while the volcanoes are active, they continue to grow, sometimes rising more than 4 km above the sea surface; but once they become inactive, they shrink from

Table 6.1   Some biogeographically interesting islands classified according to their modes of origin. Modified from Brown & Gibson 1983.

---

1. Fully oceanic
Totally volcanic islands of fairly recent origin (mostly < 10 million years) that have emerged from the ocean floor and have never been connected to any continent by a land-bridge.

Examples: Austral–Cook Island chain, Carolines, Clipperton Island, Galapagos Islands, Hawaiian Islands, Kodiak–Bowie Island Chain, Marquesas, Society–Phoenix Island Chain, Comoros, Lesser Antilles, Lesser Sunda Islands, Marianas, New Hebrides, Solomons, Tonga, Kermadec, Ascension and St Helena Islands, Faeroes, Iceland, Gough Island, Tristan da Cunha, Mascarenes.

2. Continental Islands
Formed as part of a continent and since separated from the main land-mass. Some have added oceanic material since their formation. Two types:

(a) **Continental plate (land-bridge) islands,** most recently connected with mainland by land or ice in the Pleistocene.
British Isles with Europe, Ceylon with India, Falklands with South America, Greater Sunda Islands (Java–Sumatra–Borneo) with Southeast Asia, Japan and Sakhalin with Asia, Taiwan with Asia, Greenland with Newfoundland, New Guinea with Australia, Tasmania with Australia.

(b) **Continental drift islands,** permanently separated from the mainland since an ancient split (final separation in parenthesis).
Greater Antilles (*ca.* 80 million years ago), Kerguelen Island (Upper Cretaceous), Madagascar (*ca.* 160 million years ago from Gondwanaland, *ca.* 80–85 million years ago from India), New Caledonia (*ca.* 85 million years ago), New Zealand (*ca.* 85 million years ago), Seychelles (*ca.*60 million years ago), South Georgia (*ca.* 45 million years ago).

Others
Islands formed from two sources.

Azores based partly on volcanic and partly on continental rocks.
Canary Islands, with some islands of volcanic origin, and others with past connections to mainland Africa.

---

compression and erosion, and eventually disappear beneath the waves. The majority remain above sea-level for less than ten million years (though some for much longer). The youngest of the present Hawaiian Islands (Hawaii) has existed for only 0.4 million years and the oldest (Kaui) for about five million years, but further along the chain are some older islands mostly now submerged. Thus the numbers, sizes and locations of volcanic islands are likely to have changed greatly in the last several million years and some now submerged islands could have influenced the current distinctiveness and distribution patterns of extant island birds. For example, the Mascarene Islands of Réunion, Mauritius and Rodriguez in the southern Indian Ocean were once discontinuously linked to India by a string of volcanic islands; the Chagos, Maldive and Laccadive Islands are the surviving remnants, many others having disappeared beneath the waves. These ancient islands could have formed a stepping stone route for Asian birds to the Mascarenes, thus explaining why many ancestral Mascarene birds came from Asia

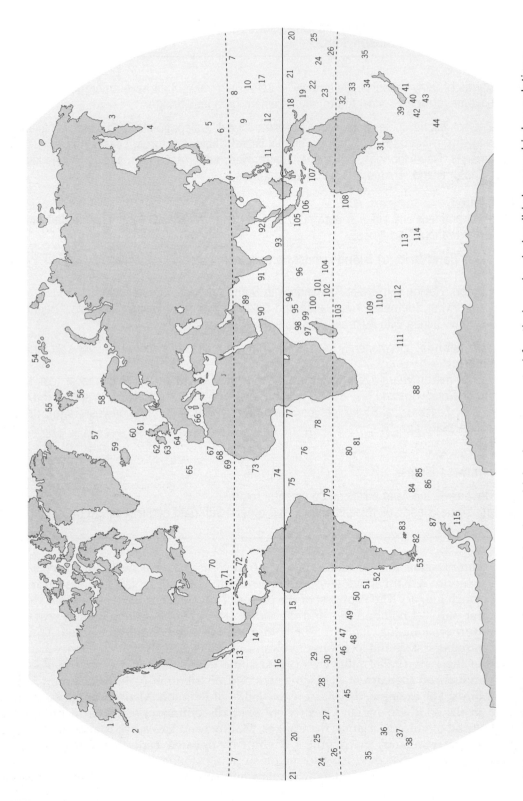

Figure 6.1 Locations of various islands and island groups, noted for their endemic landbird or seabird populations.

North Pacific Ocean
1. Pribilof Islands
2. Aleutian Islands
3. Kommandu Island
4. Kunle Islands
5. Bonin Islands
6. Volcano Island
7. Hawaiian Islands
8. Wake Island
9. Mariana-Guam
10. Marshall Islands
11. Palau Island
12. Caroline Islands
13. Revilla Gigedo Island
14. Clipperton Island
15. Galapagos Islands
16. Line Islands-Christmas Islands
17. Gilbert Islands

South Pacific Ocean
18. New Britain
19. Solomon Islands
20. Phoenix Islands
21. Elice Islands
22. New Hebrides
23. New Caledonia
24. Fiji Islands
25. Samoan Islands
26. Tonga Islands
27. Cook Islands
28. Society Islands
29. Marquessa Islands
30. Tuamotu Archipelago
31. Bass Straits Islands
32. Lord Howe Island
33. Norfolk Island
34. Three Kings Islands
35. Kermadec Islands
36. Chatham Islands
37. Bounty Islands
38. Antipodes
39. Solander Rock

40. Snares Islands
41. Stewart Island
42. Auckland Islands
43. Campbell Island
44. Macquarie Island
45. Austral Islands
46. Oneno Island
47. Henderson Island
48. Pitcairn Island
49. Ducie Island
50. Easter Island
51. Juan Fernandez Islands
52. Mocha Island
53. Cape Horn

Arctic and North Atlantic
54. Franz Joseph Land
55. Spitzbergen
56. Bear Island
57. Jan Mayen
58. Lofoten Islands
59. Iceland
60. Faroes
61. Shetland Island
62. South West Ireland Islands
63. Skokholm and Skomer Islands
64. Channel Islands
65. Azores
66. Balearic Islands
67. Madeira
68. Salvage Islands
69. Canary Islands
70. Bermuda
71. Bahama Islands
72. Hispaniola Island
73. Cape Verde Islands
74. St Pauls Rocks

South Atlantic Ocean
75. Fernando de Noronha
76. Ascension Island
77. São Tomé Island
78. St Helena

79. Trinidade–Martin Vas
80. Tristan Da Cunha
81. Gough Island
82. Staten Island
83. Falkland Islands
84. South Georgia
85. South Sandwich Islands
86. South Orkney Islands
87. South Shetland
88. Bouvet Island

Northern Indian Ocean
89. Kuria Muria Islands
90. Socotra Island
91. Laccadive Islands
92. Andaman Islands
93. Maldive Islands

Southern Indian Ocean
94. Seychelles
95. Amirantes
96. Chagos Archipelago
97. Comoro Islands
98. Aldabra Island
99. Providence Island
100. Alegela Island
101. Cargados Carajos Island
102. Mauritius
103. Réunion
104. Rodriguez
105. Cocus Keeling Islands
106. Christmas Island
107. Lacapede Islands
108. Abrolhos Islands
109. Amsterdam Island
110. St Paul
111. Marion–Prince Edward Islands
112. Crozet Islands
113. Kerguelen Islands
114. Heard Island
115. Antarctic Peninsula

and only some from Africa and Madagascar, which are closer. Long after their heyday, many submerged islands reappeared through lowered sea-levels for periods of 40,000–60,000 years at a time during the Pleistocene glaciations. This was long enough for them to acquire afresh each time their own floras and faunas. The important point is that the opportunities for colonisation of existing oceanic islands are likely to have changed repeatedly through geological time, as are the major source areas for colonists.

Any volcanic island in a warm region is likely to hold its greatest range of species while it still has a large elevational range, offering a wide range of habitats, including montane forest. Once it has shrunk to a small, low-lying plateau, it is likely to support a much narrower range of species, and towards the end of its life chiefly seabirds. Hence, on a time scale of 5–10 million years, the avifaunas of volcanic oceanic islands are likely to increase rapidly and then decline progressively in diversity of species. Low coralline islands generally have much shorter lives, often having been formed on long-dead volcanoes, and repeatedly exposed and submerged with fluctuations in sea-level. Without the fluctuations in sea-level, they could last much longer, as in pre-Pliocene times[1].

In contrast to oceanic islands are continental islands severed from continents (**Table 6.1**). Most such islands still lie close to continents and many were connected as recently as the last glaciation, only 10,000–12,000 years ago when sea-levels were lower. They are often called 'continental plate' or 'land-bridge' islands. Examples include Britain in Europe or Tasmania in Australia. The connection ensures that the plants and animals of land-bridge islands are normally a more or less representative subset of the mainland ones, though some may now be subspecifically distinct (such as the Red Grouse *Lagopus l. scoticus* in Britain). The distinction between continental and oceanic islands is not always clear cut, however, for while most continental islands were split from continents only about 10,000 years ago, others were severed many millions of years ago, and have since moved long distances from their parent continent, such as New Caledonia and New Zealand from Gondwanaland (82 million years ago) (**Table 6.1**). These older 'continental drift' islands tend to be larger than oceanic islands, and further from continents than continental plate islands. They now hold mainly modern birds that have arrived by cross-water colonisation, but most also hold some relict species that reflect their past connections. Likely avian examples include the Kagu *Rhynochetos jubatus* of New Caledonia, the moas of New Zealand and the elephant-birds of Madagascar, all of which could be of Gondwanan heritage (Chapter 2).

The situation with Sulawesi and the Philippines is somewhat intermediate. Both geological and biological evidence suggests that these islands had past land connections via other islands to southeast Asia, but not during the late Pleistocene. They are now separated by deeper water and contain much more differentiated plants and animals than other islands on the Sunda shelf, such as

---

[1]*In the tropics, corals grow around the margin of an island as it sinks and carry on growing long after the bedrock has submerged. Thus an island can survive as an atoll by coral limestone being continuously added on top long after it would otherwise have vanished. On some atolls, coral depth reaches over 1,500 m, having been growing for tens of millions of years.*

Sumatra and Java. Clearly, land-bridge islands vary greatly in the time they have been isolated, and in the proportions of their current species that have derived from cross-water colonists. Also, whether oceanic or continental in origin, the numbers, shapes and sizes of islands have changed greatly through time with geological events and with the rises and falls in sea-levels, all of which could have affected their biotas.

ISLAND BIRDS

Of the 1,627 landbird species found only on islands, some 36% relate to the Australasian Region, 24% to the Indomalayan Region, 15% to the Afrotropical Region (mainly Malagasy subregion), 12% to the Neotropical Region, 10% to Oceania, 2% to the Palaearctic and 0.5% to the Nearctic (**Table 6.2**). The contrast in endemic species numbers between high- and low-latitude regions can be attributed to the greater numbers of islands in tropical waters, the greater average distances of these islands from continental land areas, and the generally greater richness of tropical biota.

In any one region, the numbers of landbird species found on particular islands are usually correlated with island area (with more species on larger islands), and also with isolation (with fewer species on more remote islands) **(Box 6.1; Figure 6.2)**. These two relationships led to the equilibrium theory of MacArthur & Wilson (1967), according to which the rate at which new species reach oceanic islands declines with increasing remoteness of the island, while the rate at which established species die out declines with increasing size of the island. The balance point between continuing colonisations and continuing extinctions determines the equilibrium number of species on each island, which therefore varies with island size and remoteness, as described above (**Figure 6.3**). Provided that ecological conditions remain approximately constant, the theory says, so does the number of species, but the types of species may change as new ones arrive and established

Table 6.2   Number (%) of extant landbird species restricted to the islands of different biogeographical regions. Percentages are of the total world landbird species, taken as 9,416. From Newton & Dale 2001.

Region	Total species
Palaearctic	34 (0.4)
Afrotropical	236 (2.5)
Indomalayan	438 (4.7)
Australasia	666 (7.1)
Nearctic	9 (0.1)
Neotropical	200 (2.1)
Oceania	187 (1.9)
Overall	1,627 (17.3)

Note that the above figures are lower than those in some other published lists because they exclude recently extinct species, seabirds and species that are also found on continents. The figures also depend to some extent on precisely where the borders between different biogeographical regions are drawn (see **Figure 5.1**), as well as on whether some distinct taxa are classed as species or subspecies. The list above follows Sibley 1996.

> **Box 6.1**   Species numbers on islands.
>
> Whenever species on different-sized islands in the same region are compared, two trends emerge: namely that species numbers increase with island size and with proximity to a colonisation source. In other words, the smallest and most remote islands hold the fewest species. The 'species–area' relationship is evident almost everywhere: on real islands from the British Isles to the West Indies and from Galapagos to Indonesia, and on virtual islands such as mountain tops, forest fragments and lakes (**Figure 6.2**). Within each island group, it has followed a consistent pattern, with the number of native species approximately doubling with every ten-fold increase in area. Thus, if an average of about 50 landbird species is found on islands of 1,000 km^2, then about twice as many, 100 species, are found on islands of 10,000 km^2.
>
> In more exact terms, the number of species in any particular taxonomic group increases by the species-area equation $S = CA^z$, where S is the number of species and A is the island area. C is a proportionality constant that depends on the dimensions in which it is measured and on the type of organism, whether birds, butterflies, grasses and so on. The parameter z (the slope of the regression line in a log S versus log A plot) reflects the rate at which species are added with increasing area. It too depends on the type of organism and on the region — in particular, on whether the islands are close to source areas, such as Indonesia, or very remote, such as Hawaii. For different island groups and organisms, z usually falls in the range 0.20 to 0.30, being lower for more remote island groups. The rule of thumb, that a ten-fold increase in area doubles the species numbers, is the same as saying that $z = 0.30$, or that a plot of log species numbers against log area gives a linear relationship with a slope of 0.30. This relationship is called the species–area curve, because the best fit line would be a curve if the data were plotted using linear rather than logarithmic axes. As a rough guide, this rule of thumb works quite well at the spatial scale involved. It can also be applied to habitat destruction: if nine-tenths of a habitat area are destroyed while one-tenth is preserved, then eventually half the species that were originally present are likely to be lost.
>
> The second generalisation, if one compares islands of similar area, is that species numbers (S) decrease with increasing distance (D) from a colonisation source. The probability of crossing a sea gap is unlikely to decline linearly with distance, but according to some exponential function, so in analyses distances are often transformed to log values (Williamson 1981). Whether transformed or not, distance is the only variable so far shown to affect immigration, though wind strength and direction could be locally important, as could other variables such as island area (MacArthur & Wilson 1967).

ones die out. Island size is important partly because it reflects topographic and habitat diversity (Lack 1976, Whittaker 1998). However, islands hold far fewer indigenous species now than before human colonisation, as shown by bone and fossil remains (see later). To some extent, therefore, the current distributional

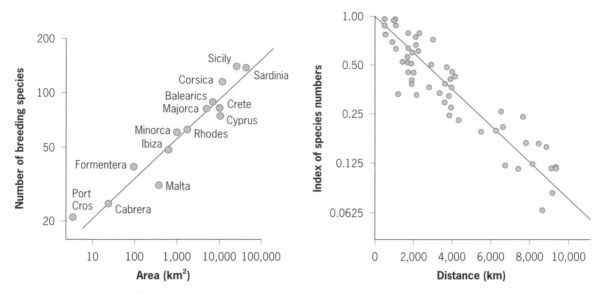

Figure 6.2   Examples of the relationships between species numbers, island sizes and isolation.

Left. Relationship between bird species numbers and land area for various Mediterranean islands. On these islands, species richness is significantly correlated with island size (r_{10} = 0.88, P<0.09) and maximum altitude (reflecting habitat diversity, r_{10} = 91, P<0.005). Together, these two variables explain 91% of variation in species numbers (r_{10} = 91, P<0.005). Neither distance to the nearest mainland nor distance to the nearest other island is significantly correlated with species numbers. These distances are evidently too small to influence species numbers. Redrawn from Blondel 1985.

Right. Relationship between species numbers and island distances from a colonisation source, for birds on tropical islands in the southwest Pacific. Ordinate (log scale): number of resident non-marine lowland bird species on islands more than 500 km from the larger source island of New Guinea, divided by the number of species on an island of equivalent area close to New Guinea. The approximately linear relation means that species numbers decrease exponentially with distance, by a factor of two per 2,600 km. Redrawn from Diamond 1972a.

patterns of island faunas, on which equilibrium and other theories of island biota were developed, are artifacts of human action, so need reassessing, as discussed in Chapter 7.

Despite a more restricted range of plant species, most of the habitat types found on continents also occur on oceanic islands, including different types of forest and grassland, although wetlands tend to be scarce and small. The range of habitats found on any one island depends on its global position, together with its size and topography (elevation). Because of their habitat and food needs, landbirds may be dependent on particular types of plants and animals being present on an island before they can survive there themselves. Avian diversity thus depends on overall biodiversity, and on any island the number of bird species present may depend as much on the dispersal powers of other organisms as on those of the birds themselves. In addition, because landbirds vary in their dispersal abilities, some

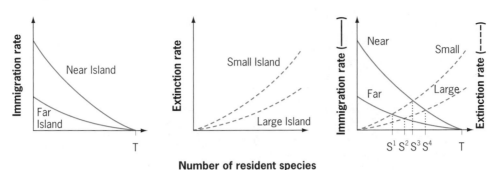

Number of resident species

Figure 6.3   The equilibrium theory of island biogeography, showing how species numbers on an oceanic island are hypothesised to depend on the balance between extinctions and colonisations, and hence on the size and isolation of the island. Based on MacArthur & Wilson 1967.

For any particular island, the net extinction rate will be zero when no species are present, and is assumed to increase as the total number of species present increases; conversely, the net rate at which new species are added — the immigration rate — will decrease as the total number of species present increases (largely because of a decline in the number of new species available for colonisation from a maximum of T). The equilibrium number of species (S) is that at which extinction and immigration rates are equal.

Species immigrate to an island as a result of dispersal from continents or other islands, and the more remote the island the lower the immigration rate (lower left curve). Species established on the island risk extinction because their numbers fluctuate; the smaller the island, the smaller the populations and the higher the extinction risk (upper right curve). All in all, the larger and less isolated the island, the higher is the species number at which it should equilibrate. S^{1-4} = equilibrium species numbers on small, far islands (S^1), large, far islands (S^2), small, near islands (S^3), and large, near islands (S^4).

Immigration lines are curved because some species disperse more readily than others (and get there early), and extinction lines are curved because, as species numbers rise, the more difficult it becomes for additional species to establish and persist (through competition).

species are unlikely to reach islands naturally (Chapter 16), and among those that do, some species can reach more remote islands than others. This is apparent in the current distribution patterns of birds. For example, as one moves eastwards from New Guinea through various Pacific islands, different types of birds drop out at different distances, producing a general decline in species diversity from west to east (**Figure 6.4**). Similar attenuation with distance from a source area is also evident on islands in other parts of the world, wherever it has been studied.

The majority of oceanic islands seem to have received their birds from the nearest continental source (judged by the distributions of their closest relatives), but some islands far from land have received them from more than one source, like the Mascarene Islands mentioned above. Even within an archipelago, individual islands vary in their avifauna according to their position relative to the colonising source (for New Hebridean Islands, see Diamond & Marshall 1976). Some island species have changed so much from their ancestral colonising form that it is difficult, without DNA analyses, to tell where the original colonists came from.

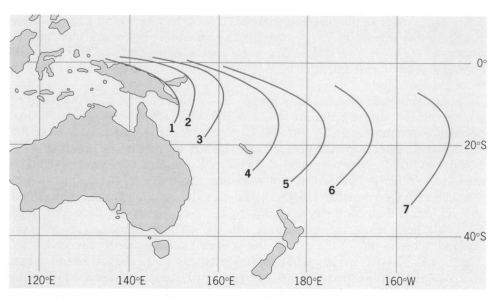

Figure 6.4   Decline in the occurrence of families and subfamilies of breeding landbirds from New Guinea eastwards on various Pacific Islands. The decline in taxa is fairly smooth and reflects both differences in dispersal ability and a general eastward decline in island sizes. (1) Not beyond New Guinea: pelicans, storks, larks, pipits, birds-of-paradise and nine others. (2) Not beyond New Britain and the Bismarck Islands: cassowaries, quails and megapodes. (3) Not beyond the Solomon Islands: strigid owls, rollers, hornbills, drongos and six others. (4) Not beyond Vanuatu and New Caledonia: grebes, cormorants, ospreys, crows and three others. (5) Not beyond Fiji and Niufo'ou: hawks, falcons, turkeys and woodswallows. (6) Not beyond Tonga and Samoa: ducks, thrushes, waxbills and four others. (7) Not beyond the Cook and Society Islands: barn owls, swallows and starlings. Beyond (7), the Marquesas and Pitcairn Islands: herons, rails, pigeons, parrots, cuckoos, swifts, kingfishers, warblers and flycatchers. From Williamson 1981 (after Firth & Davidson 1945).

The colonisation of an island can be remarkably rapid. The volcanic eruption on the Indonesian Island of Krakatoa in 1883 covered it and the two neighbouring islands with a deep blanket of ash, obliterating all life. However, various organisms soon colonised the denuded islands, probably from Java or Sumatra, 40 km and 80 km away (Bush & Whittaker 1993, Thornton *et al.* 1993). By 1933, only 50 years after the eruption, Krakatoa (renamed RaKata) was once again covered with dense tropical rain forest. By the 1990s, hundreds of insect and other invertebrate species, at least 271 plant species and 31 bird species were present on this island. The cumulative numbers of resident bird species increased from 15 in 1908 to 29 in 1919–21, 32 in 1932–34, 38 in 1951 and 43 in 1983–93 (Thornton *et al.* 1993). However, some species both came and went in that period, and no more than 31 bird species were recorded at one time (in 1983–93). These changes could be attributed partly to plant succession and forest formation, and further changes may yet occur. The case of Krakatoa is unusual only insofar as virgin land was created so quickly and dramatically. Its colonisation could therefore be documented, and the sources of the settlers identified. The same processes may be presumed to occur on every newly formed oceanic island, but would be expected to take longer on more remote islands.

Once established on an island, some species are more likely to die out than others. Species that live at naturally low densities, for example, can have only small populations on most islands. This puts them at greater risk than other species that can achieve higher numbers per unit area. Hence, for reasons of differential dispersal and local extinction risk, the bird communities of islands would be expected to differ in fairly consistent ways from those of continents.

In line with their restricted biotas, islands usually hold fewer species per genus and per family than continental areas (Järvinen 1982, Newton & Dale 2001; **Table 6.3**). This holds even on some large islands, such as Madagascar and New Zealand, where some of the genera and species have apparently evolved *in situ* (**Figure 6.5**). These are average figures, however, and, following an adaptive radiation, the numbers of species in some genera may be much higher, as in the genus *Coua* on Madagascar which has ten species (average 1.5). Niches occupied by members of different genera on mainlands are often occupied by members of the same genus on islands. On islands closer to continents, immigration is more frequent, and species per genus (or per family) ratios are not necessarily lower than expected on a random sample of species drawn from the source pool (Simberloff 1970). In fact, some genera may be exceptionally well represented because they are more effective dispersers and colonisers than others. In general, therefore, while species per genus (and per family) ratios are, on average, lower on islands than on continents, some conspicuous exceptions result from local adaptive radiation, or from the exceptional dispersal abilities of some types of birds.

Table 6.3 Number of species per genus and species per family in different zoogeographical regions. From Newton & Dale 2001.

	All species		Island species	
	Mean number of species per genus	Mean number of species per family	Mean number of species per genus	Mean number of species per family
Palaearctic	3.25	16.12	1.17	1.71
Indomalayan	3.93	23.26	2.60	3.60
Afrotropical	4.12	26.00	1.95	2.88
Australasian	3.48	21.80	2.97	4.57
Nearctic	2.43	14.06	1.00	1.13
Neotropical	3.77	47.52	1.92	3.59
Oceania	–	–	2.28	3.57

Species divergence and evolution

The isolation of islands provides ample opportunities for independent evolution, so that over time species diverge from the original colonists to become distinctive endemic forms, subspecies, species or genera. Broadly speaking, the chances of

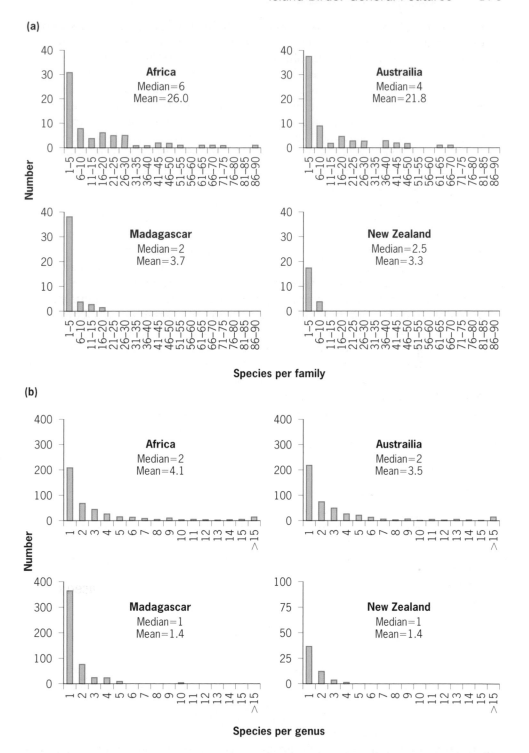

Figure 6.5   Numbers of species per family and per genus: on Madagascar compared with Africa (left), and on New Zealand compared with Australia (right).

divergence are greater the further an island lies from a mainland source. Islands close to continents receive a continual influx of immigrants which interbreed with the already established birds, preventing significant evolutionary change. In contrast, islands far from continents receive immigrants so rarely and at such long intervals that the local stock has time to change in response to local conditions. Generally, therefore, islands that are close to mainlands hold fewer endemic taxa than those that are further away. The supposed effect of invasion frequency is sometimes evident within the same island group. For example, the Common Kestrel *Falco tinnunculus* shows a stepped cline running through the Canary and Cape Verde Islands; in each archipelago, the birds of the easternmost islands, which presumably receive most immigrants, are closer in appearance to the continental form, while the birds of the outer islands are most distinct (Bourne 1986).

Speciation processes are therefore rather different on islands and continents (Chapter 3). On islands, the necessary isolation is produced by oversea dispersal, leading to the founding of new populations which can then evolve independently from the parent stock. This process occurs in one species after another, each in its own time. On continents, by contrast, the necessary isolation arises mainly when the populations of many widespread species are simultaneously fragmented by changes in geology or climate (affecting habitats) that produce new barriers to gene flow and then enable the different parts of the original population to evolve independently. The first process depends on dispersal and range expansion over existing barriers, one species at a time, while the second depends on range fragmentation through the imposition of new barriers, affecting many species at the same time in the same place (Chapters 3 and 10).

Only continents and the larger islands, such as Madagascar, are large enough to allow one bird species to evolve into two or more within the land-mass itself (Diamond 1977a). The majority of islands are too small (relative to the dispersal distances of birds) for isolaton to develop within an island, however fragmented the habitats. Hence, hybrid zones, that are common among continental birds, are unknown among oceanic island species. On islands, new colonists that are similar to an existing form either die out or are completely absorbed by hybridisation. This sometimes gives rise to a new stable form, such as the subspecies *Pachycephala pectoralis christophori* on San Cristobal Island, Melanesia (Mayr & Diamond 2001). This process in turn reduces the chance of sympatry developing on islands, compared with continents, and may help to explain the generally lower number of species per genus on islands, mentioned above. It is as though, within island groups, relatively more species than on continents remain stuck at the evolutionary stage of allospecies.

Some endemic species are restricted to particular isolated islands, while others may occur on several islands in an archipelago (**Table 6.4**). The latter provide further opportunities for speciation, giving some of the most spectacular examples of adaptive radiation in birds, as species can evolve on particular islands and then move to other islands where they undergo further divergence in competition with local forms, giving the variety of species found today among the Galapagos finches and the Hawaiian honeycreepers (Chapter 3). On single islands, any such radiation is much more limited because it mostly depends on multiple invasions at long intervals of the same or similar species, usually from the same continent. Examples of supposed double invasions, with the earliest colonist listed first,

Table 6.4 Numbers of different categories of land and freshwater bird species on different groups of Pacific Islands. Calculated from data in Pratt et al. 1987.

	Extinct[1]	Endemic species		Non-endemic natural colonist species		Introduced species[3]	
		On one island[2] or island group	On more than one island or island group	On one island or island group	On more than one island or island group	On one island or island group	On more than one island or island group
Hawaii	15	23	10	0	6	12	40
Micronesia	3	26	16	5	9	6	6
Central Pacific Islands	2	3	3	0	1	2	0
Central Polynesia	1	5	20	2	10	2	3
Southeastern Polynesia	3	30	6	3	4	4	6
Fiji	0	9	27	1	21	3	8

[1] In last 200 years only.
[2] Many species now found on one island once occurred on more than one.
[3] From continents.

include the Blue Chaffinch *Fringilla teydea* and Common Chaffinch *F. coelebs* on the Canary Islands, and the Takahe *Porphyrio mantelli* and Pukeko *P. porphyrio* on New Zealand. Some island endemics show features so extreme that they are quite atypical of the group from which they are presumed to have evolved.

On the Hawaiian Islands, which are more than 4,000 km from North America and 1,600 km from the nearest other island group, 83% of the native plant and animal genera, and 97% of native species, are endemic. These islands hold the most spectacular variety of endemic island landbirds on earth (Freed *et al.* 1987, Fleischer & McIntosh 2001). The majority are honeycreepers, which on DNA evidence evolved from a single species of cardueline finch. Species from at least two other families of passerines and five of non-passerines also radiated on Hawaii, but most are now extinct (see later). On the Galapagos Islands, 900 km from South America, some 39% of the native plant and animal genera, and 80% of the native species, are endemic. On archipelagos that are closer to other land-masses, such as Fiji (450 km) or the Philippines (350 km), relatively few of the plant and animal genera, but many of the species, are endemic (44% on Fiji and 49% on the Philippines). Again these percentages are underestimates, because they do not allow for the recent extinctions of other endemic species at the hands of human colonists.

High levels of endemism imply low natural extinction rates, because populations must survive a long time. It takes at least thousands of years to differentiate to the level of an endemic subspecies, and perhaps hundreds of thousands or millions of years to differentiate to the level of an endemic species or genus (Chapter 3). It should not surprise us, therefore, that some islands have few endemics, including those that recently rose from the sea (e.g. Surtsey) or that until recently were largely covered with glaciers (Iceland) or connected to a continent (Trinidad). There are also large differences in endemism between different families of birds, with songbirds, doves and rails providing many endemic species and herons fewer (even though representatives occur on almost every island group). Almost certainly, this is linked with the frequency with which the different types of birds reach particular islands, herons being regular cross-water travellers, thus preventing divergence of most local populations (exceptions being the three Mascarene islands each of which had an endemic *Nycticorax* heron, and Madagascar which still has three endemic species, each in a different genus).

Degree of endemism also differs between populations on the same island. For example, New Caledonia in the southwest Pacific is a geologically old island, which broke away from the ancient southern continent of Gondwanaland more than 80 million years ago. It has many distinct plants and animals. Its present breeding landbird species are of various ages, as reflected in their degrees of endemism. One New Caledonian species, the Kagu *Rhynochetos jubatus,* belongs to an endemic family; four belong to endemic genera; 14 are endemic at the species level; 26 are endemic only at the subspecies level; and 26 are no different from the parental species in Australia and New Guinea. Thus, despite the antiquity of New Caledonia itself, 52 of its 71 present bird species evidently arrived so recently or so often that they are not even specifically distinct. The avifauna is a mixture of a few old lineages, notably that of the Kagu, which have persisted for many millions of years, together with many younger lineages that have differentiated little or not at all. Some may still be in genetic contact with parental populations through the continuing influx of individuals.

Some endemic island taxa (subspecies and species) may have differentiated recently in the isolation of the island and still have close relatives on the mainland; such products of *in situ* evolution are often called 'neo-endemics'. Other endemics may be relicts of former widespread families. On the basis of morphological and DNA evidence, the nearest relative of the Kagu is the Sunbittern *Eurypyga helias* of Neotropical America. The two species are now separated by 10,000 km of ocean but, because New Caledonia and South America were both derived from Gondwanaland, the common ancestor of both species could have lived there. They may thus represent the sole survivors of a once widespread group of birds found across the southern land-masses. Similarly, members of the family Todidae are now restricted to the West Indies, but what appear to be the same type of birds are represented by fossils in both North America and Europe (Olson 1976, 1988; **Table 5.9**). Examples of other relictual bird taxa now confined to Madagascar are given in Chapter 5, and examples from among other organisms include the lizard-like Tuatara *Spenodon punctatus* (in its own order) of New Zealand, the plant *Lactoris fernandeziana* (in its own family) in the Juan Fernandez archipelago off Chile and the plant *Medusagyne oppositifolia* (in its own family) on Mahé in the Seychelles. Such 'palaeo-endemics' have no other living relatives; they represent the last vestiges of lineages that have vanished elsewhere, their taxonomic isolation being due to *ex situ* extinction (of ancestral taxa and of all continental taxa derived from these). Some palaeo-endemic plants form on some islands distinct palaeo-habitats, such as the laurel forests of the Canary Islands and the araucaria forests of New Caledonia. They provide a glimpse of continental habitats as they were millions of years ago.

We expect the percentage of old, endemic populations in an island avifauna to be a function of the island's age, but also of its area and isolation. Because of differential colonisation and extinction, it is not only the total numbers of species that vary with the size and remoteness of islands but also the proportion of endemics. This is illustrated for various Pacific islands and archipelagos in **Figure 6.6**, in which the percentage of endemic species among living birds increases with increasing area (supposedly because extinction rates decrease with increasing area) and with increasing isolation (supposedly because immigration rates decrease with distance).

Some types of birds are such good colonists that they have developed large numbers of island species or races. Among Pacific islands, the genus *Zosterops* has 33 island forms, *Halcyon* has 28, *Pachycephala* 25, *Ptilinopus* 23 and so on. However, while some species in these genera have colonised single islands or island groups, others have become much more widely spread (for *Zosterops*, see Mees 1969, Kikkawa 1973), or have colonised the same island on successive occasions (Moritz *et al.* 1998). Thus *Zosterops* has invaded Lord Howe Island off Australia at least twice and Norfolk Island at least three times, providing examples of speciation by double and triple invasion. Interestingly, the greatest extremes in divergence are found among taxa living in sympatry, again implying that competition is one of the factors promoting divergence.

Sometimes endemism is high even on islands close together. In the Solomons, for instance, endemic subspecies of *Zosterops* are separated by straits only 2 km wide, and full species by straits 5 km wide (Mayr 1942). Birds can fly such distances in a few minutes, but if the differences between these forms arose from

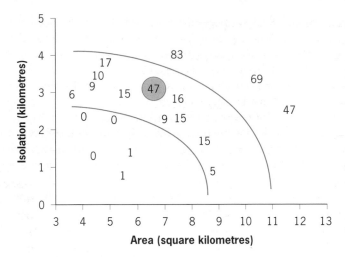

Figure 6.6 Percentages of surviving endemics among bird species on various Pacific islands in relation to island area and distance from a source. The percentage of endemic species increases with increasing area (from left to right in the figure) and with increasing isolation (from bottom to top in the figure). Curved lines separate islands with low, medium and high numbers of endemic species, and the circled number refers to the Galapagos Islands, which have a higher proportion of surviving endemics than do southwest Pacific islands. Redrawn from Diamond 1980a.

adaptations to the local conditions, there would be strong selection in favour of each individual keeping to its own island. The sedentariness of some island birds could then be as much a result as a cause of their endemism.

Another feature of the more extreme insular endemics is that they are evolutionary dead ends, their lineages perhaps doomed to extinction on much shorter time scales than most continental endemics. This results from the apparent unidirectional movement of species from continents to islands over evolutionary time. Island biotas have clearly been derived by the colonisation of taxa that came from larger land-masses either directly or via other islands. While some island species are formed by speciation within an island or archipelago, and some may survive beyond the time when their continental ancestors have died out, continents remain the ultimate source of island forms. In contrast, no oceanic island bird species is known to have colonised a continent and, because many of the more extreme island endemics are flightless, this effectively puts an end to their colonising ability. They can only change further on the island itself.

All the undifferentiated island species, which look like their continental ancestors, give us some idea of what is currently moving around in the world as it is now. In fact, the more an island species resembles its relatives elsewhere, the easier it is to see where it originated (unless a species is so widespread that it could have come from any of several source areas). The more different an island endemic is from living birds elsewhere, the harder it becomes to trace its past. This is most true for some of the ancient endemics that lack close relatives anywhere.

Some may represent relics that colonised from other now gone islands where they had already become distinct from their own continental ancestors. In other words, not all the differentiation of island species from their ancestors necessarily occurred on the islands where they now occur.

The Mauritius Dodo and Rodriguez Solitaire have diverged so far from their pigeon ancestor that they have been allocated a family of their own, the Rhaphidae. For the dodo, a DNA study singled out the Nicobar Pigeon *Caloenas nicobarica* as a close living relative, followed by the crowned pigeons *Goura* (Shapiro *et al.* 2002). The Nicobar Pigeon, which is found in the East Indies and New Guinea, differs from fruit pigeons in being able to digest the hard seeds of fruit as well as the pulp. It has in its gullet, not grit like other pigeons, but a single stone. This gives another clear link to the dodo and solitaire (Cheke & Hume, unpublished).

It seems, then, that the main evolutionary patterns on islands are: (1) rapid adaptation to local conditions, followed in suitable circumstances by adaptive radiation to produce a range of different species; and (2) prolonged status of relict forms. Islands can thus be viewed as factories for neo-endemics and as museums for palaeo-endemics. Depending partly on their isolation, islands vary in the proportions of their endemics that are products of *in situ* evolution or *ex situ* extinction. Evolution on islands can be so rapid that unusual morphology can hide a close relationship with a continental group, as in the Galapagos Finches (Chapter 3). Analyses of DNA have already helped to untangle the relationships of some strange-looking island birds, and may well reveal other cases of cryptic neo-endemism in the future.

Because palaeo-endemics, including some flightless birds, have persisted on islands for many millions of years, the islands concerned are assumed to have offered relative stability of habitats. Having an oceanic climate, islands in tropical and subtropical latitudes may well have been buffered against the extremes of climate change. It is this type of stability, coupled with low invasion rates, that has presumably enabled some species to persist for many millions of years. It is hard to imagine any continental areas that could have remained so climatically stable, or so resistant to invasion, for so long.

Features of island bird species

Species on islands often differ in particular ways from their mainland ancestors. Many have reduced or non-existent powers of flight. This is particularly notice-able among island rails. Derived from mainland forms, all these birds must have lost their powers of flight secondarily after colonising the island. This has led to speculation that, because most individual landbirds that leave remote islands are likely to die, selection has favoured the removal of long-distance dispersal ability. Other factors are likely to have been involved, however, such as the lack of native mammalian predators on many islands, and the advantages that would then have accrued from reducing unnecessary appendages and muscle mass that are ener-getically expensive to maintain (McNabb 1994). Generally, reduction in wing size is associated with reduced pectoral musculature and bone area on the sternum, and with increased leg size, so that flightlessness can be recognised even in fossil birds. Lack of mammalian predators on oceanic islands is often also associated with unusual tameness on the part of the native birds.

Secondarily derived flightlessness is shown by one or more bird species on most oceanic islands, including representatives of at least eight orders of non-marine birds, namely the Struthioniformes, Anseriformes, Psittaciformes, Strigiformes, Columbiformes, Gruiformes, Ciconiiformes and Passeriformes. It thus provides a striking example of convergent evolution. On New Zealand some 25–30% of the original native land and freshwater birds were flightless (or nearly so), including 12 out of 34 of the extant species, and more than 20 other species that have died out in the past 1,000 years, following human colonisation (**Table 6.5**). These flightless species included 11 moas, three or four kiwis, four ducks, two geese, seven rails, a parrot, an owlet-nightjar, several acanthosittid wrens, a fernbird and one or two wattlebirds (Atkinson & Milliner 1991, Bell 1991). On New Caledonia, two flightless kagus (*Rhynochetos*), a large galliform bird and at least two flightless rails occurred (Balouet & Olson 1989). On the Hawaiian Islands, at least 20 (24%) of the original terrestrial and freshwater bird species were flightless, including four duck-like birds (moa-nalos), one or two geese, three ibises and 12 rails (Olson & James 1982, 1991).

Other examples of flightless birds on islands, many eliminated following human arrival, include the Flightless Cormorant *Phalacrocorax harrisi* of the Galapagos Islands, the Dodo *Raphus cucullatus* of Mauritius, the solitaires

Table 6.5   Flightless and weakly flying species of New Zealand birds that survived beyond AD 1800. Modified from Baker 1991.

Family	Species	Distribution
Flightless		
Apterygidae	Little Spotted Kiwi *Apteryx owenii*	South Island
	Brown Kiwi *A. australis*	North, South, and Stewart Islands
	Great Spotted Kiwi *A. haastii*	South Island
Rallidae	Weka *Gallirallus australis*	North, South, Stewart, and Chatham Islands
	Takahe *Porphyrio mantelli*	South Island
Anatidae	Flightless Teal *Anas aucklandica*	Auckland Island and Campbell Island
Weak Flyers		
Psittacidae	Kakapo *Strigops habroptilus*	South and Stewart Islands
Acanthisittidae	Bush Wren *Xenicus longipes*	South Island
	South Island Wren *X. gilviventris*	South Island
	Stephens Island Wren *X. lyalli*[1]	Stephens Island
Muscicapidae	New Zealand Fernbird *Megalurus punctatus*	North, South, Stewart, Codfish, Snares and Chatham islands
Callaeidae	Saddleback *Philesturnus carunculatus*	Hauraki Gulf and offshore from Stewart Island
	Kokako *Callaeas cinerea*	North and possibly South and Stewart Island
	Huia *Heteralocha acutirostris*[1]	North Island

[1] Now extinct.

Pezophaps solitaria and *Theskiornis solitarius* of Réunion and Rodriguez, the elephant-birds (Aepyornithidae) of Madagascar, and many different rails from islands around the world. Some of these flightless birds converged on niches filled on continents by grazing mammals, while the kiwis act as specialist worm-eaters, as probably did the terrestrial ibises of the Hawaiian and Mascarene Islands. The evolution of a flightless condition can occur rapidly, as inferred from the fact that some species of flighted birds have flightless subspecies on islands. Examples include the Common Moorhen *Gallinula chloropus* which is flightless on Gough and Tristan da Cunha, and the extinct White-throated Rail *Dryolimnas cuvieri* which was flightless on Aldabra (McNabb 1994). Moreover, the endemic New Zealand Brown Teal *Anas chlorotis* varies in flight capability between islands (Bell 1991). The New Zealand mainland subspecies, *A. aucklandica chlorotis*, is more reluctant to fly than the closely related Chestnut Teal *A. castanea* of Australia. Two sub-antarctic subspecies are both totally flightless, namely *A. a. aucklandica* on the Auckland Islands and *A. a. nesiotis* on Dent Island off Campbell Island. A further subfossil form which occurred on the Chatham Islands also showed wing reduction. Hence, marked differences in dispersal ability can soon develop between different subspecies and between different allospecies.

Unusual dimensions and extreme specialisation are other features of island forms. Gigantism is shown mostly by species now extinct, such as the moas of New Zealand and the elephant-birds of Madagascar, but even recent colonists are often larger than mainland conspecifics. For example, Silvereyes *Zosterops lateralis* that have reached many islands in recent times show that change in size can occur well within 200 years (Degnan *et al.* 1998). Wrens also tend to be larger on islands, while ducks and rails tend to be smaller than their mainland ancestors.

Island birds are also often duller in colour and less sexually dimorphic than their mainland equivalents, especially those with no close relatives on the island. In most species of mainland ducks, the male is much more brightly coloured than the female, but in many island ducks, sexual dimorphism has been reduced, and the sexes look more similar. With no congeners, confusion in mate choice and consequent hybridisation become unlikely. This could reduce the strength of sexual selection which is presumed to favour colourful males, relative to natural selection which is presumed to favour dull ones. Reduced sexual dimorphism is apparent in most island ducks even though they have evolved from different parental species. They have also all produced medium-sized (generalist) species (Lack 1976). Because these features have been evolved independently by each of the duck species involved, they are presumably adaptations to conditions on remote islands most of which offer only one niche for an aquatic duck (as opposed to the terrestrial ducks (moa-nalos) of the Hawaiian Islands, Chapter 10).

Melanic or darkened plumages occur in many island birds, especially in rails, in normally black-and-white shorebirds which are black on some islands (e.g. stilts and oystercatchers in New Zealand, and extinct oystercatchers on the Canary Islands), and in some passerines, such as fantails, tits and robins in New Zealand. Uniform dark plumages have been attributed to reduced predation which allowed some species to abandon the patterns of countershading favoured in most parts of the world. Some of the most specialised of island endemics are more 'K-selected' than their continental ancestors, with lower reproductive and mortality rates, and longer life spans. This is sometimes evident in comparisons between older endemics and more recent related colonists on the same island, such as the Takahe *Porphyrio mantelli* and Pukeko *P. porphyrio* on New Zealand (Bell 1991).

Many other peculiarities are evident in the island forms of certain taxa only. For example, the various medium-sized owls, whose ancestors colonised islands that lack rodents, have developed longer legs than their continental counterparts. This is presumably an adaptation to catching alternative prey, such as birds or reptiles. Many island passerines have simpler songs than their mainland ancestors, which could again be attributed to reduced sexual selection in an environment with fewer (if any) congeners.

Niche breadths and densities

Species that occur on an island, as well as on nearby mainland, often have broader niches on the island, exploiting a wider range of habitats and food sources. This is usually attributed to the reduced number of food-types on islands, so that any one species can survive only by exploiting a wider variety of resources. This in turn reduces the possibility of close competitors becoming established there. Thus, while habitats on mainlands or on large islands may be occupied by two congeneric species, equivalent habitats on small islands may hold only one (Lack 1971).

Almost every species that has been studied has shown niche expansion on islands compared to mainland, often associated with morphological divergence, sufficient in some to warrant subspecific status. For example, where two species occur in forest on mainland, and only one of them on an island, the island one is often intermediate in character between the mainland two, exploiting both their niches. On the Swedish island of Gotland, Coal Tits *Parus ater* are larger than on the mainland, associated with the absence from the island of the larger Crested Tit *P. cristatus* and Willow Tit *P. montanus* (Alatalo *et al.* 1986). The change in morphology coincides with a shift in foraging niche, as the Coal Tits of Gotland feed more within the canopy of trees, while on the mainland they feed mostly on the edges of the canopy and on needles. In other tit communities elsewhere in Europe, the absence of any one species is commonly associated with a morphological change in another, with small species becoming larger than elsewhere (Lack 1971, Dhondt 1989). Seemingly, in the restricted ecological conditions of islands, one generalised species is liable to prevail over two specialists, thus providing another reason for the reduced number of species on islands and for the reduced numbers of species per genus.

Which of several mainland species is likely to occupy the island niche? Interestingly, while many European islands support only a single tit species occupying a generalist niche, it is not always the same species that fills this niche: different species occur on different islands. Much could depend on chance, on which species gets there first and is able to adapt to local conditions, and exclude subsequent colonists. Chance may explain why, for example, the Canary Islands and Madeira have the Firecrest *Regulus ignicapillus*, while the Azores have the almost identical Goldcrest *R. regulus*. Similar examples of so-called 'checkerboard distributions' are given in Chapter 14.

Because of their wider niches, co-existing island species often differ more markedly than would be expected if the island biota had been drawn at random from a mainland species pool. For example, among hummingbirds in the West Indies, species that co-exist on the same island tend to be more different in bill and wing measurements than would be expected if the various species associated at random (Lack 1976). The finches of the Galapagos Islands are even more instructive (Grant & Abbot 1980): two such islands have a solitary ground finch with a bill depth of 9–10 mm, but on one of the islands, the solitary species is *Geospiza fortis* and on the other it is *G. fuliginosa*. On three other islands these species co-exist, and there *G. fuliginosa* has a bill depth of about 8 mm, while *G. fortis* has a bill depth of about 12–13 mm. Thus each species has diverged in opposite directions where they occur together. This phenomenon was labelled in various kinds of animals as 'character divergence' by Darwin (1859) and as 'character displacement' by Brown & Wilson (1956). It again implies that competition is important in influencing the ecology of different species, and the numbers and types that can co-exist on different islands.

Another common feature of some island birds, associated with their wider niches, is that they occur at much higher densities than their mainland equivalents (Wright 1980). Crowell (1962) compared bird populations on Bermuda with those of similar habitats on the North American mainland, 900 km to the west. He found that the ten species of small passerines on Bermuda together maintained a population about 1.5 times greater than the combined densities of 20–30 species on the mainland. Other studies have shown similar patterns elsewhere (e.g. Lack 1947, Diamond 1975). This phenomenon, called density (over-) compensation, is often attributed to more plentiful food-supplies on islands, in turn resulting from broader niches and lower levels of inter-specific competition, not just from birds but also from other animals; but in some species high densities could also result from scarcity of predators or from other local conditions that have nothing to do with insularity (Blondel *et al.* 1988). On many islands, bird densities become much reduced after the introduction of alien predators.

To conclude, at least four features of island birds are thought to result from altered levels of inter-specific competition compared to mainlands: (1) increases in niche breadth; (2) mutually exclusive distributions of ecologically similar, closely related species among islands; (3) the apparent tendency of species from the same island to differ in ecology or taxonomic relatedness more than would be expected in a random assemblage of species drawn from the archipelago as a whole or from the mainland species pool; and (4) increases in population densities. The first three findings are taken as evolutionary consequences of greater competition on islands (resulting from a narrower range of resources), and the greater densities in certain species are taken to result from reduced competition in present conditions.

Patterns of colonisation and divergence

In some species that have reached islands in the last 200 years, the precise route of colonisation is known from historical records (e.g. the White-eye *Zosterops lateralis* from Australia to New Zealand and other islands, Clegg *et al.* 2002). In other species, established for much longer, the likely route of colonisation has been deduced from comparison of the mt DNA from parental and derived populations. On this evidence, the Common Chaffinch *Fringilla coelebs* spread from Iberia to the

Azores, on to Maderia and then to the Canary Islands (**Figure 6.7**; Marshall & Baker 1999: Chapter 3). Also on DNA evidence, the New World Yellow Warbler *Dendroica petechia* spread from Central America to the Greater Antilles and from Venezuela to the Lesser Antilles. Birds on some adjacent islands were usually nearest relatives, but not always, indicating that stepping stone colonisation was frequent but not invariable (Klein & Brown 1994).

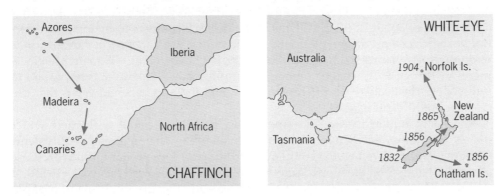

Figure 6.7   Sequence of island colonisation from a mainland source.
Left. Common Chaffinch *Fringilla coelebs*, as deduced from comparison of the mt DNA from parental and derived populations and dated as beginning 0.6 million years ago. From Marshall & Baker 1999.
Right. White-eye *Zosterops lateralis*, as known from historical records of arrival dates. From Mees 1969, Clegg *et al.* 2002.

The divergence of newly established isolates could develop through the random processes of founder events and genetic drift, as well as from natural and sexual selection (Chapter 3). Observations have confirmed the action of both types of selection on wild bird populations, but the evidence for divergence in phenotype from random genetic effects is much less firmly based. It is in island populations that divergence due to founder effects is most likely to be apparent. Any such effects should be evident immediately following colonisation, and not the result of gradual divergence, as expected under selection. They should also be more apparent in successive colonisations, where a species passes from one to another along a string of islands, with each population representing only a genetic subset of the one before (for an example in *Zosterops*, see Clegg *et al.* 2002). Moreover, in any series of colonisations, phenotypic changes due to founder effects would not all be expected to be in the same direction: rather than progressive increase in body size, say, random effects should give rise to some populations with larger, and others with smaller, body sizes than the source population. To my knowledge, no systematic tests of this sort have yet been made on bird populations and the role of founder effects in the evolution of island birds remains largely unknown.

LAND-BRIDGE ISLANDS

Land-bridge islands were once connected to mainland, but have since become isolated, mostly by rise in sea-level. During the most recent ice age, which ended about 10,000 years ago, enough water was locked up in glaciers to make the sea-level about 120 m lower than now (most estimates vary between 100 m and 150 m, according to region). Consequently, islands that are now separated from continents or from larger islands by water less than about 120 m deep were once attached. Examples include: Britain off Europe, Ceylon off India, Java–Sumatra–Borneo off Asia, Fernando Po (Bioko) off Africa, Tasmania off Australia, and Trinidad and the Falklands off South America. These are among the largest of land-bridge islands, but off some continental coastlines innumerable smaller islands were also joined to the mainland or to one another in times past.

Land-bridge islands invariably hold fewer landbird species than equivalent areas of nearby mainland, the difference being most marked on the smallest islands (**Table 6.6**; Terborgh & Winter 1980). Because all such islands, when they were part of the mainland, could be expected to support the full range of mainland species (habitats permitting), the implication is that habitat fragmentation resulting from rise in sea-level caused the loss of some of the original species, more from small islands than from large ones. Such studies gave the first indication that habitat fragmentation caused reductions in bird distributions that were greater than expected merely from the amount of habitat lost (Chapter 13). They also indicated that some species were more vulnerable to the effects of fragmentation than were others, namely those of low density, specialised requirements or poor dispersal ability (Diamond 1984). Most such species were likely to have been present in low numbers anyway, and it is not hard to imagine that small amounts of remaining habitat might have been insufficient to support them long term, especially in the absence of further immigration. Following the formation of land-bridge islands, therefore, a process of 'relaxation' is envisaged, involving the progressive loss of species until some lower level is reached, more appropriate to the habitat areas remaining, and in line with the species–area relationship. This is akin to the opposite situation in newly formed oceanic islands, where species numbers rise from zero to reach a plateau, lower on small islands than on large ones. On both kinds of islands, local losses might have been reversed by subsequent recolonisations so that their present avifaunas reflect the net outcome of species loss over immigration.

Sometimes the process of species loss has been observed directly, as in studies in forest habitats fragmented by human action. For example, early in the twentieth century, the Chagros River was dammed to create the Panama Canal and Gatun Lake, flooding lowland areas and turning forested hilltops into islands. Repeated surveys quantified the process of progressive species loss to a new lower level commensurate with the reduced area of habitat. On the largest island, Barro Colorado, surveys in the 1970–80s revealed that about 45 of the original 108 species of breeding birds had disappeared (Willis 1974, Karr 1982, 1990). This figure is minimal because intensive surveys were not made until well after the island was formed, so additional species may have gone unrecorded. Local losses and recolonisations appear to have occurred continually on all the islands studied, but turnover was lowest on the largest and most isolated ones. Studies at other

Table 6.6 Some examples of avian endemism in land-bridge islands, once connected to nearby mainland. Modified from Ridpath & Moreau 1966.

Island	Minimum breadth of strait (km)	Approximate period of isolation (years)[1]	Approximate number of breeding species	Number (%) of	
				Endemic subspecies	Endemic species
Tasmania	224[2]	12,000	104	27(26)	14(13)
Britain	34	10,000	151	27(18)[3]	–
Ireland	43	10,000	126	3(3)	–
Zanzibar	48	9,000	105	3(3)	–
Balearic Group	96		80	5(6)	–
Fernando Po	40	17,000	147	37(25)	1(1)
Japan group	192[2]	10,000	173	35(20)	7(4)
Ceylon	112[2]	9,000	225	69(31)	19(8)

Those land-bridge islands with endemic species all have substantial montane massifs that would have been isolated long before the physical separation of the islands by rising sea-levels.

[1] Calculation from the date at which the rise in sea-level that followed the end of the last glaciation severed the island from the mainland. The period depended largely on depth of intervening channel (earlier in deeper channel).

[2] Interrupted by smaller islands.

[3] Some regard two taxa, the Scottish Crossbill *Loxia scotica* and Red Grouse *Lagopus l. scoticus* as distinct species.

sites indicate that species were lost more quickly the smaller the island, but in any case were rapid initially and slower later (Diamond 1972b, Burkey 1995, Terborgh *et al.* 1997).

Interesting differences in avifaunas are seen on islands around New Guinea (Diamond 1984). Some islands were joined to New Guinea until they became cut off by rising seas, while others nearby are oceanic, never having been connected. All the large land-bridge islands now have many fewer species than equal-sized pieces of the New Guinea mainland, but they have about twice as many lowland species as oceanic islands of similar size and proximity to New Guinea (**Figure 6.8**). The difference presumably arose because the land-bridge islands started off with almost the same complement of plant and animal species as the nearby mainland, the birds benefiting from the diversity of other life; whereas the oceanic islands received all their species by overwater flight.

Some 134 (41%) of 325 bird species of lowland New Guinea seem incapable of crossing water gaps, because they occur only on land-bridge islands and not on oceanic ones. However, their numbers decrease sharply with area of land-bridge

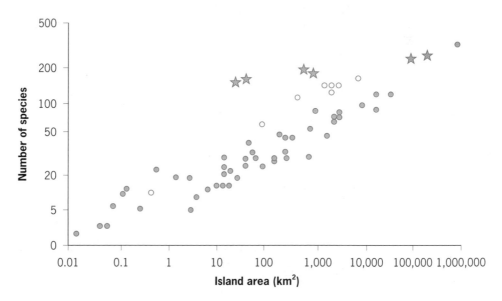

Figure 6.8   Species–area relationship for bird species on oceanic islands (⬤) and land-bridge islands (○) of the New Guinea region and for areas of various sizes on the New Guinea mainland (★). Land-bridge islands have fewer species than equivalent-sized areas of the New Guinea mainland. This is presumed to result from net excess of local extinctions over local recolonisations since the end of the Pleistocene period, when the islands were formed. Oceanic islands have fewer species than equivalent-sized land-bridge islands, presumably because only the latter were once connected to mainland. However, very small land-bridge islands do not differ from oceanic islands, supposedly because of very high post-Pleistocene extinction rates. From Diamond 1984.

island: some 36–45 such species occur on land-bridge islands of 1,600–7,800 km^2 in area, with 26 on an island of 450 km^2, three on an island of 145 km^2, and none on an island of 0.93 km^2 (which nevertheless held 14 species that regularly cross water). If extinction rates increase with decreasing island area, as expected, perhaps all the original species have disappeared in the past 10,000 years from very small land-bridge islands. Their avifaunas consist now entirely of overwater colonists, and their species numbers and composition are indistinguishable from those of like-sized oceanic islands (Diamond 1984).

Similar comparisons between land-bridge and oceanic islands can be made elsewhere. For example, in the Caribbean the continental island of Tobago lies only 120 km from the oceanic island of Grenada. The two are of similar size and climate, but Tobago has twice as many landbird species. This might be partly attributed to the longer flights needed to reach Grenada, but is probably also linked with Tobago having a much richer forest than Grenada, providing niches for more bird species (Lack 1976).

Differentiation

In general, birds on recently formed land-bridge islands, even if not connected by movements to their continental counterparts, have not been isolated for long enough to form distinct species, but many have differentiated to the level of subspecies (**Table 6.6**). On some islands, such as Tasmania and Ceylon, both endemic species and subspecies occur, but this may be because the islands concerned have long held unique habitats, not represented on the mainland, or habitat areas (such as mountains) well separated from their equivalents on mainland (Chapter 3). Their species are likely, therefore, to have been isolated for longer and even at times when the island was connected to mainland. Old land-bridge islands that have been separated for many tens of millions of years (such as Madagascar, New Caledonia and New Zealand) all contain long-established endemics in unique families that appear to be of ancient lineage. Examples include the Aepyornithidae (elephant-birds), Mesitornithidae (mesites), Leptosomidae (cuckoo-rollers), Brachypteraciidae (ground-rollers) and Philepittidae (asities) of Madagascar; the Rhynochetidae (kagu) of New Caledonia; and the Dinornithidae (moas), Apterygidae (Kiwis), Acanthisittidae (wrens) and Callaeatidae (wattlebirds) of New Zealand. Some of these families may trace back to Gondwanan times, but their member species may well have changed in this time.

Some parallels with mountain tops

The same warming process that led to the formation of land-bridge islands at the end of the last glaciation also led to the fragmentation of some mainland habitats. During cool periods of the Pleistocene, habitats that are today confined to mountains in some regions extended to lower elevations and to lower latitudes (Chapter 9). Birds and other organisms characteristic of these habitats then occupied lowland. Rising temperatures then drove the habitats and their fauna up the mountains, creating the discontinuous distributions that we see today. We may presume that the montane populations were subjected to a process of differential extinction to give the present patterns in which the numbers of montane species are greatest on the most extensive mountain ranges (Chapter 12). With at least some of the bird species concerned, movements between ranges are probably frequent, but it is not known to what extent species turnover occurs, as envisaged for islands in the sea.

The findings for birds on land-bridge islands and mountain tops are paralleled in mammals, most of which cannot move between islands through sea water, or between mountain summits through the alien habitats of lowlands. However, mammals are better represented in the fossil record than are birds, and provide many concrete examples of extinctions on small land-bridge islands and on mountain tops (Diamond 1984, Patterson 1984). To date, all known or inferred bird and mammal extinctions on land-bridge islands or on mountain tops have shown that populations in small areas are at greater risk than those in large areas. Clearly, differential local extinctions in the past could account for the patterns found today of greater species numbers in larger habitat areas, whether on land–bridge islands, mountains or other fragmented habitats (see also Chapter 13).

SUMMARY

Individual islands hold fewer plant and animal species than equivalent areas on continents; but added together, they hold many more. This is because many bird species are endemic to only one or a few oceanic islands, with different species on different islands. In general, species numbers increase with island area and decrease with distance from mainland. The small numbers of resident landbird species on individual oceanic islands are due partly to difficulties of getting there, and partly to ecological limitations, which enable fewer species with broader niches to exclude a greater number of specialists. However, some island species achieve much greater densities than their mainland equivalents.

Oceanic islands, never having been connected to mainland, have received their natural flora and fauna entirely by cross-water colonisation. Because some types of birds disperse more easily than others, islands contain an unrepresentative cross-section of mainland types. This distortion becomes more extreme the further the island lies from a continental source area.

On oceanic islands, endemism develops in response to new selection pressures because of genetic isolation from the ancestral population. Non-endemic species may be assumed to have arrived recently (in evolutionary time) or to be in continual genetic contact with the parental population through fresh arrivals. Hence, endemism is dependent primarily on long-term isolation and poor dispersal. In the absence of mammalian predators, many island endemics became flightless, as a secondary development from their flying ancestors. Other common features of island birds include melanistic plumages, tameness and evidence of well-separated double or triple colonisations by the same species.

Land-bridge islands were once connected to a nearby continent, from which they inherited their flora and fauna directly. Such islands typically have fewer species than equivalent areas on the adjacent continent, implying a net loss of species since severance. Again certain types of species are more often lost than others. Land-bridge islands connected to mainland during the last glaciation mostly contain no unique bird species, endemism being apparent only at the level of subspecies. Where endemic species exist, as on Tasmania, they might have been confined to that area for a much longer period, even when it was joined to mainland. Old continental islands, last connected to a continent many millions of years ago and now far removed from it, contain large proportions of endemic species, some of them of ancient lineage.

Extinct Dodo *Raphus cucullatus* of Mauritius.

Chapter 7
Island birds: losses and gains

Many of the original bird species are known to have disappeared from oceanic islands following human settlement. Some of these species are now extinct, while others have gone from some islands but remain on others. Human colonisation of remote oceanic islands has been occurring now for at least 4,000 years, but mainly in the last 1,500 years, depending on location. More recently, many continental bird species have been introduced to oceanic islands that they could not be expected to reach naturally. This has occurred mainly in the last 150 years. In this chapter, I shall discuss these losses and gains, and the extent to which they have affected ideas on the development of island avifaunas.

Compared with continents, greater rates of natural extinction might be expected among the plants and animals of islands because their populations and ranges are smaller, and species are therefore more vulnerable to catastrophic events. Some oceanic islands in warmer regions may have been buffered from the more extreme climatic changes experienced on continents, providing suffi-cient constancy of conditions to enable some palaeo-endemics to persist for millions of years (Chapter 6). However, islands did not escape the effects of the Pleistocene climatic and sea-level changes altogether, as confirmed by geomorphological and palaeontological evidence. Some islands were completely submerged by rises in sea-level, and others were partially submerged and fragmented, while climate change affected the extent and distribution of particular habitats. On some islands, the fossil record reveals substantial changes. For example, on the Bahamas, several endemic birds of grassland were present in dry times but disappeared with their habitats in wetter times (Olson 1982); while on St Helena a gradual change from temperate to tropical species of seabirds occurred during part of the Pleistocene, presumably in response to changing conditions (Olson 1975).

Two types of natural disaster have much more devastating effects on small oceanic islands than on larger land areas. In the space of a few hours volcanic eruptions can destroy most or all of an island's biota (as exemplified by Krakatoa in 1846, Chapter 6)[1], and in the space of seconds tsunamis can totally submerge low-lying islands, obliterating at least the non-flying animals. Caused by sea-bed earthquakes, tsunamis are high-energy tidal waves, up to 15 m high, which can travel rapidly for hundreds of kilometres across an ocean, flowing over all low-lying land in their path. In about 160 years of recording, some 130 tsunamis have occurred somewhere or other in the Pacific (Cornutt & Pimm 2001). Low islands, such as the Tuamotu archipelago, would be expected to be especially vulnerable. Forests on islands that lie in the hurricane belts, which extend some 10–20° north and south of the equator, can also suffer periodic destruction. The impact on birds is seldom devastating, and recovery is usually rapid (e.g. Askins & Ewart 1991, Wunderle *et al.* 1992), but this is not to say that some endemic species on small islands might not on rare occasions be wiped out in this way.

RECENT EXTINCTIONS

In addition to natural extinctions, many island birds have disappeared in the last 1,500 years, probably almost all through human action. The evidence for human involvement is that the extinctions occurred at widely different dates on different groups of islands, but in each case followed soon after human colonisation. Moreover, bones found at archaeological sites have revealed that many of the extinct species were frequently eaten (Duncan *et al.* 2002). In contrast, on islands that were not colonised by prehistoric people (such as Galapagos), subfossil remains have failed to reveal any extinct bird species from the Holocene period that pre-date the arrival of modern Europeans (Milberg & Tyrberg 1993). The last 1,500 years have been relatively stable in climate, and no natural factor is likely to have caused so many extinctions in this time.

Comparing different islands from around the world, the proportion of extinct species among the total known avifauna is inversely related to island area, with

---

[1] *Other examples of islands that were completely denuded of life by volcanic eruptions include Ritter (1888), Long, Crown and Tolokiwa (late seventeenth century) and Vuatom (600 AD), all in northern Melanesia (Mayr & Diamond 2001).*

relatively more extinctions on the smaller islands (Case 1996). In addition, allowing for island area, avifaunas with high levels of endemism have lost the greatest proportions of species and, on the same island, endemics have more often died out than native non-endemic forms. There are at least two likely reasons for this difference. Not only were endemics likely to have been on their islands for longer than non-endemics and lost their defences against non-endemic predators and pathogens, but they also lacked recolonisation sources, so that local extirpation meant extinction. Many of them were also highly specialised and included numerous flightless forms.

Such losses attest to the extreme vulnerability of island birds to human impact. Of the 127 named bird species that have died out in the last 400 years (the 'historic period'), 116 (91%) were island species, which therefore disappeared at a rate 40 times faster than continental species (Newton 1998a). The difference is partly because island birds have small populations and ranges, as mentioned above, but also because many such species evolved in the absence of mammalian predation so had no natural defences. Even to this day, some island birds are very tame, and allow a close approach by people, rats or other introduced predators. These recently extinct species, which are known from their remains or from written descriptions or drawings, include representatives of 26 non-passerine and 15 passerine families (Newton 1998a). Almost certainly the list is incomplete, for other species may have disappeared in the last 400 years before they could be recorded, and others not seen for many years may now be extinct too.

In addition to the 116 island endemics known to have become extinct in the last 400 years, hundreds (possibly thousands) of others were lost from islands in the prehistoric period (pre-AD 1600), following early human colonisation. The evidence is based on bones from archaeological and palaeontological sites. Each bone collection provides only a subset of the prehistoric avifauna of an island, and at least four factors influence the numbers and kinds of species found: (1) the number of bones identified; (2) the agent (human or otherwise) that accumulated the bones; (3) the age of the bone sample in relation to the date of human colonisation (most species disappeared at an early stage); and (4) the conditions of preservation and methods of field collection, both of which may reduce the representation of bones from small species relative to large ones. More than 200 such species known from their bones were listed by Milberg & Tyrberg (1993), but the number is growing all the time as new discoveries are made, and the implications become clearer. In addition, 160 other fossils listed by these same authors refer to living species (mainly petrels and pigeons) on islands where they no longer occur. Evidently many species, considered today as endemic to single islands or island groups, had much wider distributions in the past.

Pacific Islands

In recent years, evidence has emerged from bone remains of massive destruction of Pacific Island landbirds by the first human colonists, long before the coming of Europeans (Olson & James 1982, 1991, Steadman & Olson 1985, Steadman 1989, 1991, 1995, James 1995). Relevant deposits were laid down from 8,000 years ago to the present, covering the arrival of Polynesian people, 4,000–1,000 years ago, depending on location, west to east. These people not only hunted the local birds, but also introduced rats and pigs, and destroyed much of the lowland forest for agriculture (Pimm *et al.* 1994). In general, islands occupied by people for the longest periods have lost the greatest proportions of species, and on some islands all the original species have gone.

On the Hawaiian Islands, bone deposits reveal that at least 35 species identified with certainty, and up to 22 others, were eliminated by the Polynesian settlers, who arrived about AD 300 (Olson & James 1982, 1991, James & Olson 1991, James 1995). Bones are currently available from five of the main Hawaiian Islands, but are thought to provide a reasonably complete record for only two, Oahu and Maui. On these islands, bone remains include at least two extinct and one extant species of geese, together with four species of extinct moa-nalos, which were heavy-bodied, flightless derivatives of ducks, with strangely shaped bills. These species were all herbivores, some probably browsing on understorey foliage (Olson & James 1991). Two species of flightless ibises of the genus *Apteribus* probably occupied similar niches to kiwis in New Zealand. Flightless rails were represented by at least ten species, probably from at least three separate colonisations. Raptors included four species of long-legged owls (*Grallistrix*), each from a different island, a large eagle (*Haliaetus*), and a strikingly modified harrier (*Circus*) with very shortened wing bones, as well as the surviving Hawaiian Hawk *Buteo solitarius*. Two species of extinct crows (*Corvus*) were added to the one surviving crow species, and around 17 extinct honeycreepers were added to the still impressive array of living forms (James & Olson 1991, Olson 1991, Olson & James 1991, Steadman 1995). These birds together comprised more than half of the known native landbird fauna of the islands which, at first human contact, totalled more than 100 species (James 2001). When European settlers arrived after 1778, about 50 species of native landbirds remained, of which 24 have since disappeared, bringing the total extinctions to at least 81 species, comprising around 80% of the known native landbird species. The surviving Hawaiian birds are mainly small inconspicuous species, now mostly restricted to mountain forests. Twelve are unlikely to survive much longer without special management (Pimm *et al.* 1994).

On other Pacific Islands, bone remains have revealed many previously unknown species, and extended the range of other known species (Steadman 1989, 1991, 1995). The majority of the new species belong to families or genera still well represented among the living birds of Polynesia. They include *Megapodius* megapodes, *Gallirallus*, *Porzana* and *Porphyrio* rails, *Ducula* and *Gallicolumba* pigeons, *Vini* parrots and *Aplonis* starlings. Newly recorded birds for the Polynesian region include the pigeons, *Caloenas*, previously known only as far east as Micronesia and New Guinea, and *Macropygia* previously known east only to Vanuatu (Steadman 1992). At least 27 species of landbirds lived on Eua I, Tonga, in pre-human times, but only six have survived, while of 15 species known from Huahine in the Society Islands, only two nest there today.

Human-induced extinctions clearly had major effects on the distributions of particular groups. Before the Polynesian occupation, nearly every island group across the Pacific was probably home to one or more species of flightless rails. Through recent excavations, Steadman (1995) estimated that 1–4 endemic rail species could have lived on each of approximately 800 larger Pacific islands, all of which, perhaps in the thousands, disappeared after human colonisation. If these estimates prove valid, the rail family (Rallidae), which now holds about 144 species, could once have been one of the most species-rich bird families on earth. Today such species survive on only three or four Pacific islands, including New Zealand.

The pigeons and doves (Columbidae) still constitute one of the most widespread and conspicuous families of landbirds on tropical Pacific Islands. Yet in almost all the Polynesian Island groups studied, the pre-human columbid faunas had more species, more genera and more species per genus than current faunas from the same island (Steadman 1997). At least nine columbid species have become extinct, and remaining species have disappeared from many islands where they once occurred. If not for human impact, Steadman (1997) estimated, a typical east Polynesian island would support at least 5–6 species of columbids in 3–4 genera (compared to 0–3 species in 0–3 genera today); and a typical west Polynesian island would support at least 6–7 species in 4–5 genera (compared to 1–6 species in 1–5 genera today). Two or three species from the genera *Ducula*, *Ptilinopus* and *Gallicolumba* occurred together on many islands. The lack of any past record of a pigeon on the Hawaiian Islands or on Easter Island may be due to their isolation (though little research has been done on Easter Island). No Polynesian island now inhabited by columbids is more than 600 km from another island, whereas the Hawaiian and Easter Islands lie more than 1,000 km from other islands. Only one native pigeon lives in New Zealand today, and no other columbid has been found in the rich fossil fauna of these islands.

Despite an exceptionally good fossil record, no bird species of any kind is known to have become extinct on New Zealand during the climatic vicissitudes of the Pleistocene, but nearly half of all landbirds have disappeared in the last thousand years since human colonisation. Extinctions progressed in a wave from North Island to South Island as the Maoris spread south, affecting small birds (killed by rats) and large birds (killed by people). Not a single native landbird species weighing more than 2 kg has survived to the present (Worthy 1999).

When the Maoris arrived, New Zealand was home to the moas (Dinornithidae), large wingless birds unique to these islands. At least 11 species occurred, ranging from the size of a turkey to a 3 m giant (*Dinornis maximus*) of 270 kg (Anderson 1990, Cooper *et al.* 1993). Different species of moas lived in forest or open country, and occupied herbivore niches filled elsewhere by mammals, which were absent from New Zealand. To judge from remains, moas formed a major food of the early Maori colonists, who had effectively eliminated them by AD 1700. Their disappearance removed more than half of all ratite species on earth at the time. At least 24 other landbirds, including at least nine other flightless forms, disappeared in the same period (Steadman 1995, Worthy 1999). Bone remains document the extinction of the Apterornithidae, a family of large gruiform birds possibly related to the Kagu *Rhynochetos jubatus* of New Caledonia; a swan *Cygnus sumnerensis* allied to the Australian Black Swan *C. atratus*; the flightless geese *Cnemiornis calcitrans* and *C. gracilis* related to the Cape Barren Goose *Cereopsis novaehollandiae* of Australia; *Euryanas*, a nearly flightless duck related to the Australian Wood Duck *Chenonetta jubata*; a giant 13 kg eagle *Harpagornis moorei* that preyed on large birds; an owlet-nightjar *Aegotheles novaezealandia* with its presumed closest relative in New Caledonia; several flightless rails (Rallidae); plus at least two other ducks, a hawk, a pelican, a crow, and two species of acanthisittid wrens (Olson 1991). Some of these various birds had separate species on the North and South Islands.

Following British settlement from the 1840s, further extinctions of smaller bird species occurred, so that at least 49% (44 species) of all New Zealand's non-marine birds are known to have become extinct since the arrival of humans (Bell 1991). One of the most extraordinary and most recent extinctions was the crow-like Huia *Heteralocha acutirostris*, which belonged to the family Callaeatidae, all of which have bright orange or blue wattles at the corners of their beaks. The Huia was unique among the world's bird species in that the males and females had differently shaped beaks. Both sexes fed on beetle grubs from rotten wood but, whereas the male with his short stout beak would chip into the softer rotten wood, the female with her long thin tweezer-like beak would extract the beetle larvae from tunnels in harder wood. The birds were invariably found in pairs, and the partners may have co-operated in obtaining food. Much reduced in Maori times, the bird finally disappeared after European colonisation. In addition, many seabird species declined or disappeared from the mainland, but survived on offshore islets, with none as yet known to have become totally extinct.

Some of New Zealand's extraordinary landbirds still survive, albeit in much reduced numbers. They include the four species of flightless kiwis, the females of which produce the largest egg, relative to their body size, of any known bird. At 420 g in weight, the egg is 5–10 times as large as the eggs laid by similar-sized birds, weighing as much as a quarter of the female herself. The nocturnal habits of kiwis may have protected them from Maori hunters, while the large eggs (and resulting chicks) may have conferred protection from rats, which continue to be a major threat to New Zealand's surviving birds. Other strange survivors include the Kakapo *Strigops habroptilus*, a large green nocturnal flightless parrot, and the Takahe *Porphyrio mantelli*, a giant flightless swamphen, once assumed extinct but rediscovered in 1948.

The Chatham Islands, lying 860 km west of New Zealand, were colonised by people only about 450 years ago. Of the 36 former breeding landbirds, 13 species are now extinct, and eight others have gone from Chatham but survive elsewhere (Millener 1999). Without recent intervention to save them, three other species would probably also have gone. Similarly, on New Caledonia, bird remains from late Holocene cave deposits revealed that at least 25% (possibly up to 40%) of the resident non-passerine bird species were exterminated prehistorically, almost certainly as a result of human action. They included a giant flightless megapode *Sylviornis neocaledoniae* estimated at 10–20 kg in weight, a large swamphen *Porphyrio kukwicki*, somewhat like the New Zealand Takahe, and another larger species of Kagu *Rhynochetos orarius* in addition to the still extant one, *R. jubatus* (Balouet & Olson 1989). Because New Caledonia was colonised by people 3,500 years ago, none of the extinct species may have survived into recent centuries.

Other islands

People colonised Madagascar from Indonesia about 2,000 years ago, and in the ensuing centuries the larger animals gradually disappeared (Martin & Klein 1984). They included 6–12 species of elephant-birds (Aepyornithidae), flightless creatures that included the heaviest birds known to have lived. One species, *Aepyornis maximus*, stood almost 3 m tall, and probably weighed around 450 kg when fully grown. It survived in diminishing numbers well beyond AD 1600. Other animals obliterated from Madagascar at the same time included large lemurs and tortoises, and again human hunting seems the most likely cause.

In the Indian Ocean, the previously uninhabited Mascarene Islands were discovered by Portuguese navigators around 1500, and colonised by Europeans from around 1600. Within 200 years, following the introduction of rats, cats and pigs, and destruction of most of the forest, much of the avifauna of these small islands had disappeared, including the famous Dodo *Raphus cucullatus* of Mauritius and the solitaires *Pezophaps solitaria* and *Theskiornis solitarius* of Rodriguez and Réunion (the first two were highly modified pigeons, and the latter an ibis). When people arrived, Mauritius held at least 28 landbird species, including 23 endemics, of which at least 12 (52% of all known local endemics) are now extinct, and five others have gone from Mauritius but survive elsewhere. Réunion held at least 32 species, including 25 endemics, of which 17 (68%) are now extinct, while 4–5 are no longer present on Réunion but survive elsewhere (Mourer-Chauviré *et al.* 1999). The equivalent figures for Rodriguez are at least 14 landbird species, including 13 endemics of which 11 (85% of all endemics) are now extinct (Cheke 1987). In addition, at least two endemic seabirds on Rodriguez became extinct, while other seabirds disappeared from Rodriguez and other Mascarene Islands, but survived elsewhere (**Table 7.1**). One of the latter was Abbott's Booby *Papasula abbotti* which now breeds only on Christmas Island about 5,000 km to the east. This bird is the most divergent of all boobies, and is known from bone remains to have bred on at least four islands in the tropical Indian Ocean, and on other islands eastwards to the Marquesas in the Pacific.

Table 7.1   Minimum numbers of bird species present on the Mascarene Islands before human colonisation about 400 years ago, and numbers still surviving. From Cheke 1987, updated by A. Cheke.

	Réunion	Mauritius	Rodriguez
Landbirds[1]			
Minimum number of known species (endemics)	32 (25)	28 (23)	14 (13)
Minimum number of extinct species (endemics)	21 (17)	16 (12)	12 (11)
% extinct species (% endemics)	66 (68)	57 (52)	86 (85)
Seabirds[2]			
Minimum number of known species (endemics)	6 (2)	11 (0)	13 (2)
Minimum number of extinct species (endemics)	0 (0)	3 (0)	5 (2)
% extinct species (% endemics)	0 (0)	27 (0)	38 (100)

[1] *Gallinula chloropus* and *Butorides striatus* excluded as recent colonists. The seven species shared between Mauritius and Réunion are counted as endemics; the freshwater cormorant *Phalacrocorax africanus* is counted as a landbird. Only one non-endemic landbird *Streptopelia picturata* is known from Rodriguez.
[2] *Pterodroma neglecta* excluded as a recent colonist, *P. baraui* excluded as a temporary colonist in the 1970s; only one *Fregetta* species counted for each island.

At first human contact in 1692, the small Indian Ocean island of Amsterdam was estimated to hold 20 bird species (mostly seabirds), of which only ten remain today, including the endemic Amsterdam Island Albatross *Diomedea amsterdamensis*. The only known landbird species, a small flightless duck *Anas morecula*, is extinct. This island was never settled by people, but was periodically raided for food, wood and seal skins (Worthy & Jouventin 1999). It still supports rats and cats.

Turning to Atlantic Islands, on St Helena only one endemic bird species has survived (the plover *Charadrius sanctaehelenae*), but at least six others (including two seabirds) were present when Europeans arrived in 1502. At this time, the island was mainly forested, yet truly arboreal birds are all but absent from the fossils collected so far, implying that many are still to be found. On Ascension, no native landbirds have survived, but two are known from bone remains. The fossil faunas of other Atlantic islands are poorly known, but the Azores now hold only a single (poorly defined) endemic species, the Azores Bullfinch *Pyrrhula murina* (sometimes classed as a subspecies of *P. pyrrhula*). By analogy with other volcanic island groups, the Azores almost certainly held more landbird endemics when people arrived in the late fifteenth century. Since then, seabirds have largely disappeared from the main islands where they once occurred in large numbers (Monteiro *et al.* 1996). Similar drastic changes in the fauna of the West Indies (Hedges 1996) and Mediterranean Islands (Schüle 1993) have also been documented through dated fossils, again following human colonisation.

Some quite sizeable islands are now bereft of native landbirds. This is true of Easter in the Pacific, Ascension and Trinidade in the Atlantic, and St Paul and Amsterdam in the Indian Ocean. These islands cover areas of 165, 97, 10, 8 and 54 km², enough to support landbirds, judging by similar-sized islands elsewhere. Bone remains confirm that some of them had landbirds only a few centuries ago, together with more seabird species than they have now. Again the most likely explanation of their current state is human action, through either deforestation or the introduction of rats or other predators. Before the arrival of the Polynesians, Easter Island was covered in forest, all of which was eventually felled, making it the only sizeable island in the Pacific now lacking both native forest and native landbirds. Yet other islands hold no landbirds because they are too small to support viable populations, or in some cases even a single territory. Many islands in the arctic and antarctic are inhospitable for landbirds and support only seasonal seabirds, an obvious consequence of climate.

Overview

The recent extinctions of oceanic island birds have thus occurred on a wide scale. They can be traced through the Pacific, Atlantic and Indian Oceans, and the Mediterranean and Caribbean Seas. They seem to have affected endemic species more than more widespread ones, and large (mostly flightless) species more than small ones. The emphasis on large species would be expected from human hunting habits, but may also be a bias of the fossil record, with identifiable remains from large species being more likely to have been preserved and found.

On the basis of bone material, the proportions of species on different Pacific islands known to have become extinct since human occupation can be estimated as follows (updating Olson 1990): 49% for New Zealand; 40% of non-passerines alone for New Caledonia; more than 80% for the main Hawaiian islands; 39% for the Marquesas, reaching 69% on one island; 80% on Mangaia in the Cook Islands; 78% on Huahine in the Society Group, and 43% on Henderson Island in the Pitcairn Group, which was colonised and then abandoned by Polynesians. To these figures can be added the 57%, 66% and 86% for the three Mascarene Islands (**Table 7.1**), at least 80% for St Helena and 100% for Ascension and Easter (see above). Because of the inevitable incompleteness of the fossil record, and the difficulty of distinguishing closely related species from some bone remains, most of these figures must be regarded as minima. Together, however, they provide ample testimony to the devastating impact of people on island faunas.

If we assume, as a conservative overall estimate, that half of all oceanic island bird species have been eliminated by human activity, then only 1,000–1,500 years ago there must have been well over 3,000 endemic species on islands worldwide, instead of the current 1,627. At least three previous estimates have been made in various ways for Pacific Islands alone, ranging from 800 through 1,500 to more than 2,000 species (Pimm *et al.* 1994, Steadman 1995, Cornutt & Pimm 2000). Hence, on this basis, in the absence of human intervention, more than 13,000 bird species would live in the world today, instead of the current 9,800; that is, about one-third more. The majority of the missing species are likely to have been lost in the last thousand years. Considering that estimates are increasing all the time, as more subfossil material becomes available and that the remains of many small birds will never be found or correctly identified, this also remains very much a minimum estimate. Moreover, any figure is of course based on avian taxonomy as it is now, and like any other current estimates it may be changed as a result of future taxonomic revision, including any resulting from DNA analyses.

Human-induced causes of extinctions

On some islands, entire vegetation types have been destroyed, especially low-land forest, with the consequent loss of the associated fauna. The most startling example was Easter Island, mentioned above, which was totally deforested, and now holds no native landbirds. However, destruction of only part of the forest or other habitat on a small island would be expected (from the species–area relationship) to cause some extinctions, even though some pristine habitat remained (Diamond 1972b, 1975, Case 1996). In this type of impact, there is a time lag between habitat reduction and species loss, as species disappear progressively over a period of decades or longer until some lower level is reached, more appropriate to the habitat areas remaining (Chapter 6). In the meantime, such species constitute the 'living dead', destined to disappear unless their prospects are improved by expansion of habitat or some other conservation action. They probably form a large proportion of the island species listed as threatened in Red Data lists.

Attempts have been made to estimate the numbers of past extinctions, or to predict the numbers of impending ones, from measures of the extent of island habitat recently destroyed. For example, Brooks *et al.* (1997) mapped the percentage of deforestation on the islands of the Philippines and Indonesia and counted the number of forest bird species found only on these islands. They then used the species–area relationship to calculate the number of species predicted to become extinct in the near future. These figures were strikingly similar to the numbers of endemic species classed as threatened in the latest Red Data Book (**Figure 7.1**). It seemed that the number of doomed species could be predicted from the extent of deforestation, but not their identities or their times to extinction.

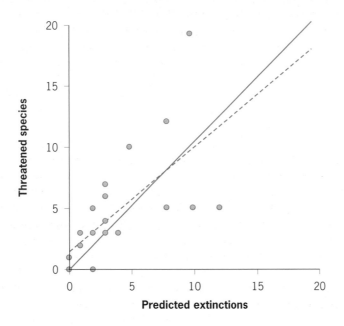

Figure 7.1   Number of bird species listed as 'threatened' in relation to the number expected to become extinct following current levels of deforestation on different southeast Asian islands. Each point refers to a different island. The regression between the points (dashed line) and the line where the two sets of numbers are equal (solid line) are not significantly different. By implication, the number of likely future extinctions correlates with the degree of forest destruction. From Brooks *et al.* 1997.

Tameness and lack of natural defences against human hunters could have accounted for the extinction of almost all large flightless birds, and many others that were good to eat. In addition, however, about half of all extinctions of island birds in the last 400 years have been attributed to introduced predatory mammals, mostly rats *Rattus* and feral cats *Felis catus*, but on some islands also Mongooses *Herpestes auropunctatus*, feral dogs *Canis familiaris*, foxes (*Vulpes* and *Alopex*), Stoats *Mustela erminea* and other mustelids, monkeys (*Cercopithecus* and *Macaca*) and feral pigs *Sus scrofa*. The more recent introductions of exotic birds to many islands may also have contributed to the extinction of some native species through competition or from the associated introduction of pathogens to which the native birds have no natural resistance (van Riper *et al.* 1986). These factors are still operating on most islands, with many remaining endemics in critically low numbers. No doubt more will disappear in the years to come.

With human help, rats have now reached more than 80% of the world's oceanic islands and island groups, though in many such groups some islets still remain rat-free (Atkinson 1985). The three species involved are the Brown Rat *Rattus norvegicus*, Black (or Ship) Rat *R. rattus* and Pacific Rat (or Kiore) *R. exulans*. All three species can climb, and eat eggs, chicks and full-grown birds up to their own weight, and sometimes larger. Together, they are held responsible for about 56% of those avian extinctions attributed to predation (King 1985). Their impact can be devastatingly rapid. For example, Black Rats reached Big South Cape Island, New Zealand, in about 1962. Within three years they had exterminated five species of native forest birds, including the last known population of the Bush Wren *Xenicus longipes* (Bell 1978). Similarly, within about 18 months of their reaching Midway Island in the Pacific in 1943, the ground-dwelling Laysan Crake *Porzana palmeri* and Laysan Finch *Telespiza cantans* disappeared (Fischer & Baldwin 1946, Johnstone 1985).

Feral cats can take larger birds than rats, and are held responsible for 26% of the avian extinctions attributed to predators (King 1985). Again, the effects can be rapid. In 1894, a lighthouse keeper took a pet cat to Stephen Island, New Zealand. Over some days the cat brought in 15 specimens of the previously unknown wren *Xenicus lyalli* which has not been seen since. This wren was notable in three respects. It may have had the smallest range of any bird, restricted to an island only 2.6 km^2 (one square mile) in area; it might have been the last surviving flight-less passerine; and thirdly, it was both discovered and exterminated by the same cat. Cats can be especially devastating to burrow-nesting seabirds that land on the ground surface at night. Soon after their introduction, cats were estimated to kill a million birds per year on the Kerguelen Isles (Pascal 1980).

Although the crucial predators of island birds are usually mammalian, this is not always so. The recent extinction of several forest species on the western Pacific island of Guam has been attributed to the accidental introduction (in the 1940s) and subsequent spread of the Brown Tree Snake *Boiga irregularis*, a mainly arbo-real, nocturnal predator which eats both eggs and adult birds, as well as small mammals and reptiles (Savidge 1987).

It is hard to assess in retrospect whether introduced predators caused all the extinctions attributed to them. For most species, the evidence is entirely circum-stantial: rapid decline in numbers following the known release of a predator, accompanied by observations of predation. Whereas on some islands the estab-lishment of a predator was the only obvious change, in others the effects of predators were confounded with other changes, such as forest clearance. How-ever, in a few species, measured predation rates were not sustainable: for example where prey breeding success was almost zero year after year (for examples, see Newton 1998a). Moreover, on several islets the control of predators by human efforts has reversed a population decline, or enabled a species previously elimi-nated from an island to thrive after reintroduction (for examples, see Forsell 1982, Veitch 1985, Griffin *et al.* 1988, Byrd *et al.* 1994; Chapter 14). These case histories can be regarded as replicated experiments, confirming the effects of introduced predators on island birds.

Introduced diseases have also had dramatic effects on some island bird species, which had no previous contact with the pathogens involved and hence no natu-ral resistance to them. On the Hawaiian Islands, mosquito-borne avian pox and malaria are thought to have eliminated roughly half the indigenous landbirds that were present in 1800, and affected the elevational distributions of others (Warner 1968). Mosquitoes were introduced to the islands in 1826, and continue to spread to even higher elevations. Pox and malaria may have always been present in the blood of migrant shorebirds wintering on the islands, and more recently in the blood of exotic birds released there by people, but without the mosquito vectors, these pathogens were very unlikely to infect native landbirds. Deductions from field observations were supported by experiments in which various Hawaiian birds in captivity were inoculated with malaria. Many infected individuals of endemic species died, whereas all the individuals of introduced species survived (van Riper *et al.* 1986). Later studies showed that honeycreepers were more susceptible than native thrushes (*Myadestes*) (Atkinson *et al.* 2001). However, levels of parasitism are not the only factor influencing the distributional patterns of Hawaiian birds, because habitat changes and introduced predators are also involved. Introduced pathogens have been found in the birds of other oceanic islands (see Peirce 1979 and Cheke *et al.* (2002) for Mascarenes; Harmon *et al.* 1987 for Galapagos), but how much they have contributed to extinctions is unknown (see Cooper 1989 for discussion of alien disease in some extant island species).

The three main causes of extinctions (habitat destruction, human hunting and predation/disease) are seldom species-specific, but vary in the proportion of an avifauna they affect at one time. Habitat destruction almost always affects entire communities; introduced predators and pathogens can affect anything from single species to whole communities; and human hunting affects mainly large species favoured as food or trophies. To these various extinctions may be added the linked ones, when species disappear as a result of the loss of others crucial to them. The main examples are mostly non-avian (e.g. the loss of plants after the demise of their specific pollinators, or the loss of parasites after the demise of their specific hosts). Such linked extinctions are little understood and almost certainly under-recorded.

Lessons from extinctions

The main lesson to be drawn from analysis of bone remains is that, before islands were colonised by people they had much richer and more diverse faunas than they have today. They contained a larger proportion of large and flightless birds than exist now, with some species occupying herbivore niches filled on continents by mammals. Human-caused extinctions seem preferentially to have included the strangest and most bizarre of island birds. Even on well-studied islands not all recently extinct species are likely to have been found, especially small ones whose bones do not readily fossilise. It is therefore impossible to know precisely how many bird species would now occur on different islands were it not for human impact.

Until recently, then, prehistoric extinctions obscured our perception of faunal development on oceanic islands, and remaining birds give a largely misleading idea of island bird communities. For example, the theory of island biogeography, which emphasises island size and remoteness in influencing species numbers, depends heavily on findings from greatly depleted faunas. It is uncertain how different these relationships would be if they could have been based on species lists compiled before human occupation. In addition, consideration of extant birds alone greatly underestimates the diversity of species that colonised islands, and the number of endemic lineages to which they gave rise. On the Hawaiian Islands, recently extinct lineages fall mainly into two categories, namely raptors (eagles, hawks, owls) or waterbirds (ibises, rails, geese, ducks), most of the latter having shifted to terrestrial habitats after colonising the islands (James 1995). In contrast, no passerine lineage is known to have become totally extinct despite the large number of species that were lost (of 67 passerine species in the known Holocene fauna of the main islands, 25 (37%) became extinct prehistorically and 15 (22%) following European settlement, **Figure 7.2**). In this way, differential extinctions downgraded the success of waterbirds and raptors in colonising the Hawaiian Islands and diversifying there, while it inflated the importance of passerines. Distribution patterns were also differentially affected, as both the surviving goose (*Branta sandwichensis*) and the surviving hawk (*Buteo solitarius*) now remain only on the largest island, whereas surviving passerines persist on all islands. On other island groups, extinctions may have had a different profile, depending on the eating habits of the people involved (whether hunters, fishers or farmers) and the types of alien animals they introduced.

Human-caused extinctions may also explain some unexpected distribution patterns. Of the largest pigeons of the genus *Ducula*, the only species to persist into the historic period in the Pacific are *D. goliath* from New Caledonia and *D. galeata* from a single island in the Marquesas. Yet bones from *D. galeata* have now been found on other islands in the Marquesas, on Mangaia in the southern Cooks, on Henderson Island and on Huahine in the Society Islands (Steadman 1989); and what was almost certainly this species was described from Tahiti on Cook's second voyage (Olson 1990). Remains of an even larger species of *Ducula* have been found on Wallis Island and this is probably the same species as on Lakeba in the Lau archipelago of Fiji; it may also be the same species as one of the two giant forms of *Ducula* recovered from midden deposits on Lifuka in the Tonga group. Evidently, the large species of *Ducula* were distributed throughout the eastern Pacific before the arrival of people. Apparent gaps in the distributions of megapodes on western Pacific islands are also being filled by palaeontological and archeological work.

Except for flightless birds, very few species were probably truly endemic to single islands, especially within archipelagos (Olson 1990). Even flightless species often had closely related representatives on other islands. Among the pigeons, distributions such as that of *Gallicolumba rubescens* found only on the small low islands of the Marquesas, and *Ducula galeata* found only on the high island of Nuku Hiva in the same group, may now be perceived as human-induced. Both of these species must once have occurred together throughout the archipelago, as shown by bones of both from archaeological deposits on Ua Huka, Hiva Oa and Tahuata. This example shows how the variable effects of extinction have altered the original biogeographical patterns; it raises questions over some of the checkerboard distribution patterns discussed in Chapter 14, which have been attributed to competition but could be due in part to differential extinctions.

Some other island birds have oddly disjunct distributions, which may also be explained by recent extinctions, but the evidence is not yet to hand. For example, the *Cyanoramphus* parrots occur only in the New Zealand region and New

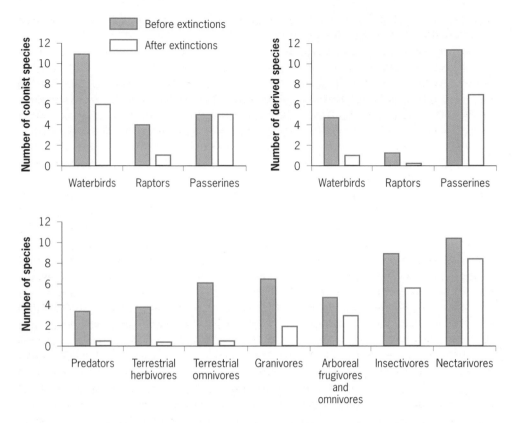

Figure 7.2 Differential extinctions among different types of birds in the Hawaiian Islands.

Upper left. Estimated number of initial colonist species, represented among living birds before human colonisation and now. Large proportions of the non-passerine lineages are now extinct.

Upper right. Estimated number of species present before human colonisation and now. A larger proportion of non-passerine than of passerine species are now extinct.

Lower. Estimated number of species of different feeding habits present before human colonisation and now. A larger proportion of species in some feeding groups (mostly of large body size) are now extinct than of species in other groups (mostly of small body size). Modified from James 1995.

Caledonia, except for two species known from the Society Islands, both now extinct. It is hard to imagine that no forms of *Cyanoramphus* have ever colonised the remaining islands of the Society Group or any of the many islands between the Society Islands and New Caledonia (Olson 1990). The current disjunct distributions of many other island birds are suspect on the same reasoning. Moreover, the realisation that many allopatric distribution patterns are probably unnatural calls for re-evaluation of the systematic relationships among some island birds. For example, knowing that *Ducula galeata* once occurred alongside *D. aurorae* on the same islands, it is no longer appropriate to consider the two as allopatric forms of a single superspecies (Steadman 1997).

The endemic or localised distributions that characterise many insular bird species may thus be more a consequence of human impact than of such factors as niche partitioning and competition which have often been assumed to influence the kinds and numbers of species on islands under natural conditions (Chapter 6). Much of the theory developed from studies of modern island avifaunas may need modification. It no longer seems sensible to try to interpret current island faunas in terms of theoretical concepts, such as species packing, niche breadths and faunal saturation, when these faunas are actually only distorted remnants of natural assemblages variously altered by human action.

Taking account of recently extinct birds also enables us to take a fresh look at speciation on islands, especially that resulting from multiple colonisation and adaptive radiation (Chapter 3). New Zealand is interesting because part of its ancestral avifauna might have been inherited direct from Gondwanaland, of which these islands once formed a part, while the rest was apparently derived by repeated cross-water colonisation mainly from Australia. Supposed examples of double invasions, with the earlier colonist mentioned first, include Weka *Gallirallus australis* and Buff-banded Rail *Gallirallus philippensis*, Takahe *Porphyrio mantelli* and Pukeko *P. porphyrio*, Black Stilt *Himantopus novaezelandiae* and White-headed Stilt *H. leucocephalus*, and New Zealand Robin *Petroica australis* and Tomtit *P. macrocephala*. Presumably because, for most of its geological history, New Zealand had two or three main islands, some *in situ* adaptive radiation occurred there, producing perhaps 2–3 species from each ancestral form, as in the wattlebirds (Callaeatidae). The main apparent exception is provided by the moas, with 11 recognised species of greatly differing body size, proportions and bill shapes (Cracraft 1976, Baker 1991), but it is uncertain how many ancestral forms might have been inherited from Gondwana. Moreover, for a period of six million years in the Oligocene, New Zealand existed as a string of small islands (Cooper & Cooper 1995), providing further opportunities for differentiation.

The Galapagos Islands where no Holocene bird species is known to have gone extinct, support three *in situ* radiations of birds, involving two species of vermilion flycatchers, four species of mockingbirds and 13 species and many local subspecies of Galapagos finches (Olson 1991, Chapter 3). These radiations appear to be very recent, the main departures from mainland forms being in the bill shapes of the finches and mockingbirds. Presumably the great number of islands in the archipelago is responsible for the diversity of finches. The fact that the subspecies, species and 'genera' of finches form essentially a continuum, with considerable hybridisation, attests to the youthfulness of this radiation which is unique among island birds (Olson 1991). There were apparently no flightless landbirds on Galapagos, the herbivore niche being occupied by tortoises.

Compared with the Galapagos, the recent avifauna of the more remote Hawaiian islands was much more diverse, with many endemic genera, and with the majority of the endemic species having diverged far more from their mainland ancestors. The most impressive radiation on the Hawaiian Islands involved the honeycreepers, of which 33 species are known from historical collections, and at least another 17 from subfossil remains, making more than 50 in total, all derived from a cardueline finch colonist species (Chapter 3). Including extinct forms, seven other Hawaiian groups underwent minor radiations, each with fewer than six species, including the thrushes (*Myadestes*), honeyeaters (*Moho* and *Chaetoptila*), owls (*Grallistrix*), crows (*Corvus*), flightless ibises (*Apteribus*) and two waterfowl lineages, the geese (*Branta*) and the dabbling duck-derived moa-nalos (three genera) (Fleischer & McIntosh 2001).

Of 21 independent lineages, the ancestors of 9–10 came from North America, four from other Pacific or Australasian Islands, 2–3 from Asia, and the rest from unknown source areas. The level of mt DNA divergence between Hawaiian and North American relatives in four lineages varies from nil to 10.3%. On these estimates, none of the Hawaiian taxa diverged from North American ancestors more than 6.4 million years ago; that is, well within the period of formation of the main islands. Compared with New Zealand, there has been much more *in situ* radiation on the Hawaiian Islands, in association with the greater number of islands.

Although large flightless herbivorous birds evolved in New Zealand, New Caledonia, the Hawaiian Islands, Madagascar and the Mascarenes, no such herbivores are known from other islands. Nor have flightless long-billed probing birds, such as kiwis and ibises, been discovered outside New Zealand, Hawaii and the Mascarenes. Aside from the Galapagos and Hawaii, Madagascar and New Zealand, few other adaptive radiations of passerine birds are recognised on islands, but this may be due to inadequate palaeontological exploration or to mistaken taxonomy through convergence (Cibois *et al.* 2001). The large high islands of Fiji and Samoa, for example, are essentially unknown in this respect, as is Vanuatu (Olson 1991). In many other island groups, the individual islands are either too small or too close to continents to support extensive *in situ* radiations. Nevertheless, modest numbers of intra-archipelagal speciations are known from the islands of Fiji, Marquesas, Societies, Philippines, West Indies and others (Diamond 1977a).

Most theories of island biogeography and speciation assume implicitly that conditions on islands have remained relatively stable over the past few million years, but it is clearly unrealistic to imagine that islands escaped the massive fluctuations in sea-level and climate of the past 2.5 million years, or that these changes had no effects on flora and fauna. The pre-human faunal changes on the Bahamas and St Helena were mentioned earlier, and following human colonisation the changes on all islands have clearly been massive.

On the question of stability, islands present us with a paradox. On the one hand, their biota are extremely vulnerable to the blitzkrieg extinctions inflicted by humanity. On the other hand, they have been stable enough to allow ancient relict endemics, such as the Kagu *Rhynochetos jubatus*, to survive for millions of years, in contrast to their disappearance from continents. The fossil evidence has indicated that similar assemblages of plants have persisted in the very small areas of many islands, such as St Helena, for many millions of years. Yet it would be practically impossible to find any equivalent area of a continent for which this was also true. This is because continents are more vulnerable to climatic effects, and any one area is continually being invaded by organisms from elsewhere. It is the isolation of islands, low arrival of outside species, low extinction due to species poverty and low competition, and the buffering against extreme climatic change provided by the ocean that favours long-term stability of the biota. Most of these features are changed by human intervention, rapidly destroying the results of millions of years of evolution.

Where does this leave the equilibrium theory of island biogeography, which has attracted so much interest in recent decades? Crucial aspects of equilibrium theory have proved hard to test, partly because of the scarcity of islands unaffected by human action, which can influence all the relevant parameters. In fact, only two of the predictions, that the numbers of species should increase with island area and isolation, have stood the test of time, albeit based on much depleted avifaunas. Evidence for an equilibrium number of species, and of turnover in species composition due entirely to natural processes, is almost non-existent. The assumptions that communities are in equilibrium, that features other than area and isolation are unimportant, that islands are equivalent with respect to the factors that influence immigration and extinction, that the identities and characteristics of species are unimportant, that colonisation and extinction can be treated as independent processes, and that *in situ* speciation can be ignored, are clearly untenable.

*   *   *   *

The main conclusion to be drawn from the work on recently extinct birds is the enormous impact of people and their commensals on island avifaunas: many island species have been eliminated altogether, while others have gone from some islands but remain on others. The second conclusion is that some previous explanations of the numbers, distributions and community structure of oceanic island birds are suspect, as they are probably based, on average, on only half the birds, and on an unrepresentative half at that. The same is also likely to be true of some other organisms. For the time being then, it seems sensible to treat with caution most of the ecological and biogeographical theory developed to explain the distribution patterns of island birds.

With such huge impact on island birds, the question inevitably arises whether early people could have had a similar impact on continental avifaunas as these people spread out of Africa over the remaining continents and nearshore islands. Few informed biologists now doubt the devastating impact that people had on the large mammal faunas of the world, and that many large scavenging and predatory birds dependent on those mammals disappeared at the same time. The loss of ostriches from southern Europe and southern Asia, where they occurred until late in the Pleistocene (Olson 1985b), could also be due to human hunting and egg eating. But could many smaller birds have been extinguished too? The fossil record from continents does not suggest that a large proportion of continental bird species died out during the Holocene (although some species did). What it does show, however, is that many large species occurred over much wider areas in the past than they do now (e.g. Tyrberg 1998), so that human impact may have greatly reduced their distributions. It is possible, therefore, that the sheer extent of their geographical ranges has so far saved most modern continental birds from extinction, coupled with the fact that they had evolved in the presence of mammalian predators, so had some natural defences against their human hunters.

RECENT INTRODUCTIONS

Human activity has not only reduced the numbers of bird species on islands; in the last 150 years many continental species have also been introduced to islands, so that their ranges have been greatly extended. Of nearly 1,200 documented successful introductions of birds from around the world, more than 70% have occurred on islands (Long 1981, Lever 1987); and on some islands, such as Hawaii and New Zealand, recently introduced landbird species (53 and 39, respectively) now outnumber native ones. Not only is it possible that introduced species hastened the demise of some indigenous species, they now form dominant elements in the avifauna of many islands, being among the commonest and most conspicuous species, especially in open lowland areas. Moreover, introduced species came from many different parts of the world, so through human agency, most oceanic islands now contain an amalgam of species, both native and exotic, which have only recently been brought into contact with one another. Both types are likely to have been exposed in the process to new competitors and new pathogens.

Islands now have much greater proportions of alien species in their flora and fauna than continents, and the effects of the aliens on native species seem to be greater. The idea that aliens establish more readily on islands than on mainlands stems partly from the facts that on islands native species are few, and certain types are lacking, leaving empty niches. Also, having evolved in isolation, native island species are assumed to be of low competitive ability, and easily displaced by aliens honed in the supposedly more exacting communities of mainlands. Neither of these ideas has been well tested, however, and the success of aliens on islands may be partly due to the prior removal of many native species by human-related action, and to the destruction and modification of native vegetation.

Questions about what factors facilitate the establishment of introduced birds on islands, and on the impact they have on native species, mostly focus on competition (for other aspects, see Chapters 14 and 16). If introduced species outcompeted and eliminated native ones or, conversely, if native species prevented introduced ones from establishing themselves, this would imply that competition was important in limiting species numbers and distributions on islands. Study is complicated because of the provision by human activity of new habitats in which most of the introduced birds live.

On the six main islands of Hawaii, more than 140 alien bird species in 14 orders have been introduced at one time or another to give the 53 species present now. These birds came from several different continents (Moulton *et al*. 2001). Like other introduced animals, introduced birds tend to follow a pattern of initial success in a restricted area, a period of quiescence with some spread, and then either a population boom with an accompanying spread to other habitats or other islands, or decrease and disappearance. On Hawaii, the Japanese Bush Warbler *Cettia diphone* remained confined to a small area for about 30 years, but then underwent a spectacular expansion in the late 1960s and 1970s, reaching several other islands. In contrast, the Varied Tit *Parus varius* was seemingly established on Hawaii until about 1960, then rapidly disappeared. The Red-billed Leiothrix *Leiothrix lutea* became abundant on all the main Hawaiian Islands in the 1940s and 1950s, but by 1995 had all but disappeared from Kauai and Oahu, and was declining elsewhere (Pratt *et al.* 1987). Some authors have speculated that competition may be involved in these changes, but again without convincing evidence.

Comparing different tropical and subtropical islands, Diamond & Case (1986) found that the success rate of bird introductions (defined as the percentage of attempted introductions that led to establishment) declined steeply with the species richness of the extant native avifauna It varied from 90% for Ascension, with no native landbirds, to 60–80% for Lord Howe, Bermuda, Rodriguez and Seychelles, which had 3–19 native landbirds, and to nil on Borneo (with about 430 species) and New Guinea (with 578 species). The chance of an invader establishing itself thus seemed to vary inversely with the numbers of species already present, as expected if competition were involved in preventing establishment. Secondly, the penetration of native forest by successfully introduced species also declined with the numbers of indigenous species present. For example, on Viti Levu, the largest island of the Fiji archipelago (with 48 native species), the introduced species are confined to human-modified open habitats. On Oahu in Hawaii, with ten native species, numerous introduced species are abundant in native forest, while Kuai is intermediate between Oahu and Viti Levu, both in species numbers (19) and penetration of forest by introduced birds (Diamond & Case 1986). Other studies revealed that a newcomer was less likely to establish itself if another species with similar morphology (and hence ecology) was already present (Moulton & Pimm 1983). All these findings would be expected if competition were important in hindering establishment, but none of these analyses took account of island area, and the degree of human-induced habitat modification or the possibilities of predation.

It would be wrong to think that no native island birds are able to live in man-made habitats. The suburban towns and large parks of tropical Pacific islands, while mainly populated by introduced species, also contain some native ones, such as the Many-coloured Fruit-Dove *Ptilinopus perousii* (Fiji, Samoa), Red-headed Parrot Finch *Erythrura cyaneovirens* (Fiji), Golden White-eye *Cleptornis marchei* (Saipan), and even the Fairy Tern *Sterna nereis* which nests in trees in downtown Honolulu (Pratt *et al.* 1987).

Almost certainly, introduced birds have had less influence on the native birds of islands than previously thought. From a study of the birds on 71 islands or island groups around the world, Case (1996) concluded the following:

(1) The number of exotic species gained by different islands was close to the number of native species lost through recent extinction. However, allowing for island area, both figures depended on the proportion of natural habitat that had been replaced by man-made habitat: the numbers of native species declined with loss of natural habitat, while the numbers of exotic species increased with expansion of man-made habitat.

(2) More than half of all known species extinctions on islands occurred long before exotics were introduced. Hence, most extinctions of native birds could not have been caused by competition with newcomers, although some might have been. Furthermore, the species lost from islands were often totally different from those gained. Hawaii, for example, lost many waterfowl, raptors, rails and forest nectar-feeders, but has gained mostly small seed-eaters and

insect-eaters of open country. Hence, most exotics have probably not greatly benefited from the loss of native species, although again some may have done so.

(3) Other factors that influence the ability of introduced species to establish themselves include the number of introduction attempts made and numbers of birds released each time, the size of the island (more exotic species established on bigger islands) and the remoteness of the island (more exotic species on remote islands that once had many endemic species).

(4) Exotics may find it more difficult to penetrate species-rich than species-poor communities, but this idea is hard to test on most islands because native species tend to remain in natural forest, while most introduced species prefer open habitats (as in their natural range). There are thus inherent differences in the habitat preferences of most species in the two groups.

In conclusion, competition has probably not been a major factor in causing the extinction of endemic island birds, or in preventing the establishment of exotic birds on islands modified by human activity. This is partly because the majority of the species introduced, in contrast to native birds, prosper in man-made habitats, which limits contact between the two groups. Competition could have played a bigger role in the replacement of one exotic species by another (Moulton & Pimm 1986), and in preventing native species from colonising man-made habitats, but this again is largely speculation. In general, it is hard to assess what, if any, effect introduced birds have had on native ones.

SUMMARY

Many endemic species have disappeared from oceanic islands in the last 1,500 years, at different dates in each island group, but in each case following human colonisation. The main likely causes include habitat destruction and overkill by human hunters, and mortality from introduced rats and other predators, or from introduced pathogens. Before humans arrived, island avifaunas often contained at least twice as many species as they do now, including many extraordinary endemics. Extinctions have been more frequent among larger and flightless forms, so that species still extant are an unrepresentative subset of the original avifauna. Surviving native species are misleading in terms of their numbers, diversity and distribution patterns. Some islands have lost all their native landbird species since human colonisation.

Several theories to explain the distributions of island birds, and the numbers and types of species on particular islands, were developed on the basis of depleted avifaunas. It is uncertain how much these same ideas would hold if recently extinct species were included.

Many continental bird species have been introduced to islands by people in the last 150 years. Such species live mainly (but not entirely) in man-made habitats, while native species live mainly (but not entirely) in remaining natural ones. Competition with introduced species may have contributed to the demise of some native species, and to the restriction of others to natural habitats, but on both counts the evidence is generally weak.

Northern Fulmar *Fulmarus glacialis*, a pelagic seabird of the North Atlantic.

Chapter 8
Seabirds

Because most seabirds do not follow the same biogeographical patterns as other birds, they are best discussed in a separate chapter. For present purposes, such birds are taken to include the 320 or so species in the families Laridae (skuas, skimmers, gulls, terns, auks), Phaethontidae (tropicbirds), Sulidae (boobies, gannets), Phalacrocoracidae (shags, cormorants), Fregatidae (frigatebirds), Spheniscidae (penguins), Gaviidae (loons) and the order Procellariiformes (petrels, shearwaters, albatrosses) **(Table 8.1)**. In five of these eight groups all the species are marine. In the Phalacrocoracidae and Laridae some species breed inland mainly beside fresh water, as do all five species of Gaviidae, but most go to sea in winter. Excluded from the list are the several species of seaducks and grebes which winter on the sea but are not usually classed as seabirds, the pelicans (Pelecanidae) because only two species are marine, and also the waders (mainly Scolopacidae), although three phalarope species winter on the sea and many other species on shorelines. It will be evident by now that the distinction between seabirds and other birds is not clear cut.

Table 8.1   Seabirds considered in this chapter, extant taxa only, following the classification of Sibley & Monroe 1990.

	Genera	Species
Family Laridae	27	129
Subfamily Larinae:	15	105
Skuas, Jaegers	1	8
Skimmers	1	3
Gulls	6	50
Terns	7	44
Subfamily Alcinae: Auks, Murres, Puffins	12	24
Family Phaethontidae: Tropicbirds	1	3
Family Sulidae: Boobies, Gannets	3	9
Family Phalacrocoracidae: Cormorants, Shags	1	38
Family Fregatidae: Frigatebirds	1	5
Family Spheniscidae: Penguins	6	17
Family Gaviidae: Loons	1	5
Family Procellariidae	23	114
Subfamily Procellariinae: Gadfly Petrels, Shearwaters, Fulmars	14	79
Subfamily Diomedeinae: Albatrosses	2	14
Subfamily Hydrobatinae: Storm Petrels	7	21
Family totals	63	320

Many authorities prefer to distinguish the Laridae and Alcidae as separate families, and four families within the Procellariiformes, namely Diomedeidae (albatrosses), Procellariidae (fulmars, gadfly petrels, prions, shearwaters and larger petrels), Pelecanoididae (diving petrels) and Hydrobatidae (storm-petrels). For other recent arrangements and species lists, see Warham 1990 (for Procellariiformes only), Bourne & Casement 1996 and Brooke 2001.

Although sea water covers more than two-thirds of the earth's surface, seabirds form only about 3% of the world's bird species. However, some of them are large and numerous, and in terms of their collective biomass they probably outweigh landbirds. They get all or much of their food from the sea, mainly fish, crustacea or cephalopods; and all have enlarged nasal glands through which they excrete brine, thereby shedding excess salt. Species vary in their dependence on land: some roost on *terra firma* each night, while others remain for months or years on end at sea, but all must return to a firm substrate to breed. Some species (notably Sooty Terns *Sterna fuscata* and frigates) hardly ever settle on water, and remain almost continuously on the wing when away from land. Most pelagic species fly by night as well as by day, and a greater proportion of seabirds than of landbirds feed mainly at night, taking advantage of the nocturnal upwards migration of plankton and its predators to near the surface.

The aims of this chapter are to outline the main distribution patterns of seabirds and the factors that influence them, and act as barriers to spread. Some seabirds have linear distributions along coastlines, obtaining their food from shorelines and from relatively shallow coastal waters over continental shelves. Many gulls, cormorants and auks are in this category. Other seabirds have wider pelagic distributions, nesting mainly on oceanic islands **(Figure 6.1)**, and foraging over large areas of open deep-water sea. The majority of petrels are in this category.

The distribution of seabirds at sea is governed by the latitudinal marine zones to which they are adapted, the distribution of food within those zones and, in the breeding season, by the locations of nesting places and distance flown from the nest. One of the most striking features of seabird distributions is the way that related forms tend to replace each other in different marine zones (e.g. Murphy 1936, Bourne 1963, Warham 1990), in the same way that related landbird species often occupy different biomes. In both groups, such allopatric distributions might now be maintained by competitive exclusion, coupled with reproductive isolation (Chapter 3). Within marine zones, the locations of food depend on oceanic conditions, notably on the whereabouts of currents, upwellings, eddies and other features that concentrate suitable prey animals within reach. In contrast, choice of breeding locations seems governed more by the availability of safe nesting cliffs, islands and promontories, which (for many species) need not necessarily be near to the feeding areas. One important distinction between pelagic seabirds and landbirds is the distance travelled for food. The majority of landbird species when breeding obtain their food from within restricted territories, or travel at most a few kilometres from the nest (although some vulture species and others travel several tens of kilometres, see later), but seabirds regularly travel at least several kilometres from the nest and some travel tens, hundreds or even thousands of kilometres.

Such long-distance feeding is necessary in many seabirds because suitable nesting places may be hundreds of kilometres from feeding areas. In addition, the nesting places sometimes hold enormous concentrations of birds which, in order to gain enough food, must range over huge sea areas. The food-sources themselves are often sporadically distributed, present at different places at different times. In contrast, landbirds nearly always find nest sites within easy reach of their feeding areas, so can breed in a more dispersed manner. Typically, in the seabirds that feed at long distances, one parent forages away for days or weeks at a time, while the other remains at the nest until the chick is large enough to be left alone. In general, larger species tend to forage further from their nesting colonies than small ones, and, within species, foraging distances vary with stage of the nesting cycle, as well as with the distribution of food. In the majority of species, individuals do not breed until they are several years of age, and, among petrels and others, birds may spend their early years at sea without making landfall. For this reason breeders and non-breeders often tend to occur in mainly different areas. There are also seasonal shifts in distribution associated with migration and with dispersal, as birds leave their nesting places after breeding.

The most pelagic of all seabirds are the Procellariiformes (petrels). Their capacity to exploit distant food-supplies is helped by: (1) their lower body temperatures (38°C versus 41°C in other birds) and thus lower energy demands; (2) their great ability to lay down subdermal fat and stomach oil, both of which can be used when food is short, enabling them to go for long periods between meals; (3) their ability to conserve energy in flight by dynamic and slope soaring and other devices; and (4) their ability to feed by day or by night, helped perhaps by their well-developed olfactory sense; and (5) the ability of their eggs to resist chilling and their chicks to become torpid when short of food (Warham 1996). Most petrel species are fast fliers and some are adept divers, but all walk with difficulty.

THE MARINE ENVIRONMENT

Sea water covers about 71% of the earth's surface, including some 61% of the northern hemisphere and 81% of the southern. This amounts to 361 million km^2 compared with 149 million km^2 taken up by land and freshwater. In contrast to land areas, marine water areas are interconnected, with continuous round-the-world seas lying both north and south of the main continents, as well as between them. They were even more connected before the closure of the Panama gap some 3.5 million years ago separated the Atlantic from the Pacific Ocean at tropical latitudes. Not unexpectedly, therefore, seabirds and other marine organisms do not follow the same geographical distribution patterns as land-based organisms. Marine faunal regions, established on the basis of submarine life, are related primarily to sea water features and are broadly latitudinal in distribution. Nine main sea water zones (or life zones) are distinguished, classed as: (1) arctic; (2) sub-arctic; (3) northern temperate (boreal); (4) northern subtropical; (5) tropical; (6) southern subtropical; (7) southern temperate; (8) sub-antarctic; and (9) antarctic (**Figure 8.1**, Bourne 1963, Briggs 1974, Rass 1986). These zones can be distinguished mainly by water temperature and density, but they also differ in salt, nutrient and oxygen content. In general, colder waters contain more nutrients and more oxygen, and therefore support more life than warmer ones, although pack-ice covers large parts of the arctic and antarctic zones, especially in winter. In places the fronts between the different zones are remarkably sharp, and within a few tens of metres water temperatures can change by 10–15°C, as warm and cold waters abut and sometimes flow in opposite directions.

Although broadly latitudinal, the distributions of these sea water zones are also affected by surface currents, which in turn are controlled by wind directions. Because the major oceanic winds flow clockwise in the northern hemisphere and anticlockwise in the southern hemisphere, owing to the Coriolis force, the different sea water zones are shifted accordingly. This is most marked in the eastern Pacific where, because South America extends far south into cold temperate and sub-antarctic waters, a cold (Humboldt) current is deflected far northwards up the western coast of that continent, veering westward in a long tongue to envelop the Galapagos Islands which lie 900 km out to sea. A similar cold (Benguela) current runs up the west side of southern Africa. Correspondingly, in the northern hemisphere, a cold (California) current runs down the west side of North America and a similar cold (Canary) current runs down the west side of northern Africa.

The broadly latitudinal pattern in the distribution of sea water zones and their associated birds parallels the latitudinal distribution of major vegetation types and their birds on continental land areas. Similarly, the local extensions of high-latitude seabirds to low latitudes that occur in association with cold currents parallel the extensions of arctic and boreal birds to lower latitudes that occur in the cold climates of high mountain ranges.

Dependence on specific sea water zones makes some seabird species extremely susceptible to changes in currents and associated water temperatures, which can suddenly cut off their food-supply, causing massive mortality, movements and breeding failure. The most significant of such events is the El Niño–Southern

Oscillation (ENSO)[1] which begins in the Pacific but whose effects spread world-wide. In particular, this event reduces the flow of the Humboldt current up the west side of South America. The rich cold waters are then replaced by warm nutrient-poor tropical ones, and seabirds die in their millions. Such events occur about every 5–7 years, but seem to have become more severe in recent times, the events of 1982–83 and 1997–98 being extremely marked. They particularly affected the seabirds of western South America and the Galapagos Islands, which had insufficient time to recover from one event before they were affected by the next. Similar periodic failures of food-supply, mostly associated with temporary changes in current systems, also affect seabirds in other cold-water regions. As seabirds are long-lived, individuals can normally expect to experience several such events during their lifetime.

Some seabird biologists, noting the association of different seabirds with particular marine zones, have used sea surface temperatures (SST) to define bird distributions. For example, in the North Pacific, Kuroda (1991) plotted the temperature preferences of 36 petrel species on a monthly basis. He found that Mottled Petrels *Pterodroma inexpectata* and Short-tailed Shearwaters *Puffinus tenuirostris* tolerate SSTs up to 2° and 3°, respectively, Sooty Shearwaters *P. griseus* up to 4°C, Flesh-footed Shearwaters *P. carneipes* and Providence Petrels *Pterodroma solandri* to 7°C, and Buller's Shearwaters *Puffinus bulleri* and Kermadec Petrels *Pterodroma neglecta* to 9°C. Warm water species include Cook's Petrels *P. cookii* to 16°C, Juan Fernanadez Petrels *P. externa* to 17–20°C, Stejneger's Petrels *P. longirostris* to 18°C, Black-winged Petrels *P. nigripennis* to 20°C and Pycroft's Petrels *P. pycrofti* to 24°C. The birds are presumably not responding to water temperature as such, but to the food-organisms found in these waters. Similar findings have emerged from studies of other seabird communities elsewhere (e.g. Ainley *et al.* 1994), and in particular regions, unseasonal changes in SST may be accompanied by unusual visitors, as when mass immigrations of Black-vented Shearwaters *Puffinus opisthomelas* off California coincide with influxes of exceptionally warm water (Helbig 1983). Similarly, during El Niño events in the Pacific, when cold water is replaced by warm, many seabirds move to higher latitudes in search of cold waters and abandon their breeding areas, the survivors returning when conditions get back to normal. These findings again emphasise how closely, in the open sea, seabirds are tied to their particular 'habitats' and associated food-supplies. In general, temperature gives stronger correlations with seabird distributions than salinity or water density but all three tend to vary in parallel.

The northern pack-ice zone forms a major feeding area for the Northern Fulmar *Fulmarus glacialis*, Ivory Gull *Pagophila eburnea* and Ross's Gull *Rhodostethia rosea*. The southern pack-ice covers a much bigger area than the northern, and forms a major feeding area for the Antarctic Petrel *Thalassoica antarctica*, Snow Petrel *Pagodroma nivea*, Southern Fulmar *Fulmarus glacialoides*, Emperor Penguin *Aptenodytes forsteri* and Adélie Penguin *Pygoscelis adeliae*. All these species depend on ice-associated prey. Except at the outer edges, both pack-ice zones lack shearwaters

---

[1] *ENSO sits on a longer term cycle of sea-temperature changes, called the Pacific Decadal Oscillation in which roughly ten years of relatively warm sea temperatures in the central ocean are followed by a similar period of cooler ones.*

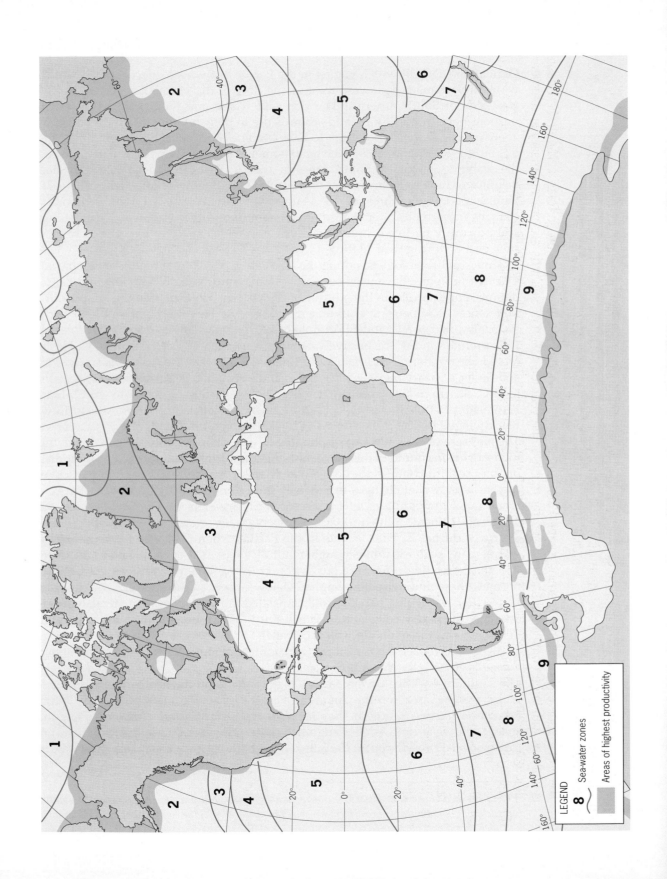

LEGEND

∿ 8 Sea-water zones

▨ Areas of highest productivity

Figure 8.1 Latitudinal biogeographical sea water zones. Areas of exceptionally high productivity are shaded. The northern and southern boundaries of tropical faunas appear to be most closely related to the 23°C isotherm for the coldest month of the year. Sea water freezes at −1.9°C. Modified from Bourne 1963, Ashmole 1971 and Rass 1986.

(1) *The arctic zone* embraces the teeming seabird stations of arctic Canada, Greenland and arctic Russia. The Arctic Ocean is shallow, with large continental shelves, and for much of the year ice-covered. Seabirds are present mainly in summer, although openings in the ice allow some to remain in winter, notably the Ivory Gull *Pagophila eburnea*.

(2) *Sub-arctic zone.* The most numerous seabirds in this region are various species of auks, and the most important seabird foods are capelin, clupeoids and sand-eels.

(3) *The northern temperate (boreal) zone* is shallow in many regions, which facilitates wind-generated mixing. Many seabirds of this zone concentrate to feed at the shelf break or at mid shelf tidal fronts. The water is cold (3–4°C), wind-swept and rich, supporting large fisheries and large seabird populations.

(4) *The northern subtropical zone* is less rich and holds fewer breeding seabirds than more northern zones.

(5) *The tropical zone* extends between the seasonally moving limits of the 23°C isotherms, north and south of the equator, rising to 29°C near the equator. Of generally low productivity, in vast areas of the tropics the main seabird foods are squid and flying fish, and the dominant seabirds are boobies, terns, frigatebirds, tropicbirds and tropical Procellariiformes.

(6) *The southern subtropical zone* contains some cold-water areas within it (notably the Humboldt and Benguela currents) which support locally endemic species and others found in colder more southerly latitudes. The warmer areas share some species with the tropics.

(7) *The southern temperate zone* supports large numbers of seabirds, especially Procellariiformes.

(8) *The sub-antarctic zone* also supports large numbers of seabirds, especially Procellariiformes and penguins.

(9) *The antarctic zone* is richer in phytoplankton than any comparable area. Under the intense and almost continuous sunlight of the short antarctic summer, productivity is enormous. Crustacea such as krill are the most numerous marine herbivores and in places average 30,000 individuals per cubic metre of surface water, reddening the sea over many square kilometres.

and albatrosses, possibly because of the difficulty for these birds of flight and take-off in calm waters. Even loose ice dampens waves, and in areas with many icebergs water can be as still as glass.

Needless to say, within each major sea water zone, areas vary greatly in their importance for seabirds.[2] In the North Sea, between Britain and continental Europe, 20 areas have been identified as especially rich in birds, on the basis of surveys during 1979–94. The top six sites, which comprised less than 5% of the total sea area, held more than 80% of the birds (Skov *et al.* 1995). These areas were already known as important to the fishing industry.

MARINE PRODUCTIVITY

Seabird numbers are greatly affected by the productivity of the seas. Almost all organic energy that sustains marine life is produced in the surface waters, and almost all organisms that live below this euphotic zone depend on sinking organic matter from above. Primary production depends on photosynthetic microscopic green algae (phytoplankton), and is limited mainly by light values and by nutrient availability (especially nitrates and phosphates), with temperature and salinity playing a lesser role. The phytoplankton are eaten by small herbivores, mostly crustacea, and these in turn by one or more tiers of carnivores up to the larger seabirds. The productivity of marine areas can be assessed in the same way as that of land areas, as 'net primary productivity', which is a measure of the rate at which solar energy is converted to plant tissue, typically expressed as dry mass produced per unit of surface area per year. On this basis, the sea emerges as generally less productive than the land; but in some sea areas, notably around estuaries and coral reefs, net primary production can be as great as in tropical rain forest (the richest of land habitats) and much greater than the most fertile farmland. Marine productivity is also measured using variants of the carbon-14 technique, and is usually expressed as grams of carbon fixed per cubic metre below the surface per year; and in recent years, it has been mapped worldwide from satellite images that reflect chlorophyll density, giving pictures of both spatial and seasonal changes. It is not that carbon or chlorophyll are important as such, but that they reflect phytoplankton density, on which the whole web of marine life ultimately depends. In some regions, productivity has also been assessed by measuring densities of plankton or other small organisms directly (e.g. Jespersen 1924).

Light penetrates the sea to a depth that varies with the turbidity of the water, seldom more than 50 m. A large standing crop of phytoplankton reduces the depth to which photosynthesis can occur. At high latitudes, seasonal variation in incident radiation causes marked seasonal fluctuations in production, but closer

---

[2]*Much research has been concerned with the distribution of seabirds at sea and, for general reviews, see Wynne-Edwards (1935), Bourne (1963, 1980, 1982), Ashmole (1971), Brown (1980), Eakin et al. (1986), and Hunt (1990, 1991). Atlases of seabird distributions in various areas include those of Watson et al. (1971), Gould et al. (1982), Powers (1983), Brown (1986), Pitman (1986), Harrison et al. (1989), Webb et al. (1990), Morgan et al. (1991) and Tickell (1993).*

to the equator radiation is more constant through the year, and is not normally limiting. Surface waters are usually poorer in nutrients than are subsurface waters, so a major factor affecting the availability of nutrients to surface plankton is the extent of vertical mixing. This is usually achieved by wind (and hence currents), but in tropical areas reduced wind action greatly lessens the upward flow of nutrients, and accounts for the general low productivity of warm tropical seas, which, although continuous through the year, is perhaps only 1% of that in temperate coastal waters (Wynne-Edwards 1985). The dearth of living organisms in most tropical seas makes the water clear and blue, deepening the euphotic layer.

In general, and in contrast to most land habitats, productivity in the sea is lowest in tropical and mid latitude regions (but with local exceptions), and highest in high latitudes and around continental shelves (including some at low latitudes), especially where rivers provide additional nutrient input. Thus, areas of high productivity occur throughout northern sea areas, especially near coasts, around antarctic coasts, and along some stretches of coast in western and southeastern South America, much of western Africa, southeast Arabia, northwest India, some southeast Asian coasts, around north Australia–New Guinea and New Zealand **(Figure 8.1)**. Among the more productive regions, the North Atlantic is generally less productive than the North Pacific, and both are less productive than parts of the Southern Ocean. Areas of lowest productivity occur mainly in the centres of the Pacific, Atlantic and Indian Oceans. These regional variations in productivity are reflected in regional variations in seabird abundance.

In some places, at both high and low latitudes, water currents flowing towards the equator along the steep margins of continents produce areas of upwelling, which bring a continuous stream of nutrients to the surface that are then dispersed by horizontal flow, creating a 'divergence'. The upwelling produces areas of very high productivity along certain westward-facing coastlines, leading in turn to an abundance of marine life, including seabirds. As mentioned above, the most famous of these areas lie off Peru (the Humboldt current), off southwest Africa (the Benguela current), off California (the California current) and off northwest Africa (the Canary current); but others occur off other western coasts that have narrow continental shelves. On some of these coasts, the upwelling that generates the high productivity occurs year-round, while on others it is seasonal, depending on winds. Off the coast of Arabia, the upwelling moves around in association with the monsoons. In addition to the main upwelling areas shown in **Figure 8.1**, other shorter-term or weaker upwellings occur locally in many other areas, especially near the edges of continental shelves and around islands and seamounts. The equatorial counter-current systems also provide important feeding areas, and are crucial for birds nesting on some islands such as Ascension (Bourne & Simmons 2001).

Together, such areas of exceptionally high productivity form only a small proportion of the total ocean area, but all have carbon production rates up to ten times greater than the average for the open ocean. They therefore have rich fisheries and some spectacular seabird concentrations. Other factors that result in vertical mixing include the shape of the sea bed, unequal salinity, convection currents around islands or ice, and tidal streaming in association with headlands, straits, shallows, estuaries and other irregularities of shorelines. These various features give rise to local upwellings, eddies and vortexes, wind streaks and other turbulence, all of which can bring nutrients and food organisms to the surface. The same occurs at fronts between different oceanic water masses.

Winds can also drive two currents together, which forces the denser (colder) water mass to flow beneath the lighter and create a 'convergence'. Such movements are vast but slow. They cause floating objects to stop when they reach the sinking front. The surface at a convergence is therefore marked by a long line of flotsam and oil, and immediately below the surface is a similar pile-up of plankton, too buoyant to sink. This concentration of food attracts fish and squid, which in turn attract birds. Many such convergences are local and transient, being produced by the swirls where currents meet. They occur in all seas, attracting many seabird species, but their importance diminishes at high latitudes where food is generally more available. In some tropical waters, such lines of convergence provide the only places in thousands of square kilometres of sea where birds can pick up an easy meal. Other feeding opportunities come when predatory fish and dolphins drive suitable food-items near to the surface (notably flying fish), or at night when plankton migrates to the surface. Spectacular feeding concentrations occur at so-called 'baitballs', swarms of tiny fish concentrated and driven to the surface by bigger fish, where they become temporarily available to birds. The sea surface appears to boil at such points, rapidly attracting birds from all directions, but within minutes the feast is over and the birds disperse.

Another factor that may increase local productivity is the enrichment associated with the seabird colonies themselves. The concentrations of birds around islands means that many faeces are dropped directly into the surrounding sea, but in rainy climates continuous run-off of guano-enriched rainwater also occurs from land to sea. In addition, many seabirds are attracted to fishing vessels, feeding on fish that escape from the nets or on offal thrown overboard — a relatively new feeding opportunity in the evolutionary history of such birds. Individual ships might attract thousands of birds during fish-cleaning operations. Nowadays some species spend so much time following fishing boats that it is hard to determine their natural feeding areas (Bourne 1982).

The marine environment may be further divided into the neritic zone, overlying continental shelves, and the oceanic zone further out. As mentioned, neritic waters nearly always have greater primary productivity, and support more life than adjacent oceanic areas. In the comparatively shallow neritic zone, nutrients are more readily returned to the surface by tidal and other currents, and are also continually added from rivers. Food for birds is often abundant over continental shelves, not only at the surface, but in mid water and on the bottom which some species can reach by diving (some penguins can dive to more than 100 m, see later). Food for some species also tends to concentrate at the shelf edge, where the sea bottom plunges abruptly downwards, deflecting deep nutrient-laden currents upwards. Other food concentrates along the tidal fronts in mid-shelf areas.

With marine productivity dependent mainly on physical processes, seabird distributions can be examined at different levels of spatial scale. Current patterns and different water masses influence bird distribution patterns on large scales of thousands of kilometres; upwellings and convergences influence distributions on scales of hundreds or tens of kilometres, while more local events influence distributions on scales of kilometres or metres.

SEABIRD DISTRIBUTIONS

Latitudinal zones

For the most part, seabird distributions match the broadly latitudinal distributions of sea water zones. This zonation is apparent worldwide, but is most clearly seen in the southern seas where no large land areas perturb it **(Figure 8.1)**. In the northern hemisphere, water tends to circulate around the oceans, to some extent disrupting the latitudinal zonation. The largest numbers of species are found in the tropical zone and the fewest in the coldest arctic and antarctic zones **(Figure 8.2)**. However, procellariiform species are much more numerous in the southern than in the northern seas, while auks are restricted to the northern seas and penguins to the southern.

It is in the cold southern ocean (zone 7), with its wide productive waters and scattered islands, that pelagic seabirds reach their maximum abundance in terms of species and individuals. The large populations of procellariiforms in this region arise partly because the shortness of their food webs allows a higher proportion of the total productivity to reach the upper trophic levels than obtains in other seas (Warham 1996). In addition, many southern seabirds have been less affected by human impact than those in the northern hemisphere. Many species eat Krill *Euphausia superba* which is supposedly the most numerous animal species on earth, with a total biomass exceeding that of the world human population. In the last 200 years, southern seabirds may have benefited from the removal by human hunting of a large proportion of the krill-feeding whale and seal populations. This reduction in competition could have enormously increased the food available to seabirds. Krill tends to be most abundant in the years when sea ice is most extensive.

Overall, some 81% of all seabird species breed in no more than two of the nine sea water zones, another 17% extend to three or four zones, and less than 2% extend to more than four zones **(Figure 8.2)**. The most widely distributed seabird species latitudinally include the White-faced Storm Petrel *Pelagodroma marina*, Roseate Tern *Sterna dougallii*, Little Tern *S. albifrons*, Caspian Tern *S. caspia* and the Great Cormorant *Phalacrocorax carbo*, all of which breed at sites in five sea water zones. In general, species of Larinae and Phalacrocoracidae show wider latitudinal distributions, and breed in a larger number of sea water zones, than species in

Figure 8.2   (See pages 224–225). Left. Numbers of seabird species known to breed in each of nine marine zones. 1—arctic; 2—sub-arctic; 3—northern temperate; 4—northern subtropical; 5—tropical; 6—southern subtropical; 7—southern temperate; 8—sub-antarctic; 9—antarctic. The largest numbers of species breed in the tropical zone, and the fewest in the arctic and antarctic zones.

Right. Numbers of seabird species known to breed in different numbers of sea water zones. Some 81% of all species are confined to one or two zones, 17% extend to three or four zones, and less than 2% to more than four zones.

Spheniscidae include all penguins, Procellariidae all petrels and other tubenose birds, Larinae all coastal and marine gulls and terns, Alcinae all auks, and along with Phalacrocoracidae (shags and cormorants) are given Phaethontidae (tropicbirds), Sulidae (gannets and boobies) and Fregatidae (frigates). Patterns calculated from distribution maps in Harrison (1983).

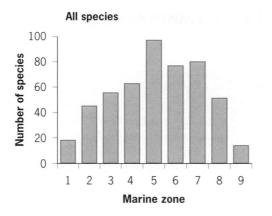

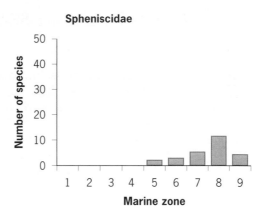

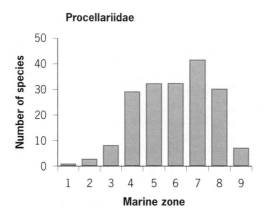

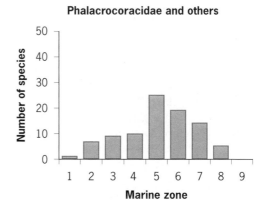

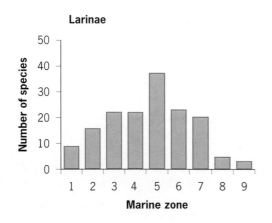

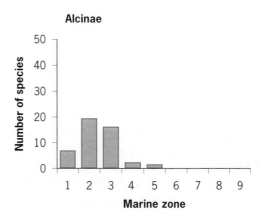

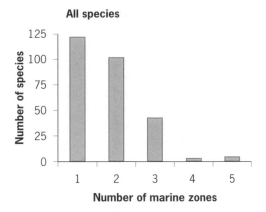

All species

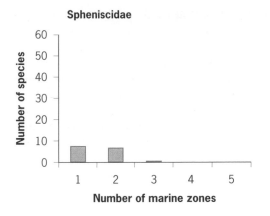

Spheniscidae

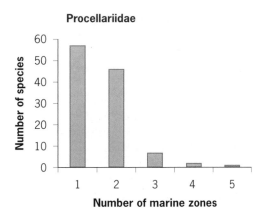

Procellariidae

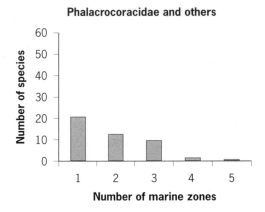

Phalacrocoracidae and others

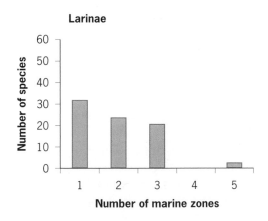

Larinae

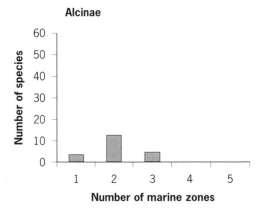

Alcinae

other families **(Figure 8.2** and **8.3)**. Some species are represented in a particular sea water zone by only a single colony, while others are represented in the same zone by a large number of colonies scattered around the world. However, because some species commute hundreds of kilometres between their nesting and foraging areas, care is needed in ascribing pelagic species to particular ocean zones solely from the locations of their breeding sites. It is where they feed that matters. Like landbirds, moreover, many seabirds migrate to spend the non-breeding season well away from their nesting areas, some in the opposite hemisphere; they may occupy a much wider geographical range then than when breeding.

Species whose breeding ranges centre near the equator have the greatest potential for wide latitudinal distributions because they could in theory extend both northwards and southwards as far as the ice. In contrast, species whose ranges centre on high latitudes are by definition constrained in latitudinal distribution. However, within different groups of seabirds, the relationship between mid latitudinal range and extent of latitudinal range is evident only in the Phalacrocoracidae and Larinae, some species of which have very wide latitudinal distributions **(Figure 8.3)**. The implication is again that factors other than simple sea space limit the latitudinal ranges of most species. In contrast to landbirds, seabirds show no obvious tendency for latitudinal range spans to increase with latitude in either the northern or the southern hemispheres (Chapter 1).

Some species that breed over a wide span of latitude show curiously staggered longitudinal distributions, breeding in one sea water zone in one part of the world and in another zone in another part of the world far to the east or west. For example, the Roseate Tern *Sterna dougallii* breeds in the northern hemisphere on both sides of the Atlantic and in southeast Asia, but in the southern hemisphere only around the Indian Ocean and Australasia. Nowhere does it extend as a breeder fully down an almost continuous coastline, such as eastern North and South America or western Europe and Africa. Similarly, the Little Tern *Sterna albifrons* breeds north of the tropics in many areas but south of the tropics only in Australia. Such longitudinally staggered breeding distributions seem rare in landbirds, but the Peregrine *Falco peregrinus*, Barn Owl *Tyto alba* and Short-eared Owl *Asio flammeus* provide examples. In both groups, they could arise because (1) ecological conditions suitable for such species are found only in widely scattered parts of the globe, which the species themselves have reached by dispersal; (2) such species were once more widespread, but have disappeared from many regions, through natural or human agency; or (3) competition from other populations prevents their establishment. Thus, some of the 'empty' areas in the southern hemisphere are used in the austral summer by migrants from the northern hemisphere and others are occupied year round by closely related forms.

In contrast to landbirds, many more seabird species breed in the southern than in the northern hemisphere **(Figure 8.2)**. This is linked partly, perhaps, with the greater total sea area in the southern hemisphere and with the many more isolated islands or island groups where nesting occurs, and where speciation could have occurred in the past. The difference is most marked in the Procellariiformes (petrels), as some 70 species breed in the southern hemisphere, compared with 30 in the northern. Moreover, some types of seabirds (notably penguins Spheniscidae and diving petrels *Pelecanoides*) are found only in the southern hemisphere while

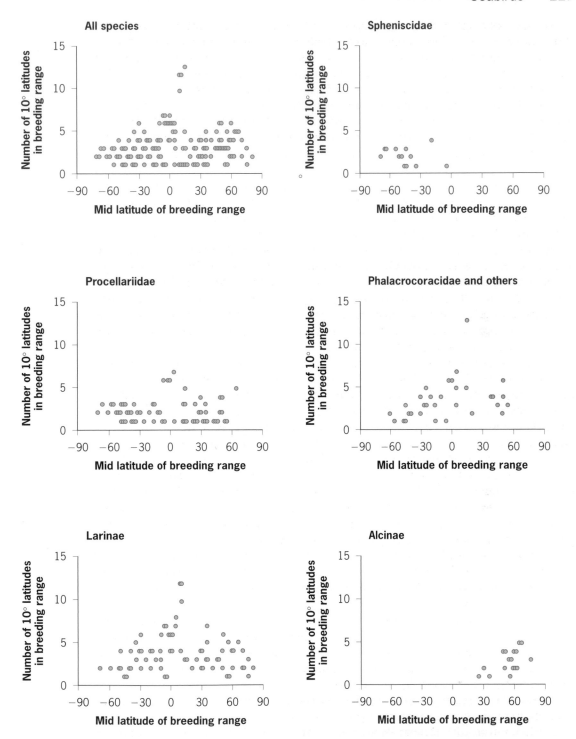

Figure 8.3   Relationship between full latitudinal extent of breeding range and mid latitude of breeding range in different seabird species. Those species whose breeding ranges centre on the equator tend to breed over the widest span of latitude. Bird families as in the caption of **Figure 8.2**. Calculated from distribution maps in Harrison (1983).

others (notably auks Alcinae) are found only in the northern hemisphere, the different types being separated by warm tropical seas which they do not cross.

Although penguins penetrate to equatorial latitudes, they do so via cold water currents from further south. Thus the Humboldt current supports Humboldt Penguins *Spheniscus humboldti* off Peru and Galapagos Penguins *S. mendiculus* on the Galapagos Islands, while the Benguela current supports African Penguins *S. demersus* off southwest Africa. Similarly, auks penetrate the subtropics mainly off California, again dependent on a cold water current. Only murrelets of the genus *Synthliboramphus* feed in subtropical waters, and a few Cassin's Auklets *Ptychoramphus aleuticus* and Rhinoceros Auklets *Cerorhinca monocerata* edge into the subtropics at the southern end of their distribution. Two of the most southerly breeding auks (Japanese Murrelet *Synthliboramphus wumizusume* and Xantus's Murrelet *S. hypoleucus*) move northwards outside the breeding season, so that they winter at higher latitudes than where they breed (Gaston & Jones 1998).

The auks, penguins and diving petrels are the only major groups of seabirds that have failed to colonise both hemispheres, and interestingly they show the most striking examples of convergent evolution in seabirds (Chapter 2). All species hunt by 'flying' underwater, and this may explain their absence from warm waters. The burst swimming speeds of most fish double between water temperatures of 5 and 15°C. If the swimming speeds of the warm-blooded birds remain unchanged, underwater pursuit of fish must become less profitable as water temperature rises. At the same time, the success of large predatory fish relative to that of birds is likely to increase with temperature, intensifying competition with diving seabirds and perhaps increasing the risk of predation from them. This may explain why the major groups of wing-propelled underwater seabirds are restricted year-round to cold water, and why none has managed to cross the tropics (Gaston & Jones 1998). Their adaptations to this mode of feeding involve reduced wing area, the penguins having become flightless and the others retaining an expensive flight mode, involving continuous rapid wing beats. This may also help to restrict them to high-latitude regions of high productivity, and make it hard for them to survive in nutrient-poor tropical waters. The same argument would apply to the divers (Gaviidae) which occur on the sea mainly in winter, but are also restricted to cold water areas of the northern hemisphere.

Two other groups of seabirds are found mainly at higher latitudes, namely the cormorants and albatrosses. Like the groups mentioned above, cormorants also hunt by swimming underwater and, although they extend through the tropics, most warm water species are found on fresh water, rather than sea water. Secondly, except for the Waved Albatross *Phoebastria irrorata* breeding on the equatorial (but cold water) Galapagos Islands off South America, the albatrosses are almost absent as breeders from equatorial waters. This may be related to food-scarcity, but also to the lack of strong and continuous winds on which albatrosses rely for gliding. Some Hawaiian colonies of Black-footed Albatrosses *Phoebastria nigripes* and Laysan Albatrosses *P. immutabilis* lie just within the tropics, but again the birds fly to feed in colder waters.

Another interesting feature of seabird distributions is the greater number of species that occur on the eastern sides of oceans than on the western sides. Thus, the eastern side of the North Atlantic (i.e. west coast of Europe) holds 37 species, while the western side (i.e. east coast of North America) holds only 26 **(Table 8.2)**.

Table 8.2   Numbers of breeding seabird species in different regions. From Bourne 1980.

Group	Southern Ocean	Indian Ocean	Tropical Pacific W	Tropical Pacific E	Tropical Atlantic W	Tropical Atlantic E	North Pacific W	North Pacific E	North Atlantic W	North Atlantic E	Total species
Penguins	13	–	–	2	–	–	–	–	–	–	15
Albatrosses	9	–	–	1	–	–	3	2	–	–	13
Petrels	41	6	9	2	3	1	6	7	2	6	56
Storm Petrels	5	–	1	5	–	1	6	8	1	4	21
Diving Petrels	3	–	–	1	–	–	–	–	–	–	4
Tropicbirds	–	3	2	1	2	2	–	–	–	–	3
Frigatebirds	–	3	2	2	3	2	–	–	–	–	5
Sulids	2	4	3	5	3	3	–	–	1	1	9
Cormorants	16	4	1	4	1	2	4	4	2	2	28
Skuas	2	–	–	–	–	–	1	1	1	2	3
Gulls	7	2	1	3	–	1	7	12	6	10	33
Terns	10	11	12	6	14	9	3	5	6	5	37
Small alcids	–	–	–	–	–	–	8	9	1	1	11
Puffins	–	–	–	–	–	–	3	3	1	1	4
Black Guillemots	–	–	–	–	–	–	2	1	1	1	3
Large auks	–	–	–	–	–	–	2	2	4	4	4
Total	108	33	31	32	26	21	45	54	26	37	249
Percent	43	13	12	13	10	8	18	22	10	15	100
Percent of same species in ocean:											
to west		27	61	28	46	68	22	56	56	69	
to east		56	29	37	58	64	68	30	93	25	

Similarly the eastern side of the North Pacific (i.e. west coast of North America) holds 54 species while the western side (i.e. east coast of Asia) holds only 45 (**Table 8.2**). In both oceans the overall population levels are also much greater on the east side than on the west. The most likely explanation relates to the greater productivity of the eastern ocean, caused by cold currents. This relationship reveals another link between marine productivity and species numbers.

High-latitude extremes

In the northern hemisphere, the Ivory Gull *Pagophila eburnea* extends further north than any other seabird, feeding at openings in the pack-ice. In the southern hemisphere, the Snow Petrel *Pagodroma nivea* and Antarctic Petrel *Thalassoica antarctica* feed in the same way, and breed further south than any other birds, along with their predator, the South Polar Skua *Catharacta maccormicki*. Both the petrels sometimes nest far inland, on snow-free ledges on rocky mountain tops protruding above the ice cap, and from one colony at 80°S Snow Petrels must fly about 350 km to reach the coastline and even further to openings in the ice (Warham 1990). They visit their antarctic nesting sites even in winter (Jouventin & Bried 2001). These birds bring life to one of the most desolate parts of the planet.

Penguins extend less far south than the petrels, but Emperor Penguins *Aptenodytes forsteri* and Adélie Penguins *Pygoscelis adeliae* penetrate to 77°S at Cape Royd. The Emperor Penguin is remarkable because it breeds in winter, nesting on sea ice, in temperatures of −10° to −20°C, and in strong biting winds. Only in this way can it raise a chick before the next winter, but this in turn requires special adaptations, including an ability in the male to incubate non-stop for 110–115 days without feeding (Chapter 14).

At present, no seabirds are known to breed beyond 83°N in the northern hemisphere, or beyond 80°S in the southern hemisphere. However, distributions may well spread to even higher latitudes in future, as ice recedes under the influence of global warming.

Bipolar distributions

A few types of seabirds are bipolar in distribution, breeding at high latitudes in both hemispheres. Closely related allospecies occur in the north and the south, as exemplified by the Northern and Southern Fulmars *Fulmarus glacialis* and *F. glacialoides*, and the Arctic and Antarctic Terns *Sterna paradisaea* and *S. vittata*. The Northern Gannet *Morus bassanus* has two counterparts in the south, the Cape Gannet *M. capensis* of South Africa and the Australasian Gannet *M. serrator* of southern Australia and New Zealand. The Great Skua *Catharacta skua* in the north has no less than four counterparts in the south, each breeding in a different region, namely the South Polar Skua *C. maccormicki* in the Antarctic and South Shetland Islands, the Chilean Skua *C. chilensis* in southern Chile, the Southern Skua *C. antarctica* in Argentina and the Falkland, Gough and Tristan da Cuhna Islands, and the Brown Skua *C. lonnbergi* on the Antarctic Peninsula and various other southern ocean islands, mainly south of New Zealand. Some of these various skuas hybridise where their ranges overlap, not having achieved full reproductive isolation.

Northern and southern types of these various species-groups are so similar to one another that most have at times been classed as subspecies, rather than as separate species, but they vary in the degree of difference between them, perhaps depending on the time since they diverged. Such split distributions, with separate populations in both hemispheres, also occur in some landbirds, in which they might have arisen from long-distance migrants remaining to breed in wintering areas, leading to the establishment of distinct breeding populations (Snow 1978b). The same explanation might hold in seabirds, but nowadays only the Arctic Tern *Sterna paradisaea* from high northern latitudes migrates to winter within the breeding range of its southern counterpart, and only the South Polar Skua *Catharacta maccormicki* from high southern latitudes migrates to winter almost within the breeding range of its northern counterpart. An alternative explanation is that the cooling and narrowing of tropical waters during the ice ages enabled some high-latitude species to pass through, possibly using cold currents, leading to the establishment of populations in both hemispheres. Not all species necessarily moved in the same direction. Many other seabirds breed in both northern and southern hemispheres but with much smaller gaps between populations, or with a more or less continuous distribution from north to south, and some northern species have started breeding in the south in recent years (e.g. Leach's Petrel *Oceanodroma leucorrhoa*, see later).

Two similar-looking gulls, one in each hemisphere, warrant closer study: the Black-headed Gull *Larus ridibundus* breeds in mid latitudes across Eurasia, and the almost identical Brown-headed Gull *L. brunnicephalus* in similar latitudes in coastal South America. The questions are whether they are really sister species, as their appearance suggests and, if so, how they achieved their present distribution, totally separated by both latitude and longitude?

Tropical distributions

Yet other seabird species are found only in tropical waters. Some extend around the world, breeding in the equatorial parts of the Atlantic, Pacific and Indian Oceans. They include one species of tropicbird *Phàëthon lepturus*, two frigates *Fregata minor* and *F. ariel*, and three boobies *Sula sula*, *S. leucogaster* and *S. dactylatra*, together with several terns and petrels. Skimmers *Rhynchops* are even more widespread, breeding in the warmer parts of all the main continents, but with different allospecies on different continents.

Barriers to spread

Pelagic seabirds are independent of land, except when breeding, and can range widely over the open seas. While each species favours particular sea water zones over others, major land areas act as the principal barriers to their dispersal and spread. Many pelagic birds have very wide geographical ranges, or are represented by different subspecies or allospecies in different regions. In contrast, some inshore species come to land to roost and are therefore restricted to the vicinity of coastlines year-round. For these species, the open ocean, as well as large land areas, act as barriers to spread, and around the world many species of gulls, terns and shags are endemic to particular coastlines, isolated from other populations associated with other land areas.

Some coastal species that depend on areas of cold water in otherwise warm seas have extremely restricted distributions. This is shown by the various species found in association with the cold Humboldt current off tropical western South America to Galapagos, or with the cold Benguela current off southwest Africa. West South American endemics include the Guanay Cormorant *Phalacrocorax bougainvillii*, Peruvian Booby *Sula variegata*; Peruvian Brown Pelican *Pelecanus (occidentalis) thagus*, Markham's Storm Petrel *Oceanodroma markhami*, Ringed Storm Petrel *Oceanodroma hornbyi*, Peruvian Diving Petrel *Pelecanoides garnotii* and the Humboldt Penguin *Spheniscus humboldti*; Galapagos endemics include the Galapagos Penguin *Spheniscus mendiculus*, Flightless Cormorant *Phalacrocorax harissi*, Swallow-tailed Gull *Creagus furcatus* and Lava Gull *Larus fuliginosus*; while both these areas are home to the Waved Albatross *Phoebastria irrorata*, Wedge-rumped Storm Petrel *Oceanodroma tethys* and White-vented Storm Petrel *Oceanites gracilis* (Murphy 1936). Southwest African endemics include the Cape Gannet *Morus capensis*, Cape Cormorant *Phalacrocorax capensi*, Bank Cormorant *P. neglectus*, Crowned Cormorant *P. coronatus* and African Penguin *Spheniscus demersus*. All these species live in cold water areas in tropical and subtropical regions that are separated by large distances from comparable habitat elsewhere. The same holds for several endemic species that breed in association with the seasonal cold current off southeast Arabia (Bailey 1966); namely Jouanin's Petrel *Bulweria phallax*, Socotra Cormorant *P. nigrogularis*, Sooty Gull *L. hemprichi*, White-eyed Gull *L. leucophthalmus*, White-cheeked Tern *Sterna repressa* and Saunders' Tern *S. saundersi*. In all these areas most of the usual pan-tropical seabirds are absent.

Other species that breed at high latitudes exploit the cold currents of tropical regions in their non-breeding season. For example, Cape Petrels *Daption capense*, Southern Fulmars *Fulmarus glacialoides* and others range north along the Humboldt current in winter; Cory's Shearwaters *Calonectris diomedea* and others range south to the Benguela current; while Sabine's Gulls *Xema sabini* and phalaropes *Phalaropus* visit both areas. They emphasise again how responsive some seabirds can be to differences in water temperature, salinity and productivity.

Because near both poles, seas circle the earth, the majority of the seabird species that breed at such high latitudes also have circumpolar ranges. The same is true for many species of tropical regions where oceans are partly interconnected, and it is chiefly in the north temperate–boreal regions that species composition differs greatly between the Atlantic and Pacific Oceans. It is at these latitudes (roughly 35–60°N), where land areas are widest and thus provide greatest separation between the seas on either side, that many different species occur in the two oceans. The North Pacific, in particular, has 17 alcid species that are found nowhere else, including two species of puffins, two guillemots and 13 species of auklets and murrelets. Speciation in this region might have been helped by the fact that at various times in the past the North Pacific was separated from the Arctic Ocean (and hence from the Atlantic) by the Bering land-bridge which joined Eurasia to North America (Chapter 10). In addition, the ice during glaciations extended much further south in the Atlantic than in the Pacific. This eliminated some Atlantic species altogether, further contributing to the difference in species numbers (and endemic species numbers) between the northern Pacific and northern Atlantic regions (Chapter 9). Contact between the North Atlantic and North Pacific oceans has been restricted to the interglacials (as now), when the two oceans were in contact via the Arctic Ocean to the north. Hence, for the most northerly distributed seabirds, on geological time scales, long periods of separation have been broken by short periods of contact.

Even at mid latitudes, however, some pelagic species breed in more than one ocean, notably Bulwer's Petrel *Bulweria bulwerii* in all three oceans, Leach's Storm Petrel *Oceanodroma leucorhoa* in the Atlantic and Pacific, and Wedge-tailed Shearwater *Puffinus pacificus* in the Indian and Pacific Oceans. This last species may once have bred in the Atlantic too, because bones of this or a very similar bird were found in Pleistocene deposits on St Helena and in North Carolina (Olson 1975).

As a generalisation, the distributions of the most pelagic seabird species are based on the geography and physical features of the oceans, and the continents act as barriers to their spread. In contrast, distributions of inshore species are based on the geography of the land, and oceans tend to act as barriers to spread. Opposite sides of each tropical ocean often have fewer inshore species in common with one another than with the adjacent parts of neighbouring oceans **(Table 8.2)**. Thus, for example, the eastern tropical Pacific shares more species with the western Atlantic than with the western Pacific, while the western Pacific shares more species with the Indian Ocean than with the eastern Pacific (Bourne 1980). This is presumably a consequence of the free communication that existed between the tropical oceans in the geologically recent past, and the reluctance of many inshore species to cross wide expanses of open sea.

In contrast, at higher latitudes, the two sides of each ocean have more species in common with one another than do the two oceans **(Table 8.2)**. This could be because the wider land areas at high latitudes, formed by North America and Eurasia, are more effective barriers to seabird movement than the narrower land areas further south. In addition, many species have stepping-stone distributions between both sides of an ocean, via Iceland and Greenland in the North Atlantic, and via the Aleutian Islands in the North Pacific. The cold circumpolar Arctic Ocean seems to act as a barrier to the movements of species that breed in association with warmer waters further south, and prevent them spreading between the Atlantic and Pacific. Among the auks, six genera are unique to the Pacific and three to the Atlantic (including the recently extinct *Pinguinus* — Great Auk), while another three are common to both oceans. The equivalent figures for species are 17, two and four.

In the southern hemisphere, where land barriers are narrower and do not extend to such high latitudes as in the northern hemisphere, the different oceans share a greater proportion of their species. Nonetheless, some species differ between oceans, the Great Shearwater *Puffinus gravis* and Short-tailed Shearwater *P. tenuirostris* being confined, respectively, to the Atlantic and Pacific (apart from strays). A strong tendency for young birds to return to breed at their natal site (philopatry), linked with wide separation of breeding places, may well have contributed to the genetic divergence in some such groups (Brooke 2001).

Foraging zones

Seabirds are often divided into three categories, according to whether they feed inshore (say 6–8 km out to sea), offshore (from there to the edge of the continental shelf), or pelagic (over the deep ocean). This is a useful general division, but even within these zones, species differ greatly in foraging range, and overlap substantially one with another, with the shelf break and tidal fronts in mid shelf areas being especially favoured. Much depends on the duration of their foraging trips and the time each bird can spend away from the nest.

Information on the potential foraging ranges of various species was obtained initially from observations of birds at sea in the breeding season, or from calculations of the maximum distances that could be travelled based on knowledge of flight speeds and absence periods. In more recent years, foraging ranges have been obtained mainly by tracking radio-marked birds. This method gives the most accurate information, because it is not biased towards areas where potential observers happen to be or on assumptions about the behaviour of absentee birds. Species that feed inshore or over continental shelves generally return to their nests up to several times each day. For example, in trips of about two hours, radio-tagged Common Terns *Sterna hirundo* in the Wadden Sea covered about 30 km (Becker *et al.* 1993); European Shags *Phalacrocorax aristotelis* off Scotland foraged an average of 7 km (up to 17 km) from the colony (Wanless *et al.* 1991); in trips of about six hours Black-legged Kittiwakes *Rissa tridactyla* off Alaska foraged up to

60 km from their colonies (Suryan *et al.* 2000); and in trips lasting 13–84 hours Northern Gannets *Morus bassanus* off Scotland extended an average of 232 km (maximum 540 km) from the colony (Hamer *et al.* 2000).

At the extreme, some pelagic species can forage at enormous distances, spending several days over each trip, before returning with food for mate or chick. Thus, satellite-based radio-tracking has revealed some spectacular journeys undertaken by various petrels, in which individuals range from the nest over vast areas of sea. Radio-marked Wandering Albatrosses *Diomedea exulans* from South Georgia ranged from 67°S off the Antarctic Peninsula to 26°S off south Brazil, and from 17°W to 85°W off the Pacific coast off southern Chile (Prince *et al.* 1998). Among individuals nesting on the Crozet Islands in the southern Indian Ocean, males tended to forage over antarctic waters (50–60°S), while their partners chose mainly tropical pelagic and offshore waters (35–40°S) (Weimerskirch & Jouventin 1987). When feeding large young from distant feeding areas, on average adults covered 6,091 km on each foraging trip, reaching 1,534 km from the colony, during an absence of 11.6 days (Weimerskirch *et al.* 1993). Waved Albatrosses *Phoebastria irrorata*, nesting on the Galapagos Islands, were found to fly 1,200 km to feed in an upwelling area off Peru (Anderson *et al.* 1997). Light-mantled Albatrosses *Phoebetria palpebrata* from Macquarie Island foraged south of the polar front, in antarctic pelagic waters at an average distance of 1,721 km from their nests. Two birds made circuits of 6,463 km and 6,975 km in ten and 15 days respectively (Weimerskirch & Robertson 1994). Similarly, White-chinned Petrels *Procellaria aequinoctialis* from South Georgia, in foraging trips lasting up to 15 days, travelled circuits of up to 8,000 km from their nests (Berrow *et al.* 2000). Foraging trips of several days' duration also occur in tropicbirds and some boobies, giving scope for hundreds of kilometres of travel. These flights have no precise parallel among nesting landbirds; the extreme is shown by some colonial *Gyps* vultures, *Apus* swifts and freshwater pelicans which when nesting can range more than 100 km from their nests.

It is not only flying seabirds that forage far from their colonies, but also some penguin species which must walk or swim. Emperor Penguins *Aptenodytes forsteri* that nest in winter on the antarctic sea ice must often walk 150–180 km to reach the open water provided by polynyas, and when feeding chicks they may swim up to 500 km for food. If they nested nearer to the ice edge, it would melt in spring before their young were ready to leave (Chapter 14). Some other penguin species swim several hundred kilometres from their nesting areas to obtain food (e.g. 300–1,000 km for King Penguin *Aptenodytes patagonicus*, see Handrich *et al.* 1997; shorter distances for smaller penguins, Williams 1995). These figures give some idea of the distances reached from colonies, but birds seldom take a direct route, and the total travel distances of some foraging trips are much greater.

The above foraging distances were obtained mainly from individual radio-marked birds from colonies that happen to have been studied. Within species, however, it is commonly observed that larger colonies are further apart than small ones (Furness & Birkhead 1984). Birds nesting in large colonies may therefore range over larger areas for their food than birds from small colonies, as in the Northern Gannet *Morus bassanus* (Lewis *et al.* 2001). Because for much of the breeding cycle, one partner stays at the nest while the other goes to sea, foraging ranges also relate to onshore fasting periods, and vary with stage of the nesting cycle, as adults can be absent for longer periods during incubation than chick rearing. Among penguins, in the pre-laying and incubation periods, inshore feeding species, such as the Yellow-eyed Penguin *Megadyptes antipodes*, Little Penguin *Eudyptula minor* and Galapagos Penguin *Spheniscus mendiculus*, usually fast ashore for periods of less than three days at a time. In contrast, long-distance foragers, such as Adélie Penguin *Pygoscelis adeliae*, Magellanic Penguin *S. magellanicus* and King Penguin *Aptenodytes patagonicus*, regularly remain ashore without feeding for 25–40 days at a time, and male Emperor Penguins *A. forsteri* for more than 110 days (Croxall & Davies 1999).

The fact that adults must return to colonies in the breeding season restricts their foraging range, compared with immatures which can be much more widely distributed (for various species around Britain, see Tasker *et al.* 1987). However, even in the non-breeding season, marine birds show different levels of aggregation. For example, in the seas around Britain, sea-ducks are usually highly aggregated on the water, Razorbills *Alca torda* moderately aggregated and Guillemots *Uria aalge* even less aggregated. Other species, such as Northern Fulmars *Fulmarus glacialis*, are usually highly dispersed and airborne, but concentrate in larger numbers at local food sources, such as fishing vessels.

Species also differ in whether they obtain their food from the surface or below, the depths to which they can penetrate, and the length of time they can remain submerged. In these respects, the auks and the penguins are perhaps the most marine of all seabirds, as they are able to 'fly' underwater for up to several minutes at a time in pursuit of prey. Comparing species, dive depth and duration increase with body size, but relative to the best human divers, some seabirds and marine mammals are outstanding performers. Few trained human divers can reach deeper than 100 m in a single-breath dive lasting four minutes, but King Penguins *Aptenodytes patagonicus* and Emperor Penguins *A. forsteri* (weighing about 12 kg and 50 kg, respectively) can dive to depths of 304 and 534 m for as long as 7.5 and 15.8 minutes respectively, with relatively short recovery times on the surface. They manage this by lowering their metabolism and body temperature during dives (Handrich *et al.* 1997). The smallest penguins, such as the Little Penguin *Eudyptula minor*, dive only to about 60 m (see Williams 1995 for summary of penguin data). Recent research has revealed that some shearwaters are also adept divers, with Sooty Shearwaters *Puffinus griseus* reaching 70 m (Weimerskirch & Sagar 1996).

About 15 species of seabirds worldwide achieve total population sizes greater than ten million individuals. Thirteen of these live at high latitudes, and most such species obtain their food by underwater pursuit. It appears that, where prey density is high enough to render underwater pursuit viable, those species adopting this lifestyle can become very numerous (Brooke 2001). They are essentially harvesting prey in three dimensions, while the surface feeders are restricted to two. Among species obtaining food at the sea's surface, those feeding offshore have a potentially greater area available in which to search for food, because of straightforward geometrical considerations, than those that feed close to shore. They might therefore have larger populations. Diamond (1978) found support for this idea at several seabird colonies. It is also notable that the surface-feeding species with populations in excess of ten million individuals (Northern Fulmer *Fulmarus glacialis*, Leach's Storm Petrel *Oceanodroma leucorrhoa*, Black-legged Kittiwake *Rissa tridactyla*, Sooty Tern *Sterna fuscata*) are all pelagic feeders.

Migrations

The above distribution patterns refer to breeding populations but, as in landbirds, many seabirds leave their high-latitude nesting areas after breeding to spend the winter elsewhere. Several northern breeding species winter in large numbers beyond the tropics in the southern hemisphere; and similarly several southern breeding species winter in large numbers beyond the tropics in the northern hemisphere. Migrant Wilson's Storm Petrels *Oceanites oceanicus* and Great Shearwaters *Puffinus gravis* in the Atlantic move up the west side, taking advantage of the flush of food off eastern North America in the northern spring, and also appear off Senegal in the eastern Atlantic in June–September, the time of upwelling, *en route* back to their southern breeding areas. Similarly, Short-tailed Shearwaters *Puffinus tenuirostris* in the Pacific are thought to perform a loop migration, most juveniles moving up the west side and down the east.

Some seabirds in the northern hemisphere go against the general trend in birds, in that they winter at latitudes higher than their breeding areas, examples including some alcid species (including the murrelets mentioned earlier), Balearic Shearwaters *Puffinus p. mauretanicus* which move from the Mediterranean to the North Atlantic, and Brown Pelicans *Pelecanus occidentalis* which in western North America move north from Mexico as far as Washington State for the winter. Other species after breeding simply disperse mainly east or west from the nesting colonies and thereby become more widely distributed but within the same sea water zone. For example, many Common Murres *Uria aalge* and Thick-billed Murres *U. lomvia* from far northern European colonies move westwards after breeding to winter mainly around southern Greenland or even Newfoundland. Similarly, many young of Antarctic Giant Petrels *Macronectes giganteus* leave their natal sites in the southern ocean by flying downwind, circling the earth in their first year of life. Some albatrosses may do the same.

LIMITATION OF NESTING PLACES

Most seabird species nest in colonies, sometimes of enormous size. A major factor limiting the breeding distributions of some species is the presence of safe colony sites on islands or on continents. Cliff-nesters nest readily on island or continental coastlines, safe from mammalian predators, but ground-nesters and burrow-nesters favour islands that lack such predators (as does Antarctica). Many islands are no longer used by certain vulnerable species, because of predation by humans or by the rats, cats, pigs and other predators that humans have introduced (Chapter 7). On some islands, introduced grazing mammals affect the vegetation cover, vital to the stability of the soil or peat in which some species nest. Hence, many seabirds now have far fewer usable colony sites than they did only a few hundred years ago.

In general, nest-sites achieve a much greater biogeographical importance in seabirds than in landbirds, and the requirement for safe sites can restrict seabird distributions locally or regionally. Large parts of all the major oceans are devoid of islands, and therefore lacking in breeding seabirds, although such sea areas are widely used by non-breeding birds. Moreover, on most oceanic islands, the different species use different types of nest sites, according to substrate, angle of slope, distance from sea, and so on, which sometimes gives a marked zonation of species from shoreline to hilltop. Different species utilise cliff ledges, cliff holes, or boulder–strewn slopes of various angles, clear or vegetated flat ground, craters, mountain tops, tree and shrub branches, ground crevices and burrows. For example, on the bare Peruvian Guano islands, the slopes are carpeted with Peruvian Boobies *Sula variegata* and Guanay Cormorants *Phalacrocorax bougainvillii*, each group clearly demarcated; the cliff ledges hold Peruvian Boobies and Inca Terns *Larosterna inca*; sea caverns hold Red-legged Cormorants *Phalacrocorax gaimardi*; lower terraces hold Peruvian Pelicans *Pelecanus occidentalis thagus*, while Peruvian Diving Petrels *Pelecanoides garnotii* burrow under the island's crust of guano (Nelson 1980). Such differences are assumed to reduce competition for sites between species, and are taken as further evidence that sites can be limiting. Bare rocky islands support fewest species. The evolution of so many flightless penguin species around the southern ocean, and their nesting on the antarctic mainland, is presumably contingent upon the lack of land-based mammalian predators in that region. Because they cannot fly, however, penguins are more restricted than most species by their need for accessible landing places and for nest sites that can be reached on foot. They may be further constrained in breeding distribution by competition with seals and sea lions which favour the same type of sites. The same may also have been true for the extinct flightless auks of the northern hemisphere.

Many species of gulls, terns, skuas and skimmers nest mainly on flat ground, on islets or sandbanks if available, but otherwise on extensive shingle beaches, sandy headlands, marshes or barren areas, including deserts. Because some of their nesting places are unstable, periodically rearranged by storms and accessible to predators, such species frequently shift their colony sites as necessity and opportunity dictate. Such sites are less often limiting than other types of site, and different larid species sometimes nest together in mixed colonies

Other species nesting at the same place that use the same type of site, such as burrows, sometimes breed at different times from one another. This gives two sets of occupants per year in the same holes, as exemplified by the winter/summer breeding of *Oceanodroma tristrami/O. matsudairae* (Volcano Island), *Pterodroma macroptera/Puffinus carneipes* (western Australia), and *Procellaria cinerea/P. aequinoctialis* (Antipodes Islands). This finding has also been attributed to competition, and taken as evidence for the limiting influence of nest sites. However, the difference in breeding seasons may be more a consequence of different feeding habits (see below) and, because of their long breeding periods, well-grown chicks of the one species are often still present when the second arrives, so the benefits, if any, of hole-sharing are unclear (Warham 1996). Moreover, among burrowing petrels, the potential for interference and competition is often present, as the excavations by some individuals may destroy the nests of other individuals (of the same or different species) already present (e.g. Ramos *et al.* 1997).

More direct evidence for the effect of nest sites in limiting seabird distributions has come from islands where introduced predators have been removed, enabling their recolonisation by nesting seabirds (for effects of fox removal on Alaskan islands, see Lensink 1984, Byrd *et al.* 1994; for effects of cat removal on Baker Island, see Forsell 1982; Chapter 7). It has also come from experiments in which nesting areas have been created artificially either by the provision of large offshore platforms or rafts, as off southwest Africa (Crawford & Shelton 1978, Crawford *et al.* 1992), or by the fencing of coastal peninsulas (to exclude mammalian predators) in Peru (Duffy 1983). In each case, the new nesting habitat was rapidly colonised, leading to an increase in regional population levels. Elsewhere, individual colonies have been expanded by the creation of extra nesting habitat, as when the felling of a coconut plantation on Seychelles led to the expansion of a Sooty Tern *Sterna fuscata* colony (Feare 1976). The rapidity with which such newly created sites are occupied signals the general shortage of nesting habitat.

Nesting seems so tight in some colonies that no more space is available. For example, the burrow-nesting Great Shearwater *Puffinus gravis* at Nightingale Island and Sooty Shearwater *Puffinus griseus* at the Snares Islands seem to occupy all available ground not taken up by other species, and many eggs are laid on the surface (Warham 1990). Moreover, shortage of nest sites has been confirmed in some species by experiments in which breeding birds were removed from nest sites, only to be replaced immediately by other (previously known non-breeding) individuals which then proceeded to nest (e.g. Cassin's Auklet *Ptychoramphus aleuticus*, Manuwal 1974). All such findings help to confirm that many seabirds are in some areas limited in both distribution and abundance by shortage of safe nesting places.

While large concentrations of breeding seabirds need rich feeding areas, nest site limitations mean that not all rich feeding areas have large seabird colonies. The upwelling season off Senegal, in February–March, attracts mainly wintering seabirds from the Palaearctic, notably Northern Gannets *Morus bassanus*, Pomarine Skuas *Stercorarius pomarinus*, Great Skuas *S. skua*, Lesser Black-backed Gulls *Larus fuscus*, Common Black-headed Gulls *L. ridibundus* and Sandwich Terns *Sterna sandvicensis*, together with Royal Terns *S. maxima* from eastern North America (which join the local breeders of this species). Relatively few African seabirds exploit this upwelling, probably because of the scarcity of secure nest sites in the region and the competing attractions of the Benguela current further south which runs year-round (Brown 1979). Likewise, the seabird population of the Banc d'Arguin, in Mauritania, which is the most important breeding area along this part of the African coast, holds only 12,000–15,000 pairs, which is trivial compared to similar concentrations elsewhere. This paucity of nesting birds is associated with a scarcity of suitable nesting sites that are inaccessible to human and other predators.

Some seabird colonies at traditional locations contain hundreds of thousands of pairs and, in a few species, occasional colonies are judged to contain more than a million pairs. Such large colonies have been recorded for Atlantic Puffin *Fratercula arctica*, Common Murre *Uria aalge* and Thick-billed Murre *U. lomvia*. In addition, one colony of Leach's Storm Petrels *Oceanodroma leucorhoa* on the 522-ha Baccalieu Island off Newfoundland was found to contain more than three million pairs (Sklepkovych & Montevecchi 1989), and various colonies of Little Auks *Alle alle* have been estimated to contain several million pairs, where food permits. Other sizeable colonies are found in pelagic species, which nest on widely spaced oceanic islands. Many procellariiform birds are in this category, as is the Sooty Tern *Sterna fuscata* which on Christmas Island in the Pacific formerly reached a staggering ten million pairs. In the southern hemisphere, Peruvian Boobies *Sula variegata* on Guañape Sur and Guanay Cormorants *Phalacrocorax bougainvillii* on

Guañape Norté off Peru have each reached about one million pairs (Nelson 1980). Further south, colonies of King Penguins *Aptenodytes patagonicus*, Macaroni Penguins *Eudyptes chrysolophus* and Chinstrap Penguins *Pygoscelis antarctica* can number several hundred thousand pairs (Williams 1995); a colony of Rockhopper Penguins *Eudyptes chrysocome* on Beauchene Island in the Falklands held an estimated 2.5 million pairs; while a colony of Chinstrap Penguins in the South Sandwich Islands held an estimated five million pairs (Nelson 1980). All these species also nest in much smaller colonies elsewhere, depending largely on nesting space and local food-supplies.

Densities of nests within colonies vary with species and terrain, but some seabirds in their colonies are remarkably concentrated. Some 50,000 pairs of Guanay Cormorants *Phalacrocorax bougainvillii* nest on little more than 2.5 ha of Guañape Norté Island off Peru (at about two pairs per m²). In various colonies, Macaroni Penguins *Eudyptes chrysolophus*, Royal Penguins *Eudyptes schlegeli* and Rockhopper Penguins *Eudyptes chrysocome* breed at densities of 2.2–2.4 pairs per m², with nests only 60–80 cm apart (Williams 1995). In the Sooty Shearwater *P. griseus* an estimated 2.75 million occupied burrows were found on 328 ha of the Snares Islands (about 1.2 burrows per m²), in the Great Shearwater *P. gravis* two million occupied burrows were found on 200 ha of Nightingale Island (about one per m²), and in the Short-tailed Shearwater *Puffinus tenuirostris* 2.86 million burrows were found in the 380 ha of Babel Island (about 1.3 per m²) (Warham 1990, 1996). Even these figures are low, however, compared with those for a Galapagos (Wedge-rumped) Storm Petrel *Oceanodroma tethys* colony, in which an estimated 200,000 pairs occupied an area of 2,500 m², giving a mean density of eight pairs per m² (Harris 1969).

Because many procellariids and others nest in underground burrows on remote islands, they are not easy to find. Their distributions may therefore still be incompletely known and extended by further exploration, but this is unlikely to alter the general conclusion that most species have restricted nesting places. This is the more striking considering that, when not nesting, some such species extend over huge areas of sea. An extreme example is the Great Shearwater *Puffinus gravis*, which nests in two main localities in the south Atlantic (on Gough and Tristan da Cunha Islands), but ranges over most of the northern Atlantic in the non-breeding season.

With many different species present in large numbers, the total seabird populations of some oceanic islands that lie within reach of good food-supplies are immense. Over 14 million birds of 18 species were estimated on the northwestern Hawaiian Islands (Fefer *et al.* 1984), over 36 million of 26 species at South Georgia (Croxall & Prince 1980, Prince & Croxall 1983), over 25 million of 37 species on the Crozet Islands, and ten million of 35 species on Kerguelen (Jouventin 1994). Nevertheless, the numbers of seabirds present on many islands today are probably only a small proportion of those present before human influence was felt, mainly through the introduction of rats and other predators.

Some species now breed on only a single island. Examples include Abbott's Booby *Papasula abbotti* and Andrews' Frigatebird *Fregata andrewsi* on Christmas Island in the Indian Ocean, the Ascension Frigatebird *F. aquila* on Boatswainbird Islet, off Ascension in mid Atlantic, and the Westland Petrel *Procellaria westlandica* on the northwest coast of South Island, New Zealand. Others breed on more than one island, but in the same limited area, such as the Flightless Cormorant *Phalacrocorax harrisi* on Galapagos. Some other species with restricted distributions are listed in **Table 8.3**. The ranges of most of these species have been reduced from former times, but the Ascension Frigatebird is not known to have nested anywhere else. Like other birds confined to a single location, they are especially vulnerable to newly introduced predators or parasites, as well as to catastrophic events, such as volcanic eruptions, tidal waves or oil spills.

The idea that many seabird species were more widely distributed before people began to visit islands than they are now rests on both direct and indirect evidence. On some islands, such as the Azores, historical accounts document massive populations on accessible parts of islands that now hold none, with reduced populations mainly restricted to small uninhabited islets still free of rats (Monteiro *et al*. 1996). For many species, remains dating from the last 1,000 years have been identified in places where those species might have been expected to breed but are now absent. On St Helena in the Atlantic, only a single petrel species still breeds, but thousands of bone remains show that, before people arrived in 1502, this island was a major breeding site for several other species. Likewise, on Ascension Island, also in mid Atlantic, abundant bone remains confirm the presence of several species, all of which survive in reduced numbers, with most confined to offshore stacks (Olsson 1977, Bourne & Simmons 2001).

The massive losses of landbirds on Pacific islands were discussed in Chapter 7, and subfossil remains confirm that many seabirds also disappeared following human colonisation. The island of Ua Huka in the Marquesas held 22 seabird species in pre-human times, compared with four now, while Easter Island held 25 species, of which only the Red-billed Tropicbird *Phaethon aethereus* remains today (Steadman 1995). Although some seabird species have disappeared from most islands studied, few have become totally extinct: examples include *Pterodroma jugabilis* on Hawaii, *Nesofregetta fuliginosa* on Cook and Henderson Islands (Olson & James 1982, Steadman & Olson 1985), and an unnamed petrel on Easter Island (Steadman 1995). Many of these remains were found at human middens, and as people may have transported recently killed birds between islands not far apart, the numbers of species attributed to some islands might have been artificially inflated.

At least one seabird species became extinct in the nineteenth century, and was already known to science, namely the flightless Great Auk *Pinguinus impennis*, once widely distributed around the North Atlantic. Another which is probably, but not certainly, extinct is the Guadalupe Storm Petrel *Oceanodroma macrodactyla*, confined to a single island off Mexico, and known into the twentieth century.

Indirect evidence of local extinctions comes from 'relict distribution' patterns, where a species is found at widely separated localities but is absent from suitable intermediate sites. Subfossil material suggests that Buller's Shearwater *Puffinus bulleri*, Westland Petrel *Procellaria westlandica* and Parkinson's Petrel *P. parkinsoni*, which are now restricted to one or two breeding places, were widespread on the New Zealand mainland until recently (Warham 1990). The same is true of the small Cook's Petrel *Pterodroma cookii* which now breeds at only two places 1,350 km apart. Similarly, Pycroft's Petrel *P. longirostris pycrofti* is now rare and breeding only on a few islands off northern New Zealand, but it formerly also bred on Norfolk Island and probably on Lord Howe. Bones of this bird from

Table 8.3   Some seabird species restricted to particular islands or island groups. Mainly from Collar & Andrew 1988.

Spheniscidae
Royal Penguin *Eudyptes schlegeli*

Macquarie Island, Southern Ocean

Procellariidae
Waved Albatross *Phoebastria irrorata*

Galapagos (Hood Island) and La Plata Island, Ecuador

Short-tailed Albatross *Phoebastria albatrus*	Torishima Island, Japan and Pelto, Taiwan
Schlegel's (Atlantic) Petrel *Pterodroma incerta*	Tristan da Cunha and Gough Islands
Providence Petrel *Pterodroma solandri*	Lord Howe and Philip Islands
Magenta Petrel *Pterodroma magentae*	Chatham Islands, New Zealand
MacGillivray's (Fiji) Petrel *Pterodroma macgillivrayi*	Gau Island, Fiji
Mascarene Petrel *Pterodroma aterrima*	Réunion and Rodriguez Islands
Fea's (Cape Verde) Petrel *Pterodroma feae*	Cape Verde and Desertas Islands
Madeira (Zino's) Petrel *Pterodroma madeira*	Madeira
Bermuda Petrel *Pterodroma cahow*	Castle Island, Bermuda
Black-capped Petrel *Pterodroma hasitata*	Hispaniola and possibly elsewhere, West Indies
Barau's Petrel *Pterodroma baraui*	Réunion and Rodriguez Islands
Chatham Islands Petrel *Pterodroma axillaris*	Rangatira, Chatham Islands
Cook's Petrel *Pterodroma cookii*	New Zealand
Mas-á-Tierra (Defilippe's) Petrel *Pterodroma defilippiana*	Juan Fernández and Desventuradas Islands, Chile
Jouanin's Petrel *Bulweria fallax*	Socotra
Westland Black Petrel *Procellaria westlandica*	NW South Island, New Zealand
Parkinson's Petrel *Procellaria parkinsoni*	Little and Great Barrier Islands, New Zealand
Buller's Shearwater *Puffinus bulleri*	Poor Knights Island, New Zealand
Pink-footed Shearwater *Puffinus creatopus*	Islands off Chile
Black-vented Shearwater *Puffinus opisthomelas*	Pacific Islands off Baja California
Fluttering Shearwater *Puffinus gavia*	New Zealand
Hutton's Shearwater *Puffinus huttoni*	New Zealand
Guadalupe Storm Petrel *Oceanodroma macrodactyla*[1]	Guadalupe Island, Mexico
Black Storm Petrel *Oceanodroma melania*	Islands off southern California
Matsudaira's Storm Petrel *Oceanodroma matsudairae*	Volcano Islands, Japan
Ashy Storm Petrel *Oceanodroma homochroa*	Islands off California and Baja California
Ringed (Hornby's) Storm Petrel *Oceanodroma hornbyi*	Coastal mountains, northern Chile

Sulidae
Abbott's Booby *Papasula abbotti*

Christmas Island, Indian Ocean

Phalacrocoracidae
Flightless Cormorant *Phalacrocorax harrisi*

Galapagos Islands

Fregatidae
Ascension Frigatebird *Fregata aquila*   Ascension Island
Andrews' Frigatebird *Fregata andrewsi*   Christmas Island, Indian Ocean

Laridae
Olrog's Gull *Larus atlanticus*   Rio Colorado River, Argentina
Japanese Murrelet *Synthliboramphus wumizusume*   Izu Island, Japan and S. Japan

[1]May be extinct

refuse dumps of early settlers on Norfolk Island suggest that they ate it, along with the Providence Petrel *Pterodroma solandri*. Likewise, Abbott's Booby *Papasula abbotti*, which is now confined to a single island (see above), once bred widely on islands across the entire Indian Ocean and east into the Pacific as far as the Marquesas (Steadman *et al.* 1988). Further extinctions are known from Pleistocene fossils, so it is uncertain whether these species disappeared during periods of climatic change, or following early human colonisation of the islands concerned.

In addition, many small seabird species are now found only on islands free of potential predators: the current nesting distribution of the European Storm Petrel *Hydrobates pelagicus* in the Mediterranean region, for example, reflects the presence or absence of the Black Rat *Rattus rattus* (Thibault *et al.* 1996). The assumptions are that, before rats were introduced, this petrel occurred on many other islands too, and while it may occasionally nest on other islets, these attempts are soon suppressed. Safe nest sites are, of course, not the only factor limiting seabird numbers, for where such sites are plentiful, seabirds may be held by food-supply at a level lower than colony sites would permit.

DIFFERENTIATION AND SPECIATION

Certain seabirds have evidently remained as separate breeding populations on different groups of islands for long enough to have evolved as separate allospecies or subspecies there. Some of these distinct forms are found on only a small number of islands. Examples from among circumpolar species include the different allospecies of *Catharacta* skuas, *Diomedia* albatrosses, *Eudyptes* penguins and *Leucocarbo* shags. Not all isolated island groups support endemic forms, however, suggesting that distance is not the only factor involved. The remote Hawaiian chain, which includes some now submerged islands, has existed for at least ten million years, and has many endemic landbirds, but no endemic seabird species. In contrast, the Galapagos Islands, only 900 km from South America, have several endemics. This may be because a cold water current around the Galapagos Islands isolates the seabird populations from the adjacent tropical waters. No such current occurs around the Hawaiian Islands, and the majority of their seabird species are widely distributed on other tropical Pacific islands. As expected, seabirds also show high levels of endemism in confined sea areas, notably the Mediterranean (Zotier *et al.* 1999).

There are four groups of seabirds where species boundaries are especially hard to define, namely the albatrosses, gadfly petrels, southern shags and northern gulls. The first three groups typically nest on islands, and a substantial proportion (3/21 albatrosses, 11/39 gadfly petrels, 7/36 shags) breeds at just a single island or archipelago (Brooke 2001). By definition, then, these species show strong natal philopatry, with all individuals returning to breed on the island where they hatched or (in archipelagos) on nearby islands. This high philopatry (despite some pre-breeding dispersal), coupled with long distances to other potential nesting sites, could reduce selection for isolating mechanisms, such as plumage divergence, which in other birds helps to prevent hybridisation. In this case, the external similarity need not indicate recent separation of the different populations, or indeed genetic similarity. Plumage and morphology would then be a poorer guide than usual to the independent evolutionary history of the birds. Only molecular studies could reveal the extent of independent history, and examination of DNA is likely to increase the number of island populations that can be separated genetically if not morphologically. It is also likely to increase the numbers of subspecies that are elevated to the level of species, whether appropriately or not.

Studies of the mt DNA cytochrome *b* gene of various albatrosses have already revealed genetic differences between subspecies of *Thalassarche cauta*, *T. chlororhynchos*, *T. melanophris* and *Diomedea exulans* (including *D. amsterdamensis*), sufficient in the view of the authors to support their separation as full species (Robertson & Nunn 1998). These authors designated 24 albatross species in place of the former 14, but for some of them the case for specific status still seems weak (for comments, see Bourne 2002). In addition, among gadfly petrels on the Pitcairn Islands, Brooke & Rowe (1996) found dark morph *Pterodroma heraldica* mating only with dark morphs and nesting away from the light morphs. On the basis of behaviour and sequence data from the mt DNA cytochrome *b* gene, they concluded that two species were involved, naming the dark morph *P. atra*, and keeping the light morph as *P. heraldica*.

The situation differs in the large gulls because they occur over wide areas of the northern continents, frequently interbreed and often produce viable hybrids. However, as the hybrids are not spreading at the expense of the parental forms, there must be some degree of reproductive isolation or some selection against the hybrids. These taxonomic problems mainly affect the north temperate species, including the Palaearctic Herring Gull *Larus argentatus*, Lesser Black–backed Gull *L. fuscus*, Yellow–legged Gull *L. cachinnans*, Armenian Gull *L. armenicus* and Slaty-backed Gull *L. schistisagus*, and the Nearctic Iceland Gull *L. glaucoides*, Thayer's Gull *L.(g.) thayeri*, Kumlien's Gull *L.(g.) kumlieni* and Glaucous-winged Gull *L. glaucescens*. These are all taxa that were probably fragmented during the glaciations, and that have re-established contact only in the last 10,000 years as the ice retreated. They are at the stage between subspecies and species. Between *Larus cachinnans*, *L. fuscus* and *L. argentatus*, mt DNA cytochrome *b* divergence was 0.69–2.04%, which is about the same magnitude as that between other subspecies of the same species (Helbig 1994).

Unlike other northern seabirds, these large gulls are spread on an east–west axis across the northern hemisphere. They would have experienced a complicated history of population fragmentation as the colonies, which now encompass the North Pacific, Great Lakes, North Atlantic, Mediterranean, Black, Caspian, Aral and Okhotsk seas, moved south and north as the ice advanced and retreated. Other northern seabirds, such as auks and Northern Fulmar *Fulmarus glacialis*, were restricted to the two major oceans, so could have moved north and south as continuous populations (Brooke 2001). Interestingly, the more marine northern gulls, such as kittiwakes *Rissa*, Ivory Gull *Pagophila eburnea*, Ross's Gull *Rhodostethia rosea* and Sabine's Gull *Xema sabini*, present clearly defined species, as do more southern gulls.

While many kinds of procellariiform birds show only slight morphological differentiation from their nearest relatives, the auks in general show much more. This is reflected in the high generic diversity in the auks, with more than half of genera containing only a single species. They lack any large genera, as occur in other types of seabirds. This suggests either little recent speciation or high extinction rates, possibly in association with the more severe and far-reaching glaciations in the northern hemisphere. In addition, their world oceanic range is smaller than that of any other seabird family, partly because few occur in waters warmer than boreal and because a greater proportion of the northern than the southern hemisphere is taken by land and is thus unsuitable for seabirds. Moreover, their almost complete restriction to continental shelf waters reduces the opportunities for isolation to occur, compared to southern hemisphere procellariiforms. An unresolved question in auk biogeography is how to explain the large numbers of mainly planktivorous species in the North Pacific, when there is only one (the Little Auk *Alle alle*) in the North Atlantic. A similar imbalance occurs among the planktivorous storm petrels.

Those auks that breed in the North Atlantic have several subspecies in the eastern (European) side but only one on the western, reflecting greater differentiation of regional populations in the east. Two possible explanations have been suggested (Gaston & Jones 1998). Firstly, auks breed and winter over a greater span of latitude in the eastern Atlantic than in the west (wintering from 45 to 75°N), and the geography is more complex in the east, favouring the development of clinal and local variation. In the colder western Atlantic, few auks winter north of 53°N, so that the total breeding range is much smaller. Secondly, owing to supposed shortage of breeding sites at appropriate latitudes during the last glaciation, auks might have been eliminated altogether from the western Atlantic and existing populations may result from subsequent colonisation from the east (Chapter 10). The same may hold for other seabirds, and several mainly European species established themselves as breeders in eastern North America in the twentieth century (e.g. Manx Shearwater *Puffinus puffinus*, Black-headed Gull *Larus ridibundus* and Little Gull *L. minutus*).

In general among closely related seabirds, allopatry is a common form of ecological segregation, as similar forms replace each other geographically. It is much rarer for closely similar species to occur sympatrically on the same island (although examples occur in most genera). This may be partly because the lifestyle of petrels and other seabirds limits the scope for morphological differentiation: species differ mainly in body size, and in proportions of black and white in the plumage. Compared with landbirds, they show much less variation in body form or colour patterns. The most striking colours, including reds, blues and yellows, occur mainly on small unfeathered areas that function in intraspecific recognition, with the small alcids showing the greatest variation between species. Where closely related, similar-sized species occur on the same oceanic island, they often forage in different sea water zones, and breed at different dates.

One way in which two closely related species might come to breed sympatrically, while maintaining ecological and reproductive isolation, is for an island to be colonized by members of a species with a breeding season different from the species already present (Fleming 1941). This is most likely to occur on islands near the boundary between two marine zones, which lie within flying range of birds adapted to conditions in different zones. Such double colonisation is especially probable in species that can feed far from their colonies even when breeding. They might then nest in one marine zone while feeding largely in another. A history of this kind may explain, for example, the sympatric occurrences of two sibling species of prions *Pachyptila desolata* and *P. belcheri* on Kerguelen Island, and of two sibling species of Giant Petrel *Macronectes halli* and *M. giganteus* at Macquarie Island and elsewhere (Bourne & Warham 1966, Warham 1990). Both petrels breed around the antarctic convergence, but *M. giganteus* feeds mainly south of the convergence, while *M. halli* feeds mainly north of it and nests six weeks earlier. Where the two species nest in close proximity, hybridisation sometimes occurs, affecting up to 1.5% of the population on South Georgia. Currently, hybridisation is known only in easily seen surface-nesting species, such as *Macronectes*, and it would be harder to detect among nocturnally active birds in burrows.

Another example concerns the White-headed Petrel *Pterodroma lessonii* and Great-winged Petrel *P. macroptera*, which are similar-sized but different coloured birds that breed sympatrically on the Kerguelen and Crozet Islands. The first is a summer breeder and the second is a winter breeder, and although their nesting seasons overlap, they are not known to interbreed (Warham 1990). Note that these various species pairs are not merely members of the same genus (there are many examples of 2–3 congeners nesting on the same islands), but are almost identical sibling species. Whenever sympatry in closely related seabirds arises in this way, the difference in laying dates (itself associated with different feeding areas) may be a prerequisite (Ashmole 1971).

Other species may represent a step *en route* to this form of speciation, for they appear to be subspecies of the same species, but nest on the same island without interbreeding because they have different nesting seasons. This occurs in two forms of Leach's Storm Petrel *Oceanodroma leucorhoa* on Guadalupe Island, one of which breeds in summer and the other in winter (Power & Ainley 1986). Even more unexpected is the situation within the Band-rumped Storm Petrel *Oceanodroma castro* on the Galapagos Islands, where two separate populations share nesting holes and breed annually, one in May–June, and the other in November–December. Non-breeders and failed breeders remained faithful to their annual cycles over four seasons (Harris 1969). The same holds for this species in the Azores, where the two populations seem to have evolved slight but consistent differences in morphology (Monteiro & Furness 1998). How the two populations of each species remain separate without giving rise to a continuous breeding regime (as *Puffinus lherminieri* does at the Galapagos Islands) is unclear, but if the temporal segregation persists, and is inherited by the young, they effectively behave as separate species. More information is needed.

The continuing operation of speciation processes in seabirds is also evident from the many species that show geographical variation in size and colour; in some species such variation follows the same eco-geographical rules as in land-birds (Chapter 2), but in others it is most readily explained by the evolution of different morphs in different areas which later came together and interbred. For example, in the Northern Fulmar *Fulmarus glacialis*, the Pacific form has a high proportion of dark phase birds, while the Atlantic form has mainly dark phase birds in the high arctic and mainly light phase birds in the low arctic and boreal regions. Van Franeker & Wattel (1982) hypothesised that the two morphs originated in different oceans — the dark in the Pacific — and that during warm interglacials with more open water at high latitudes, the two forms intermixed. Similarly, the Snow Petrel *Pagodroma nivea* has large and small morphs which occur in different proportions (along with intermediates) in different colonies, from entirely large birds on Balleny Island to entirely small birds on South

Georgia. Jouventin & Viot (1985) hypothesised that the two forms arose from a single ancestor in different glacial refuges, and subsequently came together, interbreeding to produce the present variable situation.

Continuing evolution is also evident in the marked subspeciation shown by some species, notably by petrels that are more or less isolated on different island groups. Even within subspecies, substantial genetic substructuring is sometimes apparent between the populations of different island groups. An example is Cory's Shearwater *Calonectris diomedea* in which different populations differ in mt DNA (cytochrome *b*) sequences by as much as 1.51–1.87% (Helbig 1994). This might have been expected in such a highly philopatric species, in which individuals show strong fidelity to their natal colonies. In contrast, Sooty Terns *Sterna fuscata* are scattered throughout the tropical oceans. When not breeding, individuals wander great distances across the open sea, but most tend to return faithfully to their breeding colonies. Colonies within an ocean are only weakly differentiated in mt DNA (control region), as identical haplotypes are shared across nesting sites separated by as much as 16,000 km. This provides evidence of the importance, in terms of gene flow, of the small proportion of individuals that do move long distances between their natal and subsequent breeding sites. However, slight differences are apparent between birds from Atlantic and Indo-Pacific colonies, as expected on geographical grounds (Avise *et al.* 2000).

Range expansion in coastal and inshore seabirds usually occurs as in landbirds, progressively on a rolling front. However, some petrel species have established new nesting colonies in recent decades far removed from any other known colony. The Manx Shearwater *Puffinus puffinus*, otherwise restricted to the eastern Atlantic, has established at least one colony in Newfoundland (1977) and has also bred in Massachusetts (1973). Similarly, Leach's Storm Petrel *Oceanodroma leucorhoa*, which nests primarily in the northern hemisphere, has been found breeding in South Africa and in potential nest holes on the Chatham Islands off New Zealand (Imber & Lovegrove 1982, Whittington *et al.* 1999). This range extension is remarkable because it involves a change in hemisphere; birds on the Chatham Islands were in fresh plumage and, compared to the North Pacific breeders, appeared to have switched their moult schedules by six months. In theory, such long dispersal events might give rise to isolated populations which are then free to diverge from the parental form, eventually giving a new subspecies and ultimately a new species.

CONCLUDING REMARKS

One of the most important questions to ask about seabirds is why there are so few species compared with landbirds. Although seawater covers 71% of the earth's surface, the 320 marine species represent only about 3.3% of extant bird species, while the remaining 9,416 known bird species live on the 29% of the planet that is land. One important factor concerns marine productivity. Despite impressive regional exceptions, the sea overall is one of the least productive ecosystems per unit of surface area. When all the earth's oceans are totalled, they have been estimated to account for only one-fourth of the planet's primary production, a proportion similar to that of all tropical rain forests which cover only 4.6% of the area (Whittaker & Likens 1973). A second important factor is that seabirds (as here defined) are all carnivores. Among vertebrates in general, carnivores are only about one-tenth as speciose as herbivores; in marine systems the herbivores are represented entirely by non-avian taxa, all of which live beneath the surface. Seabirds therefore represent only one component in the spectrum of food types in marine ecosystems, and the niches available for them are limited. A third factor is that relatively few bird species use the vast areas of sea surface that are far from land, the majority being confined to relatively shallow coastal areas. Moreover, all are tied to land for breeding and the numbers of island and continental areas that offer safe nesting sites are limited, despite many species being specialized cliff-nesters. The numbers of seabird species may in fact be more constrained by the restricted land areas suitable for nesting than by the vast water areas that provide their food. Fourthly, throughout the evolutionary history of birds, sea areas have remained interconnected, which must have limited the opportunities for the geographical isolation of populations necessary for speciation to occur. Likewise, coastlines of all the main continents are either connected or lie within flying distance of those of neighbouring continents, again limiting the opportunities for genetic isolation. Not surprisingly, therefore, many species have extensive breeding ranges, stretching to more than one major land-mass or ocean. Fifthly, most seabirds can fly over large distances with ease. This mobility would presumably be another obstacle to the isolation and differentiation of populations. Lastly, life at sea constrains the body designs, colours and sizes of birds to a narrower range of options than life on land, limiting the range of diversity possible. The fact that the plumage of so many seabirds is some combination of black, brown, grey or white, and lacks almost completely the vivid colours of landbirds, is almost certainly the result of convergence. It is probably some combination of these various features of the birds themselves, and of their environment, that limits the numbers of seabird species to a much smaller number than landbirds.

The numbers of seabird species alive today have been reduced to an unknown extent by recent human-induced extinctions, but we have no reason to believe that the proportion of species eliminated is greater among seabirds than among landbirds. Some seabirds could have resisted extinction partly because of their large geographical ranges, partly because of their widely separated, largely inaccessible nesting sites, and partly because younger age classes of many species tend to remain at sea away from the vulnerable nesting colonies. Japanese plume hunters are thought to have eliminated nesting Short-tailed Albatrosses *Phoebastria albatrus* from their last remaining colony on Torishima Island; but the species survived because young birds, away at sea at the time, were able to return in later years and re-establish the colony.

Notable features of modern seabird distributions include the absence, summer and winter, of albatrosses in the North Atlantic and of sulids in the North Pacific, and the far fewer species of auk and storm petrel species in the Atlantic than in the Pacific. As shown by fossil remains, however, most of these differences are relatively recent and attributable to the effects of the glaciations, followed by lack of subsequent colonisation (Chapter 9). Any southern albatross populations that wintered in the North Atlantic must also have been wiped out. The fact that the North Atlantic is not entirely inimical to albatrosses is shown by the two individual Black-browed Albatrosses *Thalassarche melanophris* that each returned annually for

30 years to particular islands in the North Atlantic in the nineteenth and twentieth centuries (Rogers *et al*. 1996, 1998).

The absence of breeding shearwaters in the North Pacific is a different matter. Shearwaters breed in the Hawaiian Islands and also in Japanese waters (Streaked Shearwater *Calonectris leucomelas*), but none is found breeding in the Pacific further north and east. This could be due to competition for food (Brooke 2001). One relevant factor is the greater species richness of auks in the north Pacific which, like most temperate zone shearwaters, catch prey underwater. The second is the huge numbers of Short-tailed Shearwaters *Puffinus tenuirostris* and Sooty Shearwaters *P. griseus* that migrate from the Antipodes into the North Pacific during the northern summer. The numbers of these two species together have been estimated at more than 45 million. Both the rich auk community and the huge influx of non-breeding shearwaters to the North Pacific almost certainly reduce prey stocks, and therefore may have contributed to the absence of breeding shear-waters in this region. Similarly in the North Atlantic, breeding shearwaters are mostly found in the northeast, occurring in trivial numbers in the northwest where transequatorial migrants, especially Great Shearwaters *Puffinus gravis*, are concentrated.

The numbers of seabirds (especially petrels) classed as distinct species, as opposed to subspecies, may be increased by further work, and conversely some taxa currently regarded as separate species might be demoted to subspecies. For example, from external morphology, Jouanin (1970) concluded that the Mascarene Petrel *Pterodroma aterrima* is the Indian Ocean representative of the larger Tahiti Petrel *P. rostrata*, a widespread form of the tropical Pacific, while Olson (1975) drew attention to the similarity in skeletal features between *P. rostrata* and the larger extinct *P. rupinarium* of St Helena. *Pterodroma* petrels need further taxonomic work, as do the various *Puffinus* shearwaters (Warham 1996). In general, mt DNA data support the five subgroups of *Puffinus* discerned on morphological and skeletal features, but also suggest a deeper split between the species that breed widely in both hemispheres (such as the polytypic *P. lherminieri* and *P. assimilis*) and the remaining four subgroups, which breed in the southern hemisphere but migrate to the northern in the austral winter (Austin 1996). It is the apparently limited range of colour options available in seabirds, together with various features of their lifestyles, that influence their speciation patterns, and makes it sometimes hard to delimit one species from another.

What have been the most important obstacles to the movements of seabirds that have promoted speciation? To judge from present distribution patterns, the most significant isolating factor in the northern hemisphere has been the continental land areas. These have led to some recent differentiation between Pacific, Atlantic and Indian Oceans, all of which hold some endemic species, as well as distinct races of widespread species. The second most important factor has been the sea water zones themselves. These constrain the north–south expansions of many species. Tropical waters have evidently provided a major barrier between northern and southern cold water species (though some taxa have crossed the tropics to give similar species in the north and south). In addition, arctic waters have per-mitted movement of arctic and boreal species between the Atlantic and Pacific Oceans, but prevented similar movement by north temperate species. The impor-tance of sea water zones is also shown by the occurrence of several endemic species in the cold Humboldt, Benguela and Arabian waters, all of which are surrounded by warmer tropical seas. The third most important factor for many island-nesting species is the sheer distance between the different island groups, many of which support endemic species and subspecies **(Table 8.3)**. However, the formation and loss of islands through volcanic activity, or through tectonic and glacial sea-level changes, implies that some islands were perhaps not always so remote as they are now. The disappearance of some islands could have provided breaks in the distributions of once-widespread species, facilitating differentiation and speciation on remaining islands.

SUMMARY

Seabirds are distributed primarily according to nine sea water zones, defined largely on the basis of water temperature, and extending mainly latitudinally around the world. The majority of species breed in only one or two of these zones and a progressively smaller number extend to three, four or more adjacent zones. Within these zones, some species are associated with the open ocean, and others with the offshore or inshore zones, or with shelf breaks and mid shelf tidal fronts.

The densities of seabirds seen in particular areas can be related to (1) food availability which is often measured indirectly from carbon levels, chlorophyll readings or plankton densities; (2) distance from the nearest land which is important for land-tied species; and (3) distances from particular nesting places. Species vary greatly in the distances they forage from the nest, with some travelling thousands of kilometres on foraging trips.

Wide land areas provide a barrier to the dispersal of many species, and large sea water areas restrict the distributions of inshore species that do not normally cross from one side of an ocean to the other. All seabirds are further constrained to breed in localities where safe nesting sites are available, which often therefore become extremely crowded. Many once-suitable nesting islands have been rendered untenable by the introduction of mammalian predators, further restricting the distributions of certain species. More now breed only on a single island than would have done so naturally.

In contrast to landbirds, latitudinal range spans of individual seabird species in the breeding season do not increase with latitude. The widest latitudinal breeding ranges are shown by the few species whose ranges are centred near the equator. Several seabirds that breed over a wide span of latitude have longitudinally staggered distributions, breeding at some latitudes on one continent, and at other latitudes far to the east or west on another continent. The gaps in breeding distributions may be exploited by young or migrants from other regions, or by related species.

Many seabirds have long migrations which enable them to exploit areas far from their breeding places, sometimes in another hemisphere. Pre-breeding age groups may also exploit different areas from breeders.

In parallel with some landbirds, some northern hemisphere seabirds have closely related southern hemisphere equivalents, with a wide latitudinal gap between them. Within hemispheres, others have distinct allospecies, subspecies and other differentiated populations in different regions. Allopatry is a much commoner form of ecological segregation among closely related seabird species than is sympatry, where related forms may differ in size, breeding season or feeding area. Procellariiformes are more species-rich in the southern than in the northern hemisphere, Alcinae (auks) are found only in the northern hemisphere and Spheniscidae (penguins) and Pelecanoididae (diving petrels) only in the southern hemisphere.

One form of speciation is possible in seabirds but not landbirds, because it depends on the long-distance foraging of seabirds. Morphologically similar populations can then exploit different (adjacent) marine zones and breed at different seasons to correspond to food-supplies in their particular zone. The two populations can then nest in the same area without interbreeding.

Part Three
Effects of Past Climate Changes

Snowy Owl *Nyctea scandiaca*, a species which was more widely distributed in glacial times.

Chapter 9
Glacial cycles in northern regions: extinctions and distributional changes

The major climatic changes of the last few million years, which included glaciations at high latitudes, inevitably affected the numbers and distributions of plants and animals. To understand how these climatic changes might have left their mark on the present distribution patterns of birds, it is necessary, as far as possible, to reconstruct the recent distributional history of the main vegetation types on which birds and other animals depend. This has been done mainly using dated plant remains, especially pollen in peat and lake sediments, but also palaeo-climatic and geomorphological evidence, particularly on glacial extents.

The past two million years (the Pleistocene era) saw many glaciations of varying severity and extent **(Box 9.1)**. Ultimately, they were caused by periodic changes in the earth's orbit and axial inclination around the sun, and by the resulting changes in solar radiation. The most recent glaciation lasted from about 90,000 to 10,000 years ago, and is the best known, mainly because its effects on the

Box 9.1   Causes and measurement of Pleistocene glacial cycles.

Some authorities prefer to set the beginning of the Pleistocene at about 1.6 million years ago and others at 2.4 million years, but two million years is often taken as a working figure. The end is generally set at 10,000 years ago around the end of the last glaciation, and beginning the present 'Holocene Period'. Estimates of the number of glaciations during the Pleistocene have increased in recent decades as fresh information has become available. Many older accounts list four, but more recent evidence suggests about 21 in the last 2.6 million years, during which time climatic cycles have increased in length and amplitude. About 900,000 years ago a shift occurred from cycles lasting about 41,000 years to longer cycles of about 100,000 years, in which glaciations lasting 60,000 to 90,000 years alternated with interglacials lasting between 40,000 and 10,000 years, respectively. These cycles, named Milankovitch cycles after their discoverer, are thought to be caused by regular changes in the ellipticity and tilt of the earth's annual orbit around the sun and associated changes in solar radiation.

The main climatic effect of these cycles is in the degree of contrast between summer and winter temperatures which, in association with changes in atmospheric composition (mainly carbon dioxide levels), causes glacial–interglacial cycles. During glacial maxima, mean world temperatures were 4–8°C cooler than in the interglacials, but the difference was much greater at high than at low latitudes. The most recent glaciation is known as the Würm or Weischelian in Europe, the Zyrianka–Sartan in Siberia and the Wisconsonian in North America. Overall, glacial conditions prevailed for most of the Pleistocene, and interglacial conditions as they are today have probably existed for only about one-tenth of the Pleistocene Period, and of the last 100,000 years. Throughout, glaciers seem to have melted much more rapidly than they formed, helped by the natural accumulation of carbon dioxide in the atmosphere.

The evidence for past global climate changes has come mainly from cores taken from sea and lake bottoms and ice sheets, which were then analysed for oxygen and carbon isotopes and for other physical and biological clues. Past temperatures over periods of hundreds of millions of years can be estimated by the oxygen isotope method. In water the two common isotopes of oxygen are ^{16}O and the heavier ^{18}O. The lighter isotope evaporates more rapidly, especially during warmer periods. Thus with some sophisticated calibrations and analyses, the ratio of ^{18}O and ^{16}O locked in the shells of marine organisms or in ice cores can serve as a palaeo-thermometer. Over shorter time periods of hundreds of thousands of years, ice cores obtained from the arctic and antarctic provide information on past precipitation (from the layering), on past temperatures (from the structure of the ice crystals) and on the composition of the atmosphere including carbon dioxide levels (from trapped air bubbles). In general, carbon dioxide and methane levels match the estimated changes in temperature. Direct evidence for past changes in plant communities has come from identified pollen remains found in cores taken from peat, lake and sea beds.

land surface have not yet been obliterated by subsequent ones. At its maximum extent, some 26,000–14,000 years ago, one-third of all land in the northern hemisphere was covered with ice, as were large areas of sea in the polar regions of both hemispheres. In Europe and North America, biomes were shifted 10–20° of latitude south from where they are today, but occupied the same relative positions north-to-south, of tundra, boreal forest, temperate forests and semi-arid scrub or grassland. This is because the climatic belts of the earth, then as now, created zonal patterns in vegetation. In addition, north–south mountain ranges permitted arctic and boreal plants and animals to extend much further south than today, as is apparent from fossil and other evidence.

As well as covering large land areas, ice formation locked up so much water that it lowered sea-levels worldwide by an average of 120–130 metres; land areas were thereby expanded, and many that are now separate became joined, thus facilitating the movements of plants and animals. Eurasia and North America became connected across the Bering land-bridge in the north Pacific, and many islands became joined to neighbouring land-masses, such as Britain to Europe and Japan to Asia (Chapter 6). Worldwide, about as much land was laid bare as was covered by ice.

During glaciations, world climates also became drier. In consequence, moisture-dependent forest and woodland habitats became restricted and fragmented, while more arid and open habitats expanded. Conversely, during the warmer and wetter interglacials, forest reached its greatest extent, and open habitats their lowest (Frenzel 1968, Frenzel *et al.* 1992, Prentice *et al.* 2000). Big reductions in habitat areas presumably led to big reductions in the population sizes of plants and animals, pushing some species through demographic bottlenecks and causing extinctions of others. In contrast, habitat expansions would have led to growth in total population sizes, giving them for a time greater security from extinction.

Such marked changes happened about 21 times during the past 2.6 million years (**Box 9.1**). Not all glaciations were equally cold, however, and not all interglacials were equally warm. The present interglacial is now past its warmest period, which occurred 7,500–4,000 years ago, depending on the region, and is less warm than the previous interglacial which peaked about 130,000–115,000 years ago. In some previous glaciations, sea-levels dropped as much as 200 m below their present levels, and in some of the strongest interglacials they may have risen as much as 70 m above. Overall, glacial conditions prevailed over nine-tenths of the last 2.6 million years and warm interglacials (as now) over only one-tenth. So the benign conditions in which most of us now live are exceptional in the context of this longer time scale.

Superimposed on these major climatic changes were more minor or more localised ones which gave temporary or regional exceptions to the overall pattern of global climatic and vegetation changes. Some areas experienced wet (pluvial) conditions when comparable areas elsewhere were dry. The western United States, for example, experienced a pluvial period between 25,000 and 10,000 years ago, at the height of the last glaciation, as did parts of Africa earlier in the same climatic cycle between 90,000 and 55,000 years ago. In addition, in the North Atlantic region very sudden climatic changes were apparently caused by the changes in the Atlantic conveyer, the current that carries the warm Gulf Stream north towards Europe. The stopping or starting of this current (in turn attributed to changes in freshwater input to the North Atlantic from run-off or ice melt) led to abrupt mean temperature changes in northern Europe of more than 10°, each change occurring within the space of a few years. These changes also had major and relatively rapid effects on flora and fauna. Events of this type occurred about 13,000 and 10,000 years ago, and one of the most recent was the 'Little Ice Age' of about AD 1650–1850.

To survive such changes, organisms must either change their geographical ranges so as to track suitable conditions, or else they must evolve so as to adapt to the new conditions. Both types of response have been confirmed in both plants and animals by fossils and other dated remains, but the most marked response in most species was change in range.

GLACIAL ADVANCE

Glaciers form from compressed snow, and spread when the weight of snow added each winter is too great to be melted the following summer. As the weight of snow accumulates over the years, the ice sheet — by then kilometres thick — creeps outwards, obliterating everything in its path. In this way, mature forests growing in a climate still hospitable to trees can be overrun and crushed by the advancing ice. Further spread is halted either by a reduction in snowfall over the glacier, perhaps hundreds of kilometres in from the edges, or when the edges reach an area where summer temperatures are high enough to melt the ice as fast as it arrives, or when the edges reach the sea and break off as icebergs. Shrinkage of an ice sheet can be brought about by a further reduction of snowfall over the central parts, or by rising temperatures which increase the melt rates at the edges. A consequence of the first process is that ice can withdraw from more than one edge simultaneously, not just from the warmest.

The climate change that led to the last ice age was apparently rapid, with estimates ranging from a few decades to a few centuries. Mean world temperatures are estimated to have fallen by 4–10°C, but most markedly at the highest latitudes. As a result, in regions with sufficient precipitation, polar and montane ice caps began to grow until they covered much of the northern land-masses. Ice reached its southernmost extent about 18,000 years ago, reaching 40°N in North America and 50°N in western Europe. In the drier climates of eastern Asia and Alaska, however, glaciers were much more restricted, and large expanses of vegetation survived at high latitudes (**Figure 9.1**).

As glaciers spread, plant communities were gradually obliterated or forced southwards, forest ahead of tundra. Wherever glaciers spread by expansion from the centre, forests could have been crushed and buried as described above, but where glaciers spread through local climatic cooling, vegetation would have been affected ahead of the ice. The less hardy plants would have gradually died off, and permafrost would have formed, inhibiting the growth of trees, but allowing the development of tundra. Further south, within the forests, imagine a process in which individual plants died at the northern boundary, and, through seed dispersal, established themselves beyond the southern edge, as changing conditions enabled them to prevail over more southern plant communities, disadvantaged by the increasing cold. In this way, over many tree generations, different plant

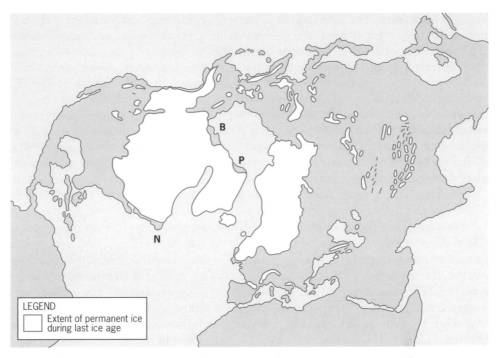

Figure 9.1   Approximate distribution of ice in the northern hemisphere at the height of the last glaciation about 14,000–26,000 years ago. Note that some shorelines were also expanded then. From various sources.

Note the areas that remained free of ice at high latitudes, namely the Angaran refuge in north–central Siberia, the Beringian refuge from eastern Siberia across the dry Beringian land-bridge into Alaska, the Banksian Refuge (Banks Island and nearby mainland), northeast Canada (parts of Labrador and Newfoundland and nearby offshore area), northeast Greenland (Peary Land). Other ice-free areas not shown included small areas of northwestern Norway, western Canada and Alaska, and possible areas now submerged, including shelf areas off Newfoundland, southwest Greenland and west Iceland. Spitzbergen appears to have been completely glaciated.

communities would have spread slowly south. However, different species of trees spread at different rates: some species might have been overtaken by the ice, unable to move fast enough, so that they disappeared altogether from the region concerned. In Europe, forest was pushed into the southern peninsulas of Iberia, Italy and the Balkans, with another area further east between the Black and Caspian Seas. The majority of tree species were apparently confined then to altitudinal forest belts on mountains, the tops being glaciated or rocky and the lower land too dry, but other trees may have occurred along rivers and lakesides as in arid lands today (Huntley 1993)[1]. Hence, the total forest area in Europe

---

[1] *A modern analogy to this situation is seen in many parts of the western United States, where forest plant species are found in a discrete altitude zone, consisting of conifer forest higher up and broad-leaved forest further down, giving way on lower ground to scrub-steppe and desert.*

became enormously reduced and fragmented, compared with earlier and later times, and the forest areas in the different southern peninsulas and islands were separated from one another by open land and sea.

By contrast, at the eastern end of the Eurasian land-mass, which remained largely free of ice, a huge area of mainly conifer forest apparently remained *in situ* and intact, extending southwards into northern China (Moreau 1955, Kurtén 1972, Frenzel *et al.* 1992, Yu *et al.* 2000). From this forest, a broken strip of conifer and broad-leaved trees ran westwards across Siberia and southern Russia to 60⁰E, at higher latitude than any forest that remained in western Europe. In far eastern China, this taiga forest joined large areas of broad-leaved deciduous and broad-leaved evergreen forest displaced to the southwest of that country (Yu *et al.* 2000). Other major forest refuges were situated on either side of the Himalayas, in Turkestan and Yunnan, respectively.

At the height of the glaciation, much of the remaining ice-free land in Europe and western Asia was covered by tundra or by 'steppe tundra', a dry windswept grass-and-shrub habitat which has no modern equivalent, but which held many plant and animal species that are now found on dry tundra or steppe. Between the forest areas in southern Europe, the predominant vegetation has been likened to some form of steppe or steppe-woodland, but in some semi-arid regions expanses of dry sclerophyllous scrub (of a type today variously called chaparral, maquis or matorral) survived in especially favourable areas (see later). In fact, all the major European vegetation types and their birds were represented in the Mediterranean region, occurring in a mosaic pattern, depending on altitude and precipitation. Moreover, in the northwestern Sahara, benefiting during glacial periods from wet summer winds off the Atlantic, Mediterranean vegetation spread south as much as 600 km from its present position, providing a wide expanse of habitat not available today, as evidenced by both plant and animal remains. This seems not to have happened in the drier eastern Sahara not reached by moisture-laden winds (Moreau 1954).

In North America, too, the advancing ice forced the forest southwards. Major refuges lay in the southeast, southwest and Mexico. Although the total forest area became reduced, and separated into at least three main regions, it was probably less fragmented than in Europe, and still covered very large areas; it was also in contact with tropical forest to the south (COHMAP 1988, Williams & Webb 1996). The main southeastern forest was mixed pine–deciduous, giving way westwards to open deciduous woodland in an increasingly grassy, open landscape in the present Gulf States. In the southwestern States where rainfall was higher than today, open conifer woodland covered areas that are now under scrub and desert (Thomson & Anderson 2000). Further north, but south of the ice, was a narrow interrupted strip of tundra-like vegetation which, on the eastern seaboard, swept north as far as Newfoundland, in a low-lying area now partly submerged as continental shelf. In southern mountain areas the treeline was depressed by 1,200–1,500 m, with alpine vegetation below the glaciated summits. Other arctic–alpine biota evidently survived on nunataks — refugia that persisted within ice sheets, as on mountain peaks or coastal cliffs too steep for ice.

In Europe and much of North America, therefore, each glaciation had four main effects on the forest: displacing its position southwards, in the process eliminating some of the original tree species, greatly reducing its total area and fragmenting it, and leaving large stretches of open land (and in Europe sea) between the remnants. During successive glaciations, as revealed by pollen and other remains, many more tree species were eliminated from Europe than from North America. This has been attributed partly to the fact that in the mid latitudes of Europe the main mountain chains run west–east rather than north–south. The mountains would have glaciated earlier than the lowlands to the north, forming in Europe a formidable barrier to the passage of many plants which, trapped against the mountains, would then have been eliminated by the advancing cold (Reid 1935). In addition, the forests were reduced to much smaller areas in Europe, which is likely to have caused further extinctions (Huntley 1993). At the glacial maximum, the European forests may have occupied less than 5% of the area they covered during an interglacial, whereas at the same time the North American forests could have occupied up to 30%.

This difference could explain why western Europe now has far fewer forest plant species than both eastern Asia and North America (where plants were able to migrate southwards ahead of the ice and persist in larger refuges), and why many of the present Asian and eastern North American tree genera are represented in Europe only by pre-Pleistocene fossils. In temperate latitudes, the forests of eastern Asia now have three times more tree species than forests in eastern North America and six times more than those in Europe (Latham & Ricklefs 1993). The number of tree genera represented in the fossil record of Europe declined abruptly around the start of the Pleistocene, as the most cold-sensitive species were lost, and then more slowly through successive glaciations (Coxon & Waldren 1997). It thus seems that a continuous forest flora once existed across the northern continents, and that this forest was subsequently broken up by the glaciations, which obliterated many species in western Eurasia, leaving disjunct populations of those species stranded in Asia and North America. Through time, some of these populations differentiated as new, but closely similar species, while others remained morphologically unchanged.

Habitats akin to modern tundra or cold steppe had a wide but disjunct distribution during the glacial maxima. They were found in several widely separated areas in the far north which for some reason escaped the ice, and in separate belts south of the ice (**Figure 9.1**). In the north, steppe-tundra formed the dominant vegetation on the dry Beringian region that for 60,000–70,000 years of the last glacial cycle linked Siberia with Alaska (**Figure 9.2**). It also occurred further west in northern Siberia (the so-called Angaran refuge), on some northern Canadian islands (the Banksian refuge) and on a small part of northeast Greenland (the Peary Land refuge) (**Figure 9.1**; Salomonsen 1972). All these latter areas are still under tundra today, and all carry some distinctive plants and animals. South of the ice, during the glaciation, tundra occurred in a narrow, interrupted belt across North America, and in a much broader belt in Eurasia where, in the form of steppe-tundra, it was the most prevalent vegetation type. All these southern areas later disappeared under forest or sea, as the ice retreated.

Because deserts in the northern hemisphere were located mostly south of 40°N latitude, they were not glaciated, but in places were pushed to the south. Moreover, because temperate zone climates were cooler than today, evaporation rates were lower, so deserts became reduced in extent and some held large freshwater or saline lakes. Many such lakes occurred in regions that are deserts today, leaving

Figure 9.2   The pattern of land, sea and ice around the Beringian refuge at the last glacial maximum, 14,000–26,000 years ago. The fine lines show the present coastlines. Modified from Pielou 1991.

Beringia was made up of three components, the ice-free parts of Alaska and the Yukon, roughly the same area of the northeastern peninsula of Siberia, and the now submerged Beringian land-bridge that united them. Thus Eurasia and North America were joined into a single supercontinent, cut in two by an enormous expanse of ice lying across North America. To the west, Beringia was largely separated from the Eurasian steppe-tundra by ice sheets on the east Siberian and Kamchatkan mountains. During the height of the glaciation, the region was covered primarily with a mosaic of different types of tundra-like vegetation, perhaps with trees in the milder spots, but as the climate warmed it became more forested, thus providing a dispersal route and refuge for both forest and tundra species for some thousands of years. In addition, the Pacific Ocean was isolated from the Arctic Ocean to the north, and hence also from the Atlantic. Little wonder that Alaska and eastern Siberia share many bird and other animal species, and that many of the seabirds of the northern Pacific are distinct from those of the Arctic and Atlantic Oceans. The Beringian land-bridge was finally broken about 14,000 years ago when rising seas flooded much of the area to create the modern Bering Strait, linking Pacific and Arctic Oceans.

remnants in the form of saline lakes or dry lake beds in all main desert regions in both hemispheres. The most notable include the Great Salt Lake in western North America which now covers 10,000 km², but once occupied more than 50,000 km² (called Lake Bonneville). The disappearance of such large lakes from all the continents since the Pleistocene must have had profound biogeographical effects.

In the southern hemisphere outside Antarctica, glaciation was much more limited than in the north, mainly because the southern tips of the southern continents lie relatively far north. Ice was confined to the highest mountains, including those within the tropics. This had the effect of pushing subalpine grassland and forest to lower elevations. South America was glaciated mainly along the Andes, with the greatest extent of ice in the south, while mainland Australia remained ice-free, except for the highest parts of Tasmania and a small mountain range in Victoria, as did Africa except for the Atlas mountains in the northwest, and some other high mountains in the east. In contrast to continents, islands at high latitudes offered only limited scope for shifts of vegetation, and many of the plant species that survived were limited to special sites such as steep slopes and gorges. The islands of New Zealand became joined together because of lowered sea-levels, and the resulting land-mass became partly ice-covered in the south, grass- and shrub-covered in the centre, with forest restricted chiefly to the west of South Island and most of North Island. The treeline on the New Zealand mountains was more than 800 m lower than now.

GLACIAL RETREAT

As the climate warmed and ice retreated, from about 14,000 years ago, forest in the northern hemisphere could spread back northwards, the fragments eventually coalescing to reform continuous belts across the Eurasian and North American land-masses (Huntley & Webb 1989). This northward spread, it may be assumed, occurred through the various plant species expanding from the northern side of their distributions and contracting from the southern, so that over the tree generations, forest crept slowly northwards in response to a warming climate, being replaced by the next vegetation community from further south. Again, to judge from the pollen record, tree species moved north at different rates, reflecting their differing needs and differing abilities to migrate into newly suitable habitats (Davis 1981, 1986, Pielou 1991, Huntley 1993).

Tree species that mature rapidly and produce wind-dispersed seeds had an obvious advantage. For example, birch *Betula* is cold-tolerant so could have survived near the glaciers; it can reproduce when only a few years old, and the seeds are so light that they can be carried by wind. It is not surprising, therefore, that birch should be among the first trees to appear abundantly in the post-glacial pollen record of northern Europe, to be followed by pine *Pinus*, elm *Ulmus*, oak *Quercus*, hazel *Corylus*, alder *Alnus* and lime *Tilia*, the latter being the least cold-tolerant of these species. The rapid rates of spread recorded in some slow-maturing species (such as oak, whose post-glacial spread reached 500 m per year in some areas) were probably dependent on their having bird-dispersed seeds (oak seeds are harvested by Eurasian Jays *Garrulus glandarius* and cached in open ground away from existing trees, which would have greatly enhanced the rate of spread). Nevertheless, the northward advance may have led to further species impoverishment, as slow-moving plant species were left behind to die out, their place being taken by more southern species.

In North America, the first tree species to appear on ice-free ground were the aspens and birches, followed by spruces, then pines, firs and a variety of hardwoods. As in Europe, different tree species spread northwards from their various refugia at different starting dates and at different speeds (Davis 1981, 1986). For example, Eastern White Pine *Pinus strobus* and Eastern Hemlock *Tsuga canadensis* are thought to have survived the last glaciation in the eastern foothills of the Appalachians and adjacent coastal plain. In contrast, Chestnut *Castanea dentata* and Sugar Maple *Acer saccharum* are thought to have survived chiefly around the mouth of the Mississippi River. The two pairs of species therefore began their northward migration from different starting points. Based on pollen remains, average annual rates of advance in the four species were calculated at 300–350 m, 200–250 m, 200 m and 100 m respectively. At these rates, these species took 2–3 thousand years to reach the furthest parts of their present ranges, and may still be on the move. As some tree species moved faster than others, woodland composition was constantly changing, and some species, now recognised as dominants, are relatively recent arrivals. Variation in the starting dates, travel rates and travel routes of different tree species is thus held partly responsible for current gradients in tree species diversity across the northern continents, with locations of high species numbers corresponding with the locations of former refuges or with areas first to be recolonised (Davis 1981, 1986, Huntley 1993).

In view of the rapidity of climate change, climate can be assumed not to have limited the rate at which trees spread north. Rather the migration rates of the trees themselves (dependent on generation time and seed dispersal) probably had most influence, modified in some places by development of suitable soil and other conditions. Hence, vegetation development was inevitably somewhat delayed relative to temperature change. Pollen recovered from lake sediments in Italy indicates that, among the numerous vegetation changes recorded in the last 100,000 years, alterations from woodland to steppe with trees (or vice versa) sometimes occurred within periods of 180–200 years (Allen *et al.* 1999). Yet the transition from glacial to post-glacial temperature could have occurred in less than 50 years (some authorities claim less than 10 years), with mean July temperatures in England, for example, rising in this time from 8°C to 17°C. This is equivalent to the present difference in July temperature between the Taimyr peninsula of Siberia and London, or between Baffin Island and Ottawa.

It is thus apparent that past climate changes did not merely involve simple spatial shifts and changes in the areal extents of modern environments. The species composition and structure of forest communities were continually changing, and unique combinations of temperature and precipitation occurred at different places at different times, with additional effects on flora and fauna. Hence, conditions in southern Europe (say) during a glaciation were not the same as those in northern Europe during an interglacial. Probably, therefore, the past bird communities of particular habitats were somewhat different from those of today, both in the numbers and types of species present and in their relative abundances. Some evidence from the fossil record, such as it is, supports this view (see later).

Another feature of de-glaciating landscapes resulted from the enormous volumes of water released from the melting ice, which gave rise to vast lakes at the southern edge (e.g. modern Lake Superior, the world's largest freshwater lake, is less than one-fourth the size of the Glacial Lake Agassiz, which covered about 350,000 km^2 of eastern North America during the early Holocene, 11–12,000 years ago). Such vast lakes would have provided, for a few thousand years, large expanses of summer habitat for northern aquatic birds much greater than those available today, and in places may have greatly slowed the northward advance of trees. In addition, under the great weight of ice, the land was pressed down, and after glacial retreat, it took hundreds or thousands of years for the downwarped crust to rebound (a process that continues in some areas). Consequently, as temperatures warmed, rising oceans inundated suppressed regions of continents, creating extensive shallow seas. Some 10–12,000 years ago, for example, the St Lawrence valley and Great Lakes of North America were flooded from the Atlantic. As the land slowly rose, the sea water receded. Shallow marine habitat was thus also much more available then than now.

The nature of refuges

The term 'climatic refuge' is used loosely for areas where plants and animals survived adverse periods and from which they could have later spread, but it is clear from the foregoing that there were at least two types of refuges:

(1) Stationary refuges — those parts of the original biome that survived *in situ* and intact, in regions that escaped the worst effects of the climatic extreme. Such refuges are likely to have held the entire original flora and fauna present in those areas, except for any species lost as a consequence of reduction in area. Examples include some large forest areas in eastern Asia and the tundra (or steppe-tundra) in eastern parts of the Beringian refuge.

(2) Displaced refuges — those parts of the original biome pushed to lower latitudes during glaciations, and whose original flora and fauna were impoverished during periods of movement, as described above, but which nevertheless provided continuously available habitat for their animal species. Examples include most of the forest areas of Europe and North America that now lie at northern and mid latitudes in formerly glaciated areas.

While both types of refuge might have lost some species because of their reduced area, displaced refuges would be expected to have lost additional species through movement, including all those plant species unable to move fast enough to escape the changing climate.

Extinctions of large mammals

During glacial changes, mobile animals would in general be expected to keep up with their moving habitat, but losses of some animal species would be expected following the loss of plant species, as well as from reductions and fragmentation of habitat areas. Surprisingly, however, many large mammal species disappeared from the northern hemisphere, not during the peaks of the glaciations as expected, but around the end of the very last glaciation, some 12,000–9,000 years ago, raising suspicion that some other factor was involved. Holarctic examples include the Mammoth *Elaphas primigeneus* and Woolly Rhinoceros *Coelodonta antiquitatis;* Eurasian ones the Giant Deer (Irish Elk) *Megaceros giganteus* and Auroch *Bos primigenius;* and North American ones the Giant Ground Sloths *Nothrotheriops* and *Megalonyx* and the Mastodon *Mammut americanum.*

Several lines of evidence point to human involvement. One is the expansion of human populations that occurred at that time, including the colonisation of the New World by hunting people from Asia (Martin 1984, Diamond 1991). Fossil evidence shows that humans and large mammals co-existed for a time in the Americas, and that the people hunted the large herbivores that became extinct. Human artifacts have been found along with the bones of the dead animals. Secondly, the extinctions of large mammals appear to have begun in the north and proceeded rapidly and systematically southwards, in line with the spread of people (Diamond 1991, Brown & Lomolino 1998). Thirdly, immigrant mammals, such as Caribou *Rangifer arcticus*, Moose *Alces alces* and Bighorn Sheep *Ovis canadensis*, which had evolved with humans in Eurasia, fared much better in North America than the native species. Fourthly, extinctions affected many more large mammal species than smaller ones, which were harder for people to hunt. It would not have been necessary for people to have killed every last individual to cause extinction, only to take more than could be offset by reproduction; and such large species would have had low breeding rates.

At the same time, the transition from late Pleistocene to Holocene was marked by widescale vegetation restructuring. This correlation has been used as the main argument for the alternative hypothesis, that many species died out because of changing conditions. But if this were so, it is hard to understand how all these animals survived previous climate cycles, which had similar effects on vegetation, and why mainly large species were affected. The two hypotheses of human over-kill and natural causes are not mutually exclusive, and it might have been some combination of the two effects that tipped the scales against so many species at the same time (for further discussion of pros and cons, see Pielou 1991). Moreover, the disappearance of the larger herbivores, which are likely to have had major effects on vegetation structure, may have predisposed the extinctions of other herbivores, as well as of their various predators and scavengers. Included among the latter were some large carrion-eating birds mentioned below.

EFFECTS ON LANDBIRDS

Fossil finds have confirmed that some former European bird species present in the Pleistocene are now extinct, while others have gone from western Eurasia but not from further east (Tyrberg 1998). Extinct species include a large vulture *Gyps melitensis*, which died out at the same time as the large mammals on whose carcasses it is presumed to have fed, a peafowl *Pavo bravardi*, a junglefowl *Gallus europaeus* related to the present Asian form *G. gallus* from which domestic fowl were allegedly derived, and possibly a duck, the Thick-legged Eider *Somateria gravipes*, the validity of which has been questioned.

Species present in western Europe in the late Pleistocene, but now restricted to more eastern areas, include Steppe Eagle *Aquila nipalensis* and a large thrush, probably Scaly (White's) Thrush *Zoothera dauma*. One of the most surprising is a supposed Mandarin Duck *Aix galericulata*, whose fossilised bones were found in mid Pleistocene deposits in southern England (Harrison 1985). If correctly identified, these bones raise the possibility that this duck, which is now restricted to eastern Asian forest, may once have extended across the whole of Eurasia in the same way that its relative, the Wood Duck *A. sponsa*, extends in forest across North America today. Many other species were much more widespread in western Europe than they are now, notably Imperial Eagle *Aquila heliaca*, Great Snipe

Gallinago media and Red-billed Chough *Pyrrhocorax pyrrhocorax* (Tyrberg 1998). Yet other species survived the Pleistocene glaciations, but became extinct in the last few thousand years, perhaps through human hunting. The most striking was a large European Crane *Grus primigenia*. Most parts of the northern hemisphere currently hold two crane species of different size, so the existence of a large extinct species, along with the smaller extant Common Crane *Grus grus*, is perhaps not surprising. Well into the Holocene, the Eagle Owl *Bubo bubo*, Black Stork *Ciconia nigra* and Hazel Grouse *Bonasa bonasia* persisted in Britain, but no longer occur here.

In addition, Guineafowl (Numididae), now confined to the tropics, were represented in Europe until the late Pleistocene or early Holocene (Bochenski 1985), and other species, now restricted to regions south of the Sahara, occurred north into Egypt, so were in much closer contact with Palaearctic species than they are now. Examples include the African Fish Eagle *Haliaeetus vocifer*, Marabou Stork *Leptoptilos crumeniferus* and White-backed Vulture *Gyps africanus* (Tyrberg 1998). In contrast, in eastern Asia where glacial effects were less severe, and where forest remained throughout the climatic cycles in a more or less continuous stretch from boreal to equatorial regions, representatives of mainly tropical Indomalayan families, such as Zosteropidae, Pycnonotidae, Timalidae and Campephagidae, still occur in the temperate zone to the north. The same is true in the New World, where some neotropical families, such as Trogonidae and Cracidae, extend into North America.

Fossil evidence has also confirmed that birds moved with their habitats during glaciations, maintaining broadly similar habitat preferences to those that they show today. Thus species associated with arctic tundra, and today found chiefly in the far north, such as Willow Ptarmigan *Lagopus lagopus* and Snowy Owl *Nyctea scandiaca*, were well represented during the last glaciation as far south as lowland France and Italy where pollen remains confirm the presence of tundra. A Snowy Owl family is also depicted in cave art from southern France in the same period (Parmelee 1992). Likewise, species now confined to the northern forests were present further south during glacial periods, with remains of Spruce Grouse *Dendragapus canadensis* found in Virginia and Georgia, of Ruffed Grouse *Bonasa umbellus* in Florida and of Northern Hawk Owl *Surnia ulula* in Tennessee (Selander 1965, Wetmore 1967). At the same time, Rock Ptarmigan *Lagopus mutus* were found in the Appalachian mountains of Virginia.

Another consistent finding, for both plants and animals, is that species that now occur in different geographical regions once lived together, apparently in the same community in the same region. Remains of species from quite disparate modern assemblages are thus found in similar-age deposits from the same place (e.g. Mourer-Chauviré 1988). In the upper Pleistocene deposits of central Europe, such northern species as Merlin *Falco columbarius*, Eurasian Dotterel *Eudromias morinellus* and Whimbrel *Numenius phaeopus* were found together with the Black Vulture *Aegypius monachus* and Alpine Swift *Tachymarptis melba* (Bochenski 1985). Similarly, at Corrèze in France, Rock Ptarmigan *Lagopus mutus* and Snowy Owl *Nyctea scandiaca* were found alongside Lesser Kestrel *Falco naumanni*, Rufous-tailed Rock Thrush *Monticola saxatilis* and *Alectoris* partridges (Mourer-Chauviré 1975).

Alternative explanations of such puzzling mixtures of species invoke (1) errors of identification, possible in some cases but not in others; (2) deposits of different ages becoming intermixed; (3) local scale juxtaposition of habitats (as on moun-tainsides); or (4) residents, summer visitors and winter visitors to the same area being intermixed in fossils (as expected). However, the same criticisms cannot be made of other kinds of animals or plants, which are not migratory and can be identified specifically with greater certainty. Such unexpected species mixtures, by today's standards, give further indication of the individualistic responses of different species to climate change, including not only plant species but birds and other animals.

In North America, too, Pleistocene extinctions affected birds, as well as mammals (Steadman & Martin 1984). Avian casualties included *Titanis walleri*, a giant flightless bird in a family otherwise known only from southern South America, various species of waterfowl among which were the scoter-like 'diving geese' *Chenodytes lawi* and *C. milleri*, two turkeys *Parapavo californicus* and *Meleagris leopoldi*, a stork *Ciconia mattla*, an unnamed owl much larger than any modern species, and many other flesh-eaters, especially carrion feeders. The latter included at least three accipitrid vultures (*Neogyps errans*, *Neophrontops vallecitrensis* and *N. americanus*), a group now represented (by different species) only in the Old World, together with several cathartid vultures, such as *Pliogyps fisheri*, *Breagyps clarki* and *Cathartarnis gracilis*, and two giant teratorns *Teratornis merriami* and *T. incredibilis*. The teratorns were huge: *T. merriami* had a wing span approaching 4 m and *T. incredibilis* had a wing span exceeding 5 m. These measurements compare to 3 m in the modern California Condor *Gymnogyps californianus* (Steadman & Martin 1984). Large teratorns and other vulture-like birds also occurred in the Plio-Pleistocene grasslands of South America and, like their North American equivalents, they disappeared at the same time as the majority of the big mammals that provided their food **(Box 9.2)**.

The fossil record for the late Pliocene and Pleistocene of Florida is especially good, enabling the history of the fauna over the last 2.5 million years to be recon-structed in exceptional detail (Emslie 1998). This low-lying area was especially susceptible to sea-level changes, decreasing to about one-half of its present size in the warmest interglacials and doubling in size in glacial times. Covering this whole period, 239 bird species have been found in Florida as fossils, of which 60 (25%) are now totally extinct. As elsewhere, the late Pleistocene extinctions included several large scavenging birds, such as teratorns, cathartid and accipitrid vultures, eagles and the California Condor *Gymnogyps californianus*, which sur-vived in the west. Of the 162 surviving species represented as fossils, 77 (48%) still breed in Florida, 64 (39%) occur as regular visitors or transients, while 21 (13%) are now rare or absent. Some species that presumably occurred in Florida during colder times now breed further north or west (e.g. Whooping Crane *Grus americana*, Northern Saw-whet Owl *Aegolius acadicus* and Band-tailed Pigeon *Columba fasciata*); while others that presumably bred in Florida during warmer times now breed further south in subtropical wetlands (e.g. Northern Jacana *Jacana spinosa*, Great Black Hawk *Buteogallus urubitinga*) or grasslands (Yellow-headed Caracara *Milvago chimachima*, Southern Lapwing *Vanellus chilensis*). Hence, these records provide further evidence of north–south range changes in birds during the past climatic cycles.

Not surprisingly, the majority of the Pleistocene fossils mentioned above refer to large species, whose bones are more readily preserved and identified. The

Box 9.2 Extinct predatory birds from the neotropical region.

The bone remains of several large predatory and scavenging birds have been found in southern South America. Like their North American equivalents, these birds probably disappeared around the end of the Pleistocene, some 12,000–9,000 years ago. Besides the surviving Andean Condor *Vultur gryphus*, they included at least four other cathartid species found in Argentina (Tambussi & Noriega 1999). They also included an even larger teratorn than the North American ones, namely *Argentavis magnificens*, which had an estimated wing span of 6–8 m and a weight of 72–79 kg (Campbell & Tonni 1980). This species was dated earlier than the others, at 8–5 million years ago (late Miocene), and occurred on the pampas, at that time a place of extremely strong winds. This could well be the largest flying bird that has ever lived.

Similarly, the remains of several species of large owls and raptors were discovered in Pleistocene deposits from Cuba and other Antillean Islands. They included a giant owl *Ornimegalonyx oteroi*, two species of giant Barn Owls (*Tyto*), a giant eagle *Aquila borrage* and a vulture (*Antillovultur*) similar in size to the Andean Condor (Arrendondo 1976). At that time, Cuba supported large populations of herbivorous mammals (rodents and edentates), but apparently no carnivorous mammal, so the birds of prey may have had this food-source to themselves. Whether these birds and their prey disappeared from natural causes in the late Pleistocene, or whether they survived until people reached the island a few thousand years ago, is an open question. Nevertheless, these and other records serve to confirm that many species of the Neotropical Region have become extinct since the late Pleistocene, and that some surviving species formerly occurred in regions in which habitats are now unsuitable for them.

weaker bones of small birds are less often preserved as fossils, and are much more difficult to identify to species. Many closely related passerine species alive today are readily distinguished on plumage but not on skeletal features, especially the fragmental remains usually available. There is no telling how many other bird species, not skeletally distinguishable from living relatives, died out too.

The figure of 25% of species found among the Florida Plio-Pleistocene avifauna that are now totally extinct compares with 17% of non-passerine species found in the Rancho La Brea Tar Pits in California, and with 23% and 17% of non-passerines in the two tar seeps in Peru and Ecuador (Campbell 1976, 1979, Feduccia 1999). If these figures are typical, we can assume that about a fifth of all non-passerines species disappeared from these New World areas during the last 2.5 million years. Some of those, like the large scavengers that died out near the end of the Pleistocene, could have been victims of human activity. Because an equivalent number of 'new' non-passerines were unlikely to have evolved in this time (Chapter 10), the total numbers of living non-passerine species must have declined during the Pleistocene. Owing to scarcity of identified material, it is impossible to say what proportion of passerines also became extinct in the last 2.5 million years, but in these birds many new species could have evolved in this time (Chapter 10).

While animals responded to climate change mainly by movement, some adapted to some extent, particularly by changes in body size. Fossil bone remains dating from cold periods of the Pleistocene were often slightly larger than are bone remains from the same species today, a difference attributed mainly to change in ambient temperature (Bergman's Rule, Chapter 2; Stewart 1999). It is known from recent studies that size variation is adaptive, and that in small birds changes in body size can evolve within a few generations (Chapter 3; Grant 1986). Body size fluctuation during the Pleistocene is thus taken as an evolutionary response to change in climate. The lack of skeletal change in other species cannot, of course, be taken to indicate lack of evolutionary change in other aspects of structure, or in behaviour or physiology.

EFFECTS ON SEABIRDS

In general, the distributions of seabirds are temperature-dependent (Chapter 8), like those of landbirds, so are also likely to have shifted southwards during glaciations, with major concentrations at mid latitudes, such as California–Oregon in western North America and the Mediterranean region in Europe (Moreau 1954). Likely east Atlantic refuges for oceanic species included the Azores, Madeira and Canary Islands, as well as various Mediterranean islands and coastlines, on many of which appropriately dated fossils have been found (Tyrberg 1999). The compression of climatic zones is likely to have led to an increase in atmospheric and oceanic circulations, thereby increasing local food-supplies through upwelling.

Species likely to have experienced big glacial range reductions include the most northerly, namely Ivory Gull *Pagophila eburnea*, Glaucous Gull *Larus hyperboreus*, Sabine's Gull *Xema sabini*, Little Auk *Alle alle* and Brunnich's Guillemot (Thick-billed Murre) *Uria lomvia*. Their present circumpolar distributions, partly along coasts concentric with the pole, would have been impossible, and at least in the North Atlantic region they are likely to have been limited to relatively short stretches of coast on the sides of the continents.

On all ice-free coasts, feeding areas are likely to have been greatly diminished, because the continental shelves over which many species feed would have been much reduced in area through lowered sea-levels. Beyond the shelves, the sea bed plunges abruptly downwards, too deep for bottom feeders. In addition, the absence of most of Iceland[2] and Greenland as breeding areas would perhaps have more effectively isolated the eastern from the western Atlantic populations of some species. In Europe, loose pack-ice extended in winter south to Spain, and in North America south to New York, compared with its present southern extent around Greenland. With much of the continental shelf laid bare, most of the existing cliff sites would have been too far inland for seabirds, and in any case land-ice would have spread out past them over the shelf. Further south, on the assumption that during glaciations cliffs existed on the exposed continental shelves, seabirds would have occupied largely different colony sites from those in

---

[2]*Iceland, being volcanic, remained partly free of ice, but at the height of the last glaciation it is likely to have lain too far removed from open sea water to serve as an important nesting site for all but a few species.*

use today. They would also have had access to any new islands exposed by lowered sea-levels.

Seabird remains have been obtained from far fewer sites than landbird remains (Tyrberg 1999).[3] Nevertheless, they are sufficient to show that some species became extinct during the Pleistocene. Associated with the greater extent of ice in the North Atlantic (**Figure 9.1**), more species were lost from the North Atlantic than from the North Pacific, which may partly account for the difference in the present numbers of seabird species between these regions. Although albatrosses are now absent as breeders from the North Atlantic, at least five species were found in Pliocene deposits in North Carolina dating from four million years ago (Olson & Rasmussen 2001). Of four *Phoebastria* albatrosses, three were very similar to living species of the North Pacific, but the fourth differed somewhat from any living species. The fifth species, *Diomedea anglica*, was larger than the others; it had also been found in Bermuda and England, and was considered most similar to the extant Short-tailed Albatross *P. albatrus* of the Pacific (Harrison 1978). To judge from these and other remains, albatrosses were prominent members of North Atlantic seabird communities until near the end of the Pleistocene (Tyrberg 1998).

Similarly, among the auks, 17 species are now confined to the North Pacific, compared with two to the North Atlantic, while four others breed in both these regions. Yet at least nine species were found in Pliocene deposits in Maryland, more than are currently found in the whole Atlantic region (S. Olson, in Gaston & Jones 1998). One flying species (*Australca*) was bigger than the flightless Great Auk *Pinguinus impennis*. Furthermore, the North Pacific now supports six cormorant species, compared with only three in the North Atlantic. On the other hand, the sulids seem to have suffered greater losses in the Pacific. The bones of both *Sula* and *Morus* species have been found in Miocene and Pliocene deposits stretching from California to British Columbia (Warheit 1992), but no sulid species breed in the North Pacific now and only one (the Northern Gannet *Morus bassanus*) in the North Atlantic. Other species lost from the Pacific about 470,000 years ago include large flightless auks *Mancalla*, the equivalents of the Great Auk *Pinguinus impennis* in the North Atlantic (Warheit 1992, 2002).

The regional extinctions of some species during the Pleistocene may have enabled other species to expand at a later date to fill the vacated niches. For example, to judge from the fossil record, *Uria* and *Cepphus* guillemots were once restricted to the North Pacific. They colonised the North Atlantic only in late Pleistocene and Holocene times, respectively, following the extinctions of many North Atlantic alcids (Olson & Rasmussen 2001).

Fossilised bone remains have also revealed marked distributional changes of seabird species that still survive (Tyrberg 1999). They confirm the north–south displacement of many species, with conditions in the west Mediterranean during much of the last glaciation similar to those around Britain today (**Figure 9.3**). In

---

[3]*The main reason for this is the lowering of sea-levels during glacial periods, which means that coastal sites of glacial age are now mostly submerged. Exceptions to this rule are found chiefly on steep coasts where the isostatic rise of formerly glaciated areas has offset the rise in sea-level, or where glacial coastal sites have been preserved through tectonic movements. In theory, interglacial sites should now be more accessible, but in practice many have been destroyed by erosion and weathering.*

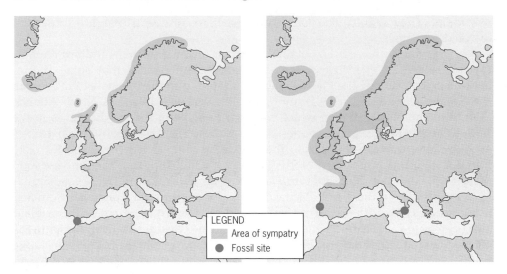

Figure 9.3   Distribution of some seabird species in glacial times and now. In each map the spots show the sites where dated fossilised bones were found, and the shading shows the area of current sympatry, where the modern distributions of those same species overlap.

Left. Remains from Gorham's Cave, Gibraltar, dated at 60–70,000 years ago, containing European Shag *Phalacrocorax aristotelis*, Little Auk *Alle alle* and Great Auk *Pinguinus impennis*.

Right. Remains from Gruta de Figueira Brava in Portugal dated at about 31,000 years ago and from Archi in Calabre, Italy, dated at about 40,000 years ago. Both contained remains of Northern Gannet *Morus bassanus* and Great Auk *Pinguinus impennis*. The Portuguese site also yielded remains of an extinct Shearwater *Puffinus holei* which disappeared before its modern distribution could be known. From Tyrberg 1999.

addition, remains of the Wedge-tailed Shearwater *Puffinus pacificus* (or a very similar bird) were found in Pleistocene deposits in North Carolina and St Helena (Olson 1975). This species is now absent in the Atlantic, but common in the Pacific and Indian Oceans.

In the southern hemisphere during glacial periods, the antarctic ice extended in summer 100–150 km out from the present coastline, making mainland nesting sites inaccessible for almost all seabird species. Forced northwards, they presumably survived by breeding mainly on the various Southern Ocean islands, which collectively may have held a greater range of species than today. Unfortunately, the fossil record for such areas is almost non-existent, and such remains as have been found are more recent than the last glaciation (Chapter 7). However, fossils have been found in southwest Africa. They indicate that the seabird communities of this region, now among the richest in the world, held even more species in the Pliocene (Olson 1983, 1985a).

Summarising, seabird remains from the Plio-Pleistocene of the northern hemisphere tell a similar story to landbird remains. Many species became extinct, especially in regions most affected by glaciations; others disappeared from parts of their range but not others; and surviving species now found at mid to high latitudes bred further south during glacial periods.

EFFECTS ON MIGRATION SYSTEMS

The Pleistocene history of most surviving Holarctic birds has thus been a series of compressions and expansions of range on an enormous scale. Such changes in the northern hemisphere must have had large impacts, not just on bird numbers and distributions, but also on migration systems. At the glacial maxima in Europe, more than half the total land area was under permanent ice, and hence incapable of supporting breeding birds, and in the remaining area the most bird-rich habitats (forests) were much restricted in distribution (**Figure 9.1**). Hence, the total bird population of the continent was probably much less than half its level during the interglacials, making the demand for winter quarters commensurately less (Moreau 1954). At the same time, over most of the habitable part of Europe winters were severe, so greater proportions of birds are likely to have crossed to Africa than do so now. The passerine migrants had to include not only insectivores but many non-insectivorous species, the group most poorly represented among the migrants entering Africa today (Moreau 1972). Because the deserts of North Africa and Arabia were more vegetated in glacial times, and have also become drier and larger through human activity, they could have supported more birds during glaciations than they can now, perhaps providing winter habitat for birds that bred on the widespread steppe-tundra habitat of glacial times (Moreau 1954). In addition, the lowering of sea-levels would have provided extensive areas of ice-free coastal lowland at suitable latitudes for wintering birds. We can only surmise that the migration systems of birds must have altered greatly in response to long-term climatic cycles, with substantial changes within the last 10,000 years (for further discussion, see Chapter 18).

CURRENT GRADIENTS IN SPECIES NUMBERS

Within any major biome, whether forest, tundra or grassland, species numbers now vary from one region to another. Typically, there are hot spots of high species numbers and high endemism, away from which species numbers progressively fall. Between such areas, hybrid zones can often be recognised, where previously separated populations have expanded and come together to interbreed (Chapter 3). To what extent, then, can these patterns in areas of continuous habitat be attributed to past climatic events?

In at least some biomes, regions of present high species numbers or endemism clearly correspond with the location of former stationary refuges, to which the flora and fauna of the biome were confined during periods of unfavourable climate. Such hotspots are discerned by superimposing the range maps of different species to find the areas where most occur together. For the northern continents, Hultén (1937) analysed 2,500 plant ranges and found that the hotspots (which he called centres of radiation) were all in regions that, on geological evidence, were not glaciated during the Pleistocene. He therefore considered that these areas more or less corresponded to the glacial refugia of the current arctic and boreal biota, and that different species were now in different stages of outward spread. Both suppositions were later supported by analyses on various animal groups (e.g. de Lattin 1956) and by additional independent palaeo-climatic evidence.

Let us look first at the boreal forest of Eurasia, which is the most extensive biome in the world, extending nearly 7,000 km from end to end. It is also one of the least human-modified biomes of the northern world. In this forest, broadly speaking, species numbers decline from east to west. According to Stegmann (1932), some 40–42 bird species breed in parts of east Siberia, falling to 30–33 species in western Siberia, then to 23–26 west of the Urals and to 15–20 in Scandinavia. Only 8–12 of these species reach the conifer forests of central Europe. Other authors have given different figures, depending on which species they class as characteristic of boreal forest, but the same east–west trend persists. Moreover, patterns of species diversity matching those for birds are also evident in other organisms, notably plants and insects (Hultén 1937, Reinig 1950). For all these organisms, therefore, a diversity gradient exists between the richer Siberian region and the relatively poor north European region. During glacial periods, the only sizeable area of boreal taiga forest to persist lay in eastern Asia, as a large refuge near the present region of maximum species diversity. Because this forest was extensive, its flora and fauna are likely to have suffered lower extinction rates than occurred in the smaller refuges further west, which were further impoverished by their continual displacements due to successive glaciations. Moreover, in eastern Asia, an almost unbroken expanse of forest extended from boreal regions southwards to the tropics, enabling easy seasonal migrations of forest birds, whereas in Europe any surviving forest areas were separated by a broad belt of sea and desert from tropical forests. All these factors are likely to have favoured the survival of eastern Asian over European forest birds.

Likewise, eastern North America holds many more forest bird species than does western Europe (95 versus 78, Mönkkönen 1994), with similar explanations. Whereas European forests were displaced, greatly reduced and fragmented during the glaciations, which led to many regional extinctions (known with certainty for some bird and many tree and mammal species), the eastern North American forest survived in a wide latitudinal swathe, extending at least 1,000 km from north to south (Huntley 1995). Conversely, open land was much more restricted in eastern North America than in Europe, which may account for the present-day difference in open-land bird species between eastern North America and western Europe (76 versus 96 species, Mönkkönen 1994).

In contrast to forest, tundra reached its smallest extent at the warmest part of the interglacials most recently some 7,500–4,000 years ago. During this period, boreal forests moved northwards to the shores of the Arctic Ocean and confined the tundras to small refuges, mainly centred on the Taimyr peninsula in northern Siberia, northeastern Siberia, Greenland and the islands of the Arctic Ocean, north of the main Eurasian and North American land-masses. In the northern hemisphere, tundra communities are unlikely to have survived anywhere further south during this period, except on some high mountain summits. However, tundra refuges also existed for much longer periods north of the ice during the glaciations (**Figure 9.1**).

The diversity of tundra-dwelling birds seems to have been influenced by the locations of both interglacial and glacial refuges (see later), but not surprisingly the last interglacial seems to have had most influence on current distributions. Thus, across the whole of the northern world, three main areas show peaks in the numbers of tundra-specialist landbirds. These are the Taimyr region with 43 species (including the endemic Red-breasted Goose *Branta ruficollis*), the northern Canadian islands with 42 species (including the endemic Ross's Goose *Chen rossii*, Thayer's Gull *Larus thayeri* and three shorebird species) and the eastern Siberian–western Alaskan Beringian region with 47 species, and no fewer than 19 endemic species, listed in **Table 9.1**. McKay's Bunting *Plectrophenax hyperboreus* is clearly derived from the more widespread Snow Bunting *P. nivalis*, but now nests only on the islands in the Bering Sea, which would have existed as mountain tops in the dry Beringian refuge of 15,000 years ago. The Beringian endemics include no fewer than ten seabird species **(Table 9.1)**, providing further testimony to the importance of the region as a centre of independent avian evolution, with the North Pacific periodically cut off by the Beringian land-bridge and hence from the Arctic and Atlantic Oceans.

Table 9.1   Endemic bird species of the Beringian Region around the North Pacific.

Landbirds	Seabirds
McKay's Bunting *Plectrophenax hyperboreus*	Spectacled Eider *Somateria fischeri*
Emperor Goose *Chen canagica*	Steller's Eider *Polysticta stelleri*
Bristle-thighed Curlew *Numenius tahitiensis*	Red-faced Cormorant *Phalacrocorax urile*
Wandering Tattler *Heteroscelus incanus*	
Long-billed Dowitcher *Limnodromus scolopaceus*	Red-legged Kittiwake *Rissa brevirostris*
	Aleutian Tern *Sterna aleutica*
Black Turnstone *Arenaria melanocephala*	Kittlitz's Murrelet *Brachyramphus brevirostris*
Surfbird *Aphriza virgata*	
Rock Sandpiper *Calidris ptilocnemis*	Parakeet Auklet *Aethia psittacula*
Western Sandpiper *Calidris mauri*	Crested Auklet *Aethia cristatella*
	Whiskered Auklet *Aethia pygmaea*
	Least Auklet *Aethia pusilla*.

Only two tundra-nesting landbirds, the Barnacle Goose *Branta leucopsis* and Pink-footed Goose *Anser brachyrhynchus*, do not occur in any of the three main centres for tundra birds. Their evolution has presumably centred somewhere in the North Atlantic, such as Greenland, Spitzbergen or Iceland where both still breed. All three islands could have offered breeding habitat in the interglacials, but in the glacial periods ice-free land remained only in extreme northeast Greenland, southwest Greenland and parts of Iceland (helped by volcanic activity), as Spitzbergen was totally glaciated. Greenland would have been too far for a goose to migrate from suitable wintering areas unless the birds staged in Iceland. Further detail on the birds of various tundra refuges is given in Chapter 10.

In conclusion, in both tundra and coniferous forest today, areas of greatest diversity are located where the largest climatic refuges are known or suspected to have lain, and where the largest numbers of species were likely to have persisted.

It is from these core areas that the bulk of the flora and fauna are likely to have dispersed to occupy their present ranges. The distributions of many bird species are still fragmented. For example, the deciduous forest avifauna of Europe (84 species altogether, according to Stegmann 1932) has 25 species (30%) in common with the deciduous forests of China. However, only 13 of these are found continuously across Eurasia (in the deciduous forests that form the boreal–steppe boundary), while the other 12 are found at both ends of the land-mass, but not in the middle. Such split distributions are shown, for example, by the Marsh Tit *Parus palustris* and Tawny Owl *Strix aluco*, and in the most extreme form by the Azure-winged Magpie *Cyanopica cyana*, found now only in southwest Europe and eastern Asia (Chapter 12). Could it that all the Eurasian species of broad-leaved woodland once occurred right across Eurasia so that the current restriction of many species to one end or the other is due to regional extinctions?

RELICT POPULATIONS

As the major vegetation zones were pushed southwards during glacial advance, many species became established in regions well to the south of both their former and their current ranges. This explains the continuing presence of relict populations of some bird species at localities within their past glacial ranges. Usually these populations occur in high mountains, where suitable habitat survives at high elevation, having retreated from the surrounding lowland. Extreme examples include the isolated populations of Black-billed Magpies *Pica pica* in the mountains of southwest Arabia and of Red-billed Choughs *Pyrrhocorax pyrrhocorax* in the mountains of Ethiopia, some 1,600 km and 2,200 km, respectively, south of the next nearest populations of their species. The many other currently disjunct distribution patterns that can be explained in terms of glacial history include the Twite *Carduelis flavirostris*, which is now found in the open grassland of various Asian mountain areas, and on the northwest seaboard of Europe, some 3,200 km to the northwest. This huge separation is explicable remembering that during glacial times suitable steppe-tundra occurred right across the Eurasian land-mass.

RECENT RANGE EXPANSIONS

Many of the bird species of boreal forests that now occur in northern Europe almost certainly spread there from eastern Asia in post-glacial times. There are two types of evidence for this view. The first, already mentioned, is that eastern Asia holds many more forest bird species than western Europe. Moreover, whereas many species in Asia do not extend as far west as Europe, few boreal bird species in Europe do not also occur in Asia. Those that do might have evolved in west European glacial refuges. Secondly, even in the past 200 years, at least 28 bird species of the Siberian taiga have spread westwards into northern Europe. Examples include the Greenish Warbler *Phylloscopus trochiloides*, Common Rosefinch *Carpodacus erythrinus*, Rustic Bunting *Emberiza rustica*, Little Bunting *E. pusilla*, Yellow-breasted Bunting *E. aureola* and Red-flanked Bluetail *Tarsiger cyanurus* (Niethammer 1937, Timoféeff-Ressovsky 1940, Sovinen 1952, Cramp &

Perrins 1994). These species still migrate east and southeast in autumn to spend the winter in southern Asia, whereas most other European birds winter in Europe or Africa. Even among the species that have long bred in western Europe, some migrate east then southeast to Africa, and may have colonised western Europe from further east (e.g. Red-backed Shrike *Lanius collurio* and Eurasian Nightjar *Caprimulgus europaeus*).

Other species have spread beyond Europe, via Iceland and Greenland, to eastern North America, including the Northern Wheatear *Oenanthe oenanthe*, Common Ringed Plover *Charadrius hiaticula* and Red Knot *Calidris canutus*, and more recently the Little Auk *Alle alle* (known on Baffin Island since 1983), Common Black-headed Gull *Larus ridibundus* (various east coast sites) and Little Gull *Larus minutus* (now breeds on the Great Lakes and at Churchill, Manitoba). Again, most of these species betray their European origins by migrating each year to winter in the Old World, but the gulls have established wintering areas on the east coast of North America while the auk winters widely in the North Atlantic.

Similar expansions have occurred from the eastern flank of the Eurasian forest and tundra zones (**Table 9.2**). At least ten species have spread across the Bering Strait from Siberia to Alaska. The majority of these species reached Alaska before ornithologists, but they reveal their recent Palaearctic origin by their lack of differentiation from Siberian populations and by their migrating every year to Old World wintering grounds, rather than to New World ones, like most other Alaskan birds. Most of these species are found nowhere else in North America but, as mentioned above, the Northern Wheatear *Oenanthe oenanthe* has also colonised the other side of North America via Greenland, and may presumably one day spread across the whole northern part of that continent. Some other species have moved the other way, from Alaska to establish populations in eastern Siberia (**Table 9.2**). The majority are still localised, but the Long-billed Dowitcher *Limnodromus scolopaceus*, Pectoral Sandpiper *Calidris melanotos* and Sandhill Crane *Grus canadensis* have spread west by hundreds of kilometres in the last 200 years. All these species return to winter with the rest of their kind in the New World.

Table 9.2   Species that have spread in recent times from Eurasia to North America, or vice versa, across the Bering Strait.

From Eurasia to North America[1]	From North America to Eurasia
A. Northern Wheater *Oenanthe oenanthe*	B. Grey-cheeked Thrush *Catharus minimus*
Bluethroat *Luscinia svecica*	Pectoral Sandpiper *Calidris melanotos*
Arctic Warbler *Phylloscopus borealis*	Baird's Sandpiper *Calidris bairdii*
Yellow Wagtail *Motacilla flava*	Buff-breasted Sandpiper *Tryngites*
White Wagtail *Motacilla alba*	*subruficollis*
Red-throated Pipit *Anthus cervinus*	Long-billed Dowitcher *Limnodromus*
Siberian Tit *Parus cinctus*	*scolopaceus*
Rustic Bunting *Emberiza rustica*	Snow Goose *Chen caerulescens*
Rufous-necked Stint *Calidris ruficollis*	Sandhill Crane *Grus canadensis*
Bar-tailed Godwit *Limosa lapponica*	

[1] In addition to the species listed, the Wood Sandpiper *Tringa glareola*, Curlew Sandpiper *Calidris ferruginea*, Lesser Sand (Mongolian) Plover *Charadrius mongolus* and other Palaearctic species have bred irregularly in Alaska at least during the last 100 years, and may eventually establish themselves there.

These recent range expansions may then be a continuation of a natural long-running process in which different bird species, like different plant species, have shown different rates of spread from former refuges, some occupying the full extent of the biome long before others. By this process, any area distant from a past refuge experiences a continual arrival of species, one after another. Such species as Northern Goshawk *Accipiter gentilis* and Northern Hawk Owl *Surnia ulula* are now so widely distributed across Eurasia and North America that it is hard to say which was their continent of origin, or whether their spread from one continent to the other followed the last glaciation (as may be assumed for the other species mentioned above) or occurred much earlier in geological time, as discussed in Chapter 5. As they are subspecifically differentiated, they may have spread much earlier. Given that the water barrier between Eurasia and Alaska is only about 115 km across and broken by islands, and that similar habitat exists on both sides, it is perhaps surprising that more species have not crossed in recent times to breed on both sides.

Rates of spread

With such highly mobile creatures as birds, the problem is to explain why gradients in species numbers still exist across each biome, and why some species have taken so long to occupy areas of apparently suitable habitat. Two possible reasons come to mind. Firstly, the habitat may remain unsuitable for some bird species for a long time, if their expansion is contingent on the prior spread of other organisms. An obvious example is the Common (Red) Crossbill *Loxia curvirostra* of the western Palaearctic. If this bird has always depended on Norway Spruce *Picea abies*, as it does now, it could have bred regularly in Scandinavia only in the last 3,000 years, following the westward spread of its food-plant (Huntley 1988). Clearly, any specialist is likely to spread to a new area only after its food-organisms have already become established there. Although most other bird species cannot be linked so tightly to particular plants or animals, it is not hard to imagine that the number of bird species that an area could hold must depend to some extent on the total diversity of plants and other organisms that occur there.

On this view, an area might at any one time hold close to the maximum number of bird species possible, with immigration of further species dependent on the prior spread of other organisms needed to provide the right conditions. In other words, each potential colonist is held back by lack of crucial resources, or by competition from species already there, but the potential colonist may co-exist with these same competing species in more diverse habitat elsewhere. This type of explanation, based on restricted floral and faunal diversity, has been used to explain the reduced numbers of bird species on islands of various sizes (Chapter 6), and elsewhere direct correlations between bird species diversity and floristic diversity have been documented for nectar-eaters and fruit-eaters (Snow & Snow 1971). Moreover, because insects are often plant-specific, insect diversity would also be expected to correlate with plant diversity.

A second hypothesis to explain the slow spread of some bird species from centres of high diversity (or former refuges) hinges on the inability of existing populations to produce enough potential colonists to occupy new areas. This inability in turn depends on conditions within the existing range (Chapter 14).

This view assumes that vacant areas within the biome are suitable for occupation, and a species is not excluded by competitors or by other environmental factors. The only limit to its rate of spread is its slow rate of increase. On this view, if a species arrived (or was released) in a new area in adequate numbers, it would be expected to establish itself, whereas on the preceding hypothesis it would not. Without experiment, it is hard to distinguish between these two explanations, and both may hold in different species.

Inability to produce a sufficient surplus of birds to expand and occupy new areas might apply particularly to migrant species which, while extending their breeding ranges, remain tied to the same limited wintering areas. The migration then becomes progressively longer as the breeding range expands, while the occupants of new areas must still compete with conspecifics from the main range on their common wintering grounds. In this situation, the new colonists could not experience the reduced competition and rapid population growth that new colonists usually enjoy. As described above, several Palaearctic species, in spreading westwards, have continued to migrate to wintering areas lying far to the southeast of their new breeding areas. However, others have suddenly acquired new wintering areas nearer their new breeding grounds, thus at once cutting the migration distance and freeing themselves from competition with conspecifics from the main range. The Shore (Horned) Lark *Eremophila alpestris* and Lapland Bunting (Longspur) *Calcarius lapponicus*, for instance, began to winter in the North Sea area only in the nineteenth century (Niethammer 1937, Jacobsen 1963), while other species, such as Red-breasted Flycatcher *Ficedula parva* and Little Bunting *Emberiza pusilla*, may now be in the process of developing a southwest migration (Jacobsen 1963, Cramp & Perrins 1994). Similarly, some European species, such as Eurasian Wigeon *Anas penelope*, Lesser Black-backed Gull *Larus fuscus* and Common Black-headed Gull *L. ridibundus*, have begun to winter regularly in eastern North America. The establishment of new wintering areas for migrants could be crucial to their further population growth and expansion of breeding range.

The likely outcome of such a pattern of westward colonisation and associated shift in wintering range would be a migratory divide, with the birds from the eastern part of the range using one wintering area, and those from the west another (Murray 1979). Such divides are common among birds that have trans-Palaearctic breeding ranges (Zink 1973–85, Schulz 1998), but they may not all have arisen in this way.

The constraint that limited wintering areas might place on the ability of migrants to extend their breeding ranges might explain why, within the Eurasian boreal region, migrants have narrower breeding ranges than residents (Bensch 1999). Among resident landbird species (excluding waterbirds and waders), 48 species breed right across the Eurasian land-mass, while 13 found at the west end do not extend to the east end, and 19 found at the east end do not extend to the west end. The equivalent figures for migrant species are 45 right across, 47 at the west end and 77 at the east **(Figure 9.4)**. Hence, migrant species from both west and east ends are significantly less likely to have breeding ranges extending all the way across than are residents (P<0.0001). This difference remains significant after controlling for potentially confounding variables, such as body mass, abundance and latitudinal range.

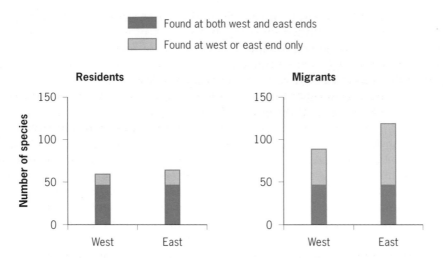

Figure 9.4   Comparison of the landbird faunas at the west and east ends of the Eurasian land-mass. More species breed at the east end than at the west end, but the discrepancy is greater in migrants than in residents. From Bensch 1999.

Although the arguments have been presented above mainly for Eurasian boreal birds (Udvardy 1958), or for other species expanding within any other natural biome. The arguments hold whether the biome was at its most restricted during glacial or interglacial periods and whether it was temperate or tropical in distribution (see later).

The main message is that localised peaks in species numbers within biomes and the restricted ranges of many taxa largely correspond to known Pleistocene refugia, established on palaeo-botanical and other evidence, so that the current ranges of many species reflect events that occurred thousands of years ago. Such findings highlight the likely role of past climatic history in influencing present distributions and range expansions. At the same time, however, it is difficult to exclude completely other factors in promoting recent range expansions, notably human influence, either in the breeding or wintering areas, which could affect the rate of population growth (Chapter 14). Whatever the reason for recent range expansions within continuous habitat, however, they would be expected to run mainly from regions of high to low species diversity, and to correlate with changes in climate, habitat and other environmental features that influence bird survival and reproductive success. The variable rates of eastward range expansion noted in the Greenish Warbler *Phylloscopus trochiloides*, for example, were related to short-term climatic fluctuations, with the most marked expansions in the most favourable periods (Valikangas 1951). A second important point to emerge is that, like the trees, bird species have responded individually to climate change, spreading at different dates and at different rates. The same has been found for other types of organisms (Huntly *et al.* 1997).

CURRENT CLIMATE WARMING

The warmest part of the current interglacial occurred about 7,500–4,000 years ago, depending on the region, at a time when ice sheets were still large, but shrinking rapidly. At that time, treelines extended to higher latitudes and to higher altitudes than today, as revealed by pollen and other evidence. If glacial cycles proceed inexorably, and the peak of the current interglacial has passed, we could now be embarked on the next glacial cycle. This likelihood is often forgotten amid the current concern over human-induced climatic warming. Recent warming is evident on short time scales of a few hundred years, and sits upon the long-term glacial cycles of around 100,000 years. Nonetheless, with the recognition of current temperature trends, knowledge of the effects of past climatic changes on flora and fauna has now achieved special significance.

Current rise in global temperature results from increased concentrations of carbon dioxide and other greenhouse gases[4] in the atmosphere, which reduce heat radiation from the earth. Through the burning of fossil fuels and other human activity, carbon dioxide concentrations have risen by 32% from about 280 ppm in pre-industrial times to about 370 ppm at present. If this trend continues, they are expected to exceed 400 ppm by the year 2100, causing a mean global temperature rise of 1–4°C in the coming century. The higher rate is not dissimilar to the global average that marked the end of the last glaciation, but as then, the rise is likely to be much greater at high than at low latitudes, and greater over land than sea. The resulting warming is likely to bring increased frequency and intensity of climatic extremes — of floods in some places and of droughts in others. Sea-levels are forecast to rise at around 30–50 cm per century owing to the thermal expansion of sea water and additional ice melt, flooding many fertile delta regions and low-lying islands. In the oceans, we can also expect shifts in water temperature, surface salinity, currents and nutrient availability, through altered patterns of rainfall. Uncertainty surrounds the direction of future change in northwest Europe, whether warmer like much of the northern hemisphere, or colder as could happen if the Gulf Stream current waned or shifted. Whatever changes materialise, interest centres on predicting the effects of global warming on fauna and flora, and on minimising its impact.

Over much of the world, to judge from past events, zones of similar climate will tend to move towards higher latitudes and altitudes, and plant and animal species will tend to follow. Past mean rates of spread of various tree species, estimated from the pollen record, mostly lie in the range 200–400 m per year, though much faster rates occurred occasionally (e.g. Davis 1981, King & Herstrom 1997). These rates lagged behind the rate of climate change, because they were constrained by other factors, as discussed above. The lag is likely to be much greater under the rapid changes predicted for the future. Under global warming, July isotherms are expected to advance northward at 4–5 km per year. If trees were to keep up with this rate, it would require migration rates 10–25 times faster than mean post-glacial

---

[4]*Including methane, nitrous oxide, chlorofluorocarbons, hydrofluorocarbons, perfluoromethane and sulphur hexafluoride.*

rates. Because this discrepancy is so much greater than past ones, it is hard to predict its effects on plant and animal distributions.

Another problem in predicting future biotic responses is that large parts of the landscape have now been converted to human use and are effectively closed to the majority of wild plants and animals. Many regions therefore provide no broad-front dispersal route, but at best a series of narrow interrupted corridors and stepping stones. This means that, for many organisms, response by movement will be impeded or prevented. The situation is likely to favour those plants with short response times, namely herbs, shrubs and fast-growing trees, while slow-growing trees with long generation times and poor dispersal could be at a particular disadvantage. Although many animals are more mobile than plants, they are still tied to areas of suitable vegetation. Those on oceanic islands, near mountain tops or on the northern edges of land-masses may have nowhere to go.

Some researchers have attempted to predict how distributions might change by use of 'climatic envelopes'. The method consists of calculating the span of climatic conditions embraced by the current geographical range of a species (its climatic envelope), and then using climate models to predict where this same range of climatic conditions might lie at some future date, perhaps 100–200 years from now, and assuming that the species will move accordingly. However, many other factors than climate also influence the range of a species, including its interactions with other species. In consequence, the current ranges of many species may be much smaller than climate alone would permit (giving a current climatic envelope that is smaller than the potential), and if species interactions also alter as a result of climate change, this could further modify distributions within the climatic envelope. So, in forecasting individual species' reactions to climate change, such climate-based models may prove to have only limited predictive power.

CONCLUDING REMARKS

Throughout this chapter, I have emphasised the fluctuating nature of distributions, especially over the last 2.5 million years. The ranges of most plant and animal species at mid to high latitudes are likely to have expanded and coalesced, contracted and fragmented or shifted in geographical position many times during this period. Distributional change was the main response to climate change, and during times of expansion different species moved at widely different rates. This resulted in continual change in the composition of communities, with one colonising species after another being added through time. In the process, many organisms became extinct, or much restricted in distribution. At some localities, the whole sequence of vegetation changes during a climatic cycle can be reconstructed from preserved pollen remains, and in other localities for shorter periods, but the numbers of sites studied are still too few to give any more than very generalised maps of the vegetation existing at one time.

For at least the past 2.5 million years, climate has always been changing between warm and cold periods. This raises the question whether plant and animal assemblages ever catch up with prevailing conditions to reach a state of equilibrium. They seem rather to exist in a state of perpetual imbalance, continually adjusting to climate, but always lagging behind and failing to achieve equilibrium before the onset of a new climatic trend. Many tree species seem still to be in the process of spreading from glacial refuges, so before they could all reach the limits of their potential range, climate trends could reverse, causing further retreat. Western Europe has still not gained many of the forest bird species that are currently spreading westwards from Asia. In North America, many bird species, isolated in forest on both sides, are spreading towards the middle but some have still not met (Chapter 10). The view that the natural world is balanced, delicately attuned to prevailing conditions, is clearly at odds with the facts, for on long time scales the geographical ranges of plants and animals, and the composition of communities, are in the process of continual change.

The maximum rates of range spread are probably much slower for trees than for other organisms, but trees limit the rates of spread of many other plants and animals, including birds, that depend on them for food or habitat. Some organisms clearly must co-exist and co-evolve, such as specific parasites and their hosts. Such partners are obliged to respond in tandem to changing conditions. They comprise some of the most closely coupled of interspecific interactions, and are thus least susceptible to disruption by environmental change. Apart from such closely coupled systems, with one species solely dependent upon another, the evidence of individual response and of distinct palaeo-communities suggests that such apparent co-evolution and interdependence is relatively rare. Current bird communities, like plant communities, are probably no more than temporary aggregations of species brought together under the influence of historical and prevailing conditions.

Although many bird and other species are known to have become extinct during the Pleistocene glaciations, the total number of extinctions is impossible to judge. The fossil record is biased towards larger species, which are more likely than small ones to have left identifiable remains, but larger species were probably also more prone to extinction when confined to climatic refuges. This is because, whereas smaller animals, such as insects and rodents, could survive generation after generation in favourable areas of only a few hectares, larger animals, such as deer and their predators, need much larger areas. Of course, they might survive on a total of many small areas, if they could move easily from one to another, but not if the small areas were too far apart to permit such movements. The fauna of the steppe-tundra zone is likely to have been especially vulnerable at the end of the last glaciation because this rich, formerly widespread habitat has no present-day equivalent. The individual plant species have persisted, as have some of the most characteristic animals, including birds, but these survivors are now distributed among different modern biomes, and nowhere come together as a distinct steppe-tundra community.

One might guess that many more species survived the climate vicissitudes of the last 2.5 million years than were eliminated by them, although the distribution patterns of the survivors may have greatly changed. Throughout, however, individual species seem to have maintained the same broad habitat preferences, in that the remains of forest species were associated with a prevalence of tree pollen, and the remains of open-country species were associated with a prevalence of grass or sedge pollen. As a result of preserved pollen, the precise distributional history of trees is much better known than that of birds, but because trees provide the habitat for so many birds, they provide important indicators of bird distributional changes, over and above the evidence from fossils.

In conclusion, reconstruction of likely climatic and vegetational history provides a plausible explanation for aspects of the current distribution patterns of birds, and of some regional variations in species numbers, patterns of endemism, hybrid zones and directions of spread. It is striking how the patterns for birds are matched by those for other animals and plants, which show the same centres of distribution and similar contact zones between related forms. This would be expected if they shared the same history. The examples mentioned above are only a small proportion of those described for animals and plants in general. Nevertheless, because such refuge-based explanations of current distributions cannot be tested, they must remain no more than hypotheses. If we ignore them, simply because they are not amenable to experiment, we risk greatly underestimating the role of climatic and vegetation history in influencing the regional numbers and distribution patterns of all land plants and animals today. We are at risk of discarding the most plausible explanation of some existing distribution and diversity patterns yet proposed.

SUMMARY

The ice sheets of polar regions are known to have fluctuated in size over at least the last 2.5 million years, associated with massive changes in world climates, affecting both temperature and precipitation. These climatic fluctuations have caused continual changes in the distributions of plants and animals, and left their imprint on current distribution and diversity patterns. A state of equilibrium, when the distributions of all plants and animals were stable, is unlikely ever to have been reached.

In the last glaciation, which peaked some 26,000–14,000 years ago, ice sheets came much further south in Europe and North America than in the drier Siberian–Beringian–Alaskan region. During this glaciation, many species of plants and animals were eliminated, either totally or regionally, while others were confined to southern refuges. The greater severity of glacial effects in Europe may partly explain why the present flora and fauna are much poorer in Europe than in eastern Asia and North America and why many forest bird species are still in the process of spreading from east to west across Eurasia.

Across any major biome, gradients in species numbers are apparent, with occasional 'core areas' of high species numbers or endemism. Many such core areas correspond with known past climatic refuges, where habitat persisted intact through a period of unfavourable climate. Current contours in species numbers across a biome can be explained on the assumption that different species (plants and animals) have spread at different rates, and hence to different distances from the former refuge. The same phenomenon is apparent in one region or another among birds of all major habitat types, and also in other animals and plants.

For seabirds, as for landbirds, glacial advance rendered uninhabitable many high-latitude areas, telescoping distributions. Lowered sea-levels laid bare much of the continental shelf, with consequent loss of feeding areas; they would also have exposed new coastal cliffs and islands, as some previously used ones were high and dry, far inland.

The majority of plant and animal species survived the climatic vicissitudes of the Pleistocene by the passive responses of geographical shifting (secular migration) and vicariance (fragmentation into separated populations), while maintaining the same broad habitat preferences. Some species became extinct regionally or totally. To judge from fossil remains, extinctions were more frequent during the earlier glaciations than during the later ones, perhaps because the most vulnerable species were eliminated at an early stage. However, a burst of mammalian extinctions occurred at the end of the Pleistocene, 12,000–9,000 years ago, which may have been due partly to human activity. It was accompanied by the loss of some bird species, especially large predators and scavengers.

At least in the last 2.5 million years, natural communities probably never existed in a state of regulated and balanced equilibrium, but were in the process of continual change. They were no more than temporary and loose assemblages of species brought together by a combination of historical circumstance and prevailing conditions. While consistent in their habitat needs, the same bird species associated and disassociated through time.

Saw-whet Owl *Aegolius acadicus* and Boreal Owl *A. funereus*, whose distributions in North America reflect glacial history.

Chapter 10

Glacial cycles in northern regions: differentiation and speciation

Major climatic change is not an entirely negative process, leading inexorably to loss of species. Paradoxically, on long time scales, it is also a major force in the evolution of new species, through promoting the fragmentation of habitats, and hence of populations. When the separated populations of a single species become far enough apart to prevent the interchange of individuals, they become genetically isolated from one another. They can then evolve independently, in response to local conditions, gradually changing from the parental form, producing separate subspecies, and eventually separate species (Chapter 3). However, this can only happen when surviving patches of habitat are each large enough to sustain their populations long-term, far enough apart to prevent the interchange of significant numbers of individuals, and isolated for long enough to enable differentiation and eventually speciation to occur.

If the isolated populations later expand, in response to spreading habitat, they may come into contact again, interbreeding if they are not very different from one another, or remaining as separate species if they have become sufficiently distinct. As separate species, they may continue to occupy separate areas, as allopatric (or parapatric) species pairs, or if sufficiently distinct in their habitat or food needs that they do not compete too much, they may spread through one another's ranges, becoming sympatric. All stages in this process of isolation, spread and subsequent contact are represented in Eurasian and North American birds, and are evident in current distribution patterns.

Climatic refuges thus provided havens where the plants and animals of a biome could survive a difficult period and also, helped by their isolation, places where divergence and speciation could occur, and from which new and emerging species could later spread (Chapter 3). Because of the sparseness of the fossil record, the range changes of most bird species can only be inferred from their present distributions, together with knowledge of the locations of past refuges. They contrast with the range changes of trees and other plants which can be mapped fairly precisely from dated pollen remains. As mentioned in Chapter 9, at the height of the last glaciation, the European forest was confined to the three southern peninsulas of Iberia, Italy and the Balkans, but other isolated forest areas remained on some Mediterranean islands and in North Africa. The North American forest was similarly confined to three main areas in the southeast, southwest and Mexico. There was thus ample opportunity on both continents for geographical differentiation in forest birds. In neither continent did such differentiation necessarily evolve in the last glaciation alone, however, for successive glaciations over the past 2.5 million years are likely to have repeatedly displaced and fragmented major habitats.

DISTRIBUTIONAL EVIDENCE FOR PAST CLIMATIC EFFECTS

European suture zones

The lines of contact that have developed where spreading sister populations have met are marked by range overlap, hybridisation or parapatry. Such lines between different pairs of taxa tend to be clustered in the same general areas that lie between former refuges. Collectively, they form a 'suture zone', defined as a 'band of geographical overlap between major biotic assemblages' (Remington 1968). At least four major suture zones are evident among the fauna and flora of Europe (**Figure 10.1**). The most obvious runs northeast–southwest across central Europe, falling mainly within Germany. It is supposedly where populations spreading from the Iberian peninsula met those spreading from the Balkans or further east (the Italian refuge could have contributed to either eastern or western populations, depending on species). Various pairs of bird taxa meet in this zone, including the Common Nightingale *Luscinia megarhynchos* and Thrush Nightingale *L. luscinia*, and the Short-toed Treecreeper *Certhia brachydactyla* and Eurasian Treecreeper *C. familiaris* (**Table 10.1**). All these various sister taxa are presumed to have split from a common ancestor into distinct western and eastern forms, and still show west-centred and east-centred distributions in Europe, but with varying degrees

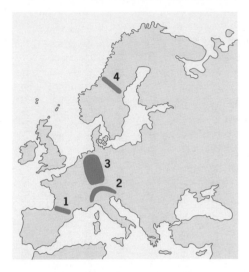

Figure 10.1   Major suture zones in Europe.

1. The Pyrenean suture zone represents the juncture between populations from the Iberian and more eastern refuges (e.g. Iberian Chiffchaff *Phylloscopus brehmii* and Eurasian Chiffchaff *P. collybita*, Spotless Starling *Sturnus unicolor* and Common Starling *Sturnus vulgaris*).
2. The Alpine suture zone represents the juncture between populations from the Italian and more eastern refuges (e.g. Italian Sparrow *Passer domesticus italiae* and House Sparrow *P.d. domesticus*).
3. The central European suture zone represents the juncture between populations from the Iberian and more eastern refuges (e.g. Common Nightingale *Luscinia megarhynchos* and Thrush Nightingale *L. luscinia*, Carrion Crow *Corvus c. corone* and Hooded *Crow C. c. cornix*).
4. The central Scandinavian suture zone represents the juncture between populations from the south and northeast (e.g. Common Chaffinch *Fringilla coelebs* and Brambling *F. montifringilla*; Common Buzzard *Buteo b. buteo* and Steppe Buzzard *B. b. vulpinus* and Willow Warblers *Phylloscopus trochilus* and *P. acredula*).
    Partly from Taberlet *et al.* 1998.

of overlap. Where they occur alone, sister taxa usually occupy similar habitats to one another, but where they overlap, some sister taxa occupy somewhat different habitats. The two nightingales, for example, occupy wet and dry woodland, respectively, while the two treecreepers occupy lowland and highland, respectively. The implication is that, where species overlap, they co-exist by using different habitats. They represent a step in the development of sympatry from allopatry.

Only a small proportion of European bird taxa have differentiated sufficiently to form separate sister species, but some others have formed obvious subspecies which have since come together, producing a mixed population in the contact zone. In the Long-tailed Tit *Aegithalos caudatus*, a northern and eastern type *A. c. caudatus* has a white head, while a southern and western type *A. c. europaeus* has

Table 10.1 Examples of European birds in which distinct western and eastern species are likely to have arisen from a common ancestor in separate western and eastern forest or scrub refuges during glaciations. Which refuges were involved can only be surmised, but for some species eastern refuges were probably in southern Asia rather than Europe. Mainly from Salomonsen 1931 and Voous 1960.

	Western form	Eastern form	Hybridisation
1.	Eurasian Green Woodpecker *Picus viridis*	Grey-faced Woodpecker *Picus canus*[1]	Very extensive overlap, no known hybridisation
2.	Great Spotted Woodpecker *Dendrocopos major*	Syrian Woodpecker *Dendrocopos syriacus*	Extensive overlap, occasional hybridisation
3.	Common Nightingale *Luscinia megarhynchos*	Thrush Nightingale *Luscinia luscinia*	Extensive overlap, no known hybridisation
4.	Melodious Warbler *Hippolais polyglotta*	Icterine Warbler *Hippolais icterina*	Little or no overlap, no known hybridisation
5.	Western Bonelli's Warbler *Phylloscopus bonelli*	Eastern Bonelli's Warbler *Phylloscopus orientalis*	Little or no overlap, no known hybridisation
6.	Firecrest *Regulus ignicapillus*	Goldcrest *Regulus regulus*	Extensive overlap, no known hybridisation
7.	European Pied Flycatcher *Ficedula hypoleuca*	Collared Flycatcher *Ficedula albicollis*[2]	Some overlap, some hybridisation
8.	Blue Tit *Parus caeruleus*	Azure Tit *Parus cyanus*	Extensive overlap, some hybridisation
9.	Short-toed Treecreeper *Certhia brachydactyla*	Eurasian Treecreeper *Certhia familiaris*	Extensive overlap, no known hybridisation
10.	Cirl Bunting *Emberiza cirlus*	Yellowhammer *Emberiza citrinella*	Extensive overlap, no known hybridisation
11.	Carrion Crow *Corvus (corone) corone*	Hooded Crow *Corvus (corone) cornix*	Stable overlap zone with hybridisation
12.	Spanish Imperial Eagle *Aquila (heliaca) adalberti*	Imperial Eagle *Aquila heliaca*	Not in contact

Note. In species pairs 2, 7 and 8, hybridisation has been recorded occasionally in the overlap zone, and in pair 11 it is regular along a narrow zone of overlap. In most of the pairs, one species has been recorded as spreading into the range of the other in the last 200 years, increasing the zone of common occurrence. Only pair 12 now occupy totally different areas, separated with no overlap.

[1]A third species, the Levaillant's Green Woodpecker *Picus vaillanti* occurs in North Africa, and is sometimes considered as a race of the Green Woodpecker.
[2]A third species, the Semi-collared Flycatcher *Ficedula semitorquata*, also occurs in a mainly different part of Europe and a newly proposed fourth species, the Atlas Flycatcher *F. speculigera*, in North Africa.

dark head markings; where the two types meet they show great variability. The geographical variation does not follow a gradual cline, but is more readily explained, Stresemann (1919) suggested, by the hybridisation of two forms which had differentiated in isolation and later re-met. The same holds for the white-throated and red-throated forms of the Bluethroat *Luscinia svecica* and for the red-breasted and white-breasted forms of the Wood Nuthatch *Sitta europaea*. Many other transition zones between sister species and subspecies in continuous habitat have been explained in the same way. The positions of the hybrid zones or bound-aries between sister species and subspecies all lie somewhere between known former refuges. They differ in position to some extent from one pair to another, presumably because not all species were found in the same refuges and because the rates of spread from these refuges is likely to have varied.

The majority of sister taxa show only slight morphological differentiation, but some show marked plumage differences. For example, the Carrion Crow *Corvus c. corone* is entirely black, while its sister form, the Hooded Crow *C. c. cornix*, has a light grey back and belly. So distinct are these two crows that, despite their inter-breeding, many taxonomists prefer to regard them as separate species. Interestingly, they show two hybrid zones, one across Europe from Italy to Schleswig-Holstein and across Scotland, and another in Asia from the Aral Sea to the Altai and then on to near the mouth of the Yenesei (**Figure 10.2**). The contact zones extend to more than 5,000 km in length, and their width varies within 25–150 km. One

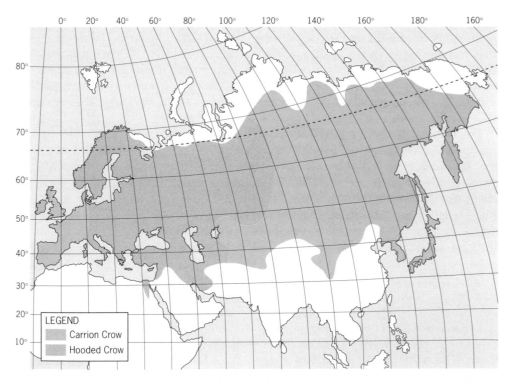

Figure 10.2   Distributions of the Carrion Crow *Corvus c. corone* and the Hooded Crow *C. c. cornix* in Eurasia, between which lie hybrid zones.

plausible explanation of this pattern is that Eurasia was originally populated entirely by black crows which during glaciations became isolated in southwest Europe, southeast Europe and southern Asia, respectively. If we assume that the central group evolved the grey and black plumage, the present distributional pattern could be explained by the expansion and re-meeting of the different populations after glacial retreat. This pattern, in which two populations very similar in appearance are separated from each other by a third, different one, is apparent in other species elsewhere (Chapter 11). It is known as the leapfrog pattern, and is purportedly caused by the central population diverging more markedly in isolation than the two end ones (Remsen 1984).

Some European birds might have been confined to Iberia and Italy by the extensive glaciers that covered the Pyrenees and the Alps, enabling the ice-free lowlands to the north to be colonised from further east. Some species, such as Spotless Starling *Sturnus unicolor* and Spanish Sparrow *Passer hispaniolensis*, remain confined to one or more of the southern peninsulas, while the rest of Europe is occupied by closely related species, namely Common Starling *Sturnus vulgaris* and House Sparrow *Passer domesticus*, which presumably came from further east. Similarly, the Pyrenees mark the juncture between the Iberian Chiffchaff *Phylloscopus brehmii* and Eurasian Chiffchaff *P. collybita* (Helbig *et al.* 1996), and the Alps the juncture between the Italian Sparrow *Passer italiae* and House Sparrow *P. domesticus*. The contact zones between these different populations now lie along the Pyrenean and Alpine mountain chains (**Figure 10.1**). Even within some species, genetic differentiation indicates that the middle latitudes of Europe were colonised by a different stock from that which survived in the Iberian or Italian peninsulas. Studies of mt DNA (control region) of Great Bustards *Otis tarda* reveal that the Iberian population differs substantially from the German–Russian one (Pitra *et al.* 2000). The Iberian stock may therefore have remained isolated within that peninsula, while the rest of Europe was colonised from further east. Similar differentiation is apparent in some other birds and in various other animals (Bilton *et al.* 1998).

The fourth cluster of contact zones occurs at about 62°N in Norway–Sweden, marking the meeting point between taxa that spread up from the south and others that spread in from the east, over the northern edge of the Baltic Sea. Some species pairs, such as Common Chaffinch *Fringilla coelebs* and Brambling *Fringilla montifringilla* meet around this zone, as do some subspecies, such as the Buzzards *Buteo b. buteo* and *B. b. vulpinus* and the Willow Warblers *Phylloscopus p. trochilus* and *P. p. acredula*. Yet another zone of secondary contact occurs further east, in Iran, but mainly affects open-country species, as exemplified by the hybrid zones between the Black-eared Wheatear *Oenanthe hispanica* and Pied Wheatear *O. pleschanka*, and between the Black-headed Bunting *Emberiza melanocephala* and Red-headed Bunting *E. bruniceps* (Haffer 1989). This suture zone extends north into Siberia, affecting different species.

The existence of contact zones implies that the presence of one population in a region can prevent another from moving in, giving a stable boundary between them. The distribution of contact zones across Europe further implies that the present bird populations of deglaciated parts of the continent were not all derived from the same refuges. In particular, the Balkans or more eastern refuges probably contributed disproportionately to north European birds, owing to the difficulty for the Iberian and Italian populations of crossing the still-glaciated Pyrenees and Alps. Future analyses of avian DNA may help to assess the importance for individual species of the various possible post-glacial colonisation routes shown in **Figure 10.3**.

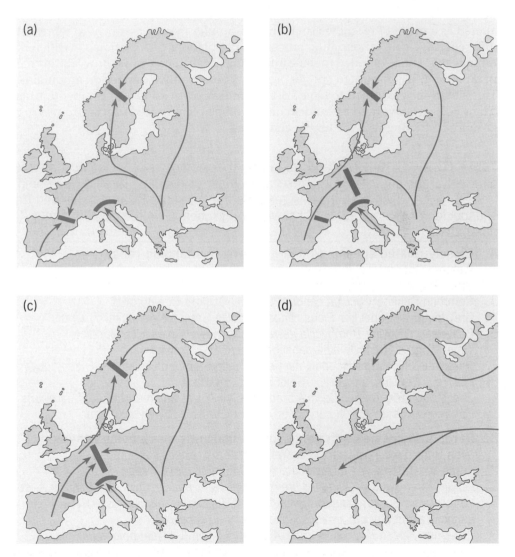

Figure 10.3   Potential routes of colonisation of mid and high latitudes of Europe from different glacial refuges, with expected contact zones: (a) colonisation from the Balkans only; (b) from Iberia and the Balkans; (c) from Iberia, Italy and the Balkans; and (d) from further east. All four patterns are evident in the morphological variation of birds (see text), and some also in the patterns of genetic variation in birds and other organisms (for specific variants of these patterns in non-avian taxa, see Hewitt 1999).

Eurasian overview

At the scale of the whole Palaearctic, Haffer (1989) listed: (1) ten pairs of distinct sister species that have mainly different ranges, but with wide overlap zones and little or no hybridisation; (2) 78 sets of 2–5 sister species that show parapatric distributions, some of which show limited hybridisation in the contact zones; and (3) nine sets of 23 subspecies that show wide contact zones and extensive hybridisation (**Table 10.2**). Although the taxonomic status of some of these birds has been altered as a result of recent research, these examples may be assumed to represent different stages in the speciation process, as populations differentiated in different refuges, and later expanded to come into contact again. Some now show 'altitudinal parapatry' in which related species occupy different altitude zones on mountains. This situation is assumed to have arisen by multiple invasion and competition, as a new colonist, favoured by milder climate, displaced a related species from lower ground, restricting it to the higher elevations.

Table 10.2   Closely related Palaearctic species and subspecies assumed to have arisen in different areas and later spread, so as to meet or overlap. The list was compiled by Haffer (1989), since which time the designated status of some taxa has changed from species to subspecies or vice versa. H = known hybridisation in limited contact zone; A = altitudinal parapatry.

A.   Closely related species with large partly overlapping ranges
1. Western Capercaillie *Tetrao urogallus*/Black-billed Capercaillie *T. parvirostris*
2. Stock Pigeon *Columba oenas*/Pale-backed Pigeon *C. eversmanni*
3. Common Cuckoo *Cuculus canorus*/Oriental Cuckoo *C. saturatus*
4. Grey-faced Woodpecker *Picus canus*/Eurasian Green Woodpecker *P. viridis*
5. Bohemian Waxwing *Bombycilla garrulus*/Japanese Waxwing *B. japonica*
6. White Wagtail *Motacilla alba*/Japanese Wagtail *M. grandis*
7. Narcissus Flycatcher *Ficedula narcissina*/Yellow-rumped Flycatcher *F. zanthopygia*
8. Firecrest *Regulus ignicapillus*/Goldcrest *R. regulus*
9. Short-toed Treecreeper *Certhia brachydactyla*/Eurasian Treecreeper *C. familiaris*
10. Common Crossbill *Loxia curvirostra*/Parrot Crossbill *L. pytyopsittacus*

B.   Parapatric species with contiguous or narrowly overlapping ranges
1. Little Bittern *Ixobrychus minutus*/Yellow Bittern *I. sinensis* (A)
2. Eurasian Griffon *Gyps fulvus*/Himalayan Griffon *G. himalayensis* (A)
3. Levant Sparrowhawk *Accipiter brevipes*/Shikra *A. badius*/Chinese Goshawk *A. soloensis*
4. Long-legged Buzzard *Buteo rufinus*/Upland Buzzard *B. hemilasius* (A)
5. Lanner Falcon *Falco biarmicus*/Saker Falcon *F. cherrug*
6. Peregrine Falcon *Falco peregrinus*/Barbary Falcon *F. pelegrinoides*
7. Red-legged Partridge *Alectoris rufa*/Rock Partridge *A. graeca*/Chukar *A. chukar*/Rusty-necklaced Partridge *A. magna* (H, A)
8. Grey Partridge *Perdix perdix*/Daurian Partridge *P. dauurica* (H)
9. Common Quail *Coturnix coturnix*/Japanese Quail *Coturnix japonica*
10. Western Tragopan *Tragopan melanocephalus*/Satyr Tragopan *T. satyra*/Blyth's Tragopan *T. blythii*/Temminck's Tragopan *T. temminckii*/Cabot's Tragopan *T. caboti*
11. Himalayan Monal *Lophophorus impejanus*/Sclater's Monal *L. sclateri*
12. Kalij Pheasant *Lophura leucomelanos*/Silver Pheasant *L. nycthemera* (H, A)

13. Imperial Pheasant *Lophura imperialis*/Edwards' Pheasant *L. edwardsi*
14. White Eared-pheasant *Crossoptilon crossoptilon*/Blue Eared-pheasant *C.auritum*
15. Golden Pheasant *Chrysolophus pictus*/Lady Amherst's Pheasant *C. amherstiae*
16. Herring Gull *Larus argentatus*/Yellow-legged Gull *L. cachinnans* (H)
17. European Turtle Dove *Streptopelia turtur*/Oriental Turtle Dove *S. orientalis*
18. European Bee-eater *Merops apiaster*/Chestnut-headed Bee-eater *M. leschenaulti*
19. Eurasian Green Woodpecker *Picus viridis*/Scaly-bellied Woodpecker *P. squamatus*
20. White-winged Woodpecker *Dendrocopos leucopterus*/Great Spotted Woodpecker *D. major*/Syrian Woodpecker *D. syriacus*/Sind Woodpecker *D. assimilis*/Himalayan Woodpecker *D. himalayensis* (H, A)
21. White-backed Woodpecker *D. leucotos*/subspecies *D. l. li llfordi* (H)
22. Rufous-tailed Lark *Ammomanes phoenicurus*/Bar-tailed Desert Lark *A. cincturus*
23. Calandra Lark *Melanocorypha calandra*/Bimaculated Lark *M. bimaculata* (A)
24. White-winged Lark *Melanocorypha leucoptera*/Mongolian Lark *M. mongolica*
25. Greater Short-toed Lark *Calandrella brachydactyla*/Hume's Lark *C. acutirostris* (A)
26. Lesser Short-toed Lark *Calandrella rufescens*/Asian Short-toed Lark *C. cheleensis*
27. Crested Lark *Galerida cristata*/Thekla Lark *G. theklae* (A)
28. Eurasian Skylark *Alauda arvensis*/Oriental Skylark *A. gulgula*
29. Horned Lark *Eremophila alpestris*/Temminck's Lark *E. bilopha* (A)
30. Rock Martin *Hirundo fuligula*/Eurasian Crag Martin *H. rupestris*/Dusky Crag Martin *H. concolor*
31. Northern House Martin *Delichon urbica*/Asian House Martin *D. dasypus*
32. Tawny Pipit *Anthus campestris*/Blyth's Pipit *A. godlewskii*
33. Water Pipit *Anthus spinoletta*/American Pipit *A. rubescens* (A)
34. Yellow Wagtail *Motacilla flava*/subspecies *M. f. flavissima*/*M. f. lutea*/*M. f. taivana*
35. White Wagtail *Motacilla alba*/White-browed Wagtail *M. madaraspatensis* (A)
36. Common Nightingale *Luscinia megarhynchos*/Thrush Nightingale *L. luscinia*
37. White-tailed Rubythroat *Luscinia pectoralis*/Siberian Rubythroat *L. calliope* (A)
38. Common Stonechat *Saxicola torquata*/White-tailed Stonechat *S. leucura* (A)
39. Black-eared Wheatear *Oenanthe hispanica*/Pied Wheatear *O. pleschanka* (H)
40. Black Wheatear *Oenanthe leucura*/White-tailed Wheatear *O. leucopyga*
41. Eyebrowed Thrush *Turdus obscurus*/Pale Thrush *T. pallidus* (A)
42. Dusky Thrush *Turdus naumanni*/Dark-throated Thrush *T. ruficollis* (H)
43. Eurasian River Warbler *Locustella fluviatilis*/Lanceolated Warbler *L. lanceolata*
44. Common Grasshopper Warbler *Locustella naevia*/Pallas's Grasshopper Warbler *L. certhiola*/Middendorf's Grasshopper Warbler *L. ochotensis* (H)
45. Great Reed Warbler *Acrocephalus arundinaceus*/Clamorous Reed Warbler *A. stentoreus*
46. Melodious Warbler *Hippolais polyglotta*/Icterine Warbler *H. icterina*
47. Olive-tree Warbler *Hippolais olivetorum*/Upcher's Warbler *H. languida*
48. Lesser Whitethroat *Sylvia curruca*/Hume's Whitethroat *S. althaea*/Small Whitethroat *S. minula* (H, A)
49. Sardinian Warbler *Sylvia melanocephala*/Ménétries's Warbler *S. mystacea*
50. Eurasian Chiffchaff *Phylloscopus collybita*/Caucasian Chiffchaff *P. lorenzii* (A)
51. Greenish Warbler *Phylloscopus trochiloides*/subspecies *P. t. nitidus*
52. Tickell's Leaf Warbler *Phylloscopus affinis*/Buff-throated Warbler *P. subaffinis* (A)
53. European Pied Flycatcher *Ficedula hypoleuca*/Collared Flycatcher *F. albicollis*/Semi-collared Flycatcher *F. semitorquata* (H)
54. Rusty-cheeked Scimitar Babbler *Pomatorhinus erythrogenys*/Spot-breasted Scimitar Babbler *P. erythrocnemis*
55. Three-toed Parrotbill *Paradoxornis paradoxus*/Brown Parrotbill *P. unicolor*
56. Giant Laughingthrush *Garrulax maximus*/Spotted Laughingthrush *G. ocellatus* (A)

57. White-throated Tit *Aegithalos niveogularis*/Black-browed Tit *A. iouschistos*
58. Black-throated Tit *Aegithalos concinnus*/White-cheeked Tit *A. leucogenys*
59. Rufous-vented Tit *Parus rubidiventris*/subspecies *P. r. beavani*
60. Blue Tit *Parus caeruleus*/Azure Tit *P. cyanus* (H)
61. Great Tit *Parus major*/Turkestan Tit *P. bokharensis* (H)
62. Black-lored Tit *Parus xanthogenys*/Yellow-cheeked Tit *P. spilonotus*
63. Chestnut-vented Nuthatch *Sitta nagaensis*/Chestnut-bellied Nuthatch *S. castanea* (A)
64. Western Rock Nuthatch *Sitta neumayer*/Eastern Rock Nuthatch *S. tephronota* (A)
65. Red-backed Shrike *Lanius collurio*/Rufous-tailed Shrike *L. isabellinus*/Brown Shrike *L. cristatus* (H)
66. Grey-backed Shrike *Lanius tephronotus*/Long-tailed Shrike *L. schach* (A)
67. Northern Shrike *Lanius excubitor*/Chinese Grey Shrike *L. sphenocercus*
68. Eurasian Jackdaw *Corvus monedula*/Daurian Jackdaw *C. dauricus* (H)
69. Carrion Crow *Corvus corone*/Collared Crow *C. torquatus*
70. Common Raven *Corvus corax*/Brown-necked Raven *C. ruficollis*
71. Spotless Starling *Sturnus unicolor*/Common Starling *S. vulgaris*
72. House Sparrow *P. domesticus*/*P. d. italiae*/*P. d. indicus* (H)
73. Common (Mealy) Redpoll *Carduelis flammea*/Arctic (Hoary) Redpoll *C. hornemanni* (A)
74. Common Crossbill *Loxia curvirostra*/Scottish Crossbill *L. scotica*
75. Trumpeter Finch *Rhodopechys githaginea*/Mongolian Finch *R. mongolica* (A)
76. Rock Bunting *Emberiza cia*/Godlewski's Bunting *E. godlewskii*
77. Ortolan Bunting *Emberiza hortulana*/Cretzschmar's Bunting *E. caesia*/Grey-necked Bunting *E. buchanani* (A)
78. Black-headed Bunting *Emberiza melanocephala*/Red-headed Bunting *E. bruniceps* (H)

C. Subspecies with obvious hybrid zones
1. White Wagtail *Motacilla alba*/*M. a. lugens*/*M. a. leucopsis*
2. Dusky Thrush *Turdus naumanni naumanni*/*T. n. eunomus*
3. Dark-throated Thrush *Turdus r. ruficollis*/*T. r. atrogularis*
4. Booted Warbler *Hippolais caligata*/*H. c. raina*
5. Azure Tit *Parus cyanus*/*P. c. flavipectus*
6. Coal Tit *Parus ater melanolophus*/*P. a. aemodius*
7. Sombre Tit *Parus lugubris*/*P. l. hyrcanus*
8. Willow Tit *Parus montanus*/*P. m. songarus*
9. European Goldfinch *Carduelis carduelis*/*C. c. caniceps*

To judge from their current distributions and overlap zones, many species were formerly isolated at both west and east ends of the Eurasian land-mass where they differentiated. They spread inwards but the zone of contact varies in position from one pair to another. With the western form listed first, examples include the European Turtle Dove *Streptopelia turtur* and Rufous Turtle Dove *S. orientalis*, the European Honey Buzzard *Pernis apivorus* and Crested (Oriental) Honey Buzzard *P. ptilorhyncus*, the Blue Tit *Parus caeruleus* and Azure Tit *P. cyanus*, and western and eastern forms of the Red-breasted Flycatcher *Ficedula parva* **(Figure 10.4)**. All these pairs meet around the middle sector of the land-mass, but others meet nearer the west end (e.g. Lesser Spotted Eagle *Aquila pomarina* and Greater Spotted Eagle *A. clanga*) or nearer the east end (e.g. Bohemian Waxwing *Bombycilla garrulus* and Japanese Waxwing *B. japonica*). Other allospecies pairs are found one at each end of the land-mass, but are still separated by a substantial gap, such as the

Figure 10.4
Examples of closely related bird species with western and eastern distribution patterns in Eurasia.

(a) Ranges of European Greenfinch *Carduelis chloris* and Oriental Greenfinch *C. sinica* currently separated by a gap.

(b) Ranges of European Honey Buzzard *Pernis apivorus* and Crested Honey Buzzard *P. ptilorhyncus*, with slight overlap.

(c) Ranges of Blue Tit *Parus caeruleus* and Azure Tit *P. cyanus*, with extensive overlap and hybridisation.

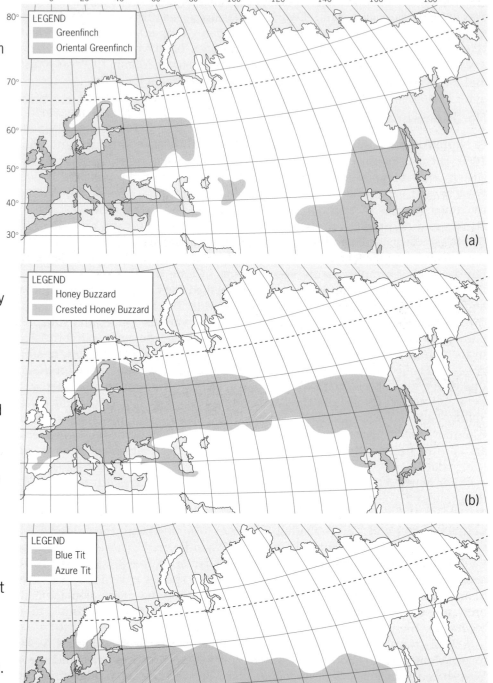

Red-footed Falcon *Falco vespertinus* and Amur Falcon *F. amurensis*, the Redstart *Phoenicurus phoenicurus* and Daurian Redstart *P. auroreus* and the European Greenfinch *Carduelis chloris* and Oriental (Grey-capped) Greenfinch *C. sinica* (**Figure 10.4**). The same holds within some species that have separate populations at each end, including the Lesser Kestrel *Falco naumanni*, Marsh Tit *Parus palustris* and Azure-winged Magpie *Cyanopica cyana* (see also Chapter 12).

North American suture zones

In North America, as in Europe, several distinct suture zones are apparent (**Figure 10.5**). The majority lie in the western half of the continent, running along the drier eastern side of various mountain chains, and marking the contacts between western and eastern taxa (**Figure 10.5**). An area in the centre of the continent remained treeless throughout the glacial cycles, and many east–west pairs of species and subspecies are thought to have arisen from a common ancestor in distinct southeast and southwest forest refuges; having diverged and spread, they now show varying degrees of overlap and hybridisation (**Tables 10.3 and 10.4, Figure 10.6**). Examples at the species level include the Red-naped Sapsucker *Sphyrapicus nuchalis* in the west and Yellow-bellied Sapsucker *S. varius* in the east, and at the subspecies level the Audubon's Warbler *D. c. auduboni* in the west and the Myrtle Warbler *D. c. coronata* in the east. These various species and subspecies pairs have maintained their distinctiveness despite, in some cases, hybridisation in the contact zone. The members of each are separated at one or more suture zones that collectively run northwest–southeast across the western half of the continent. Yet other species with western and eastern distributions are still separated by a gap. Examples include the Pacific-slope Flycatcher *Empidonax difficilis* and Yellow-bellied Flycatcher *E. flaviventris*, the Vaux's Swift *Chaetura vauxi* and Chimney Swift *C. pelagica* and, further south, the Pygmy Nuthatch *Sitta pygmaea* and Brown-headed Nuthatch *S. pusilla*. These species pairs may meet at some future date.

Other North American birds have distinct eastern and western forms, and fit the same east–west divide, but also have a third counterpart in the southwestern States or Mexico. Examples of this group include the Red-shafted Flicker *Colaptes a. cafer* and Yellow-shafted Flicker *C. a. auratus* (with the Gilded Flicker *C. a. chrysoides* further south), the Western Wood-Pewee *Contopus sordidulus* and Eastern Wood-Pewee *C. virens* (with the Greater Pewee *C. pertinax* further south), and others in **Table 10.3**. In general, these species have a more southerly distribution than those mentioned above, and some may have come together only recently as a result of human activities (such as tree-planting and fire control on the prairies).

A third group has an eastern and southern member, but lacks a distinct western form (Mengel 1970). With the southern form listed first, examples include the Long-billed Thrasher *Toxostoma longirostre* and Brown Thrasher *T. rufum*, the Great-tailed Grackle *Quiscalus mexicanus* and Boat-tailed Grackle *Q. major*, the Tropical Parula Warbler *Parula pitiayumi* and Northern Parula Warbler *P. americana*, and the Yellow-green Vireo *Vireo flavoviridis* and Red-eyed Vireo *V. olivaceus*. These different species groups again show varying degrees of differentiation, overlap and hybridisation.

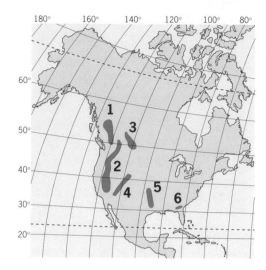

Figure 10.5 Major suture zones in North America. See **Table 10.4** for examples from relevant taxa, and Remington (1968) and Mengel (1970) for some additional minor suture zones.

1. The Northern–Cascade suture zone, where Pacific coastal–Rocky Mountain and northeast boreal populations meet.
2. The Pacific–Rocky Mountain suture zone, where coastal Pacific and Great Basin–Rocky Mountain populations meet.
3. The Northern–Rocky Mountain suture zone, where Rocky Mountain and north–northeastern populations meet.
4. The Rocky Mountain–eastern suture zone, where western and eastern populations meet.
5. The Central Texas suture zone, where western and eastern populations meet.
6. The Northern Florida suture zone, where Florida and more northern populations meet.

Because similar patterns of distribution are shown by many groups of birds (along with many other animals and plants), they are unlikely to have arisen by chance. They are again most plausibly explained by the existence, in times past, of a continuous distribution, which later split into distinct east and west forest refuges, plus a southern refuge (mainly in Mexico) for the most southern taxa. Since that time, the habitats and the animal species they contain have spread to varying extents.

On the eastern side of the continent, a minor suture zone runs across northern Florida **(Figure 10.5)**. At times of high sea-level, part of Florida became severed from the rest of North America, enabling differentiation of its plants and animals. Moreover, for much of the time that Florida was joined, a forest barrier across the top of the peninsula may have limited the north–south movement of open land species. At least six pairs of bird subspecies meet and hybridise in this zone, along with other animals and plants **(Table 10.4)**.

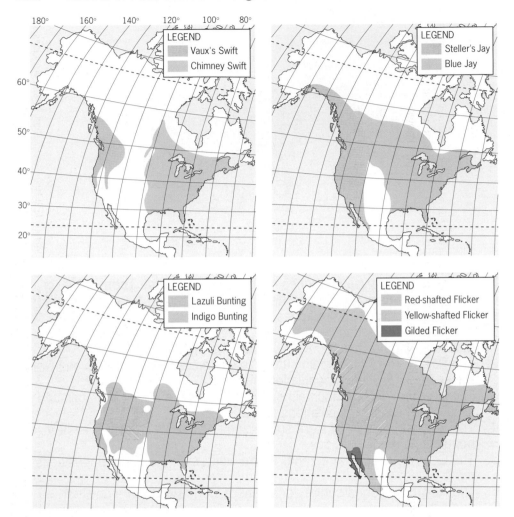

Figure 10.6   Examples of closely related bird taxa with western and eastern distribution patterns in North America.

(a) Ranges of Vaux's Swift *Chaetura vauxi* and Chimney Swift *C. pelagica* currently separated by a gap.

(b) Ranges of Steller's Jay *Cyanocitta stelleri* and Blue Jay *C. cristata* formerly separated but, following northwestward expansion of the Blue Jay, now touching in places, leading to hybridisation.

(c) Ranges of Lazuli Bunting *Passerina amoena* and Indigo Bunting *P. cyanea* overlapping with hybridisation, and in places the Indigo Bunting replacing the Lazuli Bunting westward. A third species, the Varied Bunting *P. versicolor*, occurs further south.

(d) Ranges of Red-shafted (Northern) Flicker *Colaptes auratus cafer* and Yellow-shafted (Northern) Flicker *C a. auratus* overlapping with an extensive, and apparently stable, hybrid zone. A third form, the Gilded Flicker *C. a. chrysoides*, occurs in the southwest, with additional subspecies of the Yellow-shafted Flicker on Cuba and Grand Cayman Islands (not shown).

Table 10.3   North American birds that have distinct western and eastern forms. Only species and the most distinctive subspecies are listed, but similar west–east differentiation is shown by other less-distinct subspecies (see **Table 10.4**). Mainly from Mengel 1970 and Rising 1983.

Western form	Eastern form	Hybridisation
1. Northern Spotted Owl *Strix occidentalis*	Barred Owl *Strix varia*	Occasional, recent contact
2. Western Screech Owl *Otus kennicottii*	Eastern Screech Owl *Otus asio*	In limited contact zone
3. Vaux's Swift *Chaetura vauxi*	Chimney Swift *C. pelagica*	Not in contact
4. Red-shafted (Northern) Flicker *Colaptes auratus cafer*	Yellow-shafted Flicker *C. a. auratus*	Extensive hybrid zone
5. Gila Woodpecker *Melanerpes uropygialis*	Red-bellied Woodpecker *Melanerpes carolinus*	Not in contact
6. Red-breasted Sapsucker *Sphyrapicus ruber*	Red-naped Sapsucker *Sphyrapicus nuchalis*[1]	In limited contact zone
7. Ash-throated Flycatcher *Myiarchus cinerascens*	Great-crested Flycatcher *M. crinitus*	None known
8. Western Wood-Pewee *Contopus sordidulus*	Eastern Wood-Pewee *C. virens*	None known
9. Steller's Jay *Cyanocitta stelleri*	Blue Jay *C. cristata*	None known, recent contact
10. Mountain Chickadee *Poecile gambeli*	Carolina Chickadee *P. carolinensis*[2]	Limited, hardly meet
11. Juniper Titmouse *Baeolophus griseus*	Tufted Titmouse *B. bicolor*[3]	?
12. Western Bluebird *Sialia mexicana*	Eastern Bluebird *S. sialis*	Rare
13. Bullock's Oriole *Icterus galbula bullockii*	Baltimore Oriole *I. g. galbula*	Regular, Baltimore replacing Bullock's
14. MacGillivray's Warbler *Oporornis tolmiei*	Mourning Warbler *O. philadelphia*	?
15. Western Warbling Vireo *Vireo swainsonii*	Eastern Warbling Vireo *Vireo gilvus*	None known
16. Grace's Warbler *Dendroica graciae*	Yellow-throated Warbler *D. dominica*	Not in contact
17. Audubon's Warbler *Dendroica coronata auduboni*	Myrtle Warbler *D. c. coronata*	Frequent but localised
18. Black-headed Grosbeak *Pheucticus melanocephalus*	Rose-breasted Grosbeak *P. ludovicianus*	Regular
19. Lazuli Bunting *Passerina amoena*	Indigo Bunting *P. cyanea*	Extensive hybrid zone
20. Spotted Towhee *Pipilo maculatus*	Eastern Towhee *P. erythrophthalmus*	Intergrade
21. Oregon Junco *Junco hyemalis oreganus*	Slate-coloured Junco *Junco h. hyemalis*	Intergrade
22. Western Meadowlark *Sturnella neglecta*	Eastern Meadowlark *S. magna*	Rare

[1] Also hybridises occasionally with the more eastern Yellow-bellied Sapsucker *S. varius* in the north of its range.

[2] The Carolina Chickadee *Poecile carolinensis* and Mountain Chickadee *P. gambeli* also have a northern counterpart, the Black-capped Chickadee *P. atricapillus*, with which they hybridise.

[3] Hybrid zone runs NE–SW. The eastern form is itself divided into Grey-crested *P. b. bicolor* and Black-crested *P. b. atricristatus* forms which hybridise in a few places of contact. They thus show evidence for a former disjunction, which is also displayed in other groups.

Most of the more southern species listed also have additional more southern counterparts, found in the extreme southwestern States or Mexico. These include, numbered as above, (1) Mexican Spotted Owl *Strix occidentalis*, (2) Whiskered Screech Owl *Otus trichopsis*, (3) no southwestern equivalent, (4) Gilded Flicker *Colaptes auratus chrysoides*, (5) Golden-fronted Woodpecker *M. aurifrons*, (6) no southwestern equivalent, (7) Dusky-capped Flycatcher *Myiarchus tuberculifer*, (8) Greater Pewee *Contopus pertinax*, (9) no southwestern equivalent, (10) Mexican Chickadee *Poecile sclateri*, (11) Oak Titmouse *B. inornatus*, (12) Mountain Bluebird *Sialia currucoides*, (13) Altamira Oriole *Icterus gularis*, (14–17) no southwestern equivalents, (18) Yellow Grosbeak *Pheucticus chrysopepus*, (19) Varied Bunting *Passerina versicolor*, (20) Brown (Canyon) Towhee *Pipilo fuscus*, (21) no southwestern equivalent, (22) no southwestern equivalent. In many other species, different subspecies conform to the same pattern.

Table 10.4 Examples of bird taxa (species and subspecies) that meet at particular suture zones in North America (see **Figure 10.5**). *Hybridisation recorded. Mainly from Remington 1968.

1. Northern Florida suture zone
   Bobwhite Quail *Colinus v. virginianus*   – *C. v. floridánus**
   Turkey *Meleagris gallopavo silvestris*   – *M. g. osceola**
   Brown-headed Nuthatch *Sitta p. pusilla*   – *S. p. caniceps**
   Carolina Wren *Thryothorus l. ludovicianus*   – *T. l. miamensis**
   Eastern Towhee *Pipilo erythrophthalmus rileyi*   – *P. e. alleni**
   Bachman's Sparrow *Aimophila a. aestivalis*   – *A. a. bachmani**

2. Central Texas suture zone
   Black-crested Titmouse *Baeolophus b. atricristatus*   Tufted Titmouse *P. b. bicolor**
   Varied Bunting *Passerina versicolor*   Painted Bunting *P. ciris**
   Black-chinned Hummingbird *Archilochus alexandri*   Ruby-throated Hummingbird *A. colubris*
   Western Wood-Pewee *Contopus sordidulus*   Eastern Wood-Pewee *C. virens*
   Ash-throated Flycatcher *Myiarchus cinerascens*   Great-crested Flycatcher *M.crinitus*
   Eastern Bluebird *S. sialis*   Western Bluebird *Sialis mexicana*

3. Rocky Mountain–eastern suture zone
   Western Scops Owl *Otus kennicottii*   Eastern Scops Owl *O. asio**
   Red-naped Sapsucker *Sphyrapicus nuchalis*   Yellow-bellied Sapsucker *S. varius**
   Red-shafted Flicker *Colaptes a. cafer*   Yellow-shafted Flicker *C. a. auratus**
   Western Wood-Pewee *Contopus sordidulus*   Eastern Wood-Pewee *C. virens**
   Scarlet Tanager *Piranga olivacea*   Western Tanager *P. ludoviciana**
   Bullock's Oriole *Icterus g. bullockii*   Baltimore Oriole *I. g. galbula**
   Audubon's Warbler *Dendroica c. auduboni*   Myrtle Warbler *D. c. coronata**
   Black-headed Grosbeak *Pheucticus melanocephalus*   Rose-breasted Grosbeak *P. ludovicianus**
   Lazuli Bunting *Passerina amoena*   Indigo Bunting *P. cyanea**
   Oregon Junco *Junco h. oreganus*   Dark-eyed Junco *J. h. hyemalis**
   Spotted Towhee *Pipilo maculatus arcticus*   Eastern Towhee *P. e. erythrophthalmus**
   Black-chinned Hummingbird *Archilochus alexandri*   Ruby-throated Hummingbird *A. colubris*
   Steller's Jay *Cyanocitta stelleri*   Blue Jay *C. cristata*
   Eastern Bluebird *Sialia sialis*   Western Bluebird *S. mexicana*

4. Northern–Rocky Mountain suture zone
   Franklin's Grouse *Canachites c. franklinii*   Spruce Grouse *C. c. canadensis*
   Oregon Junco *Junco h. oreganus*   Dark-eyed Junco *J. h. hyemalis**
   MacGillivray's Warbler *Oporornis tolmiei*   Connecticut Warbler *O. agilis*
   MacGillivray's Warbler *Oporornis tolmiei*   Mourning Warbler *O. philadelphia*

5. Northern–Cascade suture zone
   Red-breasted Sapsucker *Sphyrapicus ruber*   Red-naped Sapsucker *S. nuchalis**
   Red-shafted Flicker *Colaptes cafer*   Yellow-shafted Flicker *C. auratus**
   Hermit Warbler *Dendroica occidentalis*   Townsend's Warbler *D. townsendi**
   Oregon Junco *Junco h. oreganus*   Dark-eyed Junco *J. h. connectus**

6. Pacific–Rocky Mountain suture zone
   Red-breasted Sapsucker *Sphyrapicus ruber*   Red-naped Sapsucker *S. nuchalis**
   Nashville Warbler *Vermivora ruficapilla ridgwayi*   Virginia's Warbler *V. virginiae**

Eurasian birds in northwestern North America

There are yet other species pairs whose current geographical ranges may indicate their past history (Mengel 1970). The members of these pairs are usually regarded as distinct, but closely related species. They include the Northern Shrike *Lanius excubitor* and Loggerhead Shrike *L. ludovicianus*, the Bohemian Waxwing *Bombycilla garrulus* and Cedar Waxwing *B. cedrorum*, the Northern Three-toed Woodpecker *Picoides tridactylus* and Black-backed Woodpecker *P. arcticus*, and the Boreal Owl *Aegolius funereus* and Northern Saw-whet Owl *A. acadicus* (**Figure 10.7**). In each of these four pairs, the first named (larger) species has a more northerly breeding range than the second, although in three pairs the overlap is great, and the ranges do not fit together neatly like those of the west–east pairs mentioned above. The proposed course of events producing these patterns was roughly as follows. The members of each species pair are descended from a common ancestor, but they became separated and have evolved independently of each other for a longer period than the taxa already described. During glaciation, the southern species of each pair survived south of the ice, while the northern species, which all inhabit the northern coniferous forests, survived in Eurasia, where they still live. Then as the ice sheets shrank, the southern species of each pair spread northwards, while the northern species spread eastwards from Siberia into North America by way of Beringia, and now occupy extensive North American ranges.

Other species of forest or other habitats that may fit this pattern are mentioned in the footnote to **Figure 10.7**, along with some mammalian examples. Several populations of these and other species meet and hybridise along an east–west zone passing through Prince George in northern British Columbia. They include not only different species but also subspecies, such as Spruce Grouse *Dendragapus c. canadensis* and Franklin's Grouse *D. c. franklinii*.

Birds of sclerophyllous scrub and savannah

The broken nature of Mediterranean matoral vegetation during earlier periods in Europe is reflected in the geographical differentiation shown by many of the resident bird species. Good examples include the *Alectoris* partridges, which exist as several allospecies, with the Barbary Partridge *Alectoris barbara* in North Africa, the Red-legged Partridge *A. rufa* in Iberia and southern France, the Rock Partridge *A. graeca* in Italy and the Balkans, and the Chukar *A. chukar* in Turkey eastwards, all of which are themselves polytypic (Blondel 1985). Similarly, the three main groupings of *Sylvia* warblers, as revealed from a DNA-based phylogeny, correspond to the three main regions of Mediterranean endemism as defined from the distributions of various other animals and plants (**Table 10.5**; Blondel 1988, Blondel *et al.* 1996, Covas & Blondel 1998). These major warbler lineages were estimated on genetic evidence to have diverged 5.5–8.5 million years ago. This was the time of the so-called Messian Crisis, a long drought period that struck the entire Mediterranean basin east to the Caspian, when scrub habitats are likely to have shrunk to isolated areas for long periods, facilitating divergence of their occupants. Some of these taxa later came together as their habitats spread, and matoral vegetation may now be more extensive in the Mediterranean region than at any time in the past, as a result of human deforestation.

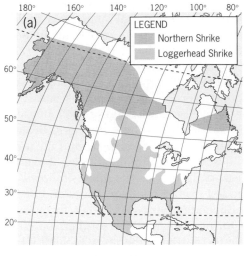

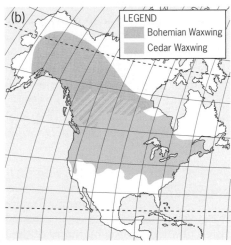

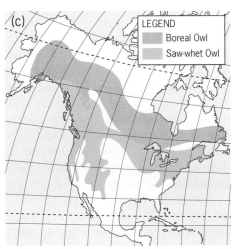

Figure 10.7 Breeding ranges of species pairs with one northern (Holarctic) species, and one more southern (Nearctic) species.

(a) Northern Shrike *Lanius excubitor* and Loggerhead Shrike *L. ludovicianus*.
(b) Bohemian Waxwing *Bombycilla garrulus* and Cedar Waxwing *B. cedrorum*.
(c) Boreal Owl *Aegolius funereus* and Northern Saw-whet Owl *A. acadicus*.

Other species that may fit this pattern and have a similar history include the Bar-tailed Godwit *Limosa lapponica* and Marbled Godwit *L. fedoa*, the Red-necked Phalarope *Phalaropus lobatus* and Wilson's Phalarope *P. tricolor*, the Three-toed Woodpecker *Picoides tridactylus* and Black-backed Woodpecker *P. arcticus*, the Black-billed Magpie *Pica pica* and Yellow-billed Magpie *P. nuttalli*, the Northern Goshawk *Accipiter gentilis* and Cooper's Hawk *A. cooperii*, the Rough-legged Hawk *Buteo lagopus* and Ferruginous Hawk *B. regalis*, the American Pipit *Anthus rubescens* and Sprague's Pipit *A. spragueii*, the Siberian Tit *Poecile cincta* and Boreal Chickadee *P. hudsonia*, the Black-capped Chickadee *P. atricapilla* (if conspecific with the Eurasian *P. montanus*) and Carolina Chickadee *P. carolinensis* (together with *P. gambeli* and *P. sclateri*), the Red-breasted Nuthatch *Sitta canadensis* and White-breasted Nuthatch *S. carolinensis*, and the Winter Wren *Troglodytes troglodytes* and House Wren *T. aedon*.

Mammalian examples include the Brown (Grizzly) Bear *Ursus arctos* and Black Bear *U. americanus*, the Collared Pika *Ochotona collaris* and American Pika *O. princeps*, the Arctic Ground Squirrel *Citellus parryi* and Columbian Ground Squirrel *C. columbianus* and the Thinhorn (Dall and Stone) Sheep *Ovis dalli* and Bighorn Sheep *O. canadensis*.

Table 10.5 The three major clades of Mediterranean *Sylvia* warblers, the evolution of which has been attributed to repeated episodes of isolation and re-expansion of maquis-type habitats in the three peninsulas. From Blondel *et al.*1996.

West Mediterranean (Iberian)	Central Mediterranean (Italian)	East Mediterranean (Balkan–Anatolian)
Marmora's Warbler *Sylvia sarda*	Sardinian Warbler *Sylvia melanocephala*	Lesser Whitethroat *Sylvia curruca*
Dartford Warbler *Sylvia undata*	Subalpine Warbler *Sylvia cantillans*	Red-Sea Warbler *Sylvia leucomelaena*
Tristram's Warbler *Sylvia deserticola*	Rueppell's Warbler *Sylvia rueppelli*	Orphean Warbler *Sylvia hortensis*
	Cyprus Warbler *Sylvia melanothorax*	Desert Warbler *Sylvia nana*
	Ménétrie's Warbler *Sylvia mystacea*	
	Spectacled Warbler *Sylvia conspicillata*	

The list excludes four more widespread species which could not be assigned (from their DNA) to the above three clusters, namely Blackcap *Sylvia atricapilla*, Garden Warbler *Sylvia borin*, Greater Whitethroat *Sylvia communis* and Barred Warbler *Sylvia nisoria*.

During glacial periods in North America, sclerophyllous scrub and savannah extended through Central America and the southern United States, giving continuous habitat from the west along the emerged Gulf coast to Florida, but since then eastern and western areas have become separated. At least two bird species of this habitat still reflect this history, occurring mainly in the west, but with outlying populations in Florida, namely the Burrowing Owl *Speotyto cunicularia* and the Scrub Jay *Aphelocoma coerulescens*, the latter now classed as a distinct species from its western counterpart *A. californica*. The same disjunct pattern of distribution is shown by several other animals, including reptiles (Woolfenden & Fitzpatrick 1996). Yet again, present distribution patterns reflect past climatic and vegetation conditions.

Birds of tundra and grassland

In contrast to forest which became most fragmented during glaciations, open habitats became most restricted and fragmented during glacial minima. It is therefore during the interglacials that most speciation in open-land birds is likely to have occurred. Overall, tundra was at its most extensive during the glacial maxima, but is unusual among open habitats in that it became highly fragmented both during interglacials and during glacials, when it persisted in some parts of the far north and also formed as separate belts south of the ice (**Figure 10.8**). Although tundra extends today all around the northern hemisphere, many of its avian inhabitants, both species and subspecies, are associated with particular segments: a distributional legacy of past refuges (**Table 10.6**).

The existence during glaciations of tundra both north and south of the ice may explain the origin of species pairs such as the Yellow-billed Loon *Gavia adamsii*, whose present distribution suggests a Beringian–Siberian origin, and

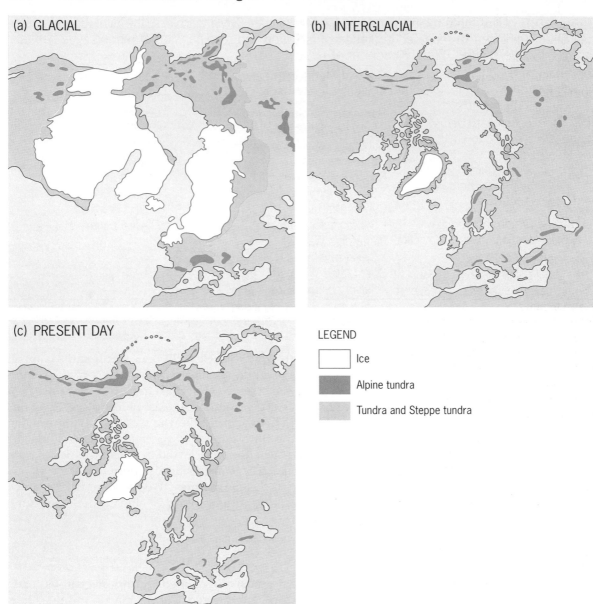

Figure 10.8   Purported distribution of tundra or steppe-tundra: (a) at the height of the last glacial period; (b) at the height of the last interglacial period, and (c) present. From various sources, especially Frenzel *et al.* 1992.

the Common Loon *G. immer*, which could have survived in the tundra and boreal forests of the south and spread north with these habitats in post-glacial times (Rand 1948). On this view, their distributional histories resemble those of the shrikes and others mentioned above. The two loon species now have almost mutually exclusive breeding ranges (**Figure 10.9**). The presence of tundra north

Table 10.6   Distribution patterns of tundra-nesting waterfowl and shorebirds. The term 'endemic' covers both species and subspecies, but some species are considered as endemic to more than one area. From Ploeger 1968.

North Atlantic region, comprising western Greenland, northeast Greenland, Iceland, northwest Norway, Spitsbergen Bank, Barents Sea, Kara Sea – 14 forms (8 endemics*)

*Anser fabalis brachyrhynchus**	*Bucephala islandica**
*Chen caerulescens atlanticus**	*Histrionicus h. histrionicus**
*Branta leucopsis**	*Clangula hyemalis*
Branta bernicla hrota	*Calidris alpina*
*Branta bernicla bernicla**	*Calidris m. maritima**
*Somateria mollissima borealis**	*Calidris alba*
Somateria spectabilis	*Calidris c. canutus*

Bering Sea region, comprising northern east Siberia, eastern Siberia, all parts of the submerged Bering Sea shelf, Aleutians, western Alaska – 25 forms (19 endemics*)

*Anser albifrons frontalis**	*Calidris ruficollis**
*Anser fabalis serrirostris**	*Calidris subminuta**
Chen c. caerulescens (white phase)*	*Calidris temminckii*
*Chen canagica**	*Calidris mauri**
*Branta canadensis leucopareia**	*Calidris ferruginea**
*Branta canadensis asiatica**	*Calidris alpina*
*Branta canadensis minima**	*Calidris ptilocnemis**
*Branta bernicla orientalis**	*Calidris acuminata**
*Polysticta stelleri**	*Calidris melanotos**
*Somateria mollissima v-nigra**	*Calidris alba*
*Somateria fischeri**	*Calidris tenuirostris**
Somateria spectabilis	*Calidris c. canutus*
Clangula hyemalis	

Canadian Arctic Archipelago – 14 forms (10 endemics*)

*Cygnus c. columbianus**	*Somateria spectabilis*
*Anser albifrons gambelli**	*Clangula hyemalis*
Chen c. caerulescens (blue phase)*	*Calidris pusilla**
*Chen rossii**	*Calidris bairdii**
*Branta canadensis hutchinsi**	*Calidris fuscicollis**
Branta bernicla hrota	*Calidris alba*
*Branta bernicla nigricans**	*Calidris canutus rufus**

The Taimyr Region, the area west of the Middle Siberian ice sheets, comprising western Siberia and Russia – 7 forms (4 endemics*)

Cygnus columbianus bewickii	*Branta ruficollis**
*Anser a. albifrons**	*Calidris temminckii*
Anser erythropus	*Calidris minuta**
*Anser fabalis rossicus**	

Rest of Europe, comprising west and southwest Europe – 5 forms (3 endemics*)

*Anser albifrons flavirostris**	*Somateria m. mollissima**
*Anser f. fabalis**	*Calidris alpina*
Anser f. brachyrhynchus	

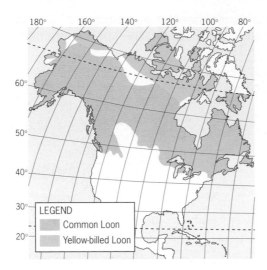

Figure 10.9   Current breeding ranges of two species of loons, the Yellow-billed Loon *Gavia adamsii* presumed to have arisen in the Beringian refuge of the northwest and the Common Loon *G. immer* presumed to have survived south of the glacial ice in mid North America. As the ice retreated, the Yellow-billed Loon is presumed to have spread eastward across the tundra, and the Common Loon northwards with the boreal forest and onto the newly exposed tundra of the northeast. Today the Common Loon occupies a wider range of habitat than the Yellow-billed Loon, but their tundra ranges are almost mutually exclusive, a situation supposedly maintained by competition between them. Their supposed distributional histories are similar to those of the forest birds in **Figure 10.7**. From Rand 1948.

and south of the glacial ice might also explain why some tundra birds exist today as both northern and southern forms. Thus the streaky grey Common (Mealy) Redpoll *Carduelis flammea* breeds in the transition zone between forest and tundra and the paler white-rumped Arctic (Hoary) Redpoll *C. hornemanni* in the bush tundra further north; similarly, the Canada Goose *Branta canadensis* exists as large-bodied forms in the south and as distinct small-bodied forms in the north. The Lapland Longspur *Calcarius lapponicus* may have evolved north of the ice and Chestnut-collared Longspur *C. ornatus* and McCown's Longspur *C. mccownii* south of the ice in North America, as they now live in tundra and prairie, respectively, as do the American Pipit *Anthus rubescens* and Sprague's Pipit *A. spragueii*. Yet other species now occur in open land north and south of the boreal forests, but their populations show only subspecific differentiation (e.g. Merlin *Falco columbarius* and Horned Lark *Eremophila alpestris*).

The Snowy Owl *Nyctea scandiaca* is especially interesting. Although placed in its own genus, recent DNA work has shown it to be very close to the big brown *Bubo* owls, represented by the Great Horned Owl *Bubo virginianus* in America and by the Eagle Owl *B. bubo* in Eurasia (Wink & Heidrich 2000). The evolution of the

Snowy Owl can thus be most readily explained by the isolation of a brown *Bubo* north of the ice sheets which, in response to local conditions, was then able to undergo independent evolution to produce the white owl of today. The divergence of the Snowy Owl was dated on mt DNA (cytochrome *b* gene) analysis at four million years ago, in the Pliocene, its closest living relative being *B. virginianus* (Wink & Heidrich 2000). The same type of differentiation, dependent on separate evolution north and south of the ice sheets, has been suggested for other organisms, notably the Polar Bear *Ursus maritimus* which is thought to have arisen in the Pleistocene from a coastal population of the Brown Bear *U. arctos* that became isolated north of the ice from the main population south of the ice (Kurtén 1968). These two species produce viable offspring if interbred in captivity, and mt DNA evidence suggests that they may have diverged only in the last few tens of thousands of years (Talbot & Shields 1996), much more recently than the owls.

Those tundra refuges that remained north of the ice during the glacials lasted for much longer than the interglacial refuges, and evidently had a major impact on the differentiation of tundra species. The distributions of tundra-nesting waterfowl were examined in detail by Ploeger (1968), who distinguished 38 morphologically distinct forms of 18 species (**Table 10.6**). These species and subspecies are all today centred on high-latitude areas that, on geomorphological and other evidence, remained ice-free through the last glaciation. From these areas, species have spread to varying extents, and breaks in their present breeding distributions all lie in or near areas that were formerly ice-covered. Various shorebirds, especially *Calidris* sandpipers, show similar patterns of distribution and differentiation (see also later).

In contrast to tundra, the grasslands of mid North America seem not to have been subdivided during glacial cycles, but remained as a single unit, expanding and contracting in area. Not surprisingly, therefore, they have not provided an important arena for speciation. The prairies support only a small avifauna, including 37 grassland-adapted species, of which 12 are endemic, plus other more widespread species. The average number of passerine species per genus is only 1.1, compared with 2.9 in the forests and other habitats surrounding the prairies (Mengel 1970). Five species are shared with South American grasslands and one with the Old World steppe (which is also poor in indigenous bird species). Whatever their past history, however, grasslands would be expected to support fewer species than forest through lack of vertical structure. Their main role in speciation in North America and elsewhere has been as an isolating agent for forest birds reluctant to cross the wide expanse of open land.

Another important region in the evolution of North American aridland birds comprises the various intermontane basins of the southwest. These lower-lying areas are largely isolated from one another and from the mid continental prairies by mountains. The various *Callipepla* quail have differentiated within this region, as have the *Toxostoma* thrashers (Hubbard 1973). Several species have distributions based around the central valley of California which is the largest low-flying area in the western States. Such localised endemics include the Lawrence's Goldfinch *Carduelis lawrencei*, Tricoloured Blackbird *Agelaius tricolor*, California Thrasher *Toxostoma redivivum*, Yellow-billed Magpie *Pica nuttalli* and Nuttall's Woodpecker *Picoides nuttalli*, but the majority of these birds are found in the

Mediterranean-type chaparral and other scrubby areas, rather than in open grassland.

Other patterns

Several other, more complex patterns of distribution are discernible among both Eurasian and North American birds. Each pattern is shown by several pairs or groups of taxa and reveals what seem to be zones of secondary contact after periods of separation. All such patterns, like those discussed above, have been interpreted in terms of past climatic and vegetation changes. All emphasise the importance of the fragmentation of habitats; and in the histories of some taxa, three or more different refugia could have been involved at one time (Rand 1948, Mengel 1970). Some distribution patterns in North America have been explained in terms of multiple invasions, in which the same (usually western) area has been colonised more than once from the same (usually eastern) source, but in the interval between successive invasions the original colonists had diverged sufficiently not to merge with later colonists (Mengel 1970). For example, this process could account for the several species of similar 'black-throated green warblers' found today in western North America (Mengel 1964). Attempts to reconstruct the phylogeny of these warblers using mt DNA (restriction sites) lent some support to this view, but also suggested that some modern species resulted from isolation in intermontane regions (Bermingham *et al.* 1992).

More on suture zones

The existence of suture zones, where a wide range of plant and animal species come together, provides strong evidence that whole communities were spatially separated in the past. We would not expect contact zones to show in all species, however, as some taxa may not have diverged in isolation, while others may have displaced their sister taxa, either completely, or from particular habitats so that they now occur sympatrically. Because taxa spread at different rates, they are unlikely all to meet at the same time or in exactly the same place, and human impact on habitats has probably increased the proportion of sister taxa now in contact (e.g. on the Great Plains of North America).

Within suture zones, the ranges of some sister taxa abut, while those of others overlap; some sister taxa hybridise, while others do not. Where hybrids are inferior in some way to the parental forms, hybridisation may be temporary, until reproductive isolation has developed (for example by mate preference). Frequent hybridisation may, therefore, be more a reflection of recent contact than of genetic similarity.

More sister species show extensive range overlap without hybridisation in Europe than in North America (compare **Tables 10.1 and 10.3**). It is as though more such species have progressed further on the path to sympatry in Europe. Perhaps European sister taxa are older, on average, than North American ones, but in addition, more frequent overlap and more marked phenotypic divergence in Europe might have been facilitated by the generally poorer avifauna, and reduced competition levels, compared with North America.

North–south movements

Consideration of glacial history has also provided an explanation of why, in the movements between continents, more species have colonised from north to south in the northern hemisphere than from south to north (as evident in movements between North and South America, Europe and Africa, and the Beringian Refuge into North America). The majority of species were likely to cross only when their particular habitat occurred across a land-bridge from one side to the other. As habitats were pushed southwards during glaciations, the general flow of species would have been from north to south, and cold periods, favouring northern species, lasted most of each climatic cycle. During the warm interglacials, southern habitats would have pushed northwards but interglacials lasted only a small part of each cycle. Moreover, for most of an interglacial, land-bridges would have been narrowed or closed by rising sea-levels, further reducing the opportunities for south–north colonisation (Vrba 1997). Darwin (1859) was aware of the greater colonisation success in the northern hemisphere of northern species moving south, and suggested that northern species were competitively superior to southern ones; however, he was probably unaware of the difference in colonisation opportunities that climatic cycles provided for the two groups.

Influence of migration

Many bird species required not only suitable breeding habitat to survive a climatic extreme, but also suitable wintering habitat at lower latitudes. For the majority of birds, which form pairs in their breeding areas, the summer refuges were crucial in allowing differentiation and speciation to occur. Just as today, different populations and subspecies might have wintered in the same areas as one another but, provided that they returned each year to their own breeding areas, they could remain genetically isolated. However, the same was not necessarily true for many waterfowl species which, unusually among birds, form pairs on their wintering areas, the males accompanying the females back to their breeding areas (Chapter 17). Hence, wherever we find genetically distinct populations of winter-pairing waterfowl, we must assume long separation of both breeding and wintering areas (for an example of different goose populations interbreeding through recent contact on wintering areas, see Chapter 3).

GENETIC EVIDENCE FOR PAST CLIMATIC EFFECTS

The history of any population, it is often said, is locked in its genes. The challenge is to unlock the salient parts of that history in order to test the kinds of inference about distributional history drawn above. One obvious question is whether current populations show genetic evidence of past distributional splits, and of bottlenecking in refuges, at dates and localities that fit the geological and palaeobotanical evidence. In theory, some of the sister taxa mentioned above might have achieved their present level of differentiation during a single climatic cycle in which two populations, originally derived from the same parental form, were kept apart for long enough to form different subspecies or species in one continuous process, but other sister taxa may have resulted from an intermittent process, extending over several climatic cycles. Periods of separation and divergence would then have been broken by intervening periods of re-contact and genetic mixing, as habitats waned and waxed. In this way, the whole speciation process could have been extended by periodic set-backs, being dependent on the frequency of isolating events, as well as on their duration. Because the cold part of each recent climatic cycle lasted an average of nine times longer than the warm part, periods of isolation for forest species would have greatly exceeded periods of mixing.

Interest has therefore centred on whether existing sister taxa arose from their common ancestor mainly during the last glacial cycle (say 100,000 years), or over multiple cycles extending over a large part of the Pleistocene or even longer. The approximate divergence dates of different sister taxa (whether species, subspecies or other phylogroups) can be estimated from their DNA (Chapter 2). This is done by measuring the 'genetic distance' between different populations and then working back on the assumption of a constant rate of nucleotide sequence divergence. For the mitochondrial cytochrome b gene, this rate has been estimated in passerines and geese from geological and fossil evidence at about 2% per million years (Chapter 2). So if two related taxa showed a 4% difference in their mt DNA cytochrome b structure, two million years is presumed to have elapsed since they first became isolated from one another, and embarked on independent evolutionary tracks. However, for other parts of the mitochondrial and nuclear genomes, and in different types of birds, different rates of sequence divergence have been estimated (Chapter 2).

Population divergence in passerines

On this basis, Klicka & Zink (1999) compared the estimated divergence dates for 74 pairs of sister taxa of North American passerines (**Figure 10.10**). They found divergence dates that ranged between less than 0.2 million years ago and 5.5 million years ago (modal value around 2.5 million years ago). In general, morphologically well-differentiated genera and species gave separation dates that preceded the Pleistocene, while morphologically less differentiated species and subspecies gave separation dates within the Pleistocene, mostly within the last 1.5 million years. On this evidence, then, divergence–speciation processes occurred throughout the Pleistocene and earlier, but few (if any) populations reached their present level of mt DNA differentiation during the last glacial cycle alone.

Many protracted speciations, considered individually, probably extended through time from Pliocene origins to Pleistocene completions. When avian speciation is viewed as an extended process, rather than as a point event, Pleistocene conditions appear to have played a role both in initiating phylogeographic separations within species, and in completing speciations that had been started earlier (Avise *et al.* 1998). Although consistency in the locations of refuges during successive climatic cycles is not essential to the prolonged speciation process envisaged, climatic refuges probably did lie in roughly the same geographical regions during each climatic extreme. The general shape and disposition of land areas has not altered greatly in the last few millions of years (except through regular changes in sea-levels); and each successive glaciation pushed the flora and

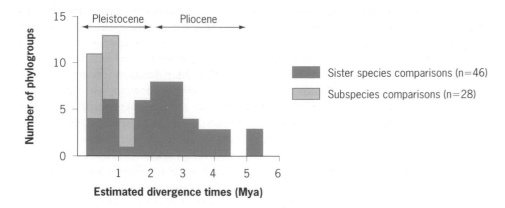

Figure 10.10   The distribution of mt DNA divergence estimates (million years ago, Mya) for North American songbird taxa including some currently recognised as subspecies and others recognised as species. Together they represent taxa at different stages of evolutionary divergence. From Klicka & Zink 1999.

fauna of forests to lower latitudes, from which they spread during interglacials. Moreover, areas that were not glaciated at any time in the last few million years, even at high latitudes, are recognisable as such from geomorphological features.

The fact that most of the North American sister taxa considered lie at the left end of **Figure 10.10** cannot be taken to imply that divergences were more frequent in the last million years than earlier. Even if the rates of origin and extinction of species were constant, most taxa would still lie at the left end, in the younger age classes, with successive age groups declining in frequency to give fewer in the older groups, in the manner shown (Nee *et al.* 1994). This would be expected firstly because extinctions would lead to progressive loss of some members of sister taxa, which would therefore not appear in the analysis, and secondly because subsequent speciation events would result in the gradual replacement of older taxa by new (younger) daughter taxa. In addition, the sister taxa examined in **Figure 10.10** are not a random sample drawn from North American passerines, but are weighted towards those showing the greatest morphological differentiation, and hence in general the oldest divergence dates (Klicka & Zink 1999). The members of most pairs are still allopatric, however, and sympatry would be expected to take longer to evolve.

One of the shortest recorded divergence events so far documented for North American passerines from mt DNA concerns the Timberline Sparrow *Spizella taverneri*, which has been classed as a subspecies of Brewer's Sparrow *S. breweri*, from which its range is completely separated. These birds differ in mt DNA cytochrome *b* sequences by an average of only 0.1%, which corresponds to a divergence date of 35,000 (range 40–80,000) years ago (Klicka *et al.* 1999). This date falls well within the last glacial period, but whether the two forms are sufficiently different for both to warrant specific status, as suggested by the authors, has been debated (Mayr & Johnson 2001). Similarly, western and eastern types of the

Yellow Warbler *Dendroica petechia* differ in mt DNA (control region), with western apparently derived from eastern types within the last climatic cycle, and possibly as recently as 7,000 years ago (Milot *et al.* 2000). Yet other species show little morphological differentiation, but significant differences in mt DNA in different parts of their range. For example, the Great Reed Warbler *Acrocephalus arundinaceus* is morphologically uniform across Europe, but shows distinct east and west genotypes, estimated from mt DNA (control region) to have diverged 70,000 years ago, which again falls in the last glaciation (Bensch & Hasselquist 1999).

The fact that pairs of taxa supposedly separated by the same events should appear to have widely different divergence dates has worried some researchers. However, even with refuges in the same places in each climatic cycle, and the same divergence rates, we would not necessarily expect the taxa evolved in them to be of similar ages. Imagine a series of range expansions and contractions based on two refuges throughout the Plio-Pleistocene. During a contraction phase, one of a pair might have become extinct, its range at the next expansion phase being filled by its partner from the other refuge. At the next contraction, the two subpopulations of the one surviving form could then start diverging afresh. This type of process would produce a phylogeographic–mitochondrial record that showed spatial concordance but temporal variation in divergence dates across species.

Suppose next that, in each of several species, population extinctions in one or other refuge were not followed by recolonisations. Then, the range of each of these species would be confined to one or other area. In both North America and Eurasia today are many bird species that are restricted to one or other side of the land-mass and do not have a counterpart species on the opposite side. Some of the gaps might be due to past extinctions, and, in addition, an unknown number of taxa could have died out in all their refuges, leaving no descendants, as discussed in Chapter 9.

Population divergence in non-passerines

So far, relatively few analyses of this type have involved non-passerines. However, studies of mt DNA have given divergence dates for various types of geese, again on the assumption of 2% divergence per million years consistent between lineages. On this basis, Canada Geese *Branta canadensis* and Brent Geese *B. bernicla* are estimated to have diverged from a common ancestor about four million years ago, only 0.5 million years after *Branta* diverged from the grey geese *Anser* (Shields & Wilson 1987a, van Wagner & Baker 1990). The two main assemblages of living Canada Geese (small and large) apparently diverged about 1.2 million years ago, and the different populations within these assemblages all diverged from one another within the last 0.5 million years (Shields & Wilson 1987b, van Wagner & Baker 1990, Quinn 1992). Atlantic Brant *B. bernicla* and Black Brant *B. nigricans* are estimated to have diverged about 0.8 million years ago (Van Wagner & Baker 1990). Another population of Pale-bellied Brant that breeds on Melville Island is estimated to have diverged from Black Brant about 0.4 million years ago, as is Ross's Goose *Chen rossii* from the Snow Goose *Chen caerulescens* (Shields & Wilson 1987a, Shields 1990). The various grey geese and snow geese are estimated to have diverged about 1.5 million years ago (van Wagner & Baker 1990, Quinn

1992). The majority of these dates lie within the Pleistocene, but well before the last glaciation.

The preceding studies were mostly based on the mt DNA cytochrome *b* gene which supposedly mutates at too slow a rate to record reliably population splits that occurred in the last glacial period or subsequently. However, studies on various shorebirds were based on the mt DNA control region, segment I of which was estimated to diverge ten times as fast as the cytochrome *b* gene (at 20.8% per million years from studies on geese, Quinn 1992) (Chapter 2). This enabled more recent, as well as older, population splits to be dated in three arctic-nesting species.

The Dunlin *Calidris alpina* breeds widely in northern Eurasia and North America, mainly on tundra in low arctic regions. It has been divided into several different subspecies mainly on size of body and bill. Genetically, the species is highly variable, and shows pronounced phylogeographic structure, with five major lineages recognised, each from a different part of the range (Wenink *et al.* 1996). Three of these genetic lineages correspond to previously designated subspecies (*C. a. hudsonia, C. a. centralis, C. a. sakhalina*), but one genetic lineage had been previously classed as two subspecies (*C. a. pacifica, C. a. arcticola*), while another genetic lineage had been split into three subspecies (*C. a. alpina, C. a. artica,* and *C. a. shinzii*).

The most genetically distinct, and therefore the oldest, form is the one breeding in central Canada (*C. a. hudsonia*). It is estimated to have split from the common ancestor of the remaining groups about 223,000 years ago, which coincides with the warm Holstein interglacial, when tundra was restricted and fragmented. The next split, between the birds of Europe and Siberia–Alaska, was dated to about 117,000 years ago, which coincides with the Emian interglacial, when tundra was again restricted. The remaining three groups (*C. a. centralis, C. a. sakhalina, C. a. pacifica*), from central and eastern Siberia and Alaska, are all dated at between 71,000 and 80,000 years, which falls within the last glaciation, when tundra habitat was widespread but broken by glaciated mountain ranges in eastern Siberia and Kamchatka. Thus the five different types of Dunlin appear to have originated at widely different times. In those lineages that contain more than one morphological subspecies that cannot be separated on mt DNA, it is assumed that the subspecies have diverged recently, perhaps within the latest interglacial (Wenink *et al.* 1996, Kraaijeveld & Nieboer 2000). All these Dunlin lineages thus appear to have evolved in isolation in different refugia within the latter half of the Pleistocene, and maintained their differences as a result of strong philopatry; they have expanded to colonise new breeding areas and, to some extent, mixed with other populations to give the patterns seen today[1].

The closely related Red Knot *Calidris canutus* also has a circumpolar distribution, but breeds only in the high arctic where it occurs as five subspecies, each breeding in a separate region and recognised mainly on bill and wing measurements. Their mt DNA (control region) shows only slight variation, however, and

---

[1] *To judge from patterns of morphological variation, some other shorebirds of low artic breeding areas, such as Golden Plover* Pluvialis apricaria/fulva/dominica *and Bar-tailed Godwit* Limosa lapponica, *may have shared a similar history to the Dunlin* Calidris alpina *(Kraaijeveld & Nieboer 2000).*

no obvious geographical structuring, as only seven haplotypes were found world-wide, all closely related and differing by only 1–3 nucleotide substitutions. The implication is that Knots were reduced to very small numbers in the late Pleistocene, presumably at a time when high arctic refuges were extremely restricted, and have expanded to their present wide distribution only in the last 10,000 years or so (Baker *et al*. 1994).

We thus have two closely related *Calidris* species, both of which now extend across the arctic, with different subspecies in different regions, yet they show different degrees of genetic differentiation, indicating markedly different phylogeographic histories, with different divergence dates. Within each species, different subspecies vary greatly in their respective ages, but all the Knot subspecies appear younger than some of the Dunlin ones. Of the two species, the high arctic Knot would have suffered the greatest restriction in range; it shows the lowest variability in mt DNA, possibly because it survived for a time as a small population in only one region.

The third species, the Ruddy Turnstone *Arenaria interpres*, breeds across the arctic, yet shows only limited geographical variation in morphology and mt DNA structure. The two subspecies (in North American and Eurasia, respectively) are estimated to have diverged only within the last 10,000 years (Wenink *et al*. 1994). This species may also have expanded recently from a refuge in which its population was bottlenecked, thus again accounting for the paucity of phylogeographic structure. The alternative explanation, that dispersal may be so great in Turnstones that the entire world population is panmixic, preventing local variants from arising, is unlikely, considering the wide geographical range and high philopatry of this species (Chapter 17).

Unlike shorebirds, Rock Ptarmigan *Lagopus mutus* stay on the tundra year-round and have a geographical range that encompasses all known high-latitude refugia. They show considerable subspecific and genetic differentiation across the arctic. Birds from 26 localities scattered from Iceland, across Greenland and North America to eastern Siberia were examined by Holden *et al*. (1999). Mitochondrial DNA (control region) and nuclear intron sequences revealed strong phylogeographic structure, with distinct lineages largely matching designated subspecies. These lineages were based around six known glacial refuges, five in North America and one in Iceland. Divergence dates varied between lineages and also according to the particular molecular clock calibration employed, but in any case they all fell within the last glacial period. The findings were thus consistent with the hypotheses that Rock Ptarmigan survived and diverged in several high-latitude refuges during the last glaciation and that current geographical subspecific and genetic variation reflects the locations of these refuges and the patterns of subsequent range expansion.

In summary, all these various studies, whether of passerines or of non-passerines, point to population divergences within the Pleistocene. Some could be dated to particular glacial or interglacial periods, including the most recent. Few studies examined the relationship between the phylogeographic structure of populations and the locations of known refuges. Those that did revealed that particular phylogroups (or subspecies) were centred on regions known to have contained glacial refuges. Such results thus add to the evidence that past glacial conditions contributed to the structuring of modern populations.

On the question of timing, much the same picture has emerged in North American mammals, in which 52 (72%) of 72 inferred phylogroup separations date to the Pleistocene, and most of the remainder to the Pliocene (Avise 2000). Among other vertebrates, estimates are complicated by suspected slower rates of mutation in mt DNA, but even on the slowest rates proposed, substantial proportions of divergence dates fell within the Pleistocene (Avise 2000). For a range of different vertebrates, therefore, the climatic vicissitudes of the Plio-Pleistocene could have been a major factor promoting the evolutionary differentiation of populations. However, two notes of caution concern the date estimates. Firstly, they are clearly sensitive to the clock calibrations employed, and to possible variations in the rates of mt DNA mutation between lineages, yet few empirical baseline data on these rates are yet available (Chapter 2). Secondly, in the use of the cytochrome *b* gene, severe bias exists against the detection of separations that occurred less than about 200,000 years ago. Such a small proportion of nucleotides are likely to have changed in this time that any estimates of divergence dates are well within the limits of error and carry large confidence intervals. Thus more taxa may have separated and diverged in the last 200,000 years than the data indicate.

Patterns of colonisation

Many bird species now live in areas that remained favourable throughout the glaciations, as well as in more northern areas that could have been colonised only since glacial retreat. The majority of such species show more mt DNA variation in the refugial than in the recently colonised areas. They are assumed to have colonised from the northern edge of their refugial ranges, and are like other founder populations which, because of the small number of initial colonists, show less genetic variation than the source population. Examples include the European Greenfinch *Carduelis chloris* in Europe (Merilä *et al.* 1997) and Macgillivray's Warbler *Oporornis tolmiei* in Central and North America (Milá *et al.* 2000). Latitudinal trends in genetic variance, associated with post-glacial leading edge spread, have also been described in plants, insects and mammals from both continents (Hewitt 2000). They represent another way in which genetic diversity is reduced from south to north in the northern continents, in addition to reductions in the numbers of subspecies and species (Hewitt 1999).

Several other widespread species examined from different parts of their deglaciated North American range have shown unexpectedly low levels of mt DNA differentiation, despite, in some species, high levels of phenotypic differentiation (for Red-winged Blackbird *Agelaius phoeniceus*, see Ball *et al.* 1988; for Song Sparrow *Melospiza melodia* see Zink & Dittman 1993a). Yet again, the implications are that large deglaciated areas were colonised by individuals from limited areas, and that such species underwent morphological differentiation only after they had spread to their current ranges, in the last 10,000 years or so (Chapter 2). However, not all widespread bird species occupying recent deglaciated areas show low genetic differentiation (see Yellow Warbler *Dendroica petechia*, mentioned above, which shows clear eastern and western types from mt DNA, Milot *et al.* 2000).

In some species, estimates have been made of the routes and dates of spread. Thus, based on diversity patterns in mt DNA (four sections), Common Chaffinches *Fringilla coelebs* appear to have expanded from North Africa into southern Europe within the last 100,000 years, and spread further north through Europe only during the past 15,000 to 3,000 years, behind the retreating ice sheets (Marshall & Baker 1999). The species colonised the Canary Islands from North Africa twice: the first colonists, about one million years ago, gave rise to a different species, the Blue Chaffinch *F. teydea;* and the second wave from Africa to the Azores, then Madeira and the Canary Islands, gave rise to additional forms, classed as subspecies of the Common Chaffinch. The common features of appearance that the different island chaffinches share could therefore be due more to a common colonisation history on a single wave from Africa than to convergent evolution in a common island environment. Similarly, studies of several North American bird species, including the Song Sparrow *Melospiza melodia* and various parids, have revealed basal haplotypes in populations on Newfoundland, suggesting that these species may have spread from there to other parts of the continent in post-glacial times (Gill *et al.* 1993, Zink 1997). These findings are consistent with other evidence of a glacial refuge on Newfoundland or in nearby areas now submerged.

Joining of populations following range expansion

The joining of previously separated populations is usually obvious when the taxa concerned have diverged in colour patterns; they often form a distinct contact zone, in which the two forms may hybridise (see above). In other species which show no plumage differentiation, contact zones are morphologically seamless, but can be recognised genetically. For example, Blue Tits *Parus caeruleus* from the same area in France all looked the same but showed two distinct mt DNA (restriction fragments) lineages (with 1.23% divergence). This dichotomy was attributed to the recent mixing of two long isolated populations (Taberlet *et al.* 1992).

Sampling of Common Ravens *Corvus corax* from various sites throughout the northern hemisphere revealed a deep genetic break between birds from the western United States and those from the rest of the northern hemisphere. These two groups differ by more than 4% in mitochondrial cytochrome *b* sequences, but come into contact over a huge area in the western United States, where they are in process of merging (Omland *et al.* 2000). Similar findings came from examination of mt DNA (control region) in Snow Geese *Chen caerulescens* in North America (Quinn 1992). In this species, however, distributional splits and subsequent re-joining apparently occurred more than once, another finding that could not have been deduced from plumage patterns.

In southeastern North America, gradual (clinal) change in genetic structure has been described in the Carolina Chickadee *Poecile carolinensis*. It was interpreted as resulting from the fusion of two populations that, at some time (estimated at 1.5 million years ago) were isolated in eastern and western refuges, but later came together (Gill *et al.* 1993, 1999). This history is reflected in the genetic structure of the current population, but not in its morphology which is uniform across the range. By contrast, conspecific populations of the Black-capped Chickadee *Poecile atricapilla* feature the same mt DNA haplotypes across the continent, as do conspecific populations of the Boreal Chickadee *P. hudsonica*. Within the last 15,000 years, each of these two species probably expanded to a continent-wide range from a single source population.

In conclusion, analysis of mt DNA can reveal past population splits that are not apparent in the morphology of re-joined populations. It thus becomes possible, given enough analyses, to calculate what proportion of species in an area have undergone past population splits, of how distinct the different forms are, and when their divergence began. Such genetic details imposed on a map may help to highlight the locations of past refuges and dispersal routes more precisely and in greater detail than can morphological data alone. They provide an independent source of evidence from that derived from studies of species numbers, distributions and morphology, discussed earlier. However, genetic analyses are still too few in number to make such avifaunal comparisons yet worthwhile, although they are sufficient to confirm that the genetic patterns found in birds are replicated in various other animals and plants (Hewitt 2000).

Genetic evidence of species responses to barriers

The avifaunas of Alaska and Siberia are not far apart, and as recently as 12,000 years ago could have been in direct contact across the Beringian land-bridge. Mitochondrial DNA (restriction fragment) profiles were examined in small numbers of 13 bird species that breed in both Alaska and Siberia (Zink *et al.* 1995). In the majority of species, genetic differences were detected between individuals collected on each side. In seven species these differences were marked, namely Marbled Murrelet *Brachyramphus marmoratus*, Three-toed Woodpecker *Picoides tridactylus*, Whimbrel *Numenius phaeopus*, Common Gull *Larus canus*, Black-billed Magpie *Pica pica*, American Pipit *Anthus rubescens* and Asian Rosy Finch *Leucosticte arctoa*. These species, it seems, have had long independent histories on either side of Beringia, even though they show only limited morphological differentiation between the two sides. Four other species showed weak mt DNA differences, namely Barn Swallow *Hirundo rustica*, Common Tern *Sterna hirundo*, Common Snipe *Gallinago gallinago* and Pelagic Cormorant *Phalacrocorax pelagicus*. In the Green-winged Teal *Anas crecca*, one haplotype was found on both continents, implying gene flow, while two others were highly divergent. On this and other evidence, the two forms of teal have since been re-classified as separate species, *A. crecca* in Eurasia and *A. carolinensis* in North America (Sangster *et al.* 2001). Only one species, the Lapland Longspur *Calcarius lapponicus*, showed no evidence of genetic differentiation between Siberia and Alaska. The fact that different species showed different levels of differentiation suggests either that they became genetically isolated at different times, or have shown different rates of evolution, or have maintained markedly different levels of gene flow between them. Whatever the explanation for the different degrees of mt DNA divergence, the findings again imply that species responded differently to the same climate and vegetation changes.

Molecular evidence for mountain ranges, deserts and sea channels acting as barriers to bird dispersal, and promoting the differentiation of populations on either side, has emerged for several North American bird species (Zink 1997). Deserts can act as barriers for non-desert species, while non-desert habitats can act as barriers for desert species. Birds found in different deserts in the southwestern United States show various patterns of genetic differentiation. The Canyon Towhee *Pipilo fuscus* and Curve-billed Thrasher *Toxostoma curvirostre* show considerable mt DNA differentiation across the Sonoran and Chihuahuan deserts, whereas several other species appear undifferentiated (Zink 1997).

Population divergence in seabirds

In the North Pacific, the Bering land-bridge existed until 6.4 million years ago when global temperatures warmed and sea-levels rose, creating a channel between the Pacific and Arctic Oceans, and hence through to the Atlantic. During the glaciations that followed, the land barrier repeatedly opened and closed with rise and fall in sea-levels. These dates give indications of when northern Pacific and Atlantic faunas are likely to have come into contact. Some seabirds are now found in both the North Atlantic and North Pacific, and have presumably dispersed between the two via the Arctic Ocean, around which some still breed (e.g. Northern Fulmar *Fulmarus glacialis*, Thick-billed Murre *U. lomvia*). Some types are represented by different allospecies in the North Atlantic and North Pacific (e.g. Atlantic Puffin *Fratercula arctica* and Horned Puffin *F. corniculata*), and others by different subspecies (e.g. Thick-billed Murre with *Uria l. lomvia* in the North Atlantic and *U. l. arra* in the North Pacific; Black-legged Kittiwake with *Rissa t. tridactyla* in the North Atlantic and *R. t. pollicaris* in the North Pacific). Atlantic and Pacific forms presumably diverged at times when the Arctic Ocean froze and the Beringian land-bridge re-emerged, breaking contact between the two oceans. At the same times, some species would have been pushed south into widely separated breeding areas on opposite sides of each ocean, breaking contact across the north and giving opportunities for divergence within each ocean.

On the basis of mt DNA (cytochrome *b*), the divergence of the Atlantic Puffin *Fratercula arctica* and the Horned Puffin *F. corniculata* is dated at 1.5 million years ago, in the early Pleistocene (Friesen *et al.* 1996a). On similar evidence, the Holarctic Black Guillemot *Cepphus grylle* and the Pacific Pigeon Guillemot *C. columba* are not as closely related as expected from appearance, the Pigeon Guillemot being more closely related to its congener, the Spectacled Guillemot *Cepphus carbo*. The divergence of the Pigeon Guillemot from the ancestral Black Guillemot probably took place in the early Pleistocene (Kidd & Friesen 1998a), so the presence of the Black Guillemot in the eastern Pacific now is probably the result of a more recent invasion. In the Thick-billed Murre *Uria lomvia*, mt DNA differentiation is apparent between Pacific and Atlantic colonies, which agrees with their phenotypic (subspecific) differentiation. No obvious structuring was apparent between the five Atlantic colonies sampled, suggesting expansion from a single homogeneous refugial population (Birt-Friesen *et al.* 1992).

The Common Murre *Uria aalge* has a more southerly distribution than the Thick-billed Murre. Atlantic and Pacific populations of the Common Murre show mt DNA (cytochrome *b*) divergence of 0.61%, which suggests a split at about 300,000 years ago (Friesen *et al.* 1996b). In addition, birds from different Atlantic colonies showed a greater average divergence from one another (0.097%) than birds from different Pacific colonies (0.008%), which may have resulted from differential patterns of post-glacial colonisation. Common Murre distribution in the Pacific is continuous from central California northwards through the Aleutians and southwards to Japan, which has presumably facilitated gene flow through the whole population. In contrast, distribution in the

Atlantic is discontinuous, with the western colonies widely separated from eastern ones.

Marbled Murrelets breed on both east and west sides of the North Pacific. Morphological and mt DNA differences between the populations of each side suggest that both could appropriately be classed as species rather than subspecies, with *Brachyramphus marmoratus* on the North American side and the 'Long-billed' Murrelet *B. perdix* on the Asian side (Friesen *et al.* 1996c). Moreover, those breeding on much of the coast of western North America nest in trees, whereas those in western Alaska nest on the ground. On the basis of nuclear DNA (micro-satellites and introns), weak but significant differentiation was apparent between tree-nesting and ground-nesting populations. The Aleutian Island population was even more different (Congdon *et al.* 2000). These North American populations evidently survived the glaciations in at least three separate areas.

These and other seabird findings are summarised in **Table 10.7**. The main conclusion to emerge is that species differentiated between the North Atlantic and North Pacific vary in their divergence dates. As in landbirds, currently co-distributed species may show different phylogeographic histories. In some species, populations within a single ocean showed signs of past separation, mainly between west and east sides, and with different degrees of mt DNA divergence.

In the southern hemisphere, Adélie Penguins *Pygoscelis adeliae* have had access to Ross Island for only 6,500–10,000 years, following glacial retreat. The mt DNA (control region) of birds from each of three colonies could be divided into two types, differing by an average of 5.1% in nucleotide sequences. The most likely interpretation is that the Ross Island population consists now of two populations that differentiated in allopatry (about 270,000 years ago) and have since come together (Monehan 1994). Morphologically, they look the same, providing a parallel with some of the landbirds mentioned earlier.

Effects of past isolation of populations are evident also in South America, in which the southern part of the continent was glaciated, pushing southern seabirds northwards and isolating the populations on opposite coasts. Thus, the Humboldt Penguin *Spheniscus humboldti* may have evolved on the west coast and the Magellanic Penguin *S. magellanicus* on the east coast, but the latter now also occurs on the west. Other species, such as the Rock Shag *Phalacrocorax magellanicus*, are genetically differentiated between west and east coasts (Siegel-Causey 1997a). These two forms were separated in the last glaciation, and have since met around the southern tip, where they hybridise.

*   *   *   *

Whether for landbirds or for seabirds, the DNA data thus provide further evidence for past population splits and divergence, recolonisation routes and contact zones. They also give date estimates, many of which fall during the climatic cycles of the Plio-Pleistocene. Let me repeat, however, that all the estimates of divergence dates depend on the crucial assumption that the molecular clock ticked at a consistent rate between related lineages (though different rates for different parts of the mitochondrial genome). This assumption has so far been based on very few calibrations (Chapter 2), so all the estimates, and the inferences derived from them, may need revising in light of further work. Furthermore, the

Table 10.7 Intraspecific phylogeographic patterns (mostly in mt DNA) reported in seabirds. Modified from Avise et al. 2000.

Species (reference)	Number of individuals	Number of colonies	Assay material	Main phylogeographic findings
Thick-billed Murre *Uria lomvia* (Birt-Friesen et al. 1992).	219	6	cytochrome *b*	Lack of genetic structure across the Atlantic; clean separation of Atlantic from Pacific.
Common Guillemot *Uria aalge* (Friesen et al. 1996b)	160	10	cytochrome *b*	Mild clinal genetic structure across the Atlantic; clean separation of Atlantic from Pacific.
Fairy Prion *Pachyptila turtur* (Ovenden et al. 1991)	61	3	restriction sites	No appreciable genetic structure around the the island of Tasmania.
Short-tailed Shearwater *Puffinus tenuirostris* (Austin et al. 1994)	335	11	restriction sites	No appreciable genetic structure throughout southern Australia.
Black Guillemot *Cepphus grylle* (Kidd & Friesen 1998a, 1998b)	65	7	control region	Only modest genetic structure and no discernible geographical pattern throughout Holarctic region.
Pigeon Guillemot *Cepphus columba* (Kidd & Friesen 1998a, 1998b)	54	3	control region	Moderate population genetic structure along the west coast of North America.
Marbled Murrelet *Brachyramphus marmoratus* (Friesen et al. 1996c)	47	9	cytochrome *b*	No appreciable genetic structure, Alaska to Oregon; highly divergent form in west Pacific
Marbled Murrelet *Brachyramphus marmoratus* (Congdon et al. 2000)	120	9	nuclear introns	Slight differentiation between Aleutian and northwest North American populations; very slight differentiation between ground-nesting Alaskan and tree-nesting British Columbian birds.
Cory's Shearwater *Calonectris diomedea* (Randi et al. 1989)	145	5	allozymes (36 loci)	Only modest geographical structure and moderately high gene flow in Mediterranean and eastern Atlantic.
Cory's Shearwater *Calonectris diomedea* (da Silva & Granadeiro 1999)	148	8	DNA fingerprinting	Only a 'small degree of population structure'.

confidence limits on the estimates are considerable (for the cytochrome *b* gene at 2% per million years, the range is 1.6–2.25% per million years from studies on geese, Shields & Wilson 1987a). Such wide limits could result in further errors in the estimated divergence dates for particular taxa. In other words, it would be premature to take the estimates of divergence dates too seriously, and use them to draw more than the most general of conclusions: namely that many population splits occurred in the Plio-Pleistocene, but at different dates, and gave rise to phylogeographic patterns that fit with the locations of known climatic refuges.

A major question remaining for some high-arctic birds, whether of land or sea, is where they lived before the arctic froze about 2.4 million years ago. The icy habitats in which they now live would have been non-existent, at least in summer. It is hard to imagine that species as distinct as the Ivory Gull *Pagophila eburnea* evolved within the last few million years, and the origin of this and some other high-arctic species cries out for study at the molecular level.

SUMMARY

By providing the necessary isolation, climatic refuges during the Plio-Pleistocene were probably a major source of new subspecies and species, offsetting to varying extents the loss of species caused by habitat shrinkage and displacement. Many aspects of current distribution patterns, and of morphological variation, can be interpreted in terms of the past fragmentation of populations, their isolation and differentiation in climatic refuges, and their subsequent expansion and (in some cases) re-contact. In both North America and Eurasia, suture zones are apparent, representing clusters of contact zones between pairs of taxa, following their spread from former refuges.

In one pattern, pairs of closely related species occur in North America; and in each pair one species in the northwest is conspecific with the Eurasian form, and another to the east and south is a distinct American species. Each pair is thought to have evolved from a common ancestor, with the northwestern species colonising in post-glacial times from Eurasia via Beringia and the more southern form having developed south of the ice in glaciated North America. They have since come together, with overlapping ranges.

On the basis of 'molecular clocks', analyses of DNA suggest that population splits leading to subspeciation and speciation have occurred throughout the Pleistocene and earlier. In general, well-differentiated species gave supposed separation dates that long preceded the Pleistocene, while less well-differentiated species and subspecies gave separation dates within the Pleistocene. Speciation in many birds was seemingly a long process, which may have extended through more than one climatic cycle, and involved several periods of isolation alternating with periods of contact.

The main contribution of recent genetic work on the effects of past climatic changes on bird speciation is to suggest that: (1) many closely related species, formerly thought to have diverged in different refuges during the last glaciation, are in fact much older, dating from the late Pliocene or early Pleistocene; (2) some less well-differentiated taxa, including some subspecies, may have diverged during or since the last glaciation; (3) species now living together in the same region may have had markedly different biogeographical histories, and arrived there by different colonisation routes; and (4) populations occupying recently deglaciated regions often show less genetic variance than other populations of their species in areas that were never glaciated. In addition, some populations isolated within Iberia and Italy may never have broken out of these areas, their conspecifics now living in central and northern Europe having colonised those regions from further east.

Budgerigar *Melopsittacus undulatus*, a widespread monotypic species of the Australian central desert.

Chapter 11
Dry–wet cycles in tropical regions

During glacial periods, world climates were not only cooler, but generally drier than today. Glacial advance had the effect of compressing the tropical zone, reducing temperatures in equatorial lowlands by 2–6°C and rainfall by up to 30% (Bonnefille *et al.* 1990), with major effects on vegetation. Yet despite these changes, the tropics maintained a great diversity of plant and animal life, unparalleled by that of any other region on earth. This chapter explores how this was possible.

In the tropics, wet (pluvial) periods generally corresponded with glacial minima and dry periods with glacial maxima. In the warm wet periods, forests reached their maximum extent, spreading over savannahs and grasslands, but in cooler dry periods forests contracted to form isolated patches in more open landscape (Prance 1982, Hamilton & Taylor 1991, Elenga *et al.* 2000). Some habitat refuges in which species survived through climatic extremes were thus most restricted during glacial maxima (mainly lowland forest and woodland) and others during glacial minima (mainly open habitats and montane habitats). Like their high-latitude equivalents, therefore, tropical bird species were exposed to alternating expansions and contractions of habitat; but geographical displacement of lowland habitat was probably much less marked in the tropics than at higher latitudes.

The apparent effects of past vegetation changes on bird populations have been studied on all three southern continents, and also in southeast Asia, where centres of species richness and endemism are just as apparent (Stattersfield *et al.* 1998). However, the limited palaeo-botanical work in tropical regions means that habitat changes can be traced in much less detail than for the northern continents, and the poorer fossil record means that former distributional changes and extinctions of birds are largely unsupported by direct evidence (but see **Box 9.2** for South America). In consequence, research on the distributional and phylogenetic history of tropical birds has taken a somewhat different course, based primarily on the analysis of existing distribution patterns. It is this aspect, and its indication of the likely locations of past refuges, that forms the bulk of this chapter, though independent palaeo-botanical and geomorphological evidence is increasing all the time (e.g. Simpson & Haffer 1978, Hamilton 1982, Maley 1991, Elenga *et al.* 2000).

Tropical habitats

In order of decreasing rainfall and tree cover, major tropical habitats include rain forest, deciduous woodland, savannah, grassland and desert. In addition, montane forest is often different in character from lowland forest, becoming more divergent with increasing elevation and giving way, on the highest summits, to open treeless moorland (the alpine zone). In the tropics, perhaps more than in northern regions, each type of habitat has its own indigenous birds, which rarely stray into other habitats, let alone breed there (although some species migrate within habitat types, or between equivalent habitat north and south of the equator). This high degree of habitat specificity partly accounts for the species richness of tropical regions. As everywhere, however, human activities are increasingly affecting bird distribution patterns, sometimes creating new habitats, such as farmland, and at other times destroying habitats, such as forest, and breaking down former barriers to dispersal.

The main areas of humid rain forest thought to have existed at the time of maximum glacial extent at high latitudes, some 26,000–4,000 years ago, are shown in **Figure 11.1** for all three southern continents. In addition to the areas shown, other forest probably survived along rivers and lakesides in open land, as well as in montane areas. Whereas temperate forests over much of the world were pushed to lower latitudes, tropical forests were not displaced latitudinally, but shrank to smaller areas, where conditions remained favourable. As at higher latitudes, some bird species might have been lost in this process, while the isolation of remaining species in forest refuges gave opportunities for further differentiation and speciation to occur. With the return of wetter conditions, the forest islands expanded over the savannahs and grasslands, taking their birds with them, and in many places, previously isolated forest taxa came into contact again. As in the northern hemisphere, these contact zones are marked today by distributional overlaps, and by habitat differences and character displacement, or hybridisation between closely related taxa.

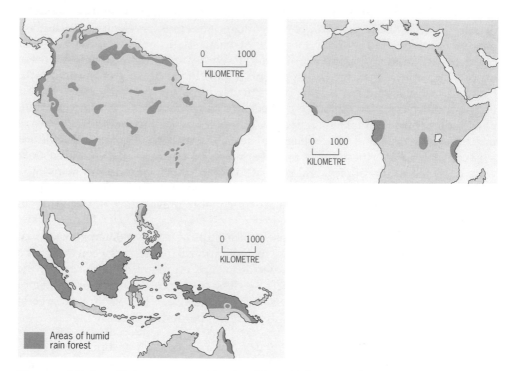

Figure 11.1   Main purported lowland humid forest refuges at the height of the latest dry period in the three southern continents. In addition, much of the exposed sea bed between Malaysia, Sumatra, Java and Borneo, and also between New Guinea and Australia, is likely to have been forested. Modified from Tallis 1991, but opinions remain divided on the exact locations and extents of these refuges.

As in high-latitude areas, the different montane vegetation zones moved downslope during cold dry periods and upslope during warm wet periods. Elevational shifts in montane forests of up to 1,500 m are revealed by pollen evidence from equatorial Africa, South America, Indomalaysia and New Guinea (Brown & Lomolino 1998). In some regions during cold periods, montane forest could have spread widely across the lowlands, replacing or intermixing with the original lowland forest; but in other regions drought may have curtailed the growth of any kind of forest. Whatever the local situation, montane bird species could be expected to have experienced conditions conducive to speciation during warm wet periods, when their habitats were most restricted and fragmented, and conditions conducive to spread and mixing during cold periods, when their habitats were most continuous and widespread.

The importance of superspecies

In research on biogeography and speciation, perhaps the most informative taxonomic unit is the superspecies, composed of a series of allospecies, each occupying separate areas, and thereby illustrating the first step in the multiplication of species through isolation and divergence (Chapter 3). The distribution patterns of superspecies, and their component allospecies, thus reveal information on nearest relatives, on the likely locations of past refuges and potential zones of contact, and on any deviation in habitat shown by specific members. All these aspects are much less readily apparent if one looks at species individually. Hence, to understand the origins of particular birds, we need to consider their distribution patterns in light of the superspecies concept (Hall & Moreau 1970). Each type of bird then falls more readily into place, reflecting the particular stage in its distributional and phylogenetic history. This is perhaps more possible for tropical than for northern areas because, when most of the distributional data were collected in the tropics (and preserved on the labels of museum skins), the vegetation and bird populations were much less modified by human action than they are now. In addition, allospecies are much more prevalent in tropical than in northern areas.

The term 'species group' is often used for groups of closely related species that do not qualify as superspecies because of geographical overlap between the group members. Being largely sympatric, species groups represent a later stage in the speciation process, and their members are less closely related than are those of superspecies. Because this process is ongoing, no clear line separates species groups from allospecies, and inevitably there are borderline cases.

AFRICA

Overview

The first ever attempt to assess continental distributional data in terms of the superspecies concept was made for Africa (Hall & Moreau 1970, Snow 1978a). Mainly using museum specimens, Hall & Moreau (1970) mapped the ranges of 962 species of African passerines, grouped when appropriate into superspecies, and superimposed these ranges on to maps of the major vegetation types. This gave 'the first attempt to show a continuing process of evolution in a continental avifauna by means of plotting on one map the distribution of species, believed to be immediately descended from a common ancestor'. Subspecies were also shown, when they were 'so well marked that they might be considered as incipient species'. The results were startling. One could see at a glance the precise range of every allopatric passerine species in Africa, together with the habitat it frequents. One could also see where gaps in ranges occur, reflecting past or present barriers, make informed guesses on the location of the former refuges where differentiation began, and pinpoint areas of contact or potential contact where hybridisation might occur. The 962 African passerine species included 486 members of superspecies (about 51% of the total), 300 members of species groups (about 31% of the total) and 176 independent species (about 18% of the total).

Because about half of all African passerines belong to superspecies, with an average of nearly three species in each, this implies much recent speciation. Few superspecies show precisely the same patterns of break-up and speciation, but the patterns are sufficiently similar between superspecies to indicate the main centres of evolution. Some 66 out of 169 superspecies[1] are confined to a single vegetation type, with 25 in acacia savannah and 19 in lowland forest, 12 in woodland and ten in montane vegetation. The low proportion in woodland, despite its enormous extent (**Figure 11.2**), may be due to its present continuity, in contrast to the more fragmented distribution of the other habitats. In the driest vegetation type, the acacia steppe, the avifauna is dominated by a few forms, notably larks. In contrast to typical forest families, the larks show little distributional overlap among close relatives, more than four-fifths belonging to superspecies, the highest proportion in any African passerine family.

Turning to species groups that show wide sympatry, 21 such groups are confined to lowland forest, with only two in woodland, one in acacia and three in montane forest. Hall & Moreau (1970) attributed this variation between habitats to the richness and structure of forest, which allows more species to live in association, co-existing in the same areas but exploiting different niches. It is not obvious why the remaining 176 species do not fall readily into superspecies or species groups: either they have undergone no recent speciation or their closest relatives have died out. Alternatively, they may have diverged so much that their closest (allopatric) relatives are no longer recognisable as such. In any large continental avifauna, cases of marked divergence could go largely unrecognised, but they may still exist (for possible examples, see Hall 1974).

In general, non-passerines have fewer allospecies within Africa than have passerines. Non-passerines are generally larger than passerines, and have longer life spans and generation times, and greater dispersal distances (Chapters 2 and 17). These features would work against the development of distinct regional forms. Where non-passerines have obvious allospecies, these are more often found on different continents than are the allospecies of passerines. Some 143 (37%) of 391 mapped African non-passerines have allospecies on other continents (mainly Eurasia), compared with only 69 (16%) of 439 mapped African passerines (Hall & Moreau 1970, Snow 1978a). Evidently superspecies and their component allospecies occur over much larger areas in the non-passerines than in the smaller, less dispersive passerines. By implication, population fragmentation and divergence occur on widely different spatial scales in the two types of birds (see also Chapter 17).

Historical factors

In Africa at the present time, rain forest occurs mainly in one continuous block straddling the equator on the west side of the continent, with isolated patches elsewhere, remaining as relicts from a time when this forest was more widely spread (**Figure 11.2**). The rain forest is surrounded by woodland and then to the north and south woodland gives way to progressively more arid habitats, through savannah to desert. These more arid habitats therefore occur mainly as separate belts in the northern and southern parts of the continent.

---

[1]*In terms of species, some 394 (41%) of 962 breeding passerine species occur only in a single vegetation type and, while the remaining species occur in more than one, hardly any occur in both forest and open land (Hall & Moreau 1970).*

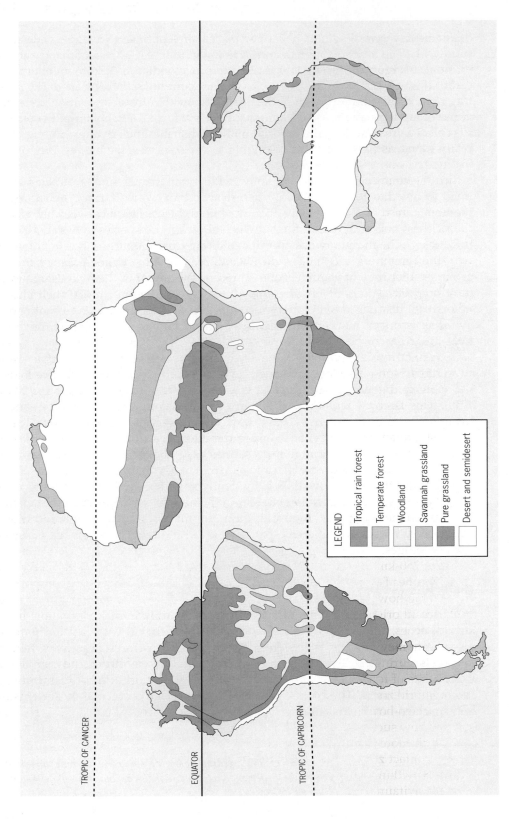

Figure 11.2 The three southern continents, showing relative sizes, latitudinal positions and major vegetation types. Redrawn from Keast 1981a and other sources.

LEGEND

- Tropical rain forest
- Temperate forest
- Woodland
- Savannah grassland
- Pure grassland
- Desert and semidesert

TROPIC OF CANCER

EQUATOR

TROPIC OF CAPRICORN

During dry periods, the African lowland forests were smaller and are thought to have become restricted to five main areas, three in west–central Africa and two in the east (**Figure 11.1**). Their locations are reflected in the current numbers and distribution patterns of forest birds (Diamond & Hamilton 1980, Crowe & Crowe 1982, Hamilton 1982, Prigogine 1984, 1988, Mayr & O'Hara 1986). Several species are confined to one or more of these purported refuges, even though their forests have since spread and joined to form larger areas of continuous forest. Parts of this newer forest, as well as neighbouring woodland, lie on Kalahari sand, confirming that in dryer times parts of these areas were desert.

In west–central Africa, the three main purported refuges include the Upper Guinea, West Lower Guinea and East Lower Guinea forests, respectively (**Figure 11.3**). Four bird taxa are endemic to the Upper Guinea Forest, four to the West Lower Guinea Forest, four to the East Lower Guinea Forest, and 17 to both East and West parts of the Lower Guinea Forest. The distributions of these endemics are clearly not random, as they centre on the three main areas, while none occurs in the present forest areas that lie in-between (**Figure 11.3**; see also Diamond & Hamilton 1980, Crowe & Crowe 1982). The fact that these species are not found in all three main areas implies that they became extinct in one or two of them, presumably during the period of greatest forest contraction.

Up to 48 other forest taxa in west–central Africa appear to have gaps in their present distributions between the three main centres (examples in **Figures 11.4 and 11.5**). In some taxa, the different populations have speciated or subspeciated, but in others they show no significant morphological divergence (Mayr & O'Hara 1986). This is consistent with evidence from elsewhere that, in the same environment, not all taxa diverged phenotypically at the same rate (Chapter 3). The simplest explanation of these localised distributions in the continuous forest of west–central Africa is again that the respective populations were isolated in forest refuges, and have not yet spread sufficiently to come into secondary contact.

For species that have spread, Mayr & O'Hara (1986) identified the range limits of 23 pairs of sister taxa. Fourteen species pairs have contact zones that lie in forest between the purported Upper and Lower Guinea refuges, and three pairs have contact zones that lie between purported West Lower Guinea and East Lower Guinea refuges. The remaining six species pairs are separated by the Dahomey Gap, the 250-km savannah barrier that still exists between the Upper and West Lower Guinea Forests (**Figure 11.5**). These latter taxa form a very small proportion of the total, however, and are not markedly different, so the gap may be of relatively recent origin. In the past, the Niger River was much wider than now in its lower reaches, and may have formed a more significant barrier, lying further east of the Dahomey Gap. This low-lying area may also have been flooded by higher sea-levels during some of the interglacials, forming an additional barrier to the movements of forest birds. These patterns of endemism, of disjunct distributions and of hybrid zones, are thus all consistent with the refuge hypothesis. With insufficient palaeo-botanical evidence, the case remains circumstantial, but it is hard to imagine how such patterns could have arisen in any other way. They provide parallels with more firmly based examples from the northern hemisphere. The fact that no contact zones occur along the Cameroon Mountains that run southwest–northeast within the Lower Guinea Forest can be attributed to the lowland and montane avifaunas being less separated in dry periods, as discussed later.

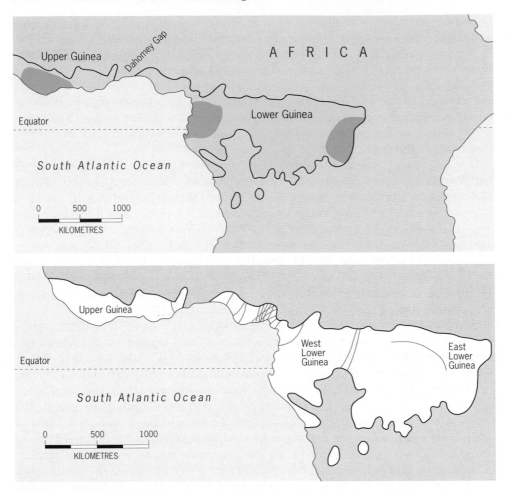

Figure 11.3   Upper. Present distribution of tropical lowland forest (light shading) in west–central Africa and postulated locations of past forest refuges (dark shading).

Lower. Locations of contact zones between 17 pairs of sister taxa (species or subspecies) within the present forest area. Six other pairs of taxa are separated by the savannah of the Dahomey Gap. From Mayr & O'Hara 1986, based largely on Hall & Moreau 1970.

The tropical forest in central Africa is surrounded to the north, east and south by a broad belt of woodland (**Figure 11.2**). Some bird species are found throughout this woodland, but others are represented as closely related allospecies that replace each other on the north and south sides. This can be explained by the forest, in wetter periods, extending right across equatorial Africa, splitting the woodland into distinct northern and southern belts. Nowadays, with the northern and southern woodland belts joined in the east of the continent, many northern and southern woodland allospecies meet without overlapping in the vicinity of Lake

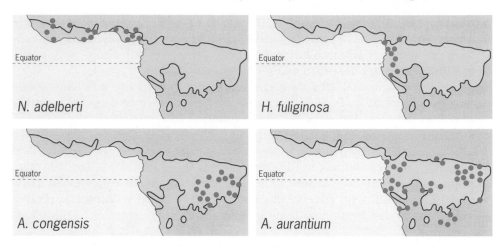

Figure 11.4 Examples of species that are endemic to specific parts of the present African lowland forest, corresponding to the locations of supposed past refuges (see **Figure 11.3**). Spots mark locations of museum specimens. Upper left: Upper Guinea Forest — Buff-throated Sunbird *Nectarinia adelberti*, Upper right: West Lower Guinea Forest — Forest Swallow *Hirundo fuliginosa*, Lower left: East Lower Guinea Forest — Congo Peacock *Afropavo congensis*, Lower right: Lower Guinea Forest, east and west — Violet-tailed Sunbird *Anthreptes aurantium*. From Mayr & O'Hara 1986, based largely on Hall & Moreau 1970 and Snow 1978.

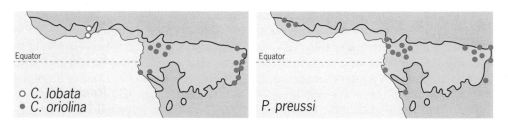

Figure 11.5 Examples of African forest species whose present discontinuous distributions are thought to centre on past forest refuges. Left: Ghana Cuckoo-shrike *Campephaga lobata* and Oriole Cuckoo-shrike *C. oriolina*, Right: Preuss's Weaver *Ploceus preussi*. Spots show collection locations of museum specimens. From Mayr & O'Hara 1986, based on Hall & Moreau 1970.

Victoria. Examples include the Abyssinian Ground-hornbill *Bucorvus abyssinicus* and the Southern Ground-hornbill *B. cafer*.

In the northern woodland some superspecies have distinct east and west forms. The main former barrier is likely to have been Lake Chad which in wetter times was much larger than its present 16,000 km². Only 8,500 years ago, at the height of the last interglacial, it covered 300,000 km² (like the present Caspian Sea) and extended over 900 km north to south, reaching to 20°N in the Sahara. However, the present dividing line between western and eastern allospecies varies greatly in position from one pair to another (**Figure 11.6**). Other extensive wetlands in northern Africa may have provided additional barriers, but in any case some variation in positions of current dividing lines between species pairs may result partly from differential rates of spread and replacement.

In the acacia savannah, the northern and southern belts are now completely separated (**Figure 11.2**), but in drier times past, a connecting corridor probably occurred in the east. Nevertheless, some different counterpart species occur in each belt: for example, the Speckle-fronted Weaver *Sporopipes frontalis* in the northern belt and Scaly Weaver *S. squamifrons* in the southern; or the Mountain Wheatear *Oenanthe monticola* in the northern belt and the White-tailed Wheatear *Oenanthe leucopyga* in the southern. Other distribution patterns are apparent in the northwest and southwest parts of this habitat, again linked with past or present barriers. Similar distribution patterns are evident in other organisms, both plants and animals.

Some of the African montane forests are separated by long distances from others, notably those in Cameroon, Angola and Ethiopia. Endemism is high in these areas at species and especially at subspecies levels, but in addition many species are shared between widely separated mountains (**Table 11.1**). Thus, well over half the species of the Cameroon Mountains are also found in the next mountains 1,800 km to the east, in the Albertine Rift; the remainder are endemic to the Cameroons, but half of these are allospecies with counterparts represented in the Albertine Rift mountains (Moreau 1966). Moreover, of the 27 shared species, six are subspecifically identical in these two areas. The Angola mountains 1,800 km to the south of the Cameroons hold about 18 forest species, including only three endemic allospecies, and nine distinct subspecies. Their avifauna is therefore even less distinct than that of the Cameroons (Moreau 1966).

Normally resident montane species are unlikely to cross regularly the 1,800 km between the western mountains of Cameroon and Angola and those in the east, and the similarity of many of their respective taxa could again be attributed to past connections. A cooler climate would have lowered the lower limit of montane forest from the present average level of 1,500 m to perhaps 600 m or less above sea-level (Moreau 1963, 1966, Elenga *et al*. 2000). This could have enabled the montane forest of western Cameroon and western Angola and eastern areas to have spread and perhaps joined, at least in the equatorial region with adequate rainfall.

Table 11.1   Affinities of African montane forest birds. From Moreau 1966.

	Numbers (and percentages) of species in different categories			
	Cameroon Mountains[1]	Tanzania–Malawi Mountains[1]	Kenyan Mountains[1]	Ethiopian Mountains[2]
Subspecies shared	6 (14)	7 (11)	20 (43)	8 (42)
Other species shared	21 (49)	27 (42)	12 (26)	7 (37)
Endemic allospecies	8 (19)	6 (9)	3 (6)	1 (5)
Other endemic species	8 (19)	12 (19)	1 (2)	3 (16)
Other species	0 (0)	12 (19)	11 (23)	0 (0)
Totals	43	64	47	19

[1] Shared with Albertine Rift mountains; [2] shared with Kenyan mountains.

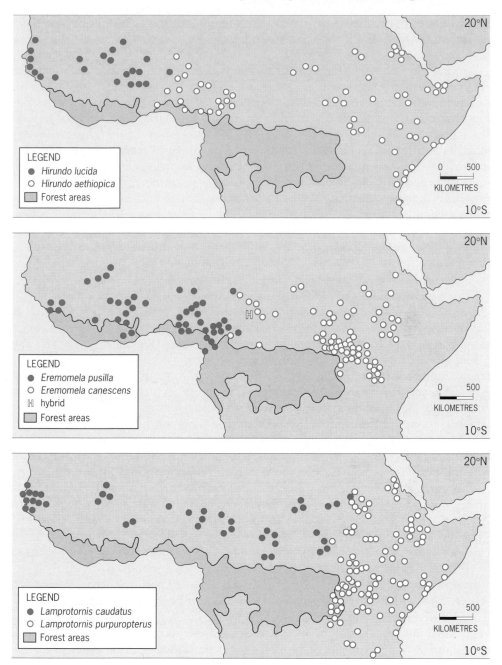

Figure 11.6   Variations in the boundaries between different species pairs in woodland in the northern tropics of Africa. Symbols show the collection locations of museum specimens.
    Upper. Red-chested Swallow *Hirundo lucida* (filled circles) and Ethiopian Swallow *Hirundo aethiopica* (open circles). Centre. Senegal Eremomela *Eremomela pusilla* (filled circles) and Green-backed Eremomela *E. canescens* (open circles). Lower. Long-tailed Glossy Starling *Lamprotornis caudatus* (filled circles) and Rueppell's Glossy Starling *L. purpuropterus* (open circles). All species shown have other allospecies further south in Africa. From Hall & Moreau 1970.

In addition, at some times in the past, montane forest species might also have occupied lowland forest (see later), and this too would have facilitated their spread.

Among the East African mountains some taxa have two allospecies, one in the Albertine Rift area and another further east. Other taxa have different forms on different mountains. These include so-called 'leapfrog patterns' where one population has diverged markedly from the flanking populations on either side (e.g. the forest greenbuls *Andropadus*, Remsen 1984). On some of the eastern mountains, speciation may be more advanced than appears from morphology, for instances exist of re-invasion of certain mountains, allowing close relatives, which have diverged only slightly in appearance, to persist either side by side or separated altitudinally (Moreau 1966). An example that would be remarkable on a marine island is afforded by the four species of *Cryptospiza* (estrildine finches) inhabiting the forests of Ruwenzori Mountain on the Congo–Uganda border. Among the closely spaced mountains of the east, movements of some species are presumably frequent (but see Chapter 12), but in any case their distributions would have been more continuous in glacial times.

The open moorlands above the treeline on the few African mountains that rise above 3,000 m also support endemic bird species (Moreau 1966). Such habitat is likely always to have been more disconnected than the forest below it, even though it spread to lower levels in dry glacial times (Elenga *et al.* 2000). The Ethiopian mountains are exceptional, with 21 of the 47 alpine species that are found there occurring nowhere else. Further south, on the mountains of East Africa, the majority of the scanty resident bird species, which vary from the alpine zone of one mountain to another, are derived from lower montane levels, without subspeciation, but three endemic species occur there, with highly discontinuous ranges. Many of the endemic species of African mountains are relatively old (Fjeldså & Lovett 1997), suggesting that they have found suitable conditions there for long periods, as their habitats moved in elevation through the climatic cycles.

Those African superspecies that extend to other continents include mainly birds of dry open country, such as larks, pipits and chats, which range from northern Africa through southwest Asia to India. In addition, several other widespread Eurasian birds are represented by one or more allospecies in Africa. In some instances, Indian and African birds are so similar in appearance as to be regarded as conspecific, such as the Wire-tailed Swallow *Hirundo smithii*.

In conclusion, current bird distribution and diversity patterns across Africa as a whole, and within particular vegetation belts, can be summarised cartographically by diversity contour maps, in which areas of high species numbers have been attributed primarily to *in situ* refuge-based allopatric speciation events, simultaneously affecting large proportions of the fauna (Crowe & Crowe 1982, Crowe & Kemp 1988). This history is well reflected in the current distribution patterns of African birds, as well as in mammals, other animals and plants.

SOUTH AMERICA

The situation in the New World tropics is more complex than in Africa, and the ranges of species are less completely known, but the distributional features are again consistent with the refuge hypothesis in explaining current patterns (Haffer 1969, 1974, 1978, 1987, Simpson & Haffer 1978, Cracraft 1985, Prum 1988). Most research has centred on the rain forest and on the Andes mountain chain, with much less on the more arid habitats. Many lowland rain forest refuges have been proposed for Amazonia, almost all on the higher ground round the edges where rainfall, even in dry periods, would have been high enough to sustain forest (**Figure 11.1**). The main areas include, from west to east, the Napo, Imerí (or Vaupés) and Guyana areas lying to the north of the Amazon and the Inambari, Rondônia and Belém-Pará areas south of the Amazon **(Figure 11.7)**. These areas were distinguished mainly from studies on about 300 species of lowland forest birds (Haffer 1978), and it is possible that other such areas might emerge if more species were examined. Moreover, other purported forest refuges have been described on the Atlantic coast of Brazil, in the Choco region in the very north of South America, and in montane areas, as discussed later. Grassy and other dry-land refuges are thought to have occurred in at least four separate areas north of Amazonia and in at least five areas to the south of it.

As on other continents, geographically restricted species and subspecies cluster in these centres of endemism (in Amazonia each comprising 10–50 such species, as well as many subspecies), and many closely related taxa come into contact,

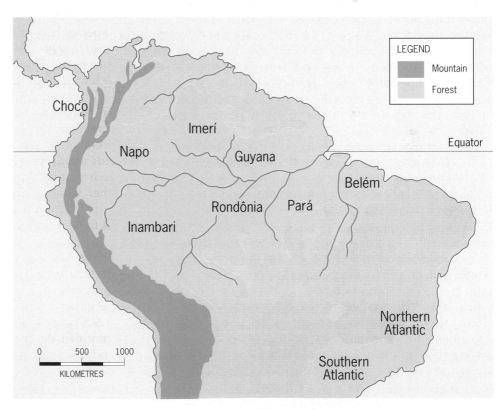

Figure 11.7   Major regions of endemism in neotropical forests. The Pará region is sometimes split into two, acknowledging the importance of the Rio Xingú as a distribution boundary for some organisms.

forming suture zones, in areas of forest between the centres. Two areas in the west are noticeably richer in species than the surrounding forest, namely the Napo centre in eastern Ecuador and the Inambari Centre in southeastern Peru (**Figure 11.7**). These two areas hold more bird species than any other similar-sized non-montane area on earth. Species numbers in lowland forest decline outwards from these centres, and many species stop at major rivers which act as barriers to further spread (Chapter 16). Other centres are described chiefly from the taxonomic distinctiveness of their birds, as they act as foci for restricted-range species and subspecies. Apart from Napo and Inambari, the total number of bird species present across lowland Amazonia shows little geographical variation, because areas between distribution centres are more or less occupied by species that have spread from neighbouring centres and by species with extensive ranges. The Guyana area in northeastern Amazonia stands out in holding a large number of endemic bird species that have not spread beyond its limits.

The role of climatic refuges in promoting bird diversification is less clear in Amazonia than in other tropical forest areas, largely because of the sparse and ambiguous nature of the geomorphological and palaeo-botanical evidence. Pollen cores confirm that 22,000–11,000 years ago, open grassy habitats occupied some areas that are now under rain forest (e.g. van der Hammen & Absy 1994, Haffer 1997), while other areas expected to be grassy remained forested throughout (Colinvaux *et al.* 1996). Hence, the extent of forest contraction remains uncertain, as does its precise past distribution. This is especially true of the Amazonian basin where much of the rain forest may have been replaced by seasonally dry forest, except along rivers and other wetter areas (Pennington *et al.* 2000). Changes in temperature and atmospheric carbon dioxide levels may have had as much effect as aridity on the tree species composition of Amazonia, with substantial changes on the higher areas designated today as refuges (Colinvaux 1998).

Some researchers have emphasised the role of prevailing conditions as possible factors in contributing to current differentiation patterns (Endler 1982, Bush 1994). In particular, many birds of the forest interior are reluctant to cross broad rivers, which could therefore have been important in fragmenting their populations (Caparella 1988). The Amazon is more than 5,000 km long, and even 1,500 km inland it is still more than 15 km across, so a bird at treetop height could not see the other side. The barrier effect of the Amazon for some birds clearly increases as it widens downstream. Among a sample of 360 forest bird species, the upper Amazon delimits the ranges of fewer than 20 species, the middle Amazon more than 50 species, and the lower reaches more than 150, decreasing at the mouth where large islands act as stepping stones to bird dispersal (Haffer 1997). Although many taxa have boundaries that correspond with rivers (**Figure 11.8**), this alone does not tell us whether the rivers provided the isolation that led to differentiation, or whether taxa that had been differentiated in climatic refuges subsequently expanded until they met rivers. Moreover, numerous contact zones are located in the headwater regions of rivers, or in areas of apparently uniform forest devoid of rivers or any other obvious discontinuities (Simpson & Haffer 1978, Haffer 1997). On the face of it, therefore, such zones cannot be attributed to anything other than competition at the line of secondary contact between expanding taxa previously separated.

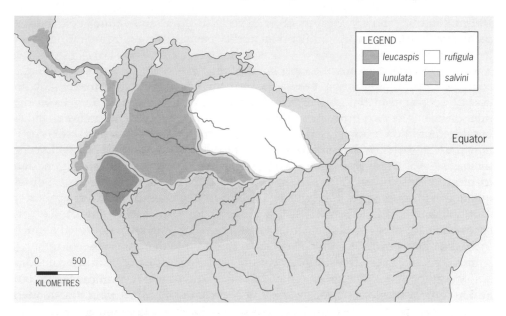

Figure 11.8   Distributions of four species of antbird in South American forest: an example of four allospecies separated by major rivers. White-cheeked Antbird *Gymnopithys leucaspis*, Rufous-throated Antbird *G. rufigula*, Lunulated Antbird *G. lunulata* and White-throated Antbird *G. salvini*. Based on allozyme differences, the split between these species is estimated to have occurred 3–5 million years ago in the Pliocene, although that between *G. leucaspis* and *G. salvini* may be more recent, at 0.7–1.0 million years ago. From Hackett 1993.

South America is unusual among the continents in the high proportion of low-lying land that could be reached by sea water. Hence, another way in which Amazonian forest could have fragmented is by a rise in sea-level in interglacials, and consequent inundation of low-lying land. This process is tantamount to river widening on a huge scale, extending more than half-way across the continent. Through a combination of eustatic and isostatic processes, sea-level was in effect up to 100 m higher at times in the Pliocene and early Pleistocene than it is now, so that Amazonian forest would have been fragmented into a number of large islands and archipelagos (Nores 1999). Current areas of high species numbers and endemism all lie in places that are now more than 100 m above sea-level, each corresponding with either a single ancient island or archipelago of islands. On this basis, both forest and non-forest bird populations would have been fragmented in a similar way at times of high sea-level, coming together between times as sea-levels fell and forest and other habitats expanded over the newly exposed ground. In support of this hypothesis, Nores (1999) pointed out that (1) many non-forest birds do indeed show the same spatial patterns of richness and endemism as forest species, and listed the various species involved; (2) different groups of animals and plants show some of the same high-diversity centres as birds and also some different ones, but all diversity centres for all groups lie above the current

100 m contour; and (3) various other aspects of floral and faunal distribution patterns fit with inundation rather than aridity being the main fragmenting agent in Amazonia.

The two hypotheses are of course not mutually exclusive, for both would have confined forest biota to more or less the same high areas around Amazonia, with aridity operating in the glacial periods and flooding to varying degrees in the interglacials. The two processes could thus have reinforced one another in causing differentiation of forest birds. However, a main period of sea-level rise is estimated to have occurred, not in the Pleistocene, but in the latter half of the Pliocene (Haq *et al.* 1987). It is thought to have lasted 800,000 years, which is long enough to produce substantial differentiation in bird populations (Chapter 10). Subsequent periods of high sea-level during the Pleistocene interglacials were of shorter duration and much less marked, reaching up to 50 m higher than today (as opposed to 100 m in the Pliocene). This flooding would still have provided a much more formidable barrier to the dispersal of forest birds than the rivers of today.

In Amazonia, then, forest blocks became separated not only during cold–dry periods when forest retreated to the high ground on the edges of the basin, but also in some warm–wet periods when low-lying parts of the basin became flooded by the sea. This dual refuge-forming process could have added to the complexity of speciation events, and contributed to the greater richness of Amazonia compared with other tropical regions. Other hypotheses proposed to account for the diversity of Amazonian birds are largely incompatible with other evidence, in particular with what is known of speciation processes elsewhere (see Haffer 1997 for discussion). Whatever the role of prevailing conditions, it seems inconceivable that climatic–vegetation changes over the past few million years have not had a major influence on the current distribution and differentiation patterns of Amazonian birds.

The limited amount of genetic work yet done on Amazonian birds implies that several lineages have shared similar distributional and phylogenetic histories, indicating that they were all exposed to the same basic set of vicariant events (Bates *et al.* 1998, Bates 2000), associated with several centres shown in **Figure 11.7**. Many apparent sister taxa, whether species or subspecies, show substantial genetic differentiation, as do other morphologically undifferentiated populations separated by the Amazon and Napo Rivers (Caparella 1988, Gerwin & Zink 1989, Hackett & Rosenberg 1990, Hackett 1993, Roy *et al.* 1997). Specific examples can be seen in northeastern Peru, where the Golden-headed Manakin *Pipra erythrocephala* occurs commonly in forest undergrowth north of the Amazon, but is replaced by the Red-headed Manakin *P. rubrocapilla* on the south side. Similarly, one of the red-crowned subspecies (*napensis*) of the Blue-backed Manakin *Chiroxiphia pareola* inhabits forest north of the river, but is replaced in similar habitat on the south side by a yellow-crowned subspecies (*C. p. regina*).

Factors that could account for such high genetic differentiation between neotropical sister taxa, compared with North America (say), include weaker dispersal tendencies, small or fluctuating long-term population sizes, social systems that discourage gene flow, as well as greater ages of taxa. Populations on either side of the Amazon River show about the same degree of genetic divergence as subspecies elsewhere (Caparella 1988, Hackett & Rosenberg 1990). The various hypotheses of factors isolating bird populations in Amazonia give broadly similar predictions on where the refuges would have lain. Only the accurate dating of divergence events between taxa may provide a means of ranking the importance of the various isolating processes involved. If calibrations are roughly correct, the molecular data suggest that various lineages began to diverge 2–6 million years ago, earlier than in many closely related North American birds. The divergence dates are mostly pre-Pleistocene, but coincide with a period of major climatic fluctuations (van der Hammen 1985), which include the prolonged period of high sea-level in the Pliocene (Haq *et al.* 1987). Again, however, Pleistocene conditions may have played a crucial role in driving the speciation process through to its present level.

Another strip of forest occurs in coastal southeastern Brazil, separated from the Amazonian forest by a stretch of dry open country. Many of the birds of this forest have closely related counterpart species and subspecies in Amazonian forest. Presumably, at some times in the past, the two forest areas were joined, allowing faunal interchange. The current distributions of open-country species suggests that this join was in the form of a forest strip which, while linking the two major forest areas, separated the open-country birds living to the north and south. The forest strip has now gone, but many open-country species occur as distinct northern and southern forms, with a suture zone between the two (for examples, see Fitzpatrick 1980).

The Andes Mountains

The Andes of South America span more than 7,000 km from north to south and cover an area of 1.8 million km². They are among the most biologically complex mountains in the world and the humid montane and premontane forests of the tropical Andes compete with the Amazon rain forest in species numbers. This richness is associated with allopatric distribution patterns, in which many species occupy only a small area, and also with dense vertical segregation of different species on the mountainsides. Many species occupy only a narrow altitudinal band, being replaced above and below by related species (Fjeldså & Krabbe 1990). Other endemic species also occur in some of the larger inter-montane valleys.

The relatively young geological age of the Andes (since the Miocene, with major upheavals in the Pliocene and early Pleistocene) implies a fairly recent evolution of their high-elevation avifaunas. In fact, avian differentiation in South America could have been prompted as much by physiographic upheavals as by climate-induced habitat changes (Cracraft 1985, Roy *et al.*1997). From the birds that occur there now, it is easy to reconstruct scenarios of colonisation of the high Andes from lower ground, followed by differentiation and diversification. Many bird groups show a substantial recent adaptive radiation within the mountains (Fjeldså & Krabbe 1990).

Many endemic species have counterparts in similar habitats on neighbouring mountains. Such patterns are found, for example, among metaltail humming-birds, thistletails, spinetails, ant-pittas, *Scylalopus* tapaculos, flowerpiercers and several brush finches (Fjeldså & Krabbe 1990). The majority of these birds are highly sedentary, and do not cross the arid tree-less valleys that separate the different forested cordilleras within the Andean range. Certain species show widely disjunct ranges, which suggests that populations in many segments of a once continuous distribution have died out. A few species occur in only one small area, and lack any near-counterparts in similar habitats elsewhere. There is a tendency for disjunct distribution patterns, and for differentiation of local populations, to increase from lowland forests to the treeline. This would be expected from the greater isolation of the high-elevation habitats.

Among the humid forest birds of the Andes, about 21% of all superspecies and species that have three or more differentiated populations show leapfrog patterns of geographical variation in which two groups of populations of similar appearance occupy disjunct ranges divided by very different populations of the same species (Remsen 1984). Leap-frog patterns are more prevalent in the Andes than anywhere else, but lack of concordance in their distribution patterns suggests a strongly random component in their variation. They are attributed not to differential dispersal and competition (like checkerboard patterns, Chapter 6), but to the chance greater divergence of the central form in isolation.

Clusters of endemic species occur at various centres in the northern tropical Andes, mostly in Venezuela and Columbia and, further south, especially in some cloud forest areas, mostly in Ecuador and Peru. The existence of such centres of endemism again suggests that many montane birds have been exposed to the same distributional histories (Vuilleumier 1969, 1980). The cold and generally dry climate that prevailed during glacial periods caused the downward shift of the forest (by up to 1,500 m near the equator), which enabled montane forest bird species to expand their ranges as their habitat moved to lower elevations (Vuilleumier & Simberloff 1980). On the higher ground, the forest was replaced by the downward spread of montane grasslands (dry puma and wet páramo). During the interglacials, the open and forested habitats shifted back upwards, causing fragmentation and isolation of the montane biota. Yet again, alternating range expansions and contractions seem to have played a crucial role in the evolution of many Andean bird families. In the hummingbirds, Trochilidae, more than 100 species have been described from the Andes range alone (see Heindl & Schuchmann 1998 for discussion of speciation patterns in *Metallura* hummingbirds). In addition to promoting their own speciation patterns, the Andes mountains also separated the lowland bird populations on either side, promoting their divergence too.

AUSTRALIA

In Australia, the periodic expansion and contraction of the arid centre during the last few million years probably had a major influence on current bird distribution and speciation patterns (Keast 1961, 1981a, Ford 1974a, b, 1988, Schodde 1982, Cracraft 1986, Schodde & Mason 1999). Its expansion would have pushed forest and woodland of all sorts into humid refuges where rainfall was still sufficient. The main refuges were mostly peripheral, in coastal regions, but others occurred inland in upland riverine areas (**Figure 11.9**). These events gave common patterns of endemism to a wide range of bird species. As on the other southern continents, forest was very much restricted at the height of the last glaciation, some 14–26,000 years ago, as documented from pollen and other evidence (references in Schneider & Moritz 1999). It then greatly expanded, reaching its peak around 8,000 years ago.

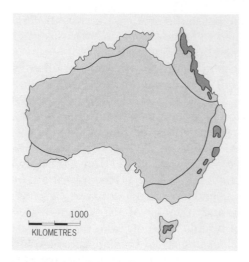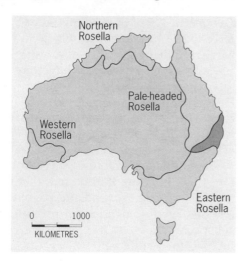

Figure 11.9   Left. The main refuge areas for birds of rain forest, sclerophyll forest and woodland in Australia. Dark blue shading – rain forest; light blue shading – sclerophyll forest and woodland. In sclerophyll forest species, there is much isolation and differentiation between the southeast, Tasmania and the southwest, and in woodland species between the northwest, southwest, northeast and southeast (to a lesser extent in the Hamersley area of the mid-west). In addition, some bird (especially parrot) distributions are centred on interior drainage basins. Partly from Keast 1981a.

Right. The current ranges of four geographically replacing, closely related parrot species in Australia which are thought to have evolved in the four main dry-period forest–woodland refuges: Northern Rosella *Platycercus venustus* (north), Pale-headed Rosella *P. adscitus* (northeast), Eastern Rosella *P. eximius* (southeast) and Western Rosella *P. icterotis* (southwest). All four show some geographical variation, and the two that overlap do not hybridise. See also **Table 11.2**.

From a study of present bird distribution patterns, Keast (1961, 1981a) proposed the following scenarios. During extreme dry periods, some forest–woodland bird species survived only in the largest southeastern refuge, while others hung on in more places. The species populations isolated in these refuges could then differentiate from each other, to produce new subspecies or species in adaptation to different local conditions. When the climate became wetter again (as now), and the desert retreated inland, flora and fauna could spread from the refuges, and secondary contact between previously separated populations became possible. Different taxa now illustrate different stages in the speciation process. We thus now see some related forms that have come into contact behaving as good species, with only slight overlap at the contact zone, as with the different fairywrens (*Malurus lamberti, elegans, pulcherrimus, amabilis*). Other forms that have come into contact show slight hybridisation (e.g. the forms of *Sitella*); and yet other forms show more pronounced hybridisation, either over a fairly narrow zone (e.g. the two magpies *Gymnorhina tibicen hypoleuca* and *G. t. tibicen*) or over a much wider zone (e.g. the striped-crowned pardalotes *Pardalotus* of southeastern Australia).

Overall, in 425 of the 531 species of breeding landbirds in Australia, Keast (1961) recorded 33 hybrid zones, which he attributed to range extensions across former barriers. Other related forms are still separated by considerable gaps.

Many superspecies now have one member in southwestern and the other in southeastern Australia (e.g. Western Spinebill *Acanthorhynchus superciliosus* and Eastern Spinebill *A. tenuirostris*). The four rosella species (*Platycercus eximius*, *P. adscitus*, *P. venustus* and *P. icterotis*) exist today as four geographically replacing forms, each centred on one of four separate refuges (**Figure 11.9**), while the races of Sitella *Daphoenositta chrysoptera* are centred on five.

The southwestern forest area in Australia is so remote, isolated from the eastern forest by 2,000 km of arid land, that it resembles an island. It has only about 66 species of forest or woodland passerines, one endemic genus (a parrot *Purpureicephalus spurius* mentioned in **Table 11.2**), and a range of well-differentiated subspecies (Keast 1981a). Apparent cases of double invasion from the southeast involve the White-tailed Black Cockatoo *Calyptorhynchus baudinii* and Slender-Billed Black Cockatoo *Calyptorhynchus latirostris*, the Red-winged Fairywren *Malurus elegans* and Blue-breasted Fairywren *Malurus pulcherrimus*, and two races of the Western Yellow Robin *Eopsaltria australis*. In other examples the western form had not attained specific status before another invasion from the east took place, resulting in widespread hybridisation, as exemplified by the Port Lincoln Parrot group *Barnardius zonarius* (Serventy 1972). In other species, the southwestern isolate has apparently spread eastwards (e.g. Western Whipbird *Psophodes nigrogularis*). The southwest also holds a number of poorly differentiated races of more widespread species.

In Australia, then, one can see in a range of species in different forest areas the whole succession from isolated populations that are identical, through those with slight, then marked differences, to clearly distinct species. The whole succession

Table 11.2 Examples of Australian parrot species, isolated in different forest–woodland refuges, and now showing different degrees of differentiation. Taxa are listed from the least to the most differentiated. From Cain 1955.

- The Purple-crowned Lorikeet *Glossopsitta porphyrocephala* appears identical morphologically in southwestern and southeastern Australia, whereas the Regent Parrot *Polytelis anthopeplus* shows slight morphological differences between east and west.
- The Ring-necks (*Barnardius*) have four different forms in different areas: the Eastern (Mallee) Ringneck *Barnardius barnardi* has a green head with red forehead, the southwestern 'Twenty-eight Parrot' *B. zonarius semitorquatus* has a blackish head and red forehead, whereas the in-between 'Port Lincoln Parrot' *B. z. zonarius* has a black head and no red forehead. These three overlap and hybridise, so it is hard to decide whether they are best called species or subspecies. The fourth is a northern subspecies which is paler and smaller with a green head, the 'Cloncurry Parrot' *B. b. macgillivrayi*.
- The Western Rosella *Platycercus icterotis* and Eastern Rosella *P. eximius* are distinct species separated by 2000 km of dry country (**Figure 11.9**). They also have northeastern (Pale-headed Rosella *P. adscitus*) and northwestern (Northern Rosella *P. venustus*) representatives. Although they differ greatly in plumage, they hybridise where they meet, and all show geographical variation.
- The Red-capped Parrot *Purpureicephalus spurius* of the southwest is so distinct from other parrots that it is placed in a genus of its own. It most closely resembles the Ring-necks (*Barnardius*).

can be illustrated by parrots, which are listed in **Table 11.2** in order of increasing differentiation. Yet again, species that have been exposed to essentially similar climatic histories have changed to different extents, presumably because, in the same time period, particular conditions exerted stronger selection pressures on some species than on others; alternatively, divergence may have begun at different dates in different species pairs. In contrast to the parrots in **Table 11.2**, which inhabit fragmented woodland and sclerophyll forest, the Budgerigar *Melopsittacus undulatus* inhabits the dry interior, a habitat not known ever to have been fragmented. It has had no known breaks in its geographical range, shows wide dispersal (being highly nomadic), and no geographical variation in morphology. It is exceptional among parrots in its lack of differentiation, and thus underlines the importance of past habitat fragmentation in influencing the distribution and speciation patterns in other Australian parrots. This situation is paralleled in other birds.

The distributional history of Australian birds has thus been worked out mainly from their morphology and distribution patterns, supported by palaeo-botanical studies on vegetation history. The limited amount of genetic work done so far has confirmed a pronounced phylogeographic divide among the rain forest birds of the northeastern mountains (Joseph & Moritz 1994, Joseph *et al.* 1995). In several species, study of mt DNA revealed a deep split between northern and southern populations. Two occurred at the Black Mountain barrier and a third at the Burdekin Gap, areas that provide obvious long-standing breaks in the distributions of rain forest species. Frogs and reptiles showed similar patterns.

COMPARISONS BETWEEN THE SOUTHERN CONTINENTS

All the southern continents provide examples of the effects of past climatic refuges on current bird distribution patterns, and of birds at different stages in the speciation process. They thus provide further evidence that speciation in birds is an overwhelmingly geographical event, requiring the isolation and divergence of populations. The distributions of the vegetation formations are strikingly different between the continents, especially between Africa and Australia. Africa has rain forest in the centre, surrounded by concentric belts of decreasing rainfall and sparser vegetation, with three separate deserts near the edges. Australia has a single central desert, with the forest in separate areas around the edge. On both continents, forest was most fragmented in dry periods while grassland and desert were most restricted in wet periods. Each major climatic phase would therefore have affected the various habitats differently, initiating isolation and differentiation in the birds of some habitats and suppressing it in the birds of others. At all times, however, we can assume that new species are evolving in one habitat or another.

In Africa, speciation processes are evident across the continent in all types of habitat, from forest to desert; but in Australia, which throughout has had a single central desert fluctuating in area, most speciation has occurred round the edges where the peripheral woodland and forest have been repeatedly fragmented, affecting species of these habitats only. It is for this reason, it may be surmised, that most woodland and forest bird species in Australia have relatively small geographical ranges, being replaced by related forms in similar habitat elsewhere on the continent.

The almost complete lack of differentiation shown by birds of the central Australian desert contrasts with the situation in Africa, where each of the three main deserts (Sahara, Somali-arid and Kalahari–Namib) has a number of distinctive species. Moreover, two of the African deserts were periodically joined to the Asian desert (or separated by narrow water gaps, as now), whereas the Australian desert has had no connection with deserts elsewhere. Little wonder that the Australian desert holds relatively few species, which themselves show little or no geographical differentiation.

Lack of opportunities for speciation in the Australian desert might also in part account for the scarcity of ground-dwelling bird species in Australia, compared with other regions (Keast 1972, 1974). The Kalahari and Australian deserts look much the same, with some scrubby *Acacia* trees and bush on red sand. However, whereas the Kalahari abounds in true ground-living birds adapted to this habitat (five bustards, five coursers, four sandgrouse, at least nine mainly ground-living larks, four pipits, some chats and warblers), the Australian centre has only one bustard, one pipit, one lark and two song-larks, none of which is common except for the pipit, a newcomer to the continent (Hall 1974). Within Africa, the Sahara has fewer species than the arid regions of the Mandeb Circle (Somalia, Ethiopia, Yemen) and the Southern African deserts.

Clearly, the numbers of species found in any continent, and in the biomes within it, depend partly on the historical opportunities in that continent to generate and maintain species, as well as on the conditions prevailing today. Past conditions have had a major influence on whether particular biomes have few species with large ranges, or many species with small ranges. About 34% of Australian passerine species fall within superspecies, compared with 51% in Africa, a difference Keast (1974) attributes to the smaller and fewer habitat areas available in Australia, and the very low levels of speciation that have occurred in Australian desert species.

SOUTHEAST ASIA

The major islands of southeast Asia, notably Sumatra, Java and Borneo, were joined to Malaysia in glacial times, forming a continuous land area that supported both forest and open habitats. As world temperatures and sea-levels rose, however, lowlands were flooded and higher ground remained as the islands we know today. It is mainly the interglacials, therefore, that led to differentiation of the forest biota of these different islands. The process is the same as envisaged in Amazonia, except that the latter may have required even higher sea-levels. In addition, climate changes caused similar vegetation changes in southeast Asia as in other tropical regions (Morley & Flenley 1986, Heaney 1991), and today hotspots of species endemism occur in forest on both the mainland and the major islands (Stattersfield *et al*. 1998, Taylor *et al*. 1999).

Uncertainty surrounds the date that the Philippines and Sulawesi were last attached to continental Asia, for they are separated by deep-water channels that would have remained through recent glacial periods. However, with much of the current sea bed dry in glacial times, they would certainly have been nearer to other land (notably Borneo) then than now. Not surprisingly, their flora and fauna are more distinctive than are those of Java, Sumatra and Borneo. The Philippines have more than 350 endemic birds (species and subspecies), with different kinds on different islands. It is possible that Sulawesi was never part of mainland Asia for, although it has land mammals, they form a depauperate selection whose ancestors could have swum there (Musser 1986).

DISCUSSION

Assessment of the refugium concept

Past refugia can be viewed as restricted areas of habitat where (1) some species became extinct, regionally or totally, and (2) others survived throughout a climatic extreme, a proportion of which (3) formed distinct species or subspecies while they were isolated. This holds whether the refuges were separated by water or by some other type of habitat. As conditions changed and habitat expanded, species subsequently spread to varying extents, interbreeding or remaining distinct when they met their counterparts spreading from other refuges. Because many species were expanding from refugia at the same time, one might expect a rough coincidence in the locations of the contact zones between faunas, which lie between former refuges and across former barriers (Endler 1982, Ford 1988). The evidence supporting the refuge concept, then, is the congruence of the distribution centres of many species, between which lie contact zones, not just of birds, but of many other animals and plants (Hamilton 1976, Simpson & Haffer 1978, Moritz *et al.* 2001). The same areas could have served repeatedly as refuges, continually isolated and rejoined during successive climatic cycles, giving what appear today as obvious areas of endemism (OAE) or centres of diversification (COD), recognised as hotspots in total species numbers, endemic species numbers or both (Nelson & Platnick 1981).

It is hard to explain this combination of distributional features on anything other than the refuge hypothesis, which has also been supported by independent geomorphological, palaeo-botanical and palaeo-climatic evidence in areas where relevant data are available. It is the general coincidence between the locations of refugia determined from these different and independent lines of evidence that has led to formulation of the theory. Criticisms have arisen mainly where uncertainty still surrounds the numbers, locations and extents of former refuges, and where factors other than refugia could account for high species numbers and high endemism. Both now and in the past, high species numbers or endemism are likely to arise in areas with high habitat diversity, rainfall or topographic relief. These are precisely the areas that could have acted as refuges in the past.

The main modification to the refuge hypothesis resulting from molecular biology stems from the dating of divergence events, which often emerge as much earlier than the last glacial or dry period and for some species date back to the climatic cycles of the Pliocene. This does not wholly undermine the refuge hypothesis, but suggests instead that young species are the result of not just one vicariant event, but of a succession of such events in the same general areas, as described in

Chapter 10. Another concern is that even those taxa (some trees, butterflies, frogs and birds) that show superficially similar patterns of endemism within the Amazon basin seem to have diverged at markedly different times in the past. This too is not inconsistent with the refuge hypothesis, for even within birds, different sister taxa diverged in mt DNA at times dated between 2.5 million and 35,000 years ago (Chapter 10). More reliable dating of divergence dates for a wider range of Amazonian forest birds should reveal whether in most species divergence began in warm wet periods in forest refuges separated by sea water, or in cool dry periods in refuges separated by grassland. Clearly, more analyses are needed, but remember that they are all based on the as yet inadequately tested concept of a consistent molecular clock. A question mark still hangs over most estimates of divergence dates.

Other uncertainties have arisen from the fact that, although many types of plants and animals show the same distribution centres as one another (as expected if they shared a common environmental history), some groups in some regions show different or additional centres (as for butterflies, lizards and certain plants in South America; Simpson & Haffer 1978, Diamond 1985, Bush 1994). This could be because different groups did in fact have different refuges, depending on their different tolerances and needs; because some groups could survive in smaller refuges than others; or because their distribution patterns are still insufficiently known to define past refuges accurately. Interestingly, the greatest criticisms and uncertainties over the refuge hypothesis concern South America, which is the continent whose palaeo-botany is least well known. Because the role of past events in shaping current distributions can seldom be known with certainty, some biologists discard the term 'refuge areas' in favour of the term 'CORE areas', an acronym for 'Centres of Richness and Endemism' (Diamond 1985).[2]

Endemism and isolation

The term endemic is applied to species that are found only in particular areas, but this restriction can occur in either of two ways. Some such species (neo-endemics) could have evolved *in situ* and failed to disperse more widely (like many island species), while others (palaeo-endemics) could have once been widespread but since contracted — so-called biogeographical relicts (in contrast to phylogenetic relicts which now lack close relatives but may be widely distributed). In addition, the term 'endemic bird areas' has been used operationally in a different sense by the conservation agency *Birdlife International* for any area that holds two or more 'restricted range species'. Such a species is defined as one with a range covering less than an arbitrary 50,000 km^2, whatever the factors that brought about such smallness in range (Stattersfield *et al.* 1998).

Isolation of populations can occur either by the fragmentation of a once wide-spread distribution, as discussed in this chapter, or by the dispersal of individuals from one area of suitable habitat to another, as in the colonisation of islands

---

[2]*At various times, such areas have also been called centres of origin (Wallace 1876, Darlington 1957), centres of radiation (Hultén 1937), centres of dispersal (de Lattin 1956), centres of diversification, evolutionary centres (Fjeldså & Lovett 1997) and hotspots (Fjeldså et al. 1997).*

(Chapter 3). Evidence for the overriding role of climatic-induced habitat fragmentation on continents lies in the facts that (1) speciation is particularly marked in areas separated by vicariance processes, and (2) members of different evolutionary lineages show the same biogeographical patterns, reflecting their similar histories. These features are apparent in the distribution patterns of birds on all the major continents. Explaining such patterns through dispersal needs not only many long-distance dispersal events into areas of eventual endemism, but also requires that each of the evolutionary lineages should show parallel cycles of dispersal and differentiation. These requirements seem beyond the limits of credibility, leaving vicariance as the main likely fragmenting agent of populations on continents (though not excluding cross-barrier dispersal in some species).

Montane and lowland forest birds

All the southern continents hold distinct montane forest bird species, which are absent from nearby lowland forest, and on the mountains themselves many species are restricted to particular altitude zones. The distributional divisions between lowland and various montane bird species often seem not to be due to altitude or vegetation differences, but to competition among the species themselves (see Moreau 1966 for Africa; Terborgh & Weske 1975 for South America; Mayr & Diamond 1976 for northern Melanesia). The evidence for this view is mainly that, on mountains that lack a particular species, its place is taken by another species that spreads in from lower or higher ground (Chapter 14). Hence, species that are at present restricted to montane forest may formerly also have occurred at lower altitudes, either because suitable forest once occurred at lower altitudes, or because the montane species themselves were once also found in lowland forest. Such species might have been forced to higher elevations as the lowlands were gradually colonised from refuges by other species better adapted to lowland conditions (Diamond & Hamilton 1980). Some montane endemics may thus have resulted from multiple invasions of species evolved in lowland or in other mountain ranges, each one pushing earlier colonists to a higher or narrower elevation range, and giving the marked zonation of species seen today (for *Scytalopus* tapaculos in the Andes mountains, see Arctander & Fjeldså 1995). The process, as envisaged, is analogous to the multiple invasions of islands by the same species, with differentiation of the colonists occurring in the long intervals between each successive influx.

On this idea of the origin of montane forest birds, montane forests are viewed as museums, containing mainly relict species generally older than their lowland congeners, but the same process could presumably also occur the other way round, with new species continually generated in the isolation of mountains, some of which then gradually invade the lowlands, where they accumulate through time. On this system, the mountains would hold a greater proportion of young species than the lowlands. This is a matter on which DNA research has thrown interesting light for, whether mountains or lowlands, regions of intensive recent speciation are likely to hold a high proportion of young species. It seems, for example, that the Andes mountain region currently holds a greater proportion of young species than the Amazon lowland, and that the African mountains contain a greater proportion of young species than the African lowland rain forests (Fjeldså 1994, Fjeldså & Lovett 1997, Roy *et al.*, 1997).

These findings have raised the possibility that isolated montane forest areas have generated more new species in the last few million years than have lowland forest refuges. In both studies, taxa were aged using the DNA hybridisation data from Sibley & Ahlquist (1990) to estimate the relative ages of several bird clades. High densities of neo-endemics and palaeo-endemics were found in the same areas, suggesting that these areas had remained stable over long periods, as expected of stationary refuges. Further evidence for tropical mountains acting as centres of bird speciation has emerged from studies of mt DNA in particular groups, such as greenbuls *Andropadus* in East Africa (Roy 1997) and spinetails *Cranioleuca* in South America (Garcia-Moreno *et al.* 1999). These findings have highlighted the importance of tropical mountains not just as centres of endemism, but as centres of speciation.

The richness of tropical regions

While various factors might contribute to greater biodiversity in tropical than in high-latitude regions, the opportunities for speciation produced by large-scale habitat fragmentation have been much greater in tropical regions, and might also have been coupled with lower extinction rates there. Tropical forest in Amazonia, west Africa and Malaysia–Indonesia sits symmetrically astride the equator, so climatically similar forest refuges could have survived at a number of sites within the original forest areas, both north and south of the equator, wherever conditions remained suitable. In contrast, the forests of northern regions were displaced long distances southwards, and survived in only three main refuges in Europe and three in North America (Chapter 9). One might imagine, therefore, that the potential for speciation was probably greater in the tropics, while the potential for extinctions was greater at higher latitudes.

Another habitat factor contributing to the greater richness of tropical biotas relates to mountains. In hotter climates, high mountains can be vegetated to much greater altitudes than in high-latitude areas, giving more vegetation zones, including several types of forest. As mentioned above, many bird species on tropical mountains, especially in forest, are confined to narrow elevational zones, being replaced above and below by related species (Moreau 1966, Terborgh & Weske 1975, Mayr & Diamond 1976). Although such 'vertical parapatry' also occurs on mountains at high latitudes, it is less marked and, with fewer vegetation bands, the opportunities are fewer. Hence, for this, as well as for other reasons, tropical mountains hold more species than high-latitude ones of similar elevation. The important point, however, is that different vegetation zones persisted through climatic cycles on tropical mountains, moving up and down the slopes, and creating further opportunities for speciation, whereas at high latitudes mountains in glacial periods were largely de-vegetated.

A third factor increasing bird species numbers in the tropics, especially in rain forest regions, may be the greater sedentariness of the bird populations themselves. Most such species do not migrate, and are reluctant even to leave the cover of the forest to cross relatively narrow tracts of open land (Chapter 16). In effect, this behaviour could increase the efficiency of topographical and habitat barriers, thus accelerating the rate of speciation, and increasing the number of allopatric species still further (Mayr 1969a).

It is not only the difference in species numbers between the northern and southern continents that can be explained partly in terms of past history and speciation opportunities, but also the differences in species numbers between the southern continents themselves. Whether created by aridity or flooding, no fewer than 14–21 forest refugia have been described in the extensive lowland forest areas of South America. Accordingly, the Amazonian basin has the richest lowland avifauna in the world, with 600–650 species, about 400 of which are endemic (Haffer 1987). This is equivalent to a density of around 90 species (with 57 endemics) per million km^2. Many species are widespread within the basin, while others are very restricted in distribution. A high proportion (70–80%) of all species are allospecies of superspecies, testimony to the recent fragmentation of their habitat.

Tropical Africa experienced a similar climatic history but, because of its different physiography, the forest area has always been smaller, and in dry periods restricted to no more than five main refuges (**Figure 11.1**). Accordingly, the African rain forest bird fauna has far fewer species than the South American, including endemics. The forested areas of the Australo-Papuan region are even smaller than the African, but the repeated joining and separation of two major land-masses, Australia and New Guinea, allowed a separate evolution of two avifaunas. Many types of birds have their equivalents on both land-masses. Moreover, distributional centres are recognisable today within the lowland forest of New Guinea, as they are in Australia, reflecting several past refuges on each land-mass (Keast 1972). This may explain why the forest avifauna of the Australian–New Guinea region is richer than the African, and includes more endemics, associated with its longer isolation. Australia itself has fewer forest species than comparable areas elsewhere, possibly because its forest was reduced to very small areas during the arid maxima, resulting in many extinctions, followed by only limited recolonisation from New Guinea (Brereton & Kikkawa 1963).

Although the processes of vicariance and differentiation have been described mainly with reference to forest, the same processes operated in the savannahs and grasslands over which the forest expanded and contracted. Moreover, the fact that there are, on average, many more species per genus, and far more allopatric sister species, in the tropics than in temperate areas further implies that much of the speciation has been relatively recent, having occurred in the past few million years, largely corresponding with the Plio-Pleistocene climatic vicissitudes. Some 65–85% of the members of certain tropical families and subfamilies are classed as allospecies of superspecies: African Laniidae (69%), Timalinae (65%), Nectariniidae (68%), Cracidae (75%), Ramphastidae (85%), Galbulidae (75%) and Piprinae (75%) (Haffer 1980). This high degree of allopatry contributes to the larger numbers of species per genus and per family in tropical regions than temperate ones (Chapter 5).

The various explanations discussed above to account for the great diversity of tropical species all concern the 'ultimate' factor of habitat vicariance, but this can only act through the 'proximate' mechanisms of species multiplication and extinction. Tropical bird families might be older, on average, than high-latitude families, and have therefore had longer in which to produce species, or tropical families might have produced more species per unit time. Phylogenetic analyses have provided support for both these explanations. To examine the question of age for the New World avifauna, Gaston & Blackburn (1996b) used the ages of bird tribes as determined from the DNA–DNA hybridisation data of Sibley & Ahlquist (1990). The mean age of tribes was found to be greatest at the equator and to decline towards the poles. The pattern resulted from the progressive loss of older tribes towards higher latitudes. Of several explanations considered, the authors concluded that tribes lasted longer, on average, in the tropics because extinction rates were lower there.

The second question of diversification (i.e. speciation minus extinction) rates was examined by sister group analysis (Barraclough *et al.* 1998). By definition, sister clades are the same age, so a clade with more species than its sister clade has diversified at a faster rate. Again using the DNA–DNA hybridisation data of Sibley & Ahlquist (1990), but at the levels of tribe, subfamily and family, Cardillo (1999) compared species numbers in 11 pairs of passerine sister clades. In each pair, one clade contained predominantly tropical species and the other predominantly extra-tropical species. In ten of the 11 comparisons, the tropical clade contained the most species, a statistically highly significant result ($P<0.01$), repeated in a similar analysis of butterflies. The implication was that, in the same period, the tropical clades produced more species. This does not necessarily imply that rates of evolutionary change were faster in the tropics, only that speciation-to-extinction ratios were greater. Thus, both the greater mean age of tropical families, and their greater diversification rates, could be explained in terms of more favourable speciation-to-extinction ratios, as suggested at the start of this section.

Allopatry and sympatry

In all parts of the world, the numbers of species per unit area increase with the structural complexity of the habitat, particularly with the vertical structuring of the vegetation. Bare deserts with sparse vegetation hold fewest species, but numbers increase through grasslands, savannah and woodland to high rain forest. It is in the structurally complex habitats of woodland and forest that we most often find several species from the same genus living side by side, each exploiting different parts of the habitat. Such sympatry becomes progressively less common from forest through woodland to savannah and desert, while allopatry becomes relatively more common. Especially in Africa, the deserts are occupied mainly by closely related species that separate by area rather than by foraging habits. Distributional overlap among congeners occurs mainly where desert adjoins other habitats. It is as though the evolutionary step from allopatry to sympatry is much more likely to occur in structurally complex habitats than in simple ones. Allopatry depends mainly on past history, on the frequency with which populations have fragmented and diverged, whereas sympatry depends much more on the current structure of habitats and the number of species they could support. On this view, the numbers of species found in particular regions is very much the result of both past and present conditions. This may seem like stating the obvious, yet some biologists have attempted to explain current biodiversity in terms of only one or the other. The numbers of species in different habitats can obviously not increase indefinitely, for each species requires a distinct niche, and resources can be subdivided only to a limited extent. Competition would be increasingly likely to eliminate species, as their numbers grew.

Constraints to speciation

Taking account of the number of times that habitats, such as rain forest, are likely to have been fragmented during the past few million years, and if full speciation had occurred each time in each refuge, the number of bird species alive now would greatly exceed their present levels. For example, imagine that a forest area was split into two separated units in which one original widespread species could give rise to two others. When the forests re-joined, these species in turn became widespread, so that at the next climatic extreme, two would occur in each refuge, able to diverge to give up to four new species. By this process of repeated isolation followed by sympatry, species could multiply geometrically, the one original giving rise to 2, 4, 8, 16, 32 and so on. Clearly, this does not happen and several factors might limit this process of species multiplication: (1) in any one climatic cycle, many species are likely to die out in one or more refuges; (2) not all the surviving species may differentiate phenotypically; (3) differentiation to species level seems to take not one but multiple climatic cycles; and (4) the expansion of species, following the re-joining of their refuges is likely to eliminate many incipient species through competition or hybridisation, only the most differentiated ones surviving. All these processes act to ensure that only a small proportion of long-term population isolations are likely to lead to speciation.

CONCLUDING REMARKS

From studies on all the continents, certain generalisations about species relationships and distributions can be drawn:

(1) Every new species, in its emergent state, has been a member of a superspecies, because it arose in one fragment of a discontinuous distribution. We should therefore always look for the original pattern, even though it might not always still be recognisable.
(2) The most likely place to find the nearest relative of any species is in the nearest block of similar vegetation rather than in the same block. It is in neighbouring blocks of similar vegetation type that we can expect to find obvious allospecies, and also cryptic taxa, recognisable only on genetic grounds.
(3) For continental birds, most active speciation occurs at times of extreme climate, sufficient to break big expanses of habitat into smaller patches, but not so extreme as to wipe out the habitat completely.

Considering the years that have elapsed since the distribution patterns of Australian birds were first examined by Keast (1961), those of African birds by Hall & Moreau (1970) and Snow (1978a), and those of South American birds by Haffer (1974, 1978, 1987) and others, it is surprising that no extensive biogeographical analyses have been conducted for the better known avifaunas of Eurasia and North America (but see Haffer 1989 for Eurasian paraspecies). Application of the superspecies concept to the full avifaunas of these regions could be extremely revealing on matters of distributional and phylogenetic history. The patterns would be more difficult to discern on the northern than on the southern continents, because of the many more substantial distributional changes likely to have occurred on the northern continents, through the greater human impacts.

SUMMARY

During glaciations at high latitudes, global climates became both cooler and drier, causing in tropical areas the contraction and fragmentation of forests and the expansion of more open habitats. During intervening pluvial periods, forests spread and joined again at the expense of open habitats.

The most useful taxonomic category for biogeographical-speciation studies is the superspecies, with its component allospecies. Sympatry, with two or more closely related species breeding in the same area, can be regarded as the next evolutionary step from allopatry. Sympatry more often develops in structurally complex habitats, such as forest, than in structurally simple habitats, such as grassland.

Within expanses of similar tropical habitat lie centres of high species diversity and/or endemism. These centres are known or supposed to correspond to past refuges, where habitats were confined during past climatic extremes (or times of high sea-level) and from which habitats (and most of their occupants) have since spread to reach their present distributions. Between such centres lie contact (or suture) zones, where differentiated populations spreading from neighbouring refuges have met.

As in the northern continents, all the southern ones provide examples of species that have apparently expanded from known or suspected refuges and, on meeting their counterparts from other refuges, have (1) interbred freely and extensively, (2) interbred along a narrow hybrid zone, or (3) behaved as distinct species, stopping where they meet, or (4) spread to greater or lesser extent through one another's ranges. Sometimes, one species has apparently pushed back another through competitive exclusion.

Refugia date not just from the last glacial–dry period, but have been formed and re-formed through a succession of climatic cycles extending back into the Pliocene. Similarly, divergence dates between closely related species and subspecies, estimated from their DNA, usually extend much further back than the last glacial–dry period, often into the Pliocene.

Although some areas of high species diversity and endemism can be explained in terms of past events, others are dependent on present conditions, or on a combination of past and present conditions. Tropical montane areas, because of the wide range of habitats they support, generally hold large numbers of bird species, many of them endemic; through fragmentation of habitats during climatic extremes, they are also likely to generate large numbers of new species.

Differences in species numbers and composition between continents can be explained partly in terms of past conditions. Thus the greater number of lowland forest species in South America than in Africa can be attributed partly to a greater number of lowland forest refuges in South America during dry periods. Likewise, the greater number of desert species in Africa than in Australia can be attributed to the greater number of well-separated desert refuges in Africa during wet periods.

Phylogenetic analyses of New World birds have indicated that, in proximate terms, the greater species richness of tropical than temperate regions is due to (1) the greater mean age of tropical families (or tribes), and (2) the greater diversification of tropical families per unit time. Both findings can be explained by the greater ratio of speciation-to-extinction events in tropical than in higher latitude areas.

Yellow-billed Chough *Pyrrhocorax graculus*, distributed patchily on high mountains.

Chapter 12
Disjunct ranges

The breeding ranges of some bird species fall in two or more discrete areas, separated from one another by hundreds or thousands of kilometres. In many species these distances greatly exceed normal dispersal distances. The different populations are therefore isolated from one another, functioning independently, with little or no exchange of individuals. This means that they are also genetically isolated, poised for independent evolution. Of course, breaks in the distribution of birds occur on various spatial scales, from the few metres that might separate the territories of neighbouring individuals in the same patch of habitat, through the unsuitable stretches that break the distributions of species in fragmented habitats, to the hundreds of kilometres that separate the different populations of certain species on the same land-mass or on different land-masses. These major discontinuities could arise either by the splitting of a once-continuous distribution (vicariance) or by long-distance dispersal and colonisation. One way or another, they reflect past events.

Modern range maps reveal that many species of birds now have markedly discontinuous ranges **(Figure 12.1)**. In the Palaearctic region about one-fifth of all

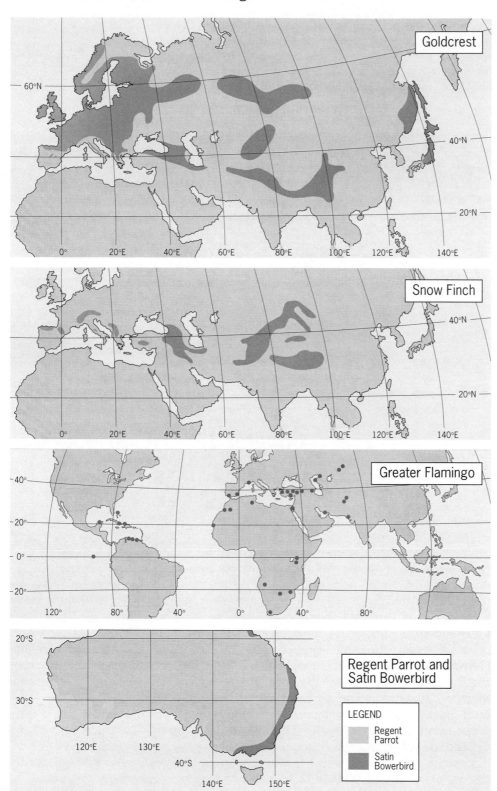

landbird species (excluding pelagic birds) have at least one population that is separated by more than 1,000 km from other populations of their species. The proportion is higher in wetland birds (23%) than in dryland birds (12%), and higher in non-passerines (22%) than in passerines (8%). Many other species have populations separated by lesser distances, and similar discontinuities can be found among the birds on every continent (for other examples, see Chapter 9).

Species with markedly discontinuous breeding distributions fall into three main categories, namely: (1) those that throughout their history have probably been confined to rare and patchily distributed habitats, relying on continual dispersal to find and colonise the suitable patches; (2) those whose habitats were once widespread, but which now have disjunct distributions because their habitats have become fragmented; and (3) those that now have disjunct distributions within an expanse of continuous and apparently suitable habitat. It is not always easy to separate these arbitrary categories or to assess the extent of human involvement, but examples of each type are given below. Discussion of a fourth process leading to discontinuous distribution, namely cross-barrier colonisation, resulting from natural or human-assisted dispersal, is left for Chapter 16.

SPECIES OF RARE OR PATCHY HABITATS

Some bird species of rare or patchily distributed habitats must always have had patchy distributions, even though at any one time they may occupy almost every piece available. Perhaps the most extreme examples are the flamingos, which are found on saline lakes with rich algal food supplies, where colonies often number many thousands or millions of individuals. The Greater Flamingo *Phoenicopterus ruber*, the range of which covers large parts of the world, is nonetheless restricted for breeding in the Old World to only about three dozen lakes or groups of lakes, separated from one another by up to hundreds or thousands of kilometres, only some of which are suitable in any one year **(Figure 12.1)**. The main concentrations of birds are centred on a series of soda lakes in the African Rift Valley, but others occur at sites in southern Africa and southern Eurasia. In recent decades in western Europe, no more than three sites have been used in any one year (Johnson *et al.* 1991). Food levels depend on salinity and water levels, and as conditions become unsuitable on any one lake, birds move to another (Cramp & Simmons 1977). Some other species of flamingos occur on even fewer lakes, which are again

Figure 12.1 Examples of discontinuous distribution patterns. Goldcrest *Regulus regulus*: found only in parts of the forested area of Eurasia, despite the presence of apparently suitable habitat in between; assumed to once have occurred right across. White-winged Snowfinch *Montifringilla nivalis*: a European montane species forced by its habitat needs into a naturally discontinuous distribution. Greater Flamingo *Phoenicopterus ruber*: confined for breeding to food-rich saline lakes in about two dozen widely separated localities. The map shows all sites known to have been used since 1950. Based on data in Cramp & Simmons 1977 and Johnson 1997. Australian Regent Parrot *Polytelis anthopeplus* and Satin Bowerbird *Ptilonorhynchus violaceus*: populations are separated by dry areas. Based mainly on data in Blakers *et al.* 1984.

often widely separated from one another; three species occupy the soda lakes of the high Andes.

The habitats of many other wetland bird species are likely always to have been patchily distributed, and often ephemeral, changing in distribution from year to year. It is not surprising, then, that wetland species in general have good dispersive powers and often turn up at suitable sites remote from other sites. The Mediterranean Gull *Larus melanocephalus* is a good example. Its main range occurs in the eastern Mediterranean and Black Sea areas, but in recent decades it has bred irregularly in various widely scattered sites in Europe as far north as Sweden and as far west as Britain, most sites lying more than 100 km from their nearest occupied neighbours. Similarly, the Little Egret *Egretta garzetta* has recently started to nest in the British Isles at localities lying more than 100 km from its nearest colonies in continental Europe, and the Eurasian Spoonbill *Platalea leucorodia* breeds in the Netherlands more than 400 km from the next nearest sites. An even more extreme example is provided by the Little Gull *Larus minutus* which has established several nesting colonies in North America, around the Great Lakes and at Churchill in Manitoba, some 5,000 km from its next nearest breeding areas in western Europe.

Long-distance movements are frequent in some aridland species that depend on ephemeral wetlands. In Australia, the Banded Stilt *Cladorhynchus leucocephalus*, Grey Teal *Anas gracilis*, Pink-eared Duck *Malacorhynchus membranaceus* and others tend to concentrate near the coast in dry periods, and spread inland to temporary floodlands during wet periods (Chapter 1; Frith 1967). They may be present in a region in one year and then absent for several further years until conditions again become suitable.

It is perhaps partly because of the dispersive powers of wetland birds that all families are now represented on every major land-mass **(Table 5.1)**, and that several taxonomically undifferentiated species breed on 2–4 different continents. This is true of various herons, ibises and waterfowl, among which the Great Egret *Ardea alba* is the most widely distributed species. Transatlantic flights by some wetland species are by no means unusual, having been recorded from ringing in species that breed on both sides or that breed on one side but not on the other (Chapter 16).

SPECIES OF ONCE-CONTINUOUS BUT NOW DISJUNCT HABITATS

Species confined to high mountain ranges would be expected to show naturally disjunct distributions because they cannot persist in the unsuitable low ground habitat in between. Most such species are associated with montane forest, tundra and scree. The usual explanation of their current distributions is that they were more widespread in glacial times, when their habitats occupied the low ground, and then moved up the mountains with their habitats as the climate warmed again. This explanation could hold at least for species at high latitudes. In a sense, the flora and fauna of such mountains parallel those of land-bridge islands, also separated since the end of the last glaciation (Chapter 6). Eurasian examples of montane species include the White-winged Snowfinch *Montifringilla nivalis*, Alpine Accentor *Prunella collaris* and Yellow-billed Chough *Pyrrhocorax graculus*, while North American examples include the White-tailed Ptarmigan *Lagopus leucurus*, Clark's Nutcracker *Nucifraga columbiana* and Rosy Finch *Leucosticte arctoa* **(Table 12.1)**. Some species that occur discontinuously on mountains occur more continuously in similar habitat elsewhere, notably in tundra or boreal forest. Examples of such arctic–alpine and boreal–montane distributions in Europe include the Rock Ptarmigan *Lagopus mutus* of tundra and the Spotted Nutcracker *Nucifraga caryocatactes* and Three-toed Woodpecker *Picoides tridactylus* of conifer forest, and in North America the Rock Ptarmigan *Lagopus mutus* of tundra and the Grey Jay *Perisoreus canadensis* and Black-backed Woodpecker *Picoides arcticus* of conifer forest.

Table 12.1   List of European and North American bird species restricted for breeding to high montane areas.

European	North American
Caucasian Black Grouse *Tetrao mlokosiewiczi*	Blue Grouse *Dendragapus obscurus*
Caucasian Snowcock *Tetraogallus caucasicus*	White-tailed Ptarmigan *Lagopus leucurus*
Caspian Snowcock *Tetraogallus caspius*	Clark's Nutcracker *Nucifraga columbiana*
Alpine Accentor *Prunella collaris*[1]	Mountain Chickadee *Poecile gambeli*[1]
Eastern Rock Nuthatch *Sitta tephronota*[1]	Mountain Bluebird *Sialia currucoides*[1]
Alpine (Yellow-billed) Chough *Pyrrhocorax graculus*[1]	Hermit Warbler *Dendroica occidentalis*[1]
White-winged Snowfinch *Montifringilla nivalis*	Olive Warbler *Peucedramus taeniatus*
Citril Finch *Serinus citrinella*[2]	Rosy Finch *Leucosticte arctoa*
Great Rosefinch *Carpodacus rubicilla*[1]	Cassin's Finch *Carpodacus cassinii*[1]

[1]Species that have obvious low-elevation equivalents, so could result from double invasions in which they were restricted to high elevation by their low-elevation competitor.
[2]Related form found on low ground on Corsica.

It could have been during the glaciations that other Palaearctic species spread south to establish populations in Africa, now marooned in montane areas. Examples include the Black-billed Magpie *Pica pica* in southern Arabia and the Red-billed Chough *Pyrrhocorax pyrrhocorax* in Ethiopia, 1,600 km and 2,000 km, respectively, from the next nearest contemporary populations (Chapter 9). Likewise, the Horned Lark *Eremophila alpestris* is found widely across North America and Eurasia, but has isolated populations more than 2,000 km to the south on the Andes mountains of Colombia, and more than 2,000 km to the south in the Atlas mountains of northwest Africa. Similar explanations have been applied to the presence of high-latitude species among the montane avifaunas of some tropical areas, as exemplified by the presence of Hairy Woodpeckers *Picoides villosus* and Rufous-collared Sparrows *Zonotrichia capensis* as far south as Panama.

Not all montane distributions necessarily result from the upslope retreat of former lowland habitat; dispersal between mountain ranges could also be involved. The isolated mountains of northeastern South America may be in this category, having received many of their bird species by dispersal from the Andes in the west (see later). They provide a parallel with oceanic islands never connected to a mainland and hence receiving their avifauna by dispersal. An alternative possibility, however, is that some of the bird species now confined to montane vegetation once also occurred in different lowland vegetation in between and have since died out in all but montane areas (Chapter 11).

Whether the populations on particular mountain ranges have been isolated since the last glacial or dry period, or whether they are connected by movement to other populations, presumably depends on the distances involved, as well as on the dispersal capabilities of the species concerned, and the nature of the intervening habitat. Most resident bird species are more likely to move the short distances from summit to summit within the same mountain range than the longer distances to more remote mountain ranges. Birds appear to colonise distant mountains more readily than oceanic islands, presumably because of the difference in intervening habitat (Vuilleumier 1970). A mountain bird that gets tired could perhaps rest and feed in the lowland, whereas a landbird that tires over the sea is doomed.

As on oceanic islands, however, the numbers of species on different mountain ranges vary with the area of the range (more species on larger ranges) and its distance from other mountain ranges (fewer species on more remote ranges). Such relationships are evident, for example, in mountains in western North America (**Figure 12.2**; Johnson 1975, Behle 1978, Brown 1978, Kratter 1992). In this region, isolated ranges, some rising to more than 3,000 m, provide islands of conifer forest and scrub within a sea of mainly desert habitats. The mountains are inhabited by a number of boreal bird and other species that are restricted to the high-elevation forests.

In the Great Basin, lying between the Sierra and Rocky Mountain ranges, plant remains have confirmed that some 10,000–12,000 years ago the climate was colder and wetter than now; and that vegetation zones were shifted several hundred metres below their present elevations, connecting several presently isolated montane habitats across the entire region. As the climate warmed and dried, the forest retreated to higher elevations, isolating on the mountain tops those forest plants and animals that are unable (or reluctant) to cross the low deserts.

Collectively, these various mountains hold fewer species than the nearest areas of extensive conifer forest, in the Sierra and Rocky Mountain ranges. The numbers of resident bird species now found on different mountains are strongly correlated with both forest area (and associated habitat diversity) and also with distance to extensive boreal forest (**Figure 12.2**; Johnson 1975, Kratter 1992). In the Great Basin, where most mountains fall well within the normal dispersal distances of birds, the effect of distance is hardly apparent, but in California, where the mountains are much further apart, species numbers decline with increasing isolation. The decline with distance is much stronger in resident species than in summer migrants (**Figure 12.3**). This implies that residents have been more affected by extinction–colonisation events than have migrants which, not surprisingly, more readily cross the intervening terrain. Among resident species, the ability or inclination to cross the desert varies between species.

So whatever the situation at the start of the present interglacial period about 10,000 years ago, subsequent local extinctions, colonisations and recolonisations have evidently influenced the present distribution patterns of non-migratory montane birds, with a greater net loss of species from smaller, more isolated

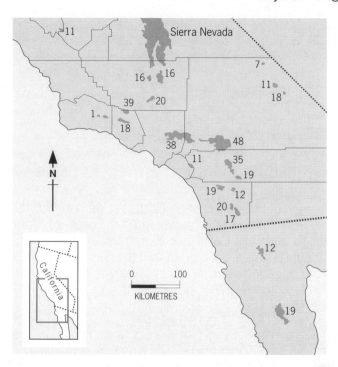

Figure 12.2 Locations of mountain conifer forest areas in California and Baja California, and number of bird species in each. Species numbers decline with habitat area and with distance from the main potential source area, the Sierra Nevada range in the north (see **Figure 12.3**). Based on data in Kratter 1992.

ranges. This is even more evident among resident montane mammals, most species of which do not cross the lowland between mountains. In consequence, species–area and species–distance relationships are stronger for mammals than for birds in the same mountains (Brown 1978), even allowing for some unrecorded species (Grayson & Livingston 1993). In the absence of recolonisation, then, montane mammal faunas have lost species in numbers proportional to area, fossils confirming their earlier presence on mountains from which they are now lacking (as well as on the intervening lowland in glacial times). Similar relationships between species numbers, habitat area and isolation have been found among the birds of European boreal and montane conifer forests (Haila *et al.* 1987) and among the forest birds of isolated Argentinian mountains (Nores 1995).

Dispersal

Mountain birds vary greatly in their dispersal abilities. Among African montane forest birds, Moreau (1966) listed several that had subspeciated (implying little or no recent immigration) on mountains only 15–30 km from their nearest neighbours. At the other extreme, other species must have colonised the mountain forests of the Yemen (Arabia) from Africa across a distance that had never been

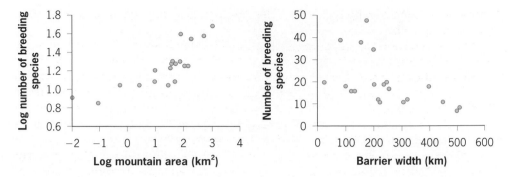

Figure 12.3 Relationships between (left) species numbers and forest area, and (right) species numbers and barrier width for birds of mountain conifer areas in California and Baja California. Barrier width reflects the distance (km) between the mountain area concerned and the Sierra Nevada, which is the nearest area of contiguous conifer forest. Regression equations, with MA = mountain area (km²), BW = barrier width (km), *P<0.05. Log resident species numbers = 1.031 + 0.280 log MA, r^2 = 0.81* Log summer migrant numbers = 0.851 + 0.109 log MA, r^2 = 0.32* Log total breeding species numbers = 1.031 + 0.156 log MA, r^2 = 0.68* Resident species numbers = 13.71 − 0.0232 BW, r^2 = 0.507* Summer migrant numbers = 16.13 − 0.0192 BW, r^2 = 0.12, n.s. Total breeding species numbers = 29.91 − 0.0427 BW, r^2 = 0.26*. From Kratter 1992.

less than 300 km, whether in glacial or interglacial times. For mountains in west Africa (in the Cameroons and Angola) that are more than 1,800 km from their nearest neighbours to the east, he considered recent colonisation as inconceivable, and all present montane species as having been marooned there since the last glacial period (Chapter 11). Montane colonisations over distances of 600 km are implied in South America, where the isolated Pantapui range in Venezuela shares about 60% of its species with the Andes or with the coastal ranges of Venezuela, the rest having been derived from lowland forms (Mayr & Phelps 1967). Again, however, one cannot exclude the possibility that some species, now restricted to mountains, once also occurred in lowland habitat from which they have since disappeared.

In the western Palaearctic, certain montane species, such as White-winged Snowfinch *Montifringilla nivalis* and Alpine Accentor *Prunella collaris*, appear as vagrants up to several hundred kilometres from their breeding areas, and also perform regular migrations. The implication is that they could easily cross most of the present discontinuities in their distributions and, as evidence that they do so, neither has subspeciated within western Europe. The situation with montane species is thus variable, with some tropical montane species apparently isolated by distances as little as 15 km, and others in tropical and temperate regions moving freely over distances of at least several hundred kilometres. It is in the latter that local extinctions could most readily be countered by recolonisations, so that distribution patterns are maintained. Regular migration may assist dispersal between mountain ranges but it is less frequent among tropical than among temperate zone birds.

While discontinuities resulting from past climatic changes are most obviously manifest in montane birds, such changes are likely to have contributed to major discontinuities in habitats on low ground too, as discussed in Chapters 9–11. In Australia, a few species have extremely disjunct distributions. The Crested Shrike Tit *Falcunculus frontatus* probably occurs in only five isolated localities, while the Yellow Chat *Epthianura crocea* is confined to some coastal rivers in northern Australia and some inland flood plains (Blakers *et al.*1984).

Mountain chains are important, not only in maintaining distinct species, but also in separating lowland avifaunas on either side, thereby promoting their separate evolution. Even in the relatively narrow land area in Costa Rica and Panama, the central mountain chain separates the lowland birds of Pacific and Caribbean slopes. We thus have several pairs of closely related species with one member on each side. With the Caribbean slope species listed first in each pair, examples include the Blue-chested Hummingbird *Amazilia amabilis* and Beryl-crowned Hummingbird *A. decora*, the Collared Aracari *Pteroglossus torquatus* and Fiery-billed Aracari *P. frantzii*, the White-collared Manakin *Manacus candei* and Orange-collared Manakin *M. aurantiacus*, the Snowy Cotinga *Carpodectes nitidus* and Yellow-billed Cotinga *C. antoniae*, and the Bay Wren *Thryothorus nigricapillus* and Riverside Wren *T. semibadius*. Similar examples of the effects of mountains as barriers to dispersal can be found on every major land-mass.

SPECIES WITH DISJUNCT DISTRIBUTIONS IN CONTINUOUS HABITAT

Some species now have discrete populations within large expanses of apparently suitable habitat. For example, the non-migratory Marsh Tit *Parus palustris* occurs at both ends of Eurasia but not in the middle, giving two populations several thousand kilometres apart. Similarly, the Goldcrest *Regulus regulus* occurs in three main regions across Eurasia, each population separated from the next by more than 500 km. Other major discontinuities affect some non-forest species, including the Twite *Carduelis flavirostris* found in northwest Europe and again in the Caucasus and central Asian Highlands, and the White Stork *Ciconia ciconia* found separately in three regions of Europe, plus Turkestan and eastern Asia. It is hard to believe that human action was involved in most of these species, and the most likely explanation is that they once occurred continuously, but have since disappeared from some regions for natural reasons.

Perhaps the most extreme example of a disjunct distribution is shown by the Azure-winged Magpie *Cyanopica cyana* which is found only in southwest Europe and southeast Asia, in regions separated by 9,000 km. It has been suggested that Asiatic birds were introduced to Europe in the sixteenth century by sailors returning from the far east, but the finding on Gibraltar of fossilised bone remains dated at 44,000 years old confirms that the species has lived in Europe for a long time (Cooper & Voous 1999). The only plausible explanation of the current distribution, therefore, is that the species once occurred right across Eurasia, but has since died out everywhere except at the two ends. Analyses of mt DNA have lent support to this view, and imply that the western and eastern populations separated about 1.2 million years ago, presumably during a glaciation (Fok *et al.* 2002). It would be especially satisfying if appropriate fossils could be found in intermediate areas, now lacking the species.

Phylogenetic study of ten species of nuthatches, based on mt DNA analyses, showed that the Corsican Nuthatch *Sitta whiteheadi* and Chinese (Snowy-browed) Nuthatch *S. villosa* are extremely closely related, even though they now occur in pine forest in far separated regions (Pasquet 1998). The degree of genetic divergence between the two species suggests a separation date as recent as one million years ago, in the mid Pleistocene. Their distributional disjunction is almost as extreme as that in the Azure-winged Magpie *Cyanopica cyana*, and their estimated separation dates are not very different.

A likely explanation for some present discontinuities is that, during a past climatic extreme, species were confined to remaining isolated refuges of habitat, and that with an improvement in climate they have spread from these refuges less rapidly than the habitat itself has spread (Chapter 9). This has left them patchily distributed in what for some species has become continuous habitat. This explanation may hold for many disjunct species, especially in tropical regions, because they remain together in some of the same areas (Chapter 11).

Not all discontinuities can be explained in this way, however, because some species show idiosyncratic patterns of patchiness. Examples are fairly numerous in pristine tropical forest, and are especially marked in the most species-rich areas. The central mountain range of New Guinea extends from west to east for 1,500 km, with elevations exceeding 5,000 m. The majority of New Guinea's 180 montane bird species extend to both west and east ends of the range. Nevertheless, each portion of the range apparently lacks some otherwise widespread species (Diamond 1980b). For example, the Papuan Treecreeper *Cormobates placens* is encountered daily by human observers in the mountains of east and west New Guinea, where it is common over a wide altitudinal range (1,500–3,300 m) and in several forest types. It is conspicuous with a loud song. Yet the species has a distributional gap of about 400 km in the middle of the central range, despite the continuous habitat and the lack of any related species likely to compete with it **(Figure 12.4)**. Twenty-one other montane species exhibit similar gaps of several hundred kilometres in one or other part of the central range. Yet other species that live in the lowlands of New Guinea have similar distributional gaps within continuous habitat. For example, the Logrunner *Orthonyx temminckii*, the sole New Guinea species in its genus, occurs in four large areas in the lowlands **(Figure 12.4)**, while the Obscure Berrypecker *Melanocharis arfakiana* is known in only two localities at opposite ends of the island, 1,500 km apart (Diamond 1980b).

Practically any area of New Guinea proves to be the site of a patch for some species, so it seems unlikely that all distributional patchiness on that island can be explained in terms of past refuges. For at least some species, other factors seem to be involved. One explanation may be habitat unsuitability: although conditions look good, some requisite may be missing or some pathogen or predator may be excluding the species from some areas, but unknown to the human observer. One can never rule out the existence of unspecified factors, but many patchily distributed species occupy such a wide range of habitats and climatic zones that this explanation seems unlikely.

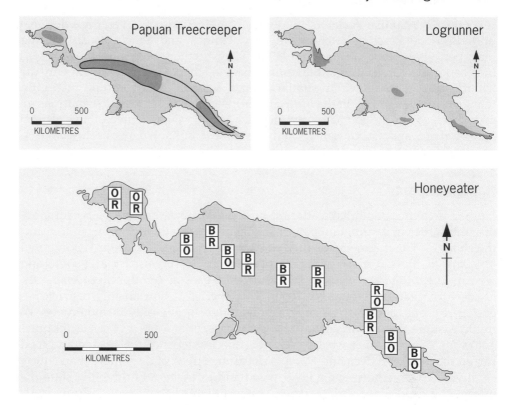

Figure 12.4 Patchy distributions of various species in New Guinea. Upper left. Distribution of the Papuan Treecreeper *Cormobates placens* in the central mountain chain, showing an area of apparently suitable habitat from which the species is unaccountably missing. Upper right. Distribution of the Logrunner *Orthonyx temminckii* in the lowlands of New Guinea. The species is known only from four far-flung areas, despite the presence of suitable forest in between. Lower. Distributions of three *Melidectes* honeyeaters in the mountains of New Guinea (O = *M. ochromelas*, B = *M. belfordi*, R = *M. rufocrissalis*). Most mountainous areas of New Guinea support two species with mutually exclusive altitudinal ranges. At each locality depicted on the map, the letters above and below indicate the species presence at higher and lower altitude, respectively. Modified from Diamond 1980a.

Another explanation is based on the presence of competitors. Checkerboard distribution patterns on islands, where species replace each other one-for-one, are familiar manifestations of apparent competitive exclusion (Chapter 14). A complex example on New Guinea involves three closely related montane honeyeater species of the genus *Melidectes* **(Figure 12.4)**. When considered individually, each species has a peculiarly disjunct range and is absent from several portions of the central cordillera. When the ranges of the three species are considered together, however, it is clear that (1) each mountainous area supports two species but not the third; (2) the identity of the locally successful combination varies in irregular checkerboard fashion; and (3) each of the three possible combinations occurs in several areas. Similar lockouts involving combinations of several species may underlie many cases of patchiness in New Guinea, and remain consistent over long periods of time (Diamond 1980b). They become increasingly likely with increasing species richness of the community, which may help to explain why distributional patchiness within similar habitat is much more common in tropical than in temperate areas. In the tropical southwest Pacific, patchy distributions are most marked on the four most species-rich islands (New Guinea, New Britain, New Ireland and Guadalcanal), and are almost unknown in the lowlands of species-poor Pacific islands (Diamond 1975).

Human impacts

Human activities could have affected bird populations, causing or accentuating discontinuities in distribution patterns, long before ornithologists began to record the changes. The fragmented nature of many bird ranges in regions lacking obvious barriers hints at much wider distributions in the recent past, and sometimes we can guess the likely causes. Take the White-backed Woodpecker *Dendrocopos leucotos* for example (Tomialojc 2000). This species is now mainly restricted to eastern Europe, and in western Europe is found locally only in mountains. We might assume that this has always been the case, or at least in post-glacial times. Yet more than any other woodpecker, this species depends for nesting and feeding on the dead and decaying wood of deciduous trees; and for more than a thousand years people have maintained woodland in western Europe in a state that offers little dead wood. In this region, then, suitable habitat remains only in the least accessible parts of mountains. The idea that this species once occurred across the whole of Europe, but has been eliminated from most western areas by human impact, is supported by occasional records in the early literature. Only the shortage of dead wood in otherwise suitable habitat seems to explain its absence from most of western Europe today.

Some wetland species are now found only or mostly in very small areas, often separated by long distances from other occupied sites. In the western Palaearctic, the Great White Pelican *Pelecanus onocrotalus* currently breeds at only about ten sites, the White headed Duck *Oxyura leucocephalus* at 21 sites, the Aquatic Warbler *Acrocephalus paludicola* at 31, the Pygmy Cormorant *Phalacrocorax pygmeus* at 34 and Audouin's Gull *Larus audouinii* at 43 (Heath & Evans 2000). The ranges of most of these species are known to have become more fragmented in recent centuries. For example, the Squacco Heron *Ardeola ralloides* was known from at least 195 sites in 1850–1900, from 115 in 1900–1920, falling to 80 sites in 1921–40 and 71 sites in 1940–60 (Jósefik 1969, 1970). Five hundred years ago, the Northern Bald Ibis *Geronticus eremita* was known from many localities in southern Europe, North Africa and the Near East. By 1950 it had become confined to two localities at either end of its range, in Morocco and Turkey, and now remains only in coastal Morocco, with a total population of less than 300 birds. The Dalmatian Pelican *Pelecanus crispus* now breeds regularly at only 22 sites, all in southeast Europe, yet Pliny writing less than 2,000 years ago mentions it as breeding at several other estuaries in western Europe, including the Elbe, the Rhine and the Sheldt. Declines in these species have been attributed mainly to habitat destruction (drainage), but in some also to killing for the plumage trade. Other apparently suitable sites remain available for most of these species, but currently remain unoccupied. Similarly, many large birds of prey are now patchily distributed in Europe because of human persecution and, if the killing stopped, some such species would almost certainly spread to areas of former habitat which still remain suitable. This is in fact happening as public attitudes are changing, as protective legislation is enforced, and as reintroduction programmes get underway.

In summary, explanations for distributional gaps in continuous habitat include the following: (1) in the past, the gap areas contained no suitable habitat and only recently has such habitat developed, giving insufficient time for colonisation (as in the glacial refuge explanation); (2) interspecific competition prevents the species from persisting in the gap area; (3) human persecution or an unknown pathogen or predator prevents the species from persisting in the gap area; (4) the species concerned has suffered serious decline, and the gaps represent areas of recent withdrawal. The first of these explanations is likely in many species but can be eliminated in others, and none of the explanations is necessarily independent of the others. In particular, many species have discontinuous distributions because they have fragmented habitats, but they have also been eliminated from some suitable habitat by human action which increases their present patchiness.

CONCLUDING REMARKS

Range fragmentation and speciation

For speciation to occur, following range fragmentation, the population concerned must be large enough to avoid extinction, isolated enough to reduce gene flow from other populations to a trivial level, and persist for long enough to allow the necessary degree of differentiation to evolve. The distributional gaps in some species resident in tropical forest are so wide and of such long standing that they could provide sufficient isolation for populations to diverge even in the absence of habitat barriers. This raises the question of whether such patchiness may contribute to the species richness of tropical areas (Chapter 11). As in the New Guinea birds mentioned above, certain groups of related species in Amazonia show peculiarly small or unique distributions, in which related forms may be separated by large gaps within apparently uniform forest habitat (**Figure 12.5**). Among some pygmy-tyrant flycatchers (subfamily Euscarthminae), current ranges seem not to reflect ancient refuges, because the distributional foci seldom coincide between species (Fitzpatrick 1980). Yet many of the species are so localised (e.g. *Hemitriccus mirandae, Todirostrum calopterum*) or discontinuous (e.g. *Hemitriccus rufigularis*) as to suggest that they have contracted from formerly more extensive distributions (**Figure 12.5**). Other characteristics of these groups include: (1) restriction to low or mid elevations on forested mountains around the Amazonian perimeter (e.g. *Hemitriccus granadensis, H. furcatus*); (2) restriction to small isolated mountain ridges, where island effects reduce the numbers of competitors (e.g. *Hemitriccus rufigularis, Todirostrum russatum*); (3) extreme rarity (e.g. *Todirostrum senex, Taeniotriccus andrei*); and (4) morphological peculiarity, as compared with the remainder of the group (e.g. *Poecilotriccus capitalis, Hemitriccus furcatus*). All these characteristics would be expected of relict species groups, in the final stages of a speciation–extinction cycle (Chapter 2).

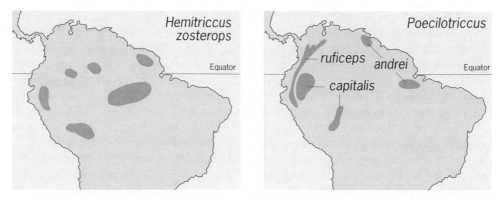

Figure 12.5   Some pygmy-tyrant flycatcher distributions. Left. The White-eyed Tody-Tyrant *Hemitriccus zosterops* purportedly shows an early stage in a taxon cycle; the distribution is patchy within continuous habitat, and each population is subspecifically different from the others. Right. *Poecilotriccus* purportedly shows a later stage in a taxon cycle; again the distribution is patchy, but differentiation is more advanced, giving rise to three distinct allospecies. Modified from Fitzpatrick 1980.

The taxon cycle within mainland forest is hypothesised to proceed as follows. As successful species disperse through continuous forest, competitively inferior forms are forced to occupy progressively smaller ranges. Their surviving populations are fragmented, enabling morphological and ecological divergence from the original, and resulting in shifts to higher elevations or to peculiar habitats. In the final stages, a species may exist only in relatively competitor-free situations, such as isolated mountain ridges. If such a remnant population can survive long enough, it could diverge sufficiently from its sister populations to form a new species, and might then even be able to re-invade a formerly occupied area. Continent-wide species diversity thus increases through the cyclic production of new species in temporary refugia, followed by their spread, which increases the competitive pressure on other species. Evidence that this process might be occurring within Amazonia lies in species that show multiple, disjunct and well-differentiated populations. An example is provided by *Hemitriccus zosterops* which has at least seven remnant populations, each recognised as a distinct race **(Figure 12.5)**. Other examples may exist in *Todirostrum calopterum* and *Poecilotriccus capitalis* (Fitzpatrick 1980).

This process of competitively induced speciation could give rise to species groups that cannot easily be distinguished from populations still restricted to former refugia (except that refugia often contain isolated populations of several species, not just one). The important points are that (1) alternative hypotheses to refugium fragmentation must be considered in attempts to explain the present-day distributions of endemic species; and (2) competition could provide a means of population fragmentation just as effective in speciation as the climatic and other physical factors that can also fragment distribution patterns. If range contraction and isolation through competition is a common mode of speciation, the process could build upon itself: the greater the number of related species that occur in a region, the greater the potential for competition, and the greater the likelihood of population shrinkage and isolation. In other words, an abundance of species in a region could promote further speciation through competition and range contraction even in the absence of any geographical or ecological barriers. The competitors themselves provide the isolation, by eliminating the species from parts of its range and isolating populations from one another.

Range fragmentation and extinction

Long-term range fragmentation can also have the opposite effect, however, increasing extinction risks, through creating smaller populations with little or no interchange, which become prone to local extinction through local events that may act on one subpopulation after another. Range fragmentation, then, sets the scene for local extinctions, which may occur tens or hundreds of years later, leading ultimately, as one population after another disappears, to range contraction and eventually to total extinction. Many species now exist as distributional relicts, surviving only in one or a few small areas that almost certainly represent a small fraction of their original ranges. In Australia, for example, the Eungella Honeyeater *Lichenostomus hindwoodi* inhabits rain forest in a single mountain range perhaps 50 km long in eastern Queensland, while the Noisy Scrub-bird *Atrichornis clamosus* occupies only a few square kilometres of coastal heath forest in southwestern Australia. In addition, many species in developed countries and on oceanic islands now exist in isolated fragments of habitat, remaining in an essentially human-modified landscape, and some may well disappear from these areas in the coming years. Species with markedly fragmented small-patch distributions therefore present major conservation problems, which have as yet hardly been addressed beyond the theoretical treatment.

SUMMARY

Some bird species have markedly discontinuous breeding ranges, through either past natural processes or more recent human action. About one-fifth of all Palaearctic bird species have at least one population separated by more than 1,000 km from the next nearest population, well beyond the normal dispersal distances of most of the species concerned. Similar examples occur on all major land-masses. They result from the splitting of once-continuous distributions (vicariance) or from distinct colonisation events.

Some species with discontinuous distributions may always have been confined to patchily-distributed habitats, such as wetlands and mountain tops. Others were once widespread, but now have disjunct distributions because their habitats have become discontinuous. Yet others now have disjunct distributions within an expanse of continuous and apparently suitable habitat. Some disjunct distribution patterns are natural, but others result from human impact, and yet others are unexplained.

Particular interest attaches to species that have patchy distributions in continuous habitat, some of which may result from species being unable to spread from past refuges, while others may result from competition, and represent the later stages in a taxon cycle. Such a process, driven by competition, is most likely to operate in tropical areas already rich in species, and could lead to further speciation and enrichment of some tropical biotas. Fragmentation of a once-widespread distribution pattern can also increase extinction risks if the remaining populations are too small and isolated to persist.

Part Four
Limitation of Species Distributions

Spectacled Eiders *Somateria fischeri*, whose entire population concentrates in about four sites for the winter.

Chapter 13
Bird distribution patterns

In this and the following chapters, we turn from the distributional histories of regional avifaunas to explore the factors that influence, here and now, the distributions of particular species. In this chapter, I discuss the measurement and mapping of bird ranges, some general patterns that have emerged from recent studies, and then explore the relationship between bird abundance and distribution, as the two are closely linked. Subsequent chapters in this section discuss the various factors that limit the spread of particular bird species, and some range changes that have occurred in recent decades.

The potential distribution of any species depends on: (1) the conditions of climate, habitat and food that the species requires; (2) the geographical extent of suitable conditions; and (3) the area of origin and dispersive powers of the species, and the historical opportunities that have enabled it to reach those areas with suitable conditions. The actual ranges of many species are probably smaller than their potential ranges, which include all the places they might live if they could get there. Some species are still restricted to the regions or islands where they evolved, while others have spread to other regions or islands far removed from their original homes. Some may even have gone altogether from their original homes, surviving only in places later colonised. In addition, many species have also been eliminated by human action from some areas that otherwise still remain suitable for them, the inability to expand and recolonise or continued persecution limiting their current distributions.

Some species have small ranges because the particular combination of conditions they require is found only over small areas, or because the species themselves are unable to disperse to other areas outside their existing range where those same conditions occur. Conversely, other species have large ranges because the particular conditions they require are found over large areas, or because the species themselves have sufficient dispersal powers to reach such areas if they are patchily distributed. Ranges are dynamic because species themselves change through time, as does the distribution of habitat conditions on which the species depend. Much the same can be said about numerical abundance, which also results from an interaction between species and their environments.

Whether they depend on terrestrial, freshwater or marine habitats, all bird species must come to land at least for breeding (though Emperor Penguins *Aptenodytes forsteri* nest on sea ice, while grebes and others build anchored floating nests). Moreover, most landbird species breed in only a single habitat type (say woodland, grassland or desert), so are inevitably restricted to where such habitat occurs. Nevertheless, some species, including many raptors, swifts and swallows, range through several habitats to obtain their food, and some aquatic species occur in wetland habitats in whatever terrestrial setting these wetlands lie.

MEASUREMENT OF RANGES

Understanding the factors that limit current bird distributions depends largely on the accurate measurement of geographical ranges. For most species this at once presents problems because of the huge spatial scale involved. We are all familiar with the hand-drawn range maps of birds in field guides and reference books, in which shading depicts the spatial extent of the range across a continent or across the globe. Typically, such maps are generated from information on the localities at which a species has been recorded over a long period of years, and inevitably involve some guesswork, notably in interpolating between records of occurrence, and in separating regular occurrence from vagrancy. In consulting such maps, we accept that some regions are better known than others, that each species may vary in abundance across its range, and may be absent altogether from areas within that range, depending on the distribution of habitat and other factors. We also accept that such maps may be out of date, in that the range of any species may have changed to some extent since the map was drawn. In other words, while the small-scale range maps typical of most bird books show an approximate (and often outdated) view of the extent of a bird range, they usually give an exaggerated view of its distribution within that range. Nonetheless, such maps are adequate for many purposes, including the majority of those addressed in this book.

Even when accurately drawn, most range maps are not depicted on equal area projection. On such a projection, all areas are drawn to precisely the same scale, so that they appear on a flat map in the same relative proportions as they occur on the earth's spherical surface. Most map projections in common use suffer distortion that results from projecting a spherical surface to a flat one; in particular, they exaggerate high-latitude land areas in relation to low ones. For the biologist interested in such matters, this complicates the measurement of range areas. By way of simplification, some authors have resorted instead to measuring indices of landbird range sizes, such as their maximum north–south or east–west dimensions (or the product of these values), expressed in kilometres or in degrees of latitude or longitude (Stevens 1989, Gaston 1994, Newton & Dale 1997). Such measures of 'extent of occurrence' are inevitably larger than the actual range size or 'area of occupancy', because species do not occupy all areas within these limits (Gaston 1994). The error could be intolerably great in species with markedly patchy distributions, like many montane ones.

Where relatively little information is available, ranges are sometimes depicted as dot maps, showing the specific location where each species has been recorded, often from verified museum skins (e.g. **Figures 11.4–11.6**; Hall & Moreau 1970, Snow 1978a). The advantages of this procedure stem from the objectivity and transparency of the underlying data-base, and the lack of extrapolation beyond it, leaving any infilling to the discretion of the reader. Such maps are produced mainly for little known areas, such as parts of South America, Africa and southeast Asia, where the numbers of records for some species are extremely limited.

More recent 'Atlas' projects, in which both survey and mapping are done on a grid basis, represent a great advance in range assessment. The surveys are usually conducted over a defined short period of years, giving a snapshot of distributions at the time, and results are usually plotted on a grid system, showing presence or absence in each square. Such maps reveal gaps in distributions across ranges and, from the sum of occupied squares, can give more precise measures of the total area of occupancy for each species. The size of the grid squares varies between schemes, and is usually set pragmatically to suit the data available, but clearly the smaller the grid, the more precise the picture and the more accurately can total area of occupancy be calculated. In general, the smaller the grid and the more refined the scale of mapping, the smaller will be the measured range size. This is because, with increasing resolution, it becomes possible to discriminate more and more localities in which the species does not occur and hence to exclude them from the measured area. We can expect, then, that on any grid system, measured range sizes will decline as resolution improves. Some recent 'Atlas' schemes have also provided measures of abundance within the range (Root 1988a, Gibbons et al. 1993, Price et al. 1995). Such measures enable contoured mapping of densities, and more objective definition of range boundaries (taken by Root (1988a) as 0.05% of maximum density). The benefits of such detailed mapping are self-evident, notably because they allow relatively small changes in distributions to be detected at future surveys.

In the mapping of geographical ranges, defining the boundaries is of paramount importance. Yet experience has shown that, unless a species stops abruptly at a coast or other barrier, range boundaries are seldom sharp like lines on a map. They appear rather as zones of progressively decreasing abundance and increasing patchiness, changing somewhat from year to year, but usually with occasional isolated groups or pairs of birds beyond the main boundary. This pattern makes boundaries hard to measure accurately. The effort required to find the outermost pairs is usually unacceptable and of little value, especially when the situation may change from year to year. An alternative procedure, when quantitative data are available, is to set the boundary as the line where population density falls to some arbitrary level, say one-hundredth of its maximum density. This kind of definition also brings difficulties but, when repeated after a period of years, it gives a quantitative measure of boundary change in which chance recording plays a smaller role. In addition, it automatically excludes vagrants.

On a world scale, knowledge of bird distributions is extremely uneven. In regions easily explored by bird-watchers, geographical ranges are based primarily on sight records, and are generally up to date, but in other parts of the world, such as many tropical regions, geographical ranges are still based mainly on localities given on museum skins. Such specimens have usually been accumulated over many years, mostly in the nineteenth century, and are far from up to date. Indeed, in many of the localities where specimens were collected, the original natural habitat might have been destroyed by subsequent human activity, with inevitable effects on the local avifauna. With the increasing development of eco-tourism and the influx of bird-watchers to previously little known areas, the museum records are increasingly being supplemented by more recent sight records. The museum material will always be useful, however, in confirming the former presence of species in localities from which they have since disappeared and, for some parts of the world, it still provides the main source of information on bird distributions.

Whatever the source of information, all mapping systems carry assumptions about the accuracy of species identifications. Among birds, this is not usually a problem, except possibly in some little known tropical avifaunas where records are based on field observation alone. However, because many bird species are migratory, we must normally separate breeding season from non-breeding season records. Most bird books, including field guides, now map breeding and non-breeding distributions separately.

Questions about current bird distributions can be addressed at several levels of scale. In discussing geographical range, we may ask what factors influence the boundaries of a range, whether the range changes with time, and why the extents and shapes of ranges vary between species. At a finer scale, we may ask about variations in abundance within a range, about why a species is numerous at some localities, scarce at other localities and absent elsewhere. At a still finer scale, we can consider how individuals distribute themselves within a local area, whether spaced out in territories or concentrated in mobile flocks. Some factors that influence the limits of a geographical range, such as an ocean barrier, are not necessarily relevant to distribution at finer scales; but other factors, such as food or nest sites, can operate at all three levels. Thus, the large-scale issues discussed in this book are not distinct, but intergrade with those concerned with local distribution and abundance, and with those concerned with individual behaviour (Newton 1998).

The search for patterns in the geographical ranges of birds and other organisms has formed a growing field of interest in recent years (called 'macroecology', Brown 1995). The main patterns that have been documented are described below, but some result from relatively obvious causes, such as the extent and shape of the various land-masses or of the habitats within them.

SMALLNESS OF GEOGRAPHICAL RANGES

Compared to the land areas that would seem to be available to them, the majority of the world's landbird species breed over remarkably small areas. Some 17% breed only on oceanic islands, mostly on a single island or island group (Chapter 6). Of the 83% that breed on continents, the vast majority (93.6%) breed on only a single continent (although some may extend to more than one on migration). Another 6.0% of species breed on two continents, 0.3% on three, only about 0.1% on four or five (counting Europe and Asia as a single continent) (Chapter 1). Moreover, in any continental avifauna most species occur over relatively small areas and few occur over large areas. Among the birds of Africa, for example, the median geographical range size corresponds to only 1% of the continental area south of the Sahara; and only about 2% of all species extend over more than half of the continent (**Figure 13.1**). Such skewed patterns in range sizes have also been well documented for the birds of North and South America and Australia (Anderson 1984, Ford 1990, Pomeroy & Ssekabiira 1990), and globally for particular families of birds (see Blackburn *et al.* 1998b for woodpeckers). On all continents, the areas (in km^2) naturally occupied by different bird species vary by six or more orders of magnitude, and the vast majority of species are not widespread.

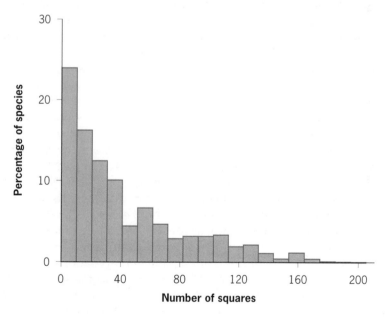

Figure 13.1   Distribution of range sizes among 1,434 African landbirds, with the range size of each species measured as the number of 2.5 × 2.5° squares in which it occurred. From a total of 422 squares available, 56 species were confined to a single square, and over 400 species to ten squares or less. Redrawn from Pomeroy & Ssekabiira 1990.

Similar patterns have emerged for the overall abundance levels of birds, taking the count data from all species over wide geographical areas: most species are relatively scarce, while a few are common or very common (e.g. Nee *et al.* 1991). It is one thing to show that such patterns in distributions and numbers exist, but it is another matter to explain them, beyond the fact that they represent patterns of ecological specialisation, with most species being habitat-specialists or food-specialists and a few being generalists. It is the latter that usually have the widest ranges and the greatest numerical abundance. The fact that similar patterns exist in other kinds of organisms, both plants and animals, suggests the operation of some underlying general cause.

Relationship between range size and body size

In those regional avifaunas that have been studied, a broad relationship is apparent between body size and geographical range size, with larger species generally occurring over larger areas than smaller ones (Brown & Maurer 1987, Nee *et al.* 1991, Gaston & Blackburn 1996a, Blackburn *et al.* 1997a). If a graph is plotted of body size against geographical range size, the points for different species fall roughly within a triangle, reflecting the fact that, while some small bird species have small ranges, others have large ranges, but almost all large bird species have large ranges (Brown & Maurer 1987, Brown 1995). The striking point is the dearth of large species with small ranges. One suggested explanation is that the minimum viable geographical range itself varies with body size. Large species generally live at low overall densities, each individual requiring a large area, so large species with small geographical ranges have low overall abundance, which makes them more vulnerable to extinction (Brown & Maurer 1987). In other words, the combination 'large body size–low density–small range' will generally not persist long through evolutionary time, accounting for its rarity in modern avifaunas. In contrast, small species, requiring less space per individual, can maintain large populations in relatively small areas, so small species with small distributions are more likely to persist long term than are large species with small distributions. This reasoning is no more than a plausible guess, but it also fits the fact that carnivores (which live at low densities) generally have larger geographical ranges than herbivores (which typically live at higher densities) (Rapoport 1982).

Another possible explanation of the dearth of large species with small ranges relates to evolutionary processes. Because of the short dispersal distances of small species, geographically separated populations are likely to achieve genetic isolation, and in the long term form separate allospecies. This is less likely to occur in large species, in which individuals disperse over longer distances, thus preventing the differentiation of geographically separated populations. Hence, small species occupying large ranges are more likely to have become split into separate species than are large species occupying large ranges (for further discussion see Chapter 17).

Relationship with geography

Not surprisingly, the range sizes of birds (like those of most other organisms) are greatly influenced by the sizes and disposition of the major land-masses, together with their topography and habitat distributions. Land covers 29% of the surface of the globe. Considered latitudinally, more than two-thirds of land lies north of the equator, and only one-third lies south of it (Chapter 18). This immediately explains why the most widespread landbird species in the northern hemisphere extend over greater areas than the most widespread species in the southern hemisphere. Considered longitudinally, about two-thirds of all ice-free land is included in the Old World and only about one-third in the New World. This allows the most widespread landbird species in the Old World to extend over greater areas than the most widespread species in the New World.

As the converse of land areas, about two-thirds of all sea areas lie south of the equator, and one-third lie north of the equator. Moreover, a broad band of open sea lies south of South America, Africa and Australia, enabling many seabird species at those latitudes to extend around the world, with no major land areas to block their spread. At comparable latitudes in the northern hemisphere, many seabird species are restricted by the northern continents to either the Atlantic or Pacific Oceans, so that their ranges are more constrained (Chapter 8).

Latitudinal trends in range size

Despite there being much more land at low than at high latitudes, another general tendency is for the range sizes (or latitudinal extents) of birds to increase from the tropics to the arctic (the latitudinal trend is called 'Rapoport's' (1982) Rule; Stevens 1989, Blackburn & Gaston 1996). In the northern hemisphere the trend is most marked above 40–50° latitude (Rohde 1992), a zone barely reached by habitable land areas in the southern hemisphere. A similar trend is apparent among birds that live on mountains, in that the altitudinal range of species tends to increase with elevation (Stevens 1992). Similar trends are found in a wide range of other organisms, and not just in birds, but again neither pattern is universal.

The net result of latitudinal and longitudinal trends in range sizes is that tropical species in general have smaller geographical ranges than their high-latitude equivalents. Tropical areas tend also to be more species rich (Chapter 11), which leads on to another broad relationship, the tendency for range size to decline geographically with increase in species richness (Anderson 1984, Stevens 1989, Rosenzweig 1995). Because of this species richness, competition is likely to be especially strong in tropical areas, and thus more important in restricting the distributions of individual species than in higher latitude areas (Chapter 11).

Coastal species

As a fourth general trend, coastal species inevitably have narrow linear ranges that follow sea coasts in whatever direction they happen to run. Many shorebirds show a narrow circumpolar distribution on their arctic inland breeding grounds, but an extensive north–south distribution on their coastal wintering areas, which in some species extend to the tips of the southern continents. Extreme examples are the Little Stint *Calidris minuta* and Bar-tailed Godwit *Limosa lapponica*, which in Europe have breeding ranges spanning less than 5% of their total annual latitudinal ranges, but outside the breeding season occur on coastlines spanning 80° and 90° of latitude, respectively (Newton & Dale 1997).

*   *   *   *

To end this section, bear in mind that all the generalisations mentioned above are no more than that. They represent overall trends, to which some species and some bird families provide conspicuous exceptions. The findings can also vary with the spatial scale on which analyses are done (Gaston & Blackburn 1996a) and sometimes also with the way the data are subdivided (Ruggiero & Lawton 1998). Other general patterns in the distribution and abundance of birds and other organisms have been documented, in addition to those mentioned above, some of which are discussed later in this book. They indicate that, to some extent, bird range sizes vary in a systematic manner along geographical gradients (smaller towards the equator) and with the type of bird involved, especially with its body size, and they tend to overlap most in particular regions, giving areas of exceptional diversity and endemism (Chapters 9 and 11). As yet, most such patterns lack generally agreed explanations (Lawton 1993), but they presumably depend on ecological features of the birds themselves, together with the history and prevailing features of their environments, including the shapes and configurations of land areas.

VARIATIONS OF ABUNDANCE WITHIN THE RANGE

Thanks to widescale bird-mapping schemes, involving hundreds or thousands of observers, it has become possible to find how the abundance of species varies across their geographical ranges (e.g. Brown 1984, Root 1988a, Blakers *et al.* 1984, Gibbons *et al.* 1993). For many bird species, population density tends to be greatest near the centre of the range and to decline towards the edges. This is evident for both breeding and wintering distributions, as exemplified in **Figure 13.2** (Brown 1984). This general pattern is attributed to the assumption that the conditions most suitable for a species are most likely to occur near the centre of its range. The pattern is not shown by all species, however, and some have multiple regions of high abundance within their ranges, presumably caused by some form of environmental patchiness, while other species remain abundant up to their range boundaries, especially where these boundaries coincide with topographical or ecotone barriers.

Two other complications affect the occupancy of particular localities, and the positions of range boundaries. Firstly, as implied by the density patterns, environmental conditions are not equally suitable for a species everywhere it occurs. Some localities may be so favourable that birth rates exceed death rates. Such localities act as sources, producing a surplus of individuals able to occupy other areas. Other localities may be so unfavourable that death rates exceed birth rates. Such localities act as sinks, whose occupancy can be maintained only by continual immigration from better areas (Pulliam 1988). This source–sink concept is applicable at various spatial scales from different territories in the same locality, through different habitats within a region, to different parts of a geographical range. Examples of apparent source and sink areas have been described for many bird species, including Sparrowhawks *Accipiter nisus* and others in different territories (Newton 1991), Eurasian Dotterel *Eudromias morinellus* on different mountain tops (Thompson & Whitfield 1993), and Barn Owls *Tyto alba* in different landscapes (de Bruijn 1994). The geographical range of any species is thus likely to contain a mixture of source and sink areas but, in general, sink areas are especially likely to

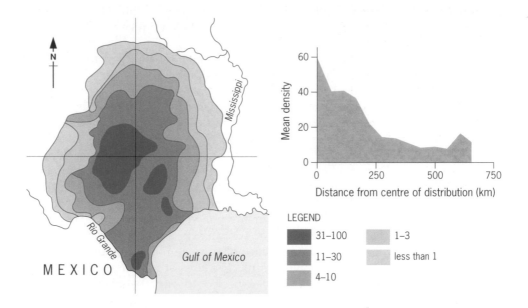

Figure 13.2   Abundance of the Scissor-tailed Flycatcher *Tyrannus forficata* over its breeding range in the United States. Shading depicts the mean population density per standardised count. On the right, abundance is shown graphically by plotting density as a function of distribution along four transects through the widest parts of the range in four major compass directions and then assessing the values. Redrawn from Brown 1984, based on data for the Breeding Bird Survey of the United States Fish and Wildlife Service.

occur near the range boundaries where conditions become unsuitable. The important point, however, is that the existence of source areas enables a species to occupy a wider distribution than would otherwise be possible. If the populations on the edge of a species range are not self-sustaining but simply result from surplus birds moving out of productive core areas, then it may be impossible to establish what limits distribution from studies on the edge of the range alone (Chapter 14).

The second complication results from habitat fragmentation. Some habitats, such as wetlands and mountain tops, have always been patchily distributed, but other, once more extensive habitats, such as forest or grassland, have been greatly fragmented by human activities. Over much of the world, therefore, most bird species now live in patchily distributed habitats. Moreover, the size and spacing of habitat patches can greatly affect the regional occurrence of birds and other organisms. This results from two simple facts: firstly, smaller patches of habitat usually support smaller populations which, in their year-to-year fluctuations, are more likely to die out than larger ones; and secondly, habitat patches that are far removed from others, relative to the dispersal powers of the species concerned, are less likely to receive immigrants from other patches, and so are less likely to be recolonised quickly after a local extinction. In other words, regardless of other limiting factors, the spatial configuration of habitat patches within a landscape can influence the persistence of populations, their overall abundance and distribution (Hanski & Gilpin 1991).

The term metapopulation is applied to any population that is composed of a number of discrete and partly independent subpopulations that live in separate habitat patches but are linked by dispersal. Not all habitat patches may be occupied at one time, and the pattern of occupancy may change continually because of local extinctions and recolonisations. Wherever local recolonisations exceed local extinctions, the metapopulation expands to fill more habitat patches, but when local extinctions exceed recolonisations, the metapopulation contracts, to nil if the process continues. The dynamics of metapopulations, based on local extinctions and recolonisations of patchy habitats, have two important implications for bird distributions. Firstly, we should not expect to find all patches of even good habitat (especially small patches) occupied in any one year; and secondly, beyond a certain threshold of habitat fragmentation, we can expect that species will decline to regional extinction, even in the presence of some remaining habitat. In other words, their geographical ranges will shrink. The level of fragmentation at which such processes come into play can be expected to vary between landscapes according to the way in which remaining habitat is distributed, and between species according to their dispersal and other features.

For some species, findings from field studies have been examined specifically in the light of metapopulation theory. In a study of the Wood Nuthatch *Sitta europaea* in the Netherlands, not all woods were occupied at one time. Over six years, occupancy varied from 14% in small (<1 ha) isolated woods to 100% in large (>10 ha) less isolated woods, where isolation was defined by the amount of woodland in the surrounding area (Opdam *et al*. 1995). From year to year, local extinctions and recolonisations occurred, giving some changes in the woods that were occupied. Local extinction rates were related to the size and quality of the woods, and colonisation rates varied according to the number of Wood Nuthatches in the surrounding area, so that the most isolated woods were least likely to be recolonised. The occurrence of this species in different woods in the same landscape could therefore be modelled explicitly as a metapopulation (Verboon *et al*. 1991).

As a second example, in western North America, the Northern Spotted Owl *Strix occidentalis caurina* is dependent on old growth forest, but lives at low density, with home ranges covering from several hundred to several thousand hectares (Thomas *et al*. 1990). This species is assumed to have once occurred continuously through extensive tracts of mature forest, but it is now restricted to clusters of old growth patches within a matrix of younger forest and clear-fells created by logging. It is likely to survive in the long term only if sufficient old growth patches are preserved in a distribution pattern that ensures their accessibility to dispersing owls. The species has been studied well enough to produce the detailed information needed for credible mathematical models. Such models have predicted a sharp threshold in the degree of fragmentation that would cause a previously viable population to decline. Specifically, they imply that this owl would disappear if the patches of old growth forest were reduced to less than 20% of the total land surface over a large region (Lande 1988, Lamberson *et al*. 1992). Metapopulation theory has thus become important in understanding the distribution patterns of some bird species, and in planning for their conservation.

Overall, the various species found in habitat patches produced by fragmentation of formerly continuous habitats often represent patterned subsets of the original pool of species. Together they may form a nested series, with the increasingly depauperate fauna in small patches made up of subsets of the more species-rich fauna of larger patches. This was true, for example, of the bird species found in isolated woodlots in Brazil, which formed a nested series, suggesting a specific sequence of local extinctions governed by habitat area (Terborgh & Winter 1980). However, nested sequences are seldom perfect: a widespread species can occasionally be missing from large species-rich patches (holes), while uncommon species may turn up in small depauperate patches (outliers). Cutler (1991) examined boreal birds and mammals among mountain tops in the Great Basin of the southwestern USA. He found a significant pattern of nested subsets in these two faunas, but whereas the mammalian patterns tended to be 'hole rich', the bird patterns were 'outlier rich'. These departures from the overall nested patterns may be associated with the mammals' background as a relict fauna, marooned on mountain tops since the Pleistocene, whereas the bird faunas have probably undergone recurring colonisation throughout their history in the region (Chapter 12). Such nested sequences, linked with the size of habitat patches, have obvious relevance in conservation, because they enable us to specify the numbers and types of species likely to be found in habitat patches of different sizes.

RELATIONSHIP BETWEEN ABUNDANCE AND DISTRIBUTION

In recent years many bird populations have been monitored by annual counts in a large sample of study sites scattered over a wide area, usually as part of some national programme (such as the Common Bird Census of the British Trust for Ornithology). The findings have revealed another general pattern, namely in comparisons between species, local abundance and spatial distribution tend to be correlated (Lawton 1994, 1996). Those bird species that have high average densities at the sites where they are counted also tend to inhabit a high proportion of sample sites within a region. Conversely, species that are rare where they occur tend to occupy small proportions of sites. Such relationships have been established from studies in various parts of the world, including Europe, North America and Australia (Fuller 1982, Hengeveld & Haeck 1982, Bock 1984, 1987, Ford 1990, Gaston & Lawton 1990, Kouki & Häyrinen 1991, Sutherland & Baillie 1993, Gaston 1996, Blackburn et al. 1997a, Newton 1997, Gregory 1998, Venier & Fahrig 1998, Gaston et al. 2000). On a graph, the points for different species mostly fall within a triangle, which reflects the fact that locally scarce species are usually found in relatively few sites, while locally abundant species can be found in few or in many sites (**Figure 13.3**).

Local abundance seems also to be correlated with geographical range. This relationship is much less firmly based, but has been shown statistically in studies of North American and Australian birds (Bock & Ricklefs 1983, Brown & Maurer 1987, Ford 1990). Again, when plotted on a graph the points for the different species fall in a rough triangle, reflecting the fact that locally scarce species usually have small ranges, while locally abundant species can have either small or large ranges. The relationship is statistically weak, because of the many exceptions to the overall trend, involving species that are common within a restricted area or scarce but widespread. Most flesh-eating species are in the latter category.

Restricting the analysis from an entire avifauna to particular trophic groups gives a clearer relationship, as shown for the winter distributions of North American seed-eating birds in **Figure 13.4** (Bock & Ricklefs 1983). Within this group, a clear tendency is apparent for species that achieve the highest mean local densities to occupy the largest geographical areas (which means that they must also have the largest total population sizes). Similar patterns have been found in other organisms, both plants and animals, but the relationship between local abundance and distribution tends to become progressively weaker the larger the geographical area involved (Gaston 1994).

Three main explanations have been proposed to explain the general link between abundance and distribution, but in any data set it is hard to discriminate between them. On one explanation, the relationship is seen as a possible sampling artifact. Because locally rare species are more difficult to detect than are locally abundant ones, the latter are more likely to be detected at a large proportion of the sites where they occur. This is less of a problem for birds than it is for other organisms, because even rare birds are relatively conspicuous. A second proposal is that species able to exploit a wide range of resources become both widespread and abundant (Brown 1984, Gaston & Lawton 1990). Alternatively, the resources used by some species are widespread and locally more abundant than are those used by other species, again yielding a positive abundance–distribution relationship. Thirdly, several metapopulation models also predict positive correlations between population density within sites and the number of sites occupied (which influence geographical range) (Hanski 1991).

Earlier, I drew attention to the greater numbers of bird species per unit area in the tropics, and to their generally smaller geographical ranges, compared with high-latitude regions. If the relationship between local density and range size, found within high-latitude regions, holds across a wider span of latitude, one would expect tropical species to occur at lower average densities than temperate ones. This is the general impression — many species occur in tropical areas, but mostly at low densities. This is true even within particular groups, such as breeding waterfowl, in which densities generally decrease towards lower latitudes (Gaston & Blackburn 1996c). More generally, in a major compendium of available data, latitude alone explained 47% of the variance in population densities, with average densities increasing linearly from equator to poles in invertebrates, cold-blooded vertebrates, birds and mammals (Currie & Fritz 1993). At any one latitude, densities ranged over four or more orders of magnitude within these groups, but the overall trend for populations, on average, to have lower densities at lower latitudes was clear and broadly consistent with a decline in range size towards the tropics. Taken at face value, both trends (towards smaller ranges and densities) must make tropical species more vulnerable to human activities.

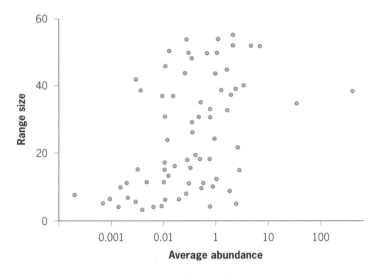

Figure 13.3 Range size in relation to average within-range abundance for 70 species of North American landbirds (Apodiformes, Piciformes, Passeriformes) in winter, as determined from Christmas Bird Counts of the National Audubon Society in North America. Abundance measured from the 10-year mean number of each species counted per party-hour of census effort ($\times$10) in each of 51 5° blocks, and range size was taken as the number of 5° blocks in which a species was detected. Correlation coefficient (r) = 0.43, P<0.01. Modified from Bock 1984.

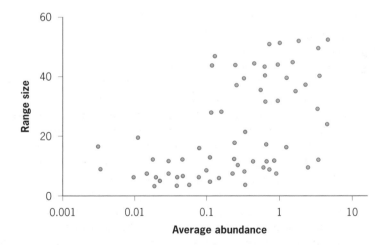

Figure 13.4 Range size in relation to average within-range abundance for 65 species of Northern American seed-eaters (Emberizinae and Carduelinae) in winter. Abundance measured as log ten-year mean number of each species counted per party-hour of census effort and range size as number of 5° blocks (T=51) in which a species was detected and which together cover much of North America. Correlation coefficient (r) = 0.54, P<0.01). Modified from Bock & Ricklefs 1983.

CONCENTRATIONS

Some highly mobile bird species, which breed over wide geographical areas, normally occupy only part of their range at any one time. This is because they depend on ephemeral habitats or food-supplies which are available in different regions in different years, so that in each year the birds have to concentrate wherever conditions are suitable at the time. Over a period of years, therefore, they may occupy a much wider area than they use in any one year. Included in this category are some seed-eaters such as crossbills that depend on variable tree-seed crops; some owls and raptors that depend on cyclic peaks in their rodent prey; some aridland and desert species whose food-supplies respond to sporadic rainfall patterns; and some wetland species dependent on temporary flood pools. All these species tend to occupy only a small part of their total potential breeding or wintering range in any one year (Chapter 17).

Other species occupy the same breeding or wintering range each year, but within their ranges they assemble in huge gatherings, whether at roosts, nesting colonies or foraging sites. Despite their large potential range, they may at times be concentrated in a small number of localities. Nightly roosts of more than a million individuals are regular in the Red-billed Quelea *Quelea quelea* in Africa, as well as in the Common Starling *Sturnus vulgaris* and Sand Martin *Riparia riparia* in Europe and North America, and in the Red-winged Blackbird *Agelaius phoeniceus* and other icterids in North America. The record for individual roosts, however, is probably held by the Brambling *Fringilla montifringilla*, whose nightly gatherings in areas of good Beech *Fagus sylvatica* crops in central Europe have sometimes numbered more than an estimated 20 million individuals, which fan out to feed over wide areas of surrounding landscape during the day (Jenni 1987).

In some species, even relatively abundant ones, the entire global population can at times be concentrated at a handful of sites. The world population of Dickcissels *Spiza americana* which breed widely over eastern North America, converge to winter in Venezuela, where they feed on cereal crops. In the 1990s, almost the entire population, numbering many millions, was thought to gather each night in only four main roosts (Attenborough 1998). Similarly, birds from the entire world population of less than 400,000 Spectacled Eiders *Somateria fischeri*, which breed on the coastal tundra of Alaska and northeast Siberia, gather to moult at only four sites far offshore in the Bering and Chukchi Seas, and then move to a similar small number of wintering sites, the biggest of which lies south of St Laurence Island in the Bering Sea (Petersen *et al.* 1999). When the sea freezes, the birds concentrate densely at breaks in the sea ice where they can dive for food. The entire sea surface is covered with birds, and from the air such sites appear as large dark marks on an otherwise white icescape.

Most species of pelagic seabirds range widely over the oceans, but while nesting gather in large colonies at traditional locations. Some species occur in concentrations of up to tens of thousands or hundreds of thousands of pairs, and in some species occasional colonies may consist of more than a million pairs (Chapter 8). Among other birds, colonies exceeding a million pairs occur in the Red-billed

Quelea *Quelea quelea* in Africa, in the Spanish Sparrow *Passer hispaniolensis* in southwest Asia, and also occasionally in flamingos that concentrate at suitable saline waters to breed. Some three million Lesser Flamingos *Phoenicopterus minor* live around the saline lakes in the African rift valley.

Some migrant waterfowl and shorebirds gather in enormous numbers at particular wintering and staging areas. For a few months in late summer, about a million Eared Grebes *Podiceps nigricollis* gather to moult on Great Salt Lake in Utah and another million on Mono Lake in California, having been drawn there from much of western North America (Storer & Jehl 1985, Jehl 1997). Similarly, the 50,000–70,000 Surfbirds *Aphriza virgata* that amass each spring in Prince William Sound are thought to include most of the North American population (Norton *et al*. 1990). The tendency of some waterfowl species to concentrate has perhaps been accentuated in recent decades by the establishment of refuges in regions in which most of the remaining wetland has been drained. Such concentrations increase the risks, for when disaster strikes it can remove a large portion of the population at one time (see Newton 1998a for examples).

CONCLUDING REMARKS

Although most of the distribution patterns described in this chapter are very general, with many specific exceptions, they carry some sobering messages for conservation. The first is that, on the scale of the earth's land areas, most bird species have small geographical ranges. The majority of species on continents occupy only a tiny fraction of the continental area concerned, and all the species confined to oceanic islands inevitably have restricted distributions. Most of the world's bird species are therefore vulnerable to small-scale events, such as local habitat destruction. Perhaps this has always been the case, but the frequency of destructive small-scale events is now very much higher than before, because of human influence. Secondly, because geographically restricted species tend also to have small local populations, they live in double jeopardy, and are likely to be more vulnerable than expected from small range or from small numbers alone (Lawton 1994). Thirdly, species occurring at low population densities are unlikely to occupy at any one time all suitable localities within their potential range. We cannot assume that habitat unoccupied at any one time is 'surplus to needs' and that it could be removed without reducing numbers in the longer term. If abundance and range size are correlated, then reducing the number of sites at which a species is found (for example by habitat destruction) might — by lowering densities at remaining undisturbed sites — reduce the overall population more than expected from the habitat lost. As in the Northern Spotted Owl *Strix occidentalis caurina* example, there may be a threshold level of habitat destruction that can drive a species regionally extinct, despite the survival of some remnant patches of habitat (Nee 1994). By the reverse argument, reducing overall population levels may lead inevitably to reduction in range size. The relationship between abundance and distribution is discussed further in the next chapter.

SUMMARY

Maps of bird ranges, as depicted in most field-guides and textbooks, give an inflated view of distributions, because they include many places where the birds concerned do not occur. Greater accuracy and information content has been achieved in recent 'Atlas' surveys, based on small-scale grid mapping, especially in those that also show variations in abundance across the range.

About 83% of all landbird species breed on continents (or on continents plus islands), while 17% breed only on oceanic islands. More than 93% of all continental landbird species breed only on a single continental land-mass, and most in only a small part of that land-mass. Most island species breed only on a single island or island archipelago.

Broad relationships are apparent between abundance and distribution in birds, especially within particular habitats and trophic groups: species that are most abundant among the birds found at particular count sites also tend to occur at the greatest proportions of count sites within a region, and often also have the largest geographical ranges. Other broad relationships exist between species range sizes on the one hand, and body size, latitude and geographical region on the other.

Species abundance varies greatly across a range. Whereas the individuals in some localities (sources) produce a surplus of young able to colonise other areas, those in other localities (sinks) produce a deficit, and are maintained by immigration. Habitat fragmentation can lead to population declines and range shrinkage much greater than expected merely from the amount of habitat lost.

Some bird species which depend on sporadic habitats or food-supplies normally occupy only part of their potential breeding or wintering range at one time. Other species gather in enormous concentrations at specific localities, mostly to nest or roost, but spread over a wider area to obtain their food.

Common Crossbill *Loxia curvirostra*, restricted to areas with fruiting conifers.

Chapter 14
Limitation of geographical ranges

Each bird species is found naturally only in certain parts of the habitable world. Each may be restricted to a particular land-mass or island, to a particular climatic or vegetation zone and to a particular type of habitat. For many bird species, apparently suitable conditions, in terms of climate and habitat, occur outside the existing range. So the question arises as to what prevents them from spreading more widely. This question is especially relevant to migrant species, which may occupy different parts of the world at different times of year, moving regularly back and forth between breeding and non-breeding ranges which may overlap or lie in completely separate regions. Yet most migrants breed in only a small part of their total annual range. Because birds are conspicuous and popular creatures, their ranges in some parts of the world are precisely known, and changes are quickly detected. Hence, the problem of what limits geographical ranges is perhaps more readily addressed in birds than in many other animals. Study is not straightforward, however, because information is needed from a wide geographical area, and also because different limiting factors may interact to determine a range boundary. Because of the nature and scale of the problem, moreover, experimentation is not generally feasible.

Broadly speaking, geographical ranges might be limited by: (1) ecological factors acting *within the existing range* to prevent a population from increasing, and spreading to adjacent areas; (2) ecological barriers (such as unsuitable climate or habitat) acting *at the boundaries of the existing range* to prevent spread; and (3) physical barriers, such as oceans or mountain ranges which curtail dispersal, and hence determine the maximum possible extent of a range. The same ecological factors that act to constrain population growth within the existing range might also, by acting more severely, prevent spread outside it.

These three types of limitation form a series of obstacles to spread. In the first place, range expansion depends on population growth within the existing range which produces the colonists needed to occupy new ground. A species could then spread outwards until constrained by ecological, or, ultimately, by physical barriers. In theory, any given species might be limited in different ways at different times and at different places, but whereas ecological factors might change through time, producing some ebb and flow in range boundaries, physical barriers are likely to remain fixed for much longer. Physical barriers do not constrain all species indefinitely, however, and from time to time occasional species manage to cross wide oceans or high mountain ranges to colonise new areas, as later examples show (Chapter 16).

DEPENDENCE OF RANGE ON POPULATION SIZE

In any species, abundance and distribution are not wholly independent attributes. In general, species with large world populations would be expected to occur over large geographical areas and vice versa, but in particular species it is often hard to separate cause from effect. Does a species that can live over a wide area thereby achieve a large population, or does a species that has a large population (for whatever reason) thereby spread over a wide area? Clearly abundance and range size are interrelated, but which has most effect on the other may vary through time, as explained below.

Associated change in abundance and range

Imagine that a species colonises a new land area and then, as it grows in numbers, it spreads to occupy the whole available area, stopping only when it meets an ocean boundary. Throughout its period of expansion, its numbers influence the size of its range: as its numbers increase, so does the area occupied. Then, as all available habitat becomes saturated and no further expansion is possible, numbers become capped by the amount of habitat or land area available. Initially numbers affect range size, and eventually range affects numbers. In the first phase, the area occupied is influenced by factors operating within the existing range (which affect growth in numbers) and in the second phase by factors outside the existing range (which prevent further growth and spread). At any one time, an avifauna is likely to contain both types of species: those whose ranges are limited by numbers, and those whose overall numbers are limited by range. The former category will include recent colonists and other species whose numbers are for various reasons below the level that the available habitat and land area would support.

To take some examples, in many seabird species, which have increased and spread in recent decades, growth of existing colonies usually preceded the establishment of the new colonies that led to range expansion. In the Northern Fulmar *Fulmarus glacialis* in the northeast Atlantic, numerical increase was first noted in Iceland in 1753, and later in the Faeroes (from about 1839), Norway (from about 1820) and then Britain (from about 1878). Within Britain, the species was once confined to St Kilda, but in 1878 it began to nest on Shetland. Once this new colony had reached about 50 pairs, the species began to establish further colonies, continuing to increase in numbers and distribution, and gradually spreading southwards round most of the British coastline. Taking the number of colonies as a measure of distribution, growth in total numbers and distribution occurred in parallel, following an exponential pattern **(Figure 14.1)**. Overall, the British population

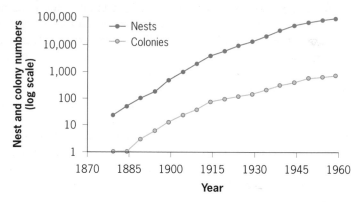

(a) Population trend, 1870–1960

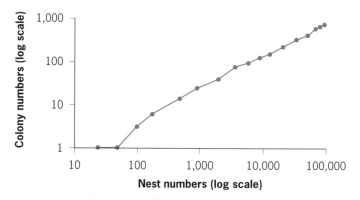

(b) Link between nest and colony numbers

Figure 14.1   Relationship between overall abundance and distribution in the Northern Fulmar *Fulmarus glacialis* in Britain during 1878–1960, based on surveys made mostly at 5-year intervals. Distribution is shown as the number of established colonies. Based on data in Fisher 1966.

(excluding St Kilda) increased from 24 pairs in 1878 to nearly 600,000 pairs in 1990, a mean rate of increase of 7–8% per year, and at the same time the range extended gradually southwards over 10° of latitude. By 1960, the species had also begun to nest in France, and by 1972 in Helgoland, off northwest Germany.

This expansion was attributed by Fisher (1966) to the growth of the fishing industry and the huge supplies of offal food that followed, first from whaling and then from fish trawling. No other obvious habitat changes occurred in this time and, without the fisheries, the Northern Fulmar might still be where it started. Substantial increases in numbers and distribution occurred over the same period in some other seabird species, in both Europe and in North America (for Northern Gannet *Morus bassanus*, see Nelson 1978; for Great Skua *Stercorarius skua*, see Furness 1987; for other species see Chapter 15). In each case, the strong implication was that range expansion depended on the availability of colonists, and hence was driven by population growth. Among landbird species, similar patterns have been described for increasing populations of the Dartford Warbler *Sylvia undata*, Bearded Tit *Panurus biarmicus* and Whooper Swan *Cygnus cygnus*, among others (Campbell *et al.* 1996, Gibbons & Wotton 1996, Nilsson *et al.* 1999).

Conversely, the White Stork *Ciconia ciconia* was once widespread in southern Sweden, but as one pair after another disappeared, the birds inevitably came to occupy a progressively smaller area, until all were gone **(Figure 14.2)**. Yet local climate and habitat remained much as before, and the region seemed to have been vacated solely through lack of takers. In any region, while some species are expanding in range, others may be contracting. Certain landbird species that in recent years colonised an island in the Baltic (= range expansion) were those whose numbers had increased on the Finnish mainland, while those that disappeared from the island (= range contraction) were those whose mainland numbers had declined (Haila *et al.* 1979). All these various species provide examples of associations between changes in numbers and changes in range, and circumstantial evidence that change in numbers drove the change in range, rather than vice versa.

Species comparisons

Further insight can be gained from comparing the numbers and distributions of closely related species at the same time in the same land area (Newton 1997). As a result of co-ordinated count and mapping schemes, fairly precise estimates have

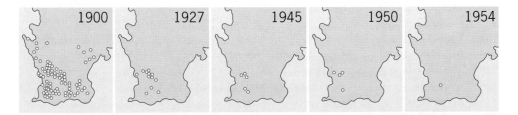

Figure 14.2   Progressive disappearance of the White Stork *Ciconia ciconia* from its breeding range in southern Sweden. Each spot represents a breeding locality. From Curry-Lindahl, in Udvardy 1969.

been made of both total numbers and total distributions of all bird species in Britain. For certain groups of related species, numbers and distributions were assessed independently by different surveys undertaken at about the same time **(Figure 14.3)**. In all these groups, a clear correlation emerged between total numbers and distribution: those species that had the greatest overall abundance bred in the greatest number of 10 km squares. In fact, overall abundance and spatial distribution were so closely coupled that any one measure could have been predicted fairly accurately from the other. In seabirds, the relationship was less tight, probably associated with their colonial nesting.

A second point to emerge was that local density (as measured by mean numbers per 10 km square) was also broadly related to distribution: those species with the highest mean local density also occupied the greatest number of squares. In other words, overall abundance, mean local abundance and distribution were all inter-correlated.

Other striking findings emerged by comparing the results for 1988 with those from similar surveys done 20 years earlier (Newton 1997). In this time interval, several species had undergone substantial changes in status. In each such species, increase or decrease in overall numbers was accompanied, at the 10 km scale, by a corresponding expansion or contraction in distribution. Moreover, these changes ran more or less parallel to the mean trend line derived from the inter-species comparisons **(Figure 14.3)**. Yet other species, whose overall numbers were similar in both surveys, also occupied approximately the same numbers of squares (but not necessarily the same squares) on each survey. This provided further evidence for the close link between abundance and distribution. It reflects the fact that the majority of the species concerned were not occupying the total land area of Britain, or even the total of suitable habitat within that land area. Most were victims of past human culling, notably various waterfowl and raptors.

How do these various relationships come about? In any species, a large total population could lead to both high local abundance and wide distribution. If density in any one locality (or 10 km square) is limited (for example by habitat or food-supplies), further increase in numbers could force birds to spread over a larger area (to more squares), and to have high local abundance within each locality (square). By way of analogy, imagine water dripping from a tap into an egg tray. Initially, the water settles in one or a few compartments close together. Then as more water drips in, some compartments begin to fill, while overspill occurs to others. As the flow of water continues, more and more compartments become filled, partly filled compartments form a smaller proportion of the total, and eventually all available compartments are filled to capacity, and no further water can be accommodated. An expanding bird population could be viewed in the same way, as areas are successively occupied and filled to capacity (Newton 1997). On this basis, total numbers, average numbers per square, and number of squares occupied, would all increase in parallel and, at whatever stages the population was examined, all three measures would be correlated. An exception occurs where birds spread from areas with high carrying capacity (supporting high densities) to areas with low carrying capacity (supporting low densities), in which case mean local densities would decline with increase in overall numbers and range size (Gregory 1998).

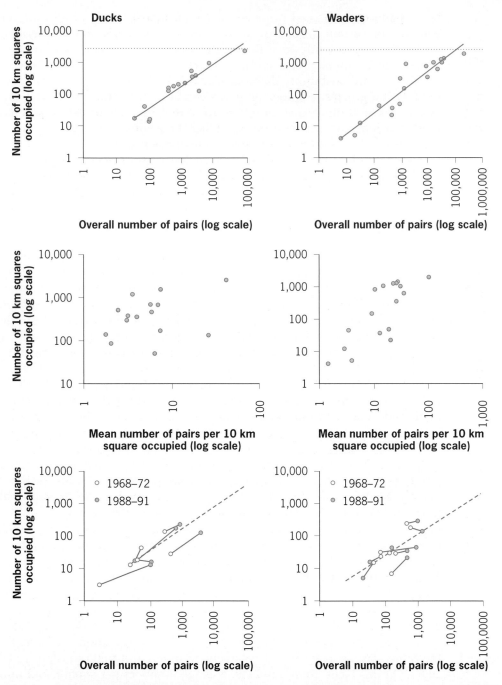

Figure 14.3 Relationship between abundance and distribution within four groups of breeding birds in Britain. For each group, upper: overall abundance and distribution, 1988–91; horizontal dotted line shows total number of 10 km squares in Britain (=2,830); centre: mean local abundance and distribution; lower: changes in overall abundance and distribution of certain species between 1968–72 and 1988–91 shown in relation to the trend line (dashed) from the upper graph.

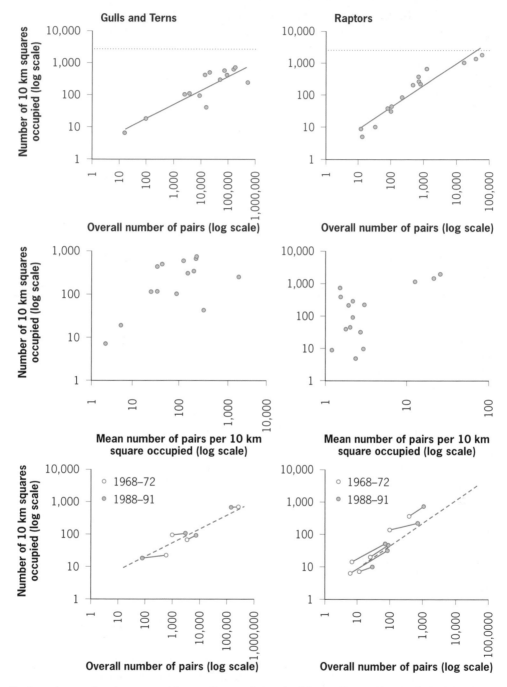

Overall abundance is shown as the estimated total of breeding pairs in Britain, distribution is shown as the number of 10 km squares in which breeding was recorded, and mean local abundance is overall abundance divided by the number of 10 km squares occupied.

From Newton 1997 — Sources: Sharrock 1976 and Gibbons *et al.*1993 for all species, plus Owen *et al.* 1986 for waterfowl, Cramp *et al.* 1974, Lloyd *et al.* 1991 for seabirds, Newton 1994a for raptors.

The expansion of overall numbers and range on a large scale has parallels with similar events at smaller scales. The proportion of any patch of habitat that is occupied by a territorial bird species depends on the numbers present. A single territory may take up a small part of a habitat patch, but as more territorial pairs settle, they occupy further parts until the whole patch is full. The same phenomenon can be seen over even smaller areas, for example in the growth of a seabird colony. African Penguins *Spheniscus demersus* recolonised Robben Island near Cape Town in 1983, when nine pairs nested, increasing to over 2,000 pairs in 1992. At the same time, the area occupied increased from only a few square metres in 1983 to some 35 ha in 1992 (Crawford *et al.* 1995). This is a familiar phenomenon in the growth of many seabird colonies which, if unchecked, can expand to occupy all suitable areas until the nesting island or cliff is replete with nests. In uniform terrain, the pairs are regularly spaced, reflecting the territorialism that limits density. Clearly, the factors that operate to give the link between abundance and distribution act over a range of spatial scales.

Similar relationships between abundance and distribution have been found in other studies of birds, as well as of other animals (Fuller 1982, Bock 1984, 1987, Bock & Ricklefs 1983, Ford 1990, Kouki & Häyrinen 1991, Lawton 1993, 1994, 1996, Gaston 1994, Blackburn *et al.* 1997a, b, 1998a, Venier & Fahrig 1998, Gaston *et al.* 2000). They are evidently a fundamental feature of animal populations at a range of spatial scales. In addition to the 'egg-box' analogy, other explanations of abundance–distribution relationships have been mooted, but have little or no empirical support (for discussion, see Gaston *et al.* 2000).

Relationships between breeding and wintering ranges

Among birds, correlations between overall abundance, local abundance and total distribution hold in winter and in summer, and in both residents and migrants. In those populations that are resident in the same areas year-round, overall numbers are determined entirely by conditions prevailing within those areas. However, in migrant species, which occupy different regions at different times of year, overall numbers may be influenced by conditions in more than one region. They may be limited in one region so severely that the species can never occupy another completely. The Bristle-thighed Curlew *Numenus tahitiensis* in North America breeds in only a small part of Alaska, in habitat that is much more widely distributed than the bird itself, but winters in the restricted coastal areas provided by certain Pacific Islands, where it could never achieve the abundance necessary to occupy a large part of North America. It provides an example where a small wintering range could limit numbers and thereby the extent of the breeding range. Short of finding new wintering areas, the species is doomed to localised rarity. If this type of situation were common in migrants, whether the main limitation was in breeding or in wintering areas, it could account for the general correlation found among birds between the sizes of breeding and wintering ranges, as discussed in Chapter 18.

Annual variations in distribution are usual in some irruptive migrants, in which the numbers of individuals reaching the furthest parts of their wintering range often fluctuate widely from year to year (Chapter 18). Birds move further than usual when their overall numbers are high, or when their food is scarce in the main range. This is evident in some northern birds of prey, such as Snowy Owl *Nyctea scandiaca* and Rough-legged Buzzard *Buteo lagopus*, which spread furthest south every 3–4 years, when their rodent prey are scarce, and in some finches, such as Eurasian Siskin *Carduelis spinus* and Common Redpoll *Carduelis flammea*, which spread furthest south when their favourite tree-seeds are scarce relative to numbers in the regular range (Lack 1954, Svärdson 1957, Newton 1972, Bock & Lepthien 1976, Köenig & Knopps 2001). In all such species, therefore, the southern limit to the wintering range varies with total population size and with food conditions further north, with birds spreading over a larger area in some years than in others. At least some of these southern areas may be suitable for occupation every year, yet are left vacant in years when, because of low numbers or abundant food, the existing population can be accommodated further north. It is thus not numbers as such that are important in these species, but numbers in relation to food-supply, for changes in either numbers or food-supply could promote year-to-year changes in distribution.

Again there is a parallel with habitat, in that many species occupy chiefly their favoured habitats in years of low numbers, and spread increasingly to poorer habitats in years of high numbers (Newton 1998a). Examples include the spread of breeding Great Tits *Parus major* or Chaffinches *Fringilla coelebs* from broad-leaved to coniferous woodland (Kluijver & Tinbergen 1953, Glas 1960), or the spread of wintering Brent Geese *Branta bernicla* from mudflat to saltmarsh to farmland as their numbers grew (Ebbinge 1992, Vickery *et al.* 1995). The interplay between population size, habitat and range is further shown in the fact that some species occupy a wider range of habitats near the centre of their range where overall densities are higher than nearer the edges where overall densities are lower.

This leads to a more general point about boundaries of breeding ranges. While populations at the periphery of a species range define the boundary, these populations may be limited in ways different from those nearer the centre, and may perish without a continual influx of immigrants (Chapter 13). In theory, some species might extend well beyond their core range, into marginal areas where they could maintain their numbers only if continually reinforced by immigration from nearer the core. The boundary may thus fluctuate in response to events happening nearer the centre and it may be impossible to discover what determines a range boundary from studies at the boundary alone. In such cases, two types of range boundary would exist: a 'functional' boundary within which reproduction equals or exceeds mortality, and an outer 'empirical' boundary, enclosing an outer zone where the reverse occurs, and where persistence is dependent on immigration (Emlen *et al.* 1986). In our present state of knowledge, mapped boundaries must usually be considered as empirical ones, liable to change on short time scales.

ECOLOGICAL BARRIERS: GENERAL SURVEYS

As a species spreads outwards, it will sooner or later meet conditions that prevent its further spread. The most interesting questions are raised by those many species whose ranges fall permanently short of obvious topographical barriers. The boreal forest runs without break from one end of Eurasia to the other, yet not all boreal bird species do so, and many are found only at one end or the other. Similarly, despite the extent of the Amazonian rain forest, most bird species have restricted distributions within it. Why do they not spread perhaps 10 km, 100 km or even 1,000 km beyond their present ranges? Many of the ecological factors involved occur not as sharp barriers, but as gradients, which interact with other factors to influence range boundaries, making it progressively more difficult for a species to persist the further it moves from its core range. Such factors include climatic and habitat conditions, together with the same factors that can limit numbers within the range, such as food and nest sites, predation and parasitism, and competition from other species.

A first step in finding which ecological factors might limit existing bird ranges is to investigate, over as wide an area as possible, whether range boundaries correlate with particular environmental features. One of the most extensive sources of information on bird distributions comes from the Christmas Bird Counts organised by the National Audubon Society in North America. These counts have been made annually since 1900 and now cover all the United States and southern Canada. Using the counts from ten years (1962–71), Root (1988c) mapped the winter distributions of 148 landbird species, in order to check how many of their range boundaries coincided with particular climatic and habitat variables. Northern, eastern and western boundaries were examined separately, but southern boundaries were ignored because most species extended south of the United States, where no counts were made.

Within this vast study area, range boundaries for the majority of bird species were associated with one or more of the six environmental variables examined (**Figure 14.4**). Of 113 species whose northern limits occurred within the study area, in 60% these limits coincided with a particular January isotherm, in 50% with a particular frost-free period and in 64% with a particular vegetation boundary. About 13% of species showed no association with any of the features examined at their northern boundary. Of 78 species whose eastern limits fell within the study area, 63% had eastern boundaries coinciding with a vegetation boundary and 40% with a precipitation measure, while 29% showed no associations with the features examined. Of 50 species whose western limits fell within the study area, 46% had western boundaries coinciding with vegetation boundaries, 36% with a precipitation measure and 38% with elevation, while 20% showed no associations. Very few associations were expected to result from chance, so the vast majority were assumed to have some biological significance to the species concerned, though not necessarily through direct causal links. Overall, the findings suggested an overwhelming importance of climate and habitat in limiting the winter distributions of North American birds, but different factors often acted along different parts of a range perimeter. Alleviation of any one such factor should allow expansion at least from that part of the boundary.

ECOLOGICAL BARRIERS: CLIMATIC CONSTRAINTS

Climate can affect bird distributions by acting directly on the birds themselves, or indirectly by acting on their habitats, food-supplies and natural enemies, or by altering the balance of competition with other species. Direct effects can result from the lowering or raising of body temperature or from dehydration. Some

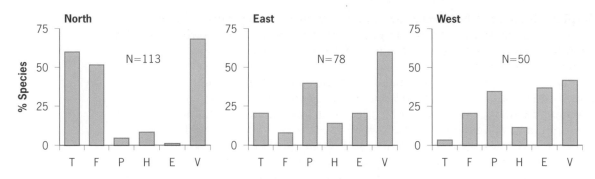

Figure 14.4   Associations between northern, eastern and western range boundaries of 148 North American landbirds and various environmental factors.

In each case an association was assumed if, along an entire boundary, the average deviation of the boundary from one environmental feature was less than 1° of latitude on either side (about 115 km) or 1° of longitude on either side (about 90 km).

N = number of species whose range boundary fell within the study area (United States or southern Canada); T = average minimum January temperature (with a contour interval of 6°C); F = mean duration of frost-free period (with a contour interval of 30 days); P = mean annual precipitation (with a contour interval of 21 cm); H = average annual general humidity (with a contour interval of 13 cm); E = elevation (with contour levels starting at 13 m and each successive level being double the previous one); V = natural vegetation boundary in the absence of human interference. The range boundary of a given species may correspond with more than one isopleth, because different climatic, vegetation and other isopleths often coincide with one another. Modified from Root 1988c.

birds cope with extremes of cold or heat, in arctic or desert regions, by remaining in the open for no more than a few favourable hours each day, spending the rest of their time in sheltered or shaded sites, but this strategy can be pursued only where sufficient high-quality food is available, so that nutritional needs can be met in a short time each day. An interaction between ambient temperature and food-supply thus influences whether a given species can persist in a climatically extreme area, and in theory a change in either temperature or food-quality could lead to an expansion or contraction of range.

Different species of birds vary greatly in their ability to cope with cold or heat. We would not expect a tropical bird, released in the arctic winter, to survive the low temperature. Nor would we be surprised if an arctic resident, released in the tropics, died from the effects of heat. The metabolic and other physiological adaptations of birds, like those of other warm-blooded animals, are adapted for the climate regimes in which those birds normally live. The term 'thermoneutral zone' is used to cover that range of outside temperatures over which the basal metabolic rate of a bird at rest remains stable. This zone covers higher temperatures in tropical than in arctic birds. To take some extreme examples, in the Spinifex Pigeon *Geophaps plumifera*, which lives in one of the hottest places on earth in the central Australian desert, the metabolic rate stays constant over the ambient temperature range 35–45°C (Withers & Williams 1990). In contrast, the Emperor Penguin *Aptenodytes forsteri*, which lives in one of the coldest places on earth in Antarctica, the thermoneutral zone spans the temperature range $-10°C$ to 20°C (Pinshow *et al.* 1976). Within the thermoneutral zone, birds allow for changing temperature mainly by adjusting their insulation, as by sleeking or fluffing their feathers.

Whatever the species, metabolic rate increases progressively as ambient temperature falls below the thermoneutral zone or rises above it. Moreover, the standard metabolic rate (SMR) of birds correlates broadly with climate in the area of origin, and allowing for body weight, tends to be higher in birds from cold climates and lower in tropical birds. SMR changes, on the average, by 1% per degree of change in latitude. This overall influence of climate is modified by lifestyle: for example, the SMR of tropical birds that forage in full sun averages less than that of other tropical birds that feed in the shade (Weathers 1979). This adaptation enables birds of all regions to maintain their body temperature under, what for them, represents an extreme of heat or cold. However, because there are limits to the extent to which individual birds can adjust for temperature extremes, we should expect that, for any species, ambient temperature might limit survival chances (of adults or chicks), and hence the potential geographical range. The question of interest is how many species are normally up against these climatic limits, in one part of their range or another, and how many are held well within these limits by other factors, such as lack of habitat.

Effects of cold

Further analysis of the North American winter bird distributions mentioned above indicated that temperature may have directly influenced the northern limits of certain species (Root 1988b). In a total of 62 species (51 passerines and 11 non-passerines), the entire length of the northern boundary was associated with a particular January isotherm, different species with different isotherms **(Figure 14.5)**. Of these 62 species, detailed studies of winter physiology had been previously made of 14 of the passerine species. In all these 14 species, the metabolic rate at the January temperature prevailing at the northern boundary (NBMR) was calculated at about 2.5 times the basal metabolic rate (BMR). Although the BMR varied greatly between species, the NBMR to BMR ratio did not, as the mean of the ratio was 2.47, with a standard error of 0.07. This strong association between NBMR and BMR implied that the winter ranges of these 14 species were restricted to areas where the extra energy needed to compensate for a colder environment was no greater than about 2.5 times their BMR. This held, even though the boundaries of the various species were separated by hundreds of kilometres, and their diets and body masses varied widely. Although no physiological studies had been made on the remaining species, their basal metabolic rates could be estimated roughly from their body weights. Even by this crude technique, the NBMR worked out, on average, at 2.64 times the BMR, a figure not dissimilar to that for the species for which precise measures were available.

Many of the species concerned were present in only low densities near the boundaries of their ranges, and reached maximum densities in warmer areas further south. In seven species that could be studied, high-density populations (defined as >80% of the maximum densities recorded) were limited to regions where individual energy needs were no more than 2.13 times the BMR (range

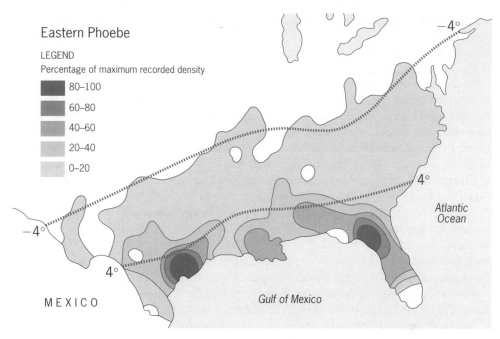

Figure 14.5 Correspondence between the northern distributional limit of the Eastern Phoebe *Sayornis phoebe* in winter, and the −4°C isotherm of average minimum January temperature. Shading depicts density as a percentage of the maximum recorded. Modified from Root 1988a, based on data from Christmas Bird Counts.

2.08 –2.34). This finding gave a further hint that the winter ranges of these North American species were influenced by the additional energy expenditure needed to keep warm. On a broad scale, then, physiological demands imposed by winter cold seemed to place constraints on birds at their northern range boundaries, and may thereby have restricted their northern limits.

Because basal metabolic rate is strongly related to body mass, large species should occur further north than small ones. According to Root & Schneider (1993), this is indeed the case among species whose northern limits are associated with temperature isotherms. Those small species that form exceptions to this generalisation all have special energy-saving mechanisms that enable them to extend further north than expected from their metabolic rates. Some such species have physiological adaptations, such as an ability to reduce their body temperature at night through torpor (as in Black-capped Chickadee *Parus atricapillus*), while others have behavioural ones, such as roosting in cavities or in clusters (Brown Creeper *Certhia familiaris* and others). Some small birds, such as Common Redpolls *Carduelis flammea* (weight 13 g), winter regularly as far north as Fairbanks in Alaska (latitude 65°N), where the mean minimum daily temperature during January is −27°C, dropping occasionally to −60°C. Redpolls have a high metabolic rate during winter and well-insulated plumage, but they also rely on a rich food-source, feed at low light values, fill their gullets with food at dusk, roost in protected sites such as holes in the snow, and can also become hypothermic at night (Repasky 1991).

In many northern species, therefore, cold seems to restrict the winter distribution. Cold can also be important in the breeding season, however, though not necessarily to the adults. The northern boundary of the Goldcrest *Regulus regulus* in the breeding season has been correlated with a particular isotherm known to be critical for the survival of nestlings (Haftorn 1978). Similarly, the ranges of several species of European ducks correlate well with the cold hardiness of day-old ducklings (Koskimies & Lahte 1964). During their first day or two of life, when they live off their yolk reserves, ducklings of different species were kept in captivity at different temperatures to find how long they could stay warm. Mallard *Anas platyrhynchos* and Eurasian Teal *Anas crecca* ducklings were unable to maintain body heat at air temperatures less than 10°C, and at 0–2°C they cooled rapidly. In contrast, more cold-tolerant species could last more than three hours at 0–2°C without cooling. Common Eider *Somateria mollissima* ducklings could endure an air temperature of −10°C with the same relative metabolic effort as Mallard would have needed to expend at 15°C. Comparing species, a clear correlation emerged between cold-hardiness and geographical range, with the most cold-resistant species having the narrowest, more northerly distribution.

The cost of cold hardiness is that, relative to body size, it requires a high metabolic rate which can be maintained only in conditions of abundant, high-quality food. It is also associated with a low degree of heat tolerance, so that high temperatures can become limiting in the southern parts of a range. Cold-sensitive species, such as Mallard, can breed in the north, but their young have to be brooded more, which restricts the feeding time, so near their northern limits such species are likely to be even more confined to areas with abundant, high-quality food. Elsewhere, they would produce too few young to maintain population levels. Once again, we see an interaction between cold tolerance and food-supplies which could restrict distribution.

In addition to adults and young, the eggs of different species also differ in their response to temperatures outside the normal range. Resistance to cold exposure is an inherent trait, and evident mainly in high-latitude species. Penguins have both a lower optimum and a broader range of acceptable incubation temperatures than other studied species. All such evidence is essentially correlative, and although it suggests a physiological constraint on range, it remains possible that some other factor limits the range, and that physiology is adjusted by natural selection to the conditions within that range, becoming limiting only in more extreme conditions.

Turning now to the indirect effects of a cold climate, the northern limits of the wintering ranges of many northern seabirds, waterbirds and waders lie just south of the line where their feeding grounds ice over, and many such species occur further north in mild winters, when feeding grounds are open, than in severe winters when they are closed. There are many examples of such birds staying north of their usual limit in special circumstances, such as the Mute Swans *Cygnus olor* that winter in Regina, Saskatchewan, where the January air temperature averages −10°C but where the occupied lake is kept open by hot water discharged from a power plant (Root 1988a). Similarly, in areas with regular snow cover, ground-feeding landbirds are forced to move south for the winter, unless they have some means of digging under the snow. In most such species, it is probably not the climatic factors acting directly on the birds themselves that limit the range, but the food-shortage that results from climate. Each year, in both Europe and

North America, occasional warm-climate bird species remain well north of their usual winter range. If these birds could not tolerate the cold they could not survive to January, yet they often do. Usually they are seen near an unusual food-source, such as a well-stocked garden feeder, hinting again that food rather than temperature limits the winter range. In both Europe and North America, a new reliable winter food-supply, provided at garden feeders, has enabled several species to winter further north than formerly (Chapter 15).

Winter weather clearly imposes a northern limit to the range of some resident species, whose numbers near the boundary fluctuate from year to year, according to winter severity. Some such species extend northwards during runs of mild winters, only to retreat again in colder periods. Examples include the Barn Owl *Tyto alba*, Little Grebe *Tachybaptus ruficollis*, Winter Wren *Troglodytes troglodytes*, Bearded Tit *Panurus biarmicus*, Stonechat *Saxicola torquata* and Razorbill *Alca torda* in northern Europe, and the Dartford Warbler *Sylvia undata*, Sardinian Warbler *S. melanocephala* and Cetti's Warbler *Cettia cetti* further south in mid-latitude Europe. While individuals of some such species may die from cold as such, most probably starve because their food becomes less available. The northern boundaries of their distributions are thus determined by winter mortality. Where the frequency of hard winters exceeds the powers of recovery, any colonists from further south are continually eliminated.

Effects of heat

The role of heat in restricting bird distributions has been examined chiefly in desert species which show varying degrees of physiological adaptation to their extreme environment. The main problems concern the extra energy and water needed to counter overheating. In the Palaearctic region, the Chukar *Alectoris chukar* has a wide distribution which extends into semi-arid areas, whereas the related Sand Partridge *Ammoperdix heyi* is found only in extremely hot, dry deserts (Frumkin *et al.* 1986). Where in some parts of Israel the two species occur together, the Chukar was found to have a lower thermoneutral zone than the Sand Partridge (28–34°C versus 30–>51°C), and could tolerate less extreme ambient heat (death at 43°C versus >51°C). The failure of the Chukar to extend into hotter areas was therefore attributed to its inferior powers of thermoregulation at high temperatures.

Further study revealed that the distribution of Chukars in Israel was influenced by limitations imposed on their activities by the physical environment (Carmi-Winkler *et al.* 1987). The birds spent twice as much time foraging in winter (7–8 hours per day), when they ate mainly natural vegetation of low energy content, than in summer (3–3.5 hours per day), when they ate high-energy foods from cultivated land. In summer, the birds were active mainly in the early morning and evening, and spent 7–8 hours in the hottest part of the day resting in shade. As outside temperatures rose over a period of weeks, the birds spent progressively more time in the shade and less in foraging. The Chukars were found in two types of habitat, either in wadis where the natural vegetation was relatively dense, thus offering substantial poor-quality food in a shaded environment, or in modified habitat around human settlements in which rich food-supplies were available. In the first habitat, Chukars could fulfil their energy needs in shaded areas, and were thus not exposed to the full sun, although they had to feed for long periods each day. In the second habitat, the presence of concentrated, high-quality food enabled Chukars to meet their energy needs within short periods before rising temperatures forced them to seek shade for most of the day. Altered habitat around human settlements thus enabled Chukars to survive and breed in areas otherwise closed to them. In general, the birds were limited to places where conditions allowed them to be active for long enough each day to obtain their food without overheating. There was thus an interaction between the physiological heat tolerance of the birds, the structure of their habitat and the quality of their food-supplies. In theory, changing any one of these components could influence survival and lead to changes in distribution.

Heat has also been suggested as directly restricting the range of several other bird species, including the Black-billed Magpie *Pica pica* in North America. Because individuals of this species die at 40°C, the species is confined to cooler areas. A related species, the Yellow-billed Magpie *Pica nuttali*, is more heat-tolerant and lives in hotter areas (Hayworth & Weathers 1984). Again, however, such findings provide no more than circumstantial evidence for a direct effect of temperature. It remains possible that the two magpies are restricted by other factors to regions of different temperature, and that the birds have evolved a temperature tolerance to match.

For eggs, the range of endurable heat above the optimal for incubation is rather limited, and overheating is more harmful to developing embryos than is cooling. Birds in hot environments have to employ both behavioural and physiological mechanisms to prevent their eggs from overheating, and attentive behaviour at the nest is especially important in this respect.

Effects of water shortage

That lack of drinking water can limit bird distributions is shown by the failure of most species to penetrate far into deserts, thus enabling the development in arid areas of special 'drought-adapted' bird species, which can obtain all the water they need from their food, or can go for long periods without drinking (Serventy 1971). Many gallinaceous birds of arid areas, which have dry seed-diets and limited powers of flight, are known to be limited in distribution by the local occurrence of drinking water (e.g. Brennan *et al.* 1987, Borralho *et al.* 1998). Moreover, when water is made available in arid areas by human action, non-desert species usually move in, and many of the true desert species decline. Such changes occurred in parts of Australia following the provision of drinking sites for cattle and sheep (Saunders & Curry 1990). These findings imply not only that many bird species are restricted in distribution by lack of drinking water within their daily flight range, but also that, in the presence of water, some desert species are restricted in distribution by competition from other, more water-dependent birds.

Often heat and water shortage act in combination. The extreme conditions of some deserts, with sizzling temperatures, intense solar radiation and scarcity of surface water, present problems for many organisms, including birds. In the Australian desert, 71 (60%) of 118 local species were never seen to drink, or visited water on fewer than half the days when temperatures exceeded 25°C (Fisher *et al.* 1972). Some carnivores and insectivores apparently obtained all their water from food, but other species, especially seed-eaters, had to drink daily and were absent from areas lacking surface water. In widespread drought, such species could die of thirst, while in extreme heat they could die even in the presence of water if their evaporative cooling mechanism could not cope (for examples, see Finlayson 1932). Normally, their distributions are continually changing in accordance with sporadic rainfall and the availability of surface water.

More often, rainfall acts on birds via the food-supply. Drought may reduce or prevent production of flowers and fruits and hence limit the numbers of insects and other animals eaten by birds. It thus affects almost the whole avian community: nectar-eaters, fruit-eaters or insectivores. It is not difficult to imagine that, with a run of dry years, too few young would be produced to maintain population levels, with inevitable decline and range contraction.

While extreme dryness can exclude some bird species, extreme wetness can do the same. In mid-to-high latitude regions, many bird species show poor breeding success in years with heavy rain in the incubation and nestling periods (Newton 1998), and Capercaillie *Tetrao urogallus* in Britain may be limited in distribution by high rainfall, which leads to poor chick survival (Moss 1986). Further evidence for the role of climate in the limitation of bird ranges comes from the long-term changes in range boundaries that have accompanied periods of climate change, as discussed in Chapter 15.

The problem of short summer seasons

Another type of physiological constraint operates at high latitudes, namely the short favourable season. This could prevent some species with long breeding cycles from nesting there, and hence restrict their breeding to lower latitudes. All species of geese feed primarily on herbaceous vegetation, even as goslings. The larger species have longer growth periods which restrict them to breeding in temperate latitudes, while smaller species can extend to progressively higher latitudes, the small Brent Goose *Branta bernicla*, Barnacle Goose *B. leucopsis* and Red-breasted Goose *B. ruficollis* nesting furthest north **(Table 14.1)**. The different races of the Canada Goose *B. canadensis* show a similar pattern in North America, with the largest races that have the longest breeding cycles nesting furthest south, and the smallest races furthest north.

The period needed to complete a breeding cycle is probably, therefore, a major feature preventing any one goose species from extending its breeding range northwards, for the further north a species breeds, the smaller the proportion of summers in which it can raise young (Newton 1977). The same constraint does not apply to moult which takes much less time than breeding, so the non-breeding individuals of many goose species migrate to moult well north of their breeding areas (Salomonsen 1968). They thus gain the advantages of the longer days and later flush of plant-growth at higher latitudes, and incidentally do not then compete for food with families of their own species further south.

In other bird species some aspects of the breeding cycle (notably nestling and post-fledging periods) are accelerated by some 10–15% in high arctic compared to more southerly populations, and in some species that moult on their breeding areas, feather replacement is also more rapid in the north (Newton 1972). Other migratory species postpone their moult until they reach a staging or wintering

Table 14.1 The duration of the breeding cycle (days) in various swans and geese of the northern hemisphere. As far as possible, species (or races) are listed according to their breeding distributions north to south. Details from Newton 1977 and Cramp & Simmons 1977.

Species	Usual clutch size	Incubation period (days)	Fledging period (days)	Total breeding period of individual[1]	Maximum breeding latitude (°N)
Bewick's Swan Cygnus columbianus	3–5	29–30	40–45	74–84	72
Whooper Swan Cygnus cygnus	3–5	35	78–96	118–140	70
Mute Swan Cygnus olor	5–8	35–41	120–150	164–205	62
Brent Goose Branta bernicla	3–5	24–26	45(?)	72–76	82
Red-breasted Goose Branta ruficollis	4–6	24	40	67–69	77
Snow Goose Chen caerulescens	4	22–23	42	68–69	82
Ross's Goose Chen rossii	4	22	–	–	73
Barnacle Goose Branta leucopsis	4–5	24–25	40–45	68–75	80
Pink-footed Goose Anser brachyrhynchus	5–6	27–28	40–43 (?)	72–77	80
Emperor Goose Chen canagica	5–6	25	–	–	71
Greater White-fronted Goose Anser albifrons	5–6	27–28	46	77–81	75
Bean Goose Anser fabalis	4–5	27–29	52	82–86	77
Canada Goose Branta canadensis[2]	5–6	28	65–70	99–104	65
Greylag Goose Anser anser	4–6	27–28	50–60	81–94	71
Lesser White-fronted Goose Anser erythropus	4–6	25–28	35–40	64–74	70

[1]Calculated assuming that swan eggs are laid at 2-day intervals and goose eggs at 1-day intervals.
[2]Calculated for one of the larger, more southern, races, B. c. moffti.

area. This allows species that breed in the high arctic to raise young in most years, but there is still a limit to how much the breeding cycle can be shortened. Many potential colonists with longer breeding cycles could thereby be excluded from high-latitude areas.

Some seabirds provide a partial exception to this generalisation, because they can establish themselves on snow-free cliffs well before the spring melt and, if necessary, travel out beyond the ice to reach their open-water feeding areas. On the antarctic peninsula, Snow Petrels *Pagodroma nivea* fly from open water over the ice to reach their breeding colonies on mountain summits up to 200 km inland, while Adélie Penguins *Pygoscelis adeliae* walk up to 100 km over the sea ice to reach their nesting colonies blown clear of snow. The strangest of all, however, is the Emperor Penguin *Aptenodytes forsteri*, which nests in winter at a few traditional sites on the sea ice itself. The sea freezes from the land outwards and melts from the outer edge inwards. Hence, because of their long breeding season, Emperor Penguins cannot nest on the edge of the ice, but have to walk inwards from the edge up to several tens of kilometres to where the ice lasts long enough for them to complete a breeding cycle (Williams 1995). This penguin, standing about a metre high, forms huddles while incubating, as protection against biting winds with temperatures down to $-70°C$. The egg is carried on the feet, heated and surrounded by a flap of belly skin that forms a pouch. Soon after the female has laid, the male takes over, while the female walks over the ice to the sea. The males incubate for the winter, lasting without food for about 115 days. As the sun returns, so do the females, walking 150 km across the ice to relieve their mates, at about the time the chicks hatch. The starving males then face a similar walk to the sea. By starting in winter, Emperor Penguins can nest every year despite their long breeding cycles, and the young can fledge before the ice breaks beneath them. Winter breeding requires a special adaptation of massive food-storage, and an extremely long fast, which probably explains why only this one large species overcomes the problem of a short summer season in this way.

Other antarctic seabirds with long breeding cycles nest only on islands that are far enough north to escape the ice, such as South Georgia, which provides breeding places for the Wandering Albatross *Diomedea exulans* and others. Similarly, King Penguins *Aptenodytes patagonicus* need continuous access to the sea, as their breeding cycles last more than a year. Pairs usually raise two young per three years, and at any one time colonies contain birds at all stages of breeding. Hence, in all these high-latitude species without special adaptations, duration of the breeding cycle seems to limit the latitude to which they can breed. Once the ice melts, they range further south to even higher latitudes to feed.

ECOLOGICAL BARRIERS: HABITAT CONSTRAINTS

Most bird species occupy fairly specific habitats and are found only where their habitats occur. Major habitats, such as deserts, grasslands or coniferous boreal forests, each have their own species which seldom or never breed elsewhere. Unlike climatic factors, which vary continuously across the globe, habitat boundaries are often fairly sharp, giving rise to clear range boundaries, as in some of the North American species studied from Christmas Bird Count data, discussed above.

The Brown-headed Nuthatch *Sitta pusilla* of the southeastern United States exemplifies this sort of boundary in its restriction to pines throughout its range (Terborgh & Weske 1975). Its northern limits coincide with the last outposts of certain species of pines (especially *Pinus tadea*) and bear no consistent relationship to any other environmental variable. Similarly the Sage Grouse *Centrocercus urophasianus* is found nowhere except in the sagebrush country of Western North America. The Giant Conebill *Oreomanes fraseri* is endemic to the high Andes of South America, being patchily distributed in islands of open woodland dominated by *Polylepis* trees (Rosaceae), which occur above the timberline in otherwise open plateaux (Vuilleumier 1984).

Some range changes have been clearly associated with the provision or removal of habitat by human action. The most obvious examples include the spread of grassland species following deforestation, and forest species following tree-planting (Chapter 15). Conversely, species that can live only in natural areas have contracted considerably in range as their habitats have been progressively destroyed. An example from North America is the Red-cockaded Woodpecker *Picoides borealis*, which lives in old pine forest. The species formerly occurred over much of the southeast, but has now become confined to small areas where its habitat remains. Almost certainly it would spread again if forests were allowed to grow to an appropriate age. Marshland species have also greatly declined and contracted, as wetlands have been drained. Such changes are unsurprising, but reaffirm the crucial importance of habitat in limiting the distributions of many birds.

The fact that many bird species are found only in a particular type of habitat raises the question why. There seem to be two main answers. Firstly, to succeed in any one habitat may require a degree of morphological or physiological specialisation that automatically excludes other options. The large bodies, webbed feet and spatulate bills of ducks are ideal for life on water, but firmly exclude other habitats, such as forest canopies. Secondly, a species may be restricted to a single habitat through the presence in other potential habitats of natural enemies or competitors, as discussed later.

ECOLOGICAL BARRIERS: FOOD AND NEST SITES

Within areas of suitable climate and habitat, birds may be limited in distribution by particular components of habitat, including food and nest sites, predators, parasites, pathogens and competitors. These are all factors that can also limit numbers within the range (Newton 1998). Some bird species have such narrow diets that their distribution can be shown to coincide with that of a particular food type. One example is the Snail Kite *Rostramus sociabilis* whose distribution in the New World is closely tied to that of its main prey, water snails of the genus *Pomacea*. These snails were introduced to Lake Gatún in central Panama in the late 1980s, and by 1994 Snail Kites had colonised the lake from sources at least 350 km away (Angehr 1999). Likewise, tick-eating Red-billed Oxpeckers *Buphagus erythrorhynchus* in Africa rely entirely on large mammals to provide their food, and are found only in areas where large species of wild game or cattle occur. Several rodent-eaters occur on tundra around the northern hemisphere, except for parts of Greenland, Iceland and Spitzbergen where rodents are lacking. Rough–legged Buzzards *Buteo lagopus* and Hen Harriers (Northern Harrier) *Circus cyaneus* are absent from the whole of Greenland, while Snowy Owls *Nyctea scandiaca* breed only in the northeast where lemmings occur. As lemmings and other rodents vary in abundance from year to year, most rodent-dependent species nest each year only in those parts of their extensive arctic range where their prey are plentiful at the time.

Some herbivores also match the distribution of their specific foods. Throughout the regions where conifers occur, different crossbill taxa breed only in places where appropriate seeds are available, changing areas from year to year, according to the fluctuations in cone crops (Newton 1972, Thies 1996). Similarly, nutcrackers *Nucifraga* species nest only in areas where large-seeded conifers grow. In Sweden and Finland where the large-seeded *Pinus cembra and P. siberica* had been planted by people, Thin-billed Nutcrackers *Nucifraga caryocatactes macrorhynchas* established new breeding populations, following an irruption from further east in 1968 (Lanner 1996). Similarly, in desert areas on all the main continents, seed-eaters settle each year wherever sporadic rainfall ensures a temporary food-supply.

Some neotropical frugivores, such as the Bearded Bellbird *Procnias averano* and Oilbird *Steatornis caripensis*, depend largely on the fruits of a relatively small number of tree species, and the Oilbird is further restricted by its need for caves for nesting. Another neotropical bird, the strange leaf-eating Hoatzin *Opisthocomus hoazin*, is restricted geographically to the range of its food plant, an arum (Araceae). The Hyacinth Macaw *Anodorhynchus hyacinthinus* eats palm nuts, has the strongest bill of any bird, and occurs only on the South American pantanal where its food-plants grow.

Another type of feeding interaction involves 'mutualism', in which each of a pair of species benefits from the other. Avian examples include various pollinating hummingbirds and their nectar-bearing food-plants. Eight hummingbird species in temperate North America feed heavily on certain plant species with red tubular flowers that they pollinate. While there is no correspondence between the ranges of particular hummingbird and particular plant species, there is collective overlap between birds and plants. Wherever hummingbirds occur they have appropriate food-plants, and wherever the plants occur there are hummingbirds to pollinate them (Kodric-Brown & Brown 1979). In tropical regions, where some hummingbirds have bills adapted to particular flower-types, close one-to-one range coincidence might be expected, but I know of no documented examples. Among other animals, many butterflies and other insects are restricted to the range of their specific food-plants. Apart from such specialists, the dependence of bird distribution on food-supplies is shown mainly by species that have spread following a known increase in food-supply (see Chapter 15).

Regarding nest-sites, birds are clearly restricted to breeding in regions where suitable sites are available. Obligate cliff-nesting or cave-nesting species may be confined to those geographical areas where cliffs or caves occur, and tree-nesters to areas where trees occur. In North America, the natural range of the cavity-nesting Wood Duck *Aix sponsa* falls within that of the Pileated Woodpecker *Dryocopus pileatus* which provides the cavities (Bellrose 1990), and in Australia, the Golden-shouldered Parrot *Psephotus chrysopterygius* relies on termite mounds for nesting and roosting, and occupies only a small range on the Cape York Peninsula where such sites are available (Pizzey & Knight 1997). The importance of nest sites in limiting bird distributions is again shown most clearly in those species for which nest sites have been provided by human activity in regions previously lacking them (see Chapter 15).

ECOLOGICAL BARRIERS: PREDATORS AND PARASITES

That predators or parasites can limit bird distributions is shown most clearly on those islands to which alien predators or disease organisms have been introduced by people, leading to extinctions or range contractions in local birds. Species of oceanic islands have often evolved in the absence of certain predators and parasites, so may be especially vulnerable. Many endemic landbirds are thought to have been eliminated from islands by predation from rats or other introduced predators (Chapter 7), and some, such as the Tuamatu Sandpiper *Prosobonia cancellata* and the Banded Rail *Rallus philippensis* in the Pacific, are now found only on those islands within their former ranges that are still free of introduced predators (Pratt *et al.* 1987). Likewise, the introduction of the Black Rat *Rattus rattus* in the western Mediterranean 2,000 years ago modified the distribution of colonial seabirds and could largely account for present patterns. Smaller species are now restricted to small rat-free islands, but larger species have proved able to persist in the presence of rats. On small islands where rat densities are highest and fairly stable from year to year, breeding by such birds is negligible, but on larger islands, where rat densities are lower and fluctuate greatly from year to year, long-lived seabird species are able to breed well in certain years. This is apparently enough to maintain their populations on both larger and smaller islands (Thibault *et al.* 1996).

On islands where introduced predators have been removed, both landbirds and seabirds have returned or increased (Chapter 7). Islands offshore from the mainland of New Zealand are now being managed to provide sanctuaries where native plants and animals can thrive in the absence of introduced mammals. After getting rid of the larger introduced mammals, such as possums *Trichosurus vulpecula*, pigs *Sus scrofa* and goats *Capra hircus*, biologists turned their attention to rats, using aerial and ground-based applications of rodenticides. To date, 41 islands around the New Zealand coastline have been cleared of rodents. In each case, this was followed by the re-establishment, improved survival and increase of native bird species (Empson & Miskelly 1999, Eason *et al.* 2001). The same has occurred on various other islands around the world that have been cleared of introduced predators (Chapter 7). These examples can be regarded as replicated experiments, and provide some of the most telling case histories of the impact of predators on bird distributions.

Predators can also restrict bird distributions in some apparently natural situations. In the Galapagos archipelago, the Red-footed Booby *Sula sula* breeds only on islands lacking Galapagos Hawks *Buteo galapagoensis* (Anderson 1991). Of 22 islands surveyed, no island had both species, seven had neither, ten had hawks but not Red-footed Boobies, while five had boobies but not hawks. This pattern was significantly different from that expected on independent distributions ($P=0.03$). Hawk distribution seemed to depend on the presence of *Tropidurus* lizards, the main prey, but hawks also took booby chicks, and Red-footed Boobies were more susceptible than other boobies on Galapagos because they foraged far from the colonies and left their chicks unattended for long periods.

A clue to the role of disease in influencing the distributions of continental species is provided by failed attempts to establish domestic poultry in areas where particular diseases (new to poultry) are endemic (Newton 1998), and by the elimination of many species from islands, or parts of islands, where alien pathogens have been introduced through human action, as in the Hawaiian Islands (Chapter 7). In time a species may develop resistance to a disease, but until it does so, the disease may limit its spread. Range restriction is most likely to occur where the pathogens have alternative, more resistant host species which can maintain pathogen numbers at sufficient levels to eliminate the more vulnerable host species (Dobson & May 1986). The same is true for predators, which, in the presence of alternative prey, can persist in sufficient numbers to limit the spread of more vulnerable prey species (Newton 1998).

ECOLOGICAL BARRIERS: COMPETITION

Acting mainly through resources, inter-specific competition can also limit the distributions of birds. It is one of the principles of ecology that, where resources are limiting, species with identical needs cannot persist together indefinitely in the same area (Gause 1934). Invariably one species would be better adapted or more efficient, and would outcompete and replace the other completely. Hence, different species of birds normally differ from one another in distribution, habitat or feeding ecology, or in more than one of these respects (Lack 1971).

Allopatric distributions

Closely related species, recently derived from a common ancestor, normally show non-overlapping distribution patterns (allopatry or parapatry; Chapter 3). The fact that the different look-alike species do not spread through one another's ranges can most plausibly be attributed to competition, with one species favoured in one region and another in the adjoining region. As Lack (1971) put it, species inhabit separate but adjoining ranges where (1) they have such similar ecology that only one of them can persist in any one area, and (2) each is better adapted than the other to part of their combined ranges. The case for competition is stronger if the zone of contact between them does not correspond with an obvious change in habitat. This type of situation is evident in superspecies around the world, where the different component species, recently derived from a single widespread parent species, now occupy non-overlapping ranges, each in similar habitat.

Various parids in North America have almost mutually exclusive ranges, and where they overlap they use different habitats (Snow 1954). The Black-capped Chickadee *Poecile atricapillus* occurs across the whole northern part of the continent, but is replaced in the southeast by the Carolina Chickadee *P. carolinensis* (**Figure 14.6**), in the Rocky Mountains by the Mountain Chickadee *P. gambeli*, on the western seaboard by the Chestnut-backed Chickadee *P. rufescens*, and in the mountains of

Mexico by the Mexican Chickadee *P. sclateri*. Other closely related bird species, at an evolutionary stage between allopatry and sympatry, also show mainly different breeding ranges, but with a broad zone of overlap. In Europe, this situation is exemplified in varying degrees by the Chaffinch–Brambling (*Fringilla coelebs* and *F. montifringilla*), the Crested Tit–Siberian Tit *(Parus cristatus* and *P. cinctus)*, the Jay–Siberian Jay *(Garrulus glandarius* and *Perisoreus infaustus)*, the Common Buzzard–Rough-legged Buzzard *(Buteo buteo* and *B. lagopus)*, the Curlew–Whimbrel *(Numenius arquata* and *N. phaeopus),* and the Red-necked–Grey Phalaropes *(Phalaropus lobatus* and *P. fulicarius)*. In each pair, the first named species breeds mainly south of the other, and, in most, the zone of equal abundance shifted north during the twentieth century, coinciding with climatic and habitat change **(Figure 14.7)**. Such boundary changes thus implicate environmental factors in influencing the competitive balance between related species (Lack 1971).

The situation in migrants is especially interesting, because some related species may be allopatric in their breeding areas yet sympatric in their wintering areas (e.g. Tawny Pipit *Anthus campestris* and Red-throated Pipit *A. cervinus*), while others show the reverse situation (e.g. Stonechat *Saxicola torquata* and Whinchat *S. rubetra* in Eurasia–Africa). Yet others may be allopatric in both breeding and wintering areas (e.g. European Pied Flycatcher *Ficedula hypoleuca* and Collared Flycatcher *F. albicollis*) or sympatric in both breeding and wintering areas (e.g. Redwing *Turdus iliacus* and Fieldfare *T. pilaris*). If competition is involved in maintaining patterns of allopatry, it would not be surprising if it varied from region to region, and from season to season, depending on the resources available and the particular mix of species exploiting them. Regional variations in conditions might well account for the same pair of species occupying a single area at one season and separate areas at another.

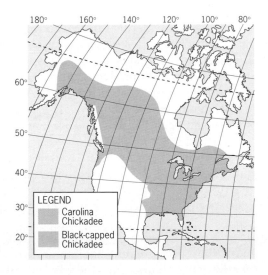

Figure 14.6   Allopatric breeding ranges of Carolina Chickadee *Poecile carolinensis* and Black-capped Chickadee *P. atricapillus* in North America. The ranges of these species abut over a wide boundary, but show little overlap.

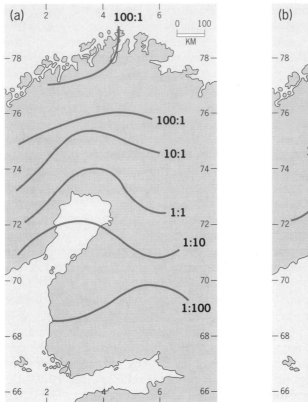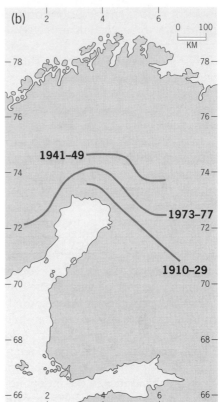

Figure 14.7   An overlap zone between closely related species. (a) The ratio of densities of Brambling *Fringilla montifringilla* to Chaffinch *F. coelebs* in Finland during the breeding seasons of 1973–77. The gradient between the species is steep, and the ratio between the two densities shows a 10,000-fold change within about 400 km. (b) Change in the line of equal abundance over a 68-year period. From Järvinen & Väisänen 1979.

Spatial segregation of species reaches a surprising degree in some tropical areas. Along the Andes range in South America, for example, many species of brush finches *Atlapetes* replace each other geographically. Species with strikingly different colours may replace each other in similar habitats on adjacent slopes, in different elevational zones on the same slopes, or in humid and adjacent rain-shadow zones (Garcia-Moreno & Fjeldså 1999). Such replacements do not always correspond to obvious physical or habitat barriers, and competition is the most plausible explanation for their mutually exclusive distributions, each being better adapted than its neighbours to the particular conditions pertaining within its range.

Species replacements in time

The invasion of an area by a new species is sometimes followed by declines among species already there that use the same resources. Forest clearance, initially to form roads, enabled non-forest birds to spread south through the Malaysian peninsula to reach Singapore (Ward 1968). Declines in eight out of ten native species recorded during the twentieth century were clearly associated with the arrival and establishment of similar species from elsewhere. Striking examples were provided by munias, with *Lonchura punctulata* replacing *L. maja*, and by the kingfishers, with *Halcyon chloris* replacing *H. smyrnensis*. Other examples of new invaders replacing earlier ones with similar ecology can be found among introduced birds on the Hawaiian Islands (Moulton & Pimm 1986).

In other cases, an invader displaced an established species from only part of its niche, so that the two persisted side by side in different habitats. More than a century ago, the Rock Sparrow *Petronia petronia* bred in both towns and country-side of the central and western Canary Islands, but from the nineteenth century the Spanish Sparrow *P. hispaniolensis* established itself there, after which the Rock Sparrow disappeared from the towns, but not from the countryside (Cullen *et al.* 1952). Similarly, wherever Spanish Sparrows and House Sparrows *P. domesticus* are sympatric, the former is found only in rural areas, but where the House Sparrow is absent, for example in Madeira and the Canary Islands, the Spanish Sparrow penetrates the towns (Summers–Smith 1988).

Not surprisingly, most recorded instances of species replacements, whether complete or partial, are in man-made habitats. Even here, all that is observed is 'replacement' and that it is due to 'competitive displacement' is an inference. Replacement might be due to some other interaction, such as disease transmission, or simply to the continuing modification of habitats by human action, or to chance coincidence in the independent loss and gain of similar species. However, it would be straining coincidence too far to suggest that in every such case the two events, involving ecologically similar species, were unconnected.

Diffuse competition

For some species, competition may not be 'direct', involving a single closely related congener, but 'diffuse', involving several species, not all of which may be closely related. Such diffuse competition is hard to detect, because we have no taxonomic or distributional clues as to who the competitors might be. The idea is that the community of species present in an area divide up the feeding opportunities in such a way that an invader would have to displace several of them from parts of their niches if it were to persist. We can only surmise the existence of diffuse competition in special circumstances, in localities where whole constellations of species are missing, and range (or niche) expansions occur in remaining species that have no clear single-species competitor. Examples are provided by the many species that occupy a wider range of habitats on islands, where competitors are lacking, than on mainlands (Lack 1971, MacArthur 1972; Chapter 6). Other examples are provided by montane bird species, some of which occupy a wider range of elevation on those mountains that hold the fewest other species (Terborgh & Weske 1975). For any one bird species, moreover, competing species increase in numbers gradually over a geographical or environmental gradient, until a threshold is reached beyond which the species of interest can no longer persist. In cases such as these, the role of competition would be practically impossible to detect, and a range boundary might be mistakenly attributed to climatic or other conditions. It is hard enough to detect the relevant competitors when they are other birds, but harder still if they happen to be other kinds of animal. For this reason, the role of competition in limiting bird ranges is almost certainly underestimated. Equally, without knowing the competitors, it would be easy to attribute to diffuse competition any boundary that did not coincide with some other environmental feature.

Better evidence for the limiting effects of competition comes from 'natural experiments' in which one species apparently by chance is absent from regions that seem suitable for it. If a second species has expanded its range to include habitat types that are normally occupied by the first species, this implies that wherever the first species is present, the second is excluded by competition. Such relationships seemed important in limiting the distributions of some Andean birds, as emerged from comparisons of bird distributions at two localities (Terborgh & Weske 1975). One locality lay within the main body of the Peruvian Andes (the Cordillera Vilcabamba) and carried what could be regarded as a complete avifauna. The other lay on an isolated massif (Cerros del Sira) that rose out of the Amazonian plain 100 km east of the main Andes. Because of the isolation of the Sira, its upper parts were exposed to invasion principally from below. About 80% of the species that might have occupied the summit zone of the Sira, had it lain in the main body of the Andes, were missing. Their absence provided openings for species living at lower elevations to move upslope to fill the gaps. Of the 24 species that were available to expand their ranges, at least 71% did so. This implied that on the main range these species had been limited in distribution by competition. In addition, 58% of 40 species that lacked close relatives at high elevation on the main range expanded upwards on the Sira. In these species diffuse competition was assumed to be the main limiting mechanism. At least two-thirds of all the distributional limits of the Andean birds that were studied could be attributed to direct or diffuse competition. Without the benefit of this special study site, the distributional limits of these species on the main range might well have been ascribed to climate or other factors that varied with elevation. At other sites, individual species also had greater elevational ranges on mountains where a close relative was absent than where one was present (Moreau 1966, Diamond 1975, Terborgh & Weske 1975, Noon 1981).

Birds on neotropical mountains have illustrated two other points (MacArthur 1972). Firstly, species higher on the mountains occupied a wider range of elevations. This was correlated with a reduction in the numbers of high-elevation species and was analogous to an island effect (island species normally occupy a wider range of habitats, Chapter 6) and to a latitudinal effect (high-latitude species have generally wider ranges, Chapter 13). Secondly, the members of a three-species group each occupied a smaller elevational range than the members of a two-species group, and these occupied a smaller range than a single species (see also Terborgh 1971). Thus, if a genus of three species with non-overlapping distributions occupied an elevation span of 3,000 m, each would occupy an average of 1,000 m; if two species subdivided this span, each would occupy an average of 1,500 m, while one species would occupy the whole span. This kind of pattern would be predicted on the basis of competition and, because it is found repeatedly, it argues against it being a chance result of independent distributions.

Checkerboard distributions

Related bird species that live side by side in the complex habitats of mainlands often occur alone in the simpler habitats of islands. Different species occur on different islands, but no more than one species occurs on any one island. The species thus have mutually exclusive, but interspersed, distributions, collectively showing a 'checkerboard' pattern. Diamond (1975) described several such patterns in the Bismarck Archipelago. In one example, the flycatcher *Pachycephala melanura dahli* is found on 18 islands and its congener *P. pectoralis* on 11, but the two species never occur together on the same island **(Figure 14.8)**.

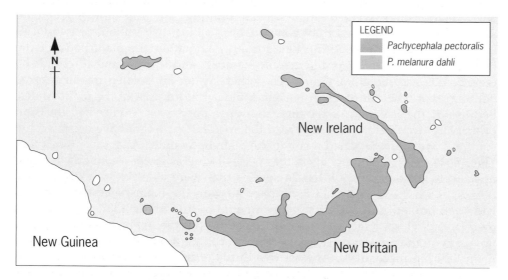

Figure 14.8 Mutually exclusive (checkerboard) distributions of *Pachycephala* flycatchers on the Bismarck Archipelago off New Guinea. Of the two species *P. pectoralis* and *P. melanura dahli*, most islands have only one, no islands have both, and a few (especially the smallest islands) have neither. From Diamond 1975.

Other negative associations may be more complex, in that species that are not found together on small islands may co-exist on larger ones where resources are more diverse. In other examples, involving several species, certain combinations occur frequently, while others seem never to occur, even though they would be expected on a random basis (Diamond 1975). Similar patterns of distribution are found among pigeons and hummingbirds in the West Indies (Lack 1976). They are also evident among tropical montane forest birds. For example, each of the isolated mountain forests in the African savannahs is occupied by only one species of woodpecker, different species on different mountains, but the several species involved co-exist in more extensive forest elsewhere (Moreau 1966).

Although such distributions imply competition, involving the local exclusion of one species by another, they do not explain why a particular species is present on certain islands (or mountains) and another species on others. The pattern could result from chance, depending on which species happened to become established first on each island, developing adaptations that made it easier to exclude close relatives. Or it could result from differences between islands, including the mix of competing species, which could tip the balance in favour of one species on certain islands and of another species on other islands. Whatever the explanation, checkerboard distributions imply that the presence of one species on an island, for whatever reason, will exclude another similar one, and thus restrict its range. However, in none of the studies cited was habitat examined in detail, and alternative explanations could not be eliminated, including the possibility of human interference (Steadman 1995). Closer inspection has sometimes revealed a more complicated situation. Thus, in the eastern Polynesian islands two species of fruit-doves (*Ducula*) occupy different islands from one another, but *D. latrans* occurs only on islands where the Fiji Goshawk *Accipiter rufitorques* is absent (Holyoak & Thibault 1978). It is possible, therefore, that *D. latrans* is excluded by predation rather than by competition, but its absence from some islands may still be what allows in its congener.

Expanding–retracting species pairs

As stronger evidence for the effects of competition in restricting bird distributions, one species can sometimes be seen in the process of expanding at the expense of another. Some examples involve related taxa coming together in secondary contact, after a period of isolation in separate regions where they diverged. In the zone of contact, some such forms remain distinct, whereas others hybridise to a greater or lesser degree (Chapter 3). In several such pairs, however, one species (or subspecies) has expanded in recent times at the expense of the other. Examples involving hybridisation include the Indigo Bunting *Passerina cyanea* replacing Lazuli Bunting *P. amoena*, and the Baltimore (Northern) Oriole *Icterus galbula* replacing the Bullock's Oriole *I. bullockii*, on the North American prairies (Emlen *et al.* 1975, Rhymer & Simberloff 1996). Other examples involve the Mallard *Anas platyrhynchos*, which is extending its range at the expense of various less dimorphic mallard-like ducks which were originally geographically separated. The latter include the Black Duck *A. rubripes* and Mottled Duck *A. fulvigula* in the New World, and the Pacific Black (or Grey) Duck *A. superciliosa* in Australia and New Zealand (where the Mallard was introduced). Replacement of the American Black Duck is partly by hybridisation and partly by competitive exclusion, for as Mallards have spread eastwards in southern Canada, they have occupied the most productive waters, leaving Black Ducks in poorer places (Merendino *et al.* 1993).

Avifaunal differences

Other evidence for competition limiting expansion occurs at the boundaries between distinct avifaunas (Chapter 5). One of the most striking occurs at the division between the Indomalayan and Australasian faunas, on some Indonesian islands. Westwards from Bali, the avifauna is predominantly Indomalayan in type, and eastwards from neighbouring Lombok predominantly Australasian. It is inconceivable that most species could not cross the short 40-km water barrier between Bali and Lombok (much wider barriers are crossed between other

Indonesian islands), and the boundary is most plausibly attributed to competition (diffuse or otherwise), which is sufficient to prevent most Indomalayan species from spreading further east and most Australasian species from spreading further west. Such competition might operate at higher taxonomic levels for an entire continent. The New World warblers (Parulidae) are replaced in Eurasia by the Old World Warblers (Sylviidae), and the New World flycatchers (Tyrannidae) are replaced by the Old World flycatchers (Muscicapidae). Australia has its own set of warblers and flycatchers (Acanthizinae, Muscicapidae), independently evolved from those in the northern hemisphere. Other examples include the four families of nectar-feeders, each from a different part of the world (Chapter 2) and the complementary distributions of the pheasants and megapodes (Chapter 16).

*   *   *   *

In conclusion, competition could be involved in limiting many bird ranges, but the evidence is largely circumstantial and alternative explanations of particular findings are sometimes hard to eliminate. Because the outcome of competition is likely to be influenced by other factors, we should expect the boundaries between apparent competitors to move as climate and other conditions change. In general, as species numbers increase from poles to tropics, one might expect a parallel increase in the role of diffuse competition and other biotic interactions in limiting bird distributions (MacArthur 1972). By the same token, if any northern hemisphere species is limited by biotic interactions, this is most likely to occur on its southern flank, and in southern hemisphere species on its northern flank, with abiotic factors more likely to operate on the opposite flank. To give a specific example, the Buff-browed Warbler *Phylloscopus humei* overwinters in the northern half of the Indian subcontinent (Gross & Price 2000). Its northern range limit coincides with the line of leaf loss associated with cool temperatures, and the associated autumn disappearance of arthropod food from relevant habitat. The southern limit is marked by high food levels, but also with increasing numbers of potential competitors, notably the Greenish Warbler *P. trochiloides* which is 40% heavier than *P. humei*.

PHYSICAL BARRIERS

Physical features, such as oceans, mountain chains or large deserts, limit species ranges because they can act as barriers to dispersal. A glance at the range maps for any set of landbird species will show how often such species extend no further than a sea coast, mountain range or other expanse of hostile terrain. Likewise, many seabird species do not extend from one side of a continent to another, the intervening land providing a barrier to their spread (Chapter 8). Such barriers can give considerable concordance and temporal stability to the range boundaries of many species, and are important in the maintenance of separate biogeographical realms, between which large proportions of the flora and fauna differ (Chapter 5).

The world's oceans are clearly major obstacles to the dispersal of most landbirds. They are largely responsible for the huge differences in avifaunas between the continents, discussed in Chapter 5, and for the high levels of endemicity among island birds. Their effectiveness as barriers is further evident from the many species that did not naturally spread to parts of the world where they have thrived when released there by people. The Eurasian Blackbird *Turdus merula* and Chaffinch *Fringilla coelebs* are obvious examples from New Zealand where they occur in both natural and man-made habitats. Birds vary enormously in their dispersal powers and, whereas some migrant landbird species regularly cross oceans with ease, others are constrained by seemingly trivial barriers (Chapter 16).

TRANSLOCATIONS BY PEOPLE

During the nineteenth century, Europeans who settled in other parts of the world often took familiar birds from home to release in their new environment. More recently, with the growing popularity of aviculture, various foreign cage birds and waterfowl have continually escaped to the wild into areas that they could not be expected to reach naturally. In addition, hunters have often introduced alien game species to add variety to their quarry. Whether deliberate or accidental, such releases have been a major factor in extending the distributions of certain birds, and at the same time provide evidence on the ability of bird species, completely new to a region, to establish themselves alongside the indigenous avifauna. They also raise questions about the effectiveness of competition in excluding species, at least in the short term.

About 5% of all bird species are known to have been introduced elsewhere, some to more than one area (Long 1981, Lever 1987). From a total of 545 introductions listed in **Table 14.2**, 29% led to a successful establishment. This proportion should be considered as a maximum, because many other unsuccessful releases are likely

Table 14.2 Number of exotic bird species introduced by human agency over a 150-year period to eight areas outside their natural range. From Long 1981.

	Number known to have been introduced[1]	Number (%) known to have become established[1]
North America	98	25 (26)
South America	15	7 (47)
Europe	42	12 (29)
Africa–Arabia	19	8 (42)
Southeast Asian mainland	15	7 (47)
Australia	71	25 (35)
New Zealand[2]	126	32 (25)
Hawaii[2]	159	43 (27)
Overall	545	159 (29)[3]

[1]Data are minima, especially the numbers of introductions which are hard to establish retrospectively.
[2]More recently Baker (1991) gave figures of 143 and 34 (24%) for New Zealand, and Moulton *et al.* (2001) gave 53 exotic species as established on the six main islands of Hawaii.
[3]This overall maximum success rate of 29% for exotics can be compared with a rate of 53% (55/104) for native species reintroduced after they had been eliminated from part of their range.

to have been made without having been recorded. Nonetheless, interesting points emerge. The majority of species died out quickly, probably without reproducing. Others persisted for a few generations and then disappeared. Yet others established themselves only in a small area, but spread no further, and only a minority increased and spread over a large area. Of 98 bird species introduced to North America in the period considered, only six became widespread, namely the House Sparrow *Passer domesticus*, Common Starling *Sturnus vulgaris*, Feral Pigeon *Columba livia*, Chukar *Alectoris chukar*, Pheasant *Phasianus colchicus* and Grey Partridge *Perdix perdix*. A few others persisted only in particular localities, such as the Skylark *Alauda arvensis* on southern Vancouver Island, the Tree Sparrow *Passer montanus* in northern Missouri and the Mute Swan *Cygnus olor* in some eastern states. More recently, the Monk Parakeet *Myiopsitta monachus* and the House Finch *Carpodacus mexicanus* have become widely naturalised, and a number of other escaped cage birds, notably parrots and doves, have become established in some large cities. All the rest died out. Moreover, most successful species settled in the man-made habitats of farmland, parks and gardens, and few successfully penetrated more natural habitats. One plausible but unproven explanation is that they were excluded from natural habitats by competition from the indigenous birds. Similar patterns are repeated in other parts of the world.

If competition is important, it is not surprising that introductions more often succeed on islands with poor avifaunas than on mainlands, and that on both mainlands and islands most newcomers have established themselves in relatively recent, man-made habitats, with which they were often already familiar (Chapter 7). In North America, for example, the House Sparrow and Starling are now extremely widespread, but only in man-made habitats. Where introduced species have established themselves in a natural habitat, such as the Eurasian Blackbird *Turdus merula* and Song Thrush *T. philomelos* in New Zealand forests and the Japanese White-eye *Zosterops japonicus* and Red-billed Leiothrix *Leiothrix lutea* in Hawaiian forests, the islands concerned had already lost many of their native birds.

While the failure of many released species to establish themselves in an apparently favourable environment may be due to competition, direct or diffuse, this may not be true of all species, because other factors influence establishment (Chapter 16). As experiments on the ability of alien birds to establish in new areas, introductions are flawed by lack of replication and randomisation, as well as by poor recording, but they still provide an unparalleled source of information on the factors that facilitate establishment, including the suitability of the local climate and habitat, and the range of potential competitors and other natural enemies present.

CONSEQUENCES OF RANGE EXPANSION AND COMPETITION

If a species spreads to a new area, or is introduced there, one of several things might happen:

(1) The species does not establish itself and dies out. The initial pioneers may be too few in number, or climatic or habitat conditions may be unsuitable, or the species may fail in the presence of local competitors or other natural enemies.

(2) The species finds a vacant, or partially vacant niche, and multiplies without causing obvious reductions in the numbers of species already there. The natural colonisation of western Europe by the Collared-Dove *Streptopelia decaocto* is perhaps in this category, as is the introduction and subsequent spread of the House Sparrow *Passer domesticus* in North America. Not surprisingly, most such species occupy a fairly distinct niche in man-made habitats. However, only a highly specialised invader, which exploits resources untouched by existing residents, is likely to have no impact whatever on other species.

(3) The invader displaces an existing species from part of its range. The invader may be better adapted than the other in the region where they meet but, after it has eliminated the less successful species from part of its range, it may come to a region where the environment is better suited to the other species. In this case the two species will come to occupy separate but adjoining geographical regions. Because environmental conditions often change gradually, there may be a zone where both species are about equally well adapted, and here their ranges could overlap. This type of situation may be exemplified by the boundary between the breeding ranges of the Chaffinch *Fringilla coelebs* and Brambling *Fringilla montifringilla* in northern Europe (**Figure 14.7**; Järvinen & Väisänen 1979).

(4) The invader may displace the native competitor from certain habitats, where it is more successful, but not from others, resulting in a new stable situation in which, through competitive exclusion, the two species occupy different habitats. The two Chaffinches of certain Canary Islands provide an example, with the European form *Fringilla coelebs* replacing the earlier-established Blue form *F. teydea* from all but the native pine forest (Lack 1971).

(5) The invader may be so much better adapted than its local counterpart that it spreads right through the range of its counterpart, and eliminates it by competitive exclusion. The chances of witnessing this are small, but the demise of certain island species, following the introduction and establishment of similar alien species, may be in this category. On some Hawaiian islands, some of the initial aliens disappeared after other ecologically similar ones were introduced, as mentioned above, and similar replacements were observed among the birds of Singapore. Where the invading and displaced species have similar ecology, it is tempting to attribute the replacement to competition, but the evidence can be no more than circumstantial, and it is usually impossible to exclude other explanations.

(6) The invader may prove better adapted for taking certain foods, the native for taking other foods. In this case, if their numbers are limited primarily by food-supply, the two species may be able to co-exist in the same habitat, dividing the available foods, but with the native species at lower density than before. The foods taken by the two species need be only partly and not wholly different. This scenario was suggested by Lack (1947) to explain the co-existence of several finch species on various Galapagos islands. The same situation might hold if the species competed for nest sites rather than food, and such competition may explain the displacement of Eastern Bluebirds *Sialia sialis* from some types of tree cavity by the introduced Common Starling *Sturnus vulgaris* in North America, and the associated reduction in Bluebird densities (Newton 1998).

(7) If the two species are closely related, an invader may replace a native species, not by direct competition, but by hybridisation. This could be regarded as a form of competition at the level of the genes. Genetic material of the original species may survive, but only in combination with that from the invader, which the hybrid stock may soon come to resemble. No species is yet known to have been lost completely through hybridisation, but some have almost disappeared from this cause, and at least three subspecies have gone (Chapter 3).

CONCLUDING REMARKS

Despite the many specific examples given above, it may not always be profitable to search for single limiting factors to the distributions of most bird species. Not only may a single species be limited by different factors in different parts of its range, but even in one area two or more factors may interact to prevent expansion. However, one widespread pattern, at least at high latitudes, is for species to be limited by abiotic factors on one range margin and by biotic interactions on another. This is especially evident in distributions that extend across environmental gradients, such as low to high latitudes or low to high altitudes.

Most of the evidence on factors limiting bird distributions is circumstantial, as it is based on correlations. However, so called 'natural experiments' have proved especially revealing, and in recent decades many species have expanded or contracted in range through specific human action. Such changes have provided further insight into the factors that can limit bird distributions, and are discussed in the next chapter.

SUMMARY

The abundance and distribution of species tend to be linked. Among closely related species, widespread ones tend to be abundant whereas narrowly distributed ones tend to be rare. Furthermore, species that decline in numbers tend also to contract in range, while those that increase in numbers tend also to expand in range. These relationships are unsurprising, but they carry implications for conservation, as well as for our understanding of distributions.

Geographical ranges may be limited in extent by shortage of potential colonists (in turn dependent on conditions within the existing range), or by conditions at range boundaries preventing further spread. Factors acting at range boundaries include ecological constraints, such as climatic and habitat features, natural enemies and competing species, or physical barriers, such as impassable oceans and mountain ranges.

Sometimes ranges end sharply at topographical or habitat boundaries, sometimes they shift back and forth along some climatic or other environmental gradient, according to fluctuations in conditions.

Climatic factors affect birds either directly through causing excess chilling, overheating or dehydration, or indirectly through causing changes in food-supplies or competition, as different limiting factors interact. Different bird species vary greatly in their ability to cope with cold or heat, each being adapted to the climatic regime in which it normally lives.

Winter weather imposes a northern limit to the ranges of some northern hemisphere resident species, whose numbers near the boundary fluctuate from year to year, according to winter severity. Many species are restricted to areas where the extra energy needed to compensate for a colder environment is no greater than about 2.5 times their normal metabolic rate. The short favourable season at high latitudes could prevent some species with long breeding cycles from nesting there, and hence restrict their breeding to lower latitudes. Other species, such as the Emperor Penguin *Aptenodytes forsteri*, have special adaptations which enable high-latitude breeding despite long breeding cycles.

Most bird species are restricted to particular habitats, within which some are further restricted by their need for specific kinds of food or nest sites, and by the need to avoid specific predators, parasites or competitors. Competition may be direct, involving a single closely similar species, or diffuse, involving several species that collectively prevent encroachment.

A species may be limited by the same factor along the entire perimeter of its existing range, or by different factors in different parts of the perimeter or by several factors acting together. Different limiting factors seem to achieve different importance in different parts of the world, and at different seasons. Winter temperatures seem particularly important at high latitudes, and competition at lower latitudes. Overall, the most frequent factor limiting bird ranges may be competition from other species, but the evidence is largely circumstantial.

Our understanding of the factors limiting bird ranges is based almost entirely on correlative evidence, because controlled experiments are usually impractical, and natural experiments are insufficiently controlled and documented, with many confounding variables.

Eurasian Collared-Dove *Streptopelia decaocto*, a species which has undergone rapid range expansion.

Chapter 15
Recent range changes

Some bird species have undergone substantial range expansions in the last 200 years, spreading over huge areas where they were formerly unknown. Such extreme changes shed light on the factors that facilitate spread, and by implication on those that restrict it. In keeping with the dynamic nature of bird ranges, most bird species that breed in well-studied areas, such as Europe and North America, are known to have expanded or contracted in range over the last 200 years. The bird faunas of particular regions have therefore changed substantially in this time and, where human activities have diversified landscapes, species numbers have increased. For example, the northern countries of Europe (Norway, Sweden, Finland and Denmark) were each colonised by an average of 2.8 species per decade during the period 1850–1970 and at the same time lost an average of 0.6 species per decade. This gave a total net gain in the region as a whole of 66 new species over this time period (Järvinen & Ulfstand 1980). Some of the newcomers, such as Eurasian Collared-Dove *Streptopelia decaocto*, are now so widespread that it is hard to believe that they were unknown in these countries only a century ago.

Because climate has changed in the last 200 years, any change in distribution recorded in this time has inevitably coincided with climate change and in some species this may indeed be the main causal factor. In other species, however, range changes have been clearly linked with human activity, which has altered the availability of habitats, foods or other resources. Some species have dispersed over barriers to reach new areas, as in the colonisation of islands, or have been translocated to new areas by people. Because of human involvement, the rate of change in bird distributions recorded in the last 200 years, and the numbers of species involved, may well have been higher than in previous centuries.

Any natural range expansion must presumably depend on two conditions, firstly on population pressure in the area already occupied, and secondly on the presence of suitable conditions in the area invaded. Population pressure in the existing range could come either from an increase in numbers there or from a rapid deterioration in local conditions, both of which could promote emigration. The conditions that enable the new colonists to succeed outside their former range might have existed for some time previously, perhaps even for hundreds of years, or they might have been created only recently through natural events or human action. Hence, range expansions could occur from changes within the existing range (encouraging birds to move out), or from changes outside, making previously unsuitable areas acceptable (Chapter 14). In theory, they could also occur from genetic or other changes in the birds themselves, enabling them to invade areas they previously could not.

DISTRIBUTIONAL CHANGES ASSOCIATED WITH CLIMATE CHANGE

The period of maximum warmth in the current interglacial was 7,000–5,000 years ago, since when the climate has cooled somewhat, with fluctuations, and with a marked human-induced rise in the last 150 years, associated with greenhouse gas emissions (Chapter 9). The upward trend was interrupted between 1945 and 1975 when a plateau in mean world temperature was apparent. Over the whole 150 years, however, the annual average temperature in many places in the northern hemisphere increased by up to a few degrees centigrade, and the May–June isotherms moved up to several hundred kilometres northwards and up to several hundred metres up mountainsides. Growing seasons lengthened by up to a fortnight and certain crop-plants could be grown at higher latitudes than before. In western Europe, summers have become wetter, while further east, on the steppes, they have become drier. It is usually the extremes in climate that have most effect on birds, and the averages are important mainly in so far as they reflect changes in the frequency or extent of extremes.

Over the same period many mid-latitude bird species increased and spread northwards, while other more northern species declined and retreated yet further north. Such range changes have usually been attributed to climate change (Burton 1995), but often without consideration of other potential causal factors. These included not only the enormous human impacts on habitat, but also in recent decades a relaxation in hunting pressure, which has enabled some species to recolonise areas from which they had been previously eliminated. Hence, some of the range changes attributed to climate might well have been due to other factors or to a combination of climate change and other factors. Almost always, uncertainty hangs over any explanation of range change, because the evidence is primarily correlative (Chapter 14).

The appeal of the climatic explanation of range change stems from the facts that: (1) many changes are latitudinal (towards the north in the northern hemisphere), and (2) in many bird-species reproductive and survival rates are clearly influenced by prevailing weather (Newton 1998). Hence, resident species that suffer high mortality in hard winters, for example, might be expected to increase and spread further north during a run of mild winters. Some might then compete with more northern species, causing them to retreat even further north.

In the first half of the twentieth century, in northern Europe major northward expansions were especially marked in Grey Heron *Ardea cinerea*, Northern Lapwing *Vanellus vanellus*, Common Starling *Sturnus vulgaris*, Wood-Pigeon *Columba palumbus*, Rook *Corvus frugilegus* and Tawny Owl *Strix aluco* **(Table 15.1)**. Similarly, marked northward withdrawals from southern range boundaries occurred in Greater Scaup *Aythya marila*, Glaucous Gull *Larus hyperboreus*, Ruddy Turnstone *Arenaria interpres*, Atlantic Puffin *Fratercula arctica* and Common Murre *Uria aalge*. In some pairs of closely related species, as the southern form pushed northwards, its northern counterpart retreated, as in the Chaffinch *Fringilla coelebs* and Brambling *F. montifringilla* and other examples mentioned in Chapter 14. Further south in Europe, other species spread north in the same way. Examples include the European Serin *Serinus serinus*, Cetti's Warbler *Cettia cetti* and Firecrest *Regulus ignicapillus* **(Table 15.1)**.

Table 15.1   Recent major range expansions by some European landbirds. Details from Kalela 1949, 1952, Voous 1960 and Burton 1995, supplemented from the additional sources quoted against each species.

1. *Westwards from Siberia into Europe*

Terek Sandpiper *Tringa cinerea*, mainly 20th century.
Grey-headed Woodpecker *Dendropicos spodocephalus* at least since 19th century.
Horned Lark *Eremophila alpestris*, mainly 19th century.
Citrine Wagtail *Motacilla citreola*, at least since 19th century (Wilson 1979).
Red-flanked Bluetail *Tarsiger cyanurus*, mainly 20th century, perhaps earlier.
Thrush Nightingale *Luscinia luscinia*, mainly 20th century.
Fieldfare *Turdus pilaris*, at least since 19th century.
Redwing *Turdus iliacus*, mainly 20th century.
Eurasian River Warbler *Locustella fluviatilis*, mainly 20th century.
Blyth's Reed Warbler *Acrocephalus dumetorum*, mainly 20th century.
Greenish Warbler *Phylloscopus trochiloides*, mainly 20th century (Valikangas 1951).
Arctic Warbler *Phylloscopus borealis*, mainly 19th century.
Red-breasted Flycatcher *Ficedula parva*, mainly 20th century.
Collared Flycatcher *Ficedula albicollis*, mainly 20th century.
Eurasian Penduline Tit *Remiz pendulinus*, mainly 20th century (Valera *et al.* 1993).

Common Rosefinch *Carpodacus erythrinus*, mainly 20th century (Stjernberg 1985).
Yellow-breasted Bunting *Emberiza aureola,* mainly 19th century.
Rustic Bunting *Emberiza rustica*, mainly 19th century.
Little Bunting *Emberiza pusilla,* mainly 20th century.

2. *Northwestwards from southeast Europe*

Stock Dove *Columba oenas*, mainly 19th century.
Eurasian Collared-Dove *Streptopelia decaocto*, mainly during 1930–70.
Syrian Woodpecker *Dendrocopus syriacus*, at least since 19th century.

3. *Northwards in northern Europe*

Great Crested Grebe *Podiceps cristatus*, mainly 20th century.
Grey Heron *Ardea cinerea*, mainly 20th century.
Tufted Duck *Aythya fuligula*, mainly 20th century.
Northern Lapwing *Vanellus vanellus*, mainly 20th century.
Eurasian Curlew *Numenius arquata*, mainly 20th century.
Common Wood-Pigeon *Columba palumbus,* mainly 20th century.
Tawny Owl *Strix aluco,* mainly 20th century.
Grey Wagtail *Motacilla cinerea*, mainly 19th century (Mayr 1942).
Eurasian Blackbird *Turdus merula*, mainly 20th century.
Mistle Thrush *Turdus viscivorus*, mainly 20th century.
Coal Tit *Parus ater*, mainly 20th century
Crested Tit *Parus cristatus*, mainly 20th century.
Blue Tit *Parus caeruleus*, mainly 20th century.
Eurasian Jay *Garrulus glandarius*, mainly 20th century.
Eurasian Jackdaw *Corvus monedula*, mainly 20th century.
Rook *Corvus frugilegus*, mainly 20th century.
Common Starling *Sturnus vulgaris,* mainly 20th century.
Chaffinch *Fringilla coelebs*, mainly 20th century.
European Greenfinch *Carduelis chloris*, mainly 20th century.

4. *Northwards in middle Europe*

Cetti's Warbler *Cettia cetti,* mainly 20th century (Bonham & Robinson 1975).
Savi's Warbler *Locustella luscinioides*, mainly 20th century.
Marsh Warbler *Acrocephalus palustris,* mainly 20th century.
Eurasian Reed Warbler *Acrocephalus scirpaceus*, mainly 20th century.
Great Reed Warbler *Acrocephalus arundinaceus*, mainly 20th century.
Olivaceous Warbler *Hippolais pallida*, mid 20th century.
Bonelli's Warbler *Phylloscopus bonelli,* mid 20th century.
Fan-tailed Warbler *Euthlypis lachrymosa*, mainly 20th century.
Firecrest *Regulus ignicapillus*, mainly 20th century.
Eurasian Golden Oriole *Oriolus oriolus*, mainly 20th century.
Crested Lark *Galerida cristata*, mainly 19th century.
European Serin *Serinus serinus*, mainly 20th century (Mayr 1926, Olsson 1969).
Black Redstart *Phoenicurus ochruros,* since mid 19th century.

5. *Northwards in southern Europe*

White-rumped Swift *Apus caffer,* 20th century.
Eurasian Crag Martin *Ptynoprogne rupestris*, 20th century.

Red-rumped Swallow *Hirundo daurica*, 20th century.
Trumpeter Finch *Rhodopechys githaginea*, 20th century.

6. *Westwards and northwestwards from eastern Europe*

Black-necked Grebe *Podiceps nigricollis*, mainly 20th century.
Red-crested Pochard *Netta rufina*, mainly 20th century.
Pochard *Aythya ferina*, from 19th century.
Ferruginous Pochard *Aythya nyroca*, mid 19th century.
Gadwall *Anas strepera*, mainly 20th century.
Little Gull *Larus minutus*, mainly 20th century.
Whiskered Tern *Chlidonias hybridus*, mainly 20th century.
White-winged Tern *Chlidonias leucopterus*, mainly 20th century.

The changes were especially striking in Iceland (Gudmundson 1951). During 1900–50, seven southern species, formerly absent, became well established as breeding birds, including the Common Starling *Sturnus vulgaris*, Short-eared Owl *Asio flammeus*, Northern Shoveler *Anas clypeata*, Tufted Duck *Aythya fuligula*, Common Black-headed Gull *Larus ridibundus*, Herring Gull *Larus argentatus* and Lesser Black-backed Gull *Larus fuscus*, while other southern species, already present, expanded their breeding ranges, including Black-tailed Godwit *Limosa limosa*, Eurasian Oystercatcher *Haematopus ostralegus* and Gadwall *Anas strepera*. Over the same period, some northern species declined, including Little Auk *Alle alle* and Long-tailed Duck *Clangula clangula*.

All these northward changes have been widely attributed to climatic amelioration, acting on the birds themselves, their habitats and food-supplies, or on the competitive balance between species. However, many such species have clearly benefited from changes in human land-use, notably the conversion of old growth forest to younger forest and farmland, a reduction of forest grazing (favouring more ground cover), and an increase in forest edge and clear-cuts, all of which began in the south of Scandinavia and spread northwards (Järvinen & Väisänen 1979). It is thus hard to tell whether climate change or habitat change has promoted the northwards spread. Moreover, while many species spread northwards in the period 1900–50, others in the same period spread south (e.g. Common Merganser *Mergus merganser*, Common Eider *Somateria mollissima*, Northern Fulmar *Fulmarus glacialis*), or retreated south (e.g. Lesser Grey Shrike *Lanius minor*, Woodchat Shrike *L. senator*). Yet others spread both north and south simultaneously (e.g. Great Skua *Stercorarius skua* and Great Black-backed Gull *Larus marinus*). It seems unlikely, therefore, that climatic amelioration could explain the range changes of all these species. It was probably only one of several factors that over the last 150 years influenced the habitats (through human land-use), food-supplies (through ice and snow periods) and competitive relationships of certain of the species listed in **Table 15.1** but by no means all. However, if climate has such all-embracing importance, we can anticipate further latitudinal changes in the range boundaries of birds in the years ahead if the predicted trend of further warming materialises.

Another reason for caution in ascribing recent range changes entirely to climatic change is that, in some bird species, interpretations have changed as fresh information has become available. In the twentieth century various European insectivorous species, which breed well in warm dry summers, greatly declined in the western parts of their range where the summers became wetter. They included Red-backed Shrike *Lanius collurio*, Eurasian Nightjar *Caprimulgus europaeus*, Eurasian Wryneck *Jynx torquilla*, Eurasian Hoopoe *Upupa epops* and Roller *Coracias garrulus*. Again, however, changes in land-use could have caused or accelerated the declines of these species by reducing insect abundance, as could changes in their African winter quarters. The fact that the Red-backed Shrike has survived in Europe wherever traditional farming practices have remained, but disappeared elsewhere, while the Eurasian Nightjar has recently increased in localities where new habitat has become available, casts doubt on the climatic explanation as the only factor.

Another group of species spread from the east into western Europe in the first half of the twentieth century. These were the waterbirds of rich shallow lakes, including Gadwall *Anas strepera*, Common Pochard *Aythya ferina*, Ferruginous Pochard *A. nyroca*, Red-crested Pochard *Netta rufina*, Black-necked Grebe *Podiceps nigricollis*, Little Gull *Larus minutus* and White-winged Tern *Chlidonias leucopterus* (Niethammer 1937). The spread of these species was linked by Kalela (1949) to increasing aridity in the steppes of the Caspian and Black Sea regions, and to the drying of shallow lakes, which caused the birds to seek new habitat in the driest years. At the same time, however, habitat in western Europe must have been suitable for these species or they could not have persisted and increased there. In fact, the man-made water-bodies on which these species are usually found in western Europe have increased greatly in the last 150 years, with the creation of reservoirs, fish ponds and gravel pits, all of which have also been subject to gradual eutrophication, giving conditions suitable for the birds concerned. Several other species of waterfowl were first recorded nesting in Britain during the late twentieth century, and some (such as Tufted Duck *Aythya fuligula*) have since become widespread. Again, however, one cannot always eliminate other explanations of spread, namely recolonisation after previous elimination by human hunting and egg-collecting. The elimination of several heron species from Britain and other parts of western Europe has been attributed to hunting and land drainage from the Middle Ages onwards (Bourne 1999), and some such species are now showing signs of returning.

With the development of increasing resolution in the mapping of bird ranges, relatively small distributional changes can now be detected. Two surveys of the distribution of breeding birds in Britain, each completed on a 10-km grid system, and compiled about 20 years apart (1968–72 and 1988–91), were used to analyse distributional changes that had occurred in this period (Thomas & Lennon 1999). The northern limits of many southern species were found to have moved north in this time, by an average of 19 km. As these changes occurred in a period of climatic warming, the authors suggested increasing warmth as the underlying causal factor. In contrast, the southern limits of more northern species had not shifted consistently in this time.

In eastern North America, marked northward expansions have occurred in several species, such as Dickcissel *Spiza americana*, Northern Cardinal *Cardinalis cardinalis*, American Wigeon *Anas americana* and Gadwall *Anas strepera*, and northward retreats in others, such as Greater Scaup *Aytha marila* and Glaucous Gull *Larus hyperboreus* (Beddall 1963, Johnson 1994, Burton 1995). In the western

United States, some species have spread south in recent decades, while others have spread westwards or southwestwards, notably Barred Owl *Strix varia*, Grey Flycatcher *Empidonax wrightii* and Summer Tanager *Piranga rubra* (Johnson 1994). At least two species (White-faced Ibis *Plegadis chihi* and American Avocet *Recurvirostra americana*) have spread from one side of their range while retreating from the other, so that the whole distribution has shifted (Johnson & Jehl 1994). The majority of these changes were attributed to climatic change (with milder winters in the north and warmer, wetter summers in the west), but it is again hard to eliminate an effect of human influence on habitats.

In addition to range changes, some bird populations in Europe and North America have shown other apparent responses to warming. In the latter half of the twentieth century, many bird species began to breed progressively earlier, though with large annual variations (Winkel 1996, Crick *et al*. 1997, Visser *et al*. 1998). Over much of North America, the laying dates of Tree Swallows *Tachycineta bicolor* advanced by about nine days during the period 1959–91 (Dunn & Winkler 1999). Many other studies have confirmed the link between spring temperatures and laying dates in a wide range of birds, as well as between spring temperatures and leafing dates of deciduous trees, spawning dates of amphibians or peak abundance dates of butterflies (Sparks & Crick 1999). The migration dates of birds have also changed (Chapter 18).

In the southern hemisphere, the Adélie Penguin *Pygoscelis adeliae* inhabits the sea ice zone that rings Antarctica, breeding only where sea ice persists well into spring. During the past 50 years, the species has been declining in the northern (warmer) part of its range, as the sea ice has begun to melt earlier in response to climate change. Winters with loose pack-ice are now too infrequent there for the population to sustain itself. By contrast, the species is increasing in southern parts of its range, where the effect of warming has been to loosen the previously hard pack-ice and make it more suitable. In effect, then, the centre of gravity of the Adélie Penguin population is shifting southwards to higher latitudes, as air temperatures rise. Interestingly, the closely related Chinstrap Penguin *Pygoscelis antarctica*, which avoids pack-ice as much as possible, is increasing and extending its range southwards along the western coast of the antarctic peninsula, apparently in response to warming and the disappearance of sea ice there. Hence, the distributions of these two species appear to be undergoing rapid and corresponding changes, with the boundary between them moving south as the climate warms (Ainley *et al*. 2001).

Rising temperatures around the antarctic peninsula, and the associated shortening of the annual period with fast sea ice, have been accompanied by a decline in Antarctic Krill *Euphausia superba*, which is a crucial food-species for many seabirds and mammals. Over the period 1980–2000, populations of several seabird species declined at South Georgia, where they have been monitored, and years of poor breeding increased in frequency (Reid & Croxall 2001). Declines in Macaroni Penguins *Eudyptes chrysolophus* and Black-browed Albatrosses *Thalassarche melanophris* exceeded 50%, although the latter may have resulted partly from mortality caused by long-line fishing operations. Further south, one of the hardiest of all birds, the Emperor Penguin *Aptenodytes forsteri*, declined by 50% in the 1970s, through increased mortality during a period of reduced sea ice (Barbraud & Weimerskirch 2001). All the ice-associated species of the high antarctic could be seriously affected by changes in the duration and persistence of polynyas (local areas kept ice-free, usually by fast currents).

It is not only in the southern seas that climate-linked changes in the numbers and distributions of seabirds have occurred. A 90% decline between 1987 and 1994 in the numbers of visiting Sooty Shearwaters *Puffinus griseus* observed in the California Current off western North America, from an initial five million individuals, was attributed to an increase in ocean surface temperatures and reduced upwelling. It remains to be seen whether this represents an overall population decline or simply a large-scale shift in distribution (Veit *et al*. 1997). Other locally breeding cold-water species also declined over the same period.

DISTRIBUTIONAL CHANGES ASSOCIATED WITH HUMAN ACTION

Species that have expanded in association with human activity fall into four main categories, namely those that have benefited from the provision of (1) new habitat; or from components of habitat, such as (2) buildings and other structures used as nest sites; (3) food-supplies resulting from agricultural procedure, garden-feeders, waste-disposal or fisheries; or (4) water in arid areas. Moreover, as the ranges of some species have expanded as a result of these changes, others have contracted. Over much of the world, widespread deforestation has enabled open-country birds to spread into regions formerly closed to them, while the tree-planting associated with human habitation has enabled some tree-dependent birds to spread through former open land. The net effect is that many species have extended their ranges as a result of human action, and birds from different formerly separated communities now live side by side in the same region. In Eurasia, most such range changes probably began more than 2,000 years ago, long before there were ornithologists to record them, but in North America and elsewhere, where major vegetation changes followed the arrival of European people, substantial range changes in birds have occurred in the last 200 years, enabling them to be more fully documented (Robbins 1985).

Changes in habitat

Throughout much of the northern hemisphere, marked range contractions have occurred in the birds of habitats vulnerable to human activity, notably old growth forest, riparian woodland, natural grassland, marshes and coastal beaches. In contrast, species that thrive in man-made habitats of parks, gardens and farmland have all expanded in range, especially in regions where these habitats have replaced continuous forest.

In North America, some of the most striking changes have occurred on the prairies, where the Bison *Bison bison* was wiped out, fires were suppressed and grassland was converted to agricultural fields (Houston & Schmutz 1999). The ploughing of the prairies led to massive declines in such species as Upland Sandpiper *Bartramia longicauda*, Chestnut-collared Longspur *Calcarius ornatus*, Baird's Sparrow *Ammodramus bairdii* and Sprague's Pipit *Anthus spragueii*, which could not thrive in seeded pasture and cropland. From the early 1970s, as agriculture intensified, declines in other species, such as Burrowing Owl *Speotyto cunicularia* and Loggerhead Shrike *Lanius ludovicianus*, have become increasingly apparent (Houston & Schmutz 1999). On the plus side, following tree-planting, some tree-dependent bird species are now so numerous on the prairies that it would be hard for a modern observer to appreciate how recent they are (Houston 1986). Common species, such as the American Crow *Corvus brachyrhynchos*, Black-billed Magpie *Pica pica*, Western Kingbird *Tyrannus verticalis*, Mourning Dove *Zenaida macroura* and others colonised the Canadian prairies only in the latter half of the nineteenth century, and other species, such as Common Grackle *Quiscalus quiscula* and Blue Jay *Cyanocitta cristata*, are still on the increase (Robbins 1985).

Other species, once mainly confined to the prairies, have spread outwards, to occupy the open land created elsewhere by forest clearance. The Brown-headed Cowbird *Molothrus ater* is a familiar example, its range having more than doubled within 100 years, so that it now occupies almost the whole continent north to the boreal forest (Brittingham & Temple 1983). Thus, as some species receded as their natural habitats were destroyed or modified, others expanded as new habitats were created. The same process is occurring in other parts of the world, wherever forests are being destroyed creating open land, with many examples from Central America (Stiles & Skutch 1989) and southeast Asia (Ward 1968). In the Malayan peninsula, following road building and associated development, several non-forest species spread southwards from Burma (now Myanmar), but not all at the same rate **(Figure 15.1)**. Such expansions provide strong circumstantial evidence that lack of suitable habitat, rather than climatic conditions, had previously restricted the ranges of the species concerned.

A species can spread through farmland or other man-made habitat only if it can accept such habitat as an alternative to natural habitat. Not all species that now live in man-made habitats moved in as soon as such habitats became available; some have done so only in recent years, or only in parts of their range. Several species, such as European Robin *Erithacus rubecula* and Dunnock *Prunella modularis*, which are now common in the gardens of western Europe, are still absent from gardens further east, occurring only in natural habitats. Other species, such as Eurasian Blackbird *Turdus merula*, occupied parks and gardens more than 100 years earlier in western than in eastern Europe. It is often hard to tell whether the sudden colonisation of a man-made habitat depends on some small change in the habitat, which turns it from unsuitable to suitable, or on some local behavioural change in the bird itself which leads it to occupy a previously unacceptable habitat. Whatever the underlying cause, however, acceptance of man-made habitat has sometimes been followed by a large expansion in range. The spread of the Mistle Thrush *Turdus viscivorus* in central Europe from the mountain forests to the lowlands followed its acceptance of cultivated land as habitat (Peus 1951). Similarly, the spread of the Black Redstart *Phoenicurus ochruros* from the mountains of central Europe through the cities of the lowlands was associated with its use of waste land with derelict buildings as nest sites (Niethammer 1937, Witherby & Fitter 1942).

Changes in a man-made habitat need only be subtle for a species to move in or, conversely, for it to disappear. Nowhere is this more apparent than on modern farmland, where seemingly minor changes in practice have caused recent declines and range contractions in several species. The disappearance of the Corn Crake

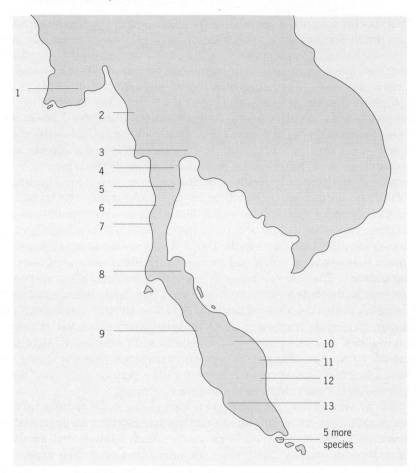

Figure 15.1   Southern limit of various birds of open country spreading south from Burma, following deforestation of the Malayan peninsula. 1. Rose-ringed Parrot *Psittacula krameri*, 2. House Sparrow *Passer domesticus*, 3. Sooty-headed Bulbul *Pycnonotus aurigaster*, 4. Barn Owl *Tyto alba*, 5. Black Kite *Milvus migrans*, 6. Black-collared Starling *Sturnus nigricollis*, 7. House Crow *Corvus splendens*, 8. Shikra *Accipiter badius*, 9. Racket-tailed Tree-pie *Crypsirina temia*, 10. Streak-eared Bulbul *Pycnonotus blanfordi*, 11. Red-whiskered Bulbul *Pycnonotus jocosus*, 12. Plain-backed Sparrow *Passer flaveolus*, 13. Coppersmith Barbet *Megalaima haemacephala*. From Ward 1968.

Crex crex from much of western Europe, which occurred around the mid twentieth century, has been attributed to mechanised grass-cutting (Norris 1945, Cadbury 1980). Traditional hayfields hand-cut in July provide ideal habitat, but an earlier cut and the use of faster (tractor-drawn) cutting equipment has killed so many Corn Crakes that the species disappeared from large areas apparently through over-kill alone. The species remains only in isolated outposts where former hay-making procedures have persisted (Green 1995). The recent disappearance of the

Grey Partridge *Perdix perdix* from large areas of farmland has been attributed mainly to herbicide use, which has destroyed the arable weeds on which the insects eaten by chicks depend (Potts 1986). For both these species, the habitat looks much the same as before, but changes in cultural practices have tipped the scales against them. Other examples of the effects of agricultural procedure on bird distributions are given later.

Some species have spread because of the creation of very specific habitats. An example is the Little Ringed Plover *Charadrius dubius*, which spread northwest in Europe during the twentieth century. Nesting naturally around lake and river shores, on areas of shingle or sand, the species has taken advantage of bare habitats provided by gravel workings, mud dredging and refuse tips. In Britain, the first known pair nested in 1938, on the dry bed of a reservoir, but in the next 50 years numbers increased to more than 400 pairs, widely scattered at suitable sites mainly in England (Parrinder 1989).

Not all attempts by birds to colonise man-made habitats have persisted. Some years ago, some Eurasian Dotterels *Eudromias morinellus*, which normally nest on the sparse alpine tundra of mountain plateaux, began to nest on arable ground in newly created drained polders, 4 m below sea-level in the Netherlands (Sollie 1961). For some years previously, the area concerned had been used as a stop-over site on migration. It is hard to imagine a more dramatic and less predictable change. Although the birds produced young, they persisted for only a few years. Similarly, isolated instances of Northern Lapwings *Vanellus vanellus* nesting on flat house roofs in England and Germany have not led to the general adoption of this habit (Cramp & Simmons 1983).

In cultivated landscapes, the Eurasian Collared-Dove *Streptopelia decaocto* has undergone one of the most rapid and spectacular range expansions ever seen in birds **(Figure 15.2)**. The species started from southern Asia, but by 1900 had extended west into Turkey, from where it later reached the Balkans. There it remained until the 1930s, after which a rapid expansion occurred northwestwards across Europe (Novak 1971, Hengeveld & Van den Bosch 1991). The first pair bred in Britain in 1955 but by 1979 numbers had increased to more than 100,000 pairs (Hudson 1965, 1972, Gibbons *et al*. 1993). Here, as in several other areas where counts were made, the population grew exponentially for the first 10–20 years, spreading by more than 40 km per year, and then levelled off, presumably as habitats became filled. The species moved mainly into villages and farms, where it could exploit the grain and other food provided for domestic livestock. It thus occupied a vacant niche, partly occupied in earlier times by the dovecote pigeon. The European expansion was only part of the total, however, for the species also spread from other parts of its range, reaching Kazakhstan and Turkmenistan, possibly from India, and down the east side of the Mediterranean through Lebanon, Israel and Egypt, presumably from Turkey **(Figure 15.2)**.

The striking feature of this unexpected range expansion was the speed with which it occurred, bringing a species from a far distant region to the western shores of Europe in 30 years. It is not obvious why the range expansion occurred, or why it occurred when it did and not earlier. Almost certainly, the impetus came from within the bird's original range, promoting a population explosion and pushing the birds beyond their existing boundaries. Other species that have spread from the southeast, but much more slowly, include the Stock Dove *Columba*

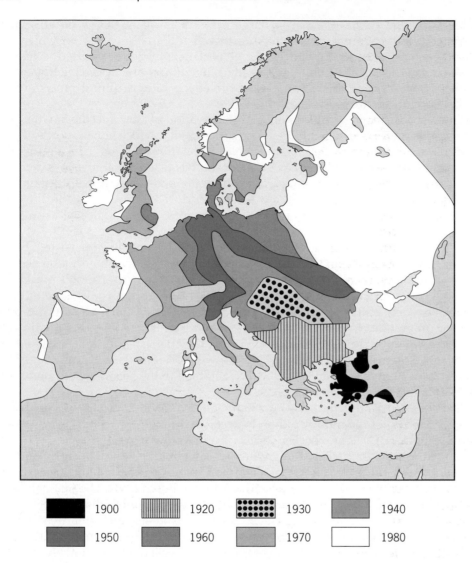

Figure 15.2   Spread of the Eurasian Collared-Dove *Streptopelia decaocto* into Europe as shown by the distribution at 10-year intervals, 1900–1980. From Hengeveld 1988.

oenas (mainly in the nineteenth century) and the Syrian Woodpecker *Dendrocopos syriacus* (mainly in the twentieth century).

The Eurasian Collared-Dove *Streptopelia decaocto* now seems set to colonise North America (Hengeveld 1993). It was introduced from the Netherlands to the Bahamas by a pet dealer. No more than 50 birds escaped in December 1974 and the first nesting was observed the next summer. Numbers increased and the species spread to other islands. In the late 1970s, birds appeared in Florida and the first nest was seen in 1982, since when it has spread across large parts of the State, and established itself in parts of Texas and Carolina (Romagosa & Labisky 2000).

Other trans-continental range expansions were shown by the House Sparrow *Passer domesticus* and Common Starling *Sturnus vulgaris*, following their introductions to North America in the late nineteenth century (Chapter 16). Like the Eurasian Collared-Dove, they depend on habitat and food provided by human activities.

Changes in food-supplies

The second category of range expansions involves bird species that have benefited from increased food-supplies. Many small birds take winter food provided by householders, enabling them to survive further north than formerly. Some seem almost entirely dependent on such handouts near the northern limits of their present ranges. European examples include European Greenfinch *Carduelis chloris* and Yellowhammer *Emberiza citrinella* and eastern North American examples include Tufted Titmouse *Parus bicolor* and Northern Cardinal *Cardinalis cardinalis* (Hildén & Koskimies 1969, Orell 1989, Robbins 1985).

By the 1980s, US citizens were spending more than \$500 million on bird-seed every year (Klinger 1982). This was enough to provide 0.5 kg of seed to each of more than a billion birds. The most abundant beneficiaries, such as Common Grackles *Quiscalus quiscula*, Red-winged Blackbirds *Agelaius phoeniceus* and Song Sparrows *Melospiza melodia*, had continental populations that then numbered tens or hundreds of millions (Terborgh 1989). However, the species that has probably benefited most from garden feeders is the House Finch *Carpodacus mexicanus* which, until the mid twentieth century, was restricted to warm regions of the southwestern States. It has since spread from its natural range, and from the New York area where it was introduced, to occupy most of the United States. In the newly colonised areas, it lives chiefly around towns, where in winter it depends almost entirely on seeds provided by householders.

In some North American species, winter survival has been further assisted by grain spilled by large-scale mechanised harvesting. Until the 1940s, maize was gathered by hand, and almost every cob was picked, but it is now harvested by machine in a process that leaves up to 10% of the crop in the field overwinter. This represents an enormous food-source, which now supports millions of birds, from songbirds to geese, over wide areas. It could have assisted the numerical increases and range expansions of several species, notably Red-winged Blackbirds *Agelaius phoeniceus* and other icterids, some of which are now able to winter further north than formerly.

In much of western Europe, where many species once fed on spilled cereal grains from stubble fields, the process has been reversed, with the switch from spring to autumn ploughing (consequent on the change from spring-sown to autumn-sown cereal varieties). This has meant that surface seeds, once available throughout the winter, are buried beyond reach soon after harvest, at once removing a large potential food-source, and thereby contributing to recent widespread declines in the numbers and distributions of Corn Buntings *Emberiza calandra*, Eurasian Tree Sparrows *Passer montanus* and other seed-eaters on European farmland (Marchant *et al.* 1990, Tucker & Heath 1994, Newton 1998).

In many seabird species, improved food-supplies and reduced persecution during the twentieth century favoured population growth and range expansion. For some species, the new food-supplies are partly land-based, provided by sewage outfalls and rubbish dumps, while for others they are largely or entirely sea-based, provided mainly by offal and other discards thrown overboard from fishing vessels and from factory ships. According to Furness *et al.* (1992), fishing boats in the waters around Britain may discharge 95,000 tonnes of offal and 135,000 tonnes of whitefish discards every year, enough to feed two million scavenging seabirds. Beneficiaries include Northern Fulmar *Fulmarus glacialis*, Northern Gannet *Morus bassanus*, Great Skua *Stercorarius skua*, Great Black-backed Gull *Larus marinus*, Lesser Black-backed Gull *L. fuscus*, Herring Gull *L. argentatus* and Black-legged Kittiwake *Rissa tridactyla*. All these species associate with fishing boats, and showed huge increases in numbers and range during the twentieth century.

Since 1880, along the western seaboard of Europe, the Northern Fulmar *Fulmarus glacialis* has spread southwards as a breeder by more than 10° of latitude (Chapter 14), the Common Gull *Larus canus* has spread southwards by about 10°, the Lesser Black-backed Gull *L. fuscus* southwards by 10° and northwards by at least 5°, the Great Black-backed Gull *L. marinus* southwards by 5° and northwards by 10° (reaching Svalbard by 1930), the Black-legged Kittiwake *Rissa tridactyla* southwards by 10°, the Great Skua *Stercorarius skua* southwards by 5° and northwards by 10° (reaching Svalbard by 1976) and the Northern Gannet *Morus bassanus* southwards by 5° and northwards by 10° (Cramp & Simmons 1977, 1983). By 1990, the Gannet bred at 21 colonies around Britain and Ireland, and at one in Brittany (France); and since then it has started nesting on Helgoland (from 1991), the Klarlov Islands off northern Russia (from 1996) and on the Mediterranean coast of France (Nelson 1997). During the twentieth century its European breeding range grew from a 15° span of latitude to more than 25°. The Lesser Black-backed Gull and Herring Gull also colonised Iceland from about 1925 (see above), and by 1980 had increased to more than 10,000 and 2,500 pairs, respectively (Cramp & Simmons 1983). Similar numerical increases and range extensions have occurred in some of these same species in North America, notably Great Black-backed Gull (by 10° south) and Herring Gull (mainly infilling) (Nisbet 1978).

Changes in nest sites

This category contains species that accept buildings as nest sites. Originally confined to breeding in areas with cliffs or caves, such species must have expanded enormously following the provision of additional sites. Striking examples are provided by swallows and other hirundines, which throughout much of the world have occupied buildings of one form or another, spreading with human settlement. Some species, such as the Barn Swallow *Hirundo rustica*, now rely almost exclusively on buildings for nest sites, so that it is hard to find them in natural sites anywhere. Other beneficiaries of man-made structures include various species of swifts and raptors which, through use of such sites, now nest widely in areas formerly closed to them (Newton 1979, 1998).

Many species that nest in trees have spread over former open grasslands, as trees have been planted and watered by people. This is evident in many North American species, which spread across the prairies following tree-planting, and by many other species in other parts of the world. In striking contrast, many hole-nesting species have almost disappeared from some forested areas, owing to modern forestry practices in which trees are harvested at a relatively young age before they can develop cavities. In such managed forests, the provision of nest boxes can lead to massive increases in the distributions and numbers of hole-nesting birds (Newton 1994a).

Changes in water supplies

In some arid regions, a major factor promoting range changes in birds has been the vastly increased availability of permanent water provided for sheep and cattle (Chapter 14). Range expansions have been noted among many species that need to drink daily, as they can now exploit food-supplies in areas previously closed to them. At the same time, true desert species have receded, possibly through competition from the newcomers (Saunders & Curry 1990). In Western Australia, one beneficiary is the Emu *Dromaius novaehollandiae*, which breeds well in wet years, but becomes short of food in subsequent dry years when many birds leave the arid areas for the neighbouring wheatlands (Davies 1977). This movement is thought to have arisen only after the start of cattle ranching in the arid zone, where provision of water enabled Emu numbers to rise above their former level. Other associated changes include the spread of grassland bird species into mallee and savannah habitats, following scrub clearance and grass growth. Species involved include the Crested Pigeon *Geophaps lophotes*, Zebra Finch *Taeniopygia guttata*, Galah *Elophus roseicapillus* and Black-shouldered Kite *Elanus axillaris*. In arid areas in various parts of the world, where water is used for irrigation of newly planted trees and crops, the avifauna has changed almost completely, as birds of cultivation have moved in and desert species withdrawn (Rustamov 1985).

Changes due to human persecution or conservation measures

Some bird species, known from bone remains or historical evidence to have formerly occurred in particular areas, are now no longer found there, even though conditions still seem suitable. The disappearance of many such species has been attributed to human persecution. Striking examples are provided by the presence, at archaeological sites on oceanic islands, of the bones of both landbird and seabird species now absent from those islands (Chapter 7). The most likely explanation is that such species were eaten to extinction on the islands concerned, thus reducing their overall ranges. Many other species were eliminated altogether, either by people directly or by predators introduced by people.

Under protection, species once reduced by human persecution have since increased to reoccupy much of their former range, with recent range expansions, evident particularly in Europe and North America, sometimes attributable to specific conservation measures. This is illustrated by various herons and ibises (formerly killed for their feathers), seabirds (killed for feathers, food, eggs collected as food), waders and waterfowl (eggs and adults taken as food), raptors (killed mainly for game preservation) and others, all of which have spread in recent decades under the benefits of protective legislation. However, these expansions represent re-occupation of former range rather than the occupation of new range.

They are different from the range expansions mentioned earlier, which are dependent on new habitats, food-supplies or nest sites, and involve the occupation of new areas.

As people have become more numerous and mobile, human presence has become a major factor preventing birds from breeding in otherwise suitable places. Obvious examples include the disappearance of plovers and other species as breeders on sand beaches used for recreation, and of ducks and other waterbirds from inland waters used for intensive boating (e.g. Cooke 1975). The disturbance caused by natural predators can have similar effects, as shown for example by the restriction of some ground-nesting birds to islands and other places secure from mammalian carnivores. In the extreme, such disturbance renders otherwise suitable habitat untenable, reducing the number of places acceptable to birds, and hence their distributions. The opposite is shown following the creation of reserves, and associated control of human disturbance, when bird usage usually increases (Madsen 1995).

OTHER CONSIDERATIONS

In most parts of the world, declines in bird numbers and distributions in recent decades have perhaps been more frequent than increases, owing almost entirely to detrimental habitat changes caused by human activities (e.g. Hildén & Sharrock 1985). The reasons for some range changes are still not understood, although they are likely to have involved one or more of the causal factors mentioned above.

The range changes described above have all occurred during the past 150 years, when they could be documented. There is no telling how many similar changes might have occurred in earlier centuries through natural or human causes. What were the distributions of species 500, 1,000 or 2,000 years ago? Historical records, in the form of unambiguous descriptions or paintings, reveal that several species were once present in particular regions from which they are now absent. Of 72 bird species recognisable in Ancient Egyptian art of 3,000 years ago, six no longer breed or winter in the region (Houlihan 1986). They include an unspecified Diver *Gavia*, the African Darter *Anhinga rufa*, Saddlebill Stork *Ephippiorhynchus senegalensis*, Bald Ibis *Geronticus calvus*, Red-breasted Goose *Branta ruficollis* and Helmeted Guineafowl *Numida meleagris*. These species include both Eurasian and African forms. The goose was last documented as present in winter as recently as the late nineteenth century, indicating that its disappearance from this area is relatively recent.

Old accounts and bone remains reveal some tantalising scraps of information, such as the unequivocal fourteenth century records of Siberian Cranes *Grus leucogeranus* in Egypt (along with the Common Crane *G. grus* and Demoiselle Crane *G. virgo* which still occur). The Siberian Crane is now known from only 2–3 discrete breeding areas: a tiny western population of less than 20 birds in the Ob-Irtysh basin which winters in Iran (10–11 birds) and India (four birds in 1995), and a larger eastern population in Yakutia (over 2,300 birds in 1993) which winters in eastern China. Hence, the nearest modern wintering site, in the Caspian lowlands of Iran, is more than 2,000 km from the Nile Delta site of the fourteenth century (Provençal & Sørensen 1998). Such records again indicate that massive changes in bird abundance levels and distributions long pre-date the main historical record. For those occurring in the last few thousand years, we cannot eliminate human activity as a possible cause.

RANGE EXPANSIONS THROUGH BREEDING IN MIGRATION AND WINTERING AREAS

Other range expansions could have apparently resulted from long-distance migrants breeding in their wintering areas (Leck 1980). Several bird species, clearly derived from European ancestors, now breed regularly in Africa. They form a continuum from long-established species, which are now specifically or subspecifically differentiated from their European counterparts, through species which appear no different from European forms, to northern migrants which nest only irregularly in Africa (Snow 1978b). Of species that are conspecific between Europe and Africa, at least five are likely to have colonised Africa in this way, namely Black Stork *Ciconia nigra*, White Stork *Ciconia ciconia*, Booted Eagle *Hieraaetus pennatus*, European Bee-eater *Merops apiaster* and Stonechat *Saxicola torquata*. This phenomenon of 'migration suspension' can also be inferred to have occurred in the New World, where several migratory species have resident insular races on various Caribbean Islands and elsewhere, and where North American Barn Swallows *Hirundo rustica* have recently nested in Argentina (Martinez 1983). It may also have led the Great Skua *Stercorarius skua*, Northern Fulmar *Fulmarus glacialis* and others to have colonised the Southern Ocean from the North Atlantic or vice versa, although long enough ago for them to have formed distinct species (Chapter 8). The same may apply to Leach's Storm Petrel *Oceanodroma leucorhoa* which breeds widely in the North Atlantic and North Pacific, and has recently been found breeding on an island off South Africa, and in potential nesting burrows on the Chatham Islands off New Zealand, within the wintering range of the northern hemisphere birds (Imber & Lovegrove 1982, Whittington *et al.* 1999). Migration suspension is a plausible explanation for such disjunct distributions, but for some species it is impossible to eliminate an alternative explanation, that they once occurred continuously between their two current ranges and have since disappeared from the intervening area.

Occasionally, expansion of breeding range has occurred through birds short-stopping on spring migration, and establishing a new population well separated from their traditional breeding area. In 1971 one pair of Barnacle Geese *Branta leucopsis* began to breed on the Baltic Island of Gotland which until then had served merely as a stop-over site on spring migration. Subsequently numbers increased rapidly to reach 1,100 breeding pairs by 1990, a mean annual rate of increase of up to 55% (Larsson *et al.* 1988, Forslund & Larsson 1991). This rate was faster than could have been achieved by reproduction alone, so must have involved immigration. The island lies in a temperate area, some 15° of latitude and 2,000 km south of the usual arctic breeding area on Novya Zemlya. In addition to the main colony on Gotland, at least nine smaller colonies have since become established around the Baltic, including some in Estonia. Similarly, a colony of Barnacle Geese of the Greenland population has recently been established in Iceland, at another stop-over site well south of the usual breeding areas.

Another example concerns the Osprey *Pandion haliaetus* which from 1984 colonised an area in central France, which lies some 800–1,000 km from the next nearest breeding areas (Thiollay & Wahl 1998). Some birds breeding in France had been ringed as nestlings in eastern Germany more than 900 km away, from which region Ospreys pass through France on migration to west Africa.

GENETIC CHANGES

Whereas in some species, the cause of range expansion has been fairly obvious, in others it has not. This has led some authors to suggest a genetic change, enabling a species to use a wider range of habitats or climes, and thereby triggering a range expansion (Wynne-Edwards 1962). This idea is almost impossible to test, but once a large range expansion is under way, genetic change has been shown to occur, as the invader adapts to its new environment by changes in morphology and physiology. Among House Sparrows *Passer domesticus* introduced to North America, changes in plumage colour and body size became apparent within 15–25 generations after birds first occupied new areas (Johnston & Selander 1964, Johnston & Klitz 1977). Populations from the north and the south of the continent now show inherent differences in morphology and in metabolic response to temperature, with birds in the north being adapted to lower temperatures than those in the south (Blem 1973). Other species, such as European Serin *Serinus serinus* and Cattle Egret *Bubulcus ibis*, developed new migratory habitats after colonising new areas, which can also be assumed to have involved genetic change (Berthold 1993). However, all these changes occurred only after the species had reached a new area, under the action of local selection. Having changed, the birds might then have been better equipped to colonise further areas where conditions were even more extreme. In this way, range expansion and genetic change could go hand in hand.

To the extent that dispersal itself is under genetic control, more dispersive genotypes might be favoured during a period of expansion. This could arise because long-dispersing individuals reaching vacant habitat, free of conspecific competitors, are likely to leave most offspring. This advantage would disappear during a period of range stability, when all areas are occupied, at least to some extent. The most obvious examples of changes in dispersal tendencies are provided by some wholly sedentary endemic island birds whose ancestors must have been highly dispersive in order to reach the islands. If over-water dispersers usually died, such dispersers would leave few, if any, descendants, and only those reluctant or unable to cross water would survive. In this way dispersers would be rapidly selected out of the island population, and traits such as flightlessness could develop. So whether or not genetic changes ever trigger range expansion, the frequency of long-distance dispersal is likely to change as a population occupies a new area, and as the trait becomes more or less advantageous.

The fact that some immigrant species have changed in response to local conditions raises the question why other species do not change continuously, enabling them to overcome one constraint after another, and undergo continual range expansion. There are several reasons why this is unlikely to happen. Firstly, in the absence of geographical isolation, peripheral populations will experience, through the movements of individuals, continual gene flow from more central parts of the range, slowing the rate of local genetic change or, in some species, preventing local adaptation altogether (Mayr 1963). Secondly, any adaptation is likely to be a compromise between conflicting selection pressures, and any change that enables a bird to cope better with one limiting factor may disadvantage it in other ways. Thirdly, most local adaptations within expanding species occur in response to climate and other physical aspects of the environment, and the scope for overcoming biotic limitations, such as competition, is probably much more limited. Within its particular niche, each species is constrained by other organisms which might themselves retaliate by genetic change. In the long term, then, the capacity of any species to continue spreading is limited by the capacity of peripheral populations to adapt to the new challenges presented by novel environments, especially when hampered by gene flow from the different environment nearer the centre of the range.

The House Sparrow *Passer domesticus*, in which the most marked genetic changes have been documented, is unusual in several respects. Firstly, the populations associated with particular settlements are often fairly isolated, and the birds themselves seldom move long distances. This would reduce gene flow from central to peripheral areas to a greater degree than in many other bird species, facilitating more rapid adaptation to local conditions. Secondly, the House Sparrow has an unusually distinct niche, being freer than most species from inter-specific competition, which may give scope for greater change. Thirdly, it often spends large parts of its time within barns and other buildings, insulated from the outside world. This could enable it to live in regions otherwise untenable, and give time for appropriate physiological and other adaptations to evolve. For these various reasons, the House Sparrow might have been able to adapt and spread to extreme areas unusually rapidly, compared with most other birds.

MODELLING RATES OF SPREAD

Mathematical models have been developed to help understand the velocity of range expansion in animals. One type, called the reaction–diffusion model, is based on a two-component process of local population increase and spatial spread (Fisher 1937, Skellam 1951). The reaction component represents the intrinsic rate of population increase of the species concerned and the diffusion component represents the measured rate of spread (Okubo 1988). This model assumes that: (1) all individuals are identical in reproduction and survival; (2) all individuals move at random throughout their lives or at least between breeding attempts; and (3) population growth is exponential. At least the first two assumptions are not realistic for birds. A second type, called the expansion model, is based on individual measures of reproduction, survival and dispersal in the species concerned (van den Bosch et al. 1990). It assumes that: (1) population growth is density independent (if not actually exponential), and (2) that individuals disperse from their natal sites but stay at the sites where they first choose to breed. As these assumptions are more realistic for many birds and other animals, the expansion model would be expected to give better predictions of the rate of range expansion than the reaction–diffusion model.

With knowledge of reproduction, survival and dispersal distances, the predicted rate of range change can be calculated for particular species and checked against observed rates (van den Bosch *et al.* 1990). Most such studies have dealt with invasions on the scale of a continent, such as the spread of the Common Starling *Sturnus vulgaris* through North America (van den Bosch *et al.* 1992) or the Eurasian Collared-Dove *Streptopelia decaocto* through Europe (Hengeveld & van den Bosch 1991). They gave a good match between predicted and observed rates. On a smaller scale, the expansion model was applied to various west European raptors that recolonised lost ground as they recovered from past persecution and organochlorine pesticide impacts. All these species had been studied in detail, and their distribution at various dates over several decades had been measured in special surveys. In several species, the square root of the area occupied increased linearly through time, showing a constant rate of range expansion (Lensink 1997). The rate of spread of different populations varied between time periods and regions, depending on persecution and other factors slowing the process. Nonetheless, the observed rates of range expansion gave a good fit to the rates predicted from published data from the same period on reproduction, survival and dispersal **(Figure 15.3)**. This held whether the range expansion involved (1) colonisation of an island (Osprey *Pandion haliaetus* re-establishment in Britain); (2) spread from a very small nucleus (Marsh Harrier *Circus aeruginosus* and Red Kite *Milvus milvus* in Britain); or (3) re-occupation of former breeding areas on a broad front from other breeding areas (Northern Goshawk *Accipiter gentilis* in the Netherlands, Common Buzzard *Buteo buteo* and Eurasian Sparrowhawk *A. nisus* in Britain and the Netherlands). Simulation modelling showed how relatively slight increases in lifetime reproduction above the required replacement rate of two young per pair could lead to rapid range expansion, and how lengthening the dispersal distances could also greatly accelerate the rate of spread.

Range contraction is more difficult to model because it is usually triggered by some environmental change (often human-induced). The pattern of range contraction therefore depends on how much of the range is affected, and populations remain in whichever parts are spared the impact involved. This mechanism has been called the 'contagion' model of range contraction by analogy with the effects of disease. Regardless of where the contagion begins, the last place affected will be the region most removed from the point of contagion, whether in the middle or on the edge of the existing range. The chief alternative to this model of range contraction is the demographic 'small population' model, which is based on two assumptions: (1) the extinction probability of a population should rise with decline in population size, and with increased variability in population size (Chapter 16); and (2) populations tend to be larger and more stable near the centre of the range, because conditions there are best suited to the species. Given these assumptions, contracting ranges should implode, with birds disappearing from edge to centre. This view makes another implicit assumption, however, namely that whatever causes the decline acts uniformly on individuals throughout the range.

Those bird populations studied during human-induced declines fit the 'contagion' hypothesis best, with populations surviving wherever (if anywhere) conditions remain suitable (Channell & Lomolino 2000). Only in the last stages of decline are such reduced populations likely to fall victim to chance demographic and genetic events that result from small population size. The Corn Crake *Crex*

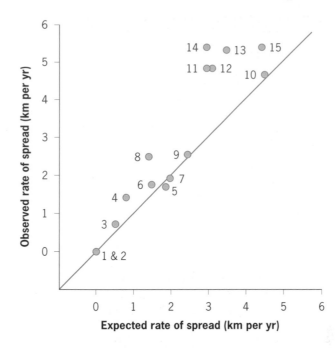

Figure 15.3 Relationship between expected rate of range expansion, based on knowledge of reproductive rates, survival rates and dispersal distances, and observed rates of range expansion in various raptor populations. The fit between predicted and observed rates is good, confirming the validity of the model and the dependence of the rate of spread on demographic values.

1. Eurasian Sparrowhawk *Accipiter nisus*, Scotland; 2. Northern Goshawk *A. gentilis*, Britain; 3. Red Kite *Milvus milvus*, Britain; 4. Common Buzzard *Buteo buteo*, Britain 1972–91; Common Buzzard, Netherlands; 6. Osprey *Pandion haliaetus*, Britain; 7. Marsh Harrier *Circus aeruginosus*, Britain; 8. Common Buzzard, Britain, 1915–54; 9. Eurasian Sparrowhawk, Netherlands; 10. Northern Goshawk, Netherlands; 11. Eurasian Sparrowhawk, eastern England; 12. Eurasian Sparrowhawk, western England; 13. Northern Goshawk, Wales; 14. Peregrine *Falco peregrinus* northern Scotland; 15. Peregrine, Wales. Modified from Lensink 1997.

crex in Britain is an example of a species that has survived in the extreme north-western edge of its range, where traditional hay-cutting methods survived, having been eliminated elsewhere because of change to other methods which kill the birds (see earlier). Similarly, on many island groups, some seabirds disappeared from the main islands after rats were introduced, surviving only on outlying islets that have so far remained rat-free. Clearly, the contraction of those bird species ranges that have been studied can best be explained in terms of the geography of extinction factors, and not simply in terms of the small population paradigm.

CONCLUDING REMARKS

Some of the above examples show how rapidly and extensively some bird species have spread over new areas that offered suitable conditions. Several species occupied whole continents within periods of 50–100 years. In each of the main examples, humans had a hand, providing suitable habitat and food-supplies and in some species also by transporting birds to new areas beyond natural barriers (see Chapters 7 and 16). Nevertheless some consistent patterns have emerged (Olsson 1969, Stjernberg 1985, Hengeveld 1989). Where it could be studied in detail, range extension occurred through occasional pairs establishing themselves outside the existing range. These pairs acted as nuclei around which local populations developed, expanding and eventually coalescing to give a continuous distribution. The initial extra-limital pairs were often first-year birds or other first-time breeders. This would be expected because the adults of most bird species tend to use the same breeding localities in successive years, so it is mostly youngsters that settle in new localities. In each area, favoured habitats were occupied first, followed by others as local numbers grew. Numbers at the front of the spreading wave at first grew rapidly, evidently from a combination of reproduction and immigration from areas already occupied, and then levelled off as the area became filled. Thus, throughout an expansion, the front appeared as a broad zone of scattered fast-growing and expanding nuclei, behind which lay an area of more continuous distribution and more stable numbers. In extreme cases, there was a clear linear relationship between the square root of the area occupied and time, suggesting unrestricted growth and little or no obstruction to onward spread (Hengeveld 1989). In some other species, however, such as the Eurasian Penduline Tit *Remiz pendulinus* or Greenish Warbler *Phylloscopus trochiloides*, the spread was slower and less regular, as periods of expansion alternated with periods of withdrawal or stability.

In theory, a species might progress in one of two main ways: by a steadily rolling continuous wave, or by spread from initial foci ahead of a spatially more contiguous region of occurrence. The first pattern (diffusion) is fed by birds dispersing short distances from previously occupied ground, and the second (called jump dispersal) is fed partly by occasional long-distance dispersers, as well as by short-distance dispersal. The distinction is partly a matter of scale, for the finer the resolution, the more a continuous wave appears like a system of isolated bridgeheads. However, when nuclei are established tens of kilometres ahead of the continuous front, they can result in a much faster rate of spread (Hengeveld 1989). The majority of the bird species that have been studied have progressed on the bridgehead system, as mentioned above.

Consideration of the historical dimension in bird ranges throws a revealing light on ecological communities. Some ecologists maintain that communities are structured in some way, mainly by competition which influences the numbers and types of species that can live together in any one habitat-area. Yet it is equally clear from the historical record, whether covering thousands, hundreds or tens of years, that communities are really no more than loose assemblages of species able to co-exist. Over time, species move individually and at different rates in response to changes in the physical environment, and come together in different combinations, producing 'communities' of varying species composition (see also Chapter 9). Communities do not persist indefinitely in balanced equilibrium.

SUMMARY

Species ranges are not static, but are continually changing in response to changing conditions. The spread of any species is likely to require both population pressure within the existing range, and suitable conditions outside it.

Many recent range changes have been associated with changing climate which may affect birds directly through influence on reproductive or survival rates, or indirectly through influence on habitats or food-supplies. Other range changes have been associated with human influence on habitats or on components of habitat, such as food-supplies or nest sites. Yet others are attributable to species recovering from past persecution, which reduced their populations to a small proportion of what the available habitat would support. Some remain poorly understood.

In many parts of the world, as forests have been replaced by farms and grazing land, birds of open country have spread while forest species have contracted. Distributional changes resulting from changes in food-supplies are evident especially in farmland species, responding to changes in cultural practices, in garden species responding to winter food-provision, and in seabirds responding to new food-supplies provided by fishing activities. Range changes resulting from nest site provision are evident, especially in swallows and other birds that make use of buildings.

When a species reaches a favourable new area, either through natural dispersal or human agency, it may spread rapidly, and several species have occupied whole continents within periods of 50–100 years. The most spectacular continent-wide range expansions witnessed in the twentieth century include (1) the natural expansions of the Eurasian Collared-Dove *Streptopelia decaocto* across Europe and of the Cattle Egret *Bubulcus ibis* through much of North and South America, southern Asia and Australia; and (2) the expansions of the House Sparrow *Passer domesticus* and Common Starling *Sturnus vulgaris* across North America and elsewhere following their localised introductions by people. These are extremes, however, and most known range expansions have occurred more slowly and over smaller areas. Some species have apparently spread by establishing breeding populations in their winter quarters, or at migration staging areas.

No evidence has emerged that genetic changes are a necessary precursor to range expansion, but genetic changes can occur during the process of expansion after the birds have reached new areas, and these changes may then facilitate further expansion. Many more avian range changes are likely to occur in the coming decades as a result of further human impact, including climate warming.

Cattle Egrets *Bubulcus ibis* over the sea.

Chapter 16
Crossing barriers

Most range expansions of landbird species involve a progressive spread through more or less continuous habitat on continents. The main requirement in such expansions is that the species should achieve a surplus of births over deaths. The advance through favourable habitat then occurs through a rolling programme of immigration and population growth, as the species spreads gradually over an ever-increasing area. In contrast, the crossing of oceans and other barriers requires a more distinct colonising event, usually by a small number of individuals that undertake a sustained long flight, the outcome of which depends largely on chance. This means that islands near to continents are likely to receive more colonists than are more remote islands, and species with good flight capabilities are more likely to achieve overseas colonisations than are others. Regular migrant birds are, of course, much more likely to turn up in unexpected places than are non-migrants, because migrants are often blown off course, but they are not necessarily more likely to stay and establish themselves. When they arrive in a new area, migrants are usually in a migratory state, programmed to continue their journeys. No doubt many, on reaching a new area, soon move on again if they are fit enough to do so.

CONSTRAINTS TO CROSS-BARRIER COLONISATIONS

Every bird-watcher delights in seeing occasional individuals of certain bird species, which normally live far away and periodically turn up as rarities. More than half of the 570 or so species on the British list are vagrants, which appear from time to time but do not regularly breed, overwinter or migrate through Britain. Some such species occur in numbers every year, while others appear less often, in extreme cases perhaps only once in several decades. The same has been noted in every well-studied region of the world, leaving no doubt that many bird species continually reach new areas. Even a land as isolated from the rest of the world as New Zealand receives a continual supply of avian vagrants, about 29% of all species recorded there being classed as 'irregular visitors' (Bell 1991). Small oceanic islands similarly receive a steady trickle of exotic birds from distant lands. Although the variety of species is somewhat restricted, dispersal is not the main obstacle to colonisation and range extension. The main problem is the difficulty of establishment. Birds that have crossed an ocean are likely to arrive so exhausted and low on body reserves that they have poor survival prospects, even in a favourable environment, and even if individuals recover from the journey, and survive long enough to breed, their low numbers render them unlikely to establish themselves.

Given favourable climate, habitat and food, factors important to overseas colonisations by birds are likely to include: (1) the chances of birds getting there, in turn dependent on distance from a source area, prevailing wind and other weather conditions, and the dispersive and flight capabilities of potential colonists; (2) the numbers of simultaneous arrivals; (3) the chances of subsequent arrivals to reinforce earlier ones; (4) the success of breeding and the rate of population growth; (5) the amplitude of population fluctuations; (6) the ultimate population size achievable; and (7) the intensity of competition, parasitism and predation from already established species. Competition with existing residents is often regarded as a major factor preventing many potential colonists from establishing themselves on islands (Lack 1976).

Probability of arrival

The importance of distance from a source area in influencing the frequency of arrivals by non-native species is shown by two types of observation. The fact that islands close to continents (such as Helgoland off Germany) receive many more vagrant birds than remote islands (such as Ascension in mid Atlantic) need hardly be stressed. On continents, the numbers of vagrant species that appear in particular localities each year decline with increasing latitude, matching the downward trend in overall species numbers, as shown for western North America in **Figure 16.1**. They also form a decreasing proportion of the local avifauna. Secondly, among closely related species, the chance of any one appearing in a given area declines with increasing distance from its regular range. This unsurprising relationship is shown for the occurrence of various warblers in California in **Figure 16.2**. These birds migrate between eastern North America and Central or South America, but species that breed nearest to California are those most often seen there. Other groups of birds, which are more or less mobile than warblers, show different levels of vagrancy, but again distance from regular range has a major influence on occurrence, as does prevailing wind direction.

The influence of wind conditions on the occurrence of birds outside their regular range is evident to any bird-watcher. Almost every autumn, on days of easterly winds, migrants that would normally pass down the western seaboard of continental Europe turn up on the east side of Britain. In some autumns with prolonged easterly winds, migrants that breed in Siberia also appear, while at times of westerly winds, occasional vagrants from North America turn up. In addition, spring migration 'overshoots' of birds that breed further south in Europe usually occur under southerly winds, especially when these are accompanied by high temperatures further north.

In addition to wind drift, other vagrants may appear because of directional or navigational 'flaws'. So-called 'reverse migration' can be illustrated by the autumn appearance in western Europe of eastern Palaearctic species, such as Pallas's

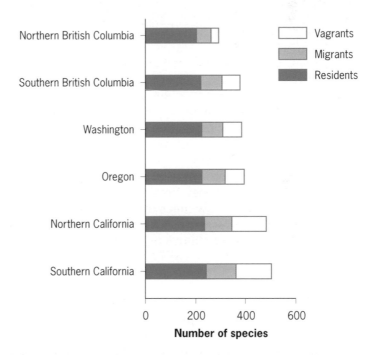

Figure 16.1   Relative contributions of resident, migrant and vagrant species to the regional bird lists of western North America. The percentage of vagrants in the total avifauna of southern California is 27%, northern California 29%, Oregon 20%, Washington 21%, southern British Columbia 20%, and northern British Columbia 9% (r = −0.84, testing percentage against latitude, P<0.05). Redrawn from Stevens 1992.

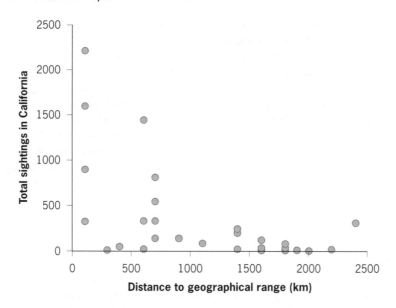

Figure 16.2   Total numbers of sightings of different species of eastern North American warblers in California in relation to the distance between California and the western edge of the breeding range (r = 0.55, P<0.01). Redrawn from Stevens 1992.

(Lemon-rumped) Warbler *Phylloscopus proregulus* and Yellow-browed (Inornate) Warbler *P. inornatus* (Rabøl 1969, Thorup 1998). These species normally migrate from Siberia to southeast Asia, but the same journey in the opposite direction on a great circle route (involving a 180° navigation error or a switch from the autumn to spring direction) would bring them to Europe. Directional tests of migrant individuals trapped in Europe are consistent with this view (Thorup 1998). If such deviation from the usual migration direction is genetically controlled (say by mutation), this is one way in which new migratory habits and range extension could arise. It is not clear, however, how much the patterns of vagrancy in migrant birds can tell us about their cross-water establishment in new areas because, as mentioned above, most migrants soon move on again.

Founder population

The importance of founder population size in the colonisation of a new area is self-evident: the more individuals that arrive at one time, the more likely is the species to become established, given favourable conditions. To state the obvious, any invading bird needs a conspecific mate before it can breed in a new area, and the resulting offspring in turn need to find mates. These requirements are more readily met if many individuals have reached an area at the same time. Because many bird species travel in flocks, multiple arrivals are perhaps more likely in birds than in most other organisms.

   In natural colonisations, the numbers of founder members are almost never known with certainty. However, the importance of numbers can be illustrated from birds released in introduction schemes. Of 133 exotic species introduced to New Zealand during the late nineteenth century, the approximate numbers released were known for 47 species. Among these, 83% of species in which more than 100 individuals were released within a 10-year period became established, compared with 35% of species in which 11–100 individuals were released, and 7% of those with 2–10 birds released. In total, 21 species still had populations surviving in the wild in 1969–79, up to 100 years later (Green 1997).

Much lower success rates were recorded for gallinaceous birds in North America, possibly because shooting hindered establishment, but the numbers released at one time were still important (Pimm 1991). Of 424 introductions of seven species that established themselves in at least one area, only 64 led to self-sustaining populations. On average, releases of 200 or more birds had a 15% chance of success, releases of 100 birds had a 10% chance of success and releases of 50 birds had a 5% chance of success. Similar relationships between numbers released and probability of establishment held for other species in other areas (Long 1981, Newsome & Noble 1986, Griffith *et al.* 1989, Beck *et al.* 1994, Veltman *et al.* 1996, Wolf *et al.* 1998). Some introduction programmes led to successful establishment only when releases were sustained over several years. Examples from Britain include the Red-legged Partridge *Alectoris rufa* and Little Owl *Athene noctua*, in both of which the initial introductions failed, but persistent releases eventually led to establishment.

The simultaneous arrival of large numbers of individuals in a new area does not necessarily result in establishment. In 1927, several flocks of Northern Lapwings *Vanellus vanellus*, totalling more than 1,500 individuals, crossed the Atlantic from northern Europe to Newfoundland (Witherby 1929). The flight occurred during a cold snap in December, when the birds would normally have moved southwest into Ireland. Instead, they became caught in a strong east–west airstream which could have carried them the 2,900 km journey in 24 hours. Some days after their arrival in Newfoundland, the weather turned cold. Many birds perished, while others scattered over a wide area, but none is known to have persisted to the spring. One of the birds found dead had been ringed as a chick in northern England. Lapwings have continued to reach North America, appearing fairly frequently on the east coast, but so far as I know they have not been recorded breeding there.

At the other extreme, some successful colonisations could indeed start with one pair. Imagine the odds against two Spotted Sandpipers *Actitis macularia* from North America reaching the same Scottish location and nesting thousands of kilometres from their normal range. Yet it happened on at least one occasion, but did not give rise to an established population (Wilson 1976). However, a single pair of Indian Peafowl *Pavo cristatus* released in 1909 led to the establishment of a sizeable population on Hawaii, which still persists (Moulton *et al.* 2001).

On matters of colonisation, records from Britain are especially revealing, not only because this species-poor island lies just off a large continental land-mass, but also because it has more skilled bird-watchers per unit area than almost any other similar-sized area. At least 27 bird species were recorded breeding either once or sporadically in Britain during a 200-year period, but without establishing a regular population (O'Connor 1986). No doubt these instances are under-estimates, considering the slim chances of such events being reported.

Nevertheless, in the same period some 30 other species did successfully establish themselves as regular breeders. Several reached Britain after a history of population increase and range expansion in continental Europe, including Cetti's Warbler *Cettia cetti*, Savi's Warbler *Locustella luscinioides* and Eurasian Collared-Dove *Streptopelia decaocto*. In addition to these European species that appear to have established themselves naturally, at least 27 non-native species (mainly waterfowl) established themselves through escapes and releases of captive birds.

The importance of a nearby population to maintain the influx to new areas is suggested by the location of establishment. Initial breeding attempts that subsequently gave rise to established British populations were almost exclusively in the east of the country, near to the source of colonists, whereas breeding attempts that failed to give rise to populations showed a much wider scatter (O'Connor 1986). Immigration over a protracted period increased the chance of success, presumably because the invasion was repeated many times in different conditions and because it could enhance the rate of population growth.

Population growth and fluctuation

Following initial breeding attempts, the factors likely to favour the establishment of a permanent population are self-evident, at least as generalisations: (1) the larger the number of founders, the greater the chance of establishment; (2) a species that in prevailing conditions can achieve a high rate of population growth will pass the vulnerable (low numbers) stages more quickly than one that can increase only slowly; (3) a species prone to large fluctuations in numbers will reach the lows that predispose extinction more often than one that is more stable; (4) a species that can achieve a large population size is more secure in the long term than one that, in the area concerned, can achieve only a small population. For any given land area, a small species that can live at high density is more likely to achieve a population size sufficient for long-term persistence than is a large species that can live only at low density.

Empirical studies have largely confirmed these predictions (e.g. Pimm *et al.* 1988, Diamond & Pimm 1993). On various islands, where annual counts of breeding birds were made, local extinction rates were related to population size and stability, and to associated factors, such as body size (large species less numerous than small ones), trophic level (carnivores less numerous than herbivores) and ecological specialism (specialists less numerous than generalists). In general, large species were more prone to local extinction than small ones because they lived at lower density; but at similar low numbers, small short-lived species were more prone to local extinction than large ones because their higher and more variable mortality rates meant that they fluctuated more from year to year, and frequently reached vulnerable low levels.

Interactions with other species

Apart from demographic problems connected with small population size, the main forces opposing establishment in an otherwise favourable area are likely to result from species already resident there, causing competition, predation or parasitism. Although these pressures can act in a density-dependent manner, being less severe at low densities, they could continually reduce a population to levels where random fluctuations could lead to local extinction (MacArthur & Wilson 1967).

The evidence that competition can prevent establishment is entirely circum-stantial, namely that invading or introduced birds have more often established themselves in areas with impoverished avifaunas (such as New Zealand and other remote islands) and in areas of new man-made habitats (such as cultivated land), than in continental areas of climax vegetation (Chapter 7). It is in the latter, with long-established species-rich communities, that a new invader is likely to meet the strongest competition and be least able to establish itself, especially if it comes up against a native species with similar ecology.

Any new invader is likely to bring some of its parasites with it, but in a new area it may in addition be exposed to new parasites and to new predators, which are again likely to be most numerous in more species-rich communities. Moreover, any new colonist may be more susceptible to infection or attack than the local species, which have evolved some measure of resistance. Examples of predation and parasitism limiting bird distributions are given in Chapter 14.

Another factor thought to contribute to success in establishment is the nature of the land-mass from which the colonist comes. Widely distributed organisms from continental areas with diverse communities tend to be relatively successful in establishing populations in small isolated areas containing few species, such as oceanic islands. Darwin (1859:299) put it this way: 'widely ranging species abounding in individuals, which have already triumphed over many competitors in their own widely extended homes, will have the best chance of seizing on new places, when they spread into new countries'. This tendency could explain why bird colonisations are normally from mainlands to islands, rather than vice versa, and why colonisations between certain continental areas have been much more one way than the other, for example from Asia to Australia (Chapter 5).

SPECIES DIFFERENCES

Whether through dispersal or establishment ability, certain bird families are more successful as colonisers than others. In particular, species vary greatly in their ability to cross water or other barriers. Individuals of some species travel thousands of kilometres between continents each year on their seasonal migra-tions, whereas those of other species spend their entire lives within the same few hectares. **Figure 6.4** gives some idea of how far eastwards into the Pacific Ocean various island colonists have reached, starting from the Australasian region. The fact that some bird families have reached more distant islands than others can be attributed partly to their superior powers of over-water dispersal. Further east, the islands of Hawaii in the central Pacific are some of the most remote, lying at 4,000 km from North America and 6,500 km from Japan. The difficulty of getting there is illustrated by the fact that the entire endemic avifauna (including extinct forms) could have been derived from only 21 colonisations (Cornutt & Pimm 2001).

In general, birds of wetland habitats are exceptional in their abilities to disperse and locate areas of suitable conditions, for they often have to change quarters as some wetlands dry up and others appear. It is perhaps not surprising, therefore, that herons, ibises, pelicans and waterfowl have become some of the most widely distributed birds in the world, with some species present on almost every continent and many oceanic islands. Six species of heron are found on New Zealand and eight on various Pacific islands. The fact that most are not subspecifically differentiated from continental forms suggests that they are not genetically isolated, and are subject to continual immigration. Good dispersal abilities are apparent in other species of unstable and successional habitats, which continually disperse to colonise new places. Such species are more likely to strike out on long flights across unsuitable terrain than are species restricted to more stable environments, such as rain forest.

Stretches of open water offer insuperable physical or psychological barriers to the movements of some bird species. For example, the Ruffed Grouse *Bonasa umbellus*, which is found across the boreal region of North America, once occurred naturally on only three of the Michigan islands in the Great Lakes, all three within 0.7 km of the mainland. It was absent from all other islands lying at greater distances from shore. To examine whether poor dispersive powers explained this distribution, Palmer (1962) tested the flight capacity of several grouse over water. None could fly as much as 700 m, so Palmer concluded that Ruffed Grouse are not capable of crossing water barriers of this distance. When more distant islands were stocked artificially with Ruffed Grouse, self-sustaining populations became established (Moran & Palmer 1963). Clearly, different species of birds vary greatly in what for them constitutes a water barrier.

Species that are capable of flight, but apparently unable or unwilling to cross water, are fairly frequent in the tropics. Large numbers of such species live in coastal lowlands, but have never been recorded from any island lacking a Pleistocene land-bridge to the mainland: not even from islands only a few kilometres offshore (Diamond 1981). In tropical southeast Asia, such species include all broadbills (Eurylaimidae), malkohas (*Phaenicophaeus*), pheasants, partridges, forktails, whistling thrushes and song babblers, and in the neotropics they include most or all species of the families Furnariidae, Formicariidae, Dendrocolaptidae, Pipridae, Bucconidae, Ramphastidae and Galbulidae. In addition, the major Amazonian rivers are such sufficient barriers to dispersal that they have promoted (or maintained) the genetic differentiation of bird populations on opposite sides (Chapter 11; Caparella 1988). The numbers of taxa separated by rivers are greatest nearest the sea, where the rivers are widest, and decline upstream where the rivers become narrower and less effective as dispersal barriers (Chapter 11). Barriers deemed inconsequential for temperate zone forest birds may thus be substantial for tropical forest birds.

The dispersal abilities of some groups of birds may incidentally affect the distributions of others. For example, the Phasianidae and Megapodiidae families have mutually exclusive and complementary distributions. This probably results from the ability of the Phasianidae, based in southeast Asia, to displace megapodes, while being unable to cross water barriers. Megapodes, on the other hand, are excellent over-water colonists and persist in areas inaccessible to phasianids, mainly in Australasia and western Pacific Islands. Despite their differences in morphology and behaviour, the distribution patterns of megapodes and phasianids seem to indicate that they act as ecological counterparts which cannot co-exist (Olson 1980). This may not be the whole story, however, for pheasants exist alongside predatory cats and civet-cats, but megapodes do not (Dekker 1989). Megapodes may therefore be more vulnerable to these predators.

Species also vary in their response to land barriers. While some species on migration regularly cross deserts as large as the Sahara, or mountains as high as the Himalayas, the fact that avifaunas often differ markedly on opposite sides of such barriers implies severe constraints on the dispersal and cross-barrier establishment of many species. At the extreme, some montane forest birds have subspeciated (implying little or no immigration) across gaps in habitat only 15–30 km wide, so closely are they tied to their particular types of forest (Moreau 1966). Equally revealing are the Rift Valley systems in eastern Africa (Benson *et al.* 1962). The bottoms of these valleys, sometimes only a few kilometres wide, have a hot, dry climate and desert-like vegetation avoided by the birds of the surrounding plateau woodlands and forest. Some species are confined to one side of the valley or the other, and some of those that occur on both sides are subspecifically different, indicating at most only limited gene flow **(Table 16.1)**.

Table 16.1   Species and subspecies differences between the two sides of the Rift Valley systems of East Africa, implying their effectiveness as barriers to bird dispersal. From Benson *et al.* 1962.

Valley	Number of species confined to		Number of subspecies differences between the two sides
	South or west side	North or east side	
North–south division			
Limpopo	15	31	15
Zambesi	8	35	30
East–west division			
Luangwa[1]	22	8	9
Nyasa/Shire	19	4	13

[1] Dowsett (1980) points out that in the Luangwa area there are other barriers to the dispersal of forest birds besides the rift.

Barriers and bridges

Studies of past large-scale colonisation events have led to the division of dispersal routes into three types: (1) corridors which provide unimpeded travelways for all species, because they contain all necessary habitats and so can support the same animals from one end to the other; (2) filter routes which connect regions with only a limited range of habitats, so favour the passage of only parts of a regional fauna; and (3) so-called sweepstake routes across hostile habitat, such as sea, which only a proportion of species succeed in crossing and only in chance favourable circumstances (Chapter 5; Simpson 1965). The three types of dispersal route allow the passage of increasingly restricted subsets of species.

It would be wrong to assume that, because two large land-masses are connected by a land-bridge, all bird species can move freely between the two. North and South America are joined by a narrow land-bridge covered mainly by tropical forest, which offers conditions suitable for only some species from each continent.

This bridge is of no more use to the birds of boreal forest or tundra than was the former Beringian land-bridge (linking Alaska and Siberia) to tropical forest birds from further south. Such bridges thus act as filter routes, allowing certain species to pass but not others. If the two ends of the corridor are occupied by different habitats, the flow of species is even further reduced. In the extreme situation, the major oceans, deserts and mountain ranges permit only sweepstake dispersal for most landbird species. This term reflects the enormous odds against crossing, and the occasional rare successes. Which of various species of similar dispersal powers make it across such barriers, and in what order, may be largely matters of chance.

The very landscape features that act as barriers for most bird species provide unimpeded passage for others. Some pelagic birds wander freely over oceans but are stopped by major land areas; some montane birds use mountain chains as corridors of dispersal, while extreme desert specialists move freely through some of the most inhospitable areas on earth. Hence, despite almost universal powers of flight, bird species differ in what constitutes for them a barrier to dispersal, and in their ability and willingness to cross different types of substrate.

As one last complication, the above generalisations apply to the extension of breeding ranges, but on migration some bird species have evolved adaptations that enable them to cross some of the most hostile of barriers (Chapter 18). For example, several North American shorebird species migrate to winter on various Pacific Islands, notably the Pacific Golden Plover *Pluvialis fulva* and Bristle-thighed Curlew *Numenius tahitiensis*, which twice each year cross the ocean from Alaska. Even more impressively, Bar-tailed Godwits *Limosa lapponica* migrate the 11,000 km journey from Alaska across the Pacific Ocean to New Zealand. Other migrant landbirds regularly cross the Atlantic between eastern North America and Brazil (up to 2,500 km), while others cross the Mediterranean Sea and Sahara desert in a single flight (up to 2,000 km). The dispersal that leads to extension of breeding range is evidently a very different process from regular seasonal migration (Chapter 17).

SUCCESSFUL CROSS-WATER COLONISATIONS

While large stretches of ocean present an effective barrier to the range extension of most birds, this is not true for all species all the time. Without long-distance cross-water colonisation, most oceanic islands would be bereft of native birds. Moreover, the fact that island birds now show different levels of distinctiveness from their continental ancestors implies that cross-water colonisations have been occurring for at least many millions of years.

Several long-distance overseas colonisations have been recorded in the last 200 years. The Cattle Egret *Bubulcus ibis*, so named because of its association with large grazing mammals, was once confined to warmer regions of the Old World, but had from time to time appeared on mid Atlantic islands, such as St Helena. Around 1880 it is thought to have crossed the Atlantic naturally, because small numbers appeared in Surinam. The species prospered, and in the next 100 years it spread to occupy grassland habitats through much of South and North America and all the major Caribbean Islands **(Figure 16.3)**. In the opposite direction, the species spread from Asia via New Guinea (1941) to western Australia (1948) and

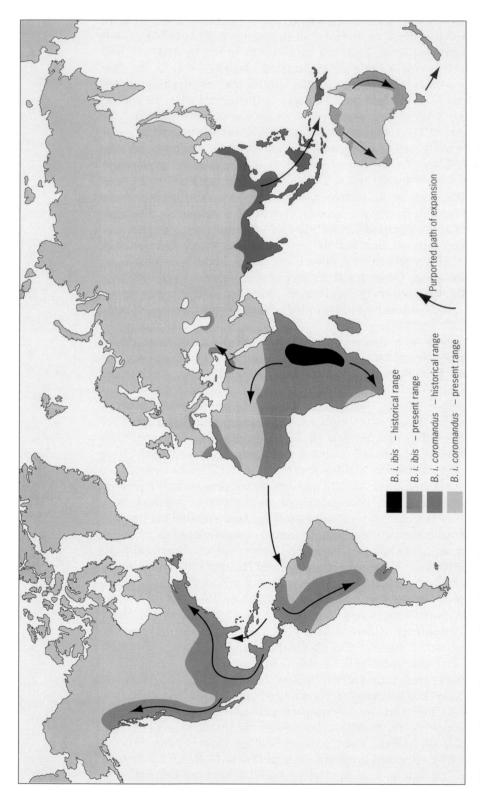

Figure 16.3 Spread of the Cattle Egret *Bubulcus ibis*. First reported sightings in westward expansion of *B. i. ibis*: Surinam 1877, British Guiana 1911, Florida 1941, Venezuela 1943, Aruba 1944, Colombia 1946, Bermuda 1953, Bolivia 1953, Costa Rica 1954, Barbados 1956, Panama 1956, Cuba 1957, Galapagos 1960. In eastward expansion of *B. i. coromandus*: New Guinea 1941, Australia 1948, New Zealand 1963, Turkey 1968. Mainly from Handtke & Mauersberger 1977 and Maddock & Geering 1994.

B. i. ibis — historical range
B. i. ibis — present range
B. i. coromandus — historical range
B. i. coromandus — present range

Purported path of expansion

then to New Zealand (1963). This species thus provides another example of natural large-scale range extension in birds, all the more impressive because it involved long sea journeys: some 2,900 km from Africa to South America, and 1,600 km from Australia to New Zealand (Maddock & Geering 1994). In the time since it arrived in new areas, the species has also established regular migrations, including one between Australia and New Zealand. In these expansions, it was no doubt helped by the widespread presence of grassland with cattle on land formerly under forest, by the fact that it occupies a distinct niche, and in some areas can breed twice in a year, achieving a rapid initial rate of population growth.

Other species may have crossed the Atlantic near its widest part long before there were ornithologists to record them. This is evident in the presence of several other Old World species in the warmer parts of the New World, notably the Striated Heron *Butorides striatus*, two species of whistling ducks *Dendrocygna bicolor* and *D. viduata* and a pochard *Netta erythrophthalma*. All these species may once have had a continuous distribution across the northern continents in regions from which they are absent now, but a transatlantic flight seems the most likely mode of colonisation. Other landbird and seabird species that breed on both sides of the Atlantic have occasionally been found from ring recoveries to have crossed it, though mostly at higher latitudes where sea distances are shorter **(Table 16.2)**.

Since the occupation of New Zealand by European people, and the resulting habitat transformations, several bird species have crossed the 1,600 km of sea and colonised the country from Australia (Baker 1991, Bell 1991). They include the Silvereye *Zosterops l. lateralis* (since 1855), Grey Teal *Anas gracilis* (re-established since 1916), Spur-winged Plover *Vanellus spinosus* (since about 1932), Masked Lapwing *Vanellus miles* (since 1940), White-faced Heron *Egretta novaehollandiae* (since 1940), Royal Spoonbill *Platalea regia* (since 1950, still rare), Eurasian Coot *Fulica atra* (since 1954), Black-fronted Dotterel *Elseyornis melanops* (since 1954), Welcome Swallow *Hirundo neoxena* (1920s) and Cattle Egret *Bubulcus ibis* (since 1958). Other species have bred occasionally, and some may now be in the process of establishing themselves. Such species almost certainly also reached New Zealand in earlier times but did not persist, possibly because suitable habitat was lacking then or because endemic species (now extinct) outcompeted them. Most have occupied new man-made habitats, in which lack of competition could have favoured their establishment. Three other species from Australia, the White-eyed Duck *Aythya australis*, Red-necked Avocet *Recurvirostra novaehollandiae* and the Dusky Woodswallow *Artamus cyanopterus*, flourished in New Zealand for a while and then died out. In contrast, only one species, the Kelp Gull *Larus dominicanus*, has colonised parts of southern Australia from New Zealand. Fewer movements this way would be expected, considering the smaller number of species in New Zealand, and the much richer established avifauna in Australia.

Some 30–40 years after their introduction to New Zealand (mainly in 1860–70), some species began to appear naturally on the small islands that lie up to 1,200 km away (Williams 1953). Immigrant species on particular islands vary from one or two on Macquarie, which has the harshest climate and the smallest range of habitats, to 11 on the Chatham Islands. The Common Starling *Sturnus vulgaris* is the most widespread, having appeared in all nine islands **(Table 16.3)**. Such extensive dispersal is unsurprising, but such high rates of establishment are unusual, and may again have been helped on some islands by human impacts on the natural vegetation, creating conditions more suited to European birds.

Since 1932 about one landbird species per decade has colonised southern Florida from the Caribbean Islands to the south. Before breeding in Florida, all these species appeared with increasing frequency as vagrants, following population increase in their homeland. This again implies that range expansion tends to be driven by population increase in source areas, rather than by habitat changes in the colonised areas. Recent immigrants include the Shiny Cowbird *Molothrus bonariensis*, Smooth-billed Ani *Crotophaga ani*, Cuban Yellow Warbler *Dendroica petechia gundlachi*, Antillean Nighthawk *Chordeiles gundlachii*, Fulvous Whistling Duck *Dendrocygna bicolor* and Cave Swallow *Hirundo fulva cavicola* (Hoffmann *et al.* 1999). Although the immigrants are presumed to have come from Cuba and other islands, most are also found widely on the South American mainland.

Other cross-sea colonisations recorded in recent times include the occupation of South Georgia by the Speckled Teal *Anas flavirostris* (population discovered 1971), of Madagascar by the Squacco Heron *Ardeola ralloides* since the 1950s (Langrand 1990) and of Greenland by the Fieldfare *Turdus pilaris* in 1937 (Salomonsen 1951). The latter followed a single invasion of birds presumed to have been blown off

Table 16.2 Transatlantic ring recoveries of various birds that breed on both sides, so could not otherwise be recorded as vagrants. From Dennis 1990, Nisbet & Safina 1996 and references therein. Regular transatlantic migrants listed in footnote.

	Recoveries from east to west	Recoveries from west to east
Little Blue Heron *Egretta caerulea*	0	1
Black-crowned Night Heron *Nycticorax nycticorax*	0	1
Caspian Tern *Sterna caspia*	0	1
Common Tern *Sterna hirundo*	0	4
Sandwich Tern *Sterna sandvicensis*	0	1
Roseate Tern *Sterna dougallii*	1	0
Gull-billed Tern *Sterna nilotica*	3	0
Black-legged Kittiwake *Rissa tridactyla*	0	1
Herring Gull *Larus argentatus*	0	1
Northern Gannet *Morus bassanus*	0	3
Leach's Storm Petrel *Oceanodroma leucorhoa*	0	3
Northern Fulmar *Fulmarus glacialis*	0	3
Green-winged Teal *Anas crecca*	0	3
Northern Pintail *Anas acuta*	0	1
Mallard *Anas platyrhynchos*	0	1
American Coot *Fulica americana*	0	1
Peregrine Falcon *Falco peregrinus*	0	1
Semipalmated Plover *Charadrius semipalmatus*	0	1

The above species are all classed as vagrants and were recovered at various sites in eastern North America or in western Europe from Britain south to Morocco and the Azores. Regular transatlantic migrants include Manx Shearwater *Puffinus puffinus*, Great Shearwater *Puffinus gravis*, Arctic Tern *Sterna paradisaea*, Sooty Tern *Sterna fuscata*, Brent Goose *Branta bernicla*, Ruddy Turnstone *Arenaria interpres*, Red Knot *Calidris canutus*, Purple Sandpiper *Calidris maritima*, Northern Wheatear *Oenanthe oenanthe* and several other species (such as Pink-footed Goose *Anser brachyrhynchus* and Barnacle Goose *Branta leucopsis*) that move between east Greenland and Europe.

Table **16.3** Birds introduced to New Zealand and Australia during 1860–70 that later established themselves naturally on other islands in the Southern and Pacific Oceans. + = breeding, (+) = present but no breeding records. From Williams 1953.

Island	Size of island (km²)	Distance in km of island from New Zealand & Australia		Arrival dates of various species	Skylark	Song Thrush	Eurasian Blackbird	Dunnock	Common Starling	Eurasian Greenfinch	Eurasian Goldfinch	Lesser Redpoll	Chaffinch	Yellowhammer	House Sparrow	Total species
Sunday (Raoul)	290	880 NNE	2,800 E	1890–1910	(+)	+	+		(+)		+			+	+	6
Chatham	906	660 ESE	2,900 ESE	1892–1900	+	+	+	?	+	+	+	+	+	+	+	10–11
Antipodes	18	800 ESE	2,100 ESE	Before 1900					(+)		+					2
Campbell	117	560 SE	1,600 E	1890–1900		+	+	+	+	+	+	(+)	(+)		+	9
Snares	1.3	96 SW	1,440 ESE	1890–1900		+	+	(+)	(+)		+	(+)	(+)		+	8
Auckland	648	320 SW	1,520 SE	1890–1900	+	+	+		+	+	+	+	+		+	8
Macquarie	260	880 SSW	1,280 SSE	1894–1912					(+)			+				2
Lord Howe	13	1,200 WNW	560 E	1913	+	+			(+)			(+)		+		4
Norfolk	31	720 NW	1,280 E	1912	+	+	+		+	+					+	5
Numbers of islands reached by each species					3	7	6	2–3	9	3	6	6	4	3	5	

Scientific names of species in table: Skylark *Alauda arvensis*, Song Thrush *Turdus philomelos*, Eurasian Blackbird *Turdus merula*, Dunnock *Prunella vulgaris*, Common Starling *Sturnus vulgaris*, Eurasian Greenfinch *Carduelis chloris*, Eurasian Goldfinch *Carduelis carduelis*, Lesser Redpoll *Acanthis flammea*, Chaffinch *Fringilla coelebs*, Yellowhammer *Emberiza citrinella*, House sparrow *Passer domesticus*.

course on migration. The population remains in southern Greenland throughout the year, free of competition from any related species. Despite being a common winter visitor to Iceland, the Fieldfare has not established itself as a breeder there, perhaps through competition with the closely related Redwing *Turdus iliacus* which breeds commonly in that country. The two species breed alongside one another in parts of Eurasia, but their co-existence may be more difficult on an island with limited food sources (Chapter 6).

TRANSLOCATIONS BY PEOPLE

About 5% of all bird species have been introduced by human action to parts of the world that they would not be expected to reach naturally (Long 1981, Lever 1987). Most are either small seed-eating or fruit-eating passerines that are commonly kept as pets, or game birds favoured by hunters. Both types adapt readily to captivity, so can survive long sea journeys. Many have thrived in their new homes, and some have undergone spectacular range expansions.

One of the most notable is the House Sparrow *Passer domesticus* which, through its association with human habitation, had already spread naturally to occupy a large part of the Palaearctic Region. Then, following introductions to other areas, it has more than doubled its geographical range since 1850 (Summers-Smith 1988). In North America, the process began in 1852 with a series of releases around New York. With the help of further introductions, notably in California in 1871 and in Utah in 1873, the species spread rapidly at about 16–24 km per year, occupying most of the continent by 1900 (Lowther & Cink 1992). Some in-filling and continued expansion have continued to the present, as people themselves have continued to occupy new areas. In the extreme north, however, where House Sparrows may be near the limits of their climatic tolerance, occupancy of settlements is still some-what tenuous, as local populations continually die out and reinvade (Johnston & Klitz 1977).

Towards the south, House Sparrows reached Mexico by 1900, spreading into Guatemala by 1966, El Salvador by 1972 and Costa Rica by 1975. They had earlier been released in Argentina (1882), Chile (1904), Brazil (1905) and Peru (1953), from which centres they spread to occupy much of South America. It seems inevitable that the birds from Central and South America will soon meet, completing their occupation of the New World. Following introductions, the House Sparrow has also occupied much of southern Africa, the eastern half of Australia, Tasmania and New Zealand, and various other islands around the world (Summers-Smith 1988). In consequence, it now occurs in every major biogeographical region and has one of the most extensive breeding ranges of any bird species. Throughout its range, moreover, it is resident year-round.

With human help, the Common Starling *Sturnus vulgaris* has also greatly expanded its range in the last 100 years. In North America, some of the earliest releases failed, and the present population stems from 80 birds released in Central Park, New York in 1890, followed by another 80 in the next year (Davis 1950, Kessel 1953). The species took about ten years to establish itself in the New York area, and then spread rapidly across the continent **(Figure 16.4)**. In 90 years it had occupied about eight million km^2, and increased to an estimated 200 million

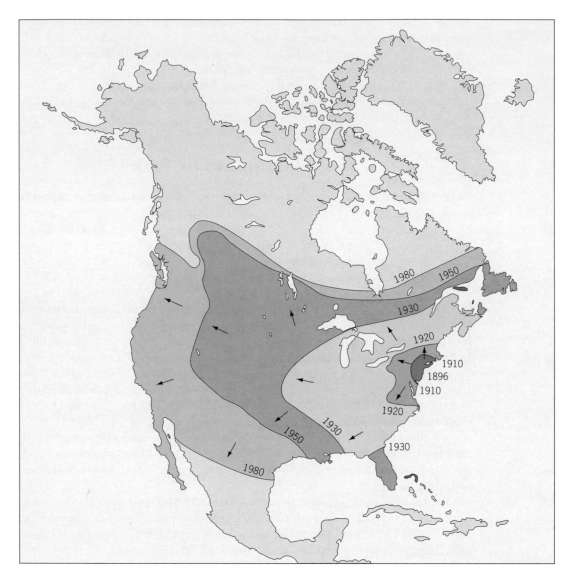

Figure 16.4   The spread of the Common Starling *Sturnus vulgaris* in North America. (Above) From the initial introduction in New York in 1890–91, the extent of the breeding range at different dates; (right) growth of population in particular areas. In each area successively occupied, numbers increased rapidly to begin with and then levelled off. The slower growth rate in New York was probably due to the lack of immigration in the founder population, compared with Ohio and Massachusetts. Modified from Davis 1950 and others.

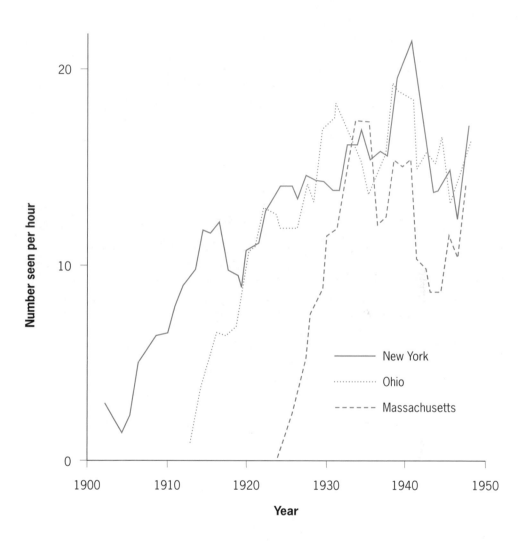

individuals. In New Zealand the species was introduced about 1890, and by 1930 had spread to suitable habitat throughout the country. Overall, through transport over ocean barriers, the Common Starling now occupies about 30% of the world's habitable land surface (Feare 1982).

Few bird species have been introduced to Europe from North America, but by far the most successful is the Canada Goose *Branta canadensis*, which is now widely established and expanding. It obtains most of its food from farmland, and may have benefited in some areas from the absence of the native Greylag Goose *Anser anser*, eliminated in earlier centuries by human hunting.

Many other examples of introductions are given in Chapter 7. The fact that they have led to the establishment of species in areas remote from their natural range implies that failure to colonise was the factor explaining their previous absence. This again underlines the importance of seas or other inhospitable areas as natural barriers to the spread of landbirds. Many introduced species are more or less restricted to man-made habitats, such as parks, gardens and farmland, and some might not have established themselves if only natural habitats were present. Other factors likely to favour establishment include the release of large rather than small numbers (as discussed above), a natural distribution that is large rather than small (assumed to reflect tolerance of a wide range of conditions, Moulton *et al.* 2001), absence of migratory behaviour, and 'a degree of adaptability' (Lever 1987).

CONCLUDING REMARKS

Much discussion has centred on what makes a 'good invader', a species able to disperse and establish itself in a new area. While all species seem able to expand through continuous habitat given favourable conditions, landbird species clearly vary in their ability to cross water and other barriers, and in the speed with which they can increase, having reached a favourable area (depending mainly on r, their intrinsic rate of natural increase). Reception areas also differ in the opportunities they offer for new colonists to establish themselves. To judge from recorded establishments, areas with impoverished avifaunas (such as New Zealand and other remote islands) and areas with simplified man-made habitats are much more open to colonisation by exotic birds than are continental areas of long-established climax vegetation, with rich existing communities. The greater array of species present in such long-established areas, and the likely resulting competition, presumably makes it more difficult for invaders to establish themselves. Hence, features both of the species themselves and of the reception areas influence the chance of establishment.

SUMMARY

Establishment of a newly arrived species on an island or other land area with suitable climate and habitat is likely to be favoured if the species: (1) arrives in large numbers or repeatedly; (2) can avoid excess competition or predation from other species; (3) can achieve a high initial rate of increase, without wide fluctuations in numbers; and (4) has a large population size. With such mobile organisms as some birds, establishment depends more on events after arrival than on the dispersal process itself. However, landbird species vary greatly in their ability to cross water and other barriers. Some flying landbird species are physically or psychologically unable to cross even 1 km of open water, while others regularly cross up to several thousand kilometres of open sea on migration. Several species have made long-distance cross-water colonisations in the last 200 years. Some highly dispersive freshwater species that depend on shallow, often ephemeral, aquatic habitats are almost cosmopolitan in their distributions.

The fact that many species have established themselves in regions where they were introduced by people indicates that barriers to natural dispersal may have accounted for their previous absence, although most occur in their new homes only or primarily in man-made habitats.

Part Five
Bird Movements

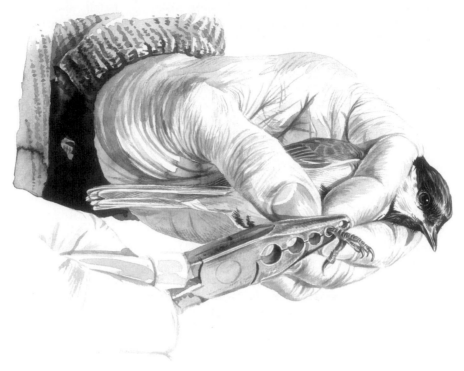

Fitting an identifying leg band to a Great Tit *Parus major*.

Chapter 17
Dispersal

The term dispersal is generally used for movements that, at the population level, have no fixed direction or distance. Such movements result in the mixing of individuals from different localities, but do not necessarily bring about any change in overall distribution. Dispersal can thus be distinguished from migration, in which individuals move in restricted directions for more or less fixed distances, only to return at a later date, thereby producing regular seasonal shifts in the centre of gravity of a population (Chapter 18). The distinction is not clear cut, however, because bird movements are so varied that they do not fall neatly into discrete categories.

In a biogeographical context, dispersal is important for two main reasons. Firstly, it can lead to range extension and to colonisation of new areas outside the existing range. Secondly, it has important genetic consequences, in helping to reduce inbreeding and promoting gene exchange between individuals derived from different localities. It thereby affects the wide-scale genetic structure of populations, and the rate at which they can differentiate. Species whose individuals show strong site fidelity, dispersing over only short distances, usually show some clinal and subspecific differentiation, as regional populations adapt to local conditions. In contrast, species whose individuals disperse over long distances, such as many waterbirds, often show little or no geographical variation in plumage or other features, supposedly because frequent long-distance gene-exchange acts against such regional adaptations arising (Chapter 2).

On a smaller spatial scale, dispersal can have a major influence on patterns of abundance and distribution within the range. It allows birds to respond rapidly to local conditions, moving from unfavourable to favourable areas, and from crowded to uncrowded localities. It can lead birds to maintain a presence in areas where they might otherwise die out, and to reoccupy vacated habitat. Dispersal is thus a crucial aspect of population ecology which affects the distribution, density and genetic structure of populations (Johnson & Gaines 1990); its study enables us to understand the dynamics of local populations, not just in terms of birth and death rates, but of immigration and emigration.

For present purposes, it is useful to distinguish: (1) natal dispersal, measured by the linear distances between natal and first breeding sites; (2) breeding dispersal, measured by the distances between the breeding sites of successive years; and (3) non-breeding dispersal, measured by the distances between the wintering sites of successive years. These three categories are relevant, not only for resident populations, but also for migratory ones which spend their breeding and non-breeding periods in different regions. In most bird species that have been studied, individuals move much greater distances between birth site and breeding site than between the breeding or wintering sites of successive years (e.g. Newton 1986, Paradis *et al.* 1998). In studying dispersal, then, the most useful measures to obtain for any population are representative samples of dispersal distances, whether natal, breeding or wintering, for it is these distances that underlie local population phenomena, and that limit the rate of range expansion and genetic interchange between populations.

In their influence on the genetic structure of populations, it is only natal and breeding dispersal that are important in most kinds of birds. This is because most species form pairs on their breeding areas, so birds choose their partners from individuals that are present in the vicinity at the time. However, some species of migratory birds, notably some waterfowl, form pairs on their winter quarters, and as birds from widely separated breeding areas may share the same wintering area, individuals from separate breeding areas can pair together. Among waterfowl, the male then accompanies the female back to her chosen breeding area. In this way, winter pairing among migrants adds an additional component to gene flow, over and above that provided by natal and breeding dispersal.

Despite their mobility, it would be wrong to think that birds can disperse anywhere within their geographical range. The majority of species show some degree of 'site fidelity', with individuals residing in the same general area throughout their lives, or returning there after successive migrations; most also show some degree of 'philopatry', with individuals remaining or returning to breed in the general area where they were hatched. There are thus behavioural constraints on the movements of individuals, which limit their dispersal distances, and hence the rate of gene flow through populations. The degree of site fidelity varies between species, however, and is often more marked in one sex than in the other.

Methods of study

Dispersal in birds has been studied primarily with the help of ringing. Most studies have been made by observers working in defined areas, recording the locations of birds ringed as nestlings or adults in one year that are found in the same area in a later year. Such records are invariably biased in favour of short-distance moves because, being confined to the study area, they are not balanced by the longer moves of other individuals which may have settled outside the area and gone undetected. Moreover, some study areas are so small compared to the natal dispersal distances of the birds themselves that only tiny proportions of ringed chicks are found breeding there in later years, the majority of survivors having settled elsewhere (see Weatherhead & Forbes 1994 for a review of passerine studies). In general, for each species, the larger the study area (up to a point), the greater the proportion of locally-raised young are later found breeding within its boundaries (see Sokolov 1997 for European Pied Flycatcher *Ficedula hypoleuca*). Some researchers, aware of the effects of size of study area on the results obtained, have attempted to devise ways of correcting for it (Barrowclough 1978, van Noordwijk 1983, Baker *et al.* 1995), or have used other means of assessing dispersal.

Less-biased information on dispersal distances comes from recoveries reported by members of the public. Even if all the birds are ringed in a particular locality, the recoveries are not confined to that locality, so can give a much more representative picture of dispersal distances. Some care is needed, however, for bird movements may be constrained by topography or landscape, and the recovery chances may vary with anthropogenic factors, such as local human population density and literacy (Baillie & Green 1987). Moreover, as most birds that are reported by members of the public are found dead, the assumption usually has to be made that individuals of breeding age recovered in the breeding season were in fact nesting, or had the potential to nest, at the localities where they were reported. Collectively, such records reflect the settling patterns of individuals with respect to ringing site, regardless of their movements in the period between ringing and recovery. In some large, long-lived species, care must also be taken to separate the immatures, for these non-breeding birds may occupy areas partly different from the breeding adults of their population (see later).

A third way of studying dispersal is by use of radio-tags, which enables the movements of individuals to be followed day to day. This method thus gives information on the actual routes travelled by the birds, and on their individual behaviour. The method is expensive, however, and with present technology can be used only on birds large enough to carry a tag for the necessary long period — ideally between fledging and first-breeding.

Costs and benefits of dispersal

Dispersal seems to be an inherent feature of birds and other animals which occurs to some extent regardless of external conditions. In this sense, it is an active prospective process. However, external conditions can have a major influence on the distances that individuals move and where they settle, and hence on whether they stay within or leave particular study areas. To some degree, then, dispersal is environmentally influenced, as the birds respond to prevailing circumstances.

The supposed advantages of site fidelity, providing conditions permit, are that the individual can benefit from local knowledge. Familiarity and prior ownership might also give a bird an advantage in competitive interactions with other individuals, making it better able to defend its feeding and breeding sites against potential takers. Moreover, through long-term residence over many generations, populations tend to become adapted through natural selection to the conditions prevailing in their particular region, so by remaining in (or returning to) the same general area, individuals occupy regions to which both they and their likely breeding partners are well adapted. It is in this way that philopatry predisposes the development of local adaptations, and once such adaptations arise, they in turn reinforce the advantages of philopatry. The benefits of local experience and of local adaptation, acting at the level of the individual, could thus be the main selective forces underlying site fidelity in birds, where this is feasible.

But there are also advantages in dispersing. One is that birds can leave areas where conditions are poor or overcrowded to find somewhere better. Some birds occupy successional habitats, which get less suitable over time, while others exploit patchy or ephemeral habitats or food-sources, which are available in different places in different years. By moving from one good area to another, as appropriate, individuals can exploit sporadic habitats or food-sources and can thereby enhance their survival and reproductive prospects.

Another advantage of dispersal is that it could reduce the chance of inbreeding. In strongly philopatric species, siblings are likely to settle close to one another and to their parents. If they have no means of kin recognition, they could end up pairing together and suffer the deleterious consequences of inbreeding (see later). In contrast, if brood-mates disperse individually, in different directions, the chances of each subsequently pairing with a close relative will decline according to the distance moved. The chances of inbreeding are reduced further if, as in many bird species, one sex tends to settle further from its natal site to breed than the other.

The avoidance of poor conditions, of competition and of inbreeding have thus all been proposed as major proximate causal factors underlying dispersal. Perhaps the main disadvantages of long-distance dispersal are the risks involved. The bird might fail to find suitable habitat and, while travelling through unfamiliar terrain, it might suffer increased risks of mortality from predation or food-shortage. It might end up in a remote region to which it, and its offspring, are not well adapted, and if it leaves the regular range completely it may not find a mate. There are clearly both benefits and costs to site fidelity and dispersal which are influenced by the ecological needs of the bird and by the conditions prevailing. In some circumstances, it pays to remain in the same general area, in others to move on elsewhere. Species vary their behaviour accordingly.

NATAL DISPERSAL

In many bird species, if the ring recoveries of young birds in a subsequent breeding season are plotted in relation to the hatch site, recoveries come from all sectors of the compass, indicating no directional preference. However, the numbers of recoveries tend to be greatest in the vicinity of the hatch site, and decline progressively with increasing distance. Such patterns are shown for several species in

Figures 17.1 and 17.2; they reflect the settling patterns of individuals with respect to natal site, regardless of their movements in the intervening period. In each species, the density of recoveries declines in approximately exponential manner in concentric circles out from the natal site.

This type of settling pattern has been found in almost all species of birds that have been studied, whether resident or migrant, the main difference between

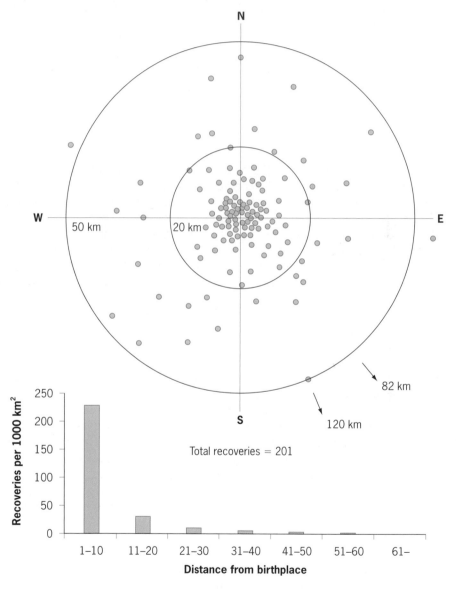

Figure 17.1 Locations of Eurasian Sparrowhawks *Accipiter nisus* ringed as chicks and recovered in a later breeding season, shown in relation to the hatch site (centre). Recoveries came from all sectors of the compass and declined in density with increasing distance from the hatch site. Based on recoveries obtained in the British ringing scheme. From Newton 1979.

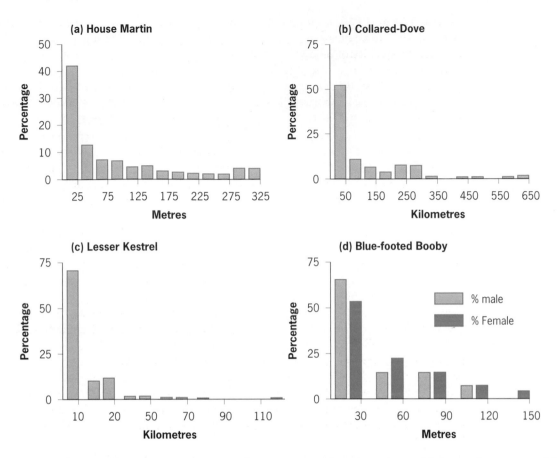

Figure 17.2 Natal dispersal patterns of several species ringed as chicks and recovered in a later breeding season. All species show a decline in numbers with increasing distance from the natal site, but the form of the relationship differs between species, and distances are greater in some species than in others. (a) Northern House Martin *Delichon urbica* (Rheinwald 1975); (b) Eurasian Collared-Dove *Streptopelia decaocto* (Hengeveld 1993); (c) Lesser Kestrel *Falco naumanni* (Negro *et al.* 1997); and (d) Blue-footed Booby *Sula nebouxii* within a colony (Osorio-Beristain & Drummond 1993).

species being one of scale: some species disperse over much longer distances than others (e.g. Paradis *et al.* 1998, Sutherland *et al.* 2000). In general, larger species tend to breed further from their natal sites than do small ones, as might be expected, but this relationship is rather loose as many other factors influence dispersal distances. In the same area of Germany, 90% of the young Blue Tits *Parus caeruleus* found had settled to breed within 4 km of where they hatched, 90% of Wood Nuthatches *Sitta europaea* had settled within 9 km, and 90% of Pied Flycatchers *Ficedula hypoleuca* within 17 km. These figures implied that 90% of the surviving nestlings from any one point source subsequently bred within circular areas centred on that point, but covering about 50 km², 250 km² and 1,000 km² respectively in the three species (Berndt & Sternberg 1968). These are big differences for similar-sized birds, but in all three species occasional individuals settled at far greater distances (and possibly some long-distance dispersers were missed altogether). Moreover, within species the settling pattern may differ somewhat from year to year or from region to region, depending on circumstances, including the density of the population and the patchiness of habitat in the region concerned (for European Pied Flycatcher *Ficedula hypoleuca*, see Sokolov 1997; for Blue Tit *Parus caeruleus* and Great Tit *P. major*, see Matthysen *et al.* 2001).

For some purposes, dispersal could be expressed in territorial species, not as metric distances, but as the number of territories traversed (Harvey *et al.* 1984). The main advantage of expressing dispersal in this way is that it could give comparable measures for different species and areas, regardless of differences in territory sizes and levels of habitat fragmentation. For example, the median dispersal distance of Wood Nuthatches *Sitta europaea* (which breed in woodland) was about 1 km in landscapes with around 30% woodland cover, but several times larger in a more open landscape within only 2% woodland. The scale of dispersal thus expanded as suitable habitat patches became further apart. However, no difference was apparent between areas in the mean number of territories traversed (Matthysen *et al.* 1995).

In a parallel situation, big differences in the natal dispersal differences of Ospreys *Pandion haliaetus* between eastern North America (median distance about 10 km) and northern Europe (median distance about 100 km) were attributed by Poole (1989) to differences between regions in the availability of nest sites. In eastern North America, many people put up platforms for nesting Ospreys, giving an unusually high density of nest sites, enabling birds to settle nearer their natal sites, and take advantage of the rich food-supplies in coastal areas. This does not happen in Europe, however, where nest sites are few and far between and where food is less plentiful.

Another indication of the influence of landscape structure on dispersal is that, within species, natal philopatry is much more marked, with higher return rates, in isolated populations (such as those on islands) than in their counterparts occupying similar areas of continuous habitat (Weatherhead & Forbes 1994). This is evident, for example, in populations of *Ficedula* flycatchers, which show much higher return rates to European study areas on islands than to similar-sized areas on the mainland. It is also evident in Blue Tits *Parus caeruleus* and Great Tits *P. major* nesting in small fragments of woodland compared to those nesting in more continuous patches (Matthysen *et al.* 2001). The implication is that the species concerned are more willing to disperse through suitable habitat than to cross areas of unsuitable habitat that presumably pose greater risks.

Although in most bird populations very few individuals make really long moves between natal and breeding sites, it is these individuals that could result in the colonisation of new areas and the greatest steps in gene flow. The extreme distances are sometimes very much larger than the average ones. For example, of 369 Northern Lapwings *Vanellus vanellus* ringed as chicks in Britain and then recovered in a later breeding season, 61% had returned to within 10 km of their natal sites and 89% had returned to within 100 km; but four others were found within the breeding season 2,000–4,000 km away in Russia and Siberia (Thompson *et al.* 1994). Not surprisingly, this species shows no subspecific differentiation across its wide Eurasian range.

Seabirds and other colonial species

In colonial species, as expected, natal dispersal (as judged from ring recoveries) is influenced by the distribution of colonies. Nevertheless, the pattern is essentially the same as in other birds, with most individuals breeding in their natal or neighbouring colonies, and fewer individuals (mostly females) moving to colonies further away. This skewed settling pattern holds for colonial landbirds, such as Sand Martin *Riparia riparia* (Mead & Harrison 1979) and Lesser Kestrel *Falco naumanni* (Negro *et al.* 1997), and for a wide range of colonial seabirds, including gulls and terns (Austin 1949, Mills 1973, Duncan & Monaghan 1977, Coulson & de Mévergnies 1992, Spendelow *et al.* 1995, Spear *et al.* 1998), auks (Swann & Ramsay 1983, Harris 1984), shags (Aebischer 1995), petrels (Fisher 1971, Brooke 1978, Thibault 1994) and others. To give one example, only 36% of surviving Black-legged Kittiwakes *Rissa tridactyla* (mostly males) bred in their natal colony, a further 43% in other colonies up to 100 km away, and the remainder moved to colonies up to 900 km away (Coulson & de Mévergnies 1992). This pattern was found from ring recoveries of birds old enough to be breeding, and from re-sightings of individually colour-marked birds.

In some seabirds, many individuals settle to breed within the same part of a colony as where they were raised. This finding is much more frequent than expected if individuals settled at random within their natal colony. It also occurs in a wide range of species, including penguins (Williams 1995), albatrosses (Fisher 1971), shearwaters (Richdale 1963), auks (Gaston *et al.* 1994, Halley *et al.* 1995), skuas (Klomp & Furness 1992), shags (Aebischer 1995) and others. An extreme example is provided by a large colony of Blue-footed Boobies *Sula nebouxii*, in which the median distance between natal site and subsequent breeding site was less than 30 m (**Figure 17.2**; Osorio-Beristain & Drummond 1993).

Such findings are more remarkable than the bland figures suggest. Take the Short-tailed Shearwater *Puffinus tenuirostris*, for example, 23 million of which breed annually in burrows and headlands around southeastern Australia, migrating to the northern Pacific for the non-breeding season (Skira 1991). On one tiny island in Bass Straight, a population of a few hundred birds has been monitored for more than 50 years. Over 40% of young hatched on this island later returned there, usually breeding for the first time in their seventh year, and a constant 45% of the breeding population consisted of locally hatched recruits (Serventy & Curry 1984). Such precision in the selection of a breeding location is extraordinary, considering the wide-ranging migration, the average seven-year period between fledging and first breeding, and the fact that less than a kilometre from the study site was a much larger island holding several hundred thousand nesting Shearwaters. Moreover, most Short-tailed Shearwaters returned not just to the island, but to that small part of the colony where they were hatched. Much the same could be said of most other pelagic seabirds that have been studied.

In any colony, then, breeders could include some individuals raised within the colony and others that have moved in from elsewhere. The proportion of immigrants recorded in nesting colonies has varied greatly with species and with circumstances at the time. During the establishment and growth phases of a colony, immigration is likely to be high, so that the majority of the occupants would have been raised elsewhere, whereas during a decline phase the reverse may be true. Some established colonies that have been studied, such as the colony of Great Skuas *Stercorarius skua* on Foula (Shetland), have appeared to be almost closed, with extremely little immigration or emigration (Klomp & Furness 1992). In contrast, in the rapidly growing colony on St Kilda off northwest Scotland, up to half of all Great Skuas that were recruited over a four-year period were immigrants from other colonies (Phillips *et al.* 1999). That some normally strongly philopatric seabirds occasionally disperse very long distances is shown by two young Common Guillemots *Uria aalge* ringed on Shetland (off Scotland) that were seen at colonies in Norway over 2,000 km away, and a Wandering Albatross *Diomedea exulans* that moved 5,000 km between its natal site and subsequent breeding site (Warham 1990).

Those gulls and terns that nest on the ground in exposed and often unstable substrates, such as sandbanks, often have to move their breeding places, as sites become flooded, accessible to mammalian predators, or infested with parasites. Hence, whole colonies can sometimes disband and re-form elsewhere, affecting the dispersal distances of young and adults alike. Sandwich Terns *Sterna sandvicensis* in a new colony in northeast Germany were drawn from as far west as Britain, and ringing records have shown that some adults moved more than 100 km between colonies in the same season (Nehls 1983). Among southern African species, entire colony shifts are frequent in King Gulls *Larus hartlaubii*, Great Crested Terns *Sterna bergii*, Roseate Terns *Sterna dougallii* and Cape Cormorants *Phalacrocorax capensis* (Crawford *et al.* 1994). These species contrast with others in the same region which occupy more stable substrates and show strong colony persistence, including African Penguins *Spheniscus demersus*, Cape Gannets *Morus capensis*, Bank Cormorants *Phalacrocorax neglectus*, Great Cormorants *P. carbo* and Great White Pelicans *Pelecanus onocrotalus*.

In many seabird species, individuals are known to visit a colony for one or more years before they attempt to breed, presumably acquiring experience and local knowledge. The number of other colonies visited by individuals during this phase seems to vary between species. In the Great Skuas *Stercorarius skua* on Foula, virtually all individuals seemed to visit only their natal colony in their pre-breeding years (Furness 1987, Klomp & Furness 1992), but in European Storm Petrels *Hydrobates pelagicus* and Atlantic Puffins *Fratercula arctica*, individuals regularly visited more than one colony before settling to breed (Mainwood 1976, Fowler *et al.* 1982, Harris 1984).

Sex differences in natal dispersal

In some bird species, both sexes show similar dispersal distances, but in many others one sex moves further than the other. The commonest pattern is for young females to disperse further between hatch site and breeding site than males (Greenwood 1980, Clarke *et al.* 1997). This is true only as a general tendency, however, and within any one population, the dispersal distances of the two sexes usually overlap greatly. Nonetheless, the tendency for greater female dispersal holds in a wide range of species, including many passerines, owls and raptors, waders and colonial seabirds. In all such species, therefore, more males than females in local populations have been raised locally.

In some group-living birds (co-operative breeders), the young remain with their parents on the natal territory for up to several years before they disperse, mostly over short distances (for Florida Scrub Jay *Aphelocoma coerulescens*, see Woolfenden & Fitzpatrick 1978; for Acorn Woodpecker *Melanerpes formicivorus*, see Köenig & Mumme 1987; for Arabian Babbler *Turdoides squamiceps*, see Zahavi 1989; for Siberian Jay *Perisoreus infaustus*, see Ekman *et al.* 1994). Delayed dispersal has developed, it is supposed, in situations where all suitable habitat is occupied by territorial groups, leaving nowhere for unattached birds to live (for experimental evidence in Seychelles Warbler *Acrocephalus sechellensis*, see Komdeur *et al.* 1995). The adults then gain by allowing their young to remain in the territory, with access to its resources, until an opening becomes available elsewhere. The young pay for their long-term accommodation by helping with defence of the territory and (in some but not all species) by feeding subsequent broods, while they forgo reproduction themselves. To become breeders, young males sometimes inherit the territory from their father, or take over another territory nearby, but young females almost always move to another territory. Despite the majority of individuals moving little or no distance from their hatch sites, they seldom mate with closely related individuals.

In a few other species, such as the Red-legged Partridge *Alectoris rufa* and Red Grouse *Lagopus l. scoticus* males seem less aggressive to their offspring than to unrelated individuals, and sons often settle near their fathers (Green 1983, Watson *et al.* 1994). In crowded populations, this may be the most effective way in which young males can acquire a territory and reproduce. As in most other birds, females disperse more widely, thereby reducing the chance of inbreeding.

Most northern waterfowl show sex-biased dispersal, but with males moving furthest, as documented in swans, geese, shelducks, and in various diving and dabbling ducks (Mihelsons *et al.* 1986, Rohwer & Anderson 1988, Clarke *et al.* 1997). In many such species, as mentioned earlier, pairing occurs in wintering areas, and the male then accompanies the female to her natal area. Because birds from different breeding areas may share the same wintering area, the males of some species have settled to breed several hundreds of kilometres from their natal sites (Udvardy 1969, Cooke *et al.* 1975, Rockwell & Cooke 1977). Moreover, because the males of some migrant duck species have a different mate each year, they often change their breeding sites substantially from year to year. Such behaviour must presumably ensure genetic mixing within the entire population, and in fact most species of ducks are monotypic across the whole of the Eurasian or North American land-masses. Subspeciation has occurred mainly between these major land-masses or in the isolated populations of some dabbling duck species that have colonised islands where they are resident. In contrast to ducks, geese and swans normally keep the same mate for several years and change their breeding sites much less often (Anderson *et al.* 1992). Most species of geese show well-marked subspecific differentiation. Although they may form pairs in wintering areas, their populations tend to remain more discrete year-round than those of ducks.

Other species in which males disperse further than females between natal and breeding sites include some shorebirds that show 'sex-role reversal', with the female defending the territory or mate, and the male doing the incubation and chick care. Examples include the Spotted Sandpiper *Actitis macularia* and various phalaropes *Phalaropus* (Oring & Lank 1982, Colwell *et al.* 1988). Like the ducks, they have a social system based on mate defence, but unlike the ducks, pair formation occurs in the breeding area.

Competition and natal dispersal

Several types of evidence point to competition in influencing dispersal distances. Firstly, the young in some studies moved further from their natal sites in years of high than low population density (for European Greenfinch *Carduelis chloris*, see Boddy & Sellers 1983; for Great Tit *Parus major*, see Greenwood *et al.* 1979, O'Connor 1981; for Marsh Tit *P. palustris*, see Nilsson 1989; for Spotted Sandpiper *Actitis macularia*, see Oring & Lank 1982), or the young moved further in successive years as a local population grew (for Eurasian Sparrowhawk *Accipiter nisus*, see Wyllie & Newton 1991; for Lesser Kestrel *Falco naumanni*, see Negro *et al.* 1997).

Secondly, the young in other studies moved further from their natal sites in years of low than high food-supply (for Coal Tit *Parus ater*, see Sellers 1984; for Northern Goshawk *Accipiter gentilis*, see Kenward *et al.* 1993; for Great Horned Owl *Bubo virginianus*, see Houston 1978; for Common Kestrel *Falco tinnunculus*, see Adriaensen *et al.* 1998; for Tengmalm's Owl *Aegolius funereus*, see Löfgren *et al.* 1986, Saurola 2002; for Tawny Owl *Strix aluco* and Ural Owl *S. uralensis*, see Saurola 2002). The last four species feed on voles, and the length of their movements fluctuated with the vole cycle in roughly three-year periodicity, with the longest natal dispersal in the low vole years. In addition, the experimental provision of supplementary food to Song Sparrows *Melospiza melodia* greatly reduced dispersal from the study area (Arcese 1989).

Thirdly, in yet other studies, young that fledged late in the season dispersed further, on average, than young that fledged earlier. In the autumn, when Eurasian Tree Sparrows *Passer montanus* occupied nest boxes as roost-sites, twice as many birds than boxes were present (Pinowski 1965). The available boxes were taken by adults and by young hatched early in the season. By the time late-hatched young attempted to obtain boxes, most were already taken, so that most late-hatched young had to move away, subsequently breeding further from their natal sites. Similar trends with fledging date were noted in other species (for Blue Tit *Parus caeruleus*, see Dhondt & Hublé 1968; for Great Tit *P. major*, see van Balen & Hage 1989; for Marsh Tit *P. palustris*, see Nilsson 1989; for Northern House Martin *Delichon urbica*, see Rheinwald & Gutscher 1969; for European Pied Flycatcher *Ficedula hypoleuca*, see Sokolov 1997; for Eurasian Sparrowhawk *Accipiter nisus*, see Newton & Rothery 2000; for Western Gull *Larus occidentalis*, see Spear *et al.* 1998). In an experimental study of Great Tits, young from late broods settled much nearer to their natal sites when most (90%) of the first-brood young had been removed, and competition was thereby reduced (Kluijver 1971).

These various findings all imply that dispersal distances may be density dependent, and influenced by levels of competition, whether for territories, food or nest sites. None of this is surprising, but it has taken detailed multi-year studies to confirm it. The biogeographical implication is that range expansions are most likely to occur with the build-up of population pressure within the existing range, as discussed in Chapter 15.

Given certain assumptions, mathematical models have shown that competition could generate skewed patterns of dispersal distances, as revealed by the decreasing densities of ring recoveries in concentric circles out from the natal site. In the simplest situation, imagine a uniform habitat filled with territories, each occupied by an adult pair. These adults are able to keep out youngsters seeking territories of their own. Imagine, too, that when young leave their natal territories, they move in random directions, but travel in a straight line and settle in the first vacant territory they meet. Vacant territories occur randomly through the area, and arise through the deaths of adults from the previous year. Given these simple assumptions, the distribution of dispersal distances would be skewed, as found in nature, and the distances themselves would depend on territory sizes and vacancy (= mortality) rates. Hence, competition for territories (or any other limiting resource) could be a major factor underlying dispersal and producing the observed patterns (Waser 1985). Moreover, as larger bird species generally have larger territories and lower mortality rates, their dispersal distances on this model would be generally longer than those of small species, as found in nature. Although migrants do not occupy their territories year-round, the adults usually return to their nesting areas before the younger first-time breeders, forcing the latter to move elsewhere, so a similar system could operate as in year-round residents. In both groups, departures from the skewed pattern could result from deviations from the assumptions mentioned above, such as habitat patchiness, a different search strategy, a non-random distribution of vacancies, and so on, or from an influence of additional factors on dispersal.

BREEDING DISPERSAL

One of the earliest findings from bird-ringing was that, once individuals had bred in an area, they tended to stay there, or to return there to breed year after year. Many individuals were found to occupy the same territories in successive years, while others moved to different territories nearby. In many species of seabirds and raptors, individuals often returned to exactly the same nest sites.

Site fidelity of adult birds has been shown mostly in specific study areas, by counting the proportion of marked individuals present in one year that returned the next. On this basis, any birds that moved outside the area would be missed, and the smaller the area the lower the expected return rate, all else being equal. Nonetheless, in some species the proportion returning to particular areas was so high that, allowing for known mortality rates, few if any individuals could have moved elsewhere (**Figures 17.3 and 17.4, Table 17.1**). In fact, as shown by general ring recoveries, the settling pattern in breeding dispersal is skewed like that of natal dispersal, but over shorter distances (e.g. Paradis *et al.* 1998). Such patterns are again evident in a wide range of species, including passerines, raptors, shorebirds, waterfowl and colonial seabirds, both resident and migratory. In the Great Tit *Parus major*, for example, juveniles in one area moved on average 4–7 territory widths during natal dispersal (from birthplace to breeding place), whereas adults in the same area usually returned to within one territory width of their previous breeding site (Greenwood *et al.* 1979, Harvey *et al.* 1979). Typically, within species, mean adult dispersal distances are about one-half of natal distances, but in some species only one-sixth as great (**Table 17.2, Figure 17.5**). The difference between age groups may arise largely because, each year when nesting begins, adults are

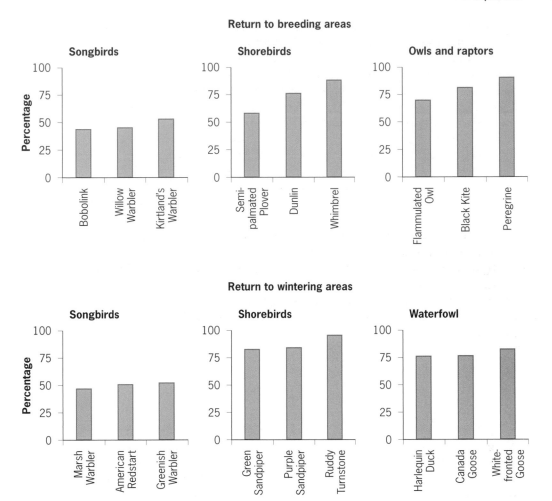

Figure 17.3 Percentage of adults in migratory bird populations that returned to the same breeding (upper) or wintering (lower) sites in successive years. Allowing for mortality, the species depicted reveal some of the most extreme examples of site fidelity in birds, in which all (or almost all) surviving individuals may be inferred to have returned to the same site in successive years. Some of the species shown bred in one continent and wintered in another. Scientific names: Bobolink *Dolichonyx oryzivorus*, Willow Warbler *Phylloscopus trochilus*, Kirtland's Warbler *Dendroica kirtlandii*, Semi-palmated Plover *Charadrius semipalmatus*, Dunlin *Calidris alpina*, Whimbrel *Numenius phaeopus*, Flammulated Owl *Otus flammeolus*, Black Kite *Milvus migrans*, Peregrine *Falco peregrinus*, Marsh Warbler *Acrocephalus palustris*, American Redstart *Setophaga ruticilla*, Greenish Warbler *Phylloscopus trochiloides*, Green Sandpiper *Tringa ochropus*, Purple Sandpiper *Calidris maritima*, Ruddy Turnstone *Arenaria interpres*, Harlequin Duck *Histrionicus histrionicus*, Canada Goose *Branta canadensis*, White-fronted Goose *Anser albifrons*. Details from **Tables 17.1** and **17.4**.

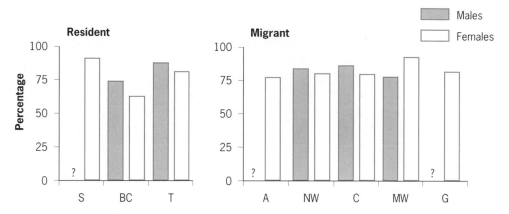

Figure 17.4 Proportions of adult Peregrine Falcons *Falco peregrinus* in resident and migratory populations that were present at nests in the same study areas in successive years. A — Alaska (Ambrose & Riddle (1988); BC — British Columbia (Nelson 1988, 1990); C — Colorado (Enderson & Craig 1988); G — Greenland (Mattox & Seegar 1988); MW — midwest United States (Tordoff & Redig 1997); NW — Northwest Territories (Court *et al.* 1989); S — Scotland (Mearns & Newton 1984); T — Tasmania (Mooney & Brothers 1993).

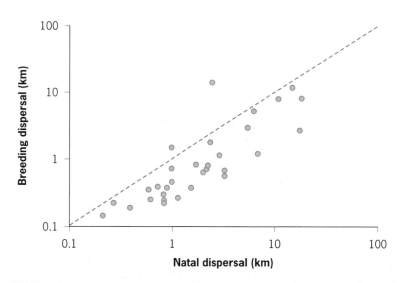

Figure 17.5 Relationship between the geometric means of the natal and breeding dispersal distances of various bird species, based on recoveries from the British ringing scheme. From data given by Paradis *et al.* 1998.

first to establish territories, or are better able to compete for them, so that the late-settling youngsters then have to search over wider areas to find vacancies.

For some species, more detailed studies have revealed the factors that underlie year-to-year territory changes by breeders, as well as the consequences of such moves for the individuals concerned. Five main patterns have emerged: (1) sex differences in site fidelity within species, with males less likely to move than females in most species (e.g. Greenwood 1980, Gavin & Bollinger 1988, Payne & Payne 1993), possibly related to sex differences in territory acquisition and defence; (2) a tendency for greater site fidelity in later life (Harvey *et al.* 1984, Newton 1993, Aebischer 1995, Morton 1997, Bried & Jouventin 1998, Forero *et al.* 1999), possibly related to increasing benefits through life of site familiarity; (3) a greater tendency to change territories after a breeding failure than after a success (Newton & Marquiss 1982, Beletsky & Orians 1991, Payne & Payne 1993, Serrano *et al.* 2001), possibly related to site quality, (4) a tendency to move to better territories through early life, possibly related to increase in status and competitive ability (e.g. Hildén 1979, Matthysen 1990, Montalvo & Potti 1992); and (5) a strong tendency for a change of territory to be associated with a change of mate (e.g. Johnston & Ryder 1987, Bradley *et al.* 1990, Payne & Payne 1993, Newton 2001). The references cited are meant only as examples: all these patterns have been confirmed in a wide range of studied species (but not all).

The three frequent findings of (1) greater tendency to move after a breeding failure; (2) greater tendency to move from a poor territory than from a good one; and (3) greater tendency to stay with increasing age, may not be wholly independent of one another, for young birds are more likely to get poor territories and to fail in their breeding than are older ones. In addition, because in many species a change of site also involves a change of mate, it is often hard to decide which event depends on the other. The advantage for birds moving after a breeding failure is that causes of breeding failure are often site-related or mate-related, in which case the failed bird may benefit by moving elsewhere. Nonetheless, while some birds may change territories voluntarily, others may change because they are driven from their territories by other individuals. Whatever the underlying cause of territory changes, or the consequences for the individuals concerned, their biogeographical importance is that they could result in the further mixing of individuals hatched in different localities, and in further gene flow through populations.

DISPERSAL WITHIN A BREEDING SEASON

Many birds raise more than one brood per season, or attempt a second nest if the first fails, normally close by in the same territory. Some even use the same nest sites for repeat broods, especially some cavity nesters where sites are scarce. In most species, the same partners stay together for successive nests in a season but in other species some individuals change partners (for Yellow-breasted Chat *Icteria virens*, see Thompson & Nolan 1973; for House Sparrow *Passer domesticus*, see Sappington 1977).

A contrasting situation is shown by species in which some individuals nest in widely separated localities within the same breeding season. Examples include some cardueline finches which feed on sporadic food-supplies, settling temporarily wherever suitable seeds are plentiful at the time (for Eurasian Bullfinch *Pyrrhula pyrrhula*, see Newton 2000; for Common Redpoll *Carduelis flammea* and Eurasian Siskin *C. spinus*, see Chapter 18). Other examples include the few species that habitually stop and breed at two or more points on a migration route, attempting to raise a brood at one site before moving on to the next (Chapter 18). Other records of long-distance dispersal within a breeding season include the

Table 17.1 Annual return rates of migrant juvenile and adult birds to specific study areas in successive breeding seasons. The data are drawn from studies in which attempts were made to identify all individuals present. Records for different years are pooled. N = number marked, % = percentage recovered. Note the lower return rates for juveniles returning to their natal areas than for adults returning to their former breeding areas. This is due partly to higher mortality rates of juveniles, which allow fewer to return, and partly to their longer dispersal distances, which lead fewer to settle in the study area, compared with returning adults.

Species	Location (Size of study area)	Natal area Unsexed young N	%	Breeding area Male N	%	Female N	%	Source
Tree Pipit *Anthus trivialis*	Switzerland (?)	114	12.5	102	51	74	32	Meury 1989
House Wren *Troglodytes aedon*	Illinois (128 ha)	6,299	2.8	643	38	1,468	23	Drilling & Thompson 1988
Whinchat *Saxicola rubetra*	Germany/Switzerland (<100 ha)		7.9	88	74	59	57	Bastian 1992
Redwing *Turdus iliacus*	Norway (?)	727	0.7	63	25	74	16	Bjerke & Espmark 1988
Great Reed Warbler *Acrocephalus arundinaceus*	Sweden (200 ha)	477	9	59	54	66	57	Bensch & Hasselquist 1991
Marsh Warbler *Acrocephalus palustris*	S. England (5.4 ha)			41	44	–	–	Kelsey 1989
Willow Warbler *Phylloscopus trochilus*	Finland (25 ha)			122	41	97	17	Tiainen 1983
Willow Warbler *Phylloscopus trochilus*	England (47 ha)	232	5.0	104	36	57	23	Lawn 1982
Willow Warbler *Phylloscopus trochilus*	England (8.8 ha)			176	30	133	17	Pratt & Peach 1991
Wood Warbler *Phylloscopus sibilatrix*	England (35 ha)			29	28			Norman 1994
Garden Warbler *Sylvia borin*	Finland (27 ha)	201	0	80	25	84	6	Solonen 1979
European Pied Flycatcher *Ficedula hypoleuca*	Germany (20 km²)	2,406	2.5	160	28	340	27	Creutz 1955
European Pied Flycatcher *Ficedula hypoleuca*	Germany (?)	229	0.4	37	35	40	38	Trettau 1952
European Pied Flycatcher *Ficedula hypoleuca*	Germany (?)	710	10.4	121	45	120	30	Curio 1958
European Pied Flycatcher *Ficedula hypoleuca*	England (24 ha)	4,086	3.9	374	24	646	23	Campbell 1959
European Pied Flycatcher *Ficedula hypoleuca*	Finland (?)	2,842	1.8	378	36	576	14	von Haartman 1960
Prairie Warbler *Dendroica discolor*	Indiana (50 ha)	272	4	55	60	105	19	Nolan 1978

Species	Location						
Kirtland's Warbler *Dendroica kirtlandii*	Michigan (47 ha)	296	2.7	51	53	110	31	Berger & Radabough 1968
Black-throated Blue Warbler *Dendroica caerulescens*	New Hampshire (55 ha)			49	39	50	36	Holmes & Sherry 1992
Prothonotary Warbler *Protonotaria citrea*	Michigan (36 ha)			18	50	59	20	Walkinshaw 1953
Yellow-breasted Chat *Icteria virens*	Indiana (18 ha)			18	11	29	0	Thompson & Nolan 1973
Bobolink *Dolichonyx oryzivorus*	New York State (43 ha)			85	44	86	25	Gavin & Bollinger 1988
American Redstart *Setophaga ruticilla*	New Hampshire (34 ha)			134	50	48	19	Holmes & Sherry 1992
Northern Lapwing *Vanellus vanellus*	England (877 ha)	801	28.1	171	74	129	70	Thompson et al. 1994
European Golden Plover *Pluvialis apricaria*	Scotland (100 ha)	100	26	77	78	35	77	Parr 1980
Semipalmated Plover *Charadrius semipalmatus*	Manitoba (384 km^2)	445	1.6	127	59	126	41	Flynn et al. 1999
Snowy (Kentish) Plover *Charadrius alexandrinus*	Utah (ca 50 km^2)			224	40	278	26	Paton & Edwards 1996
Semipalmated Sandpiper *Calidris pusilla*	Manitoba (200 ha)	802	4.0	415	61	401	56	Sandercock & Gratto-Trevor 1997
Temminck's Stint *Calidris temminckii*	Finland (12 ha)	170	21	112	79	61	70	Hildén 1978
Burrowing Owl *Speotyto cunicularia*	Manitoba (?)	538	3.5	87	40	78	24	De Smet 1997
Flammulated Owl *Otus flammeolus*	Colorado (452 ha)			21	71	16	81	Reynolds & Linkhart 1987
Black Kite *Milvus migrans*	Spain (1,000 km^2)			142	83	143	90	Forero et al. 1999
Osprey *Pandion haliaetus*	Eastern US (>1,000 km^2)			460	88[1]			Poole 1989
Lesser Kestrel *Falco naumanni*	Spain (<1,000 km^2)	997	34.1	262	71[1]			Hiraldo et al. 1996

[1] Both sexes included; no difference between them.

Table 17.2 Dispersal distances (km) calculated from recoveries of birds ringed and found dead in Britain, 1909–94. N = number of recoveries, AM = arithmetic mean distance, GM = geometric mean distance, SD = standard deviation of distances[1], SD = standard deviation of distances. All figures refer only to species for which at least 50 recoveries were available for each age group analysed. From Paradis et al. 1998.

Species	N	Natal dispersal (ND)			N	Breeding dispersal (BD)			Ratio ND:BD
		AM	GM	SD		AM	GM	SD	
Mute Swan *Cygnus olor*	49	34.3	16.772	35.9	497	18.0	2.719	48.0	6.17
Canada Goose *Branta canadensis*	173	7.0	0.969	10.6	365	8.9	1.503	10.8	1.50
Mallard *Anas platyrhynchos*	666	19.9	6.058	21.6	328	18.6	5.192	21.6	1.17
Common Black-headed Gull *Larus ridibundus*[2]	1,478	47.0	10.527	69.2	110	44.5	7.968	72.5	1.32
Lesser Black-backed Gull *Larus fuscus*[2]	1,882	28.2	2.384	40.7	190	38.2	13.805	37.4	0.17
Common Wood-Pigeon *Columba palumbus*	718	10.7	2.277	19.3	233	10.9	1.805	24.1	5.93
Sand Martin *Riparia riparia*[2]	70	20.9	6.650	22.8	144	7.7	1.221	13.4	5.45
Barn Swallow *Hirundo rustica*	395	14.1	3.194	28.4	76	4.8	0.564	9.4	5.66
Northern House Martin *Delichon urbica*	72	10.4	3.185	12.2	191	4.2	0.688	8.3	4.63
Dunnock *Prunella modularis*	237	2.1	0.380	7.2	190	1.4	0.191	8.31	1.99
European Robin *Erithacus rubecula*	409	6.0	0.571	20.2	147	8.0	0.359	35.9	1.59
Eurasian Blackbird *Turdus merula*	2,189	3.3	0.264	20.3	1,806	3.2	0.224	20.6	1.18
Song Thrush *Turdus philomelos*	779	7.0	0.591	21.6	397	4.0	0.253	21.8	2.34
Mistle Thrush *Turdus viscivorus*	92	8.3	1.490	17.4	89	2.3	0.384	5.8	3.88
Eurasian Reed Warbler *Acrocephalus scirpaceus*	77	47.0	5.215	68.6	53	32.4	2.935	61.6	1.78
Greater Whitethroat *Sylvia communis*	89	14.4	2.815	19.0	51	11.1	1.145	19.0	2.46
Blackcap *Sylvia atricapilla*	74	41.2	17.539	37.9	64	27.5	8.027	32.0	2.19
Willow Warbler *Phylloscopus trochilus*	79	20.8	2.172	46.3	58	16.9	0.816	39.6	2.66

Species								
European Pied Flycatcher *Ficedula hypoleuca*	1,551	20.6	14.272	16.5	238	20.6	11.668	17.7	1.22
Blue Tit *Parus caeruleus*	703	5.3	0.796	15.2	201	2.3	0.232	10.2	3.43
Great Tit *Parus major*	560	5.3	0.797	17.9	173	2.5	0.246	12.3	3.24
Eurasian Jackdaw *Corvus monedula*[2]	51	8.6	2.127	11.6	51	6.0	0.721	12.8	2.95
Rook *Corvus frugilegus*[2]	84	8.5	1.964	13.0	96	3.1	0.650	4.7	3.02
Common Starling *Sturnus vulgaris*	401	9.5	1.100	28.1	1,672	3.4	0.273	19.1	4.03
House Sparrow *Passer domesticus*	531	1.7	0.206	6.9	526	1.9	0.147	22.4	1.40
Chaffinch *Fringilla coelebs*	64	3.6	0.787	5.6	120	2.8	0.302	9.9	2.46
European Greenfinch *Carduelis chloris*	99	4.2	0.954	6.4	283	7.5	0.732	22.1	1.30
European Goldfinch *Carduelis carduelis*	85	11.1	1.663	18.2	63	10.6	0.835	20.8	1.99
Eurasian Linnet *Carduelis cannabina*	147	4.4	0.694	8.8	110	3.5	0.393	8.3	1.77
Eurasian Bullfinch *Pyrrhula pyrrhula*	195	4.6	0.852	9.8	194	2.5	0.382	5.2	2.23
Reed Bunting *Emberiza schoeniclus*	58	5.4	0.952	13.1	79	3.8	0.468	9.3	2.03

In these analyses, ringing and recovery occurred in the same area (Britain) for all species, so findings should have been comparable between them. This large-scale study covered a large number of ringing and recovery sites, thereby reducing to a minimum any systematic regional bias in reporting probability. All the birds were ringed in one breeding season and found dead at breeding age in a later one, excluding live recaptures by ringers which would have biased the records towards short moves. Records for the two sexes were pooled, even though they may have differed in dispersal distances. An assumption was that birds found dead in the breeding season were near their breeding sites.

[1]Geometric means calculated as the arithmetic mean of the log$_e$ distances, back-transformed.
[2]Colonial.

following: 208 km for Mountain Bluebird *Sialia currucoides* (Scott 1974), up to 210 km for Eurasian Penduline Tit *Remiz pendulinus* (Franz 1988), up to 400 km for Eurasian Dotterel *Eudromias morinellus* (Mead & Clarke 1988), and up to 1,140 km for Snowy Plover *Charadrius alexandrinus* (Stenzel *et al.* 1994). Presumably, birds that move long distances between successive nesting attempts in the same season also change mates, but I know of no information on this point.

LONG-DISTANCE DISPERSERS

The precise year-to-year homing behaviour shown by the species in **Figure 17.3** is not universal in birds. It may be favoured only where habitats remain fairly stable from year to year, and where returning birds could expect to survive and reproduce. It would not be expected in species that depend on unpredictable habitats or food-supplies, which are available in different areas in different years: for example, in some tundra-nesting species affected by variable patterns of spring snow conditions (Tomkovitch & Soloviev 1994), in some waterbirds affected by fluctuating water levels (Frith 1967, Johnson & Grier 1988), in desert species affected by irregular rainfall (Davies 1988, Dean 1997), in some boreal finches that exploit sporadic tree-seed crops (Svärdson 1957, Newton 1972), or in some predatory birds that exploit locally abundant rodents (Newton 1979, 2002a, Saurola 2002). The local population densities of such species often fluctuate greatly from year to year, in line with fluctuating habitat or food-supplies. The speed with which local numbers increase in response to improving conditions has led to the view that such species are nomadic, with individuals concentrating in different areas in different years, wherever conditions are good at the time (**Figure 17.6**). For some such species, ring recoveries have now confirmed that some individuals do indeed breed at long distances from their natal sites and in widely separated areas in different years.

Among various duck species nesting on the North American prairies, the extent and depth of wetlands varies greatly from year to year, according to previous rain and snowfall. Diving duck species, such as Redhead *Aythya americana*, Canvasback *A. valisineria* and Lesser Scaup *A. affinis*, that occupy the deepest and most stable wetlands, show the greatest degree of site fidelity; while dabbling species, such as Northern Pintail *Anas acuta* and Blue-winged Teal *A. discors*, which use the shallowest and most ephemeral wetlands, tend to settle wherever conditions are suitable at the time (Johnson & Grier 1988). The opportunist behaviour of the dabbling ducks is reflected in their long natal and breeding dispersal distances and by their low return rates to particular areas, as measured by ringing (**Table 17.3**). In years of extreme drought on the prairies, many Pintail and other ducks continue their spring migration northwards and settle to breed on the tundra. This means that many young Pintail, while raised on the prairies where they might normally have bred, moved much longer distances in drought years to breed further north (Smith 1970). Similarly, among waders, species which nest in habitats that normally remain stable from year to year, such as Eurasian Oystercatcher *Haematopus ostralegus* and Common Sandpiper *Actitis hypoleucos*, show strong philopatry and site fidelity, while species that nest in patchy and ephemeral habitats, or are affected by variable patterns of snow melt, such as Curlew Sandpiper *Calidris*

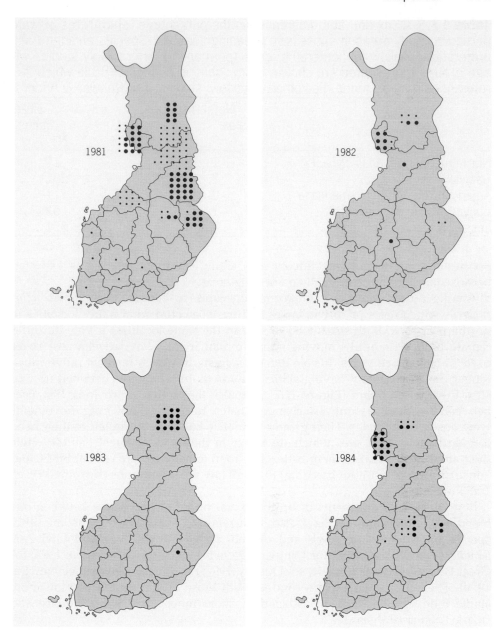

Figure 17.6   Annual variations in the breeding distribution of Great Grey Owls *Strix nebulosa* in Finland. Small dots — territorial pairs; large dots — nests. From Solonen 1986.

ferruginea and Little Stint *C. minuta*, show much less natal philopatry and less adult site fidelity (Evans & Pienkowski 1984, Tomkovich & Soloviev 1994). In such species, the numbers nesting in any one locality may fluctuate greatly from year to year, depending on how many young and adults move in from other areas.

Table 17.3   Sex and age differences in the percentage return rates of waterfowl to particular study areas in successive breeding seasons. Species arranged according to habitat preference, shallow (ephemeral) to deep (permanent) waters. Only studies with both sex and age groups represented are shown. From Johnson & Grier 1988 in which the original references may be found. For other waterfowl studies, see Rohwer & Anderson 1988.

| | Natal dispersal | | | | Breeding dispersal | | | |
| | Males | | Females | | Males | | Females | |
	N	%	N	%	N	%	N	%
Northern Shoveler *Anas clypeata*	134	1	116	3	19	11	20	15
Gadwall *Anas strepera*	42	2	28	7	236	9	54	41
American Wigeon *Anas americana*	6	0	3	33	11	9	21	38
Canvasback *Aythya valisineria*	206	1	101	27	52	10	75	76
Lesser Scaup *Aythya affinis*	91	4	76	49	351	9	58	66

Species that show low site fidelity also show low mate fidelity, as expected because pairing occurs annually on nesting areas.

Among rodent-eaters, most information concerns Tengmalm's Owl *Aegolius funereus* which nests readily in boxes and has been studied at many localities in northern Europe. Both sexes tend to stay in the same localities if vole densities remain high, with adults moving no more than about 5 km between nest boxes used in successive years. If vole densities crash, however, females move much longer distances, many having shifted hundreds of kilometres between nesting sites in different years (**Figure 17.7**). Females that moved such long distances between breeding seasons also changed mates. In contrast, few long movements were observed in males. Their greater residency has been attributed to their need to guard cavity nest sites which are scarce in their conifer forest habitat, while their smaller size makes them better able than females to catch small birds, and hence to survive (without breeding) through low vole conditions (Korpimäki *et al.* 1987).

Long-distance movements of up to several hundred kilometres between one breeding site and another have also been reported in both sexes of some other species of rodent-eating owls and diurnal raptors (for Short-eared Owl *Asio flammeus*, see Village 1987; for Long-eared Owl *A. otus*, see Marks *et al.* 1994; for Great Grey Owl *Strix nebulosa*, see Duncan 1997; for diurnal raptors, see Newton 1979). Their movements again contrast with those of species that exploit more stable food-supplies and, in association, show much greater site fidelity and smaller dispersal distances.

Within species, regional or annual differences in behaviour can sometimes be linked with regional or annual differences in the stability of food-supplies. For example, among Common Kestrels *Falco tinnunculus* in the Netherlands, fidelity to the study area was as high as 70% in years when numbers were on the increase, and as low as 10% when they were decreasing. These trends were in turn related to vole densities, and fidelity was lowest in the poor vole years (Cavé 1968). In Fennoscandia, where voles were even more markedly cyclic, Kestrel numbers fluctuated even more from year to year. Over several years in one study, Kestrels were entirely migratory and only 13% of nesting males and 3% of nesting females returned to the same area in a subsequent year (Korpimäki & Norrdahl 1991).

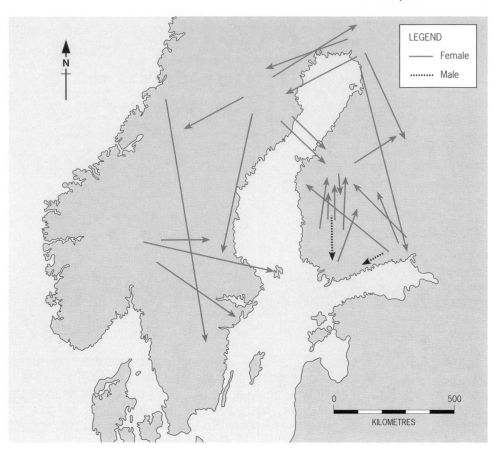

Figure 17.7 Movements of individual adult Tengmalm's Owls *Aegolius funereus* between nesting sites in different years. Compiled from information in Löfgren *et al.* 1986, Korpimäki *et al.* 1987 and Sonerud *et al.* 1988.

Most surviving Kestrels were therefore assumed to settle in different areas each year. This is in marked contrast to other raptors with more stable food-supplies, in which individuals normally return year after year to the same territory (for Peregrine Falcon *Falco peregrinus* see **Figure 17.4)**.

Among tree-seed eaters, extraordinarily long natal and breeding dispersal distances have been recorded from Common Crossbills *Loxia curvirostra* in western Eurasia, where the main food plant is Norway Spruce *Picea abies*. The seed-crops of this tree species fluctuate greatly from year to year, but poor crops in one region may coincide with good crops in another (Svärdson 1957). Crossbills therefore make one major movement each year, in June–August, leaving areas where previous crops are coming to an end, and concentrating where developing crops are good (Newton 1972). Some recorded movements between birth-place and subsequent breeding place, or between breeding places in successive years, were more than 2,000 km, and one adult was found in consecutive breeding seasons at localities 3,750 km apart (**Figure 17.8**). Again, however, because the records are based mainly on ring recoveries supplied by members of the

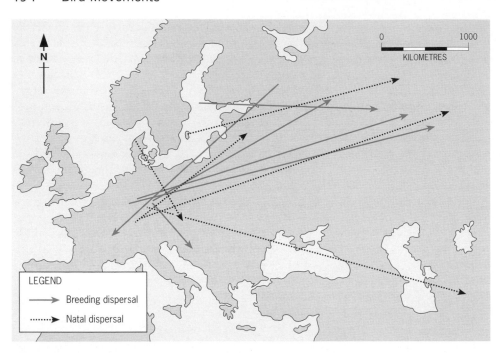

Figure 17.8   Movements of individual Common Crossbills *Loxia curvirostra* between presumed natal and breeding sites (dashed lines) and between the presumed breeding sites of different years (continuous lines). Arrows indicate direction of movement. Recoveries were all in January–April which is the main breeding season of this species in Norway Spruce *Pica abies* forest. Compiled from information in Schloss 1984, and from records in the Swiss and Finnish Bird Banding Schemes.

public, they carry the assumption that birds found in the breeding season were in fact nesting at the locality concerned. As spruce seldom crops well in the same area two years running (e.g. Svärdson 1957), it is perhaps not surprising that no records have emerged of Crossbills staying more than one year at a time. This behaviour contrasts with that of Crossbills in pine areas, where many ringed individuals were proved to remain in consecutive years (Senar *et al.* 1993). Other finches that depend on tree seeds have also shown displacements of more than 500 km between their breeding sites in successive years, or between their wintering sites in successive years (Newton 1972). They contrast with closely related finch species which feed on the more consistent seed-crops of herbaceous plants, and show relatively short natal and breeding dispersal (Newton 1972, Wernham *et al.* 2002).

In both the raptors and the seed-eaters, ring recoveries give some idea of the huge areas over which individuals may roam in order to find food-supplies sufficient for reproduction. With their ability to move rapidly over long distances, these species can exploit a specialist niche in a way that other, more sedentary animals could not, but because of the sporadic nature of their habitats or food-supplies, they occupy only parts of their vast geographical ranges at one time.

In conclusion, high site fidelity and short dispersal distances are characteristic of bird populations whose habitats or food-supplies stay reasonably stable in distribution from year to year, whereas low site fidelity and long-distance dispersal are shown by populations whose habitats or food-supplies vary in distribution from year to year. The adaptive value of moving when conditions deteriorate is obvious, because this is likely to improve individual survival and breeding success in the conditions prevailing. Because of the moves, numbers in particular localities fluctuate greatly from year to year, in line with changes in habitat or food conditions, and some localities that support high densities of breeding birds in one year may be wholly devoid of birds the next. Differences in dispersal patterns between species have led some authors to divide birds into two types: those with limited natal dispersal, and even more limited breeding dispersal (i.e. strong site fidelity) and those 'nomadic species' with wide-ranging natal and breeding dispersal (i.e. little or no site fidelity). In practice, however, variation between these extremes is continuous and, as indicated above, the same species may lie at different points on the continuum in different parts of its range, depending on the stability of local habitats and food-supplies.

NON-BREEDING DISPERSAL

In migrating to wintering areas, young birds do not have the benefits of prior experience as the adults do. The young of some birds, such as swans, geese and cranes, accompany their parents on their first autumn migration, so in this way the young could learn the locations of specific wintering sites. This behaviour could be important in maintaining the integrity of particular breeding populations, most of which also keep to distinct wintering areas, separated from those of other populations of their species (**Figure 18.17**). In most bird species, however, the young do not accompany their parents, and may migrate earlier or later in the season than adults. Many young therefore migrate to wintering areas unaided by experienced birds. This is likely to produce much more scatter in the wintering sites of birds from particular breeding localities, although the young may well return to their first wintering sites in subsequent years. The same issue does not arise with breeding sites, because both young and adults are returning to localities they already know.

In some bird species, once individuals have wintered in an area, they return from year to year with the same regularity as they return to their breeding localities. Although less information is available, site fidelity in winter seems to show the same pattern as in the breeding season, with ring recoveries in subsequent winters being centred on the first-recorded site, and declining sharply with increasing distance. Such patterns have been documented in the European Greenfinch *Carduelis chloris* (Boddy & Sellers 1983), Common Bullfinch *Pyrrhula pyrrhula* (Newton 2002b), Ruddy Turnstone *Arenaria interpres* (Clapham 1979), Eurasian Curlew *Numenius arquata* (Bainbridge & Minton 1978), Grey Plover *Pluvialis squatarola*, Dunlin *Calidris alpina* and Common Redshank *Tringa totanus* (Rehfisch *et al.* 1996) among others, but again the distance scale varies greatly between species.

Most studies of winter site fidelity, like breeding site fidelity, have been conducted by individual researchers working in the same small areas year after year.

So again, while such records are useful in confirming site fidelity, most give little idea of the proportion of surviving birds that settle elsewhere. However, in some species, the proportion of individuals that returned to the same area in successive winters was close to the expected annual (or over-summer) survival, implying that almost all individuals were site faithful. Thus, annual return rates of 37–52% have been recorded from various small songbirds, and of 77–95% for various longer-lived shorebirds (**Table 17.4, Figure 17.3**). More generally, return to the same specific localities has been recorded for a wide range of passerines, raptors, gulls, waders, waterfowl and others, including many intercontinental migrants (Moreau 1972, Holmes & Sherry 1992, Sauvage *et al.* 1998).

The fact that birds have greater freedom to move around in the non-breeding than in the breeding season gives problems in measuring site fidelity, especially in migrants. In some species that show high site fidelity, such as the Blackcap *Sylvia atricapilla*, the numbers of birds at particular sites have varied greatly from year to year, depending on the size of local fruit crops. This fluctuation was apparently due largely to the behaviour of young birds which settled in their first year mainly in sites where fruit was plentiful, and partly to individuals (young or old) varying their length of stay from year to year, moving on each year when the local crops were eaten (Cuadrado *et al.* 1995). Site fidelity with variable lengths of stay has been noted in some other species that exploit annually variable food-supplies, such as Asian Rosy Finch *Leucosticte arctoa* (Swenson *et al.* 1988) and Eurasian Siskin *Carduelis spinus* (Senar *et al.* 1992).

Typically at any one locality, some individuals seem merely to pass through, others stay for days or weeks and yet others for months, with the proportions varying from year to year in line with food-supplies. It fits with the fact that many northern seed-eaters take the same migration route each year, but pass particular latitudes much earlier in some years than in others, and reach the limits of their potential wintering range only in occasional years of widespread food shortage (Chapter 18). Species can therefore vary markedly in distribution from one winter to the next, in association with regional variations in food-supplies. This is true not only of the finches that depend on sporadic tree-seed crops, but also of some ground-feeding finches whose food may be covered by snow, of some raptors and owls which depend on temporarily and locally high rodent populations, and of shorebirds of soft substrates, whose prey varies in spatial distribution from year to year. Moreover, in any one year, many species show age-related differences in wintering range (Ketterson & Nolan 1976, Gauthreaux 1982, Dolbeer 1991). The inference that individuals of such species must use widely separated areas in different years has been increasingly supported by ring recoveries (e.g. Newton 1972, 2002a).

Where birds do not remain all winter at particular sites, but vary in their lengths of stay, many more individuals could use a site than are present at one time, and the chance of recording particular individuals increases with their length of stay. Because duration of stay is often linked with food-supplies and social status, greater return rates are recorded in good than in poor food years, and also in adults than in first-year birds (e.g. 76% versus 48% in Great Cormorant *Phalacrocorax carbo*, Yésou 1995). The potential for observational bias gives uncertainty in how much the recorded variations in return rates between years and localities, or between sex and age groups, are due to differential survival, to differential site

Table 17.4 Annual return rates of migrant birds to specific study areas in successive winters. The data are drawn from detailed studies in which attempts were made to identify all individuals present. Records for different years are pooled. N = number marked, % = percentage known to return the next winter in the same area. Sizes of study areas are given in hectares or as length of coastline.

Species	Location (Size of study area)	Unsexed N	Unsexed %	Source
Siberian Blue Robin Luscinia cyanea	Malaysia (15 ha)	156	46	Wells 1990
Great Reed Warbler Acrocephalus arundinaceus	Malaysia, area 1 (10 ha)	62	50	Nisbet & Medway 1972
	Malaysia, area 2 (2 ha)	83	37	Nisbet & Medway 1972
Marsh Warbler Acrocephalus palustris	Zambia (7.6 ha)	17	47	Kelsey 1989
Melodious Warbler Hippolais polyglotta	Ivory Coast (?)	69	6	Salewski et al. 2000
Greenish Warbler Phylloscopus trochiloides	India (200 ha)	25	52	Price 1981
Willow Warbler Phylloscopus trochilus	Ivory Coast (?)	110	0	Salewski et al. 2000
European Pied Flycatcher Ficedula hypoleuca	Ivory Coast (?)	94	23	Salewski et al. 2000
Prairie Warbler Dendroica discolor	Puerto Rico (8 ha)	25	40	Staicer 1992
Black-throated Blue Warbler Dendroica caerulescens	Jamaica (?)	57	46	Holmes & Sherry 1992
Cape May Warbler Dendroica tigrina	Puerto Rico (8 ha)	8	38	Staicer 1992
Northern Parula Warbler Parula americana	Puerto Rico (8 ha)	65	49	Staicer 1992
American Redstart Setophaga ruticilla	Jamaica (?)	111	51	Holmes & Sherry 1992
Blue Tit Parus caeruleus males	England (one garden)	69	48	Burgess 1982
females	England (one garden)	67	19	Burgess 1982
Eurasian Oystercatcher Haematopus ostralegus, 5-year-olds	England (?)	734	89	Goss-Custard et al. 1982
2–4-year-olds	England (?)	475	83	Goss-Custard et al. 1982
Grey Plover Pluvialis squatarola	England (<100 km²)	71	80	Evans 1981
Pacific Golden Plover Pluvialis fulva				
territorial (1–2 years)	Hawaiian Islands (<10 km²)	34	90	Johnson et al. 2001
territorial (older)	Hawaiian Islands (<10 km²)	78	80	Johnson et al. 2001
non-territorial (1–2 years)	Hawaiian Islands (<10 km²)	16	82	Johnson et al. 2001
non-territorial (older)	Hawaiian Islands (<10 km²)	35	67	Johnson et al. 2001
Eurasian Curlew Numenius arquata	England (<100 km²)	119	82	Evans 1981
Common Redshank Tringa totanus	Wales (4 km)	61	86	Burton 2000
Green Sandpiper Tringa ochropus	England (306 km²)	115	84	Smith et al. 1992
Purple Sandpiper Calidris maritima, adult	Helgoland (150 ha)	117	85	Dierschke 1998
Purple Sandpiper Calidris maritima, first-year	Helgoland (150 ha)	30	63	Dierschke 1998
Purple Sandpiper Calidris maritima	England (13 km)	61	66	Burton & Evans 1997
Sanderling Calidris alba	England (<100 km²)	93	91	Evans 1981
Ruddy Turnstone Arenaria interpres	England (13 km)	71	86	Burton & Evans 1997
Ruddy Turnstone Arenaria interpres	Scotland (6 km)	42	95	Metcalfe & Furness 1985

Table 17.4 continued

Species	Location (Size of study area, ha)	Unsexed N	%	Source
Bewick's (Tundra) Swan *Cygnus columbianus*	England (<100 ha)	690	67	Rees 1987
Canada Goose *Branta canadensis*	Minnesota (<100 ha)	271	78	Raveling 1979
Barnacle Goose *Branta leucopsis*	Scotland (<10 km²)	540	76	Percival 1991
Barnacle Goose *Branta leucopsis*	Scotland (<100 km²)	531	85	Wilson *et al.* 1991
Barnacle Goose *Branta leucopsis*	Netherlands (many areas)	576	90	Ebbinge *et al.* 1991
Snow Goose *Chen caerulescens*	Texas-Louisiana (<100 km²)	77	86	Prevett & MacInnes 1980
Harlequin Duck *Histrionicus histrionicus*, males	British Columbia (?)	82	77	Robertson & Cooke 1999
Harlequin Duck *Histrionicus histrionicus*, females	British Columbia (?)	66	62	Robertson & Cooke 1999
Bufflehead *Bucephala albeola*, males	Maryland	91	26	Limpert 1980
Bufflehead *Bucephala albeola*, females	Maryland	37	11	Limpert 1980

Other examples for neotropical migrants in Rappole *et al.* 1983, for waterfowl in Robertston & Cooke 1999.

fidelity or merely to different durations of stay. All three factors could contribute in some degree to recorded return rates.

In addition, some migratory species show more than one form of social behaviour in winter quarters with some individuals holding territories and others (usually females and juveniles) not. Among Black Redstarts *Phoenicurus ochruros* wintering in southern Spain, 36% of 33 territorial birds were seen in the study area the next winter, compared to 11% of 112 non-territorial ones, a highly significant difference (P=0.001, Fisher's Exact Test) (Cuadrado 1995). The non-territorial individuals may have suffered greater mortality, or may have moved elsewhere in greater proportion. A similar difference was noted in Pacific Golden Plovers *Pluvialis fulva* wintering in Hawaii, but in this species return rates were also higher in first–second-year birds than in older ones, possibly reflecting lower over-summer survival of breeders as opposed to younger non-breeders (**Table 17.4**; Johnson *et al.* 2001).

Many species show greater fidelity to breeding than to wintering sites. This reflects not only spatial variation in wintering habitat from year to year, but also the fact that individuals of many short-distance migrants migrate in their first but not in subsequent years, or migrate different distances in different years (see below). Other species show the opposite tendency, with less site fidelity to their breeding areas than to their wintering areas. In the American Redstart *Setophaga ruticilla* and Black-throated Blue Warbler *Dendroica caerulescens*, lower site-fidelity in summer reflected a breeding habitat that varied strongly in suitability from place to place, and probably also from year to year (Holmes & Sherry 1992). The same holds for some arctic-nesting shorebirds, such as Curlew Sandpiper *Calidris ferruginea*, whose breeding sites may change from year to year according to snow-melt patterns (Tomkovitch & Soloviev 1994), but whose wintering sites are more consistent and are used by the same individuals year after year (Elliott *et al.* 1977).

Other age-related differences

Outside the breeding season, when many birds move away from their nesting areas, the adults often move less far, or stay away for less long, than the young. This is true whether the species performs a fixed-direction migration or a multi-directional dispersal. Some long-lived species, in which individuals do not breed until they are several years old, show a progressive change to shorter moves or to shorter periods away from the breeding areas, with increasing age. Such patterns have been noted in a wide range of bird species, including Grey Heron *Ardea cinerea* (Olssen 1958), Black Kite *Milvus migrans* (Schifferli 1967), Eurasian Oystercatcher *Haematopus ostralegus* (Goss-Custard *et al.* 1982), Herring Gull *Larus argentatus* (Coulson & Butterfield 1985), Common Guillemot *Uria aalge* (Birkhead 1974), Great Cormorant *Phalacrocorax carbo* (Coulson 1961) and Northern Fulmar *Fulmarus glacialis* (Macdonald 1977). This means that some individual birds occupy different areas in successive winters, or that they spend progressively less time on their wintering areas as they age. Studies on other species have indicated increasing winter site fidelity with increasing age, greater in males than in females (for Mallard *Anas platyrhynchos* see Nichols & Hines 1987), and matching the findings from breeding areas (see above).

Some of the smaller arctic-nesting shorebirds that migrate in their first autumn of life to the southern hemisphere also stay in their non-breeding areas throughout the following calendar year, or return only part way towards their breeding areas, not making the full journey until they are two or more years old. Such species include Curlew Sandpiper *Calidris ferruginea* and Rufous-necked Stint *C. ruficollis* in Australia, Semipalmated Sandpiper *C. pusilla* in South America, and Curlew Sandpiper and Common Greenshank *Tringa nebularia* in South Africa (Evans & Pienkowski 1984). The same holds for some raptor and seabird species which have similar long migrations. In the Osprey *Pandion haliaetus*, for example, most young from Europe remain in Africa for their first year, and those from North America remain in Central and South America (Henny & van Velzen 1972, Saurola 1994). In this way, the age groups are separated geographically in the breeding season, and the non-breeding young avoid the risks of an unnecessary journey.

Sex-related differences

Few data are available to compare winter site fidelity between the sexes, partly because, in many species, the sexes are hard to separate at that time of year. However, among non-migratory Blue Tits *Parus caeruleus* ringed in a garden in winter, a much greater proportion of males than females (48% versus 19%) were caught there the following winter. The difference in return rates was much greater than expected from survival differences, suggesting that many females had moved elsewhere, or had spent so little time at the site that they had not been re-caught (Burgess 1982). Sex differences were also apparent in the return rates of Harlequin Ducks *Histrionicus histrionicus* to a wintering site in British Columbia (77% males versus 62% females, Robertson & Cooke 1999), and of Buffleheads *Bucephala albeola* to a site in Maryland (26% males versus 11% females, Limpert 1980). In contrast, ring recoveries gave no indication of sex or age differences in winter site fidelity in European Greenfinches *Carduelis chloris* in Britain, or of sex differences in Bewick's Swans *Cygnus columbianus* in Britain (Rees 1987), or in American Black Ducks *Anas rubripes*, Canvasbacks *Aythya valisineria* and American Woodcock *Scolopax minor* in eastern North America (Nichols & Haramis 1980, Diefenbach *et al.* 1990). Clearly, while sex differences in winter site fidelity were apparent in some species, they were absent in others.

GENETIC RESEARCH AND DISPERSAL

Studies based on ringing give a snapshot picture of contemporary dispersal patterns and (by inference) of associated gene flow, but genetic studies can give additional insight into past dispersal events that could have influenced present population structure. Genetic studies have usually compared relatedness within and among populations, using measures of DNA structure. Mitochondrial DNA has often been used for this purpose, but nuclear microsatellites and minisatellites evolve (mutate) more rapidly, so they generally show more variation than mt DNA between subpopulations (Chapter 2). Such comparisons can yield estimates of the degree of genetic differentiation between populations that can then be used, given some assumptions, to assess rates of gene flow between them. Low genetic variability is typical of inbred populations, resulting from founder or bottleneck events, or from limited dispersal and gene flow (Chapter 2).

The effects of dispersal distances on the genetic structure of populations is clearly evident in North American jays of the genus *Aphelocoma*, which contains both pair-breeding species (*A. californica*) and co-operative group-breeding species (*A. ultramarina* and *A. coerulescens*). The pair-breeders disperse over long distances (sometimes exceeding 50 km), whereas the group-breeders disperse over very short distances (mostly less than 2 km). In both types, genetic (DNA microsatellite) differences between populations increased with distance, so that populations were more divergent the further apart they lived, but the rate of genetic change per unit distance was three times greater in group-breeding than in pair-breeding species. In the group-breeding Florida Scrub Jay *A. coerulescens*, genetic differences were apparent between birds living on different sandy ridge systems, only a few kilometres apart and separated by unsuitable terrain. In this species, habitat continuity seemed more important than distance in permitting gene flow. The faster rates of local molecular differentiation in the group-breeding species could perhaps in time translate to higher speciation rates (Peterson 1992, McDonald *et al.* 1999). These findings thus confirm the role of dispersal distances in influencing the genetic structure of populations, but they also highlight in closely related species the role of habitat distribution and social system in influencing the course of evolution. Of course, the absence of long movements by individuals does not necessarily prevent all long-distance gene flow, which could still occur, but more slowly via a succession of short movements by individuals from different generations.

The degree of genetic difference between two isolated populations of a species could give some measure of the extent of past movement (and gene exchange) between them, and thus throw light on their population histories. Much recent work has concerned seabirds nesting in discrete colonies. In the Fairy Prion *Pachyptila turtur*, which breeds across the southern ocean, 21 birds taken from one colony off Australia all had identical mitochondrial (restriction fragment) haplotypes (Ovenden *et al.* 1991). The authors suggested that this uniformity reflected, not only strong philopatry, but also a past population bottleneck or founder event, following which a very small number of individuals gave rise to the present colony of 10,000 pairs, with insufficient subsequent immigration to diversify the gene pool. In contrast, two other colonies of Fairy Prions were much more heterogeneous, but showed no consistent differences, suggesting that they had experienced no such bottleneck or founder effect.

The habit of some seabirds of settling to breed in the same small part of the colony where they themselves were raised sometimes results in genetic substructuring within colonies that exceeds the structuring between colonies (for Common Murre *Uria aalge* see Friesen *et al.* 1996a; for Cory's Shearwater *Calonectris diomedea* see Rabouam *et al.* 2000). These species thus provide an extreme example of the effects of site fidelity on the genetic substructure of bird populations, although this structuring may be short-lived.

In another study, Austin *et al.* (1994) examined mt DNA in 335 Short-tailed Shearwaters *Puffinus tenuirostris* from 11 colonies across southeastern Australia. There was low diversity among individuals and the most common haplotypes were present in all the colonies. The authors suggested that this species had been greatly restricted in numbers in the past, but that subsequent range expansion, through the founding of new colonies and their subsequent growth, had probably involved the immigration of a large number of individuals to each colony. Despite high natal philopatry revealed by ringing (see earlier), this species may still show genetically significant levels of inter-colony exchange.

In conclusion, because the contemporary genetic structure of populations reflects both past and present patterns of gene flow, genetic studies are best regarded as a supplement to observational studies of marked individuals, rather than as a substitute for them.

DISCUSSION

Two main conclusions to be drawn about bird dispersal are firstly that distances vary with the body size and other features of the species themselves; and secondly that, within species, dispersal distances vary with prevailing conditions, including the structure of the landscape, the year-to-year stability in habitats and food-supplies, and the level of competition in different parts of the population. In other words, whatever the inherent components of dispersal behaviour, there is also a large facultative component that varies with circumstance.

The genetic component

The idea that aspects of dispersal can be under genetic influence comes partly by extrapolation from studies of migration, in which seasonal timing, direction and distance can all be altered by natural selection (Chapter 18). More direct evidence for inheritance of dispersal distances comes from sibling and parent–offspring comparisons. In some species, the dispersal distances of siblings were found to be similar to one another, after making allowance for any sex difference (for Spruce Grouse *Dendragapus canadensis* see Keppie 1980, Schroeder & Boag 1988; for Eurasian Sparrowhawk *Accipiter nisus* see Newton & Marquiss 1983; for Great Tit *Parus major* see Greenwood *et al.* 1979; for Spotted Sandpiper *Actitis macularia* see Alberico *et al.* 1992). Because most such records involved movements in different directions, they could not have resulted from brood-mates travelling together. Such similarities are consistent with the view that dispersal distances have a genetic component, but they are also open to other explanations. For example, they might have been influenced by the conditions of early life which brood-mates would have shared, or they could have arisen as artifacts of the study situation, such as the small size of the study area or the fixed distribution of nest sites within it (van Noordwijk 1983). The same could apply to parent–offspring comparisons.

In some species, such as swans, geese and cranes, family groups remain intact through the first winter and, when the parents begin to breed again, the siblings leave but may remain together as a group (Prevett & MacInnes 1980, Scott 1980). In some co-operative breeders, sibling groups sometimes disperse together and attempt to obtain territories (for Acorn Woodpecker *Melanerpes formicivorus*, see Köenig & Mumme 1987). In all such species, therefore, some siblings might be expected to end up close together. Hence, while dispersal distances may well have a genetic component, the evidence is equivocal (Johnson & Gaines 1990). Other aspects of dispersal, however, such as its timing and lack of directional bias, may be presumed to be under genetic influence, by analogy with comparable aspects of migration (Chapter 18).

Dispersal and inbreeding avoidance

Harmful effects of inbreeding have long been known in domestic animals and humans, and have also been demonstrated in some wild bird species, such as the Great Tit *Parus major* and Song Sparrow *Melospiza melodia* (Greenwood *et al.* 1978, Keller *et al.* 1994). In all these species, mating between close relatives produced fewer viable offspring than did other matings. Thus natural selection should favour any behaviour that reduced inbreeding, including dispersal itself, and also a difference in dispersal distances between the sexes, as well as kin recognition and avoidance. It might also favour divorce (and associated change of breeding territory in one or both partners) following poor breeding success, if one of the factors promoting poor breeding is genetic similarity of the partners. On the information available, however, it is impossible to judge whether inbreeding avoidance is merely an incidental consequence of dispersal, or whether the benefits of inbreeding avoidance have contributed to the evolution of dispersal and any associated sex difference.

Birds could also avoid inbreeding if they could recognise their close relatives and not pair with them. Recognition of family members seems to operate in swans, geese and cranes, in which families stay together for months, and young normally select breeding partners from outside the family group. It also seems to operate in species that defend group territories, and in which sexually mature, closely related individuals can occur in the same social group. In the Acorn Woodpecker *Melanerpes formicivorus*, for example, which breeds in groups of up to several dozen individuals, many birds within each group are closely related, yet only very rarely do they mate with one another (Köenig & Pitelka 1979). The males and females that join to form new groups tend to be unrelated, and sexually mature offspring seldom breed within their natal group as long as both parents are alive. This implies the existence in these species of both the recognition and avoidance of family members. However, it would be stretching the system too far to expect young to identify siblings from other years that they might never have met within the family group. So neither dispersal nor individual recognition is likely to give total protection against inbreeding, especially in small or confined populations where mate choice is limited.

Despite its rarity in most bird species, inbreeding has occurred with surprising frequency in some. For example, in the New Zealand Pukeko *Porphyrio porphyrio* (a group-breeding gallinule), incestuous matings were frequent, and inbreeding appeared commoner than outbreeding (Craig & Jamieson 1988). Some other group-breeding species, such as the Splendid Fairywren *Malurus splendens*, also revealed relatively high levels of inbreeding associated with low dispersal (Russell & Rowley 1993). Any population with a long history of inbreeding would in time rid itself of recessive lethal genes, so that incestuous matings could be less of a problem than in species in which such matings are rare. Inbreeding is usually taken to refer to matings between siblings or between offspring and parents, but it is clearly a continuous variable that could be defined more broadly.

Dispersal and population differentiation

The main advantage of philopatry is that individuals are likely to remain in, or return to, areas where their survival and reproductive chances are high. Their ancestors did well there, and so on balance so should they. Through philopatry, individuals are more likely to mate with partners from the same locality as themselves than with partners from more distant localities. This could in turn facilitate the development of local adaptations and, on a larger scale, of clines in morphology or other characters across the range (Chapter 2). In contrast, species in which individuals move long distances between natal and breeding sites would be expected to show more uniform genetic structure across the range. In fact, dispersal distances are clearly reflected in the degree of clinal or subspecific differentiation within species, as shown for various west Palaearctic birds in **Figure 17.9** (see also Belliure *et al.* 2000).

Some closely related species, which differ from one another in their movement patterns, show strikingly different degrees of subspeciation. For example, the Chaffinch *Fringilla coelebs*, which is a resident or a short-distance migrant throughout its range, is highly polytypic, with 19 subspecies, including 11 on mainland Eurasia and eight on different islands. The closely related Brambling *Fringilla montifringilla* is an irruptive migrant, in which individuals breed and winter in widely separated areas in different years; it is monotypic, showing no distinct races.

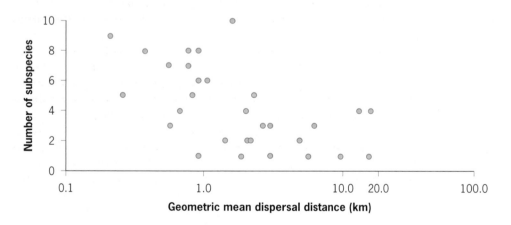

Figure 17.9   Number of west Palaearctic subspecies of different bird species shown in relation to their geometric mean natal dispersal distances. Species in which individuals move short distances between their natal and subsequent breeding sites show much subspecific variation, whereas those that move longer distances show less (r = −0.588, P<0.001). Subspecies (excluding island forms) from Cramp & Simmons 1977–83, Cramp 1985–92, Cramp & Perrins 1993–94, and dispersal distances from Paradis *et al.* 1998.

Although it breeds only on mainland Eurasia, its total range area is about twice that of the Chaffinch. Similarly, the European Goldfinch *Carduelis carduelis* is polytypic (11 black-headed races and four grey-headed ones), while the closely related irruptive Eurasian Siskin *C. spinus* is monotypic, even though its total range is greater. Overall, among west Palaearctic finches, nine out of ten widespread species that are resident or regular migrants are polytypic within this area, with an average of 6.8 races per species, while only two out of 11 irruptive species are polytypic within this area, with an overall average of 1.8 races per species (Cramp & Perrins 1994). Similarly, among the west Palaearctic owls, all six widespread species that are resident are also polytypic within this area, with an average of 6.3 races per species, while only two out of seven irruptive species are polytypic within this area, with an average of 1.3 races per species (Cramp 1985). The tendency for individuals of irruptive species to breed in widely separated areas in different years presumably ensures rapid gene flow across the range, and acts against the development of locally adapted forms.

Geographical variation is also distinctly less in long-distance migrants than in partially migrant or resident birds. Among Palaearctic passerines, for example, migratory species can each be divided into an average of 3.2 subspecies, while sedentary ones each have an average of 7.2 subspecies (**Table 2.3**, Rensch 1933). Thus, the sedentary tits and woodpeckers, which occur over much of the Palaearctic, each has several subspecies, while the highly migratory Eurasian Wryneck *Jynx torquilla* and Willow Warbler *Phylloscopus trochilus*, which breed over a similar wide area, have far fewer. This may be partly because both natal and breeding dispersal distances are in general greater in migrants than in closely related residents (as found by Paradis *et al.* 1998). Both resident and migrant species may have occupied their northern breeding areas from localised glacial refuges only in the last ten thousand years, so many local subspecies may have formed within that time (Chapter 10).

In Chapter 1, I mentioned that large species of birds generally have larger geographical ranges than small species. Study of dispersal has provided a possible explanation of this relationship, namely that small species occupying large areas are more likely than large species to split into different subspecies, and eventually different species. This is because individuals of small species typically disperse smaller distances between natal and breeding sites than individuals of large species. It is therefore easier for spatially separated populations of small species to become genetically isolated from one another, and eventually to form new species, each occupying a different part of the range of the original parental species. This process is abundantly clear in the current numbers and distributions of subspecies and allospecies of small compared with large bird species (Chapter 2).

Dispersal and range expansion

Although dispersal is a continually occurring process, most does not result in significant change in geographical range, because at the edges most species are up against barriers to further spread (Chapter 14). When expansion does occur, it may happen through progressive outward spread, or from the crossing of barriers to establish new populations some distance from the old, as in the colonisation of islands (Chapter 16). Both types of range expansion have been of immense biogeographical importance in shaping present distribution patterns, and both types have been recorded in many bird species in the last 100 years (see Chapters 15 and 16). Cross-barrier movements are assumed to be infrequent and chance events, but without them most oceanic islands would be devoid of birds, and the total variety of bird life would be much less than it is.

Some remarkably rapid rates of range expansion have been recorded in birds (Chapter 15), but they might in some species have been even faster if birds had been less faithful to their natal localities. Site fidelity can be beneficial in normal circumstances when habitats everywhere are stable and saturated, but it can greatly slow the rate of spread during a period of expansion, because in effect it makes the expansion largely dependent on birds raised near the edge of the existing range, some of which settle in the adjoining vacant ground. Surplus production further inside the range plays little or no part in the expansion process. In several raptor species that have spread in recent years, the advance occurred almost on a territory-by-territory basis, while nearer the centre of the range where habitats were saturated, potential breeding adults accumulated as non-territorial non-breeders, seemingly tied to the vicinity of their birth by strong philopatry (for Red Kite *Milvus milvus*, see Newton *et al.* 1994; for Common Buzzard *Buteo buteo*, see Walls & Kenward 1995; for Imperial Eagle *Aquila heliaca*, see Ferrer 2001). Such behaviour presumably evolved because, in the absence of human action, all parts of a geographical range would normally be occupied, so birds would gain no advantage by dispersing far. Social factors may also be important, because birds may benefit from moving to existing concentrations rather than to areas that lack conspecifics but are suitable in other respects. The biggest problem for individuals that behave independently of others is that they may not find a mate, while individuals of colonial and flock-feeding species could be disadvantaged in other ways too. There are clearly risks as well as benefits involved in range extension.

SUMMARY

In a biogeographical context, dispersal has important consequences because it facilitates gene flow and influences the genetic structure of populations, and because it can lead to range expansion. At the level of individual birds, dispersal can be measured by the distances between natal sites and subsequent breeding sites (natal dispersal), between the breeding sites of successive years (breeding dispersal), and between the wintering sites of successive years (non-breeding dispersal). In most bird species, which form pairs on their breeding areas, only natal and breeding dispersal affect rates of gene flow.

In many bird species, both resident and migratory, individuals tend to settle and breed in the neighbourhood where they were raised, and the numbers of dispersed individuals decline with increasing distance from the natal site. However, individuals of large bird species move generally longer distances than those of small species. In addition, individuals of species that depend on ephemeral habitats or food-sources tend to disperse longer distances than those of species with more predictable habitats and food-sources. This is apparent in comparisons between related species and within different populations of the same species, whose habitats or food-sources differ in spatial reliability. Dispersal distances also vary within species according to population density and other factors that promote competition, as well as with gender and other features of the individual. In many bird species, females move on average further between natal and breeding sites than males, but in waterfowl and some others, males move furthest.

In general, adults move over much shorter distances between the breeding attempts of successive years than young of their species move between their natal and first breeding sites. Where conditions remain fairly stable from year to year, adults of many species use the same nesting territories in successive years, though some individuals move to nearby territories, females more often than males, and especially after a breeding failure. Where local conditions vary from year to year, adults can move long distances between their nesting sites of successive years. Long natal and breeding dispersal distances (of up to hundreds or thousands of kilometres) are frequent, for example, in some waterbirds that depend on ephemeral wetlands, in seed-eaters that depend on sporadic tree-seed crops, and in rodent-eaters that depend on localised cyclic peaks in their prey. All such species tend to concentrate in different areas in different years, wherever conditions are suitable at the time.

Study of the genetic structure of populations can give insight into their biogeographical history, particularly in revealing past founder effects, bottlenecks or dispersal patterns. In general, species that disperse short distances show more subspecific variation than those that disperse long distances.

Barnacle Geese *Branta leucopsis.*

Chapter 18
Migration biogeography

Migration can be defined as a large-scale return movement of a population, which occurs each year between regular breeding and wintering (or non-breeding) areas. Involving seasonal shifts of millions of individuals, it produces a massive re-distribution of birds over the earth's surface twice each year. High-latitude regions, such as the arctic, support birds mainly in the breeding season, while lower latitude regions support wintering birds from higher latitudes, as well as year-round residents. At intermediate latitudes, some species leave for the winter, while others move in. Migration thus increases the numbers of species that occur in particular areas, even though some are present for only part of the year. The different aspects of migration form an enormous subject area, and in this chapter I can do no more than discuss the biogeographical aspects that relate most closely to the theme of this book.

Throughout the world, migration is most apparent in seasonal environments, in which the contrast between summer and winter (or wet season and dry season) conditions is greatest. Migration thus allows individual birds to exploit different areas at different seasons. In fact, some migratory birds exploit habitats for over-winter survival that they could not use for breeding, and then occupy breeding areas that would not support them over winter. This applies to all arctic-nesting shorebirds which spend the winter on coastal mudflats where, owing to tidal flooding, nesting would be impossible, and then migrate north to breed on the arctic tundra which is frozen and snow-covered for the rest of the year. Thus some bird species exist, over much or all of their range, only by virtue of exploiting widely separated habitats at different seasons. In the process, they make twice-yearly journeys of up to thousands of kilometres, some crossing seas, deserts or high mountain ranges (**Figure 18.1**).

Although most marked at high latitudes, migration also occurs in the tropics, especially in the savannahs and grasslands, exposed to regular wet and dry seasons. Within the northern Afrotropics, for example, many species move south for the non-breeding season, some crossing the equator, while in the other half of the year many southern ones move north. In any tropical region, the amount and timing of movement seems to be related to rainfall, and migrants appear in particular localities whenever food becomes available there. As a group, birds confined to equatorial rain forest are probably least migratory, especially the small insectivores of the understorey where conditions remain stable and suitable year-round. This consistency of environment removes any advantage in moving long distances, and many individuals may remain within the space of the same few hectares throughout their lives. In the same forests, however, some nectar-eaters and fruit-eaters move within small latitudinal or altitudinal bands in response to flowering and fruiting patterns, while other birds from higher latitudes move in for their 'winter'.

Most migrant species thus form parts of different communities at different seasons, and may interact with different species in their breeding and non-breeding homes. Each must be compatible with these different species, as well as with different climatic and dietary regimes.

EVOLUTION OF MIGRATION

Migration might be expected to occur wherever individuals survive in greater numbers if they leave their breeding areas for the non-breeding season than if they remain there for the whole year (Lack 1954). The usual reason why breeding areas become unsuitable during part of the year is lack of food. Such food-shortages occur for many birds because plant-growth stops for part of the year, and many kinds of invertebrates die or hibernate or become inaccessible under snow or ice. Hence, the purpose of the post-breeding exodus from nesting areas is fairly obvious.

The reason why birds leave their wintering areas to return in spring is less obvious, because many such areas seem able to support the birds during the rest of the year. However, if no birds migrated north in spring, the northern lands would be almost empty of many species, and a large seasonal surplus of food would go largely unexploited. Under these circumstances, any individuals that moved to higher latitudes, with increasing food and long days, might raise more young than

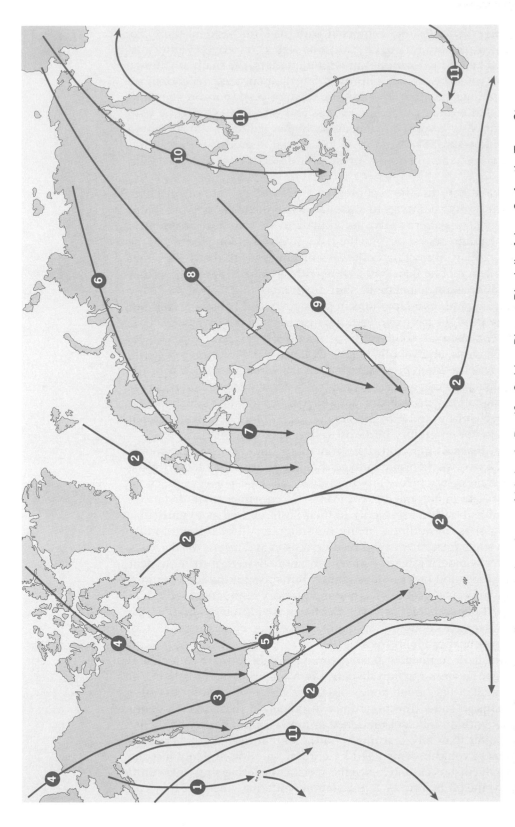

Figure 18.1 Some long-distance migrations of birds. 1. Pacific Golden Plover *Pluvialis fulva*; 2. Arctic Tern *Sterna paradisaea*; 3. Swainson's Hawk *Buteo swainsoni*; 4. Snow Goose *Anser caerulescens*; 5. Many North American breeding species that cross the Gulf of Mexico; 6. Ruff *Philomachus pugnax*; 7. Many European breeding species that cross the Mediterranean Sea and Sahara Desert; 8. Northern Wheatear *Oenanthe oenanthe*; 9. Amur Falcon *Falco amurensis*; 10. Arctic Warbler *Phylloscopus borealis*; 11. Short-tailed Shearwater *Puffinus tenuirostris*. Partly after Berthold 1993.

if they stayed at lower latitudes and competed with the birds resident there. So, whereas the advantage of autumn migration can be seen as improved winter survival, dependent on better food-supplies in winter quarters, the main advantage of spring migration could be seen as improved breeding success, dependent on food-supplies in summer quarters. In addition, because predation on eggs and chicks is a major factor reducing breeding success and is thought to decline with latitude, this might also favour movement to higher latitudes for breeding (Fretwell 1980, Pienkowski 1984). Compared to survival, reproduction often has more stringent requirements, in terms of specific food needs and predation avoidance.

The migratory habit can thus be regarded as a product of natural selection, ensuring in the long term that species in seasonal environments adopt and maintain whatever movement patterns lead individuals to survive and breed most effectively. For the habit to persist despite the risks involved in long journeys, the net benefits to individuals of moving both ways must outweigh the costs. Other animals, which are less mobile than birds, cope with seasonal shortages in other ways, notably by hibernation, a metabolic shut-down involving torpor.

Although migration requires adaptations in physiology and behaviour, it is not difficult to see how it might have evolved, because transitional or intermediate stages still occur. For instance, some bird species do not migrate at all, others travel only short distances, and yet others long distances. The full range of variation can be found among different populations of the same species, and in the intermediate populations some individuals may migrate while others do not. Further, the main adaptations needed for long-distance migration, such as fat reserves, timing mechanisms and navigation skills, are all found with somewhat different functions in non-migratory birds, as well as in other animals.

The main features that set migration apart from other long-distance movements are that it involves a two-way journey in fixed directions. An ability to perform a return movement is necessary before any migration pattern can evolve. Such an ability is already present in non-migratory birds but occurs over small distances. Thus all birds are able to return repeatedly to their nests, as well as to particular feeding or roosting sites. In addition, many non-migratory birds move locally away from their nesting areas after each breeding season and return for the next. Even juveniles, after wandering widely in winter, normally return to settle near their natal sites to breed, and other non-migratory birds re-visit the same wintering sites in successive years (Chapter 17). In non-migratory birds, such individual movements are short-distance and localised, but they provide a basis from which the longer return movements of migration might evolve, step by incremental step, each extension being beneficial in its own right.

Similarly, it is not hard to imagine how, given a sense of location, directional preferences could evolve from random dispersal movements. Any birds living in seasonal environments that move long distances are more likely to meet favourable conditions in some directions than in others. In the northern hemisphere, individuals with an inherent tendency to move south after breeding are likely to survive better than any that move north, so that over the generations directional preferences could become fixed by natural selection. It is not just the directions, but also the distances, and hence the specific wintering sites, that could be fixed in this way, the birds from each population wintering wherever they can reach and survive best, taking account of the mortality costs of getting there and back. Suppose that the birds from a certain breeding area have heritable tendencies to fly particular directions and distances at migration time and back again in spring, but that these directions and distances differ from bird to bird. Some birds will then reach suitable areas and survive to breed again, others will reach less suitable areas and survive in small numbers, and yet others will reach unsuitable areas and die. Thus those individuals with the most beneficial migratory behaviour will perpetuate themselves, and in this way the migratory habits of a population could become fixed, including the seasonal timing, the routes taken and the destinations. Only in populations (like some nomadic ones) that on balance are as likely to find food in one direction as in any other, is no directional preference likely to become fixed by natural selection.

For many years, the idea that natural selection shaped bird migration patterns was based on little more than surmise, but in recent decades the role of genetic factors in the control of migration has been shown experimentally, mainly by Peter Berthold and his colleagues, who found that they could breed Blackcaps *Sylvia atricapilla* and other songbirds in large numbers in aviaries (Berthold & Helbig 1992, Berthold 1995, 1999). In these species, populations varied from completely migratory to completely resident in different parts of the range; the migratory populations also differed in their migration dates and in the directions and distances travelled. The majority of these traits could be measured in captivity by assessing the timing and amounts of 'migratory restlessness', a specific behaviour involving fluttering and wing-whirring which appears in caged birds only at migration seasons. The test species normally migrate at night, and the number of evenings in spring and autumn on which the birds showed migratory restlessness reflected the usual duration (and distances) of migration for each population. The directional preferences of individuals could be assessed by placing them in circular 'orientation cages', from which they could see the sky, and checking where they headed.

The main finding was that, when birds from different populations were crossbred, the offspring showed migration features that were intermediate between those of their parents. This held for all aspects, whether timing, duration, direction or migrancy versus residency. Migratory or sedentary behaviour could be selected to phenotypic uniformity within as few as 3–6 generations. The overriding implication was that all these aspects of migration were under genetic control, and could thus be changed rapidly in the wild by natural selection, depending on the survival and reproductive benefits of different behaviours in the conditions prevailing.

As it happens, a new migration route among wild Blackcaps *Sylvia atricapilla* has developed in recent decades. Thousands of individuals now winter in Britain, a habit unknown before 1960, and ring recoveries have revealed that these birds are not part of the British breeding population (which continues to leave in autumn), but come from central European breeding areas. This new habit involved a marked change in the autumn direction of central European Blackcaps from southwest (towards Spain) to west–northwest (towards Britain), and a one-third shortening in the length of journey (Berthold 1995).

To see whether this new migration had a genetic basis (as opposed to being due to increased wind drift or other prevailing conditions), some wintering Blackcaps were caught in England, bred in captivity in Germany, and then tested for directional preferences in standard conditions. Both the adults and their offspring showed a west–northwest direction, indicating that the new migration route represented an evolutionary change that seemed to have occurred within a few decades (Berthold *et al*. 1992).

Such a change presumably started from one or more individuals with a new directional tendency that found themselves in a new area where they could survive the winter. It would not be enough merely for them to have been blown off course, because the new route could not have been inherited by their offspring. Once started, the selection pressures that may have favoured wintering in Britain rather than in the western Mediterranean region could have included: (1) factors acting in the new wintering area, such as milder winters, or improved food-supply provided at garden feeders and at berry bushes planted in recent decades; or (2) factors related to the location of the new wintering areas, which enable a shorter migration distance (by up to 1,500 km) and possibly also an earlier return to the breeding areas. An early return could enable British-wintering Blackcaps to pair preferentially with one another (assortative mating based on differential arrival times), which would in turn speed the evolution of the new habit. In view of these findings, it is surprising that British-breeding Blackcaps have not become partially resident, with some individuals remaining all year in their breeding areas. If such a change has started, it has not yet been confirmed by ringing, as all the recoveries point to the continued wintering of British-breeding birds in Iberia and Africa.

Many other bird species have changed their migratory habits during the last 200 years in response to changing conditions. The following examples illustrate the different types of change recorded.

Migratory to sedentary. Prior to 1940, the Lesser Black-backed Gull *Larus fuscus* was almost entirely migratory in Britain, only a few individuals remaining year-round, but nowadays large numbers stay in winter, feeding mainly on recently created refuse dumps which have increased the winter food-supply (Hickling 1984). Another example is the Eurasian Blackbird *Turdus merula*, in which the British and mid European populations became progressively more sedentary during the last century (Berthold 1990, Main 2000).

Sedentary to migratory. The European Serin *Serinus serinus* was once restricted to the south of Europe where it is resident, but in the early twentieth century it spread north, where it became migratory. In more recent years with milder winters, this migratory population has become partially resident (Berthold 1999). Another example is the Cattle Egret *Bubulcus ibis* which has developed new migrations in some newly colonised parts of the New and Old Worlds (Maddock & Geering 1994).

Shortening of migration routes. So called 'short-stopping' has occurred as more food has become available in the northern parts of the wintering range. Several populations of Canada Geese *Branta canadensis* in North America have responded in this way to agricultural changes or to the creation of refuges (e.g. Hestbeck *et al*. 1991), as have Common Cranes *Grus grus* in Europe (Alonso *et al*. 1991), and other species elsewhere. Some Barnacle Geese *Branta leucopsis* are unusual in having shortened their migration by 1,300 km by establishing nesting colonies well south of the historical range, nearer to wintering areas (Chapter 15; Larsson *et al*. 1988).

Lengthening of migration routes. In the Red-breasted Goose *Branta ruficollis*, most individuals now winter much further from their breeding areas than in the 1950s, as former wintering sites have been altered by land-use changes (Sutherland & Crockford 1993). In even earlier times, the species was found in winter even further from its breeding areas (Chapter 15).

Migration timing. Under the presumed influence of long-term global warming, many birds species are now arriving earlier and departing later from their breeding areas than in the past, and are thus spending longer each year on their breeding areas (Moritz 1993, Vogel & Moritz 1995, Sparks 1999, Jenkins & Watson 2000, Sparks & Mason 2001). However, arrival dates still fluctuate from year to year in line with local temperatures.

At present in the northern hemisphere, then, many bird species are changing their migratory behaviour with: (1) earlier arrival in spring, (2) later departure in autumn, (3) shortening or lengthening of migratory distances, (4) directional changes, and (5) reduced or enhanced migratoriness, reflected in changes in ratios of residents to migrants in particular populations and in the occurrence of wintering birds at latitudes previously lacking them. Almost all these changes are associated with changes in food-availability, or with climatic conditions that are likely to affect food-supplies, such as milder winters. The only examples of shifts to increasing migratoriness involve species (such as those mentioned above) that have extended their breeding ranges into areas where overwintering is not possible. A different type of change is shown by those northern hemisphere species introduced to the southern hemisphere, which have reversed the direction of their spring and autumn journeys, as appropriate.

These various observations, along with the experiments, all serve to confirm that migration is a dynamic phenomenon, subject to continual change in response to prevailing conditions. Some aspects, such as the development of migration in previously resident populations, imply rapid evolutionary change, but other aspects could represent facultative (non-genetic) change in response to changing conditions. Without case-by-case study, it would be hard to separate the genetically controlled changes from the non-genetic ones. Whatever their basis, however, it is presumably in these ways that birds continually adjusted their migration patterns to the massive climate changes of the past, and through which we can expect them to respond in future. This does not mean, however, that present migration patterns are completely free of the influence of past events, or that all species could alter their migration patterns so readily.

LATITUDINAL TRENDS

What effects do seasonal migrations have on the wide-scale distributions of birds? As the breeding and non-breeding ranges of birds have become increasingly well mapped in recent years, both north and south of the equator, it has become possible to examine the seasonal distributions of birds in ways not previously possible, and to draw inferences about the ecological factors underlying migration and geographical range.

As indicated above, the proportions of migrating species within the bird faunas of different regions are closely correlated with climatic seasonality, and hence with the degree of contrast between winter and summer food-supplies (Newton & Dale 1996a, b). With increasing latitude and seasonality of climate, increasing proportions of the bird fauna are migratory. Progressing northwards up the western seaboard of Europe, for example, the proportion of breeding bird species that move out totally to winter further south increases progressively from 29% of species at 35°N (North Africa) to 83% at 80°N (Svalbard), a mean increase of 1.3% of breeding species for every degree of latitude **(Figure 18.2)**.

A similar relationship between migration and latitude holds independently in the largely different avifauna of eastern North America, where the proportion of migrants among breeding species also increases with distance northwards from 12% at 25°N to 87% at 80°N, a mean increase of 1.4% per degree of latitude **(Figure 18.3)**. The difference between the two regions **(Figure 18.4)** reflects the climatic shift between east and west sides of the Atlantic: over most of the latitudinal range, at any given latitude winters are colder in eastern North America than in western Europe. Correspondingly, at any given latitude, a greater proportion (an average of about 17% more) of breeding species leaves eastern North America for the winter than leaves western Europe. The slopes of the two linear regression lines calculated for the data in **Figure 18.4** do not differ significantly, but the intercepts do ($F_{1,19}$ = 27.5, P<0.001), reflecting this climatic difference.

The few species that remain to winter in the far north (at 80°N) include the Northern Fulmar *Fulmarus glacialis*, Ivory Gull *Pagophila eburnea* and Glaucous Gull *Larus hyperboreus* in Europe, and the Common Raven *Corvus corax*, Gyrfalcon *Falco rusticolus*, Snowy Owl *Nyctea scandiaca* and Rock Ptarmigan *Lagopus mutus* in both Europe and North America. Some of these species may move south to some extent in the weeks of complete darkness. The absence of seabirds at 80°N in eastern North America is presumably due to the greater extent of sea ice there, compared with Europe. In both regions the most northerly seabirds depend on the open water provided by polynyas, and some of the gulls also scavenge the remains of seals killed by bears. Similar relationships between migration and latitude have been shown among birds of particular habitats from grassland to forests (Willson 1976, Herrera 1978), and among birds of particular families, such as flycatchers in South America (Chesser 1998). They presumably hold worldwide in both hemispheres.

The converse of these relationships for western Europe and eastern North America is shown in **Figure 18.4** as the proportions of the birds wintering at different latitudes that move north for the summer. This proportion is greatest in the south, in western Europe affecting 36% of species wintering at 35°N, and declining northwards to 8% of species wintering at 70°N (mostly seabirds) and none at 80°N. In eastern North America, the equivalent figures are 52% at 25°N to none at 70°N. Again the slopes of the regression lines to do not differ between continents but the intercepts do ($F_{1,16}$=9.9, P<0.01). Throughout the latitudinal range, the proportion of wintering species that leaves northwards for the summer averages around 10% greater in eastern North America than in western Europe, again reflecting the climatic difference between the two regions. The precise regression relationship varies somewhat between birds of different habitats. In addition, 23% of all species breeding in western Europe, and 24% of those in eastern North America, leave these areas completely in autumn for more southern climes, returning in spring (Newton & Dale 1996b).

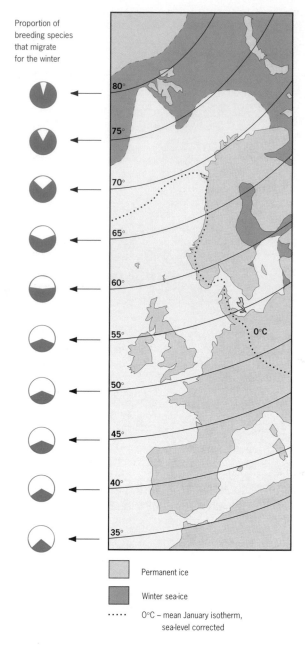

Proportion of
breeding species
that migrate
for the winter

Permanent ice

Winter sea-ice

····· 0°C – mean January isotherm,
sea-level corrected

Figure 18.2   Map of western Europe showing the proportion of breeding bird species at different latitudes that migrate south for the winter. From Newton & Dale 1996a.

The proportions of all bird species at particular latitudes that are migratory are correlated not only with latitude, but also with various environmental features that themselves correlate with latitude, such as temperatures (hottest or coldest months) or the temperature difference between the hottest and coldest months

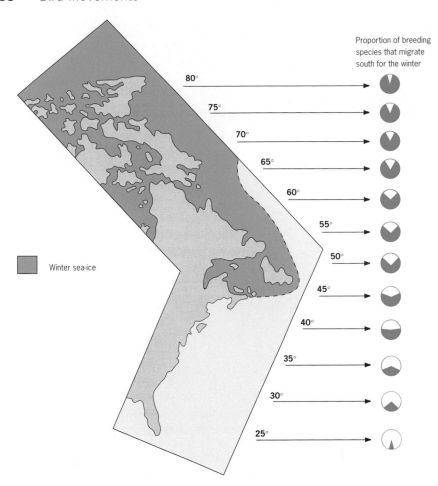

Figure 18.3 Map of eastern North America showing the proportion of breeding species at different latitudes that migrate south for the winter. From Newton & Dale 1996b.

(Appendix 18.1). All these measures are of course interrelated, but they are all surrogates for what really matters, namely the degree of climatic difference between summer and winter. It is this that, for many birds, governs the difference in food-supply between winter and summer at particular latitudes, and hence the difference in environmental carrying capacity between the two seasons.

The seasonal differential in carrying capacity may also vary from west to east across a continent (as between west and east sides of the Atlantic), according to changes in climate. Thus, as one moves from west to east across Europe, summer climates become warmer and drier, and winter climates become colder. In consequence, progressing eastwards through Europe into Asia, increasing proportions of the local breeding species become migratory. This is especially obvious in comparing populations of coastal areas that live under mild oceanic climates with those further inland that live under continental climates. For example, Common

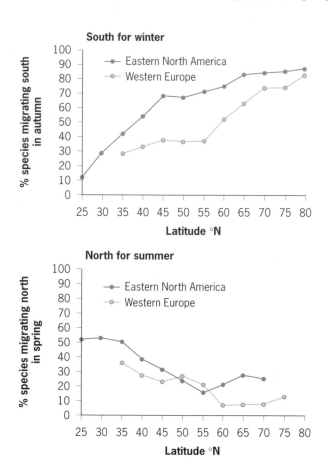

Figure 18.4 Percentage of breeding bird species (y) at different latitudes (x) in western Europe and eastern North America that migrate south for the winter (upper) or north for the summer (lower). For southward migration, on regression analysis for western Europe: $y = 41.49 - 1.03 + 0.02x^2$, $r^2 = 0.97$; for eastern North America: $y = -75.05 + 4.33x - 0.3x^2$, $r^2 = 0.98$. For northward migration, on regression analysis for western Europe: $y = 55.65 - 0.66x$, $r^2 = 0.81$; for eastern North America, $y = 123.72 - 3.22x + 0.03x^2$, $r^2 = 0.87$. For three of these relationships a quadratic equation gave a significantly better fit than a linear one. From Newton & Dale 1996b.

Starlings *Sturnus vulgaris* live year-round on the Shetland Islands at 60°N, while at the same latitude in Russia (and for 10–15° south of it) they are wholly migratory.

At some mid latitude areas, similar numbers of species may be present in summer and winter, but species composition changes somewhat between seasons, as some arrive while others leave (Gauthreaux 1982, Newton & Dale 1996a, b). In southern England, for example, insectivorous swallows and warblers arrive from the tropics for the summer, whereas fruit-eating thrushes from higher latitudes arrive for the winter. Seasonal changes in bird communities in particular regions are thus tied to seasonal changes in the types of food available. It emphasises the point that migrants often exploit seasonal abundances in both their breeding and non-breeding areas. It is a strategy that, for obvious reasons, is much more developed in birds than in most other animals.

Although seasonality in the movements of birds is evident worldwide, in warmer regions rainfall becomes more important than temperature. Thus, in different parts of Australia, the proportions of birds that are migratory decline with increase in the amount and evenness of the annual rainfall (Keast 1968). Where the annual total exceeds 125 cm, and is well distributed through the year, 70–85% of honeyeater species (Meliphagidae) are year-round residents, but in the central desert, where annual rainfall is 20–28 cm and erratic, fewer than 50% are residents.

MIGRATION AND DIET

Among species that breed in the same region, a broad relationship is apparent between migration and diet. Broadly speaking, those species that are resident year-round exploit food-supplies that are available all year, whereas those that leave after breeding exploit foods that become unavailable then. In the northern coniferous forests, for example, residents include mainly species that feed directly from trees, on bark-dwelling arthropods (tits, treecreepers), fruits and seeds (some corvids, finches, tits), buds or other dormant vegetation (grouse), or that eat mammals and other birds (some corvids, raptors and owls). Almost the entire resident landbird fauna at high northern latitudes falls into one or other of these dietary categories. In contrast, species that depart for the winter include those that eat active leaf-dwelling or aerial insects (warblers, hirundines) or foods that become inaccessible under snow or ice (ground-feeding finches and thrushes, some raptors, waterfowl and waders). Towards the equator, as winters become less severe, the range of bird dietary types that remain for the winter increases, as a wider range of food-types remains available year-round.

To illustrate in more detail the link between migration and seasonal changes in food availability, two examples should suffice. Most European songbirds feed either on (1) insects or other invertebrates, (2) seeds and fruits, or (3) a mixture of both categories (Newton 1995b). Of these food-types, seeds and fruits are clearly much more available in winter than are insects. **Figure 18.5** shows the proportions of bird species in these different winter diet categories that are migratory at different latitudes. Within each group, the proportion of migrants increases with latitude, following the general trend in birds as a whole, but at each latitude from about 35°N, a larger proportion of insect-eaters than of seed-eaters leaves, while species with mixed insect–seed diets are intermediate. Furthermore, the insect-eaters generally move longer distances, many wintering in the tropics and some south of the equator **(Figure 18.5)**. The result is that insect-eaters are concentrated at lower latitudes in winter than are seed-eaters. The same holds for New World migrants, in which most small insectivorous warblers winter in Central America (30°N–10°S) and most seed-eaters further north (mostly 40°–15°N) (Keast 1995). Among these general relationships, specific exceptions occur, such as the seed-eating European Turtle Dove *Streptopelia turtur* that winters in Africa.

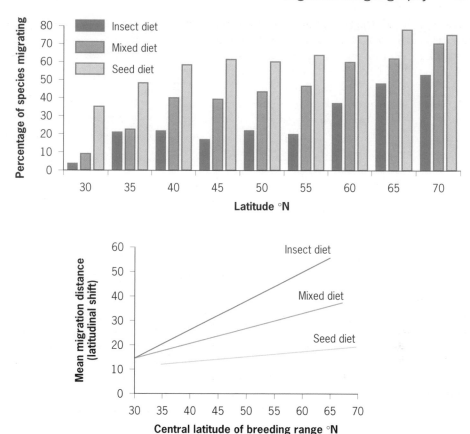

Figure 18.5   Migration in relation to diet in west Palaearctic songbirds. Upper. Percentage of species breeding at different latitudes that migrate south for the winter. Lower. Distances moved by migrants as measured by the difference between the central latitudes of the breeding and wintering ranges. Lines calculated by regression analyses. From Newton 1995b.

As a second example, consider the various European raptors, which also differ in the extents to which their foods remain available at high latitudes in winter (Newton 1998b). Species can be divided according to whether they feed primarily on warm-blooded prey (birds and mammals which remain active and available in winter at high latitudes) or on cold-blooded prey (reptiles, amphibia and insects, which become inactive and unavailable in winter). Within each group, the proportion of migrant species again increases with latitude, but at any one latitude a larger proportion of cold-blooded than of warm-blooded feeders leaves for the winter, while species with mixed diets (not shown) are intermediate **(Figure 18.6)**. The cold-blooded feeders also migrate furthest **(Figure 18.6; Table 18.1)**. The reasons for this difference are fairly obvious, in that species that feed on cold-blooded prey and breed at high latitudes must winter in the tropics or in southern hemisphere temperate areas, if they are to have access to the same types of prey year-round. Such patterns again underline the link between migration and

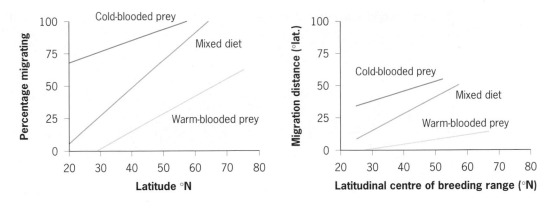

Figure 18.6   Migration in relation to diet in west Palaearctic raptors. Left. Percentage of species breeding at different latitudes that migrate south for the winter. Right. Distances moved, as measured by the difference between the central latitudes of the breeding and wintering ranges. Lines calculated by linear regression analyses.

Table 18.1   Wintering areas of west Palaearctic raptors in relation to diet. Figures show numbers of species in each category. From Newton 1998b.

Wintering area	Main prey types		
	Warm-blooded	Mixed	Cold-blooded
North of Sahara	16	0	0
North and south of Sahara	5	8	2
South of Sahara	1	4	7

Significance of variation between categories (examined by Monte Carlo randomisation test): $\chi^2_4 = 35.9$, $P<0.001$.

Of 22 species that eat mainly warm-blooded prey, only one species (Booted Eagle *Hieraaetus pennatus*) winters entirely in Africa. In contrast, seven of nine species that eat mainly cold-blooded prey winter entirely in Africa. They include six insectivores, four of which live entirely (and two largely) south of the equator, where the seasons are reversed. The 12 species with mixed diets show intermediate patterns.

the seasonal changes in specific food-sources (Newton 1979). Some authors have stressed the role of climate and competition in influencing migration patterns, but both these factors are likely to act primarily through the food-supply.

That movements are related to diet is also apparent in tropical regions, even though many such movements are relatively short, confined to narrow latitudinal or altitudinal spans. In tropical forests, fruit-eaters and nectar-eaters move in greater proportion than other birds, probably because flowers and fruit are much more seasonal in occurrence than are the insects and other small creatures eaten by birds (Morton 1980, Levey & Stiles 1992). As flowers and fruit are generally more abundant in the canopy and forest edge, it is these parts of forest environments that show the greatest seasonal variation in their bird populations, the understorey retaining most of its birds year-round (see above). However, species of open habitats, such as savannah and grassland, tend to fluctuate even more strongly through the year because such habitats tend to be more seasonal, notably in rainfall, which affects the food-supplies of a wide range of species. These patterns are apparent not just at the level of the overall avifauna, but also within particular families, where species differ in diet and habitat (e.g. Levey & Stiles 1992). There is, in fact, little evidence that the development of migration is constrained in any way by phylogeny, as many bird families contain both migratory and non-migratory species, sometimes as closest relatives.

NET LATITUDINAL SHIFTS

The net effect of bird migration is to alter the latitudinal distribution of birds between summer and winter, so that overall species numbers in the northern hemisphere are greatest in the northern summer and in the southern hemisphere in the austral summer (northern winter). Take the west European migrants as an example. Some species move relatively short distances within Europe, while others move longer distances to Africa or southern Asia, but the net result, each autumn, is a huge latitudinal shift in avifaunal distribution **(Figure 18.7)**. In summer, by definition, the whole European assemblage of breeding birds is concentrated north of 25°N, but in winter the same assemblage extends southwards as far as the southern tip of Africa (35°S) and (for some seabirds) beyond. When they are in their wintering areas, the migrants add to the local residents, increasing the total species numbers, especially in the tropics. Each year the Afrotropics receive about 177 migrant landbird species from the north (adding to the 1,951 local species), southeast Asia receives about 161 migrant species (adding to 1,697 species), Australasia receives about 36 migrant species (adding to 1,593 species), and the Neotropics receive about 147 migrant species (adding to 3,370 species).

Migration of landbirds from their breeding areas is a much more obvious phenomenon in the northern hemisphere than in the southern. This is partly because land covers three times the area in the northern hemisphere as in the southern hemisphere, and the difference is most marked at high latitudes (we can ignore Antarctica because it holds no landbirds) **(Figure 18.8)**. In North America, Greenland and Eurasia, some landbird habitat extends north of 80°N, but in the southern hemisphere, South America reaches only to 55°S, Africa to 35°S and Australia to 43°S. The net result is that, at latitudes 30–80°N, there is 15 times more land than at latitudes 30–80°S, and it is at these latitudes that migration is most developed.

Another factor is temperature which has a steeper downward gradient north of the equator than south of it. For example, the mean midwinter (January) temperature at the Tropic of Cancer is about 13°C in North America, while the mean midwinter temperature (July) at the Tropic of Cancer in South America is 16°C, a 3°C difference. Yet at 50°N in North America the mean midwinter temperature is −15°C, while at 50°S in South America it is 0°C, a 15°C difference. Similar differences are apparent in much of the Old World too. This temperature differential may explain why greater proportions of species leave from temperate latitudes in the northern hemisphere than in the southern. For example, about 29% of species leave for the winter from Morocco in North Africa, but only 6% of species leave from equivalent latitudes around the Cape in South Africa (Newton & Dale 1996a, Harrison *et al.* 1997). It is presumably largely for both these land-related and temperature-related reasons that landbird migration is much more marked in the

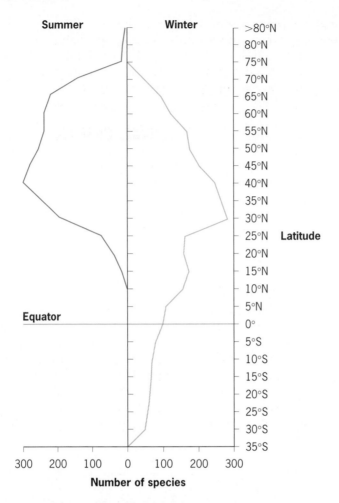

Figure 18.7   Latitudinal shift between summer and winter distributions of bird species that breed in Europe. Includes wintering areas in Europe, Asia and Africa. From Newton 1995b.

northern than in the southern hemisphere. In particular, the southern hemisphere lacks tundra-nesting shorebirds, which form a large proportion of the long-distance migrants from the northern hemisphere.

While many landbird species that breed in the northern hemisphere extend south of the tropics on migration, no landbird species that breeds in the southern hemisphere extends north of the tropics on migration. This difference in long-distance migration can also be attributed to the difference in available land areas between the two hemispheres. Birds migrating south from the northern continents encounter progressively smaller habitable land areas, which could force some individuals to travel far to the south. In contrast, birds migrating north from the southern parts of the southern continents encounter widening land areas, so may need to move less far north before they find sufficient wintering habitat. From Africa, no breeding landbird species migrates north of the Sahara; from South

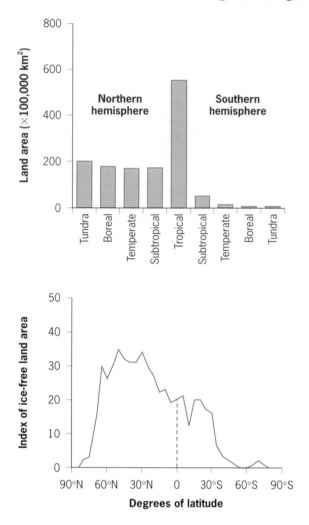

Figure 18.8   The greater land areas in the northern than in the southern hemisphere. The habitats refer to vegetation zones before human impact, only parts of which still survive intact. From Rosenzweig 1992.

America, none (apart from stragglers) penetrate north of Panama; and from Australasia, none migrate beyond Indonesia in winter. It is not that birds that breed in the southern hemisphere do not move around, but rather that most migration is short distance and partial (involving only part of a population), altitudinal (and also short distance) or nomadic (as birds concentrate locally in line with sporadic rainfall patterns). Relatively few species make regular long-distance moves. It remains to be discovered to what extent the two sets of migrants, from northern and southern regions, occupy the same niches in the tropics but at different times of year.

In extent of seasonal migration, pelagic seabirds provide a telling contrast with landbirds. Linked with the greater sea areas and large numbers of scattered island breeding sites, pelagic seabirds are much more numerous in the southern hemisphere than in the northern, both in terms of species and of individuals (Chapter 8). Correspondingly, a greater proportion of southern than of northern hemisphere breeding species make long migrations. Five (11%) of 47 species that breed north of the tropics extend to south of the tropics in the northern winter, whereas 14 (23%) of 61 species that breed south of the tropics extend to north of the tropics in the austral winter (calculated from maps in Harrison 1983), a difference that is statistically significant. The implication is again that the sheer numbers of birds, in relation to the habitat available, influence the area occupied and distances moved.

Trends within species

The foregoing analyses **(Figures 18.2–18.7)** were based on the presence or absence of species at particular latitudes in winter, ignoring partial migration, in which only a proportion of individuals of some species leave for the winter. In many species that breed over wide areas, a greater proportion of individuals migrate from higher than from lower latitudes. Thus, some bird species in the northern hemisphere are completely migratory in the north of their breeding range and completely sedentary in the south, while in intervening areas some individuals leave and others stay (partial migration). In general, therefore, the extent to which any population migrates for the winter broadly corresponds to the degree of seasonal reduction in food-supplies.

Another trend within many species that breed over a wide span of latitude is for individuals from high-latitude breeding areas to migrate further than those from low-latitude breeding areas. One pattern is illustrated by Common Grackles *Quiscalus quiscula* and Common Starlings *Sturnus vulgaris* that were ringed in the breeding season in different parts of eastern North America and recovered further south in winter **(Figure 18.9)**. In both species, birds from more northern breeding areas travelled further, on average, but wintered in the same general latitudinal band as birds from more southern breeding areas (Dolbeer 1982). A second pattern, called leapfrog migration, is illustrated by the Fox Sparrow *Passerella iliaca* in westernmost North America and the Common Ringed Plover *Charadrius hiaticula* in Europe **(Figure 18.10)**. In both these species, birds from the most northern breeding areas winter furthest south. One theory is that such leapfrog patterns arose following the last glaciation, as breeding populations spread north, and had to migrate to progressively more southern latitudes to find sufficient vacant habitat. They have persisted until now because of continuing competition.

Populations in both hemispheres

Very few migratory bird species have separate breeding populations north and south of the tropics. Examples include the Little Tern *Sterna albifrons* and Whiskered Tern *Chlidonias hybridus* in Asia–Australia, the Black Stork *Ciconia nigra* and Booted Eagle *Hieraaetus pennatus* in Europe–Africa, and the Turkey Vulture *Cathartes aura* and Black Vulture *Coragyps atratus* in North–South America. In all these species, the northern birds winter in the south when the southern birds are

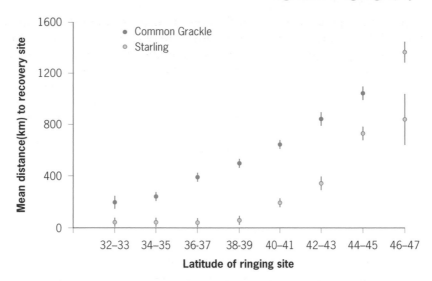

Figure 18.9 Mean distance (± SE) from ringing to recovery sites for adult Common Grackles *Quiscalus quiscula* (N=855) and Common Starlings *Sturnus vulgaris* (N=1,116). The birds were ringed on their breeding areas in eastern North America (75–100° longitude), and recovered in winter (January–February) further south. From Dolbeer 1982.

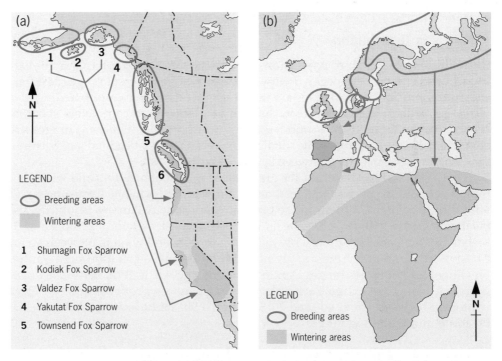

Figure 18.10 Leapfrog migration in Fox Sparrows *Passerella iliaca* (left) and Common Ringed Plovers *Charadrius hiaticula* (right). From Swarth 1920 and Salomonsen 1955.

breeding but, among the terns, the southern birds also winter in the north when the northern birds are breeding. It is as though there is a single population of terns occupying the same range but with part of the population breeding at one end of the migratory terminal and another part at the other end. In each area, terns from each population can be distinguished according to whether they are in breeding or non-breeding plumage.

The existence of such patterns raises the question, applicable to many migrants, of why the same individuals do not breed twice each year, once in each area. One reason in many species is that individuals moult while in winter quarters, a process that takes several weeks or months and could not be undertaken at the same time as breeding (the two processes are mutually exclusive in most non-migratory birds). Another reason is that many migratory species do not remain for long in the same area in winter quarters, but periodically move to other areas in response to changes in food-supplies (Jones 1995). This exploitation of temporary abundances is one way in which migrants in the southern hemisphere could avoid competition with the local residents which, breeding at that time, are tied to fixed nesting areas. Neither explanation applies to all migratory species, however, and there are still some that would seem able to breed in both hemispheres, six months apart, but do not.

RELATIONSHIP BETWEEN BREEDING AND WINTERING AREAS

Patterns in distribution

By definition, resident bird populations occupy the same geographical range year-round, while migratory species occupy partly or wholly different ranges at different times of year. The variations are in the degree of separation of breeding and wintering ranges, from coincident, through overlapping, to completely separate **(Figure 18.11; Table 18.2)**. In most species, breeding and wintering ranges are coincident or overlapping, while smaller numbers show a latitudinal gap between the two, differing in extent between species **(Figure 18.12)**.

In both Old and New Worlds, the greatest separation of breeding and wintering ranges is found in species that breed only at high latitudes in one hemisphere and winter only at high latitudes in the opposite hemisphere. Among landbirds, the Swainson's Hawk *Buteo swainsoni* is one of the most extreme examples, as it breeds between about 25° and 65°N in North America and winters between about 24° and 40°S in South America, giving a 49° latitudinal separation between the breeding and wintering ranges. Among seabirds, the Arctic Tern *Sterna paradisaea* is probably the most extreme example, breeding between about 50° and 80°N and wintering between about 40° and 70°S, giving a 90° latitudinal gap between the breeding and wintering ranges.

Comparison of sizes of breeding and wintering areas

Whatever the relative disposition of breeding and wintering ranges, those migrant species that have the largest breeding ranges also tend to have the largest wintering

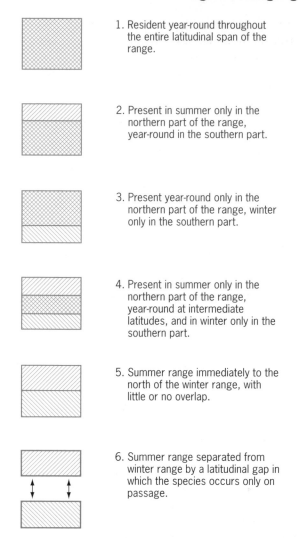

1. Resident year-round throughout the entire latitudinal span of the range.

2. Present in summer only in the northern part of the range, year-round in the southern part.

3. Present year-round only in the northern part of the range, winter only in the southern part.

4. Present in summer only in the northern part of the range, year-round at intermediate latitudes, and in winter only in the southern part.

5. Summer range immediately to the north of the winter range, with little or no overlap.

6. Summer range separated from winter range by a latitudinal gap in which the species occurs only on passage.

Figure 18.11   Main migration patterns found in northern hemisphere birds, based on the degree of separation between breeding and wintering ranges. See also **Table 18.2**.

ranges, and vice versa (Newton 1995a). This point is illustrated in **Figure 18.13** for 57 species of landbirds that breed entirely within Eurasia and winter entirely within Africa, so that their breeding and wintering ranges are completely separated. As another reflection of the same phenomenon, European landbirds and freshwater birds that breed over the widest span of latitude also winter over the widest span of latitude, and vice versa **(Figure 18.14)**. These correlations hold only as general tendencies, however, and some species provide exceptions. Moreover, because of the geographical scale involved, measures of range size can only be crude, and take no account of areas within the range that lack suitable habitat. They also take no account of the fact that the bulk of the population may occupy

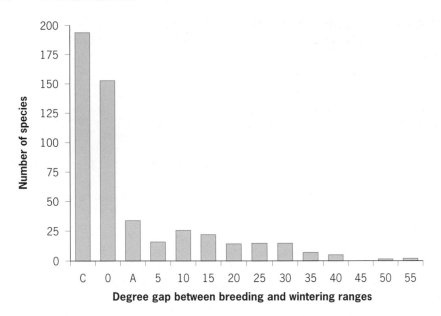

Figure 18.12   Frequency distribution of latitudinal gaps between breeding and wintering ranges, calculated for west Palaearctic breeding birds. C – coincident; O – overlapping; A – adjacent.

only part of the winter range at one time, either shifting south during the course of the northern winter or occurring at any one time only in those parts where rainfall or other factors have created suitable conditions (see later).

The correlation between sizes (or latitudinal spans) of breeding and wintering ranges may have its basis in the ecology of the species themselves, in that those that have the widest climatic and habitat tolerances may be able to spread over the largest areas, summer and winter. Alternatively, the correlation may depend on the abundance of the species concerned, in that those that have the largest populations (for whatever reason) spread over the largest areas, summer and winter (Newton 1995a). These two explanations are not mutually exclusive, and in practice are difficult to separate.

Despite the general correlation between the sizes of breeding and wintering areas, in about 69% of Eurasian–African migrants the breeding range is noticeably larger than the wintering range. This is significantly greater than the expected 50% ($\chi^2 = 4.0$, P<0.05). The most extreme example is the Lesser Grey Shrike *Lanius minor*, whose breeding range in Eurasia covers an area at least seven times greater than its known wintering range in southwest Africa, which is centred on the Kalahari basin (Herremans 1998). In contrast, in only 31% of species is the wintering range larger than the breeding range. The most extreme examples include the Olive-tree Warbler *Hippolais olivetorum* and Subalpine Warbler *Sylvia cantillans*, whose known wintering ranges cover more than twice the area of their breeding ranges. However, these low-density species are little known in Africa, and their effective wintering ranges may have been over-estimated by the inclusion of records of vagrants or of occurrences in occasional years only.

Table 18.2   Migration patterns of birds in the northern hemisphere, arranged roughly in order of decreasing segregation of breeding and wintering ranges. T = Total in each category.[1] See also **Figure 18.11**.

1. **Present year-round throughout the whole latitudinal range.**
   Old World examples: Black Grouse *Tetrao tetrix*, Eurasian Green Woodpecker *Picus viridis*, Black-billed Magpie *Pica pica*, House Sparrow *Passer domesticus* (T=195).
   New World examples: Ruffed Grouse *Bonasa umbellus*, Northern Bobwhite *Colinus virginianus*, Common Raven *Corvus corax*, Carolina Wren *Thryothorus ludovicianus* (T=64).

2. **Present only during the summer breeding season in the north of the range, year-round in the south.**
   Old World examples: Common Wood-Pigeon *Columba palumbus*, Eurasian Skylark *Alauda arvensis*, European Serin *Serinus serinus*, European Robin *Erithacus rubecula* (T=22).
   New World examples: Red-tailed Hawk *Buteo jamaicensis*, Common Moorhen *Gallinula chloropus*, Blue Jay *Cyanocitta cristata*, Common Grackle *Quiscalus quiscula* (T=47).

3. **Present year-round in the north of the range, only during winter in the south.**
   Old World examples: Eurasian Blackbird *Turdus merula*, Siberian Tit *Parus cinctus*, Willow Tit *Parus montanus*, Pine Grosbeak *Pinicola enucleator* (T=21).
   New World examples: Evening Grosbeak *Hesperiphona vespertina*, House Finch *Carpodacus mexicanus* (T=23).

4. **Present only during the summer breeding season in the north of the range, year-round at intermediate latitudes, and only during winter in the south.**
   Old World examples: Eurasian Woodcock *Scolopax rusticola*, Rook *Corvus frugilegus*, Redwing *Turdus iliacus*, Common Starling *Sturnus vulgaris* (T=111).
   New World examples: Canada Goose *Branta canadensis*, Short-eared Owl *Asio flammeus*, Cooper's Hawk *Accipiter cooperii*, Song Sparrow *Melospiza melodia* (T=52).

5. **Summer breeding range immediately to the north of the wintering range.**
   Old World examples: Ruddy Turnstone *Arenaria interpres*, Great Spotted Cuckoo *Clamator glandarius*, Bluethroat *Luscinia svecica*, Ring Ouzel *Turdus torquatus* (T=22).
   New World examples: Red-breasted Merganser *Mergus serrator*, House Wren *Troglodytes aedon*, Yellow-throated Warbler *Dendroica dominica*, Vesper Sparrow *Pooecetes gramineus* (T=34).

6. **Summer breeding range separated geographically from the wintering range by a gap in which the species occurs only on passage. Some of these species cross large stretches of inhospitable sea or desert in which they cannot feed during migration.**
   Old World examples: Arctic Tern *Sterna paradisaea*, Eurasian Dotterel *Eudromias morinellus*, Sanderling *Calidris alba*, Red-footed Falcon *Falco vespertinus*, Lesser Grey Shrike *Lanius minor*, Icterine Warbler *Hippolais icterina* (T=117).
   New World examples: Whistling Swan *Cygnus c. columbianus*, Brent Goose *Branta bernicla*, Swainson's Hawk *Buteo swainsoni*, Long-tailed Jaeger *Stercorarius longicaudus*, Iceland Gull *Larus glaucoides*, Baird's Sandpiper *Calidris bairdii*, Bristle-thighed Curlew *Numenius tahitiensis*, Snow Bunting *Plectrophenax nivalis* (T=144).

[1] The proportions in the six categories differ significantly between western Europe–Africa and eastern North America–South America (χ^2_5 = 85.7, P<0.001), reflecting mainly the smaller proportions of class 1 species, and the greater proportions of classes 5 and 6 species, in North America. The three species of southern hemisphere seabirds that spend the northern summer (austral winter) off Europe, and the eight that spend the northern summer off eastern North America, are excluded from the analysis.

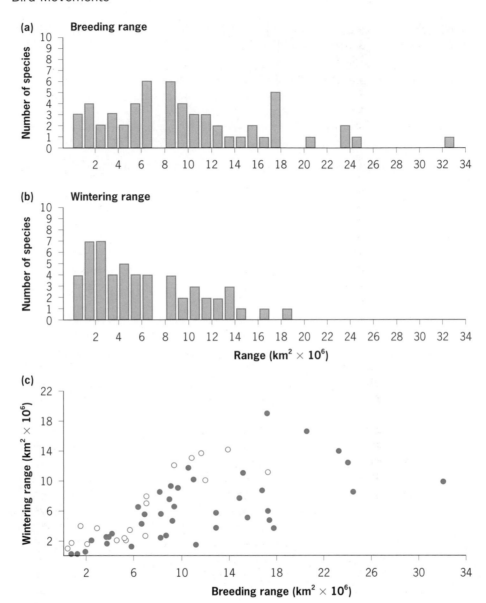

Figure 18.13   (a), (b): Sizes of breeding and wintering ranges of 57 landbird species that breed entirely in Eurasia and winter entirely in Africa. (c): Relationship between sizes of breeding and wintering ranges of the same 57 species. Warblers are shown separately as open circles. From Newton 1995a.

For the 57 Eurasian–Afrotropical landbird migrants as a whole, however, wintering ranges are, on average, about two-thirds as large as breeding ranges, and in some species only parts of the wintering range may be occupied at any one time **(Figure 18.13)**. This is because some species move further south during the course of a winter, and occupy certain areas sporadically, depending on rainfall and

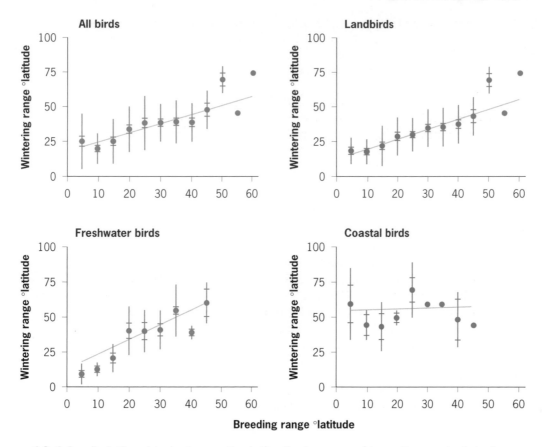

Figure 18.14   Relationship between the latitudinal spans of breeding and wintering ranges of west Palaearctic breeding birds. Some species winter entirely in the west Palaearctic and others partly or entirely in Africa. Excludes seabirds. Spots show mean values, and lines show one standard error and one standard deviation on either side of the mean. The lack of relationship in coastal birds (shorebirds) can be attributed to the fact that in winter they switch from an areal distribution in inland areas to a linear distribution along coastlines, often with a very wide latitudinal spread. From Newton & Dale 1997.

resultant food-supplies. These findings imply that most species live at greater densities in their African wintering areas than in their breeding areas, but whether this reflects differences in the per-unit-area capacities of the two regions to support the birds remains unknown.

It may be that individual birds need more space in Eurasia when they are feeding young than in Africa when they have only themselves to feed. Moreau (1972) estimated that, owing to warmer weather, the individual daily energy needs of passerines in Africa were about 60% of their breeding season needs. It is also probable that, per unit of geographical area, the amounts and productivity of suitable habitat might differ between breeding and wintering ranges and that this difference varies between species. Whatever the reason for the differences between sizes of breeding and wintering ranges, a similar phenomenon occurs in the New

World, where migrants from large parts of North America concentrate each winter in a relatively small area in the northern neotropics. On average, then, the geographical ranges of most species are smaller or more overlapping in winter than in summer. Still, however, the sizes of breeding and wintering ranges of different species are broadly correlated.

Among freshwater birds (waterfowl and waders) that winter in inland areas, the sizes of breeding and wintering ranges are also correlated, but in contrast to landbirds, wintering ranges are generally larger, with a greater latitudinal span **(Figure 18.14)**. This may be because, as wetlands become scarcer southwards (from tundra to savannah), freshwater birds have to spread over greater areas than landbirds in winter in order to find enough suitable habitat. Freshwater birds may also make longer movements within a winter than landbirds, in response to rainfall patterns (Newton & Dale 1997).

Most shorebird species switch from an areal distribution on the tundra in summer to a linear distribution on coastlines in winter (Newton & Dale 1997). In Eurasia they breed only in a narrow span of latitude, mostly between 70° and 80°N, but in winter extend southwards over 103° of latitude between about 60°N and 43°S, reaching the southern tip of Africa (35°S) or Australia (43°S). Some species, such as Ruddy Turnstone *Arenaria interpres*, can be found in winter on almost any stretch of coast within this wide latitudinal span, while others, such as Red Knot *Calidris canutus*, may be restricted to the relatively few sites where suitable conditions occur, but are found there in great numbers. Because of the seasonal switch in habitat, it is difficult to compare range sizes but, as a group, shorebirds show no relationship between the latitudinal extents of breeding and wintering ranges **(Figure 18.14)**. Overall, then, the correlation between summer and winter range sizes is most apparent in inland birds.

West–East patterns

Some European migrants show west–east patterns in distribution. Most species with wide west–east breeding ranges in western Eurasia extend from west to east across Africa. Those species that breed only in western Europe (e.g. Melodious Warbler *Hippolais polyglotta*) winter mainly in the western half of Africa; and those species that breed only in eastern Europe or Asia winter mainly in the eastern half of Africa (e.g. Masked Shrike *Lanius nubicus*). There are exceptions, however, such as Lesser Whitethroat *Sylvia curruca* and Red-backed Shrike *Lanius collurio*, in which the birds from western Europe migrate via eastern Europe to eastern Africa, where they join more eastern breeders. Likewise, European Pied Flycatchers *Ficedula hypoleuca* from Asia migrate via western Europe to reach their west African wintering areas. The situation is complicated, however, because some birds take different routes in spring from those taken in autumn, as shown for the Pied Flycatcher in **Figure 18.15** (Winkel & Frantzen 1991, Bairlein 2001). Other species show a migratory divide, with birds from western Europe crossing at the west end of the Mediterranean Sea into Africa, and birds from eastern Europe moving down the east side of the Mediterranean into Africa **(Figure 18.16)**. In this way they avoid crossing the Mediterranean Sea at its widest parts. Examples include soaring birds, such as storks and raptors, and various passerines and others.

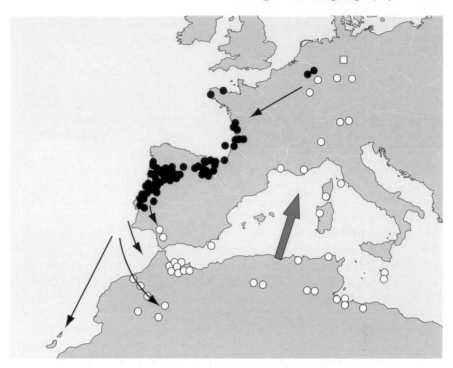

Figure 18.15   'Loop migration' of European Pied Flycatchers *Ficedula hypoleuca* ringed near Brunswick in northern Germany (open square). Black dots: recoveries during autumn migration (N=71); circles: recoveries during spring migration (N=40); arrows: main migration directions in autumn and spring. Autumn migration follows a fairly narrow westerly route, whereas spring migration follows a much broader and more easterly route. The recoveries suggest that birds stop mainly in southwest Europe on their autumn journey to west tropical Africa and mainly in North Africa on their return spring journey. From Bairlein 2001, based on Winkel & Frantzen 1991.

Similar west–east patterns are evident among New World migrants. Some species that breed only in the western states winter only on the western side of Central America (e.g. Hammond's Flycatcher *Empidonax hammondii*, Townsend's Solitaire *Myadestes townsendi*, Grey Vireo *Vireo vicinior*), while others that breed only in the eastern states winter only on the eastern side of Central America and the Caribbean Islands (e.g. Yellow-throated Vireo *Vireo flavifrons*, Northern Parula *Parula americana*, Prairie Warbler *Dendroica discolor*). As in Eurasia, however, some species with a wide longitudinal spread in breeding distribution concentrate in winter in either eastern or western parts of a potential wintering range (e.g. Rusty Blackbird *Euphagus carolinus*, which breeds from coast to coast across boreal North America and winters entirely in southeastern parts of the continent). These species form exceptions to the more or less parallel migrations shown by most species with transcontinental breeding ranges.

Within Europe an example of roughly parallel migration patterns for the Chaffinch *Fringilla coelebs* is given in **Figure 18.17**, but here the migrations run

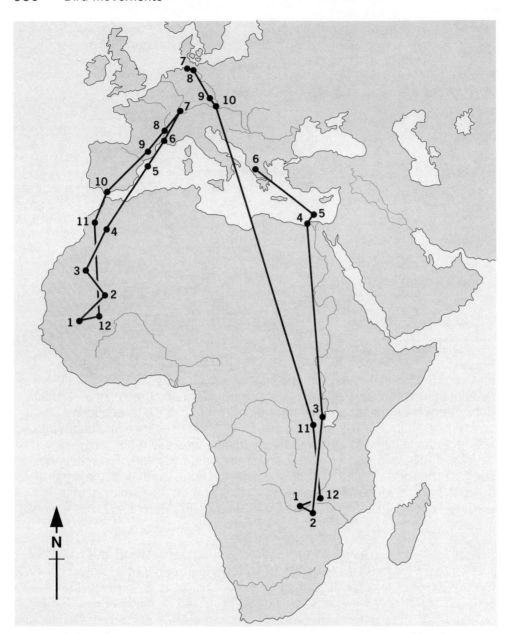

Figure 18.16   Course of first-year migration in White Storks *Ciconia ciconia* from southwestern Germany (westerly route) and northwestern Germany (easterly route), showing the migratory divide. Dots show the average monthly location of ringed birds. The lines join the dots of successive months, but do not necessarily reflect the exact routes followed. First-year storks are on the move for much of the year, and most do not reach the breeding areas until it is too late to nest that year. From Bairlein 2001.

roughly NE–SW in autumn (Bairlein 2001). Some species, notably geese, have narrow, population-specific migration routes, so that birds from different breeding areas do not mix. Examples include Barnacle Geese *Branta leucopsis* and Pink-footed Geese *Anser brachyrhynchus*, in both of which birds from three separate breeding areas have distinct migration and wintering areas **(Figure 18.18)**.

ALTITUDINAL MIGRATION

By moving a few hundred metres down the sides of a mountain, birds can achieve as much climatic benefit as by moving several hundred kilometres to lower latitude, but without the extra winter daylength. Such altitudinal (or vertical) migrations occur on mountain ranges worldwide, and can involve a large proportion of local

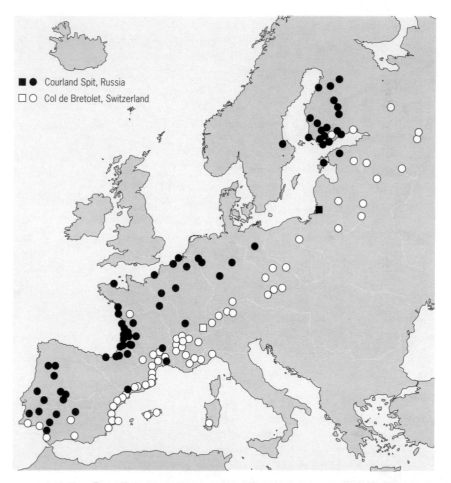

Figure 18.17   Parallel migrations within Europe, as shown from recoveries of Chaffinches *Fringilla coelebs* ringed during passage at the Courland Spit, Russia (filled square, filled dots) and at the Col de Bretolet, Switzerland (open square, open dots). From Bairlein 2001.

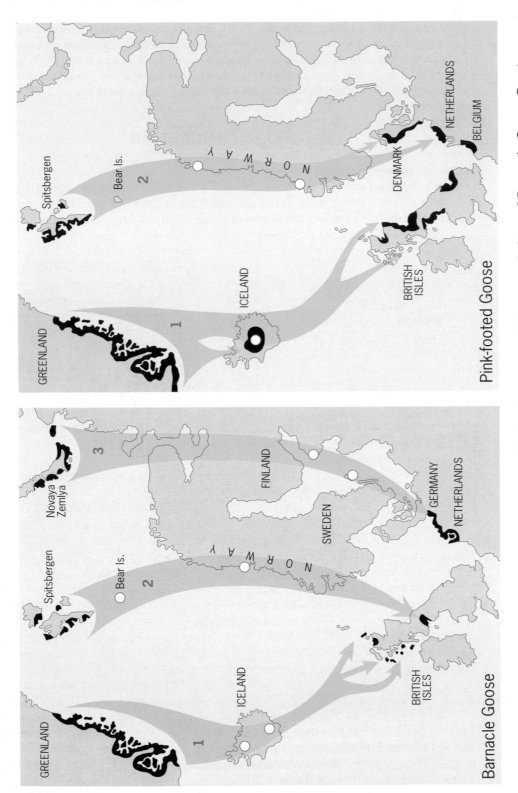

Figure 18.18 Breeding areas, migration routes and wintering areas of three populations of Barnacle Geese *Branta leucopsis* and two populations of Pink-footed Geese *Anser brachyrhynchus*. Black shading depicts the separate breeding and wintering areas, blue the migration routes, and white spots the stopover sites. From Madsen *et al.* 1999.

montane species. Examples of altitudinal migrants from different continents include the Citril Finch *Serinus citrinella* in Europe, the Rosy Finch *Leucosticte arctoa* in western North America, the Three-wattled Bellbird *Procnias tricarunculata* in Central America, the African Olive Pigeon *Columba arquatrix* in East Africa and the Olive Whistler *Pachycephala olivacea* in Australia. Seasonal altitudinal movements occur even on relatively low mountains, such as the Great Dividing Range in southeast Australia, where several montane species appear in lowland towns and farms in winter. They include many conspicuous species, such as Yellow-tailed Black Cockatoos *Calyptorhynchus funereus*, Golden Whistlers *Pachycephala pectoralis*, Regent Bowerbirds *Sericulus chrysocephalus* and others. Nectar-feeders also move in response to the progressively later flowering found at higher elevations.

Altitudinal and latitudinal movements are not mutually exclusive. For example, in western North America, the two rosy finches *Leucosticte tephrocotis dawsoni* and *L. australis* are altitudinal migrants (in any direction), two others (*L. t. wallowa* and *L. atrata*) are altitudinal migrants with small latitudinal shifts, and two others (*L. t. littoralis* and *L. t. tephrocotis*) are mainly latitudinal migrants (King & Wales 1964). The Blue Grouse *Dendragapus obscurus* in some parts of North America reverses the usual pattern, and migrates upslope in winter, as it moves from fairly open breeding areas to dense forest where it eats conifer needles. The distances are not long (perhaps up to 50 km), but much of the journey is made on foot.

MOVEMENTS WITHIN THE BREEDING SEASON

When birds are breeding, they tend to remain resident in the same areas throughout the season, so that the whole breeding range is occupied in a consistent manner from spring arrival to autumn departure. There are, however, exceptions to this general pattern, in that some individuals of some bird species breed in more than one area each year (so called itinerant breeding). In northern Europe, Common Redpolls *Carduelis flammea* breed mainly in the birch scrub that marks the boundary between forest and tundra, but in years when the spruce crop is good in southern Fennoscandia, they sometimes breed there in March–April, and then move north to the birch zone to breed again in June–July. This movement has not been confirmed by ringing, but it has been inferred from the disappearance of birds from the spruce forests in May as the seeds fall, quickly followed by the appearance of adults with young in the birch areas (Peiponen 1967, Antikainen *et al.* 1980). In years with little or no spruce seed, the birds breed only in the birch. A similar split breeding season may sometimes occur in Eurasian Siskins *Carduelis spinus*, in which adults and recently fledged juveniles have been seen migrating northeast in June over the Courland Spit in the southern Baltic (Payevsky 1971). Some of the adults that were caught were paired at the time, and many females had a well-developed brood patch. One juvenile caught in June 1959 had been ringed 25 days earlier, 760 km to the southwest, in Germany. While it could not be proved that these Siskins went on to breed elsewhere in the same year, they clearly had enough time to do so.

The raising of two or more broods, each at a different latitude, has also been suspected in the Common Quail *Coturnix coturnix* and Spanish Sparrow *Passer hispaniolensis* (Summers-Smith 1988, Aebischer & Potts 1994) in Eurasia, and in the Sedge Wren *Cistothorus platensis* in North America (Bedell 1996). Similarly, the Dickcissel *Spiza americana* is thought to breed in two different regions on spring migration through North America, first in Texas in May and again further north in Missouri in June–July (Fretwell 1980), while the Phainopepla *Phainopepla nitens* is thought to breed first in its wintering areas in the Sonoran Desert in March–April, and again in the coastal oak woodlands of southern California (Walsberg 1978).

The Common Quail *Coturnix coturnix* is unusual among these species in that the young can mature and breed at three months old, so those produced in the southern parts of Europe can then breed the same summer in the more northern parts, along with the adults on their second or third brood. Apparently the males leave southern Europe once the females have laid, and move on to establish new territories further north, where females join them later. Many females do not reach the most northern areas, so males predominate there (evidence summarised in Aebischer & Potts 1994).

Regular itinerant breeding is shown by the Red-billed Quelea *Quelea quelea*, which feeds on grass seeds on the African savannahs (Ward 1971). As the rain belts spread across equatorial Africa, they stimulate the growth and seeding of grasses, providing a sufficient temporary food-supply to enable the birds to breed. After raising their young, the birds move *en masse*, stopping again in an area where rain has recently fallen, and raise another brood (Ward 1971, Jaeger *et al.* 1986). The initial 'early rains' migration causes Quelea to leave areas where grass seed germination has occurred following the first rains, and to fly over the approaching rainfront to areas where rain fell earlier and new grass seed has already formed. The subsequent 'breeding migration' follows the rains in successive breeding attempts. Within this framework, the pattern is variable from year to year, depending on regional variations in rainfall and grass seed production. Some other birds in semi-arid regions of the world, which breed at a particular stage in the dry–wet seasonal cycle, may also be itinerant breeders (Ward 1971). Multiple breeding along a migration route, following a rain belt, has been suspected in the Eared Dove *Zenaida auriculata* in northeastern Brazil (Bucher 1982), and substantial shifts in colony sites during a single season have been suspected in the White-crowned Pigeon *Columba leucocephala* on Hispaniola (Arendt *et al.* 1979) and in the Spanish Sparrow *Passer hispaniolensis* in Kazakhstan and elsewhere (Cramp & Perrins 1994).

The raising of successive broods at different sites on a migration route could occur in a much wider range of species than recorded, but could be detected only if newly arrived adults had brood patches, recently fledged young or other signs of recent reproduction. It is clearly rare or non-existent in the vast majority of well-studied European and North American migrants. The reason for its rarity is not obvious to me, because it might be thought that many species migrating north through Eurasia or North America in spring could raise more broods by stopping at different latitudes *en route* than by breeding only for a short season in the north. The prior occupation of more southern habitat by conspecifics may be the main factor preventing this. Moreover, southern habitat remains suitable after the first broods for subsequent ones. This is not true for itinerant multi-brooded species, which are forced to move on if they are to breed again in the same season.

MOVEMENTS WITHIN THE NON-BREEDING SEASON

When they are not breeding, birds are more free to move around, so that many species do not occupy their wintering ranges in a consistent manner throughout the non-breeding season. Thus, while the individuals of some species migrate only between a single fixed breeding area and a single fixed wintering area, individuals of other species make substantial movements during the course of a non-breeding season, often extending ever further from their breeding areas as the season progresses. In some species such movements are regular, somewhat analogous to different stages in a single migration, but the birds stay in the same place for up to several weeks between each move. The separate moves occur at about the same times each year and each is preceded by fat deposition. Many species wintering in Africa perform such two-or-more-stage migrations (Jones 1995).

In other species, the movements are facultative, occurring at any time in the non-breeding season, whenever feeding becomes difficult. In some northern species, in particular, the proportion of birds that leave the breeding range, and the distance they travel, varies greatly from year to year. Most individuals stay in the north in years when food is plentiful there, wintering in, or just south of, their breeding areas, but moving further south in years when food is scarce. Such annual variation in migration is most pronounced in finches and other birds that depend on fluctuating tree-seed crops (such as Common Redpolls *Carduelis flammea* and Bohemian Waxwings *Bombycilla garrulus*), and in raptors and owls that depend on fluctuating (cyclic) prey species (such as rodent-eating Rough-legged Buzzards *Buteo lagopus* and Snowy Owls *Nyctea scandiaca*) **(Figure 18.19, Table 18.3)**. Their so-called 'invasions' or 'irruptions', in which every so often they appear in large numbers well outside their usual range, follow periodic widespread crop-failures (finches) or crashes in prey populations (raptors). Irruptions therefore occur in response to annual, as well as to seasonal, reductions in food-supplies. In some such years the birds can, in effect, be on passage for much of each winter, as they move from one area of temporary abundance to another.

Snowy Owl

Common Redpoll

Figure 18.19   North American winter ranges of Snowy Owl *Nyctea scandiaca* (left) and Common Redpoll *Carduelis flammea* (right) that winter mainly at high latitudes (shaded area), but in years of food-shortage extend far to the south (dotted line).

Table 18.3   Some irruptive migrants of the northern hemisphere.

	Preferred food[1]	Distribution
Seed-eaters		
Thick-billed Parrot *Rhynchopsitta pachyrhyncha*	Chihuahua Pine	Nearctic
Great Spotted Woodpecker *Dendrocopos major*	Spruce, Pine and other seeds	Palaearctic
Bohemian Waxwing *Bombycilla garrulus*	Berries, especially Rowan	Holarctic
Cedar Waxwing *Bombycilla cedrorum*	Berries	Nearctic
Pine Grosbeak *Pinicola enucleator*	Conifer seeds, berries	Holarctic
Fieldfare *Turdus pilaris*	Berries	Palaearctic
Coal Tit *Parus ater*	Spruce seeds, insects	Palaearctic
Black-capped Chickadee *Parus atricapillus*	Conifer seeds, insects	Nearctic
Boreal Chickadee *Parus hudsonicus*	Various tree-seeds, insects	Nearctic
Great Tit *Parus major*	Beech seeds	Palaearctic
Blue Tit *Parus caeruleus*	Beech seeds	Palaearctic
Wood Nuthatch *Sitta europaea* (Siberian populations)	Spruce seeds	Palaearctic (Siberia)
Red-breasted Nuthatch *S. canadensis*	Pine, Spruce seeds	Nearctic
Brambling *Fringilla montifringilla*	Beech seeds	Palaearctic
Eurasian Siskin *Carduelis spinus*	Birch, Alder and conifer seeds	Palaearctic
Pine Siskin *Carduelis pinus*	Conifer, Birch and Alder seeds	Nearctic
Common Bullfinch *Pyrrhula pyrrhula*	Various tree-seeds and berries	Palaearctic
Evening Grosbeak *Hesperiphona vespertina*	Maple and other tree-seeds	Nearctic
Common Redpoll *Carduelis flammea*	Birch seeds	Holarctic
Arctic Redpoll *Carduelis hornemanni*	Birch seeds	Holarctic
Purple Finch *Carpodacus purpureus*	Various tree-seeds	Nearctic
Common Crossbill *Loxia curvirostra*	Spruce and other conifer seeds	Holarctic
Two-barred Crossbill *L. leucoptera*	Larch and other conifer seeds	Holarctic
Parrot Crossbill *L. pytyopsittacus*	Pine seeds	Palaearctic (Europe)
Eurasian Jay *Garrulus glandarius*	Oak fruits	Palaearctic
Thick-billed Nutcracker *Nucifraga caryocatactes macrorhynchos*	Hazel fruits	Scandinavia
Thin-billed Nutcracker *N. c. caryocatactes*	Arolla Pine seeds	Palaearctic (Siberia)
Clark's Nutcracker *N. columbiana*	Whitebark Pine and other conifer seeds	Nearctic
Raptors and other predators		
Northern Goshawk *Accipiter gentilis*	Various grouse and hares	Holarctic
Rough-legged Buzzard *Buteo lagopus*	Lemmings, Voles	Holarctic
Great Horned Owl *Bubo virginianus*	Snowshoe Hares	Nearctic
Snowy Owl *Nyctea scandiaca*	Lemmings, Voles	Holarctic
Tengmalm's (Boreal) Owl *Aegolius funereus*	Voles	Holarctic
Hawk Owl *Surnia ulula*	Voles	Holarctic
Long-eared Owl *Asio otus*	Voles	Holarctic
Short-eared Owl *Asio flammeus*	Voles	Holarctic
Great Grey Owl *Strix nebulosa*	Voles	Holarctic
Great Grey (or Northern) Shrike *Lanius excubitor*	Voles	Holarctic
Steppe birds		
Pallas's Sandgrouse *Syrrhaptes paradoxus*	*Agriophyllum globicum*	Palaearctic
Rosy Starling *Sturnus roseus*	Locusts (in breeding season)	Palaearctic

[1] Scientific names of trees: Alder *Alnus*, Beech *Fagus sylvatica*, Birch *Betula*, Hazel *Corylus avellana*, Larch *Larix*, Maple *Acer*, Oak *Quercus*, Rowan *Sorbus aucuparia*, Scots Pine *Pinus sylvestris*, Siberian Stone Pine *Pinus sibirica*, Spruce *Picea*, Swiss Stone (Arolla) Pine *Pinus cembra*, Chihuahua Pine *Pinus leiophylla chihuahua*, Whitebark Pine *Pinus albicaulis*. Where several species in the same genus are involved, only the generic name is given.

Facultative movements are also shown by species that obtain their food from water or soft ground, and which must therefore move in response to freezing temperatures or drought. So-called 'hard weather' movements are extremely frequent and, as shown by radar, may occur almost every day and night in November–February between Britain and continental Europe (Lack 1963). The Northern Lapwing *Vanellus vanellus* and Common Starling *Sturnus vulgaris* are the most frequent participants, but many other species are involved too, including waterfowl (Ridgill & Fox 1990).

In contrast to the seed-eaters, which normally exhaust their food before moving on, such hard weather movers usually return as soon as conditions improve again, sometimes only a few days later. One probable reason for their return over several hundred kilometres is the avoidance of competition for food, which is likely to be more intense in the overcrowded hard weather refuges than in the areas they previously left. It is not that birds are necessarily driven back by competition, but this could be the ultimate factor involved. The proximate stimulus may well be temperature or wind direction, for it would be advantageous for the birds to leave before they were weakened by food-shortage. Needless to add, hard weather movements are much more pronounced in severe winters than in mild ones. It is almost as though such birds shuttle back and forth along part of their migration route, on average getting further from their breeding areas as winter advances.

Some birds seem to respond by movements not only to food-supplies, but also to predation and disturbance (which may also affect feeding rates). The banning of hunting in two coastal areas of Denmark resulted in waterfowl staying in large numbers and longer than in previous years, delaying their onward migration south (Madsen 1995). Over a five-year period, these reserves became important staging areas, increasing the national totals of several species. Hunted species in the reserves increased some 4–20-fold, while non-hunted species increased 2–5-fold. Furthermore, most quarry species stayed in the area up to several months longer each winter than in earlier years. No changes in bird numbers were noted in other areas still open to wildfowling, so the accumulation of birds in the reserves was attributed to the short-stopping of birds that would otherwise have moved on south. One consequence of this build-up of birds on the reserves was a greater depletion of the local food-supply, especially eelgrass *Zostera*. Birds presumably respond to natural predators in the same way.

A somewhat different pattern is shown by some hummingbirds in the Americas. Several species move through three different regions each year (Grinnel & Miller 1944, Stiles 1973). Anna's Hummingbird *Calypte anna*, for example, breeds in spring in the coastal chaparral of southern California, summers in the high mountains of California, and winters in the deserts of Arizona and Mexico, thereby ensuring that its need for nectar is met year-round.

Some waterfowl perform 'moult migrations' after breeding, when they move in large numbers to traditional sites, offering food and safety. They stay there long enough to moult, shedding all their flight feathers at once, remaining flightless for several weeks until the new feathers are grown. Well-known examples include the Common Shelducks *Tadorna tadorna* that gather at the German Wadden Sea, and the Eared Grebes *Podiceps nigricollis* that congregate at Mono Lake and Great Salt Lake in the western United States (Storer & Jehl 1985, Jehl 1997). The concentrations of grebes have been estimated to contain up to two million individuals. In many species of geese, the non-breeders migrate to moult at sites north of their breeding areas, starting their southwards migration soon after (Salomonsen 1968), while some shorebirds moult at stop-over sites on their southward migration (Jehl 1990).

The important point is that migration is not always a simple two-season movement between fixed breeding and wintering areas. Many species, it seems, spend much of their lives on the move, following the same route each year, pausing for a few weeks here and a few weeks there, before moving on. Their populations thus form a huge, continually shifting biomass, moving first south, then north over the earth's surface, in regular seasonal rhythm. Their mobile lifestyle enables them to exploit short-lived food-supplies at different places at different times, as they occur (Newton 1979, Lack 1990, Pearson & Lack 1992, Jones 1995). It is a lifestyle that birds more than most other kinds of animals can adopt to the full, though there are of course parallels in marine fish and reptiles, mammals and insects.

NOMADISM

Even greater flexibility in movement patterns is shown by some species that exploit sporadic habitats or food-sources. They often appear to be truly nomadic, showing little or no year-to-year consistency in their movement patterns, shifting from one area to another, and residing for a time wherever food is temporarily plentiful. The areas successively occupied may lie in various directions from one another. No one area is necessarily used every year, and some areas may be used only at intervals of several years, but for months or years at a time, whenever conditions are suitable. This kind of movement occurs on a small scale among some fruit-eating birds of tropical forests, on a larger scale among some rodent-eating owls and raptors of tundra, boreal and arid regions, among some boreal seed-eaters, and among many desert birds where infrequent and sporadic rainfall leads to unpredictable local changes in habitats and food-supplies (Chapter 17). As rainfall controls plant-growth, a wide range of desert species is affected, from songbirds to waterfowl. Australian examples include the blossom-eating Black Honeyeater *Certhionyx niger*, the seed-eating Flock Bronzewing *Phaps histrionica*, the rodent-eating Letter-winged Kite *Elanus scriptus* and the aquatic Grey Teal *Anas gracilis*, Gull-billed Tern *Sterna nilotica* and Pacific Heron *Ardea pacifica* (Frith 1961, 1982, Hollands 1977, Davies 1988). Eurasian examples include the seed-eating Pallas's Sandgrouse *Syrrhaptes paradoxus* and grasshopper-eating Rosy Starling *Sturnus roseus*; while African examples include the seed-eating Harlequin Quail *Coturnix delegorguei* and locust-eating Wattled Starling *Creatophora cinerea*. All these species may appear in enormous numbers when conditions are suitable but then disappear for several years. They all move in relation to sporadic changes in food-supplies, but without our having this knowledge their movements would appear to us random and unpredictable. The importance of food-supplies is further shown by the fact that some typical nomadic species, such as the Wattled Starling, are regular breeders in parts of their range where food is more consistently available, usually through human action. It remains a mystery how such birds find suitable localities within such vast areas: whether they search at random over appropriate range or whether they respond to climatic and other clues that indicate the most rewarding directions to fly.

Typically, species in the central Australian desert build up in numbers as a result of good breeding in occasional wet years, then move outwards to the more humid peripheral districts in the following dry years (Nix 1974). Some other Australian species, which in most years are chiefly confined to coastal localities, move inland to breed in vast numbers when rain creates suitable conditions. For example, Banded Stilts *Cladorhynchus leucocephalus* live as non-breeders on scattered briny coastal lagoons, but within days of rain falling inland, they concentrate in tens of thousands on newly formed shallow lakes, feeding on the freshly hatched swarms of brine shrimps (Burbidge & Fuller 1982). They breed while conditions last, making repeated nesting attempts, but the water evaporates rapidly, and the land soon resorts to its normal parched state. The birds then return to the coast, leaving the last thousands of eggs and young to die. Years may pass before they can breed again, and not necessarily in the same sites.

Many birds classed as nomadic still have a north–south component in their movements, in line with latitudinal–seasonal trends in temperature and rainfall. Once again we see the different kinds of bird movements grading into one another, whether migration and nomadism or migration and dispersal. In any one population, however, one kind usually prevails.

SUMMARY

Migration is associated with seasonal environments, in which food-supplies vary greatly through the year. It enables birds to exploit seasonal abundances and to avoid seasonal shortages. Broadly speaking, birds move so as to keep themselves in optimal habitat for as much of the year as possible, allowing for the fact that their requirements differ between the breeding and non-breeding seasons. Regional, annual and seasonal variations in movement patterns can all be broadly linked with variations in feeding conditions, as can the location of wintering areas.

Migration can be regarded as a product of natural selection, ensuring that species in seasonal environments adopt and maintain whatever movement pattern leads individuals to survive and breed most effectively. The advantage of autumn migration can be seen as improved winter survival, dependent on greater food-availability in winter quarters, and the advantage of spring migration can be seen as improved breeding success, dependent partly on greater seasonal food-supplies in summer quarters. From breeding experiments on captive birds, various aspects of migratory behaviour, such as timing, duration and directional preferences, were found to be under genetic control, and could be altered by selection. In the wild, many bird species have changed their migratory behaviour in the last 200 years, in response to changing conditions.

Migration causes huge seasonal latitudinal changes in the distribution of birds over the earth's surface. From low to high latitudes, a progressively greater proportion of breeding species leaves for the winter; and few species remain in winter north of the tree line. Most species occupy larger geographical ranges in summer than in winter when populations become more concentrated; however, the sizes of breeding and wintering ranges are correlated, and species that occupy the largest breeding areas also occupy the largest wintering areas, although exceptions are common. The lack of such a correlation in shorebirds can be attributed to the fact that in winter they switch from an areal distribution in inland areas to a linear distribution along coastlines, with a very wide latitudinal spread.

In montane regions, many bird species migrate altitudinally, moving from higher to lower elevations for the winter. Other species make movements within a breeding season, raising successive broods in different places, or within a winter, moving further from their breeding range as the winter progresses. Irruptive species, dependent on annually varying food-supplies, migrate further in some years than in others, concentrating wherever they encounter plentiful supplies. Nomadism occurs in some desert species which live under the influence of sporadic rainfall patterns. Typically they show little or no annual consistency in their movement patterns, but each year concentrate wherever food is available at the time. Individuals do not necessarily attempt to breed every year.

Appendix 18.1 Relationships between species numbers (or % migrants) and various environmental variables in western Europe. Upper figures show % variation accounted for by quadratic regression, and are given only where the quadratic term was statistically signifcant. All relationships were essentially monotonic over the observed range, and + or − values indicate signs of linear and (where appropriate) quadratic coefficients. *P<0.05, **P<0.01, ***P<0.001.

	Latitude	Land area	Elevation range	Mean annual temperature	Mean January temperature	Mean July temperature	Seasonal temperature range	Annual precipitation
Number of breeding species	−51* −90***	+78*** −95***	+25	+62** −88***	+75** −94***	+64** −99***	−59*	+79*** −94***
Number of wintering species	−93*** −97***	+68**	+29	+94***	+90***	+88***	−51*	+77***
% breeding species that leave for winter	+93*** +97***	−61**	−28	−95***	−87***	−85***	+50*	−75***
% wintering species that leave for summer	−85***	+32	+22	+80*** +83**	+66**	+65**	−36	+37

Relationships (Pearson correlations) between the various environmental variables listed above. *P<0.5, **P<0.01, ***P<0.001.

	Latitude	Land area	Elevation range	Mean annual temperature	Mean January temperature	Mean July temperature	Seasonal temperature range	Annual precipitation
Latitude		−0.76*	−0.60	−0.99***	−0.94***	−0.97***	0.63*	−0.80**
Land area			0.66*	0.80**	0.82**	0.77**	−0.65*	0.81**
Elevation range				0.61	0.53	0.64*	−0.16	0.66*
Mean annual temperature					0.97***	0.97***	−0.71*	0.83**
Mean January temperature						0.96***	−0.82**	0.84**
Mean July temperature							−0.61	0.83**
Seasonal temperature range								−0.54

Part Six
Conclusions

Gyrfalcon *Falco rusticolus* with Rock Ptarmigan *Lagopus mutus*.

Chapter 19
Speciation and biogeography —
a synthesis

In this book, I have discussed some of the major natural processes of bird population biology, occurring over long time periods and large geographical scales: namely the process of species formation, and the factors that have influenced the histories and geographical ranges of particular species. These various topics are inseparably linked, because new bird species arise mainly through the geographical isolation of populations, and the place where a species originates inevitably has a big influence on its subsequent distribution. If any overall conclusion can be drawn, it concerns the dynamic nature of the processes involved: change is continuous, as species evolve, expand and contract in distribution, and finally go extinct, only to be replaced by later forms. Even on the short time scales of decades, bird distributions are continually changing.

Evolutionary and distributional changes are driven mainly by changes in the physical environment, whether on the long time scales of plate tectonics and volcanism, on the medium time scales of glacial and other climatic cycles, or on the shorter time scales of recent human impact. In attempting to reconstruct any part of the history of life, therefore, we are dependent not just on biological knowledge, but on what is known of the history of the earth, as revealed by geology, climatology and palaeontology. To be valid, any explanation of the histories of species and their distribution patterns must be consistent with evolutionary theory and with the known history of the earth, including the fossil record. The relevant disciplines are developing at such a rapid rate that this ideal is unlikely to be true of all that is written in this book, and I expect that parts of it may soon become out of date.

The delimitation and formation of bird species

In considering any biogeographical problem, much depends on how we define and delimit species. To maintain their distinctiveness, species should be more or less genetically isolated from other species, and if they hybridise at all, this should not lead them to lose their genetic integrity. However, only for sympatric species, which live together in the same area, can we be sure that significant interbreeding is unlikely to occur. For allopatric populations, which live in different areas and so do not normally come into contact, we can only judge from their structure and appearance whether they would be likely to interbreed extensively if they met, and hence whether to class them as subspecies or species on the biological species concept.

In general, the more different are two taxa in their morphology and genetic (DNA) structure, the less likely they are to hybridise and merge. Nevertheless, in comparing different closely related taxa, the correlation between morphology and reproductive isolation is poor, as is the correlation between DNA structure and reproductive isolation. Divergence in the morphology of two species reflects differences in their habitat and food needs, not only in their mating preferences, so it depends on the strength of natural selection as well as on sexual selection. We thus find some closely related taxa that appear strikingly different to our eyes, yet freely interbreed, and other taxa that appear almost identical, yet live side by side as separate species. Divergence in DNA is thought to reflect the period of time since the populations were separated, and some populations have apparently been separated for hundreds of thousands of years without developing any significant divergence in morphology or mate choice. Thus some populations show two distinct mt DNA genotypes, signifying long isolation, but have recently come together, show no obvious morphological divergence and interbreed, as in the Common Raven *Corvus corax* (mean 4% divergence in mt DNA, cytochrome *b* gene, Omland *et al.* 2000), Adélie Penguin *Pygoscelis adeliae* (mean 5% divergence in mt DNA, control region, Monehan 1994) and others (Chapter 2). On the other hand, trivial genetic differences separate other taxa that look and behave like separate species, such as the Red-breasted Sapsucker *Sphyrapicus ruber* and Red-naped Sapsucker *S. nuchalis* (mean 0.4% divergence in mt DNA cytochrome *b* gene) (Cicero & Johnson 1995) or the Timberline Sparrow *Spizella taverneri* and Brewer's Sparrow *S. breweri* (mean 0.1% divergence in mt DNA cytochrome *b* gene) (Klicka *et al.* 1999). Clearly, it would be inappropriate to define related allopatric taxa as species on the degree of divergence in their mt DNA, unless their other features also differ to the extent normally expected of species.

The main value of DNA studies has been in phylogenetics, in estimating the genealogical relationships between taxa, and in reconstructing the history of lineages. The classification of birds proposed by Sibley & Ahlquist (1990) represents the biggest application to date of the methods of molecular biology to taxonomy. It has given an alternative hypothesis of avian phylogeny, at the levels of orders and families, and stimulated a flurry of new research on avian systematics. For the most part, the new phylogeny has agreed with previous ones, based mainly on anatomical structure and morphology, but has also suggested differences. Some of the new proposals have been supported by subsequent work, but others remain controversial. Molecular methods have proved especially useful in revealing examples of convergence, where unrelated birds have evolved to look like one another because they have the same way of life. The most significant example in biogeographical terms was provided by some Australasian passerine birds, previously classed with their Eurasian counterparts, but revealed from their DNA to represent a distinct radiation from unrelated ancestral forms.

With the development of biochemical and molecular technology, systematics has become increasingly grounded in genetics. Moreover, because modern genetic techniques yield information on the history of lineages that was previously unattainable, genealogy has come to play an increasing role in decisions on species delimitation and classification. This in turn has allowed phylogenetic history to be more fully represented at all levels of classification.

Speciation

The idea that new bird species arise primarily by the geographical isolation of populations, which subsequently diverge partly in response to local conditions, has stood the test of time. Such allopatric speciation can be inferred from current patterns of distribution and morphological variation, and checked by analyses of DNA. The next step in the evolutionary process from allopatry is range overlap, and the development of sympatry, in which populations formerly separated have diverged sufficiently to co-exist in the same area without excessive competition or hybridisation. Sympatry is more likely to develop in complex habitats such as some kinds of forest, which offer more opportunities for niche differentiation than structurally simpler habitats such as grassland. This provides one line of evidence that the structure of the environment itself can influence the numbers of species that can live there, and the degrees of evolutionary segregation between them.

Speciation through hybridisation, in which a new stable species may be formed from two existing ones, may well occur in birds, but has not been conclusively demonstrated. Populations are known that have apparently formed from the merging of two different taxa, but they are unlikely to have become reproductively isolated from either parental form. Similarly, sympatric speciation, in which one species splits into two in a single area, is theoretically possible in birds, but hard to prove, and again no likely cases have come to light. Species recently formed in this way would be expected to show little or no consistent difference in their mt DNA. This situation holds for at least three groups of morphologically distinct, co-occurring bird taxa, but because the different taxa are also known or suspected to hybridise, this could account for the lack of clear-cut difference in their DNA (for Galapagos finches, see Grant & Grant 1992; for redpolls, see

Seutin *et al*. 1995; for crossbills, see Piertney *et al*. 2001). So the genetic evidence for sympatric speciation is equivocal.

The geographical isolation of populations of a single species, which can promote allopatric speciation, can come about either through vicariance (fragmentation of a pre-existing range) or dispersal (giving a new population too far from the source population for regular interchange of individuals). Both processes vary in relative frequency between high- and low-latitude regions, between continents and islands, and between islands close or far from continents. Only on continents and the largest islands (such as Madagascar) has bird speciation occurred by vicariance within the land-mass itself, leading eventually to range expansion, followed by hybrid zones, overlap and sympatry. On smaller islands, dispersal between islands gives rise to separated populations; hybrid zones are unknown and sympatry among congeneric taxa on the same island is relatively uncommon. Within archipelagos, adaptive radiation can occur, giving a range of different species from a single colonisation event; but on more isolated islands radiation can come only from multiple colonisation events, which are long enough apart to allow one set of colonists to speciate before the next arrives.

The four processes hypothesised to drive the divergence of isolated bird populations to produce new species, namely founder effects and genetic drift, natural and sexual selection, have gained varying degrees of support from studies on wild populations. Both types of selection have been demonstrated from studies on wild birds, but evidence for the influence of founder effects and drift on the phenotypic features of wild birds is almost non-existent. The evidence that they can occur rests chiefly on theoretical considerations and on laboratory experiments on non-avian taxa. These aspects therefore require more study in wild bird populations.

In assessing the time taken for subspecies and species to evolve, the geological and DNA evidence has given reasonably consistent results. However, the time periods estimated vary greatly between different pairs of taxa. Divergence dates between subspecies have been estimated at periods ranging from less than 10,000 years to more than one million years. Divergence dates between sister species have been estimated at periods ranging from less than 200,000 years to more than five million years. Even longer periods are required for some sister species to diverge sufficiently to live sympatrically (Johns & Avise 1998). These estimates are surprisingly long, considering that local populations of short-lived birds can change significantly in mean body size, bill dimensions or other features within a matter of months or years (as a result of temporarily strong selection). The implication is that, while species can respond rapidly to strong selection, such changes are normally reversed by subsequent stabilising selection. The long periods of consistent directional selection required for speciation are relatively unusual.

Most recent estimates of the divergence dates of sister taxa have been based on non-coding mt DNA and depend on the assumption that equivalent genes in different related lineages mutate at the same average rate. (For the cytochrome *b* gene in the mt DNA, this rate has been estimated at 2% per million years.) However, this rate has been calibrated for very few species from geological or fossil evidence. Some geologically based estimates depend on the ages of volcanic islands, and as such islands result from not one but a series of eruptions, it is seldom certain that the age estimates were based on the oldest rocks. Similarly, in fossil-based estimates, it is seldom certain that the oldest known fossils represent the earliest divergence, as even older fossils may remain to be discovered. In effect, this means that, before we can have confidence in the estimated divergence dates, many more independent calibrations of mt DNA mutation rates are needed for a wider range of species. These rates are likely to vary more between different orders of birds than has so far been assumed.

On longer time scales, in the ageing of the different orders and families of birds, agreement between the fossil and the DNA evidence is generally poor. This has raised questions about the completeness and accuracy of some identifications in the fossil record and about the calibration and validity of molecular clocks. The fossil record suggests a major radiation of bird orders and families in the early Tertiary, but the record from before this time is relatively poor. The DNA evidence suggests that most modern bird orders and many families arose earlier, in the Cretaceous, and survived through the mass extinction event that marked the end of this period. This issue will remain unresolved until more information is available, and we can have greater confidence in one or other data base.

The avifauna of any land area, whether continent or island, typically consists of a range of species of greatly varying ages. Some belong to ancient lineages, and may have remained little changed for tens of millions of years; others may have formed within the last two million years or less. Ultimately, a major cause of extinctions is environmental change, whether geologically based (such as the joining of separate land areas or volcanic eruptions) or climatically based (such as glacial cycles, which can have profound effects on the nature and distributions of habitats). A more immediate cause of species extinctions results from interaction with other new colonist species, whether competitors, predators or parasites. Extinctions have occurred throughout the history of life, but at greatly varying frequency and, since the evolution of *Homo sapiens* less than 200,000 years ago, they have been increasingly human-driven. As an agent of extinction, no species comes close to our own in importance, the main causal factors being wide-scale habitat destruction, over-kill and translocations of species to places outside their natural range.

Continental avifaunas

On the geological time scales on which evolution occurs, the physical development of the earth is a major engine of evolutionary change. Nowhere is this more apparent than in comparisons of the avifaunas of different land-masses, each of which provides an independent platform for evolution. The earth's five main land-masses share most of the surviving orders of birds, and some of the same families, but many of the species differ from one biogeographical region to another. As expected, each such region shares most species with the closest adjacent region, and fewest species with the most distant region. This is testimony to the role of dispersal barriers in limiting bird distribution patterns. The most distinctive avifaunas are shown by the three southern regions of South America, Africa and Australasia, which have experienced the longest periods of separation. This is further testimony to the role of vicariance in bird evolution and distribution patterns. The different biogeographical regions differ from one another by more than 4.6-fold in species numbers and by more than 9.1-fold in species numbers per unit area. Such wide differences result from the past conditions on each continent that have promoted speciation or extinction, and from present conditions that support the current diversity. Those regions with tropical rain forest hold the most species. In assessing the cause of regional variations in species richness and diversity, it is important to separate the past conditions that produced the diversity from the present ones that maintain it.

The distributions of some types of birds can best be explained in terms of plate tectonics, in that land areas that are now far apart, but share related birds, were once connected as part of the same land-mass. This geological fact has been used to explain the presence of ratites, parrots and many other groups on the land areas derived from Gondwanaland, and the sharing of certain groups between South America and Africa, or between South America and Australasia. Many families that are now confined to particular continents or islands are known from fossils to have been much more widely distributed in the past. They have survived in isolated areas, having died out everywhere else. Such palaeo-endemics include the Todidae in the West Indies, represented by fossils in North America and Europe.

Island avifaunas

In discussing island avifaunas, it is important to separate continental (land-bridge) islands from oceanic (volcanic or coralline) islands, for the latter are normally much more isolated and can have received their birds only by occasional long-distance cross-water colonisation, rather than by direct inheritance and continuing immigration from the parent continent. Other ancient islands are intermediate in character in that, although they split from continents, their separation occurred many millions of years ago and they have since drifted far away (e.g. New Caledonia and New Zealand). Such islands retain endemic forms, whose ancestors were probably present at the time of the split, but also hold other species that arrived later by cross-water colonisation. Overall, their avifaunas are more like those of oceanic islands than of land-bridge islands. The range of bird types found on oceanic islands confirms that species vary greatly in dispersal ability, and that colonisation and differentiation are continuing processes. Many isolated islands hold some ancient palaeo-endemics that now lack close relatives elsewhere, other neo-endemics clearly evolved from existing continental forms, and yet other species that differ subspecifically or not at all from mainland forms.

Assessment of island avifaunas is greatly hampered by the fact that, on average, around half the species once found on oceanic islands have been eliminated by human action, mostly in the last thousand years. Many are still represented by 'subfossil' bone remains, which indicate that the most distinctive endemics have disappeared in the largest proportions, especially the flightless forms. Oceanic island birds are particularly vulnerable to human action because of their small populations (owing to small land areas) and because most have evolved in the absence of mammalian predators, so had no natural defences against people or against the dogs, cats, rats and other predators that people introduced. Theories purporting to account for the numbers and composition of species on islands were proposed before the extent of human-induced extinctions were known, and it is not clear to what extent these theories would still hold if the recently extinct species were also taken into account (Chapter 7).

Despite extinctions, islands still hold, on average, many more native species per unit area than continents, another testimony to the importance of geographical isolation in species formation. For their size, Pacific islands remain especially rich, collectively holding 20 times more species per unit area than South America, the richest of the continents. Because of the small size of the islands, however, this is still fewer than 200 native landbird species overall.

In the last century or so, many species have been introduced to islands by people, giving a 'colonisation' rate much greater than the natural, involving birds from many different parts of the world, which would not otherwise live together. It is uncertain to what extent the introduced birds have contributed (if at all) to the demise of native species, but their successful establishment (on both islands and continents) has confirmed that many bird species can live in areas they had not previously reached naturally. The implication is that inability to cross hostile habitat was a barrier to dispersal and range extension; however, most introduced species now live in man-made habitats in their new homes and may not have been able to survive in such places in the past when the vegetation was still pristine.

Marine avifaunas

Not surprisingly, seabirds show different biogeographical patterns of distribution from landbirds. Some species are found around the world at high southern latitudes, and at very high northern latitudes, where sea areas are continuous. In contrast, many different species occur in the Atlantic, Pacific and Indian Oceans separated by large land areas, so here again we see the importance of dispersal barriers in creating and maintaining separate avifaunas. Although sea water covers 71% of the globe, seabirds form only 3% of the world's bird species. This disparity can be attributed mainly to the 'design constraints' imposed on seabirds by the marine lifestyle, to their exploiting only carnivore niches, to the wide-ranging behaviour of the pelagic species, and to the more limited opportunities for speciation. Most population differentiation is allopatric, as in landbirds. In recent centuries, the nesting distributions of seabirds have been greatly reduced by human action. Subfossil bone remains and historical accounts indicate that many species were once found on more islands than they occupy now, their disappearance being linked with human colonisation, and the associated introduction of predators.

Effects of major climatic cycles

The climatic cycles of the Plio-Pleistocene clearly had major effects on bird populations. Interglacial conditions of the type we currently enjoy have prevailed periodically, but in total for only about one-tenth of the last 2.5 million years. For most of this time, the global climate was colder and drier than now. Large ice sheets covered much more of the northern hemisphere, forests were much more restricted and fragmented, while open grasslands and tundra were more extensive. Like other animals, bird species from high northern latitudes survived these changes by moving south and later north with their habitats, and by contracting to appropriate areas as their habitats fragmented. However, the continuous changes from one climatic extreme to the other eliminated some bird species altogether, but they also contributed to the production of new species through causing large-scale range fragmentation. Known extinctions and range contractions relate mainly to large species, whose bones are most readily preserved, and it is impossible to tell how many small species disappeared too. To judge from current gradients in species numbers, together with fossil remains, extinction befell greater proportions of plant and animal species in western Europe than in eastern Asia and North America where glacial effects were less severe (Chapter 9).

The climatic and vegetation changes of the last few million years, the time when the most recent species were formed, were the latest of a long series of events that contributed to the development of modern avifaunas. On all the continents, most recent speciation can be clearly linked to climatic oscillations that broke up expanses of the same vegetation type and isolated their avian inhabitants. These processes facilitated differentiation and speciation in some of the component species, thus creating areas of endemism and affecting distribution patterns. Climatically induced habitat changes continue today and, together with current bird distribution patterns, imply that speciation is ongoing.

In general, then, the present distributions of many bird species, the locations of hybrid zones, and the gradients in species numbers across biomes of fairly homogeneous vegetation, can all be explained largely in terms of the locations of known or purported past refuges, from which species have since spread to varying extents. The additional view that, in and before the Pleistocene, refuges were a major source of new species is consistent with many otherwise puzzling phenomena in bird distributions, including hybrid zones.

Past range changes had genetic consequences in surviving bird populations, and the advent of DNA technology has provided 'tests' of earlier ideas, and further insights into species formation, genetic subdivisions and colonisation patterns. From DNA evidence on the divergence dates of sister species (passerines only), few (if any) new species are likely to have been produced during the last glacial cycle alone; most were the result of a series of cycles, in which long periods of isolation and differentiation were interspersed with short periods of re-contact and interbreeding. On this view, climatically induced speciation was a prolonged and interrupted process spread over hundreds of thousands or millions of years. There was a long period of very warm conditions, lasting 800,000 years in the Pliocene, which could have been especially important in promoting speciation, because it led to raised sea-levels and the fragmentation of distributions, notably in South America (Chapter 11). The locations of climatic refuges, where new species were formed, is still evident in the distribution patterns and phylogeographic structure of current bird populations, as well as in the palaeo-botanical record. Moreover, the lowering of sea-levels that accompanied ice formation allowed many previously flooded land areas to emerge, creating many new islands, and allowing existing islands to join to mainlands. This could have been a major factor facilitating bird dispersal and range extension.

Studies of mt DNA have also proved useful in indicating significant events in the histories of particular populations, such as founder effects, bottlenecks, range expansions and colonisation histories. They have shown, for example, that many species now occupying recently deglaciated parts of North America and Europe show little or no geographical genetic structure in their populations. This is consistent with the recent colonisation of large areas from limited refuges. In general, then, the refugium hypothesis has stood the test of time, and is increasingly supported by geomorphological and palaeo-botanical evidence. Clearly, however, refugia are not the only places that promote speciation, and nor are past refugia the only places that today support large numbers of species relative to surrounding areas.

Community structure

Another major finding to emerge from the study of dated plant and animal fossils concerns the species-specific responses to climate change, which led to continual change in the composition of communities. This finding argues against the idea that plants and animals form integrated communities that remain stable through time in balanced equilibrium. Communities must instead be viewed as no more than ephemeral assemblages of species that in the past repeatedly came together and separated as climates changed. The biota at any given place were likely to follow the same basic structure, and included herbivores and carnivores and, among these, species that exploited different parts of the resource, but the actual numbers and identities of the species seem to have been in continual flux. Such change is evident, for example, in the continual addition of species to particular areas, as each arrives at a different date in spreading from a past refuge. It is also evident in the finding that some species, which once lived together at one place (as revealed by fossils), now have separate non-overlapping distributions (Huntley *et al.* 1997, Tyrberg 1999). Some introduced or expanding species often seem to have fitted easily within existing communities, so it is reasonable to expect that new communities will arise in response to future climate change. If a 'balance of nature' exists at all, it is a continually changing balance.

To what extent did species occupy the same ecological niches from one set of climatic conditions to another? We must distinguish here between the *fundamental niche* of a species, which embraces the full range of environmental conditions, habitats and food-sources that a species could potentially exploit in the absence of competitors and other constraints, and the *realised niche* which reflects the narrower range of conditions, habitats and food-sources that it actually does exploit at any specific place and time. To judge from skeletal remains, some bird species showed slight changes in body size from cold to warm periods, in the same way that they show variation in body size now from cold to warm parts of their geographical range, but their basic anatomical design remained the same throughout. This implies that they occupied much the same ecological niches throughout. They were also found consistently in the same vegetation types, whether tundra, grassland or forest, moving as necessary with their habitats. Almost certainly, therefore, particular species used their environments in much the same way in the past as they do now, occupying habitats of similar structure and foraging in the same way on similar types of food. While their fundamental niches may have remained unchanged, however, their realised niches may have fluctuated continually from time to time and from place to place, depending on the plant and animal species that they interacted with or fed upon. This is no different from the regional variation in the habitats and diets of particular species that we see today.

In tropical regions, climate change led mainly to habitat fragmentation, with much less habitat displacement than at high latitudes. Speciation was probably much greater in the tropics, and extinction lower, contributing to the greater species numbers found today in tropical compared with high-latitude regions. Greater species richness in tropical regions is manifest partly in the very large number of allospecies, compared with high-latitude regions, a legacy of past large-scale habitat fragmentation without the associated high extinction rates (Chapter 11). Differences in species numbers between the tropical regions themselves can also be linked with differences in the numbers of refuges occurring at times of climatic extremes. This is especially evident in comparisons of the numbers of forest and desert species between regions.

In the formation and maintenance of species, the importance of tropical islands and mountains is strikingly apparent, a result of the isolation and favourable habitats they provide. Tropical mountains also provided continuous habitat throughout the Plio-Pleistocene climatic cycles, as the different vegetation zones moved relatively short distances up and down the slopes, in response to changing conditions. Tropical mountains today seem to hold many old species, in addition to many young ones, suggesting high speciation and low extinction rates compared to lowlands. Mountains may thus have contributed disproportionately to the wealth of species in tropical regions, and most are important centres of endemism today. Mountains at higher latitudes could not serve such an important role, because they became largely ice-covered or barren during the long glacial periods, acquiring their flora and fauna afresh in each interglacial from surrounding lowland. Hence, at high latitudes, much smaller proportions of plant and animal species than at low latitudes are likely to have evolved on mountains.

That disproportionately high percentages of the earth's plant and animal species may be concentrated in relatively few areas, mostly recognisable as past climatic refuges, has obvious implications for conservation. With the ever accelerating destruction of natural vegetation, and current climate change, it becomes increasingly urgent to locate and protect such areas against further human impact. Areas of richness and endemism form only a small proportion of the earth's total land area, perhaps less than 10%. Their protection would probably provide the most effective way of ensuring the survival of a majority of species over the coming centuries.

Current bird distributions

The current distributions of particular bird species are limited by constraints operating either within the existing range to prevent a population increase that could fuel a spread, or just outside the existing range to prevent further expansion. Such external factors include abiotic constraints, such as climatic and topographic barriers to dispersal, or biotic constraints, such as lack of suitable habitat and food, or the presence of overwhelming competitors, predators or pathogens. Few species seem to be limited in distribution by a single factor operating around the entire range boundary, but many seem to be limited by different factors operating along different parts of a range boundary or by interactions between different limiting factors. The ranges of many bird species have contracted or expanded in area in recent centuries, associated with change in climate or with human effects on habitats and food-supplies. Such range changes have shed further light on the factors affecting bird distributions. Some species have spread over huge areas, embracing much of a continent, within periods of a few decades, again emphasising the dynamic nature of bird distributions.

The ranges of many bird species are markedly discontinuous, with two or more populations breeding so far apart that they are now genetically isolated. The distributions of some species are fragmented because their habitats are fragmented. Those of other species are fragmented because, although their habitats are now continuous, populations were fragmented in the past and have not spread with their now available habitat. In yet others, the habitat in the gap area, while appearing suitable, may be unfavourable in some respect, which is not immediately obvious. However, direct killing by people has produced large gaps in the distributions of many large bird species.

Landbird species vary greatly in their inclination or ability to cross un-favourable habitat, especially large stretches of sea. Nevertheless, the colonisation of oceanic islands is a continuing process, and even in the past 200 years many bird species have naturally established new populations after long sea-crossings. Examples include the Speckled Teal *Anas flavirostris* on South Georgia, the Fieldfare *Turdus pilaris* on Greenland, and several species on New Zealand. The Cattle Egret *Bubulcus ibis* was confined in the nineteenth century to Africa and southern Asia, but has since colonised much of the New World, much of Australasia and various oceanic islands. Presumably such movements occur all the time, but they may have more often led to range extension in recent centuries than before, owing to greater human-induced changes in habitats which for some species make previously unsuitable areas favourable. Because their ranges are influenced by climatic factors, many bird species are likely to spread to higher latitudes in the coming years, in line with global warming.

Movements

The normal dispersal movements of birds are of biogeographical importance because they can lead to range extension and influence the genetic structure of populations. In both respects, the most important movements are natal dispersal (distance between hatch site and subsequent breeding site) and breeding dispersal (distance between the breeding sites of successive years). The recoveries of ringed birds have revealed two major trends: (1) larger species generally disperse longer distances than small ones, and (2) allowing for body size, species that depend on sporadic habitats or food-supplies generally disperse longer distances than those that depend on predictable habitats or food-supplies. Hence, site fidelity is most marked in species that typically find suitable conditions in the same places year after year, and least marked in those that do not. These patterns in dispersal could explain why large birds generally show less subspeciation per unit area than small ones, and why species with predictable habitats and food-supplies show more subspecies per unit area than related species with sporadic habitats and food-supplies. They illustrate the effects of species requirements on their subsequent evolution.

Because of the mobility of birds, regular migration is more developed in birds than in most other animals. It involves large-scale seasonal movements of populations between more-or-less fixed breeding and wintering areas. Driven primarily by seasonal changes in food-supplies, migration affects a greater proportion of species with increasing latitude and, at any one latitude, it varies between species according to diet, and the availability of food-supplies in winter. Overall, it leads to huge seasonal changes in the distributions of birds over the earth's surface. Occasionally it leads to substantial extensions of breeding range, as birds remain to nest in their wintering areas or are blown far off course to other suitable areas, establishing new populations far from the original breeding range.

Human impacts

The enormous human impacts on flora and fauna need hardly be emphasised. Perhaps the most surprising finding of the past 20 years is that the destructive effects of humanity extend back for thousands of years, perhaps throughout the history of *Homo sapiens*. Of several explanations proposed for the major pulse of extinctions among large mammals (and some large birds) at the end of the Pleistocene some 12,000–9,000 years ago, over-kill by people seems the most plausible on the information available, but we can never know for sure. At a later date, from 4,000 years ago to the present, the extinctions of many island bird species so closely followed the pattern of human colonisation that no other explanation is at all plausible, whether the immediate cause of the extinctions was over-kill by people or by introduced predators and pathogens, or habitat destruction.

The impact of people on the natural world has, of course, increased greatly in the last few hundred years, as human populations have increased and spread, destroying natural plant and animal communities at an alarmingly high rate, and with increasing technological skill. One often-quoted consequence of the exponential increase in human numbers in recent centuries is that there are now more people alive on earth than have ever previously lived. Plant and other animal species are now being extinguished at rates thousands of times faster than the average background rates, and very much faster than new species can be generated. The net result is a continually and accelerating loss of biodiversity. Yet current human impacts pale to insignificance compared with changes that we can reasonably expect over the next few centuries, if human numbers and demands continue to grow. It is not the change that is new, for abundance and distribution patterns have always been changing, but the rate and spatial scale of change, and the fact that the main driving force is human impact. It is this that makes the current situation unique in the history of life on earth.

Concluding remarks

It has thus become possible to gain more understanding of the diversity and distribution patterns of birds and other organisms in terms of the past processes that have created them. The main messages are how profoundly these patterns have been influenced by major events in the history of the earth: whether geological events such as plate movements and volcanism, climatic events such as the recent glacial cycles, or biotic events such as the continuing dispersal of species or the expansion and pervasive influence of human populations. This has long been appreciated, but the recent advances of research in all the relevant sciences have provided more information and more insight on all these aspects, enabling the integration of findings from a variety of sources to produce a new and fuller picture. Further understanding is likely to emerge in the coming years from new advances in the relevant disciplines, but perhaps particularly from molecular genetics. The development of evolutionary biology and biogeography has thus reached an exciting stage; and as in previous advances, studies on birds have played a central role.

Glossary

abiotic. Pertaining to the non-living components of the environment, such as climate and soil type.

adaptation. Any feature of an organism that substantially improves its ability to survive or reproduce, compared with ancestral forms or with co-existing individuals.

adaptive radiation. The evolutionary divergence of a monophyletic taxon (from a single ancestral condition) into a number of very different forms and lifestyles.

AFLP. Amplified fragment length polymorphism; see restriction endonuclease.

Afrotropical Region. A biogeographical region, consisting of the portion of Africa south of the Sahara Desert plus Madagascar and other nearby islands.

allele. One of two or more alternative forms of a gene located at a single point (locus) on a chromosome.

Allen's Rule. Among homeotherms, the trend for limbs and extremities to become shorter and more compact in colder climates than in warmer ones.

allochthonous. Having originated outside the area in which it now occurs.

allopatric. Occurring in geographically different places; i.e. ranges that are mutually exclusive, at least in the breeding season.

allopatric speciation. The formation of new species when different populations of an ancestral species are geographically separated and thereby genetically isolated from one another.

allospecies. One of the allopatric species comprising a superspecies.

allozyme. One of several forms of an enzyme, each coded for by a different allele at a particular locus.

anagenesis. Progressive evolution of an unbranching lineage.

ancestor. The individual or population that gave rise to subsequent individuals or populations, usually with different features.

arctic. Pertaining to all non-forested areas north of the coniferous forests of the northern hemisphere, especially everything north of the Arctic Circle.

arid. Exceedingly dry; strictly defined as any region receiving less than 10 cm of annual precipitation.

austral. Pertaining to the temperate and subtemperate zones of the southern hemisphere.

Australasia. The biogeographical region including the continental fragments of the original Australian Plate, namely Australia, New Zealand, New Guinea, Tasmania, Timor, New Caledonia, and several smaller islands.

autochthonous. Having originated in the area in which it presently occurs.

avifauna. All the species of birds inhabiting a specified region.

barrier. Any abiotic or biotic feature that totally or partially restricts the movement (flow) of genes or individuals from one population or locality to another.

Bergmann's Rule. Among homeotherms, the trend for populations from cooler climates to have larger body sizes, and hence smaller surface-to-volume ratios, than related populations living in warmer climates.

Beringia. The geographical area of western Alaska, the Aleutians, and eastern Siberia that was connected by a land-bridge when the Bering Sea and adjacent shallow waters receded.

biodiversity. The total diversity of life on earth, which includes all genes, populations, species and ecosystems.

biogeography. The study of the geographical distributions of organisms, both past and present.

biological species. A group of potentially interbreeding populations that are reproductively isolated from all other populations.

biome. A major type of natural vegetation that occurs wherever a particular set of climatic and soil conditions prevail, but that may contain different taxa in different regions; e.g. tropical rain forest, savannah, desert, temperate grassland, boreal coniferous forest, tundra.

biosphere. Collectively, all the living things of the earth and the areas they inhabit.

biota. All species of plants, animals and microbes inhabiting a specified region.

biotic Pertaining to the components of an environment that are living or came from once-living forms.

bipolar. Occurring around both poles, in the cold or subtemperate zones.

boreal. Occurring in that portion of the temperate and subtemperate zones of the northern hemisphere that characteristically contains coniferous (evergreen) forests and some types of deciduous forests.

bottleneck. In evolutionary biology, any stressful situation that greatly reduces the size of a population, and usually also the genetic variance.

breeding area. In migratory birds, the area in which populations reproduce.

breeding dispersal. The movement of an individual from one breeding place to another.

carrying capacity. The total number of individuals of a species that the resources of an area support.

Cenozoic. The latest era in the history of life, spanning the last 65 million years, and including the Tertiary and Quaternary Periods.

chaparral. A type of sclerophyllous scrub vegetation occurring in the southwestern region of North America with a Mediterranean climate; see macchia.

character displacement. The divergence of the morphology of two similar species where their ranges overlap so that each uses different resources.

chromosome. A segment of a genome: a linear end-to-end arrangement of genes and other DNA, sometimes with associated protein.

clade. Any evolutionary branch in a phylogeny, especially one that is based on genealogical relationships.

cladistic biogeography. A sub-branch of vicariance biogeography that reconstructs the events that led to observed distributions of biotas, based primarily on phylogenetic methods.

cladistics. The method of reconstructing the evolutionary history (phylogeny) of a taxon by identifying the branching sequence of differentiation through analysis of shared derived character states.

cladogenesis. The production of new taxa that occurs through the branching of an ancestral lineage.

cladogram. A line diagram derived from a cladistic analysis showing the hypothesised branching sequence (genealogy) of a monophyletic taxon and using shared derived character states to estimate when each branch diverged.

climatic rules. Rules describing regular patterns of geographical variation in animals, correlated with climatic gradients; see Allen's Rule, Bergmann's Rule, Gloger's Rule.

cline. A change in one or several heritable characteristics of populations along a geographical transect, often correlated with a gradual change in the environment.

clone. A group of genetically identical individuals derived from the same parent.

cloned population. A population with no genetic input other than by mutation.

co-existence. Living together in the same local community.

colonisation. The immigration of individuals into a previously unoccupied area followed by the successful establishment of a population.

community. An assemblage of organisms that live in a particular habitat and interact with one another.

competition. Any interaction that is detrimental to one or both participants. Inter-specific competition occurs among species that share limited resources. It may affect the survival and breeding success of individuals, or the size and distribution of populations.

competitive exclusion. The principle that when two species limited by the same resources live together in the same area, one eventually outcompetes and eliminates the other.

congeners. Species belonging to the same genus.

continental drift. A model, first proposed by Alfred Wegener, stating that the continents were once united and have since become independent structures that have drifted apart. Now a part of plate tectonics.

continental island. An island that was formed as part of a continent. Some such islands (land-bridge islands) still lie close to a continent, and hold mainly or entirely continental biota. Others drifted away many millions of years ago, and hold more endemic taxa; compare with oceanic island.

convergent evolution. Unrelated species evolving to look like one another because they have the same way of life, usually in different regions.

CORE area. A centre of species richness and endemism.

Coriolis force. An apparent force, resulting from the earth's rotation, that causes oceanic and atmospheric currents to be deflected to the right in the northern hemisphere and to the left in the southern. It varies from zero at the equator to a maximum near the poles.

corridor. A dispersal route that permits the spread of many or most taxa directly from one region to another.

cosmopolitan. Occurring worldwide, as on all habitable land-masses or in all major oceanic regions.

Cretaceous. The period in the history of life, roughly spanning 144–65 million years ago, the end of which saw the extinction of the dinosaurs and many other organisms.

cryptic species. Species within a genus that are morphologically so similar that they cannot be readily distinguished by superficial features. Recognised by less obvious features, such as DNA.

deme. A local population of closely related interbreeding individuals.

density compensation. In island biogeography, an increase in the density of a species inhabiting an island habitat when one or more taxonomically similar competitors are absent.

density overcompensation. In island biogeography, when the total densities of a few species inhabiting a small island exceed the combined densities of a much greater number of species occupying similar habitats on a large island or continent.

desert. An extremely dry habitat, where water is unavailable for plant growth most of the year; in particular, a habitat with sparse coverage by plants, often with perennials covering less than 10% of the total area.

diffuse competition. A type of competition in which one species is negatively affected by numerous other species that collectively cause a significant depletion of shared resources.

dimorphic. Having two distinct forms in a population, as in colour phases.

diploid. Having a double set of chromosomes, one from each parent, as is typical in animals in which each individual is derived from a fertilised egg cell.

disjunct. A taxon whose range is geographically separated into two or more areas.

dispersal. In biogeography, range extension over a pre-existing barrier. In ecology, the movement of an individual from one place to another, usually of no fixed direction or distance. See breeding dispersal and natal dispersal.

dispersal biogeography. A branch of historical biogeography that attempts to account for present-day distributions on the assumption that they resulted from differences in the dispersal abilities of individual lineages.

dispersion. The spatial pattern of distribution of individuals within a local population.

DNA. Deoxyribonucleic acid, the double-stranded molecule that encodes and transmits genetic inheritance. It is present in the nucleus of every living cell.

DNA–DNA hybridisation. Estimation of the genetic distance between two species, by a method involving the association and disassociation of two strands of DNA, one from each species.

DNA sequencing. Determination of the order in which different nucleotide bases are arranged along a strand of DNA.

ecology. The study of the abundance and distribution of organisms and of the relationships between organisms and their environments.

ecosystem. The combined biotic and abiotic components in a given environment.

ecotone. The zone of transition between two habitats or communities.

El Niño–Southern Oscillation. Major temporary climatic change centred on the Pacific, where it is associated with failure of the cold Humboldt current, but whose effects extend over much of the world, bringing droughts in some regions and torrential rain in others.

emigration. Dispersal of organisms away from an area.

endemic. Pertaining to a taxon that is restricted to the geographical area specified, such as a continent, biome, or island.

endothermy. Ability to control body temperature by physiological means.

Eocene. The epoch spanning roughly 58–37 million years ago.

equilibrium theory of island biogeography. The theory that the number of species on an island results from a dynamic equilibrium between the opposing rates of immigration and extinction.

Ethiopian Region. See Afrotropical Region.

ethological barriers. Reproductive isolating mechanisms caused by behavioural incompatibilities of potential mates.

evolution. In the most general sense, the development of higher from lower forms of life. In the strictest sense, any irreversible change in the genetic composition of a population.

evolutionary species. A discrete cluster of individuals and populations, evolving separately from other clusters, that exhibits a clear pattern of ancestry and descent; compare with biological species.

evolutionary systematics. A method of reconstructing the evolutionary history (phylogeny) of a taxon by analysing the evolution of major features along with the distribution of both shared primitive and shared derived characteristics; compare with cladistics.

extant. Living at the present time.

extinct. No longer living.

family. A taxonomic category above the level of genus and below the level of order.

filter. A geographical or ecological barrier which blocks the passage of some species, but lets through others.

fitness. The contribution of a genotype to future generations relative to that of other genotypes.

flyway. An established air route used by large numbers of migratory birds.

forest. Any of a variety of vegetation types dominated by trees and usually having a fairly well-developed or closed canopy.

fossil. A remnant, impression, or other trace of a living organism from the past, preserved in rock after the organic material has been transformed or removed.

founder effect. The principle that the founders of a new population (as on an oceanic island) may contain only a small (and perhaps unrepresentative) subset of the total genetic variation of the parental population.

fresh water. In the strictest sense, water that has a salt concentration of less than 0.5%.

Gause's Principle. The principle stating that two species cannot co-exist at the same locality if they have identical ecological requirements.

gene. The functional unit of heredity. The small unit of a DNA molecule that codes for a specific protein (strictly one polypeptide chain).

genealogy. The study of the exact sequence of descent from an ancestor to all derived forms.

gene flow. The movement of alleles within a population or between populations caused by the dispersal of individuals (or gametes in some animals, pollen or seeds in plants).

gene frequency (allelic frequency). The proportions of gene forms (alleles) in a population.

gene pool. The totality of the genes of a given population existing at a specific time.

genetic drift. Changes in gene frequency within a population caused solely by chance, and not by selection, mutation or immigration.

genome. A full set of chromosomes; the total genetic material (DNA) of an individual.

genotype. The total genetic message found in an individual or a cell or at a particular locus.

genus. A taxonomic category above the level of species and below the level of family, comprising a group of monophyletic species.

geographical isolation. Spatial separation of two potentially interbreeding populations so that gene flow is prevented; see allopatric.

geographical speciation. See allopatric speciation.

geographical variation. The differences between spatially segregated populations.

glacial period. Periodic lowering of global temperature which causes extension of polar and montane ice caps, general drying of climates, and lowering of sea-levels worldwide.

Gloger's Rule. The trend for populations of a species living in warm and humid areas to be more heavily pigmented than those in cool and dry areas.

Gondwanaland. The southern part of the ancient supercontinent Pangaea, consisting of all the southern continental land-masses and India, which were united for at least one billion years but broke up during the late Mesozoic, around the end of the Cretaceous Period.

grassland. Any of a variety of vegetation types composed mostly of grasses and other herbaceous plants (forbs) but few if any trees and shrubs, as in the prairies, steppes or pampas.

greenhouse effect. The retention of heat in the atmosphere when clouds (water vapour), carbon dioxide and other gases absorb the infrared (heat) radiation re-radiated from the earth rather than permitting the heat to escape.

greenhouse gas. Gases that allow solar radiation to penetrate the atmosphere, but retard the outward movement of infrared radiation from the earth's surface, thereby trapping heat and increasing global surface temperatures. Such gases include carbon dioxide, methane, chlorofluorocarbons, sulphur dioxide and nitrous oxide.

guild. Groups of species that exploit similar resources in a similar manner (e.g. desert granivores or forest foliage-gleaning birds).

habitat island. A geographically isolated patch of habitat, such as mountain top, wood or lake that can be studied in the same ways as oceanic islands for patterns of colonisation and local extinction.

half-life. The amount of time needed for half of the radioactive material in a rock or other substrate to decay to a stable element.

haploid. Having only a single set of chromosomes. Gametes are usually haploid.

haplotype. A particular set of alleles at several closely linked loci that are found on a single chromosome and tend to be inherited together.

heterogametic. The sex that is heterozygous for the sex-determining chromosome; the male in mammals, the female in birds.

heterozygote. An individual with different genetic factors (alleles) at the homologous (corresponding) loci of the two parental chromosomes.

historical biogeography. The study of the relationship of present and past distributions of organisms to the physical history of the earth.

Holarctic region. The extra-tropical zone of the northern hemisphere, which includes both the Nearctic and Palaearctic regions.

Holocene. The epoch spanning the last 10,000 years or so since the last glacial retreat.

homogametic. The sex that is homozygous for the sex-determining chromosome. See heterogametic.

homozygote. An individual with identical genetic factors (alleles) at the homologous (corresponding) loci of the two parental chromosomes.

hot spot. In plate tectonics, a stationary weak point in the upper mantle that discharges magma (molten rock) as a plate passes over it, producing a narrow chain of islands, such as the Hawaiian Islands. In biodiversity, an area with a relatively high number of species or high number of endemic species.

hybrid swarm. A morphologically variable population in which introgression has occurred to varying degrees, through hybridisation and back-crossing.

hybrid zone. An area where two related taxa overlap, and where pure phenotypes of both taxa can be found alongside hybrids between them.

hybridisation. Interbreeding between individuals of distinct taxa.

immigration. The arrival of new individuals from elsewhere; in biogeography, often to a previously unoccupied area.

imprinting. A rapid learning process that occurs early in life, in which responses are acquired quickly and not usually reversed. Often involves a recognition response which, under manipulation, can be fixated on an inappropriate species or object.

inbreeding. The mating of closely related individuals, which can reduce genetic variability through increased homozygosity.

Indomalayan Region. The biogeographical region embracing tropical and subtropical parts of southeast Asia and nearby islands, eastwards to the edge of the continental plate (Wallace's Line).

interglacial. During the Quaternary, a period when glacial ice sheets retreated and the climate became milder.

introgression. The incorporation of genes of one population or species into the gene pool of another.

introns. Non-coding segments of DNA that interrupt the coding sequences of nuclear genes.

isolating mechanism. Properties of individuals that prevent successful interbreeding with individuals that belong to different populations.

Jurassic. The period in the history of life spanning roughly 213–144 million years ago, which saw the appearance of the earliest birds.

K-selection. Selection favouring a more efficient utilisation of resources, which is more pronounced when the species is at or near carrying capacity (K). Typically species with high survival and age of first breeding, and low annual reproductive rates.

land-bridge island. See continental island.

Laurasia. The northern part of the supercontinent Pangaea, including North America, Europe, and large parts of Asia.

life zones. The characteristic types of vegetation that occur along an elevational or latitudinal gradient.

limiting factor. The resource or environmental parameter that most limits the size and distribution of a population.

long-distance (jump) dispersal. The movement of an organism across inhospitable environments to colonise a favourable distant area, such as an oceanic island.

macchia. A type of sclerophyllous scrub vegetation occurring in regions of the Old World with a Mediterranean climate. Equivalent to chaparral in North America and matorral in South America.

maquis. Same as macchia.

marine. Associated with the salt water of oceans and seas.

mass extinction. A major episode of extinction for many taxa, occurring fairly suddenly in the fossil record.

matorral. A type of sclerophyllous scrub vegetation occurring in the region of Chile with a Mediterranean climate.

Equivalent to macchia in Europe and chaparral in North America.

Mediterranean climate. A semi-arid climate characterised by mild, rainy winters and hot, dry summers.

megafauna. A general term for the large terrestrial vertebrates inhabiting a specified region.

Mesozoic. The era in the history of life spanning roughly 248–65 million years ago, and including the Triassic, Jurassic and Cretaceous Periods.

Messinian crisis. The sudden drainage of the Mediterranean Basin during a period of pronounced drought in the Cenozoic.

metabolic rate. The rate at which food and body reserves are converted to energy, including heat production, and usually expressed in kcal per bird per day. Basal metabolic rate (BMR) is the minimal metabolic rate possible, when the animal is inactive in a post-absorptive state, in the thermoneutral zone, and not engaged in any physiological activity (such as moulting) that would require additional energy. Standard metabolic rate (SMR) is the same as the BMR, except that the animal is below the thermoneutral zone, usually at 0°C.

microsatellite. A short region of nuclear DNA in which a motif of a small number (often 2–5) bases is repeated, head-to-tail, a number of times at a particular locus.

migration. Regular seasonal movements between separate breeding and wintering areas.

minisatellite. Like microsatellite, but bigger, with more tandem repeats at a particular locus.

Miocene. The epoch spanning roughly 24–5 million years ago.

molecular clock. The notion that the structure of complex molecules changes (mutates) at consistent average rates through evolutionary time, enabling the ages of related lineages to be estimated. Their divergence dates are often measured from the percentage difference in nucleotide sequences in a speci-

fied segment of DNA, assuming consistent average mutation rates in both lineages through time.

monomorphic. Having only one form in a population.

monophyletic. Having arisen from one ancestral form; in the strictest sense, from one initial population.

monospecific. Having only one species in the genus.

monotypic. Having only one species in the taxon, or only one uniform subspecies in the species.

morph. Any of the genetic forms (individual variants) that account for polymorphism.

mt DNA. Mitochondrial DNA, a small haploid extra-nuclear DNA molecule, housed in the mitochondria in the cytoplasm of cells. It apparently does not recombine at fertilisation, but is instead clonally and maternally inherited, transmitted from mother to offspring of both sexes and passed unchanged (except for mutation) down the generations.

mutation. An error in the replication of a nucleotide sequence or any other alteration in the genome that is not due to recombination.

natal dispersal. The movement of an individual from birthplace to breeding place.

natural selection. The process of eliminating from a population through differential survival and reproduction those individuals of inferior fitness.

Nearctic Region. The extra-tropical biogeographical region of North America.

neo-endemic. An endemic species that evolved in fairly recent times.

Neogene. Combined Miocene, Pliocene, Pleistocene and Holocene epochs, spanning roughly the last 24 million years.

Neotropical Region. The biogeographical region from southern Mexico and the West Indies to southern South America.

niche. The total requirements of a population or species for resources and physical conditions.

niche breadth. The range of resources and physical conditions used by a particular population relative to that of other populations.

niche expansion. An increase in the range of habitats or resources used by a population, which may occur when a potential competitor is absent.

nomadic. Having no fixed spatial pattern of migration.

nucleotide. The unit structure of nucleic acids (which include DNA), and which is composed of a ribose or deoxyribose sugar bound to a purine or pyrimidine base.

nunatak. An area in a glaciated region that was not covered by an ice sheet; hence, a refuge.

oceanic. Pertaining to the open ocean with very deep water.

oceanic island. An island that was formed afresh from the floor of the ocean through volcanic activity, has never been attached to a continent, and whose biota were derived entirely by cross-water colonisation; compare with continental island.

Oligocene. The epoch spanning roughly 37–24 million years ago.

order. A taxonomic category above the level of family and below the level of class.

Oriental Region. See Indomalayan Region.

Palaearctic Region. The biogeographical region of extratropical climates in Eurasia and in northernmost Africa.

Palaeocene. The epoch spanning roughly 65–58 years ago.

palaeoclimatology. The study of past climates, as elucidated mainly through the analysis of fossils and ice cores.

palaeo-ecology. The branch of palaeontology that attempts to reconstruct the structure of ancient populations and communities.

palaeo-endemic. An endemic species that evolved in the distant past; compare with neo–endemic.

Palaeogene. Combined Palaeocene, Eocene and Oligocene epochs, roughly spanning 65–24 million years ago.

palaeontology. The field of study devoted to describing, analysing, and explaining the fossil record.

palaeotropical. Occurring in the Old World tropics and subtropics; i.e. in Africa, Madagascar, India and southeast Asia.

Pangaea. In plate tectonics, the supercontinent of the Permian that was composed of essentially all the present continents and major continental islands.

panmixis. The condition whereby interbreeding within a large population appears to be totally random.

pan-tropical. Occurring in all major tropical areas around the world.

parallel evolution. The evolution of species with strong resemblances from fairly closely related ancestors; hence the evolution of two or more taxa in the same direction from related ancestors.

paramo. Tropical alpine vegetation, characteristic of high mountains at the equator, that has a low and compact perennial cover as a response to the perpetually wet, cold and cloudy environment.

parapatric. Having contiguous but non-overlapping distributions.

parapatric speciation. A mode of speciation in which differentiation occurs when two populations have contiguous but non-overlapping ranges, often representing two distinct habitat types. See allopatric speciation.

paraphyletic. In cladistics, referring to taxa that are classified chiefly on the basis of shared primitive character states.

pelagic. Occurring in or over the open sea and away from the bottom.

peninsular effect. The hypothesised tendency for species richness to decrease along a gradient from the axis to the most distal point of a peninsula.

peripheral population. Any population of a species that occurs along or near the edges of a range, around either the perimeter or the elevational limit.

phenetics. The study of the overall similarities of organisms; compare with phylogeny.

phenotype. The totality of the characteristics of an individual that results from genetic and environmental influences.

philopatry. The tendency of an individual to return to (or stay in) its home area (birthplace).

phyletic gradualism. An evolutionary process whereby a species is gradually transformed over time into a different organism; compare with punctuated equilibrium.

phylogeny. The evolutionary relationships between an ancestor and its known descendants.

phylogeography. Study of the geographical distributions of genealogical lineages.

plankton. Small organisms (especially tiny plants, small invertebrates, and juvenile stages of larger animals) that inhabit water and are transported mainly by water currents and wave action rather than by individual locomotion.

plate. In plate tectonics, a portion of the earth's upper surface, about 100 km thick, that moves over the molten core during seafloor spreading.

plate tectonics. The study of the origin, movement and destruction of plates and how these events have influenced the evolution of the earth's crust and the distribution of continents.

Pleistocene. The epoch spanning from around two million years ago to 10,000 years ago, the start of the Holocene.

Pliocene. The epoch spanning roughly 5–2 million years ago, immediately before the Pleistocene.

pluvial. Referring to periods of high rainfall and water runoff.

polymerase chain reaction (PCR). Process for amplifying specific DNA fragments in virtually unlimited quantity, thus facilitating further analysis.

polymorphic. Having several distinct forms in a population, such as colour phases.

polynya. High-latitude area of open sea water surrounded by ice, often maintained by strong currents.

polyploid. Having more than two haploid sets of chromosomes.

polytypic. Having several types; a species with more than one subspecies.

potassium–argon dating. Estimation of the age of rock from the ratio of radio-active potassium to derived argon, which declines with age (half-life of K^{40} =1.31 million years).

primitive characters. Features low in phylogenetic information because they were inherited from a common ancestor.

provincialism. The co-occurrence of large numbers of well-differentiated endemic forms in an area; regional or provincial distinctiveness.

punctuated equilibrium. The hypothesis that evolution occurs during periods of rapid differentiation (often accompanying speciation), which are followed by long periods in which few if any characters evolve.

Quaternary Period. The period in the history of life, including the Pleistocene and Holocene epochs, spanning the last two million years.

radio-carbon dating. Estimation of the age of organic material from the ratio of two isotopes of carbon, C^{14} and C^{12}, which declines at a consistent rate with age (half-life of C^{14} = 5,730 ± 30 years).

RAPDs. Randomly amplified polymorphic DNA, used mainly for the analysis of population differentiation.

refuge. An area in which climate and vegetation type have remained relatively unchanged while areas surrounding it have changed markedly, and which has thus served as a refuge for species requiring the specific habitat it contains.

relict. A surviving taxon from a group that was once widespread (distributional relict) or diverse (taxonomic relict).

reproductive isolation. Inability of individuals from different populations to produce viable offspring.

rescue effect. On islands near continents, the tendency of a continual influx of individuals belonging to species already present to prevent the disappearance of those species. Also applied to populations in habitat patches.

resident. A species that lives year-round in the same area.

restriction endonuclease. Enzymes used to cut DNA at specific points, either as a preparative step in gene cloning or for analytical purposes, such as identification of specific restriction fragment length polymorphisms (RFLP) or amplified fragment length polymorphisms (AFLP).

reticulate evolution. Union by the horizontal transfer of genes of distinct evolutionary lineages which may themselves have been derived from a common ancestor, e.g. inter-specific hybridisation.

RFLP. Restriction fragment length polymorphism; see Restriction endonuclease.

ring species. A chain of contiguous and interbreeding populations which curves back until the terminal links overlap with each other and behave like good species (non-interbreeding).

r–selection. Selection that favours high population growth rate ('r'), which is more prominent when population size is far below carrying capacity. Typically, species with low survival and age of first breeding, and high annual reproductive rates.

savannah. Tropical and subtropical grassland with scattered trees and shrubs.

scrub. Any of a wide variety of vegetation types dominated by low shrubs; in very dry locations, scrub vegetation has few or no trees and widely spaced low shrubs, but in areas of fairly high rainfall, scrub has trees and grades into either woodland or forest.

semi-arid. Having a fairly dry climate with low precipitation, usually 25–60 cm per year, and a high evapo-transpiration rate, so that potential loss of water to the environment matches the input.

semi-desert. A semi-arid habitat characterised by low vegetation; e.g. small, widely spaced shrubs.

semi-species. Populations that have acquired some, but not yet all, attributes of species rank; borderline populations between subspecies and species.

sibling species. Morphologically similar and closely related populations that are reproductively isolated.

sister species Allopatric populations of species status derived from the geographical splitting of a single parental species.

sister taxa. The two taxa that are most closely (and therefore most recently) related.

speciation. The process in which one or more new species evolve from a single ancestral species, through the acquisition of reproductive isolation.

species. The fundamental taxonomic category for all organisms; a group of individuals that are morphologically and reproductively more similar to one another than to other populations, that share a singular ancestor-descendant heritage, and that are more or less reproductively isolated.

species-area relationship. A plot of the numbers of species of a particular taxon against the area of habitat patches, islands or other biogeographical regions. The relationship is often linearised by log-transforming one or both variables.

species composition The types of species that constitute a given sample.

species group. A group of closely related species, usually with partially overlapping ranges.

species richness. The number of species in a given sample.

stabilising selection. The elimination by selection of all phenotypes deviating too far from the population optimal, and hence also of the genes producing such phenotypes.

subfamily. A taxonomic category used for grouping genera within a family.

subfossil. Bones that have not been mineralised.

subspecies. A taxonomic category below the level of species used to designate a genetically distinct set of subpopulations, each with a discrete range, and each genetically compatible with the other subpopulations of that species.

superfamily. A taxonomic category used for grouping families within a suborder.

superspecies. A group of closely related species, each with a different geographical range, and assumed to derive from the same parental species. Too distinct to be designated as subspecies.

supertramp. A species that has excellent colonising abilities but is a poor competitor in diverse communities.

sweepstake route. A severe barrier that results in the partly stochastic dispersal of some species from a community but not others.

sympatric. In the strictest sense, living in the same local community; in the more general sense, having broadly overlapping geographical breeding ranges.

sympatric speciation. The division of one species into two or more without geographical isolation; the acquisition of isolating mechanisms by two or more parts of a single population in the same breeding area; compare with allopatric speciation; parapatric speciation.

systematics. The study of the evolutionary relationships between organisms.

taxon (pl. **taxa**). A convenient and general term for any taxonomic category, whether a species, genus, family, or order.

taxon cycle. A proposed series of ecological and evolutionary changes in a newly arrived colonist, from being indistinguishable from its relatives to becoming a highly differentiated endemic, and then to extinction and replacement by new colonists.

taxonomy. In the strictest sense, the study of the names of organisms, but often used for the entire process of classification; see systematics.

tectonic. Referring to any process involved in the production or deformation of the earth's crust.

terrestrial. Living on land.

Tertiary Period. The period in the history of life spanning roughly 65–2 million years ago.

thermoneutral zone. The range of ambient temperature in which an animal need not expend additional energy to keep its body temperature at normal level.

timberline. The uppermost limit of forest vegetation at high elevations or high latitudes.

tribe. A taxonomic category above the level of family and below the level of order.

tundra. Community of low cold-resistant plants found at high latitudes and high altitudes (= alpine tundra).

turnover. The rate of replacement of species in a particular area as some taxa become extinct but others immigrate from outside.

upwelling. The vertical movement of deep water, containing dissolved nutrients from the ocean bottom, to the surface.

vagility. The ability to move actively from one place to another.

vicariance. Disjunct distribution caused by the appearance of a barrier that splits a former continuous range into two or more separate areas.

vicariance biogeography. A branch of biogeography that attempts to reconstruct the historical events that led to current distributional patterns based largely on the assumption that these patterns resulted from the splitting (vicariance) of areas and not long-distance dispersal. Compare dispersal biogeography.

vicariants. Two disjunct taxa that are closely related to each other and that are assumed to have been created when the initial range of their ancestor was split by some past event.

Wallace's Line. The most famous biogeographical line, which marks the boundary between the Indomalayan and Australasian Regions, and runs between Borneo and Celebes and between Bali and Lombok.

wintering area. In migratory birds, the area where populations spend the cold season, but do not breed.

woodland. Any of a variety of vegetation types consisting of small, fairly widely spaced trees with or without substantial undergrowth.

zoogeography. The study of the distributions of animals, both past and present.

zygote. A fertilised egg.

Great White Heron *Egretta alba*, one of the most widely distributed of bird species.

References

Abbott, I. (1978). Factors determining the number of land bird species on islands around south-western Australia. *Oecologia* **33:** 221–223.

Ackerman, J. (1998). Dinosaurs take wing. *Nat. Geogr.* **194:** 74–99.

Adler, G. H. (1994). Avifaunal diversity and endemism on tropical Indian Ocean Islands. *J. Biogeog.* **21:** 85–95.

Adriaensen, F., Verwimp, N. & Dhondt, A. A. (1998). Between cohort variation in dispersal distance in the European Kestrel *Falco tinnunculus* as shown by ringing recoveries. *Ardea* **86:** 147–152.

Aebischer, N. J. (1995). Philopatry and colony fidelity of Shags *Phalacrocorax aristotelis* on the east coast of Britain. *Ibis* **137:** 11–18.

Aebischer, N. J. & Potts, G. R. (1994). Quail *Coturnix coturnix*. Pp. 222–223 in '*Birds in Europe. Their conservation status*' (ed. G. M. Tucker & M. F. Heath). Cambridge, UK, Birdlife International.

Ainley, D. G., Ribic, C. A. & Fraser, W. R. (1994). Ecological structure among migrant and resident seabirds of the Scotia–Weddell confluence region. *J. Anim. Ecol.* **63:** 347–364.

Ainley, D., Wilson, P. & Fraser, W. R. (2001). Effects of climate change on Antarctic sea ice and penguins. Pp. 26–27 in '*Impacts of climate change on wildlife*' (ed. R. E. Green, M. Harley, M. Spalding & C. Zöckler). Sandy, Royal Society for the Protection of Birds.

Alatalo, R. V., Gustafsson, L. & Lundberg, A. (1986). Interspecific competition and niche changes in tits (*Parus* spp) — evaluation of nonexperimental data. *Amer. Nat.* **127**: 819–834.

Alberico, J. A. R., Reed, J. M. & Oring, L. W. (1992). Non-random philopatry of sibling Spotted Sandpipers *Actitis macularia*. *Ornis Scand.* **23**: 504–508.

Allen, J. R. M., Brandt, U., Brauer, A., Hubberten, H. W., Huntley, B., Keller, J., Kraml, M., Mackensen, A., Mingram, J., Negendank, J. F. W., Nowaczyk, N. R., Oberhanski, H., Watts, W. A., Wulf, S. & Zolitschka, B. (1999). Rapid environmental changes in southern Europe during the last glacial period. *Nature* **400**: 740–743.

Alonso, J. C., Alonso, J. A. & Bautista, L. M. (1991). Carrying capacity of staging areas and facultative migration extension in Common Cranes. *J. Appl. Ecol.* **31**: 212–222.

Ambrose, R. E. & Riddle, K. E. (1988). Population dispersal, turnover and migration of Alaska Peregrines. Pp. 677–684 in '*Peregrine Falcon populations. Their management and recovery*' (ed. T. J. Cade, J. H. Enderson, C. G. Thelander & C. M. White). Boise, Idaho, The Peregrine Fund Inc.

American Ornithologists' Union. (1957). Check-list of North American birds, 5th edition. Washington, DC, American Ornithologists' Union.

Anderson, A. (1990). Prodigious birds: moas and moa-hunting in prehistoric New Zealand. Cambridge, University Press.

Anderson, D. J. (1991). Apparent predator-limited distribution of Galapagos Red-footed Boobies *Sula sula*. *Ibis* **133**: 26–29.

Anderson, D. J., Schwardt, A. J. & Douglas, H. D. (1997). Foraging ranges of Waved Albatrosses in the eastern tropical Pacific Ocean. Pp. 180–185 in '*Albatross biology and conservation*' (ed. G. Robertson & R. Gales). Chipping Norton, New South Wales, Surrey Beatty.

Anderson, M. G., Rhymer, J. M. & Rohwer, F. C. (1992). Philopatry, dispersal, and the genetic structure of waterfowl populations. Pp. 365–395 in '*Ecology and management of breeding waterfowl*' (ed. B.D. J. Batt *et al.*). Minneapolis, University of Minnesota Press.

Anderson, S. (1984). Geographic ranges of North American birds. *Amer. Mus.* **2785**: 1–17.

Andersson, M. (1994). Sexual selection. Princeton, University Press.

Andersson, M. (1999). Hybridisation and skua phylogeny. *Proc. R. Soc. Lond. B* **266**: 1579–1585.

Andrew, P. (1992). The birds of Indonesia: a checklist (Peters' sequence). Jakarta, Indonesian Ornithological Society.

Angehr, G. R. (1999). Rapid long-distance colonization of Lake Gatun, Panama, by Snail Kites. *Wilson Bull.* **111**: 265–268.

Antikainen, E., Skarén, U., Toivanen, J. & Ukkonen, M. (1980). The nomadic breeding of the Redpoll *Acanthis flammea* in 1979 in North Savo, Finland. *Orn Fenn.* **57**: 124–131.

Arcese, P. (1989). Intrasexual competition, mating system and natal dispersal in Song Sparrows. *Anim. Behav.* **38**: 958–979.

Arctander, P. & Fjeldså, J. (1995). Andean tapaculos of the genus *Scytalopus* (Aves, Rhinocryptidae): a study of modes of speciation, using DNA sequence data. Pp. 205–225 in '*Conservation genetics*' (ed. V. Loeschcke, T. Tomuick & S. K. Jain). Basel, Birkhäuser Verlag.

Arendt, W. J., Vargas Mora, T. A. & Wiley, J. W. (1979). White-crowned Pigeon: status range-wide and in the Dominican Republic. *Proc. Conf. S.E. Assoc. Fish Wildl. Agencies* **33**: 111–122.

Arrendondo, O. (1976). The great predatory birds of the Pleistocene of Cuba. Pp. 169–187 in '*Collected papers in avian palaeontology honoring the 90th birthday of Alexander Wetmore*' (ed. S. L. Olson). Washington, Smithsonian Institution Press.

Ashmole, N. P. (1971). Seabird ecology and the marine environment. Pp. 223–286 in '*Avian biology*' (ed. D. S. Farner & J. R. King). London, Academic Press.

Askins, R. A. & Ewert, D. N. (1991). Impact of hurricane Hugo on bird populations on St. John, United States Virgin Islands. *Biotropica* **23**: 481–487.

Atkinson, C. T., Lease, J. K., Drake, B. M. & Shema, N. P. (2001). Pathogenicity, serological responses, and diagnosis of experimental and natural malarial infections in native Hawaiian thrushes. *Condor* **103:** 209–218.

Atkinson, I. A. E. (1985). The spread of commensal species of *Rattus* to oceanic islands and their effects on island avifaunas. *ICBP Tech. Bull.* **3:** 35–81.

Atkinson, I. A. E. & Milliner, P. R. (1991). An ornithological glimpse into New Zealand's pre-human past. *Proc. Int. Orn. Congr.* **20:** 129–192.

Attenborough, D. (1998). The life of birds. London, BBC Books.

Audley-Charles, M. G. (1983). Reconstruction of eastern Gondwanaland. *Nature* **306:** 48–50.

Austin, J. J. (1996). Molecular phylogenetics of *Puffinus* shearwaters: preliminary evidence from mitochondrial cytochrome *b* gene sequences. *Mol. Phyl. Evol.* **6:** 77–88.

Austin, J. J., White, R. W. & Ovenden, J. R. (1994). Population–genetic structure of a philopatric, colonially nesting seabird, the Short-tailed Shearwater (*Puffinus tenuirostris*). *Auk* **111:** 70–79.

Austin, O. L. (1949). Site tenacity, a behaviour trait of the Common Tern. *Bird-Banding* **20:** 1–39.

Avise, J. C. (1989). Gene trees and organismal histories: a phylogenetic approach to population biology. *Evolution* **43:** 1192–1208.

Avise, J. C. (1994). Molecular markers, natural history and evolution. London, Chapman & Hall.

Avise, J. C. (1996). Three fundamental contributions of molecular genetics to avian ecology and evolution. *Ibis* **138:** 16–25.

Avise, J. C. (2000). Phylogeography: the history and formation of species. Cambridge, Mass., Harvard University Press.

Avise, J. C. & Ball, R. M. (1991). Mitochondrial DNA and avian microevolution. *Proc. Int. Orn. Congr.* **20:** 514–524.

Avise, J. C., Nelson, W. S. & Sibley, C. G. (1994). DNA-sequence support for a close phylogenetic relationship between some Storks and New World Vultures. *Proc. Nat. Acad. Sci. USA* **91:** 5173–5177.

Avise, J. C., Nelson, W. S., Bowen, B. W. & Walker, D. (2000). Phylogeography of colonially nesting seabirds, with special reference to global matrilineal patterns in the Sooty Tern (*Sterna fuscata*). *Mol. Ecol.* **9:** 1783–1792.

Avise, J. C., Walker, D. & Johns, G. C. (1998). Speciation durations and Pleistocene effects on vertebrate phylogeography. *Proc. R. Soc. Lond. B* **265:** 1707–1712.

Avise, J. C. & Zink, R. M. (1988). Molecular genetic divergence between avian sibling species: King and Clapper Rails, Long-billed and Short-billed Dowitchers, Boat-tailed and Great-tailed Grackles, and Tufted and Black-crested Titmice. *Auk* **105:** 516–528.

Badyaev, A. V., Hill, G. E., Stoehr, A. M., Nolan, P. M. & McGraw, K. J. (2000). The evolution of sexual size dimorphism in the House Finch. II. Population divergence in relation to local selection. *Evolution* **54:** 2134–2144.

Bailey, R. S. (1966). The seabirds of the southeast coast of Arabia. *Ibis* **108:** 224–264.

Baillie, S. R. & Green, R. E. (1987). The importance of variation in recovery rates when estimating survival rates from ringing recoveries. *Acta Orn.* **23:** 41–60.

Bainbridge, I. & Minton, C. D. (1978). The migration and mortality of the Curlew in Britain and Ireland. *Bird Study* **25:** 39–50.

Bairlein, F. (2001). Results of bird ringing in the study of migration routes. *Ardea* **89:** 7–19.

Baker, A. J. (1980). Morphometric differentiation in New Zealand populations of the House Sparrow (*Passer domesticus*). *Evolution* **34:** 638–653.

Baker, A. J. (1991). A review of New Zealand ornithology. *Curr. Orn.* **8:** 1–67.

Baker, A. J., Daugherty, C. H., Colbourne, R. & McLennan, J. L. (1995). Flightless Brown Kiwis of New Zealand possess extremely subdivided population structure and cryptic species like small mammals. *Proc. Nat. Acad. Sci. USA* **92:** 8254–8258.

Baker, A. J., Dennison, M. D., Lynch, A. & Le Grande, G. (1990). Genetic divergence in peripherally isolated populations of chaffinches in the Atlantic islands. *Evolution* **44**: 981–999.

Baker, A. J. & Moeed, A. (1979). Evolution in the introduced New Zealand populations of the Common Myna *Acridotheres tristis*. *Can. J. Zool.* **57**: 570–584.

Baker, A. J., Piersma, T. & Rosenmeier, L. (1994). Unravelling the intraspecific phylogeography of Knots *Calidris canutus*. *J. Orn.* **135**: 599–608.

Baker, M., Nur, N. & Geupel, G. R. (1995). Correcting biased estimates of dispersal and survival due to limited study area: theory and an application using Wrentits. *Condor* **97**: 663–674.

Baker, M. C. & Boylan, J. T. (1999). Singing behaviour, mating association and reproductive success in a population of hybridising Lazuli and Indigo Buntings. *Condor* **101**: 493–504.

Ball, R. M. & Avise, J. C. (1992). Mitochondrial DNA phylogeographic differentiation among avian populations and the evolutionary significance of subspecies. *Auk* **109**: 626–636.

Ball, R. M., Freeman, S., James, F. C., Bermingham, E. & Avise, J. C. (1988). Phylogeographic population structure of Red-winged Blackbirds assessed by mitochrondrial DNA. *Proc. Nat. Acad. Sci. USA* **85**: 1558–1562.

Ballmann, P. (1969). Die Vögel aus der altburdigalen Spalterfullung von Wintershaf (West) bei Eichstätt in Bayern. *Zitteliana* **1**: 5–60.

Balouet, J. C. & Olson, S. L. (1989). Fossil birds from late Quaternary deposits. *Smithson. Contrib. Zool.* **469**: 1–38.

Barbraud, C. & H. Wiemerskirch. (2001). Emperor Penguins and climate change. *Nature* **411**: 183.

Barker, F. K., Barrowclough, G. F. & Groth, J. G. (2002). A phylogenetic hypothesis for passerine birds: taxonomic and biogeographic implications of an analysis of nuclear DNA sequence data. *Proc. R. Soc. Lond B* **269**: 295–308.

Barnes, I., Mathews, P., Shapiro, B., Jensen, D. & Cooper, A. (2002). Dynamics of Pleistocene population extinctions in Beringian Brown Bears. *Science* **295**: 2267–2270.

Barraclough, T. G., Harvey, P. H. & Nee, S. (1995). Sexual selection and taxonomic diversity in passerine birds. *Proc. R. Soc. Lond. B* **259**: 211–215.

Barraclough, T. G., Vogler, A. P. & Harvey, P. H. (1998). Revealing the factors that promote speciation. *Phil. Trans. R. Soc. Lond.* **353**: 241–249.

Barrowclough, G. F. (1978). Sampling bias in dispersal studies based in finite area. *Bird-Banding* **49**: 333–341.

Barton, N. H. & Hewitt, G. M. (1989). Adaptation, speciation and hybrid zones. *Nature* **341**: 497–503.

Bastian, H. V. (1992). Breeding and natal dispersal of Whinchats *Saxicola rubetra*. *Ringr. Migr.* **13**: 13–19.

Bates, A. J., Hackett, S. J. & Cracraft, J. (1998). Area-relationships in the neotropical lowlands: an hypothesis based on raw distributions of Passerine birds. *J. Biogeog.* **25**: 783–793.

Bates, J. M. (2000). Allozymic genetic structure and natural habitat fragmentation: data for five species of Amazonian forest birds. *Condor* **102**: 770–783.

Baverstock, P. R., Schodde, R., Christidis, L., Krieg, M. & Sheedy, C. (1991). Microcomplement fixation: preliminary results from the Australasian avifauna. *Proc. Int. Orn. Congr.* **20**: 611–618.

Beck, B. B., Rapaport, L. G., Stanley Price, M. R. & Wilson, A. C. (1994). Reintroduction of captive-born arrivals. Pp. 264–286 in '*Creative conservation*' (ed. P. J. S. Olney, C. M. Mace & A. T. C. Feistner). London, Chapman & Hall.

Becker, P. H., Frank, D. & Sudmann, S. R. (1993). Temporal and spatial pattern of Common Tern (*Sterna hirundo*) foraging in the Wadden Sea. *Oecologia* **93**: 389–393.

Beddall, B. G. (1963). Range expansion of the cardinal and other birds in the northeastern States. *Wilson Bull.* **75:** 140–158.

Bedell, P. A. (1996). Evidence of dual breeding ranges for the Sedge Wren in the Central Great Plains. *Wilson Bull.* **108:** 115–122.

Beehler, B. M., Pratt, T. K. & Zimmerman, D. A. (1986). Birds of New Guinea. Princeton, University Press.

Behle, W. H. (1978). Avian biogeography of the Great Basin and Intermontane Region. *Great Basin Natur. Mem.* **2:** 55–80.

Beletsky, L. D. & Orians, G. H. (1991). Effects of breeding experience and familiarity on site fidelity in female Red-winged Blackbirds. *Ecology* **72:** 787–796.

Bell, B. D. (1978). The Big South Cape Islands rat irruption. Pp. 33–45 in '*The ecology and control of rodents in New Zealand Nature Reserves.*' (ed. P. R. Dingwall, I. A. E. Atkinson & C. Hay). New Zealand, Dept. Lands and Survey.

Bell, B. D. (1991). Recent avifaunal changes and the history of ornithology in New Zealand. *Proc. Int. Orn. Congr.* **20:** 195–230.

Belliure, J., Sorci, G., Møller, A. P. & Clobert, J. (2000). Dispersal distances predict subspecies richness in birds. *J. Evol. Biol.* **13:** 480–487.

Bellrose, F. C. (1990). The history of Wood Duck management. Pp. 13–20 in '*Proceedings of the 1988 North American Wood Duck Symposium*' (ed. L. H. Fredrickson, G. V. Burger, S. P. Havera, D. A. Graber, R. E. Kirby & T. S. Taylor). St. Louis, Missouri.

Benkman, C. W. (1987). Food profitability and the foraging ecology of crossbills. *Ecol. Monogr.* **57:** 251–267.

Benkman, C. W. (1993). Adaptation to single resources and the evolution of crossbill (*Loxia*) diversity. *Ecol. Monogr.* **63:** 305–325.

Bensch, S. (1999). Is the range size of migratory birds constrained by their migratory programme? *J. Biogeog.* **26:** 1225–1235.

Bensch, S. & Hasselquist, D. (1991). Territory infidelity in the polygynous Great Reed Warbler *Acrocephalus arundinaceus*: the effect of variation in territory attractiveness. *J. Anim. Ecol.* **60:** 857–871.

Bensch, S. & Hasselquist, D. (1999). Phylogeographic population structure of Great Reed Warblers: an analysis of mt DNA control region sequences. *Biol. J. Linn. Soc.* **66:** 171–185.

Benson, C. W., Irwin, M. P. S. & White, C. M. N. (1962). The significance of valleys as avian zoogeographical barriers. *Ann. Cape Prov. Museums* **2:** 155–189.

Berger, A. J. & Radabaugh, B. E. (1968). Returns of Kirtland's Warblers to the breeding grounds. *Bird-Banding* **39:** 161–186.

Bermingham, E. S., Rohwer, S., Freeman, S. & Wood, C. (1992). Vicariance biogeography in the Pleistocene and speciation in North American wood warblers. *Proc. Nat. Acad. Sci., USA* **89:** 6624–6628.

Berndt, R. & Sternberg, H. (1968). Terms, studies and experiments on the problems of bird dispersion. *Ibis* **110:** 256–269.

Berrow, S. D., Wood, A. G. & Prince, P. A. (2000). Foraging location and range of White-chinned Petrels *Procellaria aequinoctialis* breeding in the South Atlantic. *J. Avian Biol.* **31:** 303–311.

Berthold, P. (1990). Die Vogelwelt Mitteleuropas: Entstehung der Diversität, gegenwärtige Veränderungen und Aspekte der Zukunftigen Entwicklung. *Verh. Dtsch. Zool. Ges.* **83:** 227–244.

Berthold, P. (1993). Bird migration. A general survey. Oxford, University Press.

Berthold, P. (1995). Microevolution of migratory behaviour illustrated by the Blackcap *Sylvia atricapilla* — 1993 Witherby Lecture. *Bird Study* **42:** 89–100.

Berthold, P. (1999). A comprehensive theory for the evolution, control and adaptability of avian migration. *Ostrich* **70:** 1–12.

Berthold, P. & Helbig, A. J. (1992). The genetics of bird migration: stimulus, timing and direction. *Ibis* **134**: 35–40.

Berthold, P., Helbig, A. H., Mohr, G. & Querner, U. (1992). Rapid microevolution of migratory behaviour in a wild bird species. *Nature* **360**: 668–670.

Besse, J. & Courtillot, V. (1988). Paleogeographic maps of the continents bordering the Indian Ocean since the early Jurassic. *J. Geophys. Res.* **93(B10)**: 11791–11808.

Bibby, C. J., Collar, N. J., Crosby, M. J., Heath, M. F., Imboden, C., Johnson, T. H., Long, A. J., Stattersfield, A. J. & Thirgood, S. J. (1992). Putting biodiversity on the map: priority areas for global conservation. Cambridge, International Council for Bird Preservation.

Bilton, D. T., Mirol, P. M., Mascheretti, S., Fredga, K., Zima, J. & Searle, J. B. (1998). Mediterranean Europe as an area of endemism for small mammals rather than a source for northwards postglacial colonisation. *Proc. R. Soc. Lond. B* **265**: 1219–1226.

Birkhead, T. R. (1974). Movement and mortality rates in British Guillemots. *Bird Study* **21**: 241–253.

Birt-Friesen, V. L., Montevecchi, W. A., Gaston, A. J. & Davidson, W. S. (1992). Genetic structure of Thick-billed Murre (*Uria lomvia*) populations examined using direct sequence analysis of amplified DNA. *Evolution* **46**: 267–272.

Bjerke, T. & Espmark, Y. (1988). Breeding success and breeding dispersal in recovered Redwings *Turdus iliacus*. *Fauna Norveg.* **11**: 45–46.

Blackburn, T. M. & Gaston, K. J. (1995). What determines the probability of discovering a species?: a study of South American oscine passerines. *J. Biogeog.* **22**: 7–14.

Blackburn, T. M. & Gaston, K. J. (1996). Spatial patterns in the geographic range sizes of bird species in the New World. *Phil. Trans. R. Soc. Lond. B* **351**: 897–912.

Blackburn, T. M., Gaston, K. J., Greenwood, J. J. D. & Gregory, R. D. (1998a). The anatomy of the interspecific abundance–range size relationship for the British avifauna II. Temporal dynamics. *Ecol. Lett.* **1**: 47–55.

Blackburn, T. M., Gaston, K. J. & Gregory, R. D. (1997a). Abundance–range size relationship in British birds: is unexplained variation a product of life history? *Ecography* **20**: 466–474.

Blackburn, T. M., Gaston, K. J. & Lawton, J. H. (1998b). Patterns in the geographic ranges of the world's woodpeckers. *Ibis* **140**: 626–638.

Blackburn, T. M., Gaston, K. J., Quinn, R. M., Arnold, H. & Gregory, R. D. (1997b). Of mice and wrens: the relationship between abundance and geographic range size in British mammals and birds. *Phil. Trans. R. Soc. Lond. B* **352**: 419–427.

Blakers, M., Davies, S. J. J. F. & Reilly, P. N. (1984). The atlas of Australian birds. Melbourne, University Press.

Bleiweiss, R., Kirsch, J. A. W. & Lapointe, F.-J. (1994a). DNA–DNA hybridisation-based phylogeny for 'higher' non-passerines: re-evaluating a key portion of the avian family tree. *Mol. Phyl. Evol.* **3**: 248–255.

Bleiweiss, R., Kirsch, J. A. W. & Matheus, J. C. (1994b). DNA–DNA hybridisation evidence for subfamily structure among hummingbirds. *Auk* **111**: 8–19.

Bleiweiss, R., Kirsch, J. A. W. & Shafi, N. (1995). Confirmation of a portion of the Sibley–Alquist 'tapestry'. *Auk* **112**: 87–97.

Blem, C. R. (1973). Geographic variation in the bioenergetics of the House Sparrow. *Orn. Monogr.* **14**: 96–121.

Blondel, J. (1985). Historical and ecological evidence on the development of Mediterranean avifaunas. *Proc. Int. Orn. Congr.* **18**: 373–386.

Blondel, J. (1988). Biogéographic évolutive à différentes échelles: l'histoire des avifaunes méditerranéenees. *Proc. Int. Orn. Congr.* **19**: 155–188.

Blondel, J., Catzeflis, F. & Perret, P. (1996). Molecular phylogeny and the historical biogeography of the warblers of the genus *Sylvia* (Aves). *J. Evol. Biol.* **9**: 871–891.

Blondel, J., Chessel, D. & Frochot, B. (1988). Bird species impoverishment, niche expansion, and density inflation in Mediterranean Island habitats. *Ecology* **69**: 1899–1917.

Bochenski, Z. (1985). The development of western Palaearctic avifaunas from fossil evidence. *Proc. Int. Orn. Congr.* **18:** 338–347.

Bock, C. E. (1984). Geographical correlates of abundance vs. rarity in some North American winter landbirds. *Auk* **101:** 266–273.

Bock, C. E. (1987). Distribution–abundance relationships of some Arizona landbirds: a matter of scale? *Ecology* **68:** 124–129.

Bock, C. E. & Lepthien, L. W. (1976). Synchronous eruptions of boreal seed-eating birds. *Amer. Nat.* **110:** 559–571.

Bock, C. E. & Ricklefs, R. E. (1983). Range size and local abundance of some North American songbirds: a positive correlation. *Amer. Nat.* **122:** 295–299.

Boddy, M. & Sellers, R. M. (1983). Orientated movements by Greenfinches in southern Britain. *Ring. Migr.* **4:** 129–138.

Böhning-Gaese, K., González-Guzmán, L. I. & Brown, J. H. (1998). Constraints on dispersal and the evolution of the avifauna of the northern hemisphere. *Evol. Ecol.* **12:** 767–783.

Boles, W. E. (1995). The world's oldest songbird. *Nature* **374:** 21–22.

Bonham, P. F. & Robertson, J. C. M. (1975). The spread of Cetti's Warbler in northwest Europe. *Brit. Birds* **68:** 393–408.

Bonnefille, R., Roeland, J. C. & Guiot, J. (1990). Temperature and rainfall estimates for the past 40,000 years in equatorial Africa. *Nature* **345:** 347–349.

Borralho, R., Rito, A., Rego, F., Simões, H. & Pinto, P. V. (1998). Summer distribution of Red-legged Partridges *Alectoris rufa* in relation to water availability on Mediterranean farmland. *Ibis* **140:** 620–625.

Bourne, W. R. P. (1955). The birds of the Cape Verde Islands. *Ibis* **97:** 508–556.

Bourne, W. R. P. (1963). A review of oceanic studies of the biology of seabirds. *Proc. Int. Orn. Congr.* **13:** 831–854.

Bourne, W. R. P. (1980). The habitats, distribution, and numbers of northern seabirds. *Trans. Linn. Soc. New York* **9:** 1–14.

Bourne, W. R. P. (1982). Concentrations of Scottish seabirds vulnerable to oil pollution. *Marine Pollut. Bull.* **13:** 270–273.

Bourne, W. R. P. (1986). Recent work on the origin and suppression of bird species in the Cape Verde Islands, especially the shearwaters, the herons, the kites and the sparrows. *Bull. Brit. Orn. Club* **104:** 163–170.

Bourne, W. R. P. (1999). The past status of the herons in Britain. *Bull. Brit. Orn. Club* **119:** 192–196.

Bourne, W. R. P. (2002). The classification of albatrosses. *Australasian Seabird Bulletin* **38:** 11–12.

Bourne, W. R. P. & Casement, M. B. (1996). RNBWS checklist of seabirds. *Sea Swallow* **45:** 2–12.

Bourne, W. R. P. & Simmons, K. E. L. (2001). The distribution and breeding success of seabirds on and around Ascension in the tropical Atlantic Ocean. *Atlantic Seabirds* **3:** 187–200.

Bourne, W. R. P. & Warham, J. (1966). Geographical variation in the Giant Petrels of the genus *Macronectes*. *Ardea* **54:** 45–67.

Bouzat, J. L., Lewin, H. A. & Paige, K. N. (1998). The ghost of genetic diversity past: historical DNA analysis of the Greater Prairie Chicken. *Amer. Nat.* **152:** 1–6.

Bradley, J. S., Wooller, R. D., Skira, R. J. & Serventy, D. L. (1990). The influence of mate retention and divorce upon reproductive success in Short-tailed Shearwater *Puffinus tenuirostris*. *J. Anim. Ecol.* **59:** 487–496.

Brennan, C. A., Block, W. M. & Gutiérrez, R. J. (1987). Habitat use by Mountain Quail in northern California. *Condor* **89:** 66–74.

Brereton J. Le G. & Kikkawa, J. (1963). Diversity of avian species. *Aust. J. Sci.* **26:** 12–14.

Bried, J. & Jouventin, P. (1998). Why do Lesser Sheathbills *Chionis minor* switch territory? *J. Avian Biol.* **29:** 257–265.

Briggs, J. C. (1974). Marine zoogeography. New York & London, McGraw-Hill.

Brittingham, M. C. & Temple, S. A. (1983). Have cowbirds caused forest song birds to decline? *Bioscience* **33:** 31–35.

Brodkorb, P. (1971). Origin and evolution of birds. Pp. 19–55 in *'Avian biology'* (ed. D. S. Farner & J. R. King). Vol. I. London, Academic Press.

Brooke M. de L. (1978). The dispersal of female Manx Shearwaters *Puffinus puffinus*. *Ibis* **120:** 545–551.

Brooke M. de L. (2001). Seabird systematics and distribution: a review of current knowledge. Pp. 57–85 in *'Biology of marine birds'* (ed. E. A. Shreiber & J. Burger). London, CRC Press.

Brooke M. de L. & Rowe, G. (1996). Behavioural and molecular evidence for specific status of light and dark morphs of the Herald Petrel *Pterodroma heraldica*. *Ibis* **138:** 420–432.

Brooks, T. M., Pimm, S. L. & Collar, N. J. (1997). Deforestation predicts the number of threatened birds in insular southeast Asia. *Conserv. Biol.* **11:** 382–394.

Brown, J. H. (1978). The theory of insular biogeography and the distribution of boreal birds and mammals. *Great Basin Natur. Mem.* **2:** 209–227.

Brown, J. H. (1984). On the relationship between abundance and distribution of species. *Amer. Nat.* **124:** 253–279.

Brown, J. H. (1995). Macroecology. Chicago, University Press.

Brown, J. H. & Gibson, A. C. (1983). Biogeography. St Louis, Missouri, Mosby.

Brown, J. H. & Lomolino, M. V. (1998). Biogeography. Sunderland, Mass, Sinauer Associates.

Brown, J. H. & Maurer, B. A. (1987). Evolution of species assemblages: effects of energetic constraints and species dynamics on the diversification of the North American avifauna. *Amer. Nat.* **130:** 1–17.

Brown, R. G. B. (1979). Seabirds of the Senegal upwelling and adjacent waters. *Ibis* **121:** 283–292.

Brown, R. G. B. (1980). The pelagic ecology of seabirds. *Trans. Linn. Soc. NY.* **9:** 15–21.

Brown, R. G. B. (1986). Revised atlas of Eastern Canadian seabirds. 1. Shipboard surveys. Dartmouth, Canadian Wildlife Service.

Brown, W. L. & Wilson, E. O. (1956). Character displacement. *Syst. Zool.* **5:** 49–54.

Bucher, E. H. (1982). Colonial breeding of the Eared Dove (*Zenaida auriculata*) in northeastern Brazil. *Biotropica* **14:** 255–261.

Burbidge, A. A. & Fuller, P. J. (1982). Banded Stilt breeding at Lake Barlee, Western Australia. *Emu* **82:** 212–216.

Burgess, J. P. C. (1982). Sexual differences and dispersal in the Blue Tit *Parus caeruleus*. *Ring. Migr.* **4:** 25–32.

Burkey, T. V. (1995). Extinction rates in archipelagos: implications for populations in fragmented habitats. *Conserv. Biol.* **9:** 527–541.

Burton, J. F. (1995). Birds and climate change. London, Christopher Helm.

Burton, N. H. K. (2000). Winter site-fidelity and survival of Redshank *Tringa totanus* at Cardiff, south Wales. *Bird Study* **47:** 102–112.

Burton, N. H. K. & Evans, P. R. (1997). Survival and winter site fidelity of Turnstones *Arenaria interpres* and Purple Sandpipers *Calidris maritima* in northeast England. *Bird Study* **44:** 35–44.

Bush, M. B. (1994). Amazonian speciation: a necessarily complex model. *J. Biogeog.* **21:** 5–17.

Bush, M. B. & Whittaker, R. J. (1991). Krakatau: colonisation patterns and hierarchies. *J. Biogeog.* **18:** 341–356.

Bush, M. B. & Whittaker, R. J. (1993). Non-equilibrium in island theory of Krakatau. *J. Biogeog.* **20:** 453–457.

Byrd, G. V., Trapp, J. L. & Zeillemaker, C. F. (1994). Removal of introduced foxes: a case study in restoration of native birds. *Trans. N.A. Wildl. & Nat. Res. Conf.* **59**: 317–321.

Cadbury, C. J. (1980). The status and habitats of the Corncrake in Britain 1978–79. *Bird Study* **27**: 203–218.

Cade, T. J. & Temple, S. A. (1995). Management of threatened bird species: evaluation of the hands-on approach. *Ibis* **137**: 161–172.

Cain, A. J. (1955). A revision of *Trichoglossus haematodus* and of the Australian platycercine parrots. *Ibis* **97**: 432–479.

Campbell, B. (1959). Attachment of Pied Flycatchers *Muscicapa hypoleuca* to nest sites. *Ibis* **101**: 445–448.

Campbell, K. E. (1976). The Late Pleistocene avifauna of La Carolina southwestern Ecuador. Collected papers in avian paleontology honoring the 90th birthday of Alexander Whetmore, ed. S. L. Olson. *Smithson. Contrib. Paleobiol.* **27**: 155–168.

Campbell, K. E. (1979). The non-passerine Pleistocene avifauna of the Talara tar seeps, northwestern Peru. *Life Sci. Contrib. R. Ont. Mus.* **118**: 1–203.

Campbell, K. E. & Tonni, E. P. (1980). A new genus of teratorn from the Huayguerian of Argentina (Aves: Terathornidae). *Nat. Hist. Mus. LA Co., Contrib. Sci.* **330**: 59–68.

Campbell, L., Cayford, J. & Pearson, D. (1996). Bearded Tits in Britain and Ireland. *Brit. Birds* **89**: 335–346.

Caparella, A. P. (1988). Genetic variation in neotropical birds: implications for the speciation process. *Proc. Int. Orn. Congr.* **19**: 1658–1664.

Cardillo, M. (1999). Latitude and rates of diversification in birds and butterflies. *Proc. R. Soc. Lond. B* **266**: 1221–1225.

Carmi-Winkler, N., Degen, A. A. & Pinshow, B. (1987). Seasonal time–energy budgets of free-living Chukars in the Negev Desert. *Condor* **89**: 594–601.

Case, T. J. (1996). Global patterns in the establishment and distribution of exotic birds. *Biol. Conserv.* **78**: 69–96.

Catchpole, C. K. & Slater, P. J. B. (1995). Bird song: biological themes and variations. Cambridge, University Press.

Cavé, A. J. (1968). The breeding of the Kestrel, *Falco tinnunculus* L., in the reclaimed area Oostelijk Flevoland. *Neth. J. Zool.* **18**: 313–407.

Channell, R. & Lomolino, M. V. (2000). Trajectories to extinction: spatial dynamics of the contraction of geographical ranges. *J. Biogeog.* **27**: 169–179.

Chapman, F. M. & Griscom, M. L. (1924). The House Wrens of the genus *Troglodytes*. *Bull. Amer. Mus. Nat. Hist.* **50**: 279–304.

Cheke, A. S. (1987). An ecological history of the Mascarene Islands, with particular reference to extinctions and introductions of land vertebrates. Pp. 5–89 in '*Studies of Mascarene Island Birds*' (ed. A. W. Diamond). Cambridge, University Press.

Cheke, A. S. & Hume, J. P. (In prep.). Lost lands of the Dodo. London, Academic Press.

Chesser, R. T. (1998). Further perspectives on the breeding distribution of migratory birds: South American austral migrant flycatchers. *J. Anim. Ecol.* **67**: 69–77.

Christidis, L. (1991). Biochemical evidence for the origins and evolutionary radiations in the Australian avifauna: the songbirds. *Proc. Int. Orn. Congr.* **20**: 392–397.

Christidis, L. & Schodde, R. (1991). Relationship of the Australo-Papuan songbirds — protein evidence. *Ibis* **133**: 277–285.

Cibois, A., Slikas, B., Schulenberg, S. & Pasquet, E. (2001). An endemic radiation of Malagasy songbirds is revealed by mitochondrial DNA sequence data. *Evolution* **55**: 1198–1206.

Cicero, C. & Johnson, N. K. (1995). Speciation in sapsuckers (*Sphyrapicus*): III Mitochondrial DNA sequence divergence at the cytochrome-*b* locus. *Auk* **112**: 547–563.

Clapham, C. (1979). The Turnstone populations of Morecambe Bay. *Ring. Migr.* **2**: 144–150.

Clarke, A. L., Saether, B.-E. & Røskaft, E. (1997). Sex biases in avian dispersal: a reappraisal. *Oikos* **79**: 429–438.

Clegg, S. M., Degnan, S. M., Kikkawa, J., Moritz, C. & Owens, I. P. F. (2002). Morphological divergence of island silvereyes (Zosteropidae). Is random drift a plausible mechanism? Submitted.

COHMAP Members (1988). Climatic changes of the last 18,000 years: observations and model simulations. *Science* **241**: 1043–1052.

Colinvaux, P. A. (1998). A new vicariance model for Amazonian endemics. *Global Ecol. & Biogeog. Letters* **7**: 95–96.

Colinvaux, P. A., De Oliveira, P. E., Moreno, J. E., Miller, M. C. & Bush, M. B. (1996). A long pollen record from lowland Amazonia: forest and cooling in glacial times. *Science* **247**: 85–88.

Collar, N. J. & Andrew, P. (1988). Birds to watch. Cambridge, ICBP.

Colwell, M. A., Reynolds, J. D., Gratto, C. L., Schamel, D. & Tracy, D. M. (1988). Phalarope philopatry. *Proc. Int. Orn. Congr.* **19**: 585–593.

Congdon, B. C., Piatt, J. F., Martin, K. & Friesen, V. L. (2000). Mechanisms of population differentiation in Marbled Murrelets: historical versus contemporary processes. *Evolution* **54**: 974–986.

Cooke, A. (1975). The effects of fishing on waterfowl on Grafham Water. *Cambridge Bird Club* **48**: 40–46.

Cooke, F., MacInnes, C. D. & Prevett, J. P. (1975). Gene flow between breeding populations of the Lesser Snow Goose. *Auk* **92**: 493–510.

Cooke, F., Rockwell, R. F. & Lank, D. B. (1995). The Snow Geese of La Pérouse Bay. Oxford, University Press.

Cooper, A., Atkinson, I. A. E., Lee, W. G. & Worthy, T. H. (1993). Evolution of the Moa and their effect on the New Zealand flora. *Trends Ecol. Evol.* **8**: 433–437.

Cooper, A. & Cooper, R. A. (1995). The oligocene bottleneck and New Zealand biota: genetic record of a past environmental crisis. *Proc. R. Soc. Lond. B* **261**: 293–302.

Cooper, A., Mourer-Chauviré, C., Chambers, G. K., Vonhaeseler, A., Wilson, A. C. & Paabo, S. (1992). Independent origins of New Zealand moas and kiwis. *Proc. Nat. Acad. Sci. USA* **89**: 8741–8744.

Cooper, A. & Penny, D. (1997). Mass survival of birds across the Cretaceous–Tertiary boundary: molecular evidence. *Science* **275**: 1109–1113.

Cooper, A., Rhymer, J., James, H. F., Olson, S. L., McIntosh, C. E., Sorenson, M. D. & Fleischer, R. C. (1996). Ancient DNA and island endemics. *Nature* **381**: 484.

Cooper, J. E. (1989). The role of pathogens in threatened populations: an historical review. Pp. 51–61 in '*Disease and Threatened Birds*' (ed. J. E. Cooper). Cambridge, ICBP.

Cooper, J. H. & Voous, K. H. (1999). Iberian Azure-winged Magpies come in from the cold. *Brit. Birds* **92**: 659–665.

Cooper, R. A. & Millener, P. (1993). The New Zealand biota: historical background and new research. *Trends Ecol. Evol.* **8**: 429–433.

Cornutt, J. & Pimm, S. (2001). How many bird species in Hawaii and the central Pacific before first contact? *Stud. Avian Biol.* **22**: 15–30.

Coulson, J. C. (1961). Movements and seasonal variation in mortality of Shags and Cormorants ringed on the Farne Islands, Northumberland. *Brit. Birds* **54**: 225–235.

Coulson, J. C. & Butterfield, J. (1985). Movements of British Herring Gulls. *Bird Study* **32**: 91–103.

Coulson, J. C. & Mévergnies, G. N. (1992). Where do young Kittiwakes *Rissa tridactyla* breed, philopatry or dispersal? *Ardea* **80**: 187–197.

Court, G. S., Bradley, D. M., Gates, C. C. & Boag, D. A. (1989). Turnover and recruitment in a tundra population of Peregrine Falcons *Falco peregrinus*. *Ibis* **131**: 487–496.

Covas, R. & Blondel, J. (1998). Biogeography and history of the Mediterranean bird fauna. *Ibis* **140:** 395–407.

Coxon, P. & Waldren, S. (1997). Flora and vegetation of the Quaternary temperate stages of NW Europe: evidence for large-scale range changes. Pp. 227–238 in *'Past and future rapid environmental changes: the spatial and evolutionary responses of terrestrial biota'* (ed. B. Huntley, W. Cramer, A. V. Morgan, H. C. Prentice & J. R. M. Allen). London, Springer.

Cracraft, J. (1972). Continental drift and Australian avian biogeography. *Emu* **72:** 171–174.

Cracraft, J. (1973). Continental drift, paleoclimatology, and the evolution and biogeography of birds. *J. Zool. Lond.* **169:** 455–545.

Cracraft, J. (1976). The species of Moas (Aves, Dinornithidae). *Smithson. Contrib. Paleobiol.* **27:** 189–205.

Cracraft, J. (1981). Towards a phylogenetic classification of the recent birds of the world (class Aves). *Auk* **98:** 681–714.

Cracraft, J. (1983). Species concepts and speciation analysis. *Curr. Orn.* **1:** 159–187.

Cracraft, J. (1985). Historical biogeography and patterns of differentiation within the South American avifauna: areas of endemism. *Ornithol. Monogr.* **36:** 49–84.

Cracraft, J. (1986). Origin and evolution of continental biotas: speciation and historical congruence within the Australian avifauna. *Evolution* **40:** 977–996.

Cracraft, J. (2001). Avian evolution, Gondwana biogeography and the Cretaceous–Tertiary mass extinction event. *Proc. R. Soc. Lond. B* **268:** 459–469.

Cracraft, J. & Feinstein, J. (2000). What is not a bird of paradise? Molecular and morphological evidence places *Macgregoria* in the Meliphagidae and the Cnemophilinae near the base of the corvid tree. *Proc. R. Soc. Lond. B* **267:** 233–241.

Craig, J. L. & Jamieson, I. G. (1988). Incestuous mating in a communal bird: a family affair. *Amer. Nat.* **131:** 58–70.

Cramp, S. (1985–92). Handbook of the Birds of Europe, the Middle East and North Africa, Vols 4–6. Oxford, University Press.

Cramp, S., Bourne, W. R. P. & Saunders, D. (1974). The seabirds of Britain and Ireland. London, Collins.

Cramp, S. & Perrins, C. M. (1993–94). Handbook of the Birds of Europe, the Middle East and North Africa, Vols 7–9. Oxford, University Press.

Cramp, S. & Simmons, K. E. L. (1977–83). Handbook of the Birds of Europe, the Middle East and North Africa, Vols. 1–3. Oxford, University Press.

Crawford, R. J. M., Allwright, D. M. & Heyl, C. W. (1992). High mortality of Cape Cormorants (*Phalacrocorax capensis*) off western South Africa in 1991 caused by *Pasteurella multocida*. *Colon. Waterbirds* **15:** 236–238.

Crawford, R. J. M., Boonstra, H. G. V. D., Dyer, B. M. & Upfold, L. (1995). Recolonisation of Robben Island by African Penguins, 1983–1992. Pp. 333–363 in *'The Penguins. Ecology and Management'* (ed. P. Dann, I. Norman & P. Reilly). Chipping Norton, Surrey Beatty.

Crawford, R. J. M., Dyer, B. M. & Brooke, R. K. (1994). Breeding nomadism in southern African seabirds – constraints, causes and conservation. *Ostrich* **65:** 231–246.

Crawford, R. J. M. & Shelton, P. A. (1978). Pelagic fish and seabird inter-relationships off the coasts of South West and South Africa. *Biol. Conserv.* **14:** 85–109.

Creutz, G. (1955). Der Trauerschnapper (*Muscicapa hypoleuca* Pall.), Eine Populationsstudie. *J. Orn.* **96:** 241–326.

Crick, H. Q. P., Dudley, C., Glue, D. E. & Thomson, D. L. (1997). UK birds are laying eggs earlier. *Nature* **388:** 526.

Cronin, M. A., Grand, J. B., Esler, D., Derksen, D. V. & Scribner, K. T. (1996). Breeding populations of northern Pintails have similar mitochondrial DNA. *Can. J. Zool.* **74:** 992–999.

Crowe, T. M. & Crowe, A. A. (1982). Patterns of distribution, diversity and endemism in Afrotropical birds. *J. Zool. Lond.* **198:** 417–442.

Crowe, T. M. & Kemp, A. C. (1988). African historical biogeography as reflected by galliform and hornbill evolution. *Proc. Int. Orn. Congr.* **19:** 2510–2518.

Crowell, K. L. (1962). Reduced interspecific competition among the birds of Bermuda. *Ecology* **43:** 75–88.

Croxall, J. P. & Davies, L. S. (1999). Penguins: paradoxes and patterns. *Mar. Orn.* **27:** 1–12.

Croxall, J. P. & Prince, P. A. (1980). Food, feeding ecology and ecology and ecological segregation of seabirds at South Georgia. *Biol. J. Linn. Soc.* **14:** 103–131.

Croxall, J. P., Ricketts, C. & Prince, P. A. (1984). Impact of seabirds on marine resources, especially krill, off South Georgia waters. Pp. 285–317 in *'Seabird Energetics'* (ed. G. C. Whittow & H. Rahn). New York, Plenum Press.

Cuadrado, M. (1995). Winter territoriality in migrant Black Redstarts *Phoenicurus ochruros* in the Mediterranean area. *Bird Study* **42:** 232–239.

Cuadrado, M., Senar, J. C. & Copeti, J. C. (1995). Do all Blackcaps *Sylvia atricapilla* show winter site fidelity? *Ibis* **137:** 70–75.

Cullen, J. M., Guiton, P. E., Horridge, G. A. & Peirson, J. (1952). Birds on Palma and Gomera (Canary Islands). *Ibis* **94:** 68–84.

Curio, E. (1958). Geburtsortstreue und Lebenser- wartung juger Traverschnapper (*Muscicapa hypoleuca* Pall.). *Vogelwelt* **79:** 135–149.

Currie, D. J. & Fritz, J. T. (1993). Global patterns of animal abundance and species energy use. *Oikos* **67:** 56–68.

Cutler, A. (1991). Nested faunas and extinction in fragmented habitats. *Conserv. Biol.* **5:** 496–505.

da Silva, M. C. & Granadeiro, J. P. (1999). Genetic variability and isolation of Cory's Shearwater colonies in the Northeast Atlantic. *Condor* **101:** 174–179.

Darlington, P. J. (1957). Zoogeography: the geographical distribution of animals. New York, Wiley.

Darwin, C. (1859). The origin of species. London, John Murray.

Davies, S. J. J. F. (1977). Man's activities and bird distribution in the arid zone. *Emu* **77:** 169–172.

Davies, S. J. J. F. (1988). Nomadism in the Australian Gull-billed Tern *Sterna nilotica*. *Proc. Int. Orn. Congr.* **19:** 744–753.

Davis, D. E. (1950). The growth of Starling populations. *Auk* **67:** 460–465.

Davis, M. B. (1981). Quaternary history and the stability of forest communities. Pp. 132–153 in *'Forest succession: concepts and application'* (ed. D. C. West, H. H.Shugart & D. B. Botkin). New York, Springer-Verlag.

Davis, M. B. (1986). Climate stability, time lags and community disequilibrium. Pp. 269–284 in *'Community Ecology'* (ed. J. M. Diamond & T. Case). New York, Harper & Row.

de Bruijn O. (1994). Population ecology and conservation of the Barn Owl *Tyto alba* in farmland habitats in Liemers and Achterhoek (The Netherlands). *Ardea* **82:** 1–109.

de Lattin, G. (1956). Die Ausbreclungsenken der Holarktischen Landtierwelt Verkhandl. *Dt. Zool. Ges. Hamburg* **1956:** 380–410.

de Menocal P.B. & Bloemendal, J. (1995). Plio-Pleistocene climatic variability in subtropical Africa and the paleo-environment of hominid evolution: a combined data–model approach. Pp. 262–288 in *'Paleoclimate and evolution, with emphasis on human origins'* (ed. E. S. Vrba, G. H. Denton, T. C. Partridge & L. H. Burckle). New Haven, Yale University Press.

De Smet, K. D. (1997). Burrowing Owl (*Speotyto cunicularia*) monitoring and management activities in Manitoba 1987–1996. Pp. 123–130 in *'Biology and Conservation of Owls of the Northern Hemisphere'* (ed. J. R. Duncan, D. H. Johnson & T. H. Nicholls). Second

International Symposium, February 5–9, 1997. Winnipeg, Manitoba, Canada, St Paul Minnesota, Forest Service–United States Department of Agriculture.

Dean, W. R. J. (1997). The distribution and biology of nomadic birds in the Karoo, South Africa. *J. Biogeog.* **24:** 769–779.

Degnan, S. M. (1993). Genetic variability and population differentiation inferred from DNA fingerprinting in Silvereyes (Aves, Zosteropidae). *Evolution* **47:** 1105–1117.

Degnan, S. M., Owens, I. P. F., Clegg, S. M., Moritz, C. C. & Kikkawa, J. (1999). Mt DNA, microsatellites and coalescence: tracing the colonisation of Silvereyes through the southwest Pacific. *Proc. Int. Orn. Congr. Durban* **22:** 1881–1898.

Dekker, R. W. R. J. (1989). Predation and the western limits of megapode distribution (Megapodiidae; Aves). *J. Biogeog.* **16:** 317–321.

del Hoyo, J., Elliott, J. & Sargatel, J. (eds.) (1999). The handbook of the birds of the world, Vol. 5. Barcelona, Lynx Edicions.

Dennis, J. V. (1990). Banded North American birds encountered in Europe: an update. *North Amer. Bird Bander* **15:** 130–133.

Dhondt, A. A. (1989). Ecological and evolutionary effects of interspecific competition in tits. *Wilson Bull.* **101:** 198–216.

Dhondt, A. A. & Hublé, J. (1968). Age and territory in the Great Tit (*Parus major* L.). *Angew. Orn.* **3:** 20–24.

Diamond, A. W. (1978). Feeding strategies and population size in tropical seabirds. *Amer. Nat.* **12:** 215–223.

Diamond, A. W. (1985). The selection of critical areas and current conservation efforts in tropical forest birds. *ICBP Tech. Publ.* **4:** 33–48.

Diamond, A. W. & Hamilton, A. C. (1980). The distribution of forest passerine birds and Quaternary climatic change in tropical Africa. *J. Zool. Lond.* **191:** 379–402.

Diamond, J. M. (1972a). Avifauna of the Eastern Highlands of New Guinea. Cambridge, Mass., Nuttall Ornithological Club.

Diamond, J. M. (1972b). Biogeographic kinetics: estimation of relaxation time for avifaunas of southwest Pacific Islands. *Proc. Nat. Acad. Sci.USA* **69:** 3199–3203.

Diamond, J. M. (1975). Assembly of species communities. Pp. 342–444 in '*Ecology and evolution of communities*' (ed. M. L. Cody & J. M. Diamond). Cambridge, Mass., Belknap Press.

Diamond, J. M. (1977a). Colonisation cycles in man and beast. *World Archaeol.* **8:** 249–261.

Diamond, J. M. (1977b). Continental and insular speciation in Pacific land birds. *Syst. Zool.* **26:** 263–268.

Diamond, J. M. (1980a). Species turnover in island bird communities. *Proc. Int. Orn. Congr.* **17:** 777–782.

Diamond, J. M. (1980b). Why are many tropical bird species distributed patchily with respect to available habitat? *Proc. Int. Orn. Congr.* **17:** 968–973.

Diamond, J. M. (1981). Flightlessness and fear of flying in island species. *Nature* **293:** 507–508.

Diamond, J. M. (1984). 'Normal' extinctions of isolated populations. Pp. 191–246 in '*Extinctions*' (ed. M. H. Nitecki). Chicago, University Press.

Diamond, J. M. (1991). The rise and fall of the third chimpanzee. London, Vintage.

Diamond, J. M. & Case, T. J. (1986). Overview: introduction, extinctions, exterminations and invasions. Pp. 65–79 in '*Community ecology*' (ed. J. M. Diamond & T. J. Case). New York, Harper Row.

Diamond, J. M. & Marshall, A. G. (1976). Origin of the New Hebridean avifauna. *Emu* **76:** 187–200.

Diamond, J. M. & May, R. M. (1976). Island biogeography and the design of natural reserves. Pp. 228–252 in '*Theoretical ecology*' (ed. R. M. May). Oxford, Blackwell Scientific Publications.

Diamond, J. M. & Pimm, S. (1993). Survival times of bird populations: a reply. *Amer. Nat.* **142:** 1030–1035.

Diefenbach, D. R., Derleth, E. L., Haegen, W. M. V., Nichols, J. D. & Hines, J. E. (1990). American Woodcock winter distribution and fidelity to wintering areas. *Auk* **107:** 745–749.

Dierschke, V. (1998). Site fidelity and survival of Purple Sandpipers *Calidris maritima* at Helgoland (SE North Sea). *Ring. Migr.* **19:** 41–48.

Dobson, A. P. & May, R. M. (1986). Patterns of invasions by pathogens and parasites. Pp. 58–76 in '*Ecology of biological invasions of North America and Hawaii*' (ed. H. A. Mooney). New York, Springer Verlag.

Dolbeer, R. A. (1982). Migration patterns for age and sex classes of Blackbirds and Starlings. *J. Field Orn.* **53:** 28–46.

Dolbeer, R. A. (1991). Migration patterns of Double-crested Cormorants east of the Rocky Mountains. *J. Field Orn.* **62:** 83–93.

Dowsett, R. J. (1980). Post-pleistocene changes in the distributions of African montane forest birds. *Proc. Int. Orn. Congr.* **17:** 787–792.

Drilling, N. E. & Thompson, L. F. (1988). Natal and breeding dispersal in House Wrens (*Troglodytes aedon*). *Auk* **105:** 480–491.

Duffy, D. C. (1983). Competition for nesting space among Peruvian guano birds. *Auk* **100:** 680–688.

Duncan, J. R. (1997). Great Grey Owls (*Strix nebulosa nebulosa*) and forest management in North America: a review and recommendations. *J. Raptor Res.* **31:** 160–166.

Duncan, R. P., Blackburn, T. M. & Worthy, T. W. (2002). Prehistoric bird extinctions and human hunting. *Proc. R. Roc. Lond. B* **269:** 517–521.

Duncan, W. N. M. & Monaghan, P. (1977). Infidelity to the natal colony by breeding Herring Gulls. *Ring. Migr.* **1:** 166–172.

Dunn, P. O. & Winkler, D. W. (1999). Climate change has affected the breeding date of Tree Swallows throughout North America. *Proc. R. Soc. Lond. B* **266:** 2487–2490.

Eakin, R. R., Dearborn, J. H. & Townsend, W. C. (1986). Observations of marine birds in the South Atlantic Ocean in the late austral autumn. *Antarct. Res. Ser.* **44:** 69–86.

Eason, C. T., Murphy, E. C., Wright, G. R. G. & Spurr, E. B. (2002). Assessment of risks of brodifacoum to non-target birds and mammals in New Zealand. *Ecotoxicology* **11:** 35–48.

Ebbinge, B. S. (1992). Regulation of numbers of Dark-bellied Brent Geese *Branta bernicla* on spring staging sites. *Ardea* **80:** 203–228.

Ebbinge, B. S. & St Joseph, A. K. M. (1992). The Brent Goose colour ringing scheme: unravelling annual migratory movements from high arctic Siberia to the coasts of western Europe. Pp. 93–104 in '*Population limitation in arctic breeding geese*' (ed. B. S. Ebbinge). University of Groningen, unpublished thesis.

Ebbinge, B. S., van Biezen, J. B. & van der Voet, H. (1991). Estimation of annual adult survival rates of Barnacle Geese *Branta leucopsis* using multiple re-sightings. *Ardea* **79:** 73–112.

Ekman, J., Sklepkovych, B. & Tegelström, H. (1994). Offspring retention in the Siberian Jay (*Perisoreus infaustus*): the prolonged brood care hypotheses. *Behav. Ecol.* **5:** 245–253.

Elenga, H. & 21 others. (2000). Pollen-based biome reconstruction for southern Europe and Africa 18,000 yr BP. *J. Biogeog.* **27:** 621–634.

Elliott, C. C. H., Waltner, M., Underhill, L. G., Pringle, G. S. & Dick, W. J. A. (1977). The migration system of the Curlew Sandpiper *Calidris ferruginea* in Africa. *Ostrich* **47:** 191–213.

Emlen, S. T., DeJong, M. J., Jaeger, M. J., Moermond, T. C., Rusterholz, K. A. & White, R. P. (1986). Density trends and range boundary constraints of forest birds along a latitudinal gradient. *Auk* **103:** 791–803.

Emlen, S. T., Rising, J. D. & Thompson, W. L. (1975). A behavioural and morphological study of sympatry in the Indigo and Lazuli Buntings of the Great Plains. *Wilson Bull.* **87:** 145–179.

Empson, R. A. & Miskelly, C. M. (1999). The risks, costs and benefits of using brodifacoum to eradicate rats from Kapiti Island, New Zealand. *N. Z. J. Ecol.* **23:** 241–254.

Emslie, S. D. (1998). Avian community, climate and sea-level changes in the Plio-Pleistocene of the Florida peninsula. *Orn. Monogr.* **50:** 1–113.

Enderson, J. H. & Craig, G. R. (1988). Population turnover in Colorado Peregrines. Pp. 685–688 in *'Peregrine Falcon populations. Their management and recovery'* (ed. T. J. Cade, J. H. Enderson, C. G. Thelander & C. M. White). Boise, Idaho, The Peregrine Fund Inc.

Endler, J. A. (1977). Geographic variation, speciation and clines. Princeton, Princeton University Press.

Endler, J. A. (1982). Problems in distinguishing historical from ecological factors in biogeography. *Am. Zool.* **22:** 441–452.

Ericson, P. G. P., Christidis, L., Cooper, A., Irestedt, M., Jackson, J., Johansson, U. S. & Norman, J. A. (2002). A Gondwanan origin of passerine birds supported by DNA sequences of the endemic New Zealand wrens. *Proc. Roy. Soc. Lond. B* **269:** 235–241.

Evans, P. R. (1981). Migration and dispersal of shorebirds as a survival strategy. Pp. 275–290 in *'Feeding and survival strategies of estuarine organisms'* (ed. N. V. Jones & W. J. Wolff). New York, Plenum Press.

Evans, P. R. & Pienkowski, M. W. (1984). Population dynamics of shorebirds. Pp. 83–123 in *'Behaviour of marine animals, Vol. 5, Shorebirds, breeding biology and populations'* (ed. J. Burger & B. L. Olla). New York, Plenum Press.

Falla, R. A., Sibson, R. B. & Turbott, E. G. (1989). Collins guide to the birds of New Zealand and Outlying Islands. Auckland, Collins.

Feare, C. J. (1976). The breeding of the Sooty Tern *Sterna fuscata* in the Seychelles and the effects of experimental removal of its eggs. *J. Zool. Lond.* **179:** 317–360.

Feare, C. J. (1982). The Starling. Oxford, University Press.

Feduccia, A. (1995). Explosive evolution in Tertiary birds and mammals. *Science* **267:** 637–638.

Feduccia, A. (1999). The origin and evolution of birds. 2nd ed. New Haven, Yale University Press.

Feduccia, A. & Olson, S. L. (1982). Morphological similarities between the Menurae and the Rhinocryptidae, relict passerine birds of the southern hemisphere. *Smithson. Contrib. Zool.* **366:** 1–22.

Fefer, S. T., Harrison, C. S., Naughton, M. B. & Shallenberger, R. J. (1984). Synopsis of results of recent seabird research conducted in the Northwestern Hawaiian Islands. Vol 1. Pp. 9–76 in *'Proceedings of 2nd Symposium on Resource Investigations in the Northwestern Hawaiian Islands'* (ed. R. W. Grigg & K. Y. Tanoue). Honolulu, University of Hawaii, Sea Grant College.

Ferrer, M. (2001). The Spanish Imperial Eagle. Barcelona, Lynx Edicions.

Finlayson, H. H. (1932). Heat in the interior of South Australia and in central Australia. *S. Aust. Orn.* **11:** 158–163.

Firth, R. & Davidson, J. W. (1945). Pacific Islands, Vol. 1. General Survey. London, Naval Intelligence Division.

Fischer, H. I. & Baldwin, P. H. (1946). War and the birds on Midway Atoll. *Condor* **48:** 3–15.

Fisher, C. D., Lindgren, E. & Dawson, W. R. (1972). Drinking patterns and behavior of Australian desert birds in relation to their ecology and abundance. *Condor* **74:** 111–136.

Fisher, H. I. (1971). Experiments on homing in Laysan Albatrosses (*Diomedea immutabilis*). *Condor* **73:** 389–400.

Fisher, J. (1966). The Fulmar population of Britain and Ireland, 1959. *Bird Study* **13:** 5–76.

Fisher, R. A. (1937). The wave of advance of advantageous genes. *Ann. Eugen.* **7:** 355–369.

Fitzpatrick, J. W. (1980). Some aspects of speciation in South American flycatchers. *Proc. Int. Orn. Congr.* **17**: 1273–1279.

Fjeldså, J. (1994). Geographical patterns of relict and young species of birds in Africa and South America and implications for conservation priorities. *Biodiv. Conserv.* **3**: 107–226.

Fjeldså, J., Ehrlich, D., Lambin, E. & Prins, E. (1997). Are biodiversity 'hotspots' correlated with current ecoclimatic stability? A pilot study using the NOAA-AVHRR remote sensing data. *Biodiv. Conserv.* **6**: 401–422.

Fjeldså, J. & Krabbe, N. (1990). Birds of the High Andes. Copenhagen, Zoological Museum, University.

Fjeldså, J. & Lovett, J. C. (1997). Geographical patterns of old and young species in African forest biota: the significance of specific montane areas as evolutionary centres. *Biodiv. Conserv.* **6**: 325–346.

Fleischer, R. C. & McIntosh, C. E. (2001). Molecular systematics and biogeography of the Hawaiian avifauna. *Stud. Avian Biol.* **22**: 51–60.

Fleischer, R. C., McIntosh, C. E. & Tarr, C. E. (1998). Evolution on a volcanic conveyor belt using phylogeographic reconstructions and K–Ar-based ages of the Hawaiian Islands to estimate molecular evolutionary rates. *Mol. Ecol.* **7**: 533–545.

Fleischer, R. C. & Rothstein, S. I. (1988). Known secondary contact and rapid gene flow among subspecies and dialects in the Brown-headed Cowbird. *Evolution* **42**: 1146–1158.

Fleming, C. A. (1941). The phylogeny of the prions. *Emu* **41**: 134–155.

Flessa, K. W. (1975). Area, continental drift and mammalian diversity. *Paleobiology* **1**: 189–194.

Flynn, L., Nol, E. & Zharikov, Y. (1999). Philopatry, nest-site tenacity, and mate fidelity of Semipalmated Plovers. *J. Avian Biol.* **30**: 47–55.

Fok, K. W., Wade, C. M. & Parkin, D. (2002). Inferring the phylogeny of disjunct populations of the Azure-winged Magpie *Cyanopica cyanus* from mitochondrial control region sequences. *Proc. R. Soc. Lond. B*

Ford, H. A. (1989). Ecology of birds. An Australian perspective. Chipping Norton, New South Wales, Surrey Beatty.

Ford, H. A. (1990). Relationships between distribution, abundance and foraging specialization in Australian landbirds. *Ornis Scand.* **21**: 133–138.

Ford, J. (1974a). Concepts of subspecies and hybrid zones, and their application in Australian ornithology. *Emu* **74**: 113–123.

Ford, J. (1974b). Speciation in birds adapted to arid habits. *Emu* **74**: 161–168.

Ford, J. (1988). Avian speciation patterns in Australia as indicated by hybrid zones and minor isolates. *Proc. Int. Orn. Congr.* **19**: 2554–2561.

Forero, M. G., Donázar, J. A., Blas, J. & Hiraldo, F. (1999). Causes and consequences of territory change and breeding dispersal distance in the Black Kite. *Ecology* **80**: 1298–1310.

Forsell, D. J. (1982). Recolonisation of Baker Island by seabirds. *Bull. Pac. Seabird Group* **9**: 75–76.

Forslund, P. & Larsson, K. (1991). Breeding range expansion of the Barnacle Goose in the Baltic area. *Ardea* **79**: 343–346.

Fowler, J. A., O'Kill, J. D. & Marshall, B. (1982). A retrap analysis of Storm Petrels tape-lured in Shetland. *Ring. Migr.* **4**: 1–7.

Fox, A. D. & Madsen, J. (1997). Behavioural and distributional effects of hunting disturbance on waterbirds in Europe: implications for refuge design. *J. Appl. Ecol.* **34**: 1–13.

Franz, D. (1988). Migration of the Penduline Tit (*Remiz pendulinus*) during the breeding period — range, frequency and ecological importance. *Vogelwelt* **109**: 188–206.

Freed, L. A., Conant, S. & Fleischer, R. C. (1987). Evolutionary ecology and radiation of Hawaiian passerine birds. *Trends Ecol. Evol.* **2**: 196–203.

Freeland, J. R. & Boag, P. T. (1999). Phylogenetics of Darwin's finches: paraphyly in the tree-finches, and two divergent lineages in the Warbler Finch. *Auk* **116:** 577–588.

Freeman, S. & Zink, R. M. 1995. A phylogenetic study of the blackbirds based on variation in mitochondrial DNA restriction sites. *Syst. Biol.* **44:** 409–420.

Frenzel, B. (1968). The Pleistocene vegetation of northern Eurasia. *Science* **161:** 637–649.

Frenzel, B., Pecsi, M. & Velichko, A. A. (1992). Atlas of palaeoclimates and palaeoenvironments of the northern hemisphere. Budapest, Research Institute, Hungarian Academy of Sciences.

Fretwell, S. (1980). Evolution of migration in relation to factors regulating bird numbers. Pp. 517–527 in '*Migrant birds in the Neotropics*' (ed. A. Keast & E. Morton). Washington DC, Smithsonian Institution Press.

Friesen, V. L., Baker, A. J. & Piatt, J. F. (1996a). Phylogenetic relationships within the Alcidae (Charadriiformes: Aves) inferred from total molecular evidence. *Mol. Biol. Evol.* **13:** 359–367.

Friesen, V. L., Montevecchi, W. A., Baker, A. J., Barrett, R. T. & Davidson, W. S. (1996b). Population differentiation and evolution in the Common Guillemot *Uria aalge*. *Molec. Ecol.* **5:** 793–805.

Friesen, V. L., Piatt, J. F. & Baker, A. J. (1996c). Evidence from cytochrome *b* sequences and allozymes for a 'new' species of alcid: the Long-billed Murrelet (*Brachyramphus perdix*). *Condor* **98:** 681–690.

Frith, H. J. (1961). Ecology of wild ducks in inland Australia. *The Wildfowl Trust 12th Ann. Rep.* **1959–60:** 81–91.

Frith, H. J. (1967). Waterfowl in Australia. Sydney, Angus & Robertson.

Frith, H. J. (1982). Pigeons and doves of Australia. Adelaide, Rigby.

Frumkin, R., Pinshow, B. & Weinstein, Y. (1986). Metabolic heat production and evaporative heat loss in desert phasianids: Chukar & Sand Partridge. *Physiol. Zool.* **59:** 592–605.

Fryer, G. (2001). On the age and origin of the species flock of haplochromine cichlid fishes of Lake Victoria. *Proc. R. Soc. Lond. B.* **268:** 1147–1152.

Fuller, R. J. (1982). Bird habitats in Britain. Calton, Poyser.

Furness, R. W. (1987). The Skuas. Calton, Poyser.

Furness, R. W. & Birkhead, T. R. (1984). Seabird colony distributions suggest competition for food supplies during the breeding season. *Nature* **311:** 655–656.

Furness, R. W., Ensor, K. & Hudson, A. V. (1992). The use of fishing waste by gull populations around the British Isles. *Ardea* **80:** 105–113.

Garcia-Moreno, J., Arctander, P. & Fjeldså, J. (1999). A case of rapid diversification in the Neotropics: phylogenetic relationships among *Cranioleuca* spinetails (Aves, Furnariidae). *Mol. Phyl. Evol.* **12:** 273–281.

Garcia-Moreno, J. & Fjeldså, J. (1999). Re-evaluation of species limits in the genus *Atlapetes* based on mt DNA sequence data. *Ibis* **141:** 199–207.

Gaston, A. J., Deforest, C. N., Donaldson, L. & Noble, D. G. (1994). Population parameters of Thick-billed Murres at Coats Island, northwest Territories, Canada. *Condor* **96:** 935–948.

Gaston, A. J. & Jones, I. L. (1998). The auks. Oxford, University Press.

Gaston, K. J. (1994). Rarity. London, Chapman & Hall.

Gaston, K. J. (1996). The multiple forms of the interspecific abundance–distribution relationship. *Oikos* **76:** 211–220.

Gaston, K. J. & Blackburn, T. M. (1996a). Range size–body size relationships: evidence of scale dependence. *Oikos* **75:** 479–485.

Gaston, K. J. & Blackburn, T. M. (1996b). The tropics as a museum of biological diversity: an analysis of the New World avifauna. *Proc. R. Soc. Lond. B* **263:** 63–68.

Gaston, K. J. & Blackburn, T. M. (1996c). Global scale macroecology: interactions between population size, geographic range size and body size in the Anseriformes. *J. Anim. Ecol.* **65:** 701–714.

Gaston, K. J., Blackburn, T. M., Greenwood, J. J. D., Gregory, R. D., Quinn, R. M. & Lawton, J. H. (2000). Abundance–occupancy relationships. *J. Appl. Ecol.* **37, suppl. 1:** 39–59.

Gaston, K. J. & Lawton, J. H. (1990). Effects of scale and habitat on the relationship between regional distribution and local abundance. *Oikos* **58:** 329–335.

Gause, G. F. (1934). The strategy for existence. Baltimore, Williams & Wilkins.

Gauthreaux, S. A. (1982). The ecology and evolution of avian migration systems. Pp. 93–167 in '*Avian Biology Vol. 6*' (ed. D. S. Farner & J. R. King). New York, Academic Press.

Gavin, T. A. & Bollinger, E. K. (1988). Reproductive correlates of breeding site fidelity in Bobolinks (*Dolichonyx oryzivorus*). *Ecology* **69:** 96–103.

Gelter, H. P., Tegelström, H. & Gustafsson, L. (1992). Evidence from hatching success and DNA fingerprinting for the fertility of hybrid Pied × Collared Flycatchers *Ficedula hypoleuca × albicollis*. *Ibis* **134:** 62–68.

Gerwin, J. A. & Zink, R. M. (1989). Phylogenetic patterns in the genus *Hediodoxa* (Aves: Trochilidae): an allozymic perspective. *Wilson Bull.* **101:** 525–544.

Gibbons, D. W., Reid, J. B. & Chapman, R. A. (1993). The New Atlas of breeding birds in Britain and Ireland: 1988–1991. London, Poyser.

Gibbons, D. W. & Wotton, S. (1996). The Dartford Warbler in the United Kingdom in 1994. *Brit. Birds* **89:** 203–212.

Gill, F. B. (1980). Historical aspects of secondary contact and hybridisation between Blue-winged and Golden-winged Warblers. *Auk* **97:** 1–18.

Gill, F. B. (1997). Local cytonuclear extinction of the Golden-winged Warbler. *Evolution* **51:** 519–525.

Gill, F. B., Mostrom, A. M. & Mack, A. L. (1993). Speciation in North American chickadees. I. Patterns of mtDNA genetic divergence. *Evolution* **47:** 195–212.

Gill, F. B., Slikas, B. & Agro, D. (1999). Speciation in North American Chickadees: II Geography of mt DNA haptotypes in *Poecile carolinensis*. *Auk* **116:** 274–277.

Gingerich, P. D. (1993). Quantification and comparison of evolutionary rates. *Am. J. Sci.* **293A:** 453–478.

Glas, P. (1960). Factors governing density in the Chaffinch (*Fringilla coelebs*) in different types of wood. *Arch. Néere. Zool.* **13:** 466–472.

Good, T. P., Ellis, J. C., Annett, C. & Pierotti, R. (2000). Bounded hybrid superiority in an avian hybrid zone: effects of mate, diet and habitat choice. *Evolution* **54:** 1774–1783.

Goss-Custard, J. D., Le V. dit. Durell, S. E. A., Sitters, H. P. & Swinfen, R. (1982). Age-structure and survival of a wintering population of Oystercatchers. *Bird Study* **29:** 83–98.

Gould, P. J., Forsell, D. J. & Lensink, C. J. (1982). Pelagic distribution and abundance of seabirds in the Gulf of Alaska and Eastern Bering Sea. Washington, US Fish Wildl. Serv.

Grant, P. R. (1986). Ecology and evolution of Darwin's Finches. Princeton, University Press.

Grant, P. R. & Abbot, I. (1980). Interspecific competiton, island biogeography and null hypotheses. *Evolution* **34:** 332–341.

Grant, P. R. & Grant, B. R. (1992). Hybridisation of bird species. *Science* **256:** 193–197.

Grant, P. R. & Grant, B. R. (1996). Speciation and hybridisation in island birds. *Phil. Trans. R. Soc. Lond. B* **351:** 765–772.

Grayson, D. K. & Livingston, S. D. (1993). Missing mammals on Great Basin Mountains: Holocene extinctions and inadequate knowledge. *Conserv. Biol.* **7:** 527–532.

Green, R. E. (1983). Spring dispersal and agonistic behaviour of the Red-legged Partridge (*Alectoris rufa*). *J. Zool. Lond.* **201:** 541–555.

Green, R. E. (1995). The decline of the Corncrake *Crex crex* in Britain continues. *Bird Study* **42:** 66–75.

Green, R. E. (1997). The influence of numbers released on the outcome of attempts to intro-
duce exotic bird species to New Zealand. *J. Anim. Ecol.* **66:** 25–35.

Greenberg, R., Cordero, P. J., Droege, S. & Fleischer, R. C. (1998). Morphological adaptation
with no mitochondrial DNA differentiation in the coastal plain Swamp Sparrow. *Auk*
115: 706–712.

Greenwood, P. J. (1980). Mating systems, philopatry and dispersal in birds and mammals.
Anim. Behav. **28:** 1140–1162.

Greenwood, P. J., Harvey, P. H. & Perrins, C. M. (1978). Inbreeding and dispersal in the
Great Tit. *Nature* **271:** 52–54.

Greenwood, P. J., Harvey, P. H. & Perrins, C. M. (1979). The role of dispersal in the Great Tit
(*Parus major*): the causes, consequences and heritability of natal dispersal. *J. Anim. Ecol.*
48: 123–142.

Gregory, R. D. (1998). An interspecific model of species' expansion, linking abundance and
distribution. *Ecography* **21:** 92–96.

Griffin, C. R., King, C. M., Savidge, J. A., Cruz, F. & Cruz, J. B. (1988). Effects of introduced
predators on island birds: contemporary case histories from the Pacific. *Proc. Int. Orn.
Congr.* **19:** 688–698.

Griffith, B., Scott, J. M., Carpenter, J. W. & Reed, C. (1989). Translocation as a species
conservation tool: status and strategy. *Science* **245:** 477–480.

Grinnel, J. & Miller, A. H. (1944). The distribution of the birds of California. *Pacif. Coast
Avifauna* **27:** 216–225.

Groombridge, J. J., Bruford, M. W., Jones, C. G. & Nichols, R. A. (2000). 'Ghost' alleles of the
Mauritius Kestrel. *Nature* **403:** 616.

Gross, S. J. & Price, T. D. (2000). Determinants of the northern and southern range limits of
a warbler. *J. Biogeog.* **27:** 869–878.

Groth, J. G. (1988). Resolution of cryptic species in Appalachian Red Crossbills. *Condor* **90:**
745–760.

Gudmundson, F. (1951). The effects of recent climatic changes on the bird life of Iceland.
Proc. Int. Orn. Congr. **10:** 502–514.

Hackett, S. J. (1993). Phylogenetic and biogeographic relationships in the neotropical genus
Gymnopithys (Formicariidae). *Wilson Bull.* **105:** 301–315.

Hackett, S. J. & Rosenberg, S. V. (1990). Comparison of phenotypic and genetic differentiation in South American antwrens (Formicariidae). *Auk* **107:** 473–489.

Haddon, M. (1984). A re-analysis of hybridisation between Mallards and Grey Ducks in
New Zealand. *Auk* **101:** 190–191.

Haddrath, O. & Baker, A. J. (2001). Complete mitochrondrial DNA genome sequences of
extinct birds: ratite phylogenetics and the vicariance biogeography hypothesis. *Proc. R.
Soc. Lond. B* **268:** 939–945.

Haffer, J. (1969). Speciation in Amazonian forest birds. *Science* **165:** 131–137.

Haffer, J. (1974). Avian speciation in tropical South America. Cambridge, Mass., Nuttall
Ornithological Club.

Haffer, J. (1978). Distributions of Amazonian forest birds. *Bonn. Zool. Beitr.* **29:** 38–78.

Haffer, J. (1980). Avian speciation patterns in upper Amazonia. *Proc. Int. Orn. Congr.* **17:**
1251–1255.

Haffer, J. (1987). Biogeography of neotropical birds. Pp. 105–150 in *'Biogeography and
quaternary history in tropical America'* (ed. T. C. Whitmore & G. T. Prance). Oxford,
University Press.

Haffer, J. (1989). Parapatrische Vogelarten der palaarktischen Region. *J. Orn.* **130:** 475–512.

Haffer, J. (1997). Alternative models of vertebrate speciation in Amazonia: an overview.
Biodiv. Conserv. **6:** 451–476.

Haftorn, S. (1978). Energetics of incubation by the Goldcrest *Regulus regulus* in relation to
ambient air temperatures and the geographical distribution of the species. *Ornis Scand.*
9: 22–30.

Hagemeijer, W. J. M. & Blair, M. J. (eds.) (1997). The EBCC Atlas of European breeding birds. Their distribution and abundance. London, Poyser.

Haila, Y., Jarvinen, O. & Raivio, S. (1987). Quantitative versus qualitative distribution patterns of birds in the western Palaearctic taiga. *Ann. Zool. Fenn.* **24**: 179–194.

Haila, Y., Järvinen, O. & Väisänen, R. A. (1979). Effects of mainland population changes on the terrestrial bird fauna of a northern island. *Ornis Scand.* **10**: 48–55.

Haldane, J. B. S. (1922). Sex ratio and unisexual sterility in hybrid animals. *J. Genet.* **12**: 101–109.

Hall, B. P. (1972). Causal ornithogeography of Africa. *Proc. Int. Orn. Congr.* **17**: 585–593.

Hall, P. (1974). Species and speciation. *Bird Study* **21**: 93–101.

Hall, P. & Moreau, R. E. (1970). An atlas of speciation in African passerine birds. London, Brit. Mus. Nat. Hist.

Halley, D. J., Harris, M. P. & Wanless, S. (1995). Colony attendance patterns and recruitment in immature Common Murres (*Uria aalge*). *Auk* **112**: 947–957.

Hamer, K. C., Phillips, R. A., Wanless, S., Harris, M. P. & Wood, A. G. (2000). Foraging ranges, diets and feeding locations of Gannets *Morus bassanus* in the North Sea: evidence from satellite telemetry. *Mar. Ecol. Prog. Ser.* **200**: 257–264.

Hamilton, A. C. (1976). The significance of patterns of distribution shown by forest plants and animals in tropical Africa for the reconstruction of Upper Pleistocene palaeo-environments. *Palaeoecol. Afr.* **9**: 63–97.

Hamilton, A. C. (1982). Environmental history of East Africa. A study of the Quaternary. London, Academic Press.

Hamilton, A. C. & Taylor, D. (1991). History of climate and forests in tropical Africa during the last 8 million years. *Climate Change* **19**: 65–78.

Handrich, Y., Bevan, R. K., Charrassin, J.-B., Butler, P. J., Pütz, K., Woakes, A., Lage, J. & Le Maho Y. (1997). Hypothermia in foraging King Penguins. *Nature* **388**: 64–67.

Handtke, K. & Mauersberger, G. (1977). Die Ausbreitung des Kuhreibers, *Bubulcus ibis* (L.) mit. Zool. Mus. Berlin, Suppl. Bd. 3. *Ann. Orn.* **1**: 1078.

Hanski, I. (1991). Single-species metapopulation dynamics: concepts, models and observations. *Biol. J. Linn. Soc.* **42**: 17–38.

Hanski, I. & Gilpin, M. (1991). Metapopulation dynamics: brief history and conceptual domain. *Biol. J. Linn. Soc.* **42**: 3–16.

Haq, B. U., Harbenbol, J. & Vail, R. R. (1987). Chronology of fluctuating sea levels since the Triassic. *Science* **235**: 1156–1167.

Härlid, A. & Arnason, U. (1999). Analyses of mitochondrial DNA nest ratite birds within the Neognathae: supporting a neotenous origin of ratite morphological characters. *Proc. R. Soc. Lond. B* **266**: 305–309.

Härlid, A., Janke, A. & Arnason, U. (1998). The complete mitochondrial genome of *Rhea americana* and early avian divergences. *J. Mol. Evol.* **46**: 669–679.

Harmon, W. M., Clark, W. A., Hawbecker, A. C. & Stafford, M. (1987). *Trichomonas gallinae* in columbiform birds from the Galapagos Islands. *J. Wildl. Dis.* **23**: 492–494.

Harris, M. P. (1969). Biology of Storm Petrels in the Galapagos Islands. *Proc. Calif. Acad. Sci. Ser 4.* **37**: 95–166.

Harris, M. P. (1970). Abnormal migration and hybridisation of *Larus argentatus* and *L. fuscus* after interspecies fostering experiments. *Ibis* **112**: 488–498.

Harris, M. P. (1984). The Puffin. Calton, Poyser.

Harrison, C. J. O. (1977). Non-passerine birds of the Ipswichian Interglacial from the Gower Caves. *Trans. Brit. Cave Res. Assoc.* **4**: 441–442.

Harrison, C. J. O. (1982). An atlas of the birds of the western Palaearctic. London, Collins.

Harrison, C. J. O. (1985). The Pleistocene birds of south-eastern England. *Bull. Geol. Soc. Norfolk* **35**: 53–69.

Harrison, C. J. O. (1998). The North Atlantic Albatross *Diomedea anglica*, a Pliocene–Lower Pleistocene species. *Tertiary Res.* **2**: 45–46.

Harrison, J. A., Allan, D. G., Underhill, L. G., Herremans, M., Tree, A. J., Parker, V. & Brown, C. J. (1997). The Atlas of Southern African Birds. Vols. 1 & 2. Johannesburg, Birdlife South Africa.

Harrison, N. M., Webb, A., Leaper, G. M. & Steele, D. (1989). Seabird distributions and the movements of moulting auks west of Scotland and in the Northern Irish Sea (late summer 1988). Aberdeen, Nature Conservancy Council.

Harrison, P. (1983). Seabirds. An identification guide. Beckenham, Kent, Croom Helm.

Harvey, P. H., Greenwood, P. J., Campbell, B. & Stenning, M. J. (1984). Breeding dispersal of the Pied Flycatcher (*Ficedula hypoleuca*). *J. Anim. Ecol.* **53:** 727–736.

Harvey, P. H., Greenwood, P. J. & Perrins, C. M. (1979). Breeding area fidelity of Great Tits (*Parus major*). *J. Anim. Ecol.* **48:** 305–313.

Hayworth, A. M. & Weathers, W. (1984). Temperature regulation and climatic adaptation in Black-billed and Yellow-billed Magpies. *Condor* **86:** 19–26.

Heaney, L.R. (1991). A synopsis of climate and vegetational change in southeast Asia. *Climate change* **19:** 53–61.

Heard, S. B. & Hauser, D. L. (1995). Key evolutionary innovations and their ecological mechanisms. *Hist. Biol.* **10:** 151–173.

Heath, M. & Evans, M. (2000). Important bird areas in Europe: priority sites for conservation. Vols. 1 & 2. Birdlife Conservation Series No. 8. Cambridge, Birdlife International.

Hedges, S. B. (1996). Historical biogeography of West Indian vertebrates. *Ann. Rev. Ecol. Syst.* **27:** 163–196.

Hedges, S. B., Hass, C. A. & Maxson, L. R. (1992). Caribbean biogeography: molecular evidence for dispersal in West Indian terrestrial vertebrates. *Proc. Nat. Acad. Sci. USA* **89:** 1909–1913.

Hedges, S. B. & Sibley, C. G. (1994). Molecules vs morphology in avian evolution — the case of the Pelecaniform birds. *Proc. Nat. Acad. Sci., USA* **91:** 9861–9865.

Hedges, S. B., Simmons, M. D., van Dijk, M. A. M., Caspers, G.-J., de Jong, W. W. & Sibley, C. G. (1995). Phylogenetic relationships of the Hoatzin, an enigmatic South American bird. *Proc. Nat. Acad. Sci., USA* **92:** 11662–11665.

Heidrich, P. & Wink, M. (1998). Phylogenetic relationships in Holarctic owls (Order Strigiformes): evidence from nucleotide sequences of the mitochondrial cytochrome *b* gene. Pp. 73–87 in '*Holarctic birds of prey*' (ed. R. D. Chancellor, B.-U. Meyburg & J. J. Ferrero). Berlin, World Working Group on Birds of Prey and Owls & Adenex.

Heindl, M. & Schuchmann, K.-L. (1998). Biogeography, geographical variation and taxonomy of the Andean hummingbird genus *Metallura* Gould, 1847. *J. Orn.* **139:** 425–473.

Helbig, A. (1983). Mass occurrence of Manx Shearwaters (*Puffinus puffinus*) at the coast of southern California coincides with exceptionally warm water temperatures. *Ardea* **71:** 161–162.

Helbig, A. J. (1994). Genetische Differenzierung von Möwen und Sturmtauchern: Ein Kommentar. *J. Orn.* **135:** 609–615.

Helbig, A. J., Martens, J., Seibold, I., Henning, F., Schöttler, B. & Wink, M. (1996). Phylogeny and species limits in the Palaearctic Chiffchaff *Phylloscopus collybita* complex: mitochondrial genetic differentiation and bioacoustic evidence. *Ibis* **138:** 650–666.

Helbig, A. J., Seibold, I., Martens, J. & Wink, M. (1995). Genetic differentiation and phylogenetic relationships of Bonelli's Warbler *Phylloscopus bonelli* and Green Warbler *P. nitidus*. *J. Avian Biol.* **26:** 139–153.

Helm-Bychowski, K. M. & Wilson, A. C. (1986). Rates of nuclear DNA evolution in pheasant-like birds: evidence from restriction maps. *Proc. Nat. Acad. Sci., USA* **83:** 688–692.

Hengeveld, R. (1988). Mechanisms of biological invasions. *J. Biogeog.* **15:** 819–828.

Hengeveld, R. (1989). Dynamics of biological invasions. London, Chapman & Hall.

Hengeveld, R. (1993). What to do about the North American invasion by the Collared Dove? *J. Field Orn.* **64:** 477–489.

Hengeveld, R. & Haeck, J. (1982). The distribution of abundance. 1. Measurements. *J. Biogeog.* **9:** 303–316.

Hengeveld, R. & van den Bosch, F. (1991). The expansion velocity of the Collared Dove *Streptopelia decaocto* population in Europe. *Ardea* **79:** 67–72.

Henny, C. J. & van Velzen, W. T. (1972). Migration patterns and wintering localities of American Ospreys. *J. Wildl. Manage.* **36:** 1133–1141.

Herremans, M. (1998). Strategies, punctuality of arrival and ranges of migrants in the Kalahari basin, Botswana. *Ibis* **140:** 585–590.

Herrera, C. M. (1978). Ecological correlates of residence and non-residence in a Mediterranean passerine bird community. *J. Anim. Ecol.* **47:** 871–890.

Hestbeck, J. B., Nichols, J. D. & Malecki, R. A. (1991). Estimates of movement and site fidelity using mark–resight data of wintering Canada Geese. *Ecology* **72:** 523–533.

Hewitt, G. M. (1999). Post-glacial re-colonisation of European biota. *Biol. J. Linn. Soc.* **69:** 87–112.

Hewitt, G. M. (2000). The genetic legacy of the Quaternery ice ages. *Nature* **405:** 907–913.

Hickling, R. A. O. (1984). Lesser Black-backed Gull numbers at British inland roosts in 1979–80. *Bird Study* **31:** 157–160.

Hildén, O. (1978). Population dynamics in Temminck's Stint *Calidris temminckii*. *Oikos* **30:** 17–28.

Hildén, O. (1979). Territoriality and site tenacity of Temminck's Stint *Calidris temminckii*. *Ornis Fenn.* **56:** 56–74.

Hildén, O. & Koskimies, J. (1969). Effects of the severe winter of 1965/66 upon winter bird fauna in Finland. *Ornis. Fenn.* **46:** 22–31.

Hildén, O. & Sharrock, J. T. R. (1985). A summary of recent avian range changes in Europe. *Proc. Int. Orn. Congr.* **18:** 716–736.

Hiraldo, F., Negro, J. H., Donázar, J. A. & Gaona, P. (1996). A demographic model for a population of the endangered Lesser Kestrel in southern Spain. *J. Appl. Ecol.* **33:** 1085–1093.

Hoffman, W., Woolfenden, G. E. & Smith, P. W. (1999). Antillean Short-eared Owls invade southern Florida. *Wilson Bull.* **111:** 303–446.

Holden, K., Montgomerie, R. & Friesen, V. L. (1999). A test of the glacial refugium hypothesis using patterns of mitochondrial and nuclear DNA sequence variation in Rock Ptarmigan (*Lagopus mutus*). *Evolution* **53:** 1936–1950.

Hollands, D. (1977). Field observations on the Letter-winged Kite, Eastern Simpson Desert, 1974–1976. *Aust. Bird Watcher, Sept* **1977:** 73–80.

Holmes, R. T. & Sherry, T. W. (1992). Site fidelity of migratory warblers in temperate breeding and neotropical wintering areas: applications for population dynamics, habitat selection, and conservation. Pp. 563–575 in 'Ecology and conservation of neotropical migrant landbirds' (ed. J. M. Hagan III & D. W. Johnston). Washington, Smithsonian Institution Press.

Holyoak, D. T. & Thibault, J.-C. (1978). Notes on the phylogeny, distribution and ecology of frugivorous pigeons in Polynesia. *Emu* **78:** 201–206.

Houde, P. (1987). *Paleotis wiegelti* restudied: a small middle Eocene ostrich (Aves Struthioniformes). *Palaeovertebrata* **17:** 27–42.

Houde, P., Cooper, A., Leslie, E., Strand, A. E. & Montaño, G. A. (1997). Phylogeny and evolution of 128 *r* DNA in Gruiformes (Aves). Pp. 121–158 in 'Avian molecular evolution and systematics' (ed. D. Mindell). London, Academic Press.

Houlihan, P. F. (1986). The birds of Ancient Egypt. Warminster, Wiltshire, Aris & Phillips.

Houston, C. S. (1978). Recoveries of Saskatchewan-banded Great Horned Owls. *Can. Field Nat.* **92:** 61–66.

Houston, C. S. (1986). Mourning Dove numbers explode on the Canadian prairies. *Amer. Birds* **40:** 52–54.

Houston, C. S. & Schmutz, J. K. (1999). Changes in bird populations on Canadian grasslands. *Stud. Avian Biol.* **19:** 87–94.

Howell, S. N. G. & Robbins, M. B. (1995). Species limits of the Least Pygmy Owl (*Glaucidium minutissimum*) complex. *Wilson Bull.* **107:** 7–25.

Hubbard, J. (1973). Avian evolution in the aridlands of North America. *Living Bird* **12:** 155–196.

Hudson, R. (1965). The spread of the Collared Dove in Britain and Ireland. *Brit. Birds* **58:** 105–139.

Hudson, R. (1972). Collared Doves in Britain and Ireland during 1965–1970. *Brit. Birds* **65:** 139–155.

Hughes, B. (1993). Stiff-tail threat. *BTO News* **185:** 14.

Hultén, E. (1937). Outline of the history of the arctic and boreal biota during the Quaternary period. Sweden, Thule.

Hunt, G. L. J. (1990). The distribution of seabirds at sea: physical and biological aspects of their marine environment. Pp. 167–171 in '*Current topics in avian biology*' (ed. R. van den Elzen, K.-L. Schuchmann & K. Schmidt-Koenig). Berlin, Deutschen Orn. Gesellschaft.

Hunt, G. L. J. (1991). Marine ecology of seabirds in polar oceans. *Amer. Zool.* **31:** 131–142.

Huntley, B. (1988). European post-glacial vegetation history. *Proc. Int. Orn. Congr.* **19:** 1061–1077.

Huntley, B. (1993). Species-richness in north-temperate zone forests. *J. Biogeog.* **20:** 163–180.

Huntley, B. (1995). Plant species' response to climate change: implications for the conservation of European birds. *Ibis* **137 (Suppl.1):** 127–138.

Huntley, B., Cramer, W., Morgan, A. V., Prentice, H. C. & Allen, J. R. M. (eds.) (1997). Past and future rapid environmental changes: the spatial and evolutionary responses of terrestrial biota. Proceedings of the NATO Advanced Research Workshop. London, Springer.

Huntley, B. & Webb, T. (1989). Migration: species' response to climatic variations caused by changes in the earth's orbit. *J. Biogeog.* **16:** 5–19.

Huxley, J. (1942). Evolution. The modern synthesis. London, Allen & Unwin.

Imber, M. J. & Lovegrove, T. G. (1982). Leach's Storm Petrels (*Oceanodroma l. leucorhoa*) prospecting for nest sites at the Chatham Islands. *Notornis* **29:** 101–108.

Immelmann, K. (1965). Objektifixierung geslechtlicher Triehandhungen bei Prachtfinken. *Naturwissenschaften* **7:** 169–170.

Irwin, D. E., Alstrom, P., Olsson, U. & Benowitz-Friederichs, Z. M. (2001a). Cryptospecies in the genus *Phylloscopus* (Old World leaf warblers). *Ibis* **143:** 233–247.

Irwin, D. E., Bensch, S. & Price, T. D. (2001b). Speciation in a ring. *Nature* **409:** 333–337.

Jablonski, D. (1995). Extinctions in the fossil record. *Phil. Trans. R. Soc. Lond. B* **344:** 11–17.

Jacobsen, J. R. (1963). The migration and wintering in northwestern Europe of Lapland Bunting (*Calcarius lapponicus lapponicus*) (L.). *Dansk. Orn. Foren. Tidskrift.* **57:** 181–211.

Jaeger, M. M., Bruggers, R. L., Johns, B. E. & Erickson, W. A. (1986). Evidence of itinerant breeding of the Red-billed Quelea *Quelea quelea* in the Ethiopian Rift Valley. *Ibis* **128:** 469–482.

James, H. F. (1991). The contribution of fossils to knowledge of Hawaiian birds. *Proc. Int. Orn. Congr.* **20:** 420–424.

James, H. F. (1995). Prehistoric extinctions and ecological changes on oceanic islands. Pp. 87–102 in '*Islands. Biological diversity and ecosystem function*' (ed. R. M. Vitousek, L. L. Loope & H. Adsersen). Berlin, Springer-Verlag.

James, H. F. (2001). Systematics — introduction. *Stud. Avian Biol.* **22:** 48–50.

James, H. F., Ericson, P. G. P., Slikas, B., Lei, F.-m., Gill, F. B. & Olson, S. L. (2003). *Pseudopodoces humilis*, a misclassified terrestrial tit (Aves: Paridae) of the Tibetan Plateau: evolutionary consequences of shifting adaptive zones. *Ibis* **145,** in press.

James, H. F. & Olson, S. L. (1991). Descriptions of thirty-two new species of birds from the Hawaiian Islands. Part II. Passeriformes. *Orn. Monogr.* **46:** 1–88.

Järvinen, O. (1982). Species-to-genus ratios in biogeography — a historical note. *J. Biogeog.* **9:** 363–370.

Järvinen, O. & Ulfstand, S. (1980). Species turnover of a continental bird fauna, northern Europe 1850–1970. *Oecologia* **46:** 186–195.

Järvinen, O. & Väisänen, R. (1979). Climatic changes, habitat changes and competition: dynamics of geographical overlap in two pairs of congeneric bird species in Finland. *Oikos* **33:** 261–271.

Jehl, J. R. (1990). Aspects of the molt migration. Pp. 102–113 in *'Bird migration physiology and ecophysiology'* (ed. E. Gwinner). Berlin, Springer-Verlag.

Jehl, J. R. (1997). Cyclical changes in body composition in the annual cycle and migration of the Eared Grebe *Podiceps nigricollis*. *J. Avian Biol.* **28:** 132–142.

Jenkins, D. & Watson, A. (2000). Dates of first arrival and song of birds during 1974–99 in mid-Deeside, Scotland. *Bird Study* **47:** 249–251.

Jenni, L. (1987). Mass concentrations of Bramblings *Fringilla montifringilla* in Europe 1900–1983: their dependence upon beech mast and the effect of snow cover. *Ornis Scand.* **18:** 84–94.

Jespersen, P. (1924). On the frequency of birds over the high Atlantic Ocean. *Nature* **114:** 281–283.

Johns, G. C. & Avise, J. C. (1998). A comparative summary of genetic distances in the vertebrates from the mitochondrial cytochrome *b* gene. *Mol. Biol. Evol.* **15:** 1481–1490.

Johnson, A. (1997). Greater Flamingo. Pp. 15–23 in *'BWP update 1'* (ed. M. A. Ogilvie). Oxford, University Press.

Johnson, A. R., Green, R. E. & Hirons, G. J. M. (1991). Survival rates of Greater Flamingos in the west Mediterranean region. Pp. 249–271 in *'Bird population studies'* (ed. C. M. Perrins, J.-D. Lebreton & G. J. M. Hirons). Oxford, University Press.

Johnson, D. H. & Grier, J. W. (1988). Determinants of breeding distributions of ducks. *Wildlife Monogr.* **100:** 1–37.

Johnson, M. L. & Gaines, M. S. (1990). Evolution of dispersal: theoretical models and empirical tests using birds and mammals. *Ann. Rev. Ecol. Syst.* **21:** 449–480.

Johnson, N. K. (1975). Controls of number of bird species on montane islands in the Great Basin. *Evolution* **29:** 545–567.

Johnson, N. K. (1994). Pioneering and natural expansion of breeding distributions in western North American birds. *Stud. Avian Biol.* **15:** 27–44.

Johnson, N. K. & Jehl, J. R. (1994). A century of change in western North America: an overview. *Stud. Avian Biol.* **15:** 1–3.

Johnson, N. K. & Johnson, C. B. (1985). Speciation in sapsuckers (*Sphyrapicus*): II. Sympatry, hybridisation, and mate preference in *S. ruber daggeti* and *S. nuchalis*. *Auk* **1021:** 1–15.

Johnson, N. K., Marten, J. A. & Ralph, C. J. (1989). Genetic evidence for the origin and relationships of the Hawaiian Honeycreepers (Aves: Fringillidae). *Condor* **91:** 379–396.

Johnson, O. W., Bruner, P. L., Rotella, J. J., Johnson, P. M. & Bruner, A. E. (2001). Long-term study of apparent survival in Pacific Golden Plovers at a wintering ground on Oahu, Hawaiian Islands. *Auk* **118:** 342–351.

Johnston, R. F. & Klitz, W. J. (1977). Variation and evolution in a granivorous bird. Pp. 15–51 in *'Granivorous birds in ecosystems'* (ed. J. Pinowski & S. C. Kendeigh). Cambridge, Cambridge University Press.

Johnston, R. F. & Selander, R. K. (1964). House Sparrow: rapid evolution of races in North America. *Science* **144:** 548–550.

Johnston, V. H. & Ryder, J. P. (1987). Divorce in larids: a review. *Colon. Waterbirds* **10:** 16–26.

Johnstone, G. W. (1985). Threats to birds on subantarctic islands. *ICPB Tech. Bull.* **3**: 101–121.

Jones, P. J. (1995). Migration strategies of Palaearctic passerines in Africa: an overview. *Israel. J. Zool.* **41**: 393–406.

Jósefik, M. (1969). Studies on the Squacco Heron. *Acta Orn., Warsz.* **11**: 103–262.

Jósefik, M. (1970). Studies on the Squacco Heron. *Acta Orn., Warsz.* **12**: 57–102, 394–504.

Joseph, L., Hugall, C. & Moritz, C. (1995). Molecular support for vicariance as a source of diversity in rainforest. *Proc. R. Soc. Lond. B* **260**: 177–182.

Joseph, L. & Moritz, C. (1994). Mitochondrial DNA phylogeography of birds in eastern Australian rain forests: first fragments. *Aust. J. Zool.* **42**: 385–403.

Jouanin, C. (1970). Le Petrel Noir de Bourbon, *Pterodroma aterrima* Bonaparte. *Oiseau Rev. Fr. Orn.* **40**: 48–68.

Jouventin, P. (1994). Les populations d'oiseaux marins des T.A.A.F.: resumé de 20 années de réserche. *Alauda* **62**: 44–47.

Jouventin, P. & Bried, J. (2001). The effect of mate choice on speciation in Snow Petrels. *Anim. Behav.* **62**: 123–132.

Jouventin, P. & Viot, C.-R. (1985). Morphological and genetic variability of Snow Petrels *Pagodroma nivea*. *Ibis* **127**: 430–441.

Juan, C., Emerson, B. C., Oromi, P. & Hewitt, G. M. (2000). Colonisation and diversification: towards a phylogeographic synthesis of the Canary islands. *Trends Ecol. Evol.* **15**: 104–109.

Kalela, O. (1949). Changes in geographic ranges in the avifauna of northern and central Europe in relation to recent changes in climate. *Bird-Banding* **20**: 77–103.

Kalela, O. (1952). Changes in the geographic distribution of Finnish birds and mammals in relation to recent changes in climate. *Fennia* **75**: 38–57.

Karr, J. R. (1982). Avian extinction on Barro Colorado Island, Panama: a reassessment. *Amer. Nat.* **119**: 220–239.

Karr, J. R. (1990). Avian survival rates and the extinction process on Barro Colorado Island, Panama. *Conserv. Biol.* **4**: 391–397.

Keast, A. (1961). Bird speciation on the Australian continent. *Bull. Mus. Comp. Zool.* **123**: 305–495.

Keast, A. (1968). Seasonal movements in the Australian honeyeaters (*Melaphagidae*) and their ecological significance. *Emu* **67**: 159–209.

Keast, A. (1972). Faunal elements and evolutionary patterns: some comparisons between the continental avifaunas of Africa, South America and Australia. *Proc. Int. Orn. Congr.* **17**: 594–622.

Keast, A. (1974). Avian speciation in Africa and Australia: some comparisons. *Emu* **74**: 261–269.

Keast, A. (1981a). The evolutionary biogeography of Australian birds. Pp. 1586–1634 in 'Ecological biogeography of Australia' (ed. A. Keast). London, Dr W. Junk Publishers.

Keast, A. (ed.) (1981b). Ecological biogeography of Australia. London, Dr W. Junk Publishers.

Keast, A. (1991). Avian evolution of southern Pacific island groups. An ecological perspective. *Proc. Int. Orn. Congr.* **20**: 435–446.

Keast, A. (1995). The Nearctic–Neotropical bird migration system. *Israel J. Zool.* **41**: 455–476.

Keller, L. F., Arcese, P., Smith, J. N. M., Hochachka, W. M. & Stearns, S. (1994). Selection against inbred Song Sparrows during a natural bottleneck. *Nature* **372**: 356–357.

Kelsey, M. G. (1989). A comparison of the song and territorial behaviour of a long-distance migrant, the Marsh Warbler *Acrocephalus palustris*, in summer and winter. *Ibis* **131**: 403–414.

Kennedy, R. S., Gonzales, P. C., Dickinson, E. C., Miranda, H. C. & Fisher, T. H. (2000). A guide to the birds of the Philippines. Oxford, University Press.

Kenward, R. E. (2001). A manual for wildlife radio tagging. London, Academic Press.

Kenward, R. E., Marcström, V. & Karlbom, M. (1993). Post-nestling behaviour in Goshawks, *Accipiter gentilis*: I. The causes of dispersal. *Anim. Behav.* **46**: 365–370.

Kenward, R. E., Walls, S. S. & Hodder, K. H. (2001). Life path analysis: scaling indicates priming effects of social and habitat factors on dispersal distances. *J. Anim. Ecol.* **70**: 1–13.

Keppie, D. M. (1980). Similarity of dispersal among sibling male Spruce Grouse. *Can. J. Zool.* **58**: 2102–2104.

Kessel, B. (1953). Distribution and migration of the European Starling in North America. *Condor* **55**: 49–67.

Kessler, L. G. & Avise, J. C. (1984). Systematic relationships among wildfowl (Anatidae) inferred from restriction endonuclease analyses of mitochondrial DNA. *Syst. Zool.* **33**: 370–380.

Ketterson, E. D. & Nolan, V. J. (1976). Geographic variation and its climatic correlates in the sex ratio of eastern-wintering Dark-eyed Juncos (*Junco hyemalis hyemalis*). *Ecology* **57**: 679–693.

Kidd, M. G. & Friesen, V. L. (1998a). Patterns of control region variation in populations of *Cepphus* guillemots: testing microevolutionary hypotheses. *Evolution* **52**: 1158–1168.

Kidd, M. G. & Friesen, V. L. (1998b). Sequence variation in the Guillemot (Alcidae: *Cepphus*) mitochondrial control region and its nuclear homolog. *Mol. Biol. Evol.* **15**: 61–70.

Kikkawa, J. (1973). The status of the Silvereyes *Zosterops* on the islands of the Great Barrier Reef. *Sunbird* **4**: 30–37.

Kimura, M. (1983). The neutral theory of molecular evolution. Cambridge, University Press.

King, G. A. & Herstrom, A. A. (1997). Holocene tree migration rates objectively determined from fossil pollen data. Pp. 91–107 in '*Past and future rapid environmental changes: the spatial and evolutionary responses of terrestrial biota*' (ed. B. Huntley, W. Cramer, A. V. Morgan, H. C. Prentice & J. R. M. Allen). London, Springer-Verlag.

King, J. R. & Wales, E. E. (1964). Observations of migration, ecology and population flux of wintering Rosy Finches. *Condor* **66**: 24–31.

King, W. B. (1985). Island birds: will the future repeat the past? *ICBP Tech. Pub.* **3**: 3–15.

Klein, N. K. & Brown, W. M. (1994). Intraspecific molecular phylogeny in the Yellow Warbler (*Dendroica petechia*), and implications for avian biogeography in the West Indies. *Evolution* **48**: 1914–1932.

Klicka, J. & Zink, R. (1997). The importance of recent Ice Ages in speciation: a failed paradigm. *Science* **277**: 1666–1669.

Klicka, J. & Zink, R. M. (1999). Pleistocene effects on North American songbird evolution. *Proc. R. Soc. Lond. B* **266**: 695–700.

Klicka, J., Zink, R. M., Barlow, J. C., McGillivray, W. B. & Doyle, T. J. (1999). Evidence supporting the recent origin and species status of the Timberline Sparrow. *Condor* **101**: 577–588.

Klinger, D. (1982). 1980 fishing and hunting survey results told. *Fish & Wild. News*: 1–7.

Klomp, N. I. & Furness, R. W. (1992). The dispersal and philopatry of Great Skuas from Foula, Shetland. *Ring. Migr.* **13**: 73–82.

Kluijver, H. N. (1971). Regulation of numbers in populations of Great Tits (*Parus m. major* L.). Pp. 507–524 in '*Dynamics of populations*' (ed. P. J. den Boer & G. R. Gradwell). Wageningen, Centre for Agricultural Publishing and Documentation.

Kluijver, H. N. & Tinbergen, L. (1953). Territory and the regulation of density in titmice. *Arch. néel. Zool.* **10**: 265–289.

Knox, A. G., Helbig, A. J., Parkin, D. T. & Sangster, G. (2001). The taxonomic status of Lesser Redpoll. *Brit. Birds* **94**: 260–267.

Kodric-Brown, A. & Brown, J. H. (1979). Competition between distantly-related taxa and the co-evolution of plants and pollinators. *Amer. Zool* **19**: 1115–1127.

Köenig, W. D. & Knops, J. M. H. (2001). Seed crop size and eruptions of North American boreal seed-eating birds. *J. Anim. Ecol.* **70:** 609–620.

Köenig, W. D. & Mumme, R. L. (1987). Population ecology of the cooperatively breeding Acorn Woodpecker. Princeton, University Press.

Köenig, W. D. & Pitelka, F. A. (1979). Relatedness and inbreeding avoidance: counterploys in the communally nesting Acorn Woodpecker. *Science* **206:** 1103–1105.

Komdeur, J., Huffstadt, A., Prast, W., Castle, G., Mileto, R. & Wattel, J. (1995). Transfer experiments of Seychelles Warblers to new islands: changes in dispersal and helping behaviour. *Anim. Behav.* **49:** 695–708.

König, C., Weick, F. & Becking, J.-H. (1999). Owls. A guide to owls of the world. Sussex, Pica Press.

Kornegay, J. R., Kocher, T. D., Williams, L. A. & Wilson, A. C. (1993). Pathways of lysozyme evolution inferred from the sequences of cytochrome *b* in birds. *J. Mol. Evol.* **37:** 367–379.

Korpimäki, E., Lagerström, M. & Saurola, P. (1987). Field evidence for nomadism in Tengmalm's Owl *Aegolius funereus. Ornis Scand.* **18:** 1–4.

Korpimäki, E. & Norrdahl, K. (1991). Numerical and functional responses of Kestrels, Short-eared Owls and Long-eared Owls to vole densities. *Ecology* **72:** 814–825.

Koskimies, J. & Lahte, L. (1964). Cold-hardiness of the newly hatched young in relation to ecology and distribution in ten species of European ducks. *Auk* **81:** 281–307.

Kouki, J. & Häyrinen, U. (1991). On the relationship between distribution and abundance in birds breeding on Finnish mires: the effects of habitat specialisation. *Ornis Fenn.* **68:** 170–177.

Kraaijeveld, K. & Nieboer, E. N. (2000). Late quaternery paleogeography and evolution of arctic breeding waders. *Ardea* **88:** 193–205.

Krajewski, C. & King, D. G. (1996). Molecular divergence and phylogeny: rates and patterns of cytochrome *b* evolution in cranes. *Mol. Biol. Evol.* **13:** 21–30.

Kratter, A. W. (1992). Montane avian biogeography in southern California and Baja California. *J. Biogeog.* **19:** 269–283.

Kurochkin, E. N. (1985). A true carinate bird from Lower Cretaceous deposits in Mongolia and other evidence of early Cretaceous birds in Asia. *Cretaceous Research* **6:** 271–278.

Kuroda, N. (1991). Distributional patterns and seasonal movements of Procellariiformes in the North Pacific. *J. Yamashina Inst. Orn.* **23:** 23–84.

Kurtén, B. (1968). The Pleistocene mammals of Europe. London, Wiedenfield & Nicholson.

Kurtén, B. (1972). The ice age. London, Rupert Hart-Davis.

Lack, D. (1944). Ecological aspects of species formation in passerine birds. *Ibis* **86:** 260–286.

Lack, D. (1947). Darwin's finches. Cambridge, University Press.

Lack, D. (1954). The natural regulation of animal numbers. Oxford, University Press.

Lack, D. (1963). Migration across the southern North Sea studied by radar. Part 5. Movements in August, winter and spring, and conclusion. *Ibis* **105:** 461–492.

Lack, D. (1969). The numbers of bird species on islands. *Bird Study* **16:** 193–209.

Lack, D. (1971). Ecological isolation in birds. Oxford, Blackwells.

Lack, D. (1976). Island biology illustrated by the birds of Jamaica. Oxford, Blackwell Scientific Publications.

Lack, P. (1986). The atlas of wintering birds in Britain and Ireland. Calton, Poyser.

Lack, P. (1990). Palaearctic–African systems. Pp. 345–356 in '*Biogeography and ecology of forest bird communities*' (ed. A. Keast). The Hague, Netherlands, SPB Academic Publishing.

Lamb, S. & Sington, D. (1998). Earth story. London, BBC Books.

Lamberson, R. H., McKelvey, R., Noon, B. R. & Voss, C. (1992). A dynamic analysis of Northern Spotted Owl viability in a fragmented forest landscape. *Conserv. Biol.* **6:** 505–512.

Lande, R. (1988). Demographic models of the Northern Spotted Owl (*Strix occidentalis caurina*). *Oecologia* **75:** 601–607.

Langrand, O. (1990). Guide to the birds of Madagascar. New Haven, Yale University Press.

Lanner, R. M. (1996). Made for each other. A symbiosis of birds and pines. Oxford, University Press.

Lanyon, S. M. & Hall, J. G. (1994). Re-examination of barbet monophyly using mitochrondrial–DNA sequence data. *Auk* **111**: 389–397.

Lanyon, W. E. (1979). Hybrid sterility in Meadowlarks. *Nature* **279**: 557–558.

Larsson, K., Forslund, P., Gustafsson, L. & Ebbinge, B. S. (1988). From the high arctic to the Baltic: the successful establishment of a Barnacle Goose *Branta leucopsis* population on Gotland, Sweden. *Ornis Scand.* **19**: 182–189.

Latham, R. E. & Ricklefs, R. E. (1993). Global patterns of tree species diversity in moist forests: energy diversity theory does not account for variation in species richness. *Oikos* **67**: 325–333.

Lawn, M. R. (1982). Pairing systems and site tenacity of the Willow Warbler *Phylloscopus trochilus* in southern England. *Ornis Scand.* **13**: 193–199.

Lawton, J. H. (1993). Range, population abundance and conservation. *Trends Ecol. Evol.* **8**: 409–413.

Lawton, J. H. (1994). Population dynamic principles. *Phil. Trans. R. Soc. Lond. B* **344**: 61–68.

Lawton, J. H. (1996). Population abundances, geographic ranges and conservation: 1994 Witherby Lecture. *Bird Study* **43**: 3–19.

Lawton, J. H. & May, R. M. (eds.) (1995). Extinction rates. Oxford, University Press.

Leck, C. F. (1980). Establishment of new population centers with changes in migration patterns. *J. Field Orn.* **51**: 168–173.

Lensink, C. J. (1984). The status and conservation of seabirds in Alaska. Pp. 13–27 in '*Status and conservation of the world's seabirds*' (ed. J. P. Croxall, P. G. H. Evans & R. W. Schreiber). Cambridge, ICBP.

Lensink, R. (1997). Range expansion of raptors in Britain and the Netherlands since the 1960s: testing an individual-based diffusion model. *J. Anim. Ecol.* **66**: 811–826.

Lever, C. (1987). Naturalised birds of the world. Harlow, Longman.

Levey, D. J. & Stiles, F. G. (1992). Evolutionary precursors of long-distance migration: resource availability and movement patterns in neotropical landbirds. *Amer. Nat.* **140**: 447–476.

Lewis, S., Sherratt, T. N., Hamer, K. C. & Wanless, S. (2001). Evidence of intra-specific competition for food in a pelagic seabird. *Nature* **412**: 816–819.

Limpert, R. J. (1980). Homing success of adult Buffleheads to a Maryland wintering site. *J. Wildl. Manage.* **44**: 905–908.

Lloyd, C., Tasker, M. L. & Partridge, K. (1991). The status of seabirds in Britain and Ireland. London, Poyser.

Löfgren, O., Hörnfeldt, B. & Carlsson, B.-G. (1986). Site tenacity and nomadism in Tengmalm's Owl (*Aegolius funereus* (L)) in relation to cyclic food production. *Oecologia* **69**: 321–326.

Löhrl, H. (1955). Beziehungen zwischen Halsband- und Trauerfliegenschnäpper (*Muscicapa albicollis* and *M. hypoleuca*) in dem Brulgebiet. *Proc. Int. Orn. Congr.* **11**: 333–336.

Long, J. L. (1981). Introduced birds of the world. New York, Universe Books.

Lowther, P. E. & Cink, C. L. (1992). House Sparrow. Birds of North America No. 12. Washington, D.C., American Ornithologists' Union.

Lydekker, R. (1896). A geographical history of mammals. Cambridge, University Press.

MacArthur, R. H. (1972). Geographical ecology. Patterns in the distribution of species. New York, Harper & Row.

MacArthur, R. H. & Wilson, E. O. (1967). The theory of island geography. Princeton, University Press.

Macdonald, M. A. (1977). Adult mortality and fidelity to mate and nest-site in a group of marked Fulmars. *Bird Study* **24**: 165–168.

Maddock, M. & Geering, D. (1994). Range expansion and migration of the Cattle Egret. *Ostrich* **65**: 191–203.

Madsen, J. (1995). Impacts of disturbance on migratory waterfowl. *Ibis* **137 (suppl.):** 67–74.

Madsen, J., Cracknell, G. & Fox, T. (1999). Goose populations of the western Palaearctic. Rönde, National Environment Research Institute, Denmark.

Main, I. G. (2000). Obligate and facultative partial migration in the Blackbird (*Turdus merula*) and the Greenfinch (*Carduelis chloris*): uses and limitations of ringing data. *Vogelwarte* **40**: 286–291.

Mainwood, A. R. (1976). The movements of Storm Petrels as shown by ringing. *Ring. Migr.* **1**: 98–104.

Maley, J. (1991). The African rain forest vegetation and palaeoenvironments during the late Quaternary. *Climate Change* **19**: 79–98.

Mallet, J. (1995). A species definition for the modern synthesis. *Trends Ecol. Evol.* **10**: 294–299.

Manuwal, D. A. (1974). Effects of territoriality on breeding in a population of Cassin's Auklet. *Ecology* **55**: 1399–1406.

Marchant, J. H., Hudson, R., Carter, S. P. & Whittington, P. (1990). Population trends in British breeding birds. Tring, British Trust for Ornithology.

Marchant, S. (1972). Evolution of the genus *Chrysococcyx*. *Ibis* **114**: 219–233.

Markgraf, V. & Kenny, R. (1997). Character of rapid vegetation and climate change during the late-glacial in southernmost South America. Pp. 81–90 in *'Past and future rapid environmental changes: the spatial and evolutionary responses of terrestrial biota'* (ed. B. Huntley, W. Cramer, A. V. Morgan, H. C. Prentice & J. R. M. Allen). London, Springer Verlag.

Marks, J. S., Evans, D. L. & Holt, D. W. (1994). Long-eared Owl. The Birds of North America No. 133. Washington, D.C., American Ornithologists' Union.

Marquiss, M. & Rae, R. (2002). Ecological differentiation in relation to bill size amongst sympatric, genetically undifferentiated crossbills (*Loxia* spp.). *Ibis* **144**: 494–508.

Marshall, H. D. & Baker, A. J. (1999). Colonisation history of Atlantic Island Chaffinches (*Fringilla coelebs*) revealed by mitochrondrial DNA. *Mol. Phyl. Evol.* **11**: 201–212.

Martin, A. J. & Palumbi, S. R. (1993). Body size, metabolic rate, generation time, and the molecular clock. *Proc. Nat. Acad. Sci. USA* **90**: 4087–4091.

Martin, P. S. (1984). Prehistoric overkill: the global model. Pp. 354–403 in *'Quarternary extinctions'* (ed. P. S. Martin & R. G. Klein). Tucson, Arizona, University of Arizona Press.

Martin, P. S. & Klein, R. G. (1984). Quaternary extinctions. Tucson, Arizona, University of Arizona Press.

Martin, T. E. & Badyaev, A. (1996). Sexual dichromatism in birds: importance of nest predation and nest location for females versus males. *Evolution* **50**: 2454–2460.

Martinez, M. M. (1983). Nidificacion de *Hirundo rustica erythrogaster* (Boddaert) en la Argentina (Aves, Hirundinidae). *Neotropica* **29**: 83–86.

Marzluff, J. M. & Dial, K. P. (1991). Life history correlates of taxonomic diversity. *Ecology* **72**: 428–439.

Matthysen, E. (1990). Behavioral and ecological correlates of territory quality in the Eurasian Nuthatch (*Sitta europaea*). *Auk* **107**: 86–95.

Matthysen, E., Adriaensen, F. & Dhondt, A. A. (1995). Dispersal distances of Nuthatches, *Sitta europaea*, in a highly fragmented forest habitat. *Oikos* **72**: 375–381.

Matthysen, E., Adriaensen, F. & Dhondt, A. A. (2001). Local recruitment of Great and Blue Tits (*Parus major, P. caeruleus*) in relation to study plot size and degree of isolation. *Ecography* **24**: 33–42.

Mattox, W. G. & Seegar, W. S. (1988). The Greenland Peregrine Falcon Survey, 1972–1985, with emphasis on recent population status. Pp. 27–36 in *'Peregrine Falcon populations. Their management and recovery'* (ed. T. J. Cade, J. H. Enderson, C. G. Thelander & C. M. White). Boise, Idaho, The Peregrine Fund Inc.

Maurer, B. A., Brown, J. H. & Rusler, R. D. (1992). The micro and macro of body size evolution. *Evolution* **46:** 939–953.

May, R. M. (1990). Taxonomy as destiny. *Nature* **339:** 104.

Mayden, R. L. (1997). A hierarchy of species concepts: the denouncement in the saga of the species problem. Pp. 381–424 in '*Species: the units of biodiversity*' (ed. M. F. Claridge, H. A. Dawah & M. R. Wilson). London, Chapman & Hall.

Mayden, R. L. & Wood, R. M. (1995). Systematics, species concepts and the evolutionary significant unit in biodiversity and conservation biology. *Amer. Fisheries Soc. Symp.* **17:** 58–113.

Mayr, E. (1926). Die Ausbreitung des Girlitz (*Serinus canaria serinus* L.). *J. Orn.* **74:** 571–671.

Mayr, E. (1942). Systematics and the origin of species. New York, Columbia University Press.

Mayr, E. (1944). Wallace's Line in the light of recent zoogeographical studies. *Quart. Rev. Biol.* **19:** 1–14.

Mayr, E. (1946). History of the North American bird fauna. *Wilson Bull.* **58:** 3–41.

Mayr, E. (1954). Change of genetic environment and evolution. Pp. 157–180 in '*Evolution as a process*' (ed. J. Huxley, A. C. Hardy & E. B. Ford). London, Allen & Unwin.

Mayr, E. (1963). Animal species and evolution. London, Oxford University Press.

Mayr, E. (1969a). Bird speciation in the tropics. *Biol. J. Linn. Soc.* **1:** 1–17.

Mayr, E. (1969b). Principles of systematic zoology. New York, McGraw-Hill.

Mayr, E. (1985). Nearctic Region. Pp. 379–381 in '*A dictionary of birds*' (ed. B. Campbell & E. Lack). Calton, Poyser.

Mayr, E. & Diamond, J. M. (1976). Birds on islands in the sky: origin of the montane avifauna of Northern Melanesia. *Proc. Nat. Acad. Sci. USA* **73:** 1765–1769.

Mayr, E. & Diamond, J. M. (2001). The birds of Northern Melanesia. Speciation ecology and biogeography. Oxford, University Press.

Mayr, E. & Johnson, N. K. (2001). Is *Spizella taverneri* a species or a subspecies? *Condor* **103:** 418–419.

Mayr, E. & O'Hara, R. J. (1986). The biogeographic evidence supporting the Pleistocene forest refuge hypotheses. *Evolution* **40:** 55–67.

Mayr, E. & Phelps, W. J. (1967). The origin of the bird fauna of South Venezuelan highlands. *Bull. Amer. Mus. Nat. Hist.* **136:** 269–327.

Mayr, G. (1999). Caprimulgiform birds from the Middle Eocene of Messel (Hessen, Germany). *J. Vert. Paleont.* **19:** 521–532.

McDonald, D. B., Potts, W. K., Fitzpatrick, J. W. & Woolfenden, G. E. (1999). Contrasting genetic structures in sister species of North American scrub jays. *Proc. R. Soc. Lond. B* **266:** 1117–1125.

McNabb, B. K. (1994). Energy conservation and the evolution of flightlessness in birds. *Amer. Nat.* **144:** 628–642.

Mead, C. J. & Clark, J. A. (1988). Report on bird ringing in Britain and Ireland for 1987. *Ring. Migr.* **9:** 169–204.

Mead, C. J. & Harrison, J. D. (1979). Colony fidelity and interchange in the Sand Martin. *Bird Study* **26:** 99–106.

Mearns, R. & Newton, I. (1984). Turnover and dispersal in a Peregrine *Falco peregrinus* population. *Ibis* **126:** 347–355.

Mees, G. (1969). A systematic review of the Indo-Australian Zosteropidae (Part III). *Zool. Verhand.* **102:** 1–390.

Meise, W. (1936). Zur systematik und Verbreitungsgeschichte der Hause- und Weidensperlinge, *Passer domesticus* (L.) und *hispaniolensis* (T.). *J. Orn.* **84:** 631–672.

Mengel, R. M. (1964). The probable history of species formation in some northern wood warblers (Parulidae). *Living Bird* **3:** 9–43.

Mengel, R. M. (1970). The North American central plains as an isolating agent in bird speciation. Pp. 279–340 in *Pleistocene and recent environments of the central Great Plains*. Lawrence, Kansas, University Press.

Merendino, M. T., Ankney, C. D. & Dennis, D. G. (1993). Increasing Mallards, decreasing American Black Ducks: more evidence for cause and effect. *J. Wildl. Manage*. **57:** 199–208.

Merilä, J., Björkland, M. & Baker, A. J. (1997). Historical demography and present day population structure of the Greenfinch *Carduelis chloris* — an analysis of mtDNA control–region sequences. *Evolution* **51:** 946–956.

Metcalfe, N. B. & Furness, R. W. (1985). Survival, winter population stability and site-fidelity in the Turnstone *Arenarea interpres*. *Bird Study* **32:** 207–214.

Meury, R. (1989). Brutbiologie und Ortstreue einer Baumpieper population *Anthus trivialis* in einem inselartig verteilten Habitat des schweizerischen Mittellandes. *Orn. Beob*. **86:** 219–233.

Mihelsons, H., Mednis, A. & Blums, P. (1986). Populaton ecology of migratory ducks in Latvia. Riga, Zinatne. In Russian.

Milá, B., Girman, D. J., Kimura, M. & Smith, T. B. (2000). Genetic evidence for the effect of a postglacial population expansion on the phylogeography of a North American songbird. *Proc. R. Soc. Lond. B* **267:** 1033–1040.

Milberg, P. & Tyrberg, T. (1993). Naive birds and noble savages — a review of man-caused prehistoric extinctions of island birds. *Ecography* **16:** 229–250.

Millener, P. R. (1999). The history of the Chatham Islands' bird fauna of the last 7000 years — a chronicle of change and extinction. *Smithson. Contrib. Paleobiol*. **89:** 85–109.

Miller, A. H. (1931). Systematic revision and natural history of the American shrikes (*Lanius*). *Univ. Calif. Publ. Zool*. **38:** 11–242.

Miller, A. H. (1941). Speciation in the avian genus *Junco*. *Univ. Calif. Publ. Zool*. **44:** 173–434.

Mills, J. A. (1973). The influence of age and pair-bond on the breeding biology of the Red-billed Gull *Larus novaehollandiae scopulinus*. *J. Anim. Ecol*. **42:** 147–163.

Milot, E., Gibbs, H. L. & Hobson, K. A. (2000). Phylogeography and genetic structure of northern populations of the Yellow Warbler (*Dendroica petechia*). *Mol. Ecol*. **9:** 667–681.

Mindell, D. P., Sorenson, M. D., Huddleston, C. J., Miranda, H. C., Knight, A., Sawchuck, S. J. & Yuri, T. (1997). Phylogenetic relationships among and within select avian orders based on mitochondrial DNA. Pp. 213–247 in *Avian molecular evolution and systematics* (ed. D. P. Mindell). San Diego, Academic Press.

Mitra, S., Landel, H. & Pruett-Jones, S. (1996). Species richness covaries with mating system in birds. *Auk* **113:** 544–551.

Miura, G. I. & Edwards, S. V. (2001). Cryptic differentiation and geographic variation in genetic diversity of Hall's Babbler *Pomatostomus halli*. *J. Avian Biol*. **32:** 102–110.

Miyaki, C. M., Matioli, S. R., Burke, T. & Wajntal, A. (1998). Parrot evolution and paleogeographical events: mitochrondrial DNA evidence. *Mol. Biol. Evol*. **15:** 544–551.

Møller, A. P. & Cuervo, J. J. (1998). Speciation and feather ornamentation in birds. *Evolution* **52:** 859–869.

Monehan, T. M. (1994). Molecular genetic analysis of Adélie Penguin populations, Ross Island, Antarctica. New Zealand, University of Auckland.

Mönkkönen, M. (1994). Diversity patterns in Palaearctic and Nearctic forest bird assemblages. *J. Biogeog*. **21:** 193–195.

Mönkkönen, M., Helle, P. & Welsh, D. (1992). Perspectives on Palaearctic and Nearctic bird migration; comparisons and overview of life-history and ecology of migrant passerines. *Ibis* **134:** 7–13.

Montalvo, S. & Potti, J. (1992). Breeding dispersal in Spanish Pied Flycatchers *Ficedula hypoleuca*. *Ornis Scand*. **23:** 491–498.

Monteiro, L. R. & Furness, R. W. (1998). Speciation through temporal segregation of Madeiran Storm Petrel (*Oceanodroma castro*) populations in the Azores? *Phil. Trans. R. Soc. Lond. B* **353**: 945–953.

Monteiro, L. R., Ramos, J. A. & Furness, R. W. (1996). Past and present status and conservation of the seabirds breeding in the Azores Archipelago. *Biol. Conserv.* **78**: 319–328.

Mooney, N. & Brothers, N. (1993). Dispersion, nest and pair fidelity of Peregrine Falcons *Falco peregrinus* in Tasmania. Pp. 33–42 in '*Australian raptor studies*' (ed. P. Olsen). Melbourne, Australasian Raptor Association, RAOU.

Moore, W. S., Graham, J. H. & Price, J. T. (1991). Mitochondrial DNA variation in the Northern Flicker (*Colaptes auratus*, Aves). *Mol. Biol. Evol.* **8**: 327–344.

Moore, W. S. & Köenig, W. D. (1986). Comparative reproductive success of Yellow-shafted, Red-shafted, and hybrid flickers across a hybrid zone. *Auk* **103**: 42–51.

Moran, R. J. & Palmer, W. L. (1963). Ruffed Grouse introduction and population trends on Michigan Islands. *J. Wildl. Manage.* **27**: 606–664.

Moreau, R. E. (1954). The main vicissitudes of the European avifauna since the Pliocene. *Ibis* **96**: 411–431.

Moreau, R. E. (1955). Ecological changes in the Palaearctic region since the Pliocene. *Proc. Zool. Soc. Lond.* **125**: 253–295.

Moreau, R. E. (1963). Vicissitudes of the African biomes in the late Pleistocene. *Proc. Zool. Soc. Lond.* **141**: 395–421.

Moreau, R. E. (1966). The bird faunas of Africa and its islands. London, Academic Press.

Moreau, R. E. (1972). The Palaearctic–African bird migration system. London, Academic Press.

Morgan, K. H., Vermeer, K. & McKelvey, R. W. (1991). Atlas of pelagic birds of western Canada. Occas. Pap. 72. Ottawa, Can. Wildl. Serv.

Moritz, C., Richardson, K. S., Ferrier, S., Monteith, G. B., Stanisic, J., Williams, S. E. & Whiffin, T. (2001). Biogeographical concordance and efficiency of taxon indicators for establishing conservation priority in a tropical rainforest biota. *Proc. R. Soc. Lond. B* **268**: 1875–1881.

Moritz, D. (1993). Long-term monitoring of Palaearctic–African migrants at Helgoland/German Bight, North Sea. *Proc. Pan-African Orn. Congr.* **8**: 579–586.

Morley, R. J. & Flenley, J. R. (1986). Late Cenozoic vegetational and environmental changes in the Malay archipelago. Pp. 50–59 in '*Biogeographical evolution of the Malay archipelago*' (ed. T. C. Whitmore). Oxford, Clarendon Press.

Morton, E. S. (1980). Adaptations to seasonal changes by migrant land-birds in the Panama Canal Zone. Pp. 437–453 in '*Migrant birds in the Neotropics*' (ed. A. Keast & E. S. Morton). Washington, D.C., Smithsonian Institution Press.

Morton, M. L. (1997). Natal and breeding dispersal in the Mountain White-crowned Sparrow *Zonotrichia leucophrys oriantha*. *Ardea* **85**: 145–154.

Moss, R. (1986). Rain, breeding success and distribution of Capercaillie *Tetrao urogallus* and Black Grouse *Tetrao tetrix* in Scotland. *Ibis* **128**: 65–72.

Moulton, M. P., Miller, K. E. & Tillman, E. (2001). Patterns of success among introduced birds in the Hawaiian Islands. *Stud. Avian Biol.* **22**: 31–46.

Moulton, M. P. & Pimm, S. L. (1983). The introduced Hawaiian avifauna: biogeographic evidence for competition. *Amer. Nat.* **121**: 669–690.

Moulton, M. P. & Pimm, S. L. (1986). The extent of competition in shaping an introduced avifauna. Pp. 80–97 in '*Community ecology*' (ed. J. Diamond & T. J. Case). New York, Harper & Row.

Mourer-Chauviré, C. (1975). Les oiseaux du Pleistocène moyen et supérieur de France. Lyon, Thèse de l'Université Claude Bernard.

Mourer-Chauviré, C. (1982). Les oiseaux fossiles des Phosphorites due Quercy (Eocène Supérieur à Oligocène Supérieur): implications paléobiogéographiques. *Géobios Mem. Spec.* **6**: 413–426.

Mourer-Chauviré, C. (1985). Les Todidae (Aves, Coraciformes) des Phosphorites due Quercy (France). *Proc. K. Ned. Akad. Wet. Ser. B* **88:** 407–414.

Mourer-Chauviré, C. (1988). Les Caprimulgiformes et les Coraciiformes de l'Eocène et de l'Oligocène des Phosphorites du Quercy et description de deux genes nouveaux de Podargedae et Nyctibiidae. *Proc. Int. Orn. Congr.* **19:** 2045–2055.

Mourer-Chauviré, C., Bour, R., Ribes, S. & Moutou, F. (1999). The avifauna of Réunion Island (Mascarene Islands) at the time of the arrival of the first Europeans. *Smithson. Contrib. Paleobiol.* **89:** 1–38.

Mundy, N. I., Winchell, C. S., Burr, T. & Woodruff, D. S. (1997). Microsatellite variation and microevolution in the critically endangered San Clemente Island Loggerhead Shrike (*Lanius ludovicianus mearnsi*). *Proc. R. Roc. Lond. B* **264:** 869–875.

Murphy, R. C. (1936). Oceanic birds of South America. New York, Am. Mus. Nat. Hist.

Murray, R. D. (1979). Colonisation of Scotland by northern birds, 1820–1977. *Scot. Birds* **10:** 158–174.

Musser, G. G. (1986). The mammals of Sulawesi. Pp. 73–93 in *'Biogeographical evolution of the Malay Archipelago'* (ed. T. C. Whitmore). Oxford, Clarendon Press.

Nee, S. (1994). How populations persist. *Nature* **367:** 123–124.

Nee, S., May, R. M. & Harvey, P. H. (1994). The reconstructed evolutionary process. *Phil. Trans. R. Soc. Lond. B* **344:** 305–311.

Nee, S., Mooers, A. Ø. & Harvey, P. H. (1992). Tempo and mode of evolution revealed from molecular phylogenies. *Proc. Nat. Acad. Sci. USA* **89:** 8322–8326.

Nee, S., Read, A. F., Greenwood, J. J. D. & Harvey, P. H. (1991). The relationship between abundance and body size in British birds. *Nature* **351:** 312–313.

Negro, J. J., Hiraldo, F. & Donázar, J. A. (1997). Causes of natal dispersal in the Lesser Kestrel: inbreeding avoidance or resource competition? *J. Anim. Ecol.* **66:** 640–648.

Nehls, H. W. (1983). In Proceedings of the 3rd conference on the study and conservation of the migratory birds of the Baltic basin. *Ornis Fenn.* **Suppl. 3**

Nelson, B. (1978). The Gannet. Berkhamsted, Hertfordshire, Poyser.

Nelson, B. (1980). Seabirds. Their biology and ecology. London, Hamlyn.

Nelson, B. (1997). The Gannet. BWP Update. The journal of birds of the western Palaearctic. Oxford, University Press.

Nelson, G. & Platnick, N. (1981). Systematics and biogeography: cladistics and vicariance. New York, Columbia University Press.

Nelson, R. W. (1988). Do large natural broods increase mortality of parent Peregrine Falcons? Pp. 719–728 in *'Peregrine Falcon populations. Their management and recovery'* (ed. T. J. Cade, J. H. Enderson, C. G. Thelander & C. M. White). Boise, Idaho, The Peregrine Fund Inc.

Nelson, R. W. (1990). Status of the Peregrine Falcon *Falco peregrinus pealei*, on Langara Island, Queen Charlotte Islands, British Columbia, 1968–1989. *Can. Field-Nat.* **104:** 193–199.

Nesje, M., Røed, K. H., Bell, D. A., Lindberg, P. & Lifjeld, J. T. (2000). Microsatellite analysis of population structure and genetic variability in Peregrine Falcons (*Falco peregrinus*). *Anim. Conserv.* **3:** 267–275.

Newsome, A. E. & Noble, I. R. (1986). Ecological and physiological characters of invading species. Pp. 1–20 in *'Ecology of biological invasions'* (ed. R. H. Groves & J. J. Burdon). Cambridge, University Press.

Newton, I. (1967). The adaptive radiation and feeding ecology of some British finches. *Ibis* **109:** 33–98.

Newton, I. (1972). Finches. London, Collins.

Newton, I. (1977). Timing and success of breeding in tundra-nesting geese. Pp. 113–126 in *'Evolutionary ecology.'* (ed. B. Stonehouse & C. M. Perrins). London, Macmillan.

Newton, I. (1979). Population ecology of raptors. Berkhamsted, Hertfordshire, Poyser.

Newton, I. (1986). The Sparrowhawk. Calton, Staffordshire, Poyser.

Newton, I. (1991). Habitat variation and population regulation in Sparrowhawks. *Ibis* **133, suppl. 1.**: 76–88.

Newton, I. (1993). Age and site fidelity in female Sparrowhawks *Accipiter nisus*. *Anim. Behav.* **46:** 161–168.

Newton, I. (1994a). Current population levels of diurnal raptors in Britain. *The Raptor* **21:** 17–21.

Newton, I. (1994b). The role of nest-sites in limiting the numbers of hole-nesting birds: a review. *Biol. Conserv.* **70:** 265–276.

Newton, I. (1995a). Relationship between breeding and wintering ranges in Palaearctic–African migrants. *Ibis* **137:** 241–249.

Newton, I. (1995b). The contribution of some recent research on birds to ecological understanding. *J. Anim. Ecol.* **64:** 675–696.

Newton, I. (1997). Links between the abundance and distribution of birds. *Ecography* **20:** 137–145.

Newton, I. (1998a). Population limitation in birds. London, Academic Press.

Newton, I. (1998b). Migration patterns in West Palaearctic raptors. Pp. 603–612 in 'Holarctic birds of prey' (ed. R. D. Chancellor, B.-U. Meyburg & J. J. Ferrero). Calamonte, Spain, ADENEX-WWGBP.

Newton, I. (2000). Movements of Bullfinches *Pyrrhula pyrrhula* within the breeding season. *Bird Study* **47:** 372–376.

Newton, I. (2001). Causes and consequences of breeding dispersal in the Sparrowhawk. *Ardea* **89:** 143–154.

Newton, I. (2002a). Population limitation in owls. Pp. 1–27 in 'Ecology and conservation of owls' (ed. I. Newton, R. Kavanagh, J. Olson & I. R. Taylor). Canberra, CSIRO.

Newton, I. (2002b). Bullfinch. In *The migration atlas: movements of the birds of Britain and Ireland*' (ed. C. V. Wernham, M. P. Toms, J. H. Marchant, J. A. Clark, G. M. Siriwardena & S. R. Baillie). London, Poyser.

Newton, I. & Dale, L. (1996a). Relationship between migration and latitude among west European birds. *J. Anim. Ecol.* **65:** 137–146.

Newton, I. & Dale, L. (1996b). Bird migration at different latitudes in eastern North America. *Auk* **113:** 626–635.

Newton, I. & Dale, L. (1997). Effects of seasonal migration on the latitudinal distribution of west Palaearctic bird species. *J. Biogeog.* **24:** 781–789.

Newton, I. & Dale, L. (2001). A comparative analysis of the avifaunas of different zoogeographical regions. *J. Zool., Lond.* **254:** 207–218.

Newton, I., Davis, P. E. & Moss, D. (1994). Philopatry and population growth of Red Kites *Milvus milvus* in Wales. *Proc. R. Soc. Lond. B* **257:** 317–323.

Newton, I., Kavanagh, R. P., Olsen, J. & Taylor, I. R. (eds.) (2002). The ecology and conservation of owls. Canberra, CSIRO.

Newton, I. & Marquiss, M. (1982). Fidelity to breeding area and mate in Sparrowhawks *Accipiter nisus*. *J. Anim. Ecol.* **51:** 327–341.

Newton, I. & Marquiss, M. (1983). Dispersal of Sparrowhawks between birthplace and breeding place. *J. Anim. Ecol.* **52:** 463–477.

Newton, I. & Rothery, P. (2000). Post–fledging recovery and dispersal of ringed Eurasian Sparrowhawks *Accipiter nisus*. *J. Avian Biol.* **31:** 226–236.

Nichols, J. D. & Haramis, G. M. (1980). Sex-specific differences in winter distribution patterns of Canvasbacks. *Condor* **82:** 406–416.

Nichols, J. D. & Hines, J. E. (1987). Population ecology of the Mallard. VIII. Winter distribution patterns and survival rates of winter-banded Mallards. Res. Publ. 162. Washington, U.S. Fish & Wildl. Serv.

Niethammer, G. (1937). Handbuch der deutschen Vogelkunde. Leipzig, Akademische Verlagsgesellschaft.

Nilsson, J.-A. (1989). Causes and consequences of natal dispersal in the Marsh Tit, *Parus palustris*. *J. Anim. Ecol.* **58**: 619–636.

Nilsson, L., Andersson, O., Gustafsson, R. & Svensson, M. (1999). Increase and changes in distribution of breeding Whooper Swans in northern Sweden from 1972–75 to 1997. *Wildfowl* **49**: 6–17.

Nisbet, I. C. T. (1978). Recent changes in gull populations in the western North Atlantic. *Ibis* **120**: 129–130.

Nisbet, I. C. T. & Medway, L. (1972). Dispersion, population ecology and migration of Eastern Great Reed Warblers *Acrocephalus orientalis* wintering in Malaysia. *Ibis* **114**: 451–494.

Nisbet, I. C. T. & Safina, C. (1996). Transatlantic recoveries of ringed Common Terns *Sterna hirundo*. *Ring. Migr.* **17**: 28–30.

Nix, H. A. (1974). Environmental control of breeding, post-breeding dispersal and migration of birds in the Australian region. *Proc. Int. Orn. Congr.* **16**: 272–305.

Nolan, V. (1978). The ecology and behaviour of the Prairie Warbler (*Dendroica discolor*). *Orn. Monogr.* **26**: 1–595.

Noon, B. R. (1981). The distribution of an avian guild along a temperate elevational gradient — the importance and expression of competition. *Ecol. Monogr.* **51**: 105–124.

Norell, M., Ji, Q., Gao, K., Yuan, C., Zhao, Y. & Wang, L. (2002). 'Modern' feathers on a non–avian dinosaur. *Nature* **416**: 36.

Nores, M. (1995). Insular biogeography of birds on mountain-tops in northwestern Argentina. *J. Biogeog.* **22**: 61–70.

Nores, M. (1999). An alternative hypothesis for the origin of Amazonia bird diversity. *J. Biogeog.* **26**: 475–485.

Norman, S. C. (1994). Dispersal and return rates of Willow Warbler *Phylloscopus trochilus* in relation to age, sex and season. *Ring. Migr.* **15**: 8–16.

Norris, C. A. (1945). Summary of a report on the distribution and status of the Corncrake (*Crex crex*). *Brit. Birds* **38**: 142–148, 162–168.

Norton, D. W., Senner, S. E., Gill, R. E., Martin, R. D., Wright, J. M. & Fukuyama, A. S. (1990). Shorebirds and herring roe in Prince William Sound, Alaska. *Amer. Birds* **44**: 367–371.

Norton, I. & Sclater, J. G. (1979). A model for the evolution of the Indian Ocean and the breakup of Gondwanaland. *J. Geophys. Res.* **84**: 6803–6830.

Novak, E. (1971). The range expansion of animals and its causes. *Zeszyty Naukowe* **3**: 1–255.

Nunn, G. B., Cooper, J., Jouventin, P., Robertson, C. J. R. & Robertson, G. C. (1996). Evolutionary relationships among extant albatrosses (Procellariiformes: Diomedeidae) established from complete cytochrome-*b* gene sequences. *Auk* **113**: 784–801.

Nunn, G. B. & Stanley, S. E. (1998). Body size effects and rates of cytochrome b evolution in tube-nosed seabirds. *Mol. Biol. Evol.* **15**: 1360–1371.

O'Connor, R. J. (1981). Comparisons between migrant and non-migrant birds in Britain. Pp. 167–195 in '*Animal migration*' (ed. D. J. Aidley). Cambridge, University Press.

O'Connor, R. J. (1986). Biological characteristics of invaders among bird species in Britain. *Phil. Trans. R. Soc. Lond. B* **314**: 583–598.

O'Donald, P. (1959). Possibility of assortive mating in the Arctic Skua. *Nature* **183**: 1210–1211.

Okubo, A. (1988). Diffusion-type models for avian range expansion. *Proc. Int. Orn. Congr.* **19**: 1038–1049.

Olson, S. L. (1975). Paleornithology of St. Helena Island, South Atlantic Ocean. *Smithson. Contrib. Paleobiol.* **23**: 1–49.

Olson, S. L. (1976). Oligocene fossils bearing on the origins of the Todidae and Momotidae (Aves: Coraciiformes). *Smithson. Contrib. Paleobiol.* **27**: 111–119.

Olson, S. L. (1977). Additional notes on the subfossil bird remains from Ascension Island. *Ibis* **119**: 37–43.

Olson, S. L. (1980). The significance of the distribution of the Megapodiidae. *Emu* **80:** 21–24.

Olson, S. L. (1982). Fossil vertebrates from the Bahamas. *Smithson. Contrib. Paleobiol.* **48:** 1–65.

Olson, S. L. (1983). Fossil seabirds and changing marine environments in the late Tertiary of South Africa. *S. Afr. J. Sci.* **79:** 399–402.

Olson, S. L. (1985a). Early Pliocene Procellariiformes (Aves) from Langebaanweg, South-western Cape Province, South Africa. *Ann. S. Afr. Mus.* **95:** 123–145.

Olson, S. L. (1985b). The fossil record of birds. Pp. 79–238 in *'Avian biology, Vol. 8'* (ed. D. S. Farner, J. R. King & K. C. Parkes). London, Academic Press.

Olson, S. L. (1988). Aspects of global avifaunal dynamics during the Cenozoic. *Proc. Int. Orn. Congr.* **19:** 2023–2029.

Olson, S. L. (1990). The prehistoric impact of man on biogeographical patterns of insular birds. Pp. 45–51 in *'Biogeographical aspects of insularity'*. Rome, Accademia Nazionale dei Lincei.

Olson, S. L. (1991). Patterns of avian diversity and radiation in the Pacific as seen through the fossil record. Pp. 314–318 in *'The unit of evolutionary biology'* (ed. E. C. Dudley). Portland, Oregon, Dioscorides Press.

Olson, S. L. & Hasegawa, Y. (1979). Fossil counterparts of giant penguins in the North Pacific. *Science* **206:** 688–689.

Olson, S. L. & James, H. F. (1982). Fossil birds from the Hawaiian Islands: evidence for wholesale extinction by man before western contact. *Science* **217:** 633–635.

Olson, S. L. & James, H. F. (1991). Descriptions of thirty-two new species of birds from the Hawaiian Islands. Parts 1 & 2. *Orn. Monogr.* **45:** 1–88; **46:** 1–88.

Olson, S. L. & Kurochin, E. N. (1988). The early radiation of birds. Introduction. *Proc. Int. Orn. Congr.* **19:** 2022.

Olson, S. L. & Rasmussen, P. C. (2001). Miocene and Pliocene birds from the Lee Creek Mine, North Carolina. Pp. 233–365 in *'Geology and paleontology of the Lee Creek Mine, North Carolina, III'* (ed. E. R. Clayton & D. J. Bohaska). Smithsonian Contributions to Paleobiology 90. Washington, Smithsonian Institution Press.

Olssen, O. (1958). Dispersal, migration, longevity and death causes of *Strix aluco, Buteo buteo, Ardea cinerea* and *Larus argentatus*. *Acta Vert.* **1:** 91–189.

Olsson, V. (1969). Die Expansion der Girlitzes (*Serinus serinus*) in Nordeuropa in den letzten Jahrzeheten. *Vogelwarte* **25:** 147–156.

Omland, K. E., Tarr, C. L., Boarman, W. I., Marzluff, J. L. & Fleischer, R. C. (2000). Cryptic genetic variation and paraphyly in Ravens. *Proc. R. Soc. Lond. B* **267:** 2475–2482.

Opdam, P., Foppen, R., Reijnen, R. & Schotman, A. (1995). The landscape ecological approach in bird conservation: integrating the metapopulation concept into spatial planning. *Ibis* **137, Suppl. 1:** 139–146.

Orell, M. (1989). Population fluctuations and survival of Great Tits *Parus major* dependent on food supplied by man in winter. *Ibis* **131:** 113–127.

Oring, L. W. & Lank, D. B. (1982). Sexual selection, arrival times, philopatry and site fidelity in the polyandrous Spotted Sandpiper. *Behav. Ecol. Sociobiol.* **10:** 185–191.

Osorio-Beristain, M. & Drummond, H. (1993). Natal dispersal and deferred breeding in the Blue-footed Booby. *Auk* **110:** 234–239.

Ovenden, J. R., Wurt-Saucy, A., Bywater, R., Brothers, N. & White, R. W. G. (1991). Genetic evidence for philopatry in a colonially nesting seabird, the Fairy Prion (*Pachyptila turtur*). *Auk* **108:** 688–694.

Owen, M., Atkinson-Willes, G. L. & Salmon, D. G. (1986). Wildfowl in Great Britain. 2nd ed. Cambridge, University Press.

Owens, I. P. F., Bennett, P. M. & Harvey, P. H. (1999). Species richness among birds: body size, life history, sexual selection or ecology? *Proc. R. Soc. Lond. B.* **266:** 933–939.

Owens, I. P. F. & Hartley, I. R. (1998). Sexual dimorphism in birds: why are there so many forms of dimorphism? *Proc. R. Soc. Lond. B* **265:** 397–407.

Padian, K. & Chappe, L. M. (1998). The origin and early evolution of birds. *Biol. Rev.* **73:** 1–42.

Palmer, W. L. (1962). Ruffed Grouse flight capability over water. *J. Wildl. Manage.* **26:** 338–339.

Paradis, E., Baillie, S. R., Sutherland, W. J. & Gregory, R. D. (1998). Patterns of natal and breeding dispersal in birds. *J. Anim. Ecol.* **67:** 518–536.

Parkin, D. T. (2002). Birding and DNA: species for the new millennium. *Bird Study* **49,** in press.

Parmelee, D. F. (1992). Snowy Owl. *Birds N. Amer.* **19:** 1–19.

Parr, R. (1980). Population study of Golden Plover *Pluvialis apricaria*, using marked birds. *Ornis Scand.* **11:** 179–189.

Parrinder, E. D. (1989). Little Ringed Plovers *Charadrius dubius* in Britain in 1984. *Bird Study* **36:** 147–153.

Pascal, M. (1980). Population structure and dynamics of feral cats on Kerguelen Islands. *Mammalia* **44:** 161–182.

Pasquet, E. (1998). Phylogeny of the nuthatches of the *Sitta canadensis* group and its evolutionary and biogeographic implications. *Ibis* **140:** 150–156.

Paterson, H. E. H. (1985). The recognition concept of species. Pp. 21–29 in '*Species and speciation*' (ed. E. S. Vrba). Transvaal Museum Monograph No.4. Pretoria, Transvaal Museum.

Paton, P. W. C. & Edwards, T. C. (1996). Factors affecting interannual movements of Snowy Plovers. *Auk* **113:** 534–543.

Paton, T., Haddrath, O. & Baker, A. J. (2002). Complete mitochondrial DNA genome sequences show that modern birds are not descended from transitional shorebirds. *Proc. R. Soc. Lond. B* **269:** 839–846.

Patterson, B. D. (1984). Mammalian extinction and biogeography in the southern Rocky Mountains. Pp. 247–293 in '*Extinctions*' (ed. M. H. Nitecki). Chicago, University Press.

Payevsky, V. A. (1971). Atlas of bird migrations according to banding data at the Courland Spit. Pp. 1–124 in '*Bird migrations. Ecological and physiological factors*' (ed. B. E. Bykhovskii). Leningrad.

Payne, R. B. (1991). Natal dispersal and population structure in a migratory songbird, the Indigo Bunting. *Evolution* **45:** 49–62.

Payne, R. B. & Payne, L. L. (1993). Breeding dispersal in Indigo Buntings: circumstances and consequences for breeding success and population structure. *Condor* **95:** 1–24.

Pearson, D. J. & Lack, P. C. (1992). Migration patterns and habitat use by passerine and near-passerine migrant birds in eastern Africa. *Ibis* **134, suppl. 1.:** 89–98.

Peiponen, V. A. (1967). Südliche Fortpflanzung und Zug von *Carduelis flammea* (L.) in Jahre 1965. *Ann. Zool. fenn.* **4:** 547–549.

Peirce, M. A. (1979). Some additional observations on haematozoa of birds in the Mascarene islands. *Bull. Brit. Orn. Club* **99:** 68–71.

Pennington, R. T., Prado, D. E. & Pendry, C. A. (2000). Neotropical seasonally dry forests and Quaternary vegetation changes. *J. Biogeog.* **27:** 261–273.

Percival, S. M. (1991). The population structure of Greenland Barnacle Geese *Branta leucopsis* on the winter grounds on Islay. *Ibis* **133:** 357–364.

Peters, J. L. (1940). Checklist of the birds of the world, Vol. 4. Cambridge, Mass., Harvard University Press.

Petersen, M., Larned, W. H. & Douglas, D. C. (1999). At sea distribution of Spectacled Eiders: a 120-year-old mystery resolved. *Auk* **116:** 1009–1020.

Peterson, A. T. (1992). Philopatry and genetic differentiation in the *Aphelocoma* jays (Corvidae). *Biol. J. Linn. Soc.* **47:** 249–260.

Petren, K., Grant, B. R. & Grant, P. R. (1999). A phylogeny of Darwin's finches based on microsatellite DNA length variation. *Proc. R. Soc. Lond. B* **266:** 321–329.

Petrie, M. (1999). Sexual selection and avian speciation. *Proc. Int. Orn. Congr.* **22**

Petrie, M., Halliday, T. & Sanders, C. (1991). Peahens prefer peacocks with elaborate trains. *Anim. Behav.* **41:** 323–331.

Peus, F. (1951). Nüchterne Analyse der Massenvermehrung der Misteldrossel (*Turdus viscivorus* L.) in Nordwesteurope. *Bonn. Zool. Beitr.* **2:** 55–81.

Phillips, R. A., Bearhop, S., Hamer, K. C. & Thompson, D. R. (1999). Rapid population growth of Great Skuas *Catharacta skua* at St Kilda: implications for management and conservation. *Bird Study* **46:** 174–183.

Pielou, E. C. (1979). Biogeography. New York, Wiley.

Pielou, E. L. (1991). After the Ice Age. The return of life to glaciated North America. Chicago, University Press.

Pienkowski, M. W. (1984). Breeding biology and population dynamics of Ringed Plovers *Charadrius hiaticula* in Britain and Greenland: nest predation as a possible factor limiting distribution and timing of breeding. *J. Zool. Lond.* **202:** 83–114.

Piertney, S. B., Summers, R. & Marquiss, M. (2001). Microsatellite and mitochondrial DNA homogeneity among phenotypically diverse crossbill taxa in the UK. *Proc. R. Soc. Lond.* B **268:** 1511–1517.

Pimm, S. L. (1991). The balance of nature? Chicago, University of Chicago Press.

Pimm, S. L., Jones, H. L. & Diamond, J. (1988). On the risk of extinction. *Amer. Nat.* **132:** 757–785.

Pimm, S. L., Moulton, M. P. & Justice, L. J. (1994). Bird extinctions in the central Pacific. *Phil. Trans. R. Soc. Lond.* B **344:** 27–33.

Pinowski, J. (1965). Overcrowding as one of the causes of dispersal of young Tree Sparrows. *Bird Study* **12:** 27–33.

Pinshow, B., Fedah, M. A., Battles, D. R. & Schmidt-Nielsen, K. (1976). Energy expenditure for thermoregulation and locomotion in Emperor Penguins. *Amer. J. Physiol.* **231:** 903–912.

Pitman, R. L. (1986). Atlas of seabird distribution and relative abundance in the eastern tropical Pacific. La Jolla, Administrative Rep. LT-86–02C, South Eastern Fisheries Centre.

Pitra, C., Lieckfeldt, D. & Alonso, J. C. (2000). Population subdivision in Europe's Great Bustard inferred from mitochondrial and nuclear DNA sequence variation. *Mol. Ecol.* **9:** 1165–1170.

Pizzey, G. & Knight, F. (1997). Field guide to the birds of Australia. Sydney, Angus & Robertson.

Ploeger, P. L. (1968). Geographical differentiation in arctic Anatidae as a result of isolation during the last glacial. *Ardea* **56:** 1–159.

Pomeroy, D. & Ssekabiira, D. (1990). An analysis of the distributions of terrestrial birds in Africa. *Afr. J. Ecol.* **28:** 1–13.

Poole, A. F. (1989). Ospreys. A natural and unnatural history. Cambridge, University Press.

Potts, G. R. (1986). The Partridge: pesticides, predation and conservation. London, Collins.

Power, D. M. & Ainley, D. G. (1986). Seabird geographic variation: similarity among populations of Leach's Storm Petrel. *Auk* **103:** 575–585.

Powers, K. D. (1983). Pelagic distributions off the northeastern United States. Washington, NOAA.

Prance, G. J. (ed.) (1982). Biological diversification in the tropics. New York, Columbia University Press.

Pratt, A. & Peach, W. (1991). Site tenacity and annual survival of a Willow Warbler *Phylloscopus trochilus* population in southern England. *Ring. Migr.* **12:** 128–134.

Pratt, H. D., Bruner, P. L. & Berrett, D. G. (1987). A field guide to the birds of Hawaii and the tropical Pacific. Princeton, University Press.

Pratt, T. K. (1991). Biogeography of New Guinea birds: a re-evaluation in light of new systematic and ecological information. *Proc. Int. Orn. Congr.* **20:** 425–433.

Pregill, G. K. & Olson, S. L. (1981). Zoogeography of West Indian vertebrates in relation to Pleistocene climatic cycles. *Ann. Rev. Ecol. Syst.* **12:** 75–98.

Prentice, C., Jolly, D. & BIOME 6000 participants. (2000). Mid-Holocene and glacial maximum vegetation geography of the northern continents and Africa. *J. Biogeog.* **27:** 507–519.

Prevett, J. P. & MacInnes, C. D. (1980). Family and other social groups in Snow Geese. *Wildlife Monogr.* **71:** 1–46.

Price, J., Droege, S. & Price, A. (1995). The summer atlas of North American birds. London, Academic Press.

Price, T. (1981). The ecology of the Greenish Warbler *Phylloscopus trochiloides* in its winter quarters. *Ibis* **123:** 131–144.

Price, T. (1996). Exploding species. *Trends Ecol. Evol.* **11:** 314–315.

Price, T. (1998). Sexual selection and natural selection in bird speciation. *Phil. Trans. R. Soc. Lond. B* **353:** 251–260.

Prigogine, A. (1984). Speciation problems in birds with special reference to the Afrotropical region. *Mitt. Zool. Berl. 60 Suppl. Ann. Orn* **8:** 3–27.

Prigogine, A. (1988). Speciation pattern of birds in the central African forest refugia and their relationship with other refugia. *Proc. Int. Orn. Congr.* **19:** 2537–2546.

Prince, P. A. & Croxall, J. P. (1983). Birds of South Georgia: new records and re-evaluation of status. *Br. Antarct. Surv. Bull.* **59:** 15–27.

Prince, P. A., Croxall, J. P., Trathan, P. N. & Wood, A. G. (1998). The pelagic distribution of South Georgia Albatrosses and their relationships with fisheries. Pp. 137–167 in 'Albatross biology' (ed. G. Robertson & R. Gales). Chipping Norton, New South Wales, Surrey Beatty.

Provençal, P. & Sørensen, U. G. (1998). Medieval records of the Siberian White Crane *Grus leucogeranus* in Egypt. *Ibis* **140:** 333–335.

Prum, R. O. (1988). Historical relationships around avian forest areas of endemism in the Neotropics. *Proc. Int. Orn. Congr.* **19:** 2562–2572.

Prum, R. O. (2002). Why ornithologists should care about the theropod origin of birds. *Auk* **119:** 1–17.

Pulliam, H. R. (1988). Sources, sinks, and population regulation. *Amer. Nat.* **132:** 652–661.

Quinn, T. W. (1992). The genetic legacy of mother goose — phylogeographic patterns of Lesser Snow Goose *Chen caerulescens caerulescens* maternal lineages. *Mol. Evol.* **1:** 105–117.

Quinn, T. W., Shields, G. F. & Wilson, A. C. (1991). Affinities of the Hawaiian Goose based on two types of mitochondrial DNA data. *Auk* **108:** 585–593.

Rabøl, J. (1969). Reversed migration as the cause of westward vagrancy by four *Phylloscopus* warblers. *Brit. Birds* **62:** 89–92.

Rabouam, C., Bretagnolle, V., Bigot, Y. & Periquet, G. (2000). Genetic relationships of Cory's Shearwater: parentage, mating assortment, and geographic differentiation revealed by DNA fingerprinting. *Auk* **177:** 651–662.

Ramos, J. A., Monteiro, L. R., Sola, E. & Moniz, Z. (1997). Characteristics and competition for nest cavities in burrowing Procellariiformes. *Condor* **99:** 634–641.

Rand, A. L. (1948). Glaciation, an isolating factor in speciation. *Evolution* **2:** 314–321.

Randi, E., Spina, F. & Marsa, B. (1989). Genetic variability in Cory's Shearwater (*Calonectris diomedea*). *Auk* **106:** 411–417.

Randler, C. (2002). Avian hybridisation, mixed pairing and female choice. *Anim. Behav.* **63:** 103–119.

Rapoport, E. H. (1982). Areography: geographical strategies of species. Oxford, Pergamon.

Rappole, J. H., Morton, E. S., Lovejoy, T. E. & Ruos, J. L. (1983). Nearctic avian migrants in the Neotropics. Washington, United States Fish & Wildlife Service.

Rass, T. S. (1986). Vicariance ichthyogeography of the Atlantic ocean pelageal. Pp. 237–241 in 'Pelagic biogeography'. UNESCO Technical Paper in Marine Science 49.

Raup, D. M. (1976). Species diversity in the Phanerozoic: an interpretation. *Paleobiology* **2:** 289–297.

Raveling, D. G. (1979). The annual cycle of body composition of Canada Geese with special reference to control of reproduction. *Auk* **96:** 234–252.

Rechner, H. (1992). The effect of extra food on fitness in breeding Carrion Crows. *Ecology* **73:** 330–335.

Rees, E. C. (1987). Conflict of choice within pairs of Bewick's Swans regarding their migratory movement to and from wintering grounds. *Anim. Behav.* **35:** 1685–1693.

Rees, E. C. (1989). Consistency in the timing of migration for individual Bewick's Swans. *Anim. Behav.* **38:** 384–393.

Rehfisch, M. M., Clark, N. A., Langston, R. H. W. & Greenwood, J. J. D. (1996). A guide to the provision of refuges for waders: an analysis of 30 years of ringing data from the Wash, England. *J. Appl. Ecol.* **33:** 673–687.

Reid, E. M. (1935). British floras antecedent to the Great Ice Age. *Proc. R. Soc. Lond. B* **118:** 197.

Reid, K. & Croxall, J. P. (2001). Environmental response of upper trophic level predators reveals a system change in an Antarctic marine ecosystem. *Proc. R. Soc. Lond. B* **268:** 377–384.

Reinig, W. F. (1950). Chorologische Voraussetzungen für die Analyse von Formenkreisen. Pp. 346–378 in 'Syllegomena Biol. (Festschr. Kleinschmidt).' Leipzig, Geest & Portig.

Remington, C. L. (1968). Suture-zones of hybrid interaction between recently joined biotas. *Evol. Biol.* **2:** 321–428.

Remsen, J. V. (1984). High incidence of 'leapfrog' pattern of geographic variation in Andean birds: implications for speciation process. *Science* **224:** 171–172.

Rensch, B. (1933). Zoologische Systematik und Artbildungsproblem. *Verh. dtsch. zool. Ges.* **1933:** 19.

Repasky, R. R. (1991). Temperature and the northern distributions of wintering birds. *Ecology* **72:** 2274–2285.

Reynolds, R. T. & Linkhart, B. D. (1987). Fidelity to territory and mate in Flammulated Owls. Pp. 234–238 in 'Biology and conservation of northern forest owls' (ed. R. W. Nero, R. J. Clark, R. J. Knapton & R. H. Hamre). Fort Collins, Colorado, U.S. Dept. Agric., Forest Service, Rocky Mountain Forest & Range Experimental Station.

Rheinwald, G. (1975). The pattern of settling distances of House Martins *Delichon urbica*. *Ardea* **63:** 136–145.

Rheinwald, G. & Gutscher, H. (1969). Dispersion und Ortstreue der Mehlschwalbe (*Delichon urbica*). *Die Vogelwelt* **90:** 121–140.

Rhymer, J. M. & Simberloff, D. (1996). Extinction by hybridisation and introgression. *Ann. Rev. Ecol. Syst.* **27:** 83–109.

Richdale, L. E. (1963). Biology of the Sooty Shearwater *Puffinus griseus*. *Proc. Zool. Soc. Lond.* **141:** 1–117.

Ricklefs, R. E. & Bermingham, E. (1999). Taxon cycles in the Lesser Antillean avifauna. *Ostrich* **70:** 49–59.

Ricklefs, R. E. & Cox, G. W. (1972). Taxon cycles in the West Indian avifauna. *Amer. Nat.* **106:** 195–219.

Ricklefs, R. E. & Cox, G. W. (1978). Stage of taxon cycle, habitat distribution, and population density in the avifauna of the West Indies. *Amer. Nat.* **112:** 875–895.

Ridgill, S. C. & Fox, A. D. (1990). Cold weather movements of waterfowl in western Europe. Spec. Publ. 13. Slimbridge, Gloucestershire, IWRB.

Ridpath, M. G. & Moreau, R. E. (1966). The birds of Tasmania: ecology and evolution. *Ibis* **108:** 348–393.

Rising, J. D. (1983). The Great Plains hybrid zones. *Curr. Ornith.* **1**: 131–157.

Robbins, C. S. (1985). Recent changes to the ranges of North American birds. *Proc. Int. Orn. Congr.* **18**: 737–742.

Robertson, C. J. R. & Nunn, G. B. (1998). Towards a new taxonomy for albatrosses. Pp. 13–19 in *'Albatross biology and conservation'* (ed. G. Robertson & R. Gales). Chipping Norton, New South Wales, Surrey Beatty.

Robertson, G. & Cooke, F. (1999). Winter philopatry in migratory waterfowl. *Auk* **116**: 20–34.

Rockwell, R. F. & Cooke, F. (1977). Gene flow and local adaptation in a colonially nesting dimorphic bird: the Lesser Snow Goose (*Anser caerulescens caerulescens*). *Amer. Nat.* **11**: 91–97.

Rogers, M.J. & the Rarities Committee (1996). Report on rare birds in Great Britain in 1995. *Brit. Birds* **89**: 481–531.

Rogers, M. J. & the Rarities Committee (1998). Report on rare birds in Great Britain in 1997. *Brit. Birds* **91**: 455–517.

Rohde, K. (1992). Latitudinal gradients in species diversity: the search for the primary cause. *Oikos* **65**: 514–527.

Rohwer, F. C. & Anderson, M. G. (1988). Female-biased philopatry, monogamy, and the timing of pair formation in migratory waterfowl. *Curr. Ornith.* **5**: 187–221.

Rolando, A. (1993). A study on the hybridisation between Carrion and Hooded Crow in northwest Italy. *Ornis Scand.* **24**: 80–83.

Romagosa, C. M. & Labisky, R. F. (2000). Establishment and dispersal of the Eurasian Collared Dove in Florida. *J. Field Orn.* **71**: 159–166.

Root, T. (1988a). Atlas of wintering North American birds. Chicago, University Press.

Root, T. (1988b). Energy constraints on avian distributions and abundances. *Ecology* **69**: 330–339.

Root, T. (1988c). Environmental factors associated with avian distributional boundaries. *J. Biogeog.* **15**: 489–505.

Root, T. L. & Schneider, S. H. (1993). Can large-scale climatic models be linked with multi-scale ecological studies. *Conserv. Biol.* **7**: 256–278.

Rosenzweig, M. L. (1992). Species diversity gradients: we know more and less than we thought. *J. Mammal.* **73**: 715–730.

Rosenzweig, M. L. (1995). Species diversity in space and time. Cambridge, University Press.

Roy, H. S., da Silva, J. M. C., Arctander, P., Garcia-Moreno, J. & Fjeldså, J. (1997). The speciation of South American and African birds in montane regions. Pp. 325–343 in *'Avian molecular evolution and systematics'* (ed. D. P. Mindell). London, Academic Press.

Roy, M. S. (1997). Recent diversification in African greenbuls (Pycnonotidae: *Andropadus*) supports a montane speciation model. *Proc. R. Soc. Lond. B* **264**: 1337–1344.

Ruggiero, A. & Lawton, J. H. (1998). Are there latitudinal and altitudinal Rapoport effects in the geographic ranges of Andean passerine birds? *Biol. J. Linn. Soc.* **63**: 283–304.

Russell, E. & Rowley, I. (1993). Philopatry or dispersal: competition for territory vacancies in the Splendid Fairy Wren *Malurus splendens*. *Anim. Behav.* **43**: 519–539.

Rustamov, A. K. (1985). Birds and man-made environmental changes in the arid zone of the USSR. *Proc. Int. Orn. Congr.* **18**: 584–587.

Ryan, P. G., Hood, I., Bloomer, P., Komen, J. & Crowe, T. W. (1998). Barlow's Lark: a new species in the Karoo Lark *Certhilauda albescens* complex of southwest Africa. *Ibis* **140**: 605–619.

Saetre, G.-P., Borge, T. & Moum, T. (2001). A new bird species? The taxonomic status of 'the Atlas Flycatcher' assessed from DNA sequence analysis. *Ibis* **143**: 494–497.

Saetre, G.-P., Král, M., Bures, S. & Ims, R. A. (1999). Dynamics of a clinal hybrid zone and a comparison with island hybrid zones of flycatchers (*Ficedula hypoleuca* and *F. albicollis*). *J. Zool. Lond.* **247**: 53–64.

Saetre, G.-P., Moum, T., Král, M., Burns, S., Adamjan, M. & Moreno, J. (1997). A sexually selected character displacement in flycatchers reinforces premating isolation. *Nature* **387**: 589–592.

Saino, N. & Villa, S. (1992). Pair composition and reproductive success across a hybrid zone of Carrion Crows and Hooded Crows. *Auk* **109**: 543–555.

Salewski, V., Bairlein, F. & Leisler, B. (2000). Recurrence of some Palaearctic migrant passerine species in West Africa. *Ring. Migr.* **20**: 29–30.

Salomonsen, F. (1931). Diluviale Isolation und Artenbildung. *Proc. Int. Orn. Congr.* **7**: 413–438.

Salomonsen, F. (1951). The immigration and breeding of the Fieldfare (*Turdus pilaris* L.) in Greenland. *Proc. Int. Orn. Congr.* **10**: 515–525.

Salomonsen, F. (1955). The evolutionary significance of bird-migration. *Dan. Biol. Medd* **22**: 1062.

Salomonsen, F. (1967). Fuglene på Grønland. Copenhagen, Rhodos.

Salomonsen, F. (1968). The moult migration. *Wildfowl* **19**: 5–24.

Salomonsen, F. (1972). Zoogeographical and ecological problems in arctic birds. *Proc. Int. Orn. Congr.* **15**: 25–77.

Sandercock, B. K. & Gratto-Trevor, C. L. (1997). Local survival in Semipalmated Sandpipers *Calidris pusilla* breeding at La Perouse Bay, Canada. *Ibis* **139**: 305–312.

Sangster, G. (2000). Genetic distance as a test of species boundaries in the Citril Finch *Serinus citrinella*: a critique and taxonomic reinterpretation. *Ibis* **142**: 487–490.

Sangster, G., Collinson, M., Helbig, A. J., Knox, A. G., Parkin, D. T. & Prater, T. (2001). The taxonomic status of Green-winged Teal. *Brit. Birds* **94**: 218–226.

Sangster, G., Hazevoet, C. J., van den Berg, A. B., Roselaar, C. S. & Slerys, K. (1999). Dutch avifaunal list: species concepts, and taxonomic changes in 1977–1998. *Ardea* **87**: 139–166.

Sappington, J. N. (1977). Breeding biology of House Sparrows in Mississippi. *Wilson Bull.* **89**: 300–309.

Sato, A., Tichy, H. O'h Uigin, C., Grant, P.R., Grant, R. and Klim, J. (2001). On the origin of Darwin's Finches. *Mol. Biol. & Evol.* **18**: 299–31.

Saunders, D. A. & Curry, P. J. (1990). The impact of agricultural and pastoral industries on birds in the southern half of western Australia: past, present and future. *Proc. Ecol. Soc. Aust.* **16**: 303–321.

Saurola, P. (1994). African non-breeding areas of Fennoscandian Ospreys *Pandion haliaetus*: a ring recovery analysis. *Ostrich* **65**: 127–136.

Saurola, P. (2002). Natal dispersal distances of Finnish owls: results from ringing. Pp. 40–53 in 'Ecology and conservation of owls' (ed. I. Newton, R. Kavannagh, J. Olsen & I. Taylor). Canberra, CSIRO.

Sauvage, A., Rumsey, S. & Rodwell, S. (1998). Recurrence of Palaearctic birds in the lower Senegal river valley. *Malimbus* **20**: 33–53.

Savidge, J. A. (1987). Extinction of an island forest avifauna by an introduced snake. *Ecology* **68**: 660–668.

Schifferli, A. (1967). Vom Zug Schweizerischer und Deutscher Schwarzer Milane nach Ringfunden. *Orn. Beob.* **64**: 34–51.

Schloss, W. (1984). Ringfunde des Fichtenkreuzschnabels (*Loxia curvirostra*). *Auspicium* **7**: 257–284.

Schluter, D. & Smith, J. N. M. (1986). Natural selection on beak and body size in the Song Sparrow. *Evolution* **40**: 221–231.

Schneider, C. & Moritz, C. (1999). Rainforest refugia and evolution in Australia's wet tropics. *Proc. R. Soc. Lond. B* **266**: 191–196.

Schneider, C. J., Cunningham, M. & Moritz, C. (1998). Comparative phylogeography and the history of endemic vertebrates in the wet tropical rainforests of Australia. *Mol. Ecol.* **7**: 487–498.

Schodde, R. (1982). Origin, adaptation and evolution of birds in arid Australia. Pp. 191–224 in 'Evolution of the flora and fauna of arid Australia' (ed. W. R. Barker & P. J. M. Greenslade). Frewville, Australia, Peacock Publications.

Schodde, R. (1991). Concluding remarks: origins and evolution of the Australasian avifauna. Proc. Int. Orn. Congr. 20: 413–416.

Schodde, R., Fullagar, P. & Hermes, N. (1983). A review of Norfolk Island birds: past and present. Canberra, Australian National Parks & Wildlife Service.

Schodde, R. & Mason, I. J. (1999). The directory of Australian birds. Passeriformes: a taxonomic and zoogeographic atlas of the biodiversity of birds in Australia and its territories. Canberra, CSIRO.

Schroeder, M. A. & Boag, R. A. (1988). Dispersal in Spruce Grouse: is inheritance involved? Anim. Behav. 36: 305–307.

Schüle, W. (1993). Mammals, vegetation and the initial human settlement of the Mediterranean islands: a palaeoecological approach. J. Biogeog. 20: 399–412.

Schulz, H. (1998). World status and conservation of the White Stork. Torgos 28: 49–65.

Schutz, V. F. (1965). Sexuelle Prägung bei Anatiden. Z. Tierpsychol. 22: 50–103.

Sclater, P. L. (1858). On the general geographical distribution of the members of the class Aves. J. Proc. Linn. Lond., Zool. 2: 130–145.

Scott, D. K. (1980). Functional aspects of prolonged parental care in Bewick's Swans. Anim. Behav. 28: 938–952.

Scott, L. (1974). Mountain Bluebird travels 130 miles to renest. Blue Jay 32: 44–45.

Scott, S. L. (ed.) (1987). Field guide to the birds of North America. 2nd ed. Washington, D.C., National Geographic Society.

Seibold, I. & Helbig, A. J. (1995). Evolutionary history of New and Old World vultures inferred from nucleotide sequences of the mitochondria cytochromic b gene. Phil. Trans. R. Soc. Lond. B 350: 163–168.

Selander, R. K. (1965). Avian speciation in the Quaternary. Pp. 527–542 in 'The Quaternary of the United States' (ed. H. E. Wright & D. G. Frey). Princeton, University Press.

Sellers, R. M. (1984). Movements of Coal, Marsh and Willow Tits in Britain. Ring. Migr. 5: 79–89.

Senar, J. C., Borras, A., Cabrera, T. & Cabrera, J. (1993). Testing for the relationship between coniferous crop stability and Common Crossbill residence. J. Field Orn. 64: 464–469.

Senar, J. C., Burton, P. J. K. & Metcalfe, N. B. (1992). Variation in the nomadic tendency of a wintering finch Carduelis spinus and its relationship with body condition. Ornis Scand. 23: 63–72.

Sepkoski, J. J. (1989). Periodicity in extinction and the problem of catastrophism in the history of life. J. Geol. Soc. Lond. 146: 7–19.

Serrano, D., Tella, J. L., Forero, R. G. & Donázar, J. A. (2001). Factors affecting breeding dispersal in the facultative Lesser Kestrel: individual experience vs. conspecific cues. J. Anim. Ecol. 70: 568–578.

Serventy, D. L. (1971). The biology of desert birds. Pp. 287–339 in 'Avian biology, Vol 1' (ed. D. S. Farner & J. R. King). New York, Academic Press.

Serventy, D. L. (1972). Causal ornithogeography of Australia. Proc. Int. Orn. Congr. 17: 574–584.

Serventy, D. L. & Curry, P. J. (1984). Observations on colony size, breeding success, recruitment and inter-colony dispersal in a Tasmanian colony of Short-tailed Shearwaters Puffinus tenuirostris over a 30-year period. Emu 84: 71–79.

Seutin, G., Ratcliffe, L. M. & Boag, P. T. (1995). Mitochondrial DNA homogeneity in the phenotypically diverse redpoll finch complex (Aves: Carduelinae: Carduelis flammea-hornemanni). Evolution 49: 962–973.

Shackleton, N. J. & 16 others. (1984). Oxygen isotope calibration of the onset of ice-rafting and the history of glaciation in the North Atlantic region. Nature 307: 620–623.

Shapiro, B., Sibthorpe, D., Rambaut, A., Austin, J., Wragg, G. M., Bininda-Emonds, O. R. P., Lee, P. L. M. & Cooper, A. (2002). Flight of the Dodo. *Science* **295**: 1683.

Sharpe, R. B. (1899–1909). A hand-list of the genera and species of birds. Vols. 1–5. London, British Museum (Natural History).

Sharrock, J. T. R. (1976). The atlas of breeding birds in Britain and Ireland. Berkhamsted, Hertfordshire, Poyser.

Sheldon, F. H. & Bledsoe, A. H. (1993). Avian molecular systematics, 1970s to 1990s. *Ann. Rev. Ecol. Syst.* **24**: 243–278.

Sheldon, F. H. & Gill, F. B. (1996). A reconsideration of songbird phylogeny, with emphasis on the evolution of titmice and their sylvioid relatives. *Syst. Biol.* **45**: 473–495.

Shields, G. F. (1990). Analysis of mitochondrial DNA of Pacific Black Brant (*Branta bernicla nigricans*). *Auk* **107**: 620–623.

Shields, G. F. & Wilson, A. C. (1987a). Calibration of mitochondrial DNA evolution in geese. *J. Molec. Evol.* **24**: 212–217.

Shields, G. F. & Wilson, A. C. (1987b). Subspecies of the Canada Goose (*Branta canadensis*) have distinct mitochondrial DNAs. *Evolution* **41**: 662–666.

Shirihai, H., Gargallo, G. & Helbig, A. J. (2001). *Sylvia* warblers. Identification, taxonomy and phylogeny of the genus *Sylvia*. London, A. & C. Black.

Sibley, C. (1996). Distribution and taxonomy of birds of the world. CD rom, Version 2. Cincinnati, Thayer Birding Software.

Sibley, C. G. & Ahlquist, J. E. (1985). The phylogeny and classification of the Australo-Papuan passerine birds. *Emu* **85**: 1–14.

Sibley, C. & Ahlquist, J. (1990). Phylogeny and classification of birds. A study in molecular evolution. New Haven, Yale University Press.

Sibley, C. & Monroe, B. (1990). Distribution and taxonomy of birds of the world. New Haven, Yale University Press.

Siegel-Causey, D. (1997a). Molecular variation and biogeography of Rock Shags. *Condor* **99**: 139–150.

Siegel-Causey, D. (1997b). Phylogeny of the Pelecaniformes: molecular systematics of a primitive group. Pp. 159–171 in '*Avian molecular evolution and systematics*' (ed. D. Mindell). London, Academic Press.

Simberloff, D. S. (1970). Taxonomic diversity of island biotas. *Evolution* **24**: 23–47.

Simpson, B. B. & Haffer, J. (1978). Speciation patterns in the Amazon forest biota. *Ann. Rev. Ecol. Syst.* **9**: 497–518.

Simpson, G. G. (1951). The species concept. *Evolution* **5**: 285–298.

Simpson, G. G. (1965). The geography of evolution. Collected essays. Philadelphia & New York, Chilton.

Skellam, J. G. (1951). Random dispersal in theoretical populations. *Biometrika* **38**: 196–218.

Skira, I. (1991). The Short-tailed Shearwater: a review of its biology. *Corella* **15**: 45–52.

Sklepkovych, B. O. & Montevecchi, W. A. (1989). The world's largest known nesting colony of Leach's Storm Petrels on Baccalieu Island, Newfoundland. *Am. Birds* **43**: 38–42.

Skov, H., Durinck, J., Leopold, M. F. & Tasker, M. L. (1995). Important bird areas for seabirds in the North Sea. Sandy, Royal Society for the Protection of Birds.

Smith, A. G., Hurley, A. M. & Briden, J. C. (1981). Phanerozoic palaeocontinental world map. Cambridge, University Press.

Smith, K. W., Reed, J. M. & Trevis, B. E. (1992). Habitat use and site fidelity of Green Sandpipers *Tringa ochropus* wintering in southern England. *Bird Study* **39**: 155–164.

Smith, R. I. (1970). Response of Pintail breeding populations to drought. *J. Wildl. Manage.* **34**: 943–946.

Snow, D. W. (1954). The habitats of Eurasian tits (*Parus* spp.). *Ibis* **96**: 565–585.

Snow, D. W. (ed.) (1978a). An atlas of speciation in African non-passerine birds. London, Brit. Mus. (Nat.Hist.).

Snow, D. W. (1978b). Relationships between European and African avifaunas. *Bird Study* **25:** 134–148.

Snow, D. W. (1997). Should the biological be superseded by the phylogenetic species concept? *Bull. Brit. Orn. Club* **117:** 110–121.

Snow, D. W. & Snow, B. K. (1971). The feeding ecology of tanagers and honeycreepers in Trinidad. *Auk* **88:** 291–322.

Sokal, R. R. & Rohlf, F. S. (1981). Biometry, 2nd ed. New York, W.H. Freeman.

Sokolov, L. V. (1997). Philopatry of migratory birds. *Phys. Gen. Biol. Rev.* **11:** 1–58.

Sollie, J. F. (1961). Dotterels nesting in the Netherlands at 4 metres below sea level. *Limosa* **34:** 274–276.

Solonen, T. (1979). Population dynamics of the Garden Warbler *Sylvia borin* in southern Finland. *Ornis Fenn.* **56:** 3–12.

Solonen, T. (1986). Breeding of the Great Grey Owl *Strix nebulosa* in Finland. *Lintumies* **21:** 11–18.

Sonerud, G. A., Solheim, R. & Prestrud, K. (1988). Dispersal of Tengmalm's Owl *Aegolius funereus* in relation to prey availability and nesting success. *Ornis Scand.* **19:** 175–181.

Sorenson, M. D., Cooper, A., Paxinos, E. E., Quinn, T. W., James, H. F., Olson, S. L. & Fleischer, R. C. (1999). Relationships of the extinct moa-nalos, flightless Hawaiian waterfowl, based on ancient DNA. *Proc. R. Soc. Lond. B.* **266:** 2187–2193.

Sovinen, M. (1952). The Red-flanked Bluetail *Tarsiger cyanurus* (Pall.) spreading into Finland. *Ornis Fenn.* **29:** 27–35.

Sparks, T. (1999). Phenology and the changing pattern of bird migration in Britain. *Int. J. Biometeorol.* **42:** 134–138.

Sparks, T. & Crick, H. (1999). The times they are a-changing? *Bird Conserv. Int.* **9:** 1–7.

Sparks, T. & Mason, C. (2001). Dates of arrivals and departures of spring migrants taken from Essex Bird Reports 1950–1998. Pp. 154–164 in *'Essex Bird Report 1999'* (ed. A. Mullins & H. Vaughan). Essex, Essex Birdwatching Society.

Spear, L. B., Pyle, P. & Nur, N. (1998). Natal dispersal in the Western Gull: proximal factors and fitness consequences. *J. Anim. Ecol.* **67:** 165–179.

Spendelow, J. A., Nichols, J. D., Nisbet, I. C. T., Hays, H., Cormons, G. D., Burger, J., Safina, C., Hines, J. E. & Gochfield, M. (1995). Estimating annual survival and movement rates of adults within a metapopulation of Roseate Terns. *Ecology* **76:** 2415–2428.

Staicer, C. A. (1992). Social behaviour of the Northern Parula, Cape May Warbler, and Prairie Warbler wintering in second-growth forest in southwestern Puerto Rico. Pp. 308–320 in *'Ecology and conservation of neotropical migrant landbirds'* (ed. J. M. Hagen & D. W. Johnston). Washington, Smithsonian Institution Press.

Stattersfield, A., Crosby, M. J., Long, A. J. & Wege, D. C. (1998). Endemic bird areas of the world: priorities for biodiversity conservation. Cambridge, Birdlife International.

Steadman, D. W. (1989). Extinction of birds of Eastern Polynesia: a review of the record and comparisons with other Pacific Island groups. *J. Archaeol. Sci.* **16:** 177–205.

Steadman, D. W. (1991). Extinction of species: past, present and future. Pp. 156–169 in *'Global climate change and life on earth'* (ed. R. C. Wyman). New York, Routledge, Chapman & Hall.

Steadman, D. W. (1992). New species of *Gallicolumba* and *Macropygia* (Aves: Columbidae) from archeological sites in Polynesia. Los Angeles, County Museum of Natural History Science Series.

Steadman, D. W. (1995). Prehistoric extinctions of Pacific island birds: biodiversity meets zooarchaeology. *Science* **267:** 1123–1131.

Steadman, D. W. (1997). The historic biogeography and community ecology of Polynesian pigeons and doves. *J. Biogeog.* **24:** 737–753.

Steadman, D. W. & Martin, P. S. (1984). extinctions of birds in the late Pleistocene of North America. Pp. 466–477 in *'Quaternary extinctions'* (ed. P. S. Martin & R. J. Klein). Tucson, University of Arizona Press.

Steadman, D. W. & Olson, S. R. (1985). Bird remains from an archaeological site on Henderson Island, South Pacific; man-caused extinctions on an 'uninhabited' island. *Proc. Nat. Acad. Sci. USA* **82**: 6191–6195.

Steadman, D. W., Schubel, S. E. & Pahlavan, D. (1988). New subspecies and new records of *Papasula abbotti* (Aves, Sulidae) from archaeological sites in the Tropical Pacific. *Proc. Biol. Soc.Washington* **101**: 487–495.

Stegmann, B. (1932). Die Lerkunft der palaarktischen Taiga-vogel. *Arch. f. Naturg.(N.F.)* **1**: 355–398.

Stenzel, L. E., Warriner, J. C., Warriner, J. S., Wilson, K. S., Bidstrup, F. C. & Page, G. W. (1994). Long-distance dispersal of Snowy Plovers in western North America. *J. Anim. Ecol.* **63**: 887–902.

Stevens, G. C. (1989). The latitudinal gradient in geographical range: how so many species coexist in the tropics. *Amer. Nat.* **133**: 240–256.

Stevens, G. C. (1992). The elevational gradient in altitudinal range: an extension of Rapoport's latitudinal rule to altitude. *Amer. Nat.* **140**: 893–911.

Stewart, J. R. (1999). Intraspecific variation in modern and Quaternary European *Lagopus*. *Smithson. Contrib. Paleobiol.* **89**: 159–168.

Stiles, F. G. (1973). Food supply and the annual cycle of the Anna Hummingbird. *Univ. Calif. Publ. Zool.* **97**: 1–109.

Stiles, F. G. & Skutch, A. F. (1989). A guide to the birds of Costa Rica. London, Christopher Helm.

Stjernberg, T. (1985). Recent expansion of the Scarlet Rosefinch *Carpodacus erythrinus* in Europe. *Proc. Int. Orn. Congr.* **18**: 743–753.

Storer, R. W. & Jehl, J. R. (1985). Moult patterns and moult migration in the Black-necked Grebe *Podiceps nigricollis*. *Ornis Scand.* **16**: 253–260.

Stresemann, E. (1919). Beiträge zur Zoogeographie der paläatischen Region, 1. Munich, Gustav Fischer.

Summers-Smith, J. D. (1988). The Sparrows. Calton, Poyser.

Suryan, R. M., Irons, D. B. & Benson, J. (2000). Prey switching and variable foraging strategies of Black-legged Kittiwakes and the effect on reproductive success. *Condor* **102**: 374–384.

Sutherland, G. D., Harestad, A. S., Price, K. & Lertzman, K. P. (2000). Scaling of natal dispersal distances in terrestrial birds and mammals. *Conserv. Ecol.* **4**: 16(online) URL: http://www.consecol.org/vol4/iss1/art16.

Sutherland, W. J. & Baillie, S. R. (1993). Patterns in the distribution, abundance and variation of bird populations. *Ibis* **135**: 209–210.

Sutherland, W. J. & Crockford, N. J. (1993). Factors affecting the feeding distribution of Red-breasted Geese *Branta ruficollis* wintering in Romania. *Biol. Conserv.* **63**: 61–65.

Svärdson, G. (1957). The 'invasion' type of bird migration. *Brit. Birds* **50**: 314–343.

Swann, R. L. & Ramsay, A. D. K. (1983). Movements from, and age of return to, an expanding Scottish Guillemot colony. *Bird Study* **30**: 207–214.

Swarth, H. S. (1920). Revision of the avian genus *Passerella* etc. *Univ. Calif. Publ. Zool.* **21**: 75–224.

Swenson, J. E., Jensen, K. C. & Toepfer, J. E. (1988). Winter movements by Rosy Finches in Montana. *J. Field Orn.* **59**: 157–160.

Taberlet, P., Fumagalli, L., Wust-Saucy, A.–G. & Cosson, J.–F. (1998). Comparative phylogeography and postglacial colonization routes in Europe. *Mol. Ecol.* **7**: 453–464.

Taberlet, P., Meyer, A. & Bouvet, J. (1992). Unusual mitochondrial DNA polymorphism in two local populations of Blue Tit *Parus caeruleus*. *Mol. Ecol.* **1**: 27–36.

Talbot, S. L. & Shields, G. F. (1996). Phylogeography of Brown Bears (*Ursus arctos*) of Alaska and paraphyly within the Ursidae. *Mol. Phylogenet. Evol.* **5**: 477–494.

Tallis, J. H. (1991). Plant community history. London, Chapman & Hall.

Tambussi, C. P. & Noriega, J. I. (1999). The fossil record of condors (Ciconiiformes: Vulturidae) in Argentina. *Smithson. Contrib. Paleobiol.* **89**: 177–184.

Tarr, C. L. & Fleischer, R. C. (1993). Mitochondrial DNA variation and the evolutionary relationships in the *Amakihi* complex. *Auk* **110**: 825–831.

Tarr, C. L. & Fleischer, R. C. (1995). Evolutionary relationships of the Hawaiian honeycreepers (Aves: Drepanidinae). Pp. 147–159 in '*Hawaiian biogeography: evolution on a hot spot archipelago*' (ed. W. L. Wagner & V. A. Funk). Washington, Smithsonian Institution Press.

Tasker, M. L., Webb, A., Hall, A. J., Pienkowski, M. W. & Langslow, D. R. (1987). Seabirds in the North Sea. Peterborough, Nature Conservancy Council.

Taylor, D., Saksena, P., Sanderson, P. G. & Kucera, K. (1999). Environmental change and rain forests on the Sunda shelf of Southeast Asia: drought, fire and the biological cooling of biodiversity hotspots. *Biodiv. Conserv.* **8**: 1159–1177.

Templeton, A. R. (1989). The meaning of species and speciation: a genetic perspective. Pp. 3–27 in '*Speciation and its consequences*' (ed. D. Otte & J. A. Endler). Sunderland, Mass., Sinauer Associates.

Templeton, A. R. (2002). Out of Africa again and again. *Nature* **416**: 45–50.

Terborgh, J. (1971). Distribution on environmental gradients. Theory and a preliminary interpretation of distributional patterns in the avifauna of the Condillera Vilcabamba, Peru. *Ecology* **52**: 23–40.

Terborgh, J. (1989). Where have all the birds gone? Princeton, University Press.

Terborgh, J., Lopez, L. & Telles, J. (1997). Bird communities in transition: the Lago Guri Islands. *Ecology* **78**: 1494–1501.

Terborgh, J. & Weske, J. S. (1975). The role of competition in the distribution of Andean birds. *Ecology* **56**: 562–576.

Terborgh, J. W. & Winter, B. (1980). Some causes of extinction. Pp. 119–33 in '*Conservation biology: an evolutionary–ecological perspective*' (ed. M. E. Soulé & B. A. Wilcox). Sunderland, Mass., Sinauer Associates.

Thibault, J. C. (1994). Nest-site tenacity and mate fidelity in relation to breeding success in Cory's Shearwater *Calonectris diomedea*. *Bird Study* **41**: 25–28.

Thibault, J. C., Zotier, R., Guyot, I. & Bretagnolle, V. (1996). Recent trends in breeding marine birds of the Mediterranean region with special reference to Corsica. *Colon. Waterbirds* **19 (spec. pub. 1):** 31–40.

Thies, H. (1996). Zum Vorkommen des Fichtenkreuzschnabels (*Loxia curvirostra*) und anderer *Loxia*-Arten im Segeberger Forst 1970–1995 mit besonderer Erörterung der Zugphänologie. *Corax* **16**: 305–334.

Thiollay, J.-M. & Wahl, R. (1998). Le Balbuzard Pêcheur *Pandion haliaetus* nicheur en France continentale. *Alauda* **66**: 1–12.

Thomas, C. D. & Lennon, J. J. (1999). Birds extend their ranges northwards. *Nature* **399**: 213.

Thomas, J. W., Forsman, E. D., Lint, J. B., Meslow, E. C., Noon, B. R. & Verner, J. (1990). A conservation strategy for the Northern Spotted Owl. Report of the Interagency Scientific Committee to address the conservation of the Northern Spotted Owl. United States Department of Agriculture, Forest Service and Fish and Wildlife Service, United States Department of the Interior, Bureau of Land Management and National Park Service, Portland, Oregon.

Thompson, C. F. & Nolan, V. (1973). Population biology of the Yellow-breasted Chat (*Icteria virens* L.) in southern Indiana. *Ecol. Monogr.* **43**: 145–171.

Thompson, D. B. & Whitfield, D. P. (1993). Research on mountain birds and their habitats. *Scott. Birds* **17**: 1–8.

Thompson, P. S., Baines, D., Coulson, J. C. & Longrigg, G. (1994). Age at first breeding, philopatry and breeding site fidelity in the Lapwing *Vanellus vanellus*. *Ibis* **136**: 474–484.

Thomson, R. S. & Anderson, K. H. (2000). Biomes of western North America at 18,000, 6,000 and 0 ^{14}C yr BP reconstructed from pollen and packrat midden data. *J. Biogeog.* **27:** 555–584.

Thornton, I. W. B., Zann, R. A. & van Balen, S. (1993). Colonisation of Rakata (Krakatau Islands) by non-migrant landbirds from 1883 to 1992 and implications for the value of island biogeography theory. *J. Biogeog.* **20:** 441–452.

Thorpe, W. H. (1961). Bird-song. Cambridge, University Press.

Thorup, K. (1998). Vagrancy of Yellow-browed Warbler *Phylloscopus inornatus* and Pallas's Warbler *Ph. proregulus* in north-west Europe: misorientation or great circles? *Ring. Migr.* **19:** 7–12.

Tiainen, J. (1983). Dynamics of a local population of the Willow Warbler *Phylloscopus trochilus* in southern Finland. *Ornis Scand.* **14:** 1–15.

Tickell, W. L. N. (1993). Atlas of southern hemisphere albatrosses. *Bull. Pacif. Seabird Group* **20:** 22–38.

Timoféeff-Ressovsky, N. W. (1940). Zur Frage über die 'Eliminationsregel': die geographische Grössenvariabilität von *Emberiza aureola* Pall. *J. Orn.* **88:** 334–340.

Tomialojc, L. (2000). Did White-backed Woodpeckers ever breed in Britain? *Brit. Birds* **93:** 453–456.

Tomkovich, P. J. & Soloviev, M. Y. (1994). Site fidelity in high arctic breeding waders. *Ostrich* **65:** 174–180.

Tordoff, H. B. & Redig, P. T. (1997). Midwest Peregrine Falcon demography, 1982–1995. *J. Raptor Res.* **31:** 339–346.

Trettau, W. (1952). Planberingung des Trauerfliegenschnappers (*Muscicapa hypoleuca*) in Hessen. *Vogelwarte* **16:** 89–95.

Tucker, G. & Heath, M. (1994). Birds in Europe: their conservation status. Cambridge, Birdlife International.

Tyrberg, T. (1998). Pleistocene birds of the Palearctic: a catalogue. Cambridge, Mass., Nuttall Orn. Club.

Tyrberg, T. (1999). Seabirds and late Pleistocene marine environments in the northeast Atlantic and the Mediterranean. *Smithson. Contrib. Paleobiol.* **89:** 139–157.

Udvardy, M. D. F. (1958). Ecological and distributional analyses of North American birds. *Condor* **60:** 50–66.

Udvardy, M. D. F. (1969). Dynamic zoogeography with special reference to land animals. New York, Van Nostrand Reinhold.

Valera, F., Rey, P., Sanchezlafuente, A. M. & Munozcobo, J. (1993). Expansion of Penduline Tit (*Remiz pendulinus*) through migration and wintering. *J. Orn.* **134:** 273–282.

Valikangas, I. (1951). The expansion of the Greenish Warbler (*Phylloscopus trochiloides viridanus* Blyth) in the Baltic area, especially in Finland, towards north and northwest, and its causes. *Proc. Int. Orn. Congr.* **10:** 527–531.

van Balen, J. H. & Hage, F. (1989). The effects of environmental factors on tit movements. *Ornis Scand.* **20:** 99–104.

van den Bosch, F., Hengeveld, R. & Metz, J. A. J. (1992). Analysing the velocity of range expansion. *J. Biogeog.* **19:** 135–150.

van den Bosch, F., Metz, J. A. J. & Diekmann, O. (1990). The velocity of spatial population expansion. *J. Mathem. Biol.* **28:** 529–565.

van der Hammen, T. (1985). The Plio-Pleistocene climatic record of the tropical Andes. *J. Geog. Soc. Lond.* **142:** 483–489.

van der Hammen, T. & Absy, M. L. (1994). Amazonia during the last glacial. *Palaeogeogr., Palaeoclimatol. Palaeoecol.* **109:** 247–261.

van Franeker, J. A. & Wattel, J. (1982). Geographical variation of the Fulmar *Fulmarus glacialis* in the North Atlantic. *Ardea* **70:** 31–44.

van Noordwijk, A. J. (1983). Problems in the analysis of dispersal and a critique on its 'heritability' in the Great Tit. *J. Anim. Ecol.* **53:** 533–544.

van Riper, C., van Riper, S. G., Goff, M. G. & Laird, M. (1986). The epizootiology and ecological significance of malaria in Hawaiian land birds. *Ecol. Monogr.* **56:** 327–344.

van Tuinen, M. & Hedges, S. B. (2001). Calibration of avian molecular clocks. *Mol. Biol. Evol.* **18:** 206–213.

van Wagner, C. E. & Baker, A. J. (1990). Association between mitochondrial DNA and morphological evolution in Canada Geese. *J. Mol. Evol.* **31:** 373–382.

Veit, R. R., McGowan, J. A., Ainley, D. G., Wahls, T. R. & Pyle, P. (1997). Apex marine predator declines ninety percent in association with climatic change. *Global Change Biol.* **3:** 23–28.

Veitch, C. R. (1985). Methods of eradicating feral cats from offshore islands in New Zealand. *ICBP Tech. Bull.* **3:** 125–141.

Veltman, C. J., Nee, S. & Crawley, M. J. (1996). Correlates of introduction success in exotic New Zealand birds. *Amer. Nat.* **147:** 542–557.

Venier, L. A. & Fahrig, L. (1998). Intraspecific abundance–distribution relationships. *Oikos* **82:** 483–490.

Verboon, J., Schotman, A., Opdam, P. & Metz, H. (1991). European Nuthatch metapopulations in a fragmented agricultural landscape. *Oikos* **61:** 149–156.

Vickery, J. A., Sutherland, W. J., Watkinson, A. R., Lane, S. J. & Rowcliffe, J. M. (1995). Habitat switching by Dark-bellied Brent Geese *Branta b. bernicla* (L.) in relation to food depletion. *Oecologia* **103:** 499–508.

Village, A. (1987). Numbers, territory size and turnover of Short-eared Owls *Asio flammeus* in relation to vole abundance. *Ornis Scand.* **18:** 198–204.

Visser, M. E., van Noordwijk, A. J., Tinbergen, J. M. & Lessells, C. M. (1998). Warmer springs lead to mistimed reproduction in Great Tits. *Proc. R. Soc. Lond. B* **265:** 1867–1870.

Vogel, C. & Moritz, D. (1995). Langjährige Ànderungen von Zugzeiten auf Helgoland. *Jber Inst. Vogelforshung* **2:** 8–9.

von Haartman, L. (1960). The Ortstreue of the Pied Flycatcher. *Proc. Int. Orn. Congr.* **12:** 266–273.

Voous, K. H. (1960). Atlas of European birds. London, Nelson.

Voous, K. H. (1985). Palaearctic Region. Pp. 429–431 in '*A dictionary of birds*' (ed. B. Campbell & E. Lack). Calton, Poyser.

Vrba, E. S. (1997). Species 'habitats' in relation to climate, evolution, migration and conservation. Pp. 275–285 in '*Past and future rapid environmental changes: the spatial and evolutionary responses of terrestrial biota*' (ed. B. Huntley, W. Cramer, A. V. Morgan, H. C. Prentice & J. R. M. Allen). London, Springer Verlag.

Vuilleumier, F. (1969). Pleistocene speciation in birds living in the High Andes. *Nature* **223:** 1179–1180.

Vuilleumier, F. (1970). Insular biogeography in continental regions. I. The northern Andes of South America. *Amer. Nat.* **104:** 373–388.

Vuilleumier, F. (1980). Speciation in birds of the high Andes. *Proc. Int. Orn. Congr.* **17:** 1256–1261.

Vuilleumier, F. (1984). Patchy distribution and systematics of *Oreomanes fraseri* (Aves, Coerebidae) of Andean *Polylepis* woodlands. *Amer. Mus. Novitates* **2777:** 1–17.

Vuilleumier, F., LeCroy, M. & Mayr, E. (1992). New species of birds described from 1981 to 1990. *Bull. Brit. Orn. Club Centenary Suppl.* **112A:** 267–309.

Vuilleumier, F. & Simberloff, D. (1980). Ecology vs. history as determinants of patchy and insular distributions in high Andean birds. *Evol. Biol.* **12:** 235–379.

Walker, C. A., Wragg, G. M. & Harrison, C. J. O. (1990). A new shearwater from the Pleistocene of the Canary Islands and its bearing on the evolution of certain *Puffinus* shearwaters. *Hist. Biol.* **3:** 203–224.

Walkinshaw, L. H. (1953). Life history of the Prothonotary Warbler. *Wilson Bull.* **65:** 152–168.

Wallace, A. R. (1876). The geographical distribution of animals. New York, Harper.

Walls, S. S. & Kenward, R. E. (1995). Movements of radio-tagged Common Buzzards *Buteo buteo* in their first year. *Ibis* **137**: 177–182.

Walsberg, G. E. (1978). Brood size and the use of time and energy by the Phainopepla. *Ecology* **59**: 147–153.

Wanless, S., Harris, M. P. & Morris, J. A. (1991). Foraging range and feeding locations of Shags *Phalacrocorax aristotelis* during chick-rearing. *Ibis* **133**: 30–36.

Ward, P. (1968). Origin of the avifauna of urban and suburban Singapore. *Ibis* **110**: 239–255.

Ward, P. (1971). The migration patterns of *Quelea quelea* in Africa. *Ibis* **113**: 275–297.

Warham, J. (1990). The Petrels. London, Academic Press.

Warham, J. (1996). The behaviour, population biology and physiology of the petrels. London, Academic Press.

Warheit, K. I. (1992). A review of the fossil seabirds from the Tertiary of the North Pacific: plate tectonics, paleoceanography, and faunal changes. *Paleobiology* **18**: 401–424.

Warheit, K. I. (2002). The seabird fossil record and the role of paleontology in understanding seabird community structure. Pp. 57–85 in *'Biology of marine birds'* (ed. E. A. Schreiber & J. Burger). London, CRC Press.

Warner, R. E. (1968). The role of introduced diseases in the extinction of the endemic Hawaiian avifauna. *Condor* **70**: 101–120.

Waser, P. M. (1985). Does competition drive dispersal? *Ecology* **66**: 1170–1175.

Watson, A., Moss, R., Parr, R., Mountford, M. D. & Rothery, P. (1994). Kin landownership, differential aggression between kin and non-kin, and population fluctuations in Red Grouse. *J. Anim. Ecol.* **63**: 39–50.

Watson, G. E., Angle, J. P., Harper, P. C., Bridge, M. A., Schlatter, R. P., Tickell, W. L. N., Boyd, J. C. & Boyd, M. M. (1971). Birds of the Antarctic and Subantarctic. New York, Amer. Geogr. Soc.

Weatherhead, P. J. & Forbes, M. R. L. (1994). Natal philopatry in passerine birds: genetic or ecological influences? *Behav. Ecol.* **5**: 426–433.

Weathers, W. W. (1979). Climatic adaptation in avian standard metabolic rate. *Oecologia* **42**: 81–89.

Webb, A., Harrison, N. M., Leaper, G. M., Steele, R. D., Tasker, M. L. & Pienkowski, M. W. (1990). Seabird distribution west of Britain. Aberdeen, Nature Conservancy Council.

Webb, T. J. & Gaston, K. J. (2000). Geographic range size and evolutionary age in birds. *Proc. R. Soc. Lond. B* **267**: 1843–1850.

Weckstein, J. D., Zink, R. M., Blackwell-Rago, R. C. & Nelson, D. A. (2001). Anomalous variation in mitochondrial genomes of White-crowned (*Zonotrichia leucophrys*) and Golden-crowned (*Z. atricapilla*) Sparrows: pseudogenes, hybridisation or incomplete lineage sorting? *Auk* **118**: 231–236.

Weimerskirch, H. & Jouventin, P. (1987). Population dynamics of the Wandering Albatross *Diomedea exulans*, of the Crozet Islands: causes and some consequences of the population decline. *Oikos* **49**: 315–322.

Weimerskirch, H. & Robertson, G. (1994). Satellite tracking of Light-mantled Sooty Albatross. *Polar Biol.* **14**: 123–126.

Weimerskirch, H. & Sagar, P. M. (1996). Diving depths of Sooty Shearwaters *Puffinus griseus*. *Ibis* **138**: 786–788.

Weimerskirch, H., Salamolard, M., Sarrazin, F. & Jouventin, P. (1993). Foraging strategy of Wandering Albatrosses through the breeding season: a study using satellite telemetry. *Auk* **110**: 325–342.

Wells, D. R. (1990). Migratory birds and tropical forest in the Sunda region. Pp. 357–369 in *'Biogeography and ecology of forest bird communities.'* (ed. A. Keast). The Hague, SPB Academic Publishers.

Wenink, P. W., Baker, A. J., Rösner, H.-U. & Tilanus, M. G. J. (1996). Global mitochondrial DNA phylogeography of holarctic breeding Dunlins (*Calidris alpina*). *Evolution* **50:** 318–330.

Wenink, P. W., Baker, A. J. & Tilanus, M. G. J. (1993). Hypervariable–control-region sequences reveal global population structuring in a long-distance migrant shorebird, the Dunlin (*Calidris alpina*). *Proc. Nat. Acad. Sci. USA* **90:** 94–98.

Wenink, P. W., Baker, A. J. & Tilanus, M. G. J. (1994). Mitochondrial control region sequences in two shorebird species, the Turnstone and the Dunlin, and their utility in population genetic studies. *Mol. Biol. Evol.* **11:** 22–31.

Wernham, C. V., Toms, M. P., Marchant, J. H., Clark, J. A., Siriwardena, G. M. & Baillie, S. R. (2002). The migration atlas: movements of the birds of Britain and Ireland. London, Poyser.

West-Eberhard, M. J. (1983). Sexual selection, social competition and speciation. *Quart. Rev. Biol.* **58:** 155–183.

Wetmore, A. (1967). Pleistocene Aves from Ladds, Georgia. *Bull. Georgia Acad. Sci.* **25:** 151–153.

Whittaker, R. H. & Likens, G. E. (1973). Carbon in the biota. Pp. 281–300 in '*Carbon on the biosphere*' (ed. G. M. Woodwell & E. V. Pecan). Springfield, National Technical Information Service.

Whittaker, R. J. (1998). Island biogeography. Oxford, University Press.

Whittington, P. A., Dyer, B. M., Crawford, R. J. M. & Williams, A. J. (1999). First recorded breeding of Leach's Storm Petrel *Oceanodroma leucorhoa* in the southern hemisphere, at Dyer Island, South Africa. *Ibis* **141:** 327–330.

Wiley, E. O. (1981). Phylogenetics: the theory and practice of phylogenetic systematics. New York, Wiley-Interscience.

Williams, G. R. (1953). The dispersal from New Zealand and Australia of some introduced European passerines. *Ibis* **95:** 676–692.

Williams, T. C. & Webb, T. (1996). Neotropical bird migration during the ice ages: orientation and ecology. *Auk* **113:** 105–118.

Williams, T. D. (1995). The Penguins. Oxford, University Press.

Williamson, K. (1975). Birds and climatic change. *Bird Study* **22:** 143–164.

Williamson, M. (1981). Island populations. Oxford, University Press.

Willis, E. O. (1974). Populations and local extinctions on Barro Colorado Island, Panama. *Ecol. Monogr.* **44:** 153–169.

Willson, M. F. (1976). The breeding distribution of migrant birds: a critique of MacArthur (1959). *Wilson Bull.* **88:** 582–587.

Wilson, E. O. (1961). The nature of the taxon cycle in the Melanesian ant fauna. *Amer. Nat.* **95:** 165–193.

Wilson, G. E. (1976). Spotted Sandpipers resting in Scotland. *Brit. Birds* **69:** 288–292.

Wilson, H. J., Norriss, D. W., Walsh, A., Fox, A. D. & Stroud, D. A. (1991). Winter site fidelity in Greenland White-fronted Geese *Anser albifrons flavirostris*, implications for conservation and management. *Ardea* **79:** 287–294.

Wilson, M. (1979). Further range expansion by Citrine Wagtail. *Brit. Birds* **72:** 42–43.

Wink, M. & Heidrich, P. (2000). Molecular systematics of owls (Strigiformes), based on DNA-sequences of the mitochondrial cytochrome *b* gene. Pp. 819–828 in '*Raptors at risk*' (ed. R. D. Chancellor & B.-U. Meyburg). Berlin, Hancock House Publishers & World Working Group on Birds of Prey and Owls.

Wink, M., Seibold, I., Lotfikhah, F. & Bednarek, W. (1998). Molecular systematics of Holarctic raptors (Order Falconiformes). Pp. 29–48 in '*Holarctic birds of prey*' (ed. R. D. Chancellor, B.-U. Meyburg & J. J. Ferrero). Berlin, World Working Group on Birds of Prey and Owls and Adenex.

Winkel, W. (1996). Langzeit–Erfassung brutbiologischer Parameter bei Höhlenbrüter (*Parus* spp., *Sitta europaea*) — gibt es signifikante varänderungen? *Verh. Dtsch. Zool. Ges.* **89:** 1–130.

Winkel, W. & Frantzen, M. (1991). Ringfun–Analyse zum Zug einer niedersächsischen Population des Trauerschnäppers *Ficedula hypoleuca*. *Vogelkdl. Beitr. Nieders*. **23:** 90–98.

Witherby, H. (1929). A transatlantic passage of Lapwings. *Brit. Birds* **22:** 6–13.

Witherby, H. F. & Fitter, R. S. R. (1942). Black Redstarts in England in the summer of 1942. *Brit. Birds* **36:** 132–139.

Withers, P. C. & Williams, J. B. (1990). Metabolic and respiratory physiology of an arid-adapted Australian bird, the Spinifex Pigeon. *Condor* **92:** 961–969.

Wolf, C. M., Garland, T. & Griffith, B. (1998). Predictors of avian and mammalian translocation success: reanalysis with phylogenetically independent contrasts. *Biol. Conserv*. **86:** 243–255.

Woolfenden, G. E. & Fitzpatrick, J. W. (1978). The inheritance of territory in group- breeding birds. *Bioscience* **28:** 104–108.

Woolfenden, G. E. & Fitzpatrick, J. W. (1996). Florida Scrub Jay. *Birds N. Am*. **228:** 1–27.

Worthy, T. H. (1999). The role of climate change versus human impacts — avian extinction on South Island, New Zealand. *Smithson. Contrib. Paleobiol* **89:** 111–123.

Worthy, T. H. & Jouventin, P. (1999). The fossil avifauna of Amsterdam Island, Indian Ocean. *Smithson. Contrib. Paleobiol*. **89:** 39–65.

Wright, S. J. (1980). Density compensation in island avifaunas. *Oecologia* **45:** 385–389.

Wright, T. F. & Wilkinson, G. S. (2001). Population genetic structure and vocal dialects in an amazon parrot. *Proc. R. Soc. Lond. B* **268:** 609–616.

Wu, C.-I. & Li W.-H. (1985). Evidence for higher rates of nucleotide substitution in rodents than in man. *Proc. Nat. Acad. Sci. USA* **82:** 1741–1745.

Wunderle, J. M., Lodge, J. M. & Waide, R. B. (1992). Short-term effects of Hurricane Gilbert on terrestrial bird populations on Jamaica. *Auk* **109:** 148–166.

Wyllie, I. & Newton, I. (1991). Demography of an increasing population of Sparrowhawks. *J. Anim. Ecol*. **60:** 749–766.

Wynne-Edwards, V. C. (1935). On the habits and distribution of birds on the North Atlantic. *Proc. Boston Soc. Nat. Hist*. **40:** 233–346.

Wynne-Edwards, V. C. (1962). Animal dispersion in relation to social behaviour. Edinburgh & London, Oliver & Boyd.

Wynne-Edwards, V. C. (1985). Oceanic birds. Pp. 403–405 in '*A new dictionary of birds*' (ed. B. Campbell & E. Lack). Calton, Poyser.

Yang, S. Y. & Patton, J. L. (1981). Genic variability and differentiation in the Galapagos Finches. *Auk* **98:** 230–242.

Yésou, P. (1995). Individual migration strategies in Cormorants *Phalacrocorax carbo* passing through or wintering in western France. *Ardea* **83:** 267–274.

Young, H. G. & Rhymer, J. M. (1998). Meller's Duck: a threatened species receives recognition at last. *Biodiv. Conserv*. **7:** 1313–1323.

Yu, G., Chen, X., Ni, J., Cheddadi, R., Guiot, J., Han, H., Harrison, S. P., Huang, C., Ke, M., Kong, Z., Li, S., Li, W., Liew, P., Liu, G., Liu, J., Liu, Q., Liu, K. B., Prentice, I. C., Qui, W., Ren, G., Song, C., Sugita, S., Sun, X., Tang, L., Van Campo, E., Y. Xia, Xu, Q., Yan, S., Yang, X., Zhao, J. & Zheng, Z. (2000). Palaeovegetation of China: a pollen data-based synthesis for the mid-Holocene and last glacial maximum. *J. Biogeog*. **27:** 635–664.

Zahavi, A. (1989). Arabian Babbler. Pp. 253–275 in '*Lifetime reproduction in birds*' (ed. I. Newton). London, Academic Press.

Zink, G. (1973–85). Der Zug Europäischer Singvögel. Lieferung, Vogelwarte Radolfzell.

Zink, G. & Bairlein, F. (1995). Der Zug Europäischer Singvögel. Wiesbaden, AULA-Verlag.

Zink, R. M. (1994). The geography of mitochondrial DNA variation, population structure, hybridisation, and species limits in the Fox Sparrow (*Passerella iliaca*). *Evolution* **48:** 96–111.

Zink, R. M. (1996a). Bird species diversity. *Nature* **381:** 566.

Zink, R. M. (1996b). Comparative phylogeography of North American birds. *Evolution* **50:** 308–317.

Zink, R. M. (1997). Phylogeographic studies of North American birds. Pp. 301–324 in '*Avian molecular evolution and systematics*' (ed. D. P. Mindell). London, Academic Press.

Zink, R. M., Blackwell-Rago, R. C. & Ronquist, F. (2000). The shifting roles of dispersal and vicariance in biogeography. *Proc. R. Soc. Lond. B*. **267:** 497–503.

Zink, R. M. & Dittman, D. L. (1991). Evolution of Brown Towhees: mitochondrial DNA evidence. *Condor* **93:** 98–105.

Zink, R. M. & Dittman, D. L. (1993a). Gene flow, refugia and evolution of geographic variation in the Song Sparrow (*Melospiza melodia*). *Evolution* **47:** 717–729.

Zink, R. M. & Dittman, D. L. (1993b). Population structure and gene flow in the Chipping Sparrow (*Spizella passerina*) and a hypothesis for evolution in the genus *Spizella*. *Wilson Bull*. **105:** 399–413.

Zink, R. M. & Hackett, S. J. (1988). Historical biogeography of the North American avifauna. *Proc. Int. Orn. Congr*. **19:** 2574–2580.

Zink, R. M. & McKitrick, M. C. 1995. The debate over species concepts and its implications for ornithology. *Auk* **112:** 701–719.

Zink, R. M., Rohwer, S., Andreev, A. V. & Dittman, D. L. (1995). Trans-Beringia comparisons of mitochondrial DNA differentiation in birds. *Condor* **97:** 639–649.

Zotier, R., Bretagnolle, V. & Thibault, J.-C. (1999). Biogeography of the marine birds of a confined sea, the Mediterranean. *J. Biogeog*. **26:** 297–313.

Zwartjes, P. W. (1999). Genetic variability in the endemic vireos of Puerto Rico and Jamaica contrasted with the continental White-eyed Vireo. *Auk* **116:** 964–975.

Index